CONTROL SYSTEMS

Prof V Krishnamurthi BE, MSc (Engg), PhD
Former Head
Department of Electronics and Communication Engineering
College of Engineering, Anna University
Chennai, India

IKON BOOKS
Publishers and Distributors
New Delhi • Hyderabad

CBS Publishers & Distributors Pvt Ltd

New Delhi • Bengaluru • Chennai • Kochi • Mumbai • Pune
Hyderabad • Kolkata • Nagpur • Patna • Vijayawada

CBS Publishers & Distributors Pvt Ltd

4819/XI, 24 Ansari Road, Daryaganj
New Delhi-110 002 (India)

IKON BOOKS, Publishers and Distributors
B-37, First Floor, Office No. 105
Street No. 14, Madhu Vihar
I.P. Extention, Delhi - 110 092
Telephone: 91-11-43618579
Email: ikonbooks@gmail.com
            cs@ikonbooks.com
            Website: www.ikonbooks.com

CONTROL SYSTEMS

First Edition: 2014

ISBN: 978-81-239-2388-8

Published by Satish Kumar Jain for
CBS Publishers & Distributors Pvt Ltd
4819/XI Prahlad Street, 24 Ansari Road, Daryaganj, New Delhi 110002, India.
Ph: 23289259, 23266861, 23266867   Website: www.cbspd.com
Fax: 011-23243014   e-mail: delhi@cbspd.com; cbspubs@airtelmail.in.
Corporate Office: 204 FIE, Industrial Area, Patparganj, Delhi 110092
Ph: 4934 4934   Fax: 4934 4935   e-mail: publishing@cbspd.com; publicity@cbspd.com

Branches

- **Bengaluru:** Seema House 2975, 17th Cross, K.R. Road,
  Banasankari 2nd Stage, Bengaluru 560 070, Karnataka
  Ph: +91-80-26771678/79   Fax: +91-80-26771680   e-mail: bangalore@cbspd.com
- **Chennai:** 20, West Park Road, Shenoy Nagar, Chennai 600 030, Tamil Nadu
  Ph: +91-44-26260666, 26208620   Fax: +91-44-42032115   e-mail: chennai@cbspd.com
- **Kochi:** 36/14 Kalluvilakam, Lissie Hospital Road, Kochi 682 018, Kerala
  Ph: +91-484-4059061-65   Fax: +91-484-4059065   e-mail: kochi@cbspd.com
- **Mumbai:** 83-C, Dr E Moses Road, Worli, Mumbai-400018, Maharashtra
  Ph: +91-9833017933   e-mail: mumbai@cbspd.com
- **Pune:** Bhuruk Prestige, Sr. No. 52/12/2+1+3/2 Narhe, Haveli
  (Near Katraj-Dehu Road Bypass), Pune 411 041, Maharashtra
  Ph: +91-20-64704058, 64704059, 32392277   Fax: +91-20-24300160   e-mail: pune@cbspd.com

Representatives

- **Hyderabad** 0-9885175004   • **Kolkata** 0-9831437309, 0-9051152362   • **Nagpur** 0-9021734563
- **Patna** 0-9334159340   • **Vijayawada** 0-9000660880

Printed at India Binding House, Noida, UP

PREFACE

This text book is the outgrowth of several years of experience in teaching the subject Control Systems to engineering students in India and Abroad. The book aims at presenting to the readers a complete understanding of the various principles and techniques involved in the analysis and synthesis of linear feedback control systems The book is intended to serve as a text for a first course in linear control systems to electrical, electronics, computer science, information technology, mechanical and chemical engineering and polytechnic courses. The book is also suitable for students of Grad. IETE, IE, HNC, UPSC and City and Guilds examinations. This book is very much useful to practicing engineers for self-study.

Chapter 1 reviews the history of automatic control systems and discusses its advantages. It also introduces the concept of open and closed loop control systems, deals with their merits and limitations and describes a number of practical feedback control systems to illustrate the diverse fields of application. The effects of feedback on system response, bandwidth, noise and distortion are also brought out clearly.

Chapter 2 deals with Laplace transform theorems and their applications. Laplace transforms of simple functions are derived. Partial fraction expansion method of obtaining inverse Laplace transforms is described and illustrated with suitable examples. The applications of Laplace transforms in solving differential equations are also explained.

Chapter 3 describes the method of formulating governing equations of electrical and mechanical translational and rotational systems and the representation of physical systems by transfer functions, block diagrams and signal flow graphs. Methods of obtaining the transfer functions of electrical, mechanical and chemical process control systems are dealt with in detail. Obtaining the overall gain by block diagram reduction technique and Mason's formula is illustrated extensively.

Chapter 4 brings out the analogies between mechanical and electrical systems and their importance. The steps in obtaining the electrical analogs of mechanical translational and rotational systems are explained and illustrated their suitable examples. Analog and digital simulations of system transfer function are also dealt with.

Chapter 5 describes the constructional features and operating characteristics of electrical, mechanical, hydraulic and pneumatic system components that are extensively used in control systems. Many practical systems using these components are also discussed in detail and their transfer functions are derived.

In Chapter 6, the standard test signals are described and the time-domain response of first-order, second-order and higher order systems to these test signals are discussed in detail. The performance specifications of systems is also defined and derived. The effects of adding poles and zeros are discussed.

In Chapter 7, the system types are defined and the methods of determining the steady-state errors for the three standard inputs are explained. The error performance indices such as ISE, ITSE, IAE and ITAE and their applications are discussed. The concept of system optimization is introduced.

Chapter 8 discusses the time-domain design of proportional, proportional and derivative, proportional and integral, PID and bridged-T controllers. In addition, minor loop and rate feedback designs are dealt with.

Chapter 9, the concept of system stability is introduced and the relationship between characteristic equations of systems and their stability is brought out clearly. The Routh-Hurwitz stability criteria are explained in detail and its applications are illustrated with suitable examples.

Chapter 10 introduces the concept of root locus and describes the rules for constructing the root locus with illustrative examples. The effect of adding poles and zeros on the root locus and system stability are discussed in detail. The root contours are also explained and illustrated with an example.

Chapter 11 describes the frequency domain response of systems. Three forms of responses-Bode plots, Polar plots and Lm-ϕ diagram are explained. The advantages of frequency domain analysis are brought out clearly. A new concept of relative stability- stability ratio- is introduced and its merits and applications are illustrated.

Chapter 12 deals with the development of Nyquist path, Nyquist stability criterion and its applications to system stability. The method of drawing the complete Nyquist plot is extensively described and illustrate with suitable examples. The relative stability is defined and the methods of obtaining the relative stability of systems from the Nyquist plot are discussed in detail and illustrated with large number of examples.

Chapter 13 describes the development of M and N loci in the frequency domain and discusses the method of deriving the closed loop frequency response of systems from M and N circles. The method of adjusting the system gain for a specified value of M_o is also discussed in detail.

In Chapter 14, the need for compensators to improve the performance of systems is explained and the various types of compensators and their design in frequency domain are discussed. The design of compensators is illustrated in detail with suitable examples. Their merits and limitations are brought out clearly.

Chapter 15 introduces the concept of state variable representation of systems and explains the derivation of state equations of systems, their transformation to phase-variable canonical forms and state diagrams. It also deals with the solution of homogeneous and non-homogeneous state equations by transfer function decomposition, Laplace transform matrix exponential, series summation, diagonal matrix, Cayley-Hamilton theorem and numerical methods. The concept of controllability and observability is introduced and its applications to systems are illustrated with suitable examples. The state variable representation of linear time-varying systems is discussed. To improve the performance of systems, the methods of designing the compensators by state feedback control and pole placement design approach are dealt with in detail. State transition matrix and general solution of state equations are also presented.

Chapter 16, MATLAB commands are introduced and the method of preparing MATLAB programs useful in the analysis and design of control systems is dealt with in detail. About 45 MATLAB programs are developed. All these programs are tested and test results are obtained.

A large number of illustrations and figures are provided for clear understanding of the concepts. Exercises, problems and end of the chapter review questions will assist the reader to understand and grasp the concepts and principles with ease. Constructive suggestions and comments are welcome.

AUTHOR

ACKNOWLEDGMENT

This Book on *Control Systems* is the outcome of teaching the subject to engineering degree students for many years and the feedback received from them. The students have played a greater part in the preparation of the manuscript by way of persuasion, encouragement and assistance. The author wishes to extend his whole-hearted thanks to all his students who have attended his lectures and the feedback he has received from them.

The author also thanks all his colleagues who were very helpful during the preparation of the manuscript and their valuable suggestions and guidance.

The author extends his sincere thanks to the publisher and their staff for bringing up this book in time. The author wishes to express his special thanks to Mr. A.K. Bharti for his continuous interaction and supervision of this project.

Finally, the author thanks his wife, Mrs. Gajalakshmi, B.Sc., B.Ed. for her forbearance, patience and assistance in the preparation of the manuscript and proof reading.

CONTENTS

Chapter 1

INTRODUCTION TO FEEDBACK CONTROL SYSTEM

1.1. INTRODUCTION

A system is a combination of interacting elements or components which act together to achieve a common objective. A few common examples of systems are transportation, educational and communication systems. Systems engineering is concerned with the analysis, design and development of systems to perform various tasks. System may be broadly classified as *controllable* and *uncontrollable*. A communication system or a power system is controllable whereas pollution of air is uncontrollable. Systems which are controllable are known as *control systems*. Control system engineering deals with the analysis, design and development of controllable systems. If a system is controlled automatically without the intervention of a human operator, it is called an *automatic control system*. A control system may be as large as a chemical plant or as small and simple as an electric toaster.

In recent years, automatic control systems have assumed an increasingly important role in all fields because of the following advantages over manual systems.

1. They relieve one of many monotonous activities so that he/she can devote his/her abilities to more productive and developmental endeavours.

2. They perform functions at a fast rate unmatched by human beings and unaffected by environmental disturbances.

3. They can function under hazardous conditions which are harmful and fatal for human life.

4. Their performances are most reliable and can be standardized.

5. The accuracy and repeatability of their performance are very high.

6. Though the initial costs of automatic control systems are high, they are cheaper in the long run.

7. Manufactured products are of uniform quality.

8. Production is increased due to less rejection of manufactured products.

9. Large savings in materials result due to low rejection.

10. Cost of production is reduced due to increased production.

Because of the many advantages, automatic control systems find extensive applications in almost all areas. In the domestic field, automatic controls are used in automatic toaster, in heating and air-conditioning systems. In defense application, it finds extensive use in anti-aircraft gun control, guided missiles and other weapon systems. In industries, it is used to control the quality of manufactured products. In machine – tool applications, it is used in numerical control machines and in computerized NC machines. In thermal power station, it is used to control the turbine speed and the frequency of generators. In nuclear power station, it is used to control the power level of the nuclear reactors. In space technology, it finds application in space travel missions and moon landings. In robotic engineering it is used in positioning and orienting the robot arms. In process industries, it is used to control the various process parameters as well as to design massive industrial process equipments. In the inertial guidance system, it is used in the design of delicate instruments.

Because of myriad application of automatic control systems, the demand imposed on their reliability, precision and speed of operation has also increased. Fortunately, many of the demands have been met with due to simultaneous advances in the various branches of engineering

1.2. BRIEF HISTORICAL REVIEW

Feedback control system is as old as human race, since the human beings are bestowed by nature with a set of biological controls. Consider a simple example of following a prescribed path. When one tries to follow this desired path, the actual path taken by him deviates from the desired path. The eyes compare the actual path (output) with the desired path (reference input) and transmit a signal to the brain (controller). The brain generates control signal and transmits it to the legs. It adjusts the actual path so as to bring it in line with the desired path in order to reduce the deviation. Thus, the eyes, the brain and legs constitute the feedback control system.

The earliest practical control system was the *float-controlled irrigation system* of the Babylonian Civilization (2000 BC). The next practical system was a fan-tail arrangement devised in Holland in 1750 to turn the large windmills into the wind for maximum operating efficiency. However, the significant milestone in the early history of control systems was the development of the fly-ball governor by James Watt in 1788. The fly-ball governor used to control the speed of steam engines is shown in Fig. 1.1.

The amount of steam admitted to the engine cylinder is proportional to the difference between the desired and actual speeds. The set point is the reference input and is set according to the desired speed. If the actual speed of the engine increases above the desired value, the centrifugal force of the governor is increased and control valve is moved downwards, thus reducing the steam supply. The speed decreases until the desired value is reached. On the other

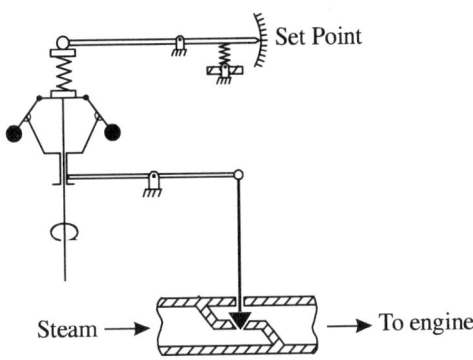

Fig. 1.1. Fly-ball governor

hand, if the speed decreases the centrifugal force of the governor decreases and the control valve is moved upward increasing the steam supply. The speed of the engine increases until it reaches the desired value. Thus the corrective action is proportional to the difference between the desired and actual speeds. This is known as *proportional control* and it gives rise to steady-state error in the engine speed. Attempts to eliminate the error led to integral control. However, when it was used with steam engines, they became unstable. Since no significant theoretical foundation was available during this period of evolution, the methods of control were mostly based on one's experience and intuition. The only exception was the classical paper entitled "On Governors" by J.C. Maxwell in 1868. He made an analytical study of the fly-ball governors and their stability. In 1877, R.J. Routh presented a theory on "Stability of Given State of Motion" which had won the Adams Essay Prize. In 1922, Minorsky, in his study of automatic ship steering system, deliberately made use of nonlinear elements in the closed-loop systems. In 1932, Nyquist published his study of stability theory providing a graphical method to determine system stability. This was followed by Black's important paper on "Stabilized Feedback Amplifiers" in 1934. In the same year, Hazen published his paper on "Theory of Servo-mechanism", thus originating the term *servomechanism* for slave mechanism. This publication created intense interest in control systems. However, the developments during the World-War II were obscured due to security reasons. In 1950s, developments in control theory and their applications were spurred by several factors. Perhaps the most significant was the emergence of analog and digital computers which promoted the design of complex controllers and interest in sample data systems.

The control theory developed through the late 1950's is categorized as *classical control theory*. This is effectively applied to many control design problems, especially to *single-input* and *single-output* systems. The control theory developed since the late fifties for the design of more complicated systems such as *multiple-input* and *multiple-output* systems is called *modern control theory*. Problem areas which are very important in space travel such as trajectory optimization, minimum – time and/or minimum-fuel problems can readily be handled by modern control theory. Because of the advent of modern control theory, it was possible to investigate the Universe by landing explorers on the moon and instrument-packages on our neighbouring planets.

In additions, the concept of feedback control finds extensive use in metallurgy, agriculture, inventory, biological, biomedical, physiological, social, economic and political processes.

1.3. OPEN-LOOP AND CLOSED-LOOP SYSTEMS

Fig.1.2 shows the block diagram of a control system. Its main function is to maintain the controlled variable or the output $c(t)$ in close correspondence with the reference input $r(t)$ which is frequently or continuously changed. If the operation of the system solely depends on the reference input $r(t)$ and independent of the output $c(t)$, then the system is called *open-loop control system*. If, on the other hand, the operation of the system depends on the reference input as well as the output, then the system is said to be *closed-loop control system*. In the closed-loop system, a fraction or a function of the output $c(t)$ is fed back into the system, compared with reference input and the difference or the *error* is used to implement necessary corrective control action. Whereas, in the open-loop system, the output is not measured and compared with the reference input, and hence no feedback exists. It relies for its action on the settings and calibration of its components.

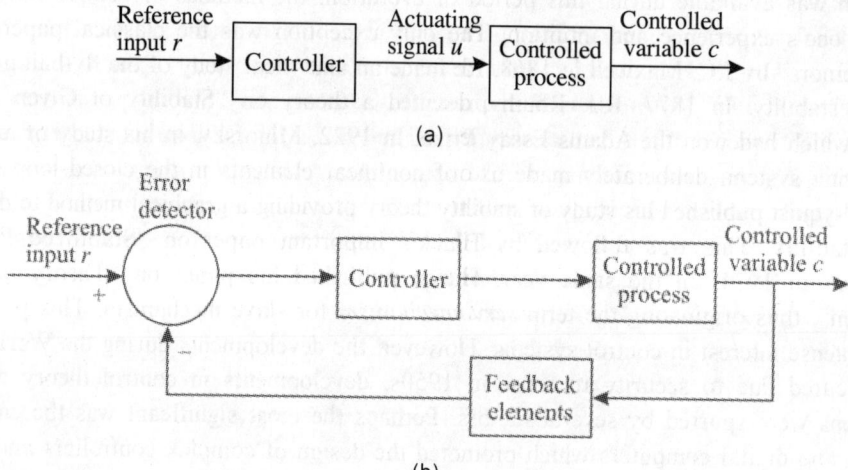

Fig. 1.2. Block diagram of (a) Open-loop control system (b) Closed-loop control system

An example of open-loop system is time-operated traffic signals as shown in Fig 1.3. It consists of a controller and the process.

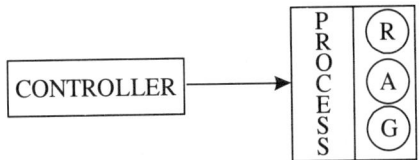

Fig. 1.3. Time-operated traffic control

The process variable is the time duration for which each lamp glows which ultimately controls the actual flow of traffic. The controller switches the set of lamps ON or OFF in a particular sequence for a preset time irrespective of the volume of traffic present in each direction. When there is heavy traffic waiting in one direction, the green lamp in that direction does not glow automatically for a longer duration to permit more traffic flow. Or when there is no or little traffic, the duration of the green lamp glow is not automatically reduced. Thus there is no device in the traffic signal system to measure the volume of traffic in each direction and control the signals accordingly. In other words, there is no feedback of the volume of the traffic present to the controller. Hence this is an open-loop control system. However, if the volume of the traffic is measured by means of vehicle-detectors and fed to the controller, the duration of the green lamp glow can be effectively controlled accordingly. Then the system becomes a closed-loop system.

As a second example, consider the operation of a dc shunt motor shown in Fig. 1.4. The system consists of a potentiometer P, a dc amplifier A, the motor M and the load. The potentiometer is calibrated in terms of the desired motor speed. This voltage is amplified by the dc amplifier whose output excites the field of the motor. If the load is changed, the speed of the motor also changes. Thus, the speed of the motor is not maintained at the desired value when the load fluctuates. This is because of the fact that there is no device in the system which measures the motor speed, and if the speed is not same as the desired speed, no control action is initiated to restore the speed to the desired value. Therefore the system is an open-loop system.

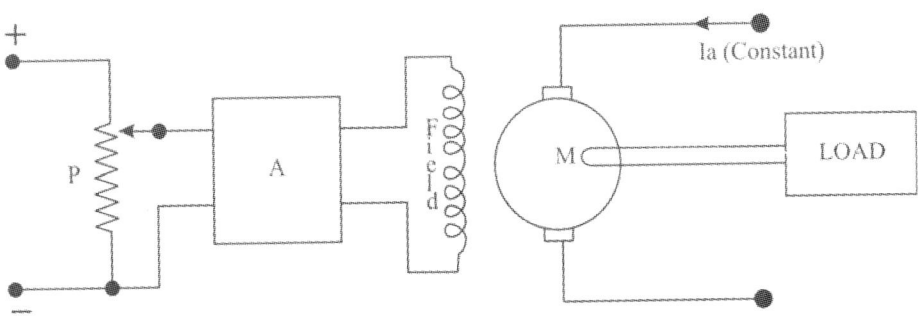

Fig.1.4. D.C. Shunt motor

Let us introduce a d.c. tachogenerator in the system as shown in Fig 1.5. It functions as a feedback element. It generates an output voltage which is proportional to the speed of the motor. This feedback voltage is compared with the input and the difference or error voltage is amplified by the dc amplifier.

When motor runs at the desired speed, the feedback voltage is such that, the error voltage supplies the required field current. Due to load fluctuation if the speed changes, corrective actions are initiated. For example, if the speed increases, the feedback voltage is also increased. This decreases the field current and the motor torque reduces. This reduction in the torque restores the speed of the motor to the desired value. If the speed decreases, a reverse process sets in and the speed of the motor is again restored. Thus this system becomes a closed-loop system.

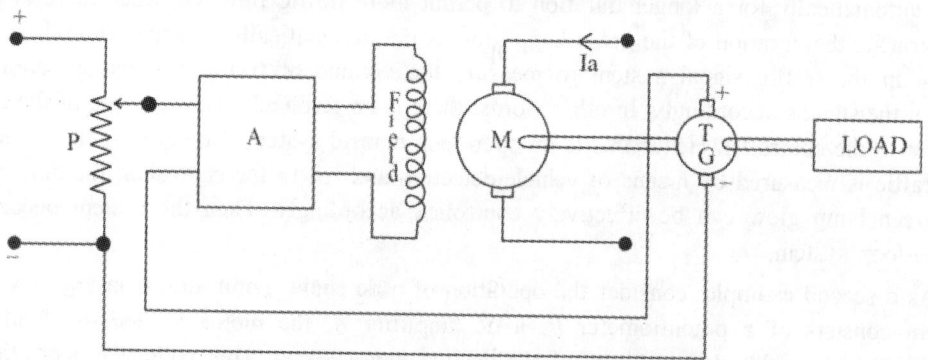

Fig. 1.5. D.C. shunt motor with feedback

1.3.1. Comparison between Open-loop and Closed-loop Systems

In the open-loop system, the control acts in accordance with the dictates or commands of some arbitrary quantity, and fidelity of action is depended wholly on the linearity of the mechanism, and/or on calibration. *In the closed-loop system, the control acts in accordance with the dictates of an arbitrary quantity, and also in accordance with what has happened as a result of the control operation.* Linearity of the mechanism and calibration occupy only a secondary role.

Open-loop systems require careful attention to insure their correct continued operation. Many factors can alter their performance as precision instruments. They must be calibrated and maintained in calibration, against wear and changes in environment such as temperature, humidity, friction or vibration. This class of control systems is called open-loop because there is nothing in the mechanism which actually measures the result of control operation and does something about it, if the result is not what is desired. On the other hand the essential requirement of the closed-loop control system is that the error between the state desired and the state existing is constantly measured and it there is any error, some thing is done about it to reduce the error. Thus *the closed-loop system is an error sensitive system.*

Open-loop systems are the simplest and most economical and are free from problems of instability. However, they are seldom used because they are usually inaccurate and unreliable. The open-loop system must be calibrated first in order to perform some specific task, and the perfection of operation depends upon the maintenance of this calibration. In other words, the open-loop system operates without the knowledge of precision it attains during operation. In contrast, the closed-loop system operates in terms of the error, and tends to minimize the error at all times. Thus such a system knows the accuracy of its performance at all times during operation and continuously tends to keep its output in agreement with its command. However, the closed loop system has a tendency to oscillate because of the inherent error sensitive (feedback) property.

The important features of open-loop and closed-loop systems are summarized below in Table 1.1.

Table 1.1. Comparison of open-loop and closed-loop systems

Open-loop system	Closed-loop system
1. Control acts in accordance with the reference input or the command.	Control acts in accordance with the reference input and also in accordance with what has happened as a result of control action.
2. There is no device in the system which measures the output and takes corrective action if the output is not in the desired state. Thus there is no feedback and the system is not error sensitive.	The output is continuously measured, and if there is any error, corrective action is taken to reduce the error. Thus there is feedback and the system is error sensitive.
3. They must be calibrated and maintained in calibration against wear and tear, and changes in environment such as temperature, humidity, friction and vibration.	External disturbances and internal parameter variations do affect the system performance but taken care of by the feedback mechanism.
4. The fidelity of action is dependent solely on the linearity of the mechanism and calibration.	Linearity of mechanism and calibration occupy a secondary role.
5. Simple and economical	Complex and costlier.
6. Free from problems of instability	Liable to become unstable because of inherent feedback.
7. System non-linearities tend to degrade the performance	Effect of non-linearities is less because of feedback.

1.4. SERVOMECHANISM AND REGULATOR

A Servomechanism is a class of feedback control system in which the controlled variable or the output is usually a *mechanical quantity* such as *position, velocity* or *acceleration*. The reference input may vary over a wide range and the servomechanism functions to hold the output at the desired value anywhere within this range. It can thus cope up with random variations of command as well as random disturbances. The system performance is a function of error. The servomechanism is thus error sensitive and eventually the error is reduced to zero. Thus the *servomechanism is an error-sensitive follow-up system permitting wide range of input, and commanding remotely located elements.*

A regulator is a feedback system in which the controlled quantity is *other than mechanical quantity* such as *voltage, temperature* or *pressure*. Its reference input remains relatively constant. It usually functions to maintain the output at a fixed level established by the setting of the reference input. Disturbances, if any, are usually due to variation in system parameters. In a regulator the error is necessary to generate the power required to produce the output. Thus, an error signal is always present to maintain the output.

1.4.1. Comparison between Servomechanism and Regulator

Comparison between servomechanism and regulator is summarized below in Table 1.2.

Table 1.2. Comparison between servomechanism and regulator

Servomechanism	Regulator
1. The output of the servomechanism is usually a mechanical quantity such as position, velocity or acceleration.	The output of the regulator is usually other than mechanical quantities such as voltage, temperature speed etc.
2. The reference input varies over a wide range	The reference input remains relatively constant.
3. The servomechanism can cope up with random input variations as well as random disturbances.	The regulator cannot cope up with random input variations and disturbances
4. The system error ultimately reduces to zero.	The system error is necessary for proper operation.

1.5. TERMINOLOGY AND DEFINITIONS

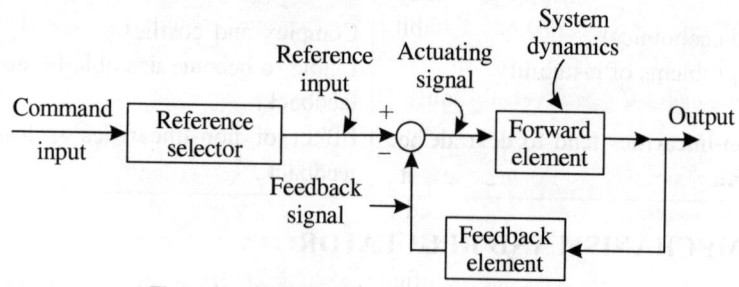

Fig. 1.6. Block diagram of a feedback system

The functional block diagram of a feedback system is shown in Fig. 1.6. Reference selector, comparator and controller are in the forward path of the energy flow, and the feedback element is in the feedback path. The direction of energy flow along with the associated signals is also indicated.

Command input: This is the motivating input signal to the system and is independent of the system output.

Reference Selector: This unit, also called reference input element, establishes the value of the reference input. It is calibrated in terms of the desired value of the system output.

Reference Input: The reference input signal is produced by the reference selector. By virtue of its relation with the command input, it constitutes, at all times, a standard or reference with which the feedback signal is compared.

Comparator (error detector): This unit compares the feedback signal with reference input. Its output is the difference of the two and is known as error signal.

Actuating or Error Signal: The error or actuating signal is the difference between the reference input and the feedback signals and is produced by the comparator. It actuates the controller in order to cause the output to have the desired value.

Controller (forward element): This is the dynamic unit of the system and is in the nature of a power device. It reacts to the actuating signal to produce the desired output.

Controlled Output Variable: It is the quantity to be maintained at prescribed value in accordance with the reference input.

Feedback Element: This is the unit that provides the means of feeding back a fraction or a function of the output to the comparator in order to compare it with the reference input.

Feedback Signal: This is the signal produced by the feedback element. This is compared with the reference input to produce an actuating signal.

1.6. FEEDBACK AND ITS EFFECTS

In the previous sections, it has been stated that the motivation for using feedback is for the purpose of reducing the error between the desired output and the actual system output. The reduction of error is merely one of the many important effects of feedback. In this section we shall discuss its effect on overall gain, stability, sensitivity and external disturbance or noise.

1.6.1. Effect of Feedback on Overall Gain

The existence of feedback in a system can be easily identified when it is introduced deliberately for the purpose of control. In some cases, it cannot be seen explicitly. In general, feedback is said to exist whenever there exists a closed sequence of cause-and-effect relationships among the system variables. For investigating the effect of feedback on system performance, let us consider the negative feedback system configuration shown in Fig.1.7. Here r in the reference input, c is the controlled variable or the output, e is the error and b is the feedback signal. G and H are forward and feedback gains respectively and are considered here as constants.

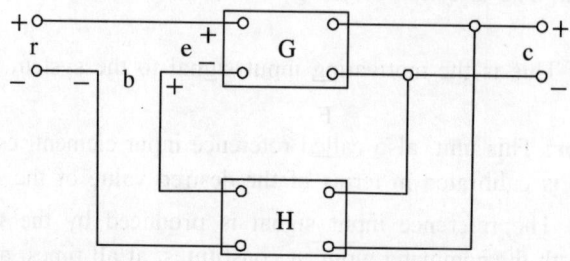

Fig. 1.7. Negative feedback system

The input-output relationship of the system is derived below.

$$b = Hc, \qquad e = (r - b) = (r - Hc)$$
$$c = Ge = G(r - Hc)$$

or,
$$\frac{c}{r} = M = \frac{G}{1 + GH} \qquad\qquad ... (1.1)$$

where M is the overall gain of the system. In the absence of feedback, $H = 0$ and hence $M = G$. With the feedback present, M is given by Eq. (1.1). *Thus the negative feedback reduces the non-feedback system gain by a factor of* $(1 + GH)$. Since the error is the difference between r and b, the system is a negative feedback system and hence the over-all gain is reduced by $(1 + GH)$. In a practical control system, G and H are functions of frequency. Therefore the magnitude of $(1 + GH)$ may be greater than 1 during one frequency range and less than 1 in another range. Depending upon this, the overall gain may decrease or increase in the frequency range of interest.

1.6.2. Effect of Feedback on Stability

In general, *a system is said to be stable if its output is bounded for bounded input, and unstable if its output is not bounded for bounded input.*

To investigate the effect of feedback on system stability, refer to Eq. (1.1). If $GH = -1$, the overall gain is infinity and the system output is unbounded. Hence system becomes *unstable*. This happens when the feedback becomes positive. This unstable system can be stabilized by introducing another feedback loop of gain F as shown in Fig. 1.8. The overall gain of this system is given by

$$M = \frac{c}{r} = \frac{G/1 + GH}{1 + \dfrac{GF}{1 + GH}} = \frac{G}{1 + GH + GF}$$

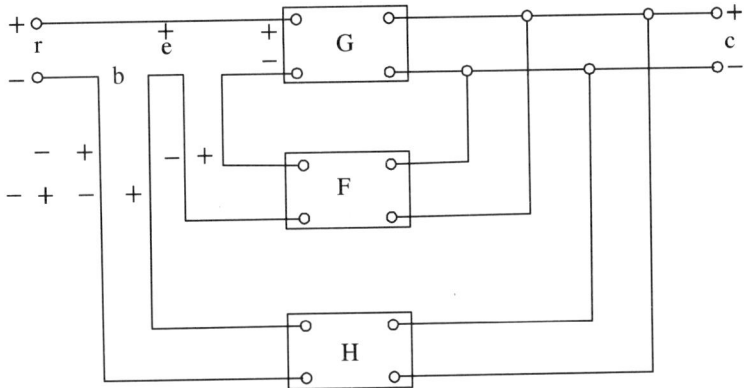

Fig. 1.8. Feedback system with a second or minor feedback loop

The system can be stabilized by properly selecting the outer-loop gain F such that $(1 + GH + GF)$ never becomes zero even though $GH = -1$.

1.6.3. Effect of Feedback on Sensitivity

The parameters of a control system may vary during the operating life of the system due to aging of components and changes in the environment. To investigate the effect of negative feedback on sensitivity of parameter variations, consider G as a variable parameter in Eq. (1.1). The sensitivity of overall system gain M due to variations in G is defined as

$$S_G^M = \frac{\partial M / M}{\partial G / G} \qquad \dots (1.3)$$

Here ∂M represents the incremental change in M due to the incremental change in G. $\partial M/M$ and $\partial G/G$ are the percentage changes in M and G respectively.

Thus the sensitivity of the system is defined as the ratio of percentage change in the system transfer function M to the percentage change in the process transfer function G.

From Eq. (1.1),

$$M = \frac{G}{1 + GH}$$

$$\frac{\partial M}{\partial G} = \frac{\partial}{\partial G}\left\{\frac{G}{1 + GH}\right\} = \frac{(1 + GH) - GH}{(1 + GH)^2} = \frac{1}{(1 + GH)^2}$$

$$S_G^M = \frac{\partial M}{\partial G}\frac{G}{M} = \frac{1}{(1 + GH)^2} \times \frac{G(1 + GH)}{G}$$

$$= \frac{1}{1 + GH} \qquad \dots (1.4)$$

Eq. (1.4) shows that, by increasing GH, the sensitivity function can be made very small provided that the system remains stable. In general, a good control system should be very insensitive to parameter variations.

1.6.4. Effect of Feedback on External Disturbance or Noise

During operation, all practical control systems are subject to extraneous unwanted signals such as thermal noise voltage in electronic amplifiers and commutator noise in electric motors.

No general conclusions can be made regarding the noise reduction by feedback because it depends greatly on where, in the system, the noise is introduced or generated. In many situations, however, negative feedback can reduce the effect of noise on the performance of the system.

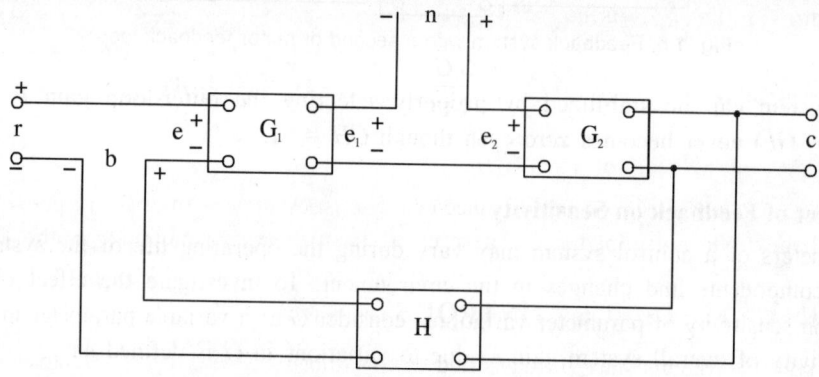

Fig. 1.9. Negative feedback system with noise

Fig. 1.9 shows a feedback system with a noise signal n. In the absence of feedback, $e = r$ and the output c is given by

$$c = G_1 G_2 r + G_2 n \qquad \text{... (1.5)}$$

The signal-to-noise ratio (SNR) of the output is defined as

$$SNR = \frac{\text{Output due to signal}}{\text{Output due to noise}} = \frac{G_1 G_2 r}{G_2 n} = G_1 \frac{r}{n} \qquad \text{... (1.6)}$$

The output in the presence of feedback is given by

$$c = \frac{G_1 G_2 r}{1 + G_1 G_2 H} + \frac{G_2 n}{1 + G_1 G_2 H} \qquad \text{... (1.7)}$$

Eq.(1.5) and (1.7) show that the noise component in the output is reduced by a factor of $(1 + G_1 G_2 H)$. But the signal-to-noise ratio remains same since the signal component is also reduced by the same factor. However, under certain conditions, the SNR can be improved.

Assume that G_1 and r are increased to G'_1 and r' respectively with all other parameters unchanged. Also assume that the output due to signal acting alone is at the same level as that when the feedback is absent. In other words,

$$c_{n=0} = \frac{G'_1 G_2 r'}{1 + G'_1 G_2 H} = G_1 G_2 r \qquad \qquad ... (1.8)$$

The output due to noise signal acting alone is given by

$$c_{r=0} = \frac{G_2 n}{1 + G'_1 G_2 H} \qquad \qquad ... (1.9)$$

Comparison of Eq. (1.9) with Eq. (1.7) shows that the noise component in the output is reduced with G'_1. The *SNR* is now

$$SNR = \frac{G_1 G_2 r}{G_2 n / [(1 + G'_1 G_2 H)]} = (1 + G'_1 G_2 H) G_1 \frac{r}{n} \qquad \qquad ... (1.10)$$

which is higher by a factor of $1 + G'_1 G_2 H$.

Feedback also has effects on performance characteristics such as transient response, frequency response, bandwidth and impedance. These effects will be discussed when the need arises.

1.7. TYPES OF FEEDBACK CONTROL SYSTEMS

Feedback control systems may be classified in a number of ways according to some special features of the system. Based on the method of analysis and design, they may be classified as linear and non-linear, time-varying and time invariant; according to the types of signal in the system as continuous, on-off and sampled-data systems; based on the type of system components as electromechanical, thermal, hydraulic, pneumatic and biological control system, and according to the output variable as position control and velocity control systems. Based on the type of controllers (computers), they may also be classified as analog and digital control system. We shall briefly outline below the classification based on the types of signal in the system.

1.7.1. Continuous-Data Control System

A continuous-data system is one in which the signals at all points are *continuous functions of time*. In this system, the input signal to the controller is zero only when the error signal is exactly zero. The greater the deviation between the desired output and the actual output, the larger the error is. The corrective control action is directly proportional to the error. Although physical systems are non-linear to some extent, they do have a definite range of linear operation as shown in Fig. 1.10.

A continuous–data system may further be classified as a.c. and d.c. control systems. In an a.c. control system, the signals in the system are modulated signals. Typical components of a.c.

control systems are: synchros, a.c. amplifiers, a.c. motors, gyroscopes and accelerometers. In a d.c. control system, the signals are un-modulated but are varying. Typical components of a d.c. control systems are: potentiometers, d.c. amplifiers, d.c. motors and d.c. tachogenerators. Servomechanisms and regulators are examples of continuous–data control systems.

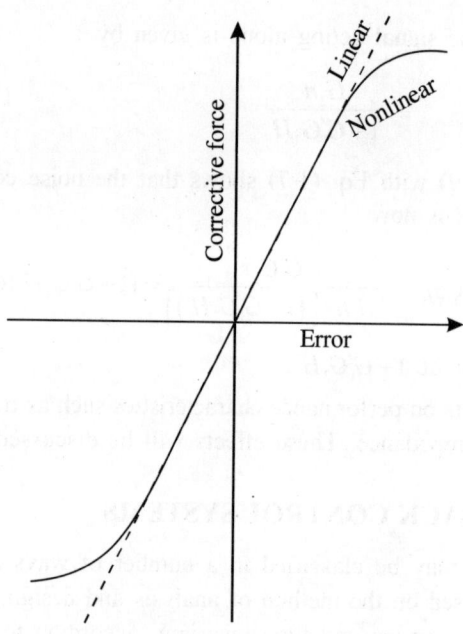

Fig. 1.10. Linear operation of physical system

1.7.2. On-Off or Relay Control System

This is a discontinuous type of control system in which the signals in parts of the system are rectangular pulses whose polarity depends on the polarity of the error signal. In this system, there is a *dead region in which no corrective control action is applied* (Fig. 1.11). However, when the amplitude of the error signal is equal to or greater than ±*E*, full corrective action is applied to the system. Because of this type of operation, they are called bang-bang control systems. Examples of bang-bang system are temperature control system and missile or space-craft control system.

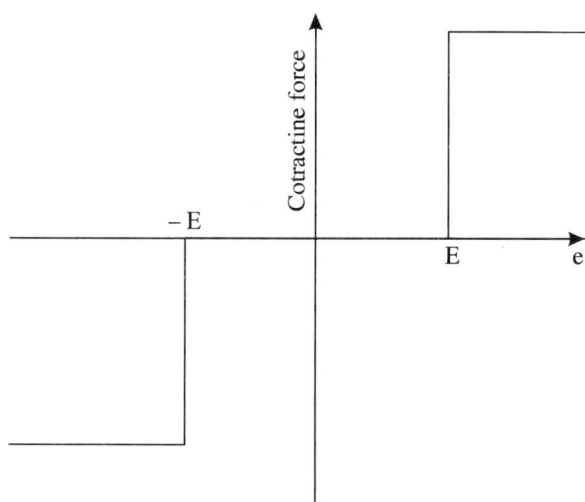

Fig.1.11. On-Off control system

1.7.3. Sampled-data Control System

This is also a discontinuous type but the *discontinuity is in time rather than amplitude*. The signals in sampled-data system are in the form of pulsed data. Fig 1.12 shows the operation of a typical sampled-data system. The error signal $e(t)$ is sampled by a sampler and the output of the sampler is a sequence of pulses. There are many advantages of incorporating a sampler. One important advantage is that the expensive equipment can be shared on time-shared basis among several control channels.

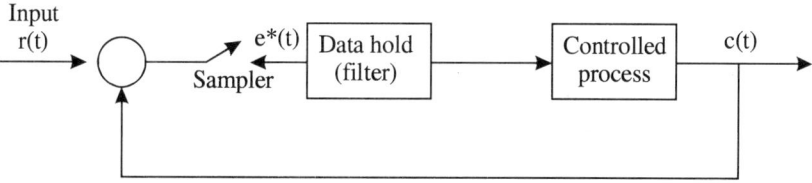

Fig. 1.12. Sampled-data control system

1.8. ILLUSTRATIVE EXAMPLES OF CONTROL SYSTEMS

In this section we shall discuss the operation of some typical control systems in order to have a better understanding of the physical concept of feedback control, their operation and their applications in diverse fields.

1.8.1. D.C. Generator Voltage Regulator

The output voltage of a d.c. generator depends on its field current and the prime mover speed. Due to fluctuations in either, the output voltage may change. In order to maintain the output voltage at the desired value, the field current must be automatically adjusted in such a way as to counter-act the fluctuations and restore the output voltage to the desired value. This must be caused by the change in the output voltage for proper operations. Fig. 1.13 shows a d.c. generator voltage regulator system. It consists of a pair of potentiometers P_1 and P_2, a reference voltage source V, an amplifier A and the d.c. generator G. The reference voltage source is the command input. P_1 functions as the reference selector and is usually calibrated in terms of the desired output voltage. P_1 is set according to the output voltage desired and this fixes the reference voltage e_r. P_2 is the feedback element through which a fraction $h = [R_2/(R_1 + R_2)]$ of the output voltage is fedback to the amplifier. The feedback voltage e_f is equal to he_o. The pair of potentiometer P_1 and P_2 together function as the error detector or comparator. The error voltage e which is the difference between e_r and e_f is amplified by the amplifier. The amplifier output voltage is used to excite the generator field.

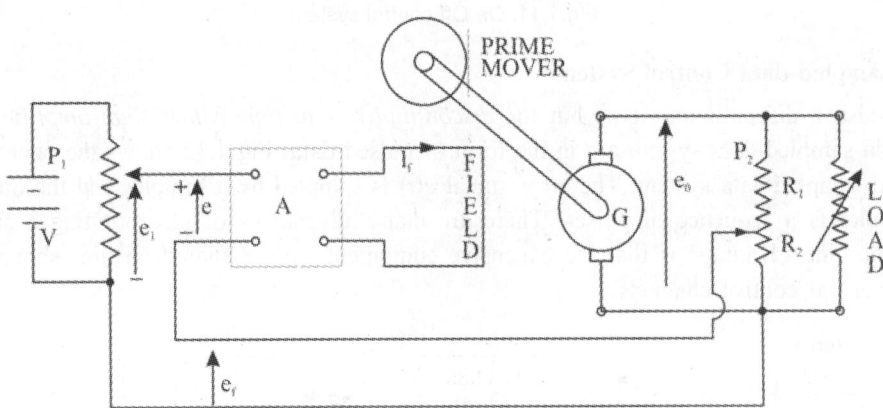

Fig. 1.13. D.C. generator voltage regulator

Due to fluctuation in load and/or speed, if the output voltage decreases, it is reflected in the feedback voltage e_f since $e_f = he_o$. Thus the error voltage is increased causing the field excitation to increase. This, in turn, increases the output voltage and restores it to the desired value. On the other hand, if the output voltage increases above the desired value, a reverse process is set up and the output voltage is restored to the desired value. Thus the output voltage of the d.c. generator can be closely regulated and maintained despite fluctuations in load and/or prime-mover speed.

1.8.2. Position Control System

In a position control system, the output position is required to follow the movement of the input shaft position as closely as possible. Fig. 1.14 shows the schematic diagram of a d.c. position control system. It consists of a pair of potentiometers P_1 and P_2 which function as transducers and error detector. The reference voltage source V is the command input. The input potentiometer P_1 is the reference selector and provides a reference input voltage e_r proportional to the angle through which the input shaft is rotated and the output shaft is required to rotate. The error voltage e is amplified by the d.c. amplifier which provides the field current to drive the d.c. motor. The armature current of the motor is assumed to be constant. A gear system is used for speed reduction. The output shaft is coupled to the output potentiometer P_2 which provides a voltage proportional to the angular position of the output shaft. This voltage e_f is the feedback voltage and is compared with reference input voltage e_r.

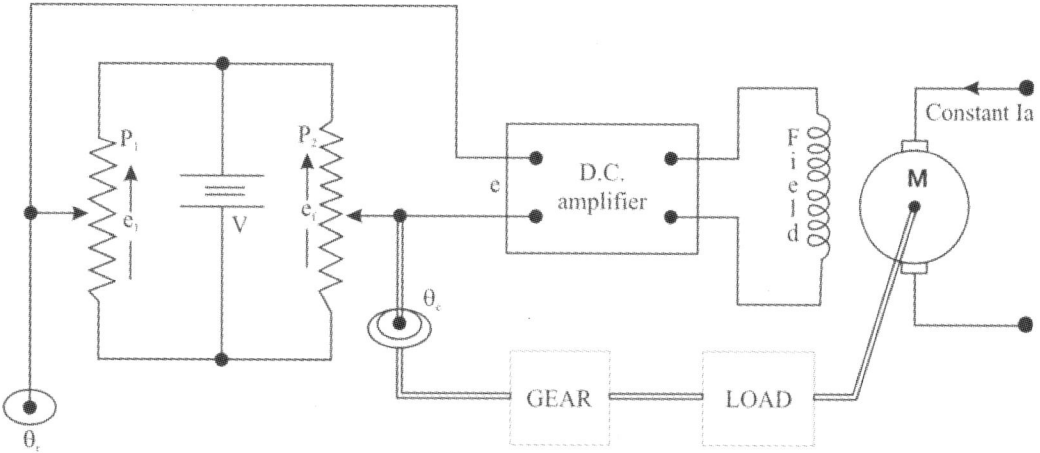

Fig. 1.14. Position control system

When the input shaft is moved through an angle θ^0, an error voltage appears at the amplifier input since the output shaft has not rotated yet. This causes the motor to rotate the output shaft until the output shaft is brought in line with the input shaft position. Since the two shafts have aligned, error voltage now becomes zero and the motor stops.

Position control system finds extensive application in defense such as radar tracking, gun directors, missile launchers and guidance systems. In industrial and commercial applications, the position control is used in machine tools, profile cutting, positioning of antennas, servo-multipliers and automatic pilots. In scientific research, it is used for automatic handling of radio-active materials, remote control of objects such as T.V. Cameras and mechanical robots.

1.8.3. Process Control System

In process control system, process parameters such as temperature, pressure, flow or liquid level are under control. An example of such a system is a chemical plant. Consider that some chemicals are to be mixed up in a tank at a constant temperature. This temperature is to be monitored and maintained constant. Fig. 1.15 shows a simplified schematic and functional diagram. The potentiometer P is usually calibrated in terms of the desired temperature and is set according to the desired temperature. This setting provides the reference input e_r. A thermocouple is used to sense the tank temperature. Its output voltage is very low and hence is amplified by a high gain amplifier A_2. The output of A_2 is proportional to the tank temperature. This feedback voltage is compared with the reference input and the difference or error voltage is amplified by the amplifier A_1. The output voltage of A_1 energizes the solenoid and the valve position is adjusted accordingly. This varies the supply of hot air to the tank, thus maintaining the temperature at the preset value.

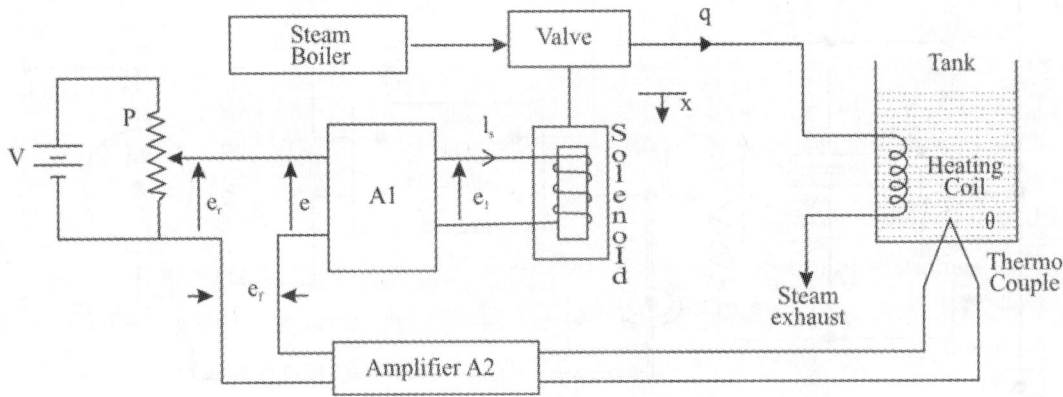

Fig. 1.15. Temperature control system

If the tank temperature increases above the desired value, the feedback voltage increases thus reducing the error voltage. This results in a reduced force on the solenoid valve and the opening of the valve is reduced and hot air supply is also reduced. The temperature of the tank decreases until it reaches the preset valve. If the tank temperature decreases, a reverse process is set in and the temperature is maintained constant.

1.8.4. Learning System

The concept of feedback control may be applied to non-engineering fields as well. For example, consider the student-teacher learning process. The teacher imparts certain amount of knowledge to the student. The student in the process of learning acquires a good amount of the imparted knowledge. The teacher measures the acquired knowledge of the student by questioning and obtaining the students response. This feedback information is compared by the teacher with the knowledge imparted and if the knowledge acquired is equal to the desired level, he is satisfied. If not, he takes necessary action so that the acquired knowledge is equal to the desired level. In this system, the objective is to reduce the system error to a minimum and generally a minimum amount of error is always present.

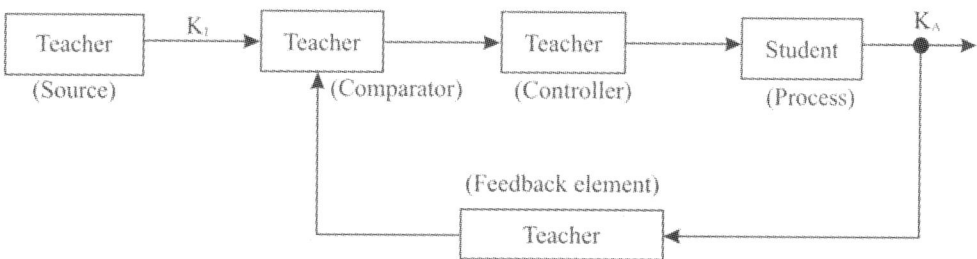

Fig. 1.16. Schematic representation of learning system

The functional block diagram of the learning process is shown in Fig. 1.16. The imparted knowledge K_I is the reference input, the teacher functions as comparator, controller and feedback element and the student as the process. The desired output is the acquired knowledge K_A.

1.9. SOLVED EXAMPLES

Example 1.1. *The forward and feedback transfer function are G = 10 and H = 1 respectively. Determine the overall gain of the system.*

Solution:

Given: $\qquad\qquad G = 10, H = 1$

Overall gain $\qquad M = \dfrac{G}{1+GH} = \dfrac{10}{1+10} = 0.91$

The overall gain is reduced by 11 times compared to non-feedback gain G of 10.

Example 1.2. *For the system in Ex. 1.1, the bandwidth is B rad/sec. without feedback. What is the bandwidth with feedback?*

Solution:

The gain-bandwidth product is constant. Since the gain is reduced by $(1 + GH)$ times, the bandwidth will be increased by the same amount to maintain the gain-bandwidth constant. The new bandwidth B′ is :

$$B' = (1 + GH) B = 11 B \text{ rad/sec.}$$

Example 1.3. *Determine the sensitivity of the closed loop transfer function M with respect to the feedback transfer function H.*

Solution:

$$S_H^M = \frac{\partial M/M}{\partial H/H} = \frac{\partial M}{\partial H} \cdot \frac{H}{M} = G \cdot \frac{-G}{(1+GH)^2} \cdot \frac{H}{G/(1+GH)} = \frac{-GH}{1+GH}$$

Example 1.4. *The forward and feedback transfer function of a position control system are given by*

$$G(s) = 10/s(s+1) \ and \ H(s) = 5$$

Determine the sensitivity of the closed-loop transfer function with respect to G and H.

Solution:

(a)
$$S_G^M = \frac{1}{1+GH} = \frac{1}{1+\dfrac{10}{s(s+1)}\times 5} = \frac{s(s+1)}{s^2+s+50}$$

(b)
$$S_H^M = \frac{-GH}{1+GH} = \frac{-50/s(s+1)}{1+50/s(s+1)} = \frac{-50}{s^2+s+50}$$

1.10. SUMMARY

Feedback is the fundamental feature of automatic control systems and feedback control is the essential feature of modern industries and present-day technological society. The application of automatic control to industrial processes leads to the increase in productivity due to increase in quantity, quality and uniformity of products as well as savings in processing materials, power consumption and plant equipments.

In this chapter, we have discussed the concept of feedback, the open-loop and closed-loop systems. Examples of closed-loop control system were presented to show the diverse field of their applications. It can be seen that the closed-loop systems are superior to open-loop systems in their reliability, performance and accuracy. However, in closed-loop systems, it is quite possible that during certain frequency range of operation the feedback may become positive resulting in an increase in the input to the system. This will set in motion a chain reaction which ultimately may result in the output becoming uncontrollable or the system as such becoming unstable. Thus the stability problem looms large in the case of closed-loop systems.

REVIEW QUESTIONS

1.1. Give four advantages of automatic control systems.

1.2. What are the application areas of the automatic control systems?

1.3. Distinguish between open-loop and closed-loop systems.

1.4. What are the advantages and disadvantages of open-loop systems?

1.5. What are the advantages and disadvantages of closed-loop systems?

1.6. Distinguish between servomechanism and regulator.

1.7. Compare open-loop and closed-loop systems.

1.8. Mention the effect of feedback on

     (i) System gain

    (ii) System Stability

   (iii) System sensitivity and

   (iv) Noise or disturbance.

1.9. What are the different types of feedback control systems?

1.10. Give four examples of feedback control systems.

EXERCISE

1.1. What are the advantages of automatic control system compared to manual control systems? Describe in detail an automatic control system.

1.2. Distinguish between open-loop and closed-loop control systems with suitable examples.

1.3. Compare the open-loop and closed loop systems.

1.4. Discuss the merits and demerits of open-loop and closed-loop systems.

1.5. Write short note on servomechanism and regulator with examples.

1.6. What is the effect of negative feedback on (a) Overall gain, (b) Stability, (c) sensitivity (d) external disturbance or noise. Derive necessary equations.

1.7. With suitable example, explain continuous, on-off and sample-data control system.

1.8. With a neat diagram, explain the operation of a d.c. generator voltage regulator.

1.9. Draw the schematic diagram of a position control system and describe its operation.

1.10. Explain, with a schematic diagram, the operation of a closed-loop thermal system.

PROBLEMS

1.1. The forward and feedback transfer function of a system used to track the sun are given by $G(s) = 100 / (3s+1)$ and $H(s) = 1$. Calculate the system sensitivity.

1.2. A digital audio system may be represented by Fig 1.9 with $G_1 = 10$, $G_2 = 2$ and $H = 1$. (a) Find the system sensitivity due to G_2. (b) Determine the effect of noise on the output voltage. (c) What value would you select for G_1 to minimize the effect of noise on the output?

1.3. A control system has two forward signal paths and its transfer function is given by

$$M = (G_1 G_2 G_3 + G_3 G_4)/(1 + G_2 G_3)$$

(a) Find system sensitivity due to G_3 and G_4. (b) Does the sensitivity depend on G_3 or G_4?

1.4. A closed-loop transfer function is $M = (G_1 + KG_2)/(G_3 + KG_4)$. Find the system sensitivity due to K.

1.5. One important objective of the paper-making process is to maintain uniform consistency of the stock-output as it progresses to drying and rolling. The closed-loop transfer function is

$$M = K/(20s^2 + 12s + 1 + K)$$

Find the system sensitivity due to K.

1.6. Two feedback control systems have the following transfer functions.

$$M_1 = K_1 K_2/(1 + 0.0099 K_1 K_2)$$

$$M_2 = K_1 K_2/(1 + 0.09 K_1 + 0.09 K_2 + 0.0081 K_1 K_2)$$

(a) Find M_1 and M_2 when $K_1 = K_2 = 100$. (b) Compare the system sensitivities due to K_1 when $K_1 = K_2 = 100$.

Chapter 2

LAPLACE TRANSFORMS

2.1. INTRODUCTION

The behaviour of dynamic system is completely described by integrodifferential equations. Solving these equations by the classical method is *laborious and time consuming.* In addition, the evaluation of the integration constants required the simultaneous solution of a number of algebraic equations equal to the order of these equations. Laplace transformation is a technique by which the integrodifferential equations are *transformed into algebraic equations* thereby facilitating their solution with ease. The Laplace transform method has the following advantages.

1. It transforms the system equations into algebraic equations the solution of which is simple.
2. The work is systematized and the initial conditions are incorporated at the time of transformation itself.
3. The use of a table of Laplace transform pairs (Appendix 1) reduces the labour required considerably.
4. The transient and steady-state solutions are obtained simultaneously.
5. This method is applicable to the solution of system equations in matrix state-variable form.
6. Discontinuous inputs can also be treated easily.

2.2. COMPLEX VARIABLES AND FUNCTIONS

The classical control theory is based on the application of complex variables and their functions. A complex variable s has a real component σ and an imaginary component $j\omega$. The real component is represented along the horizontal axis and the imaginary component along the vertical axis. The plane containing the two axes is called s-plane. In the s-plane, a complex variable is represented by a point. In Fig. 2.1, a typical point $s_1 = \sigma_1 + j\omega_1$ is shown.

A function G(s) is said to be a complex function if, for every value of s, there exits a corresponding value or values of G(s). The function G(s) may be represented by its real and imaginary components as

$$G(s) = ReG(s) + ImG(s) = G_x + G_y \qquad\qquad ... (2.1)$$

The plane containing G_x along the x-axis and G_y along the y-axis is called G-plane. $G(s)$ is said to be single-valued function if, for every value of s, there exists only one corresponding value of $G(s)$.

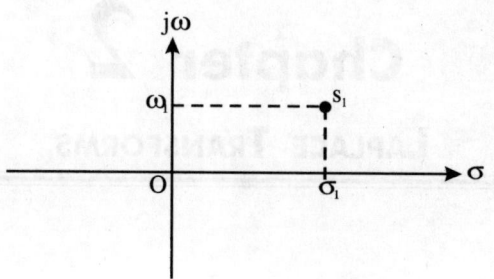

Fig. 2.1. Complex s-plane

2.3. POLES AND ZEROS

A complex function $G(s)$ is said to be *analytic* in a region if $G(s)$ and all its derivatives exist in the region. The function $G(s) = s + 1$ is analytic at every point in the s-plane whereas $G(s) = 1/(s + 1)$ is analytic at all points except at the point s = –1. Points in the s-plane at which the function $G(s)$ is analytic are called ordinary points. Points in the s-plane at which the function $G(s)$ is not analytic are known as *singular points*. Singular points at which the function or its derivatives approach infinity are called *poles*. Singular points at which the function $G(s)$ is equal to zero, are called *zeros*. For example, consider the function given in Eq. (2.2).

$$G(s) = \frac{K(s+2)}{(s+1)(s+3)} \qquad\qquad ...(2.2)$$

It has two finite poles at s = –1 and –3, and a zero at s = –2. Fig 2.2 shows the location of poles and zeros in the s-plane. A pole is represented in the s-plane by a cross and a zero by

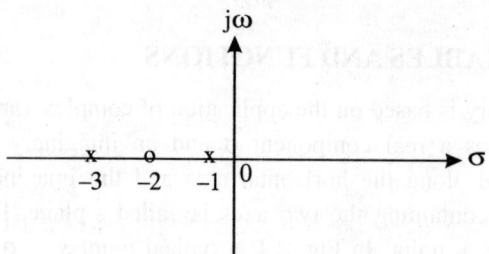

Fig. 2.2. Pole-zero diagram

a circle. However if we also consider singular points at infinity, $G(s)$ has equal number of poles and zeros. In addition to a finite zero at s = –2, $G(s)$ has one more zero at infinity, since

$$\lim_{s \to \infty} G(s) \;=\; \lim_{s \to \infty}(K\,/\,s) \to 0 \qquad\qquad ...(2.3)$$

2.4. DEFINITION OF THE LAPLACE TRANSFORM

Provided $f(t)$ is Laplace transformable, the Laplace transform $F(s)$ of a function of time $f(t)$ is defined as

$$F(s) \;=\; \mathcal{L}[f(t)] = \int_0^\infty f(t)e^{-st}\,dt \qquad\qquad ...(2.4)$$

where $\mathcal{L}\,[f(t)]$ is a short-hand notation for "Laplace transform of $f(t)$". The time function $f(t)$ is Laplace transformable if it satisfies the following conditions:

(i) It must be piecewise continuous over every finite interval $0 \le t_1 \le t \le t_2$ and

(ii) It must be of exponential order.

A function $f(t)$ is of exponential order if a real, positive constant α exists such that the function $e^{-\alpha t}/f(t)$ approaches zero as t approaches infinity. All signals that can be physically generated and all linear differential equations with constant coefficients and a finite number of terms are Laplace transformable.

2.5. DERIVATION OF LAPLACE TRANSFORMS OF SIMPLE FUNCTIONS

In this section, the method of deriving the Laplace transform of several time functions is illustrated.

Example 2.1. *Find the Laplace transform of a step function of amplitude A shown in Fig. 2.3.*

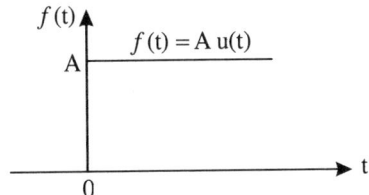

Fig. 2.3. Step function with amplitude A.

Solution:

The step function has constant amplitude of A for $t \ge 0$ and zero value for $t < 0$. It is expressed mathematically.

$$f(t) = A \qquad \text{for } t \ge 0$$
$$ = 0 \qquad \text{for } t < 0 \qquad\qquad ...\,(2.5)$$

The step function is undefined at $t = 0$ but it is immaterial since

$$\int_{0^-}^{0^+} Ae^{-st} dt \;=\; 0$$

The Laplace transform of $f(t)$ is given by

$$F(s) \;=\; \int_0^\infty Ae^{-st} dt \;=\; -\frac{A}{s} e^{-st} \Big|_0^\infty = \frac{A}{s} \qquad\qquad \ldots (2.6)$$

The step function whose amplitude A is unity is called a *unit–step function* and is denoted by $u(t)$. The function in Eq. (2.5) can therefore by denoted by A $u(t)$. The Laplace transform of the unit step function is given by

$$F(s) \;=\; \mathcal{L}\,[u(t)] = 1/s \qquad\qquad \ldots (2.7)$$

The unit step function occurring at $t = \tau$ is denoted by $u(t - \tau)$. A step function which occurs at $t = \tau$ corresponds to a constant signal such as a d.c. voltage suddenly applied to the system at a time t equals τ. Its Laplace transform is given by $F(s) = \mathcal{L}[u(t - \tau)] = e^{-\tau s}/s$.

Example 2.2. *Find the Laplace transform of the ramp function* $f(t) = At$.

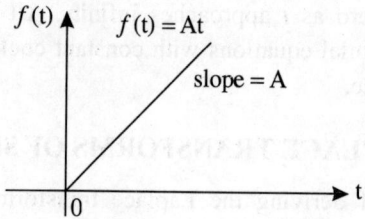

Fig. 2.4. Ramp function with slope A

Solution:

The ramp function (Fig. 2.4) is defined by

$$f(t) = At \qquad \text{for } t > 0$$
$$= 0 \qquad \text{for } t \leq 0 \qquad\qquad \ldots (2.8)$$

Here A is a constant. The Laplace transform is

$$F(s) \;=\; A \int_0^\infty te^{-st} dt$$

$$=\; At \frac{e^{-st}}{-s} \Big|_0^\infty - A \int_0^\infty \frac{e^{-st}}{-s} dt$$

$$=\; 0 + \frac{A}{s} \int_0^\infty e^{-st} dt$$

$$=\; \frac{A}{s} \int_0^\infty e^{-st} dt$$

$$= \frac{A}{-s^2} e^{-st} \Big|_0^\infty$$

$$= \frac{A}{s^2}$$

The constant A is the slope of the time function $f(t)$. If the slope is unity, $A = 1$ and the function is known as *unit ramp*. The Laplace transform of unit ramp $f(t) = t$ is given by

$$F(s) = 1/s^2 \qquad \qquad ...(2.9)$$

Example 2.3. *Find the L. T. of parabolic function shown in Fig. 2.5*

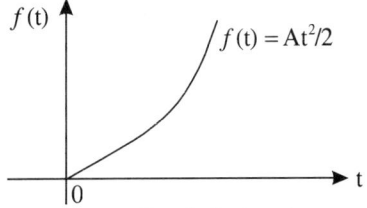

Fig. 2.5. Parabolic function

Solution:

The parabolic function is defined by

$$f(t) = \frac{At^2}{2} \qquad \text{for } t > 0$$

$$= 0 \qquad \text{for } t \le 0 \qquad \qquad ... (2.10)$$

where A is a constant. The Laplace transform is

$$F(s) = \int_0^\infty [At^2 / 2] e^{-st} dt$$

$$= \frac{At^2}{2} \frac{e^{-st}}{-s} \Big|_0^\infty - A \int_0^\infty t \frac{e^{-st}}{-s} dt$$

$$= \frac{A}{s} \int_0^\infty t e^{-st} dt$$

$$= \frac{At}{s} \frac{e^{-st}}{-s} \Big|_0^\infty - \frac{A}{s} \int_0^\infty \frac{e^{-st}}{-s} dt$$

$$= \frac{A}{s^2} \int_0^\infty e^{-st} dt$$

$$= \frac{A}{s^2} \frac{e^{-st}}{-s} \Big|_0^\infty$$

$$= \frac{A}{s^3}$$

If the constant $A = 1$, then the function is called *unit parabolic function* and its Laplace transform is given by

$$F(s) = 1/s^3 \qquad\qquad\qquad ... (2.11)$$

Example 2.4. *Find the Laplace transform of the exponential function, Ae^{-at}.*

Solution:

The exponential function is expressed as

$$f(t) = Ae^{-at} \qquad \text{for } t \geq 0$$
$$= 0 \qquad\qquad \text{for } t < 0 \qquad\qquad ... (2.12)$$

where A is a constant. The Laplace transform $F(s)$ is

$$F(s) = \int_0^\infty Ae^{-at}e^{-st}dt = A\int_0^\infty e^{-(s+a)}dt$$

$$= -A\frac{e^{(s+a)t}}{s+a}\Big|_0^\infty$$

$$= \frac{A}{s+a} \qquad\qquad\qquad ... (2.13)$$

The exponential function produces a pole at $s = -a$ in the complex plane.

Example 2.5. *Find the L. T. of $f(t) = A \sin \omega t$*

Solution:

$$f(t) = A\sin \omega t \qquad\qquad \text{for } t \geq 0$$
$$= 0 \qquad\qquad\qquad \text{for } t < 0$$

where A and ω are constants.

Expressing $\sin \omega t$ in exponential form we have

$$\sin \omega t = (e^{j\omega t} + e^{-j\omega t})/2j$$

Therefore,

$$F(s) = \frac{A}{2j}\int_0^\infty (e^{j\omega t} - e^{-j\omega t})e^{-st}dt$$

$$= \frac{A}{2j}\left(\frac{1}{s-j\omega} - \frac{1}{s+j\omega}\right)$$

$$= \frac{A}{2j}\frac{(s+j\omega)-(s-j\omega)}{s^2+\omega^2}$$

$$= \frac{A\omega}{s^2+\omega^2} \qquad\qquad\qquad ... (2.14)$$

Example 2.6. *Find L. T. of A cos ωt*

Solution:

$$f(t) = A\cos\omega t \qquad \text{for } t \geq 0$$
$$= 0 \qquad \text{for } t < 0 \qquad \text{... (2.15)}$$

where A and ω are constants.

Expressing cos ωt in exponential form we have

$$\cos\omega t = \frac{e^{j\omega t} + e^{-j\omega t}}{2}$$

Hence

$$F(s) = \frac{A}{2}\int_0^\infty (e^{j\omega t} + e^{-j\omega t})e^{-st}\,dt$$

$$= \frac{A}{2}\left(\frac{1}{s - j\omega} + \frac{1}{s + j\omega}\right)$$

$$= \frac{As}{s^2 + \omega^2} \qquad \text{... (2.16)}$$

2.6. LAPLACE TRANSFORM THEOREMS

The Laplace transform has several properties that are helpful in evaluating the transforms and solving differential equations. These properties are mentioned in the following theorems.

Theorem 1: Linearity. If $F(s)$ is the Laplace transform of $f(t)$, then the Laplace transform of $Af(t)$ is given by $AF(s)$.

$$\mathcal{L}\,[Af(t)] = AF(s) \qquad \text{... (2.17)}$$

Theorem 2: Superposition. The Laplace transform of the sum or difference of time functions is the sum or difference of their Laplace transform.

$$\mathcal{L}\,[f_1(t) \pm f_2(t)] = F_1(s) \pm F_2(s) \qquad \text{... (2.18)}$$

Theorem 3: Translation in Time Domain. If $F(s)$ is the Laplace transform of $f(t)$, then the Laplace transform of the delayed function $f(t - T)$ is $e^{-Ts}\,F(s)$.

$$\mathcal{L}[f(t - T)] = e^{-Ts}F(s) \qquad \text{... (2.19)}$$

Translation in real domain corresponds to multiplication by e^{-Ts} in s-domain

Theorem 4: Translation in s-domain. If $F(s)$ is the Laplace transform of $f(t)$, then the Laplace transform of $e^{-at}\,f(t)$ is $F(s + a)$.

$$\mathcal{L}[e^{-at}f(t)] = F(s + a) \qquad \text{... (2.20)}$$

Multiplication by e^{-at} in real domain corresponds to translation in s-domain.

Theorem 5: Complex Differentiation. If $F(s)$ is the Laplace transform of $f(t)$, then

$$\mathcal{L}[tf(t)] = -\frac{d}{ds}F(s) \qquad\qquad ... (2.21)$$

Multiplication by time in real domain corresponds to differentiation with respect to s in the s-domain.

Theorem 6: Complex Integration. If $F(s)$ is the Laplace transform of $f(t)$ and if $f(t)/t$ has a limit as $t \to 0^+$, then

$$\mathcal{L}\left[\frac{f(t)}{t}\right] = \int_0^\infty F(s)ds \qquad\qquad ... (2.22)$$

Division by time in real domain corresponds to integration with respect to s in the s-domain.

Theorem 7: Real Differentiation. If $F(s)$ is the Laplace transform of $f(t)$, then

$$\mathcal{L}\left[\frac{df(t)}{dt}\right] = sF(s) - f(0^+) \qquad\qquad ... (2.23)$$

where $f(0^+)$ is the value of $f(t)$ as the origin is approached from the right-side. In general, the Laplace transform of the n-th derivative is

$$\mathcal{L}\left[\frac{d^n f(t)}{dt^n}\right] = s^n F(s) - s^{n-1}f(0^+) - s^{n-2}f'(0^+)........f^{(n-1)}(0^+) \qquad\qquad ... (2.24)$$

The transform thus includes all the initial conditions.

Theorem 8: Real Integration. If $F(s)$ is the Laplace transform of $f(t)$, then

$$\mathcal{L}\left[\int_0^t f(\tau)d\tau\right] = F(s)/s \qquad\qquad ... (2.25)$$

The Laplace transform of the n-th order integration is $F(s)/s^n$.

Theorem 9: Initial – Value theorem. If $F(s)$ is the Laplace transform of $f(t)$ then,

$$\lim_{t \to 0} f(t) = \lim_{s \to \infty} sF(s) \qquad\qquad ... (2.26)$$

Theorem 10: Final – Value theorem. If $F(s)$ is the Laplace transform of $f(t)$, then

$$\lim_{t \to \infty} f(t) = \lim_{s \to 0} sF(s) \qquad\qquad (2.27)$$

The final value theorem is very useful in the analysis and design of feedback control system since it gives the steady state behaviour.

2.7. APPLICATION OF LAPLACE TRANSFORM THEOREMS

The application of Laplace transform (LT) theorems are illustrated with the following examples.

Example 2.7. *Find the LT of a pulse function of amplitude A.*

Solution:

The pulse function is shown in Fig. 2.6(a)

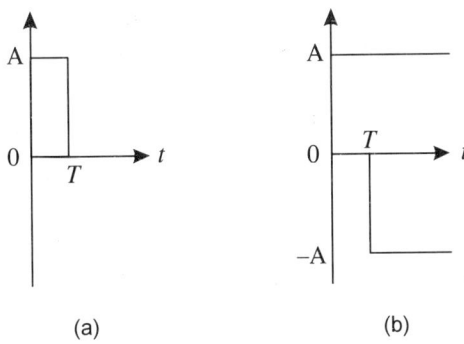

(a)   (b)

Fig. 2.6: (a) Pulse function (b) Pulse function in terms of step function

$$f(t) = A \text{ constant for } 0 < t < T$$
$$= 0 \text{ for } t < 0 \text{ and } t > T \qquad \qquad ... (2.28)$$

The pulse function may be considered as a sum of a positive step function of amplitude A from $t = 0$ and a negative step function of the same amplitude from $t = T$ as shown in Fig. 2.6(b). Thus

$$f(t) = Au(t) - Au(t - T)$$

The Laplace transform of $Au(t)$ is A/s and that of $Au(t - T)$, by applying Theorem 3, is $A e^{-Ts}/s$. Hence

$$\mathcal{L}[f(t)] = F(s) = \frac{A}{s}(1 - e^{-Ts}) \qquad \qquad ... (2.29)$$

Example 2.8. *Find the LT of an impulse function.*

Solution:

The impulse function is a special limiting case of a pulse function. Consider the impulse function $f(t)$ shown in Fig. 2.7(b). It is given by

$$f(t) = \lim_{t \to 0}(A/T) \qquad \text{for } 0 < t < T$$
$$= 0 \qquad \text{for } t < 0, t > T \qquad \qquad ... (2.30)$$

The area under the impulse is equal to A. As $T \to 0$, the amplitude of impulse A/T approaches infinity. However the area under the impulse remains same. The size of an impulse is always

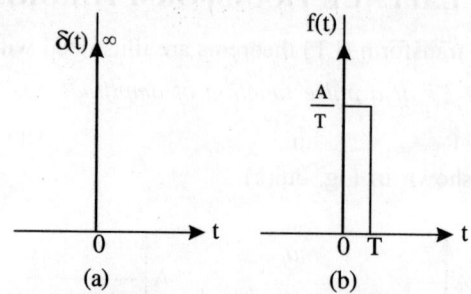

Fig. 2.7. (a) Impulse function (b) Equivalent waveform (A → ∞ as T → 0)

measured by its area. The impulse function may be considered as a pulse of amplitude A/T as $T \to 0$. Hence

$$\mathcal{L}[f(t)] = \lim_{T \to 0} \mathcal{L}\left[\frac{A}{T}u(t) - \frac{A}{T}u(t-T)\right]$$

$$= \lim_{T \to 0}\frac{A}{Ts}\left(1-e^{-Ts}\right)$$

$$= \lim_{T \to 0}\frac{d[A(1-e^{-Ts})]/dt}{dTs/dT} \qquad \text{(D' Alembert's Principle)}$$

$$= \lim_{T \to 0}\frac{Ase^{-Ts}}{s}$$

$$= A \qquad\qquad\qquad\qquad\qquad\qquad (2.31a)$$

The Laplace transform of an impulse function is equal to the area under the impulse.

Impulses do not occur in physical systems. However, the concept is useful as a tool in the system analysis and design.

The impulse function whose area is equal to unity is called *unit-impulse function* or the *Dirac-delta function*. Mathematically it is expressed as

$$\delta(t) = 0 \qquad\qquad t \neq 0$$

$$\int_0^\infty \delta(t) = 1 \qquad\qquad\qquad\qquad\qquad ... (2.31b)$$

The Laplace transform of unit impulse function occurring at $t = 0$ is 1. The unit impulse function occurring at $t = T$ is denoted by $\delta(t-T)$ and it satisfies the following conditions.

$$\delta(t-T) = 0 \qquad\qquad \text{for } t \neq T$$

$$\int_{-\infty}^{\infty} \delta(t-T)dt = 1$$

$$\mathcal{L}\,\delta(t-T) = e^{-sT} \qquad \text{... (2.32)}$$

Hence its Laplace transform is also unity but occurs at $t = T$. The unit-impulse function is the time derivative of unit step function, or, conversely, the unit step function is the integral of unit-impulse function.

Example 2.9. *Find the initial and final value of* $F(s) = 2/s(s^2 + 5s + 2)$

Solution:

By applying initial value theorem,

$$f(0) = \lim_{s \to \infty} sF(s) = 0$$

By applying final value theorem,

$$f(\infty) = \lim_{s \to 0} sF(s) = 1$$

Example 2.10. *Given the function* $F(s) = K(s + 2)/s (s + 4) (s + 5)$, *find the value of K so that the final value of* $f(t) = 1$.

Solution:

By applying final value theorem and equating the result to $f(\infty)$, we get

$$\lim_{s \to 0} s\,F(s) = \frac{K}{10} = 1$$

Hence, $\qquad\qquad\qquad K = 10$

Example 2.11. *Find the final value of* $f(t)$ *whose LT is* $\omega/(s^2 + \omega^2)$.

Solution:

$$f(\infty) = \lim_{s \to 0} s\,F(s) = \lim_{s \to 0} \frac{s\omega}{s^2 + \omega^2} = 0\,(\text{wrong})$$

This result is not correct since $f(t) = sin\ \omega t$ is not equal to zero when $t = \infty$. Since $F(s)$ has two poles on the $j\omega$-axis, the final value theorem cannot be applied. Therefore, *care should be exercised in applying the final value theorem.*

2.8. INVERSE LAPLACE TRANSFORM

Given $F(s)$, the process of obtaining $f(t)$ is called inverse Laplace transform and is denoted by $\mathcal{L}^{-1}$. The inverse LT is defined as

$$\mathcal{L}^{-1}[F(s)] = f(t) = \frac{1}{2\pi j} \int_{c-j\infty}^{c+j\infty} F(s)e^{st}\,ds \qquad \text{... (2.33)}$$

where c is a real constant greater than the real parts of all singularities of $F(s)$. Thus the path of integration is parallel to and displaced from $j\omega$ axis by the amount c, and lies to the right of all singular points.

The evaluation of complex integral given by Eq. (2.33) is tedious and time-consuming. Fortunately there are simpler methods by which $f(t)$ can be found. One such method is the use of the table of Laplace transform pairs given in Appendix 1. Quite often a particular transform $F(s)$ may not find a place in the table. In such cases, we have to expand it into partial fractions of standard forms for which inverse LT is found in the table.

2.9. PARTIAL FRACTION EXPANSION METHOD

If the Laplace transform of $f(t)$ is $F(s)$ and is expressed as

$$F(s) = F_1(s) + F_2(s) + \ldots\ldots\ldots + F_n(s)$$

then $f(t)$ is given by

$$\mathcal{L}^{-1}[F(s)] = f(t) = \mathcal{L}^{-1}[F_1(s)] + \mathcal{L}^{-1}[F_2(s)] + \cdots \cdots \mathcal{L}^{-1}[F_n(s)]$$

$$= f_1(t) + f_2(t) + \cdots \cdots f_n(t)$$

where $f_1(t)$, $f_2(t)$, $\ldots\ldots\ldots f_n(t)$ are the inverse LT of $F_1(s)$, $F_2(s)$, $\ldots\ldots\ldots$, $F_n(s)$ respectively.

In a majority of problems in control systems, $F(s)$ can be expressed as a ratio of two polynomials, $P(s)$ and $Q(s)$. Thus,

$$F(s) = \frac{P(s)}{Q(s)} = \frac{b_m s^m + b_{m-1} s^{m-1} + \ldots\ldots\ldots + b_1 s + b_0}{a_n s^n + a_{n-1} s^{n-1} + \ldots\ldots\ldots + a_1 s + a_0}, \quad m < n \quad \ldots (2.34)$$

$$= K \frac{(s+z_1)(s+z_2)\ldots\ldots\ldots(s+z_m)}{(s+p_1)(s+p_2)\ldots\ldots\ldots(s+p_n)}, \quad m < n \quad \ldots (2.35)$$

where the zeroes z_i and the poles p_j are either real or complex conjugate pairs. In Eq. (2.34), if m is greater than n, $P(s)$ must be divided by $Q(s)$ to produce a polynomial in s plus a rational polynomial.

The advantage of the partial fraction expansion method is that the terms in the expansion are very simple functions of s whose inverse L.T. can be simply obtained. The limitation of this method is that the denominator polynomial $Q(s)$ should be expressed in factored form.

2.9.1. Partial Fraction Expansion with Simple and Real Poles

When all poles of $F(s)$ are real and simple, $F(s)$ is expressed as

$$F(s) = \frac{P(s)}{(s+p_1)(s+p_2)(s+p_3)\ldots\ldots\ldots(s+p_n)} \quad \ldots (2.36)$$

By applying partial fraction expansion method, $F(s)$ is written as

$$F(s) = \frac{A_1}{s+p_1} + \frac{A_2}{s+p_2} + \frac{A_3}{s+p_3} + \ldots\ldots + \frac{A_n}{s+p_n} \qquad \ldots (2.37)$$

The coefficient A_i is found by multiplying both sides of Eq. (2.36) by the factor $(s + p_i)$ and then substituting $s = -p_i$ in the resulting function. Thus,

$$A_i = (s+p_i)\frac{P(s)}{Q(s)}\Big|_{s=-p_i} \qquad i=1,2,\ldots\ldots.n \qquad \ldots (2.38)$$

For example, A_1 is obtained by

$$A_1 = (s+p_1)\frac{P(s)}{Q(s)}\Big|_{s=-p_1} = \frac{P(-p_1)}{(-p_1+p_2)(-p_1+p_3)\ldots\ldots.(-p_1+p_n)}$$

Example 2.12. *Find f(t), if F(s) is given by*

$$F(s) = \frac{7s+2}{(s+1)(s+3)(s+4)}$$

Solution:

In the partial-fractioned form, $F(s)$ is

$$F(s) = \frac{A_1}{s+1} + \frac{A_2}{s+3} + \frac{A_3}{s+4}$$

By the application of Eq. (2.38), we have

$$A_1 = (s+1)F(s)\big|_{s=-1} = \frac{7s+2}{(s+3)(s+4)}\Big|_{s=-1}$$

$$= -\frac{5}{6}$$

$$A_2 = (s+3)F(s)\big|_{s=-3} = \frac{7s+2}{(s+1)(s+4)}\Big|_{s=-3}$$

$$= 9.5$$

$$A_3 = (s+4)F(s)\big|_{s=-4} = \frac{7s+2}{(s+1)(s+3)}\Big|_{s=-4}$$

$$= -\frac{26}{3}$$

Therefore, $\qquad F(s) = -\frac{5/6}{s+1} + \frac{9.5}{s+3} - \frac{26/3}{s+4}$

$$f(t) = \mathcal{L}^{-1}[F(s)] = -\frac{5}{6}e^{-t} + 9.5e^{-3t} - \frac{26}{3}e^{-4t}$$

The inverse Laplace transform of each term is obtained from the Laplace transform pairs shown in Appendix 1.

2.9.2. Partial Fraction Expansion with Multiple Order Real Poles

When some of the real poles are repeated, Eq. (2.38) cannot be applied. The technique followed in such cases is explained below.

Consider that the pole at $s = -p_i$ is repeated r times.

Then $F(s)$ is given by

$$F(s) = \frac{P(s)}{(s+p_1)(s+p_2)...(s+p_{i-1})(s+p_i)^r(s+p_{i+1})...(s+p_n)}$$

...(2.39)

In the partial–fractioned form, $F(s)$ is written as

$$F(s) = \frac{A_1}{s+p_1} + \frac{A_2}{s+p_2} + ... + \frac{A_{i-1}}{s+p_{i-1}} + \frac{A_{i+1}}{s+p_{i+1}} + ... + \frac{A_n}{s+p_n}$$

$$+ \frac{B_1}{(s+p_i)^r} + \frac{B_2}{(s+p_i)^{r-1}} + ... + \frac{B_i}{(s+p_i)}$$

...(2.40)

The coefficients corresponding to simple poles are obtained by Eq. (2.38). The coefficients of multiple order poles are determined by evaluating the following equations.

$$B_1 = (s+p_i)^r F(s)\big|_{s=-p_i}$$

...(2.41)

$$B_2 = \frac{d}{ds}\left[(s+p_i)^r F(s)\right]\Big|_{s=-p_i}$$

...(2.42)

$$B_3 = \frac{1}{2!}\frac{d^2}{ds}\left[(s+p_i)^r F(s)\right]\Big|_{s=-p_i}$$

...(2.43)

$$\vdots$$

$$B_r = \frac{1}{(r-1)!}\frac{d^{r-1}}{ds^{r-1}}\left[(s+p_i)^r F(s)\right]\Big|_{s=-p_i}$$

...(2.44)

In general,

$$B_k = \frac{1}{(k-1)!}\frac{d^{k-1}}{ds^{k-1}}\left[(s+p_i)^r F(s)\right]\Big|_{s=-p_i}$$

...(2.45)

Example 2.13. *Find f(t) for the transform*

$$F(s) = \frac{1}{(s+1)^3 (s+2)}$$

Solution:

In the partial–fraction form

$$F(s) = \frac{B_1}{(s+1)^3} + \frac{B_2}{(s+1)^2} + \frac{B_3}{(s+1)} + \frac{A}{(s+2)}$$

Applying Eq. (2.41)

$$B_1 = (s+1)^3 \frac{1}{(s+1)^3 (s+2)}\Big|_{s=-1} = \frac{1}{s+2}\Big|_{s=-1}$$

$$= 1$$

Applying Eq. (2.45)

$$B_2 = \frac{d}{ds}\left[\frac{1}{s+2}\right] = \frac{-1}{(s+2)^2}\Big|_{s=-1}$$

$$= -1$$

$$B_3 = \frac{1}{2!}\frac{d^2}{ds^2}\left[(s+1)^3 F(s)\right] = \frac{1}{2!}\frac{d}{ds}\left[\frac{-1}{(s+2)^2}\right]$$

$$= \frac{1}{2!}\frac{2}{(s+2)^3}\Big|_{s=-1} = 1$$

Applying Eq. (2.38)

$$A = (s+2)F(s)\Big|_{s=-2} = \frac{1}{(s+1)^3}\Big|_{s=-2}$$

$$= -1$$

Therefore,

$$F(s) = \frac{1}{(s+1)^3} - \frac{1}{(s+1)^2} + \frac{1}{(s+1)} - \frac{1}{(s+2)}$$

Hence,

$$f(t) = \frac{1}{2}t^2e^{-t} - te^{-t} + e^{-t} - e^{-2t}$$

2.9.3. Partial Fraction Expansion with Complex Conjugate Poles

The complex conjugate poles often occur in control system studies and therefore of special interest. Consider a transform function with a pair of complex poles

$$F(s) = \frac{P(s)}{(s+p_1)...(s+p_m)(s^2 + 2\alpha s + \alpha^2 + \beta^2)} \qquad ...(2.46)$$

In the partial – fractioned form, $F(s)$ is written as

$$F(s) = \frac{A_1}{s+p_1} + ... + \frac{A_m}{s+p_m} + \frac{C}{s+\alpha+j\beta} + \frac{C*}{s+\alpha-j\beta} \qquad ...(2.47)$$

The coefficients A_1 to A_m are evaluated using Eq. (2.38).

By using the same equation, C and $C*$ are also evaluated.

$$C = (s+\alpha+j\beta)F(s)\Big|_{s=-\alpha-j\beta} \qquad ...(2.48)$$

$$C* = (s+\alpha-j\beta)F(s)\Big|_{s=-\alpha+j\beta} \qquad ...(2.49)$$

This is illustrated by the following example.

Example 2.14. *Find f(t) of the transform function*

$$F(s) = \frac{\omega_n^2}{s\left(s^2 + 2\zeta\omega_n s + \omega_n^2\right)} \qquad ...(2.50)$$

Assume that roots of the quadratic factor are complex conjugate.

Solution:

Writing $F(s)$ in the partial–fraction form,

$$F(s) = \frac{A}{s} + \frac{C}{s+\alpha+j\beta} + \frac{C*}{s+\alpha-j\beta} \qquad ...(2.51)$$

where

$$\alpha = \zeta\omega_n \qquad ...(2.52)$$

and

$$\alpha^2 + \beta^2 = \omega_n^2 \qquad ...(2.53)$$

$$\beta = \omega_n \sqrt{1 - \zeta^2} \qquad \qquad ...(2.54)$$

The coefficients are evaluated as

$$A = sF(s)\big|_{s=0} = 1$$

$$C = (s + \alpha + j\beta)F(s)\big|_{s=-\alpha-j\beta}$$

$$= \frac{\omega_n^2}{-2j\beta(-\alpha - j\beta)}$$

$$= \frac{\omega_n}{2\beta} e^{-j\left(\frac{\pi}{2} + \theta\right)}$$

where

$$\theta = \tan^{-1}\left[\frac{\beta}{\alpha}\right] = \tan^{-1}\frac{\sqrt{1 - \zeta^2}}{\zeta} \qquad \qquad ...(2.55)$$

$$C^* = (s + \alpha - j\beta)F(s)\big|_{s=-\alpha+j\beta}$$

$$= \frac{\omega_n^2}{2j\beta(-\alpha + j\beta)} = \frac{\omega_n}{2\beta} e^{+j\left(\frac{\pi}{2} + \theta\right)}$$

Therefore, $F(s)$ is given by

$$F(s) = \frac{1}{s} + \frac{\omega_n}{2\beta}\left[\frac{e^{-j\left(\frac{\pi}{2} + \theta\right)}}{s + \alpha + j\beta} + \frac{e^{+j\left(\frac{\pi}{2} + \theta\right)}}{s + \alpha - j\beta}\right]$$

Taking inverse Laplace Transform,

$$f(t) = 1 + \frac{\omega_n}{2\beta}\left[e^{-j\left(\frac{\pi}{2}+\theta\right)} \cdot e^{-(\alpha+j\beta)t} + e^{+j\left(\frac{\pi}{2}+\theta\right)} \cdot e^{-(\alpha-j\beta)t}\right]$$

$$= 1 + \frac{\omega_n}{2\beta} e^{-\alpha t}\left[e^{+j\left(\frac{\pi}{2}+\theta+\beta t\right)} + e^{-j\left(\frac{\pi}{2}+\theta+\beta t\right)}\right]$$

$$= 1 + \frac{\omega_n}{\beta} e^{-\alpha t} \cos\left(\frac{\pi}{2} + \beta t + \theta\right)$$

$$= 1 - \frac{\omega_n}{\beta} \cdot e^{-\alpha t} \sin(\beta t + \theta)$$

Substituting the values of α and β we get

$$f(t) = 1 - \frac{1}{\sqrt{1-\zeta^2}} e^{-\zeta \omega_n t} \sin\left(\omega_n \sqrt{1-\zeta^2}\ t + \theta\right) \qquad \qquad ...(2.56)$$

where θ is given by Eq. (2.55). Note C and C^* are complex conjugates.

2.10. SOLUTION OF LINEAR ORDINARY DIFFERENTIAL EQUATIONS USING LAPLACE TRANSFORM

The differential equations of systems can be easily solved using the Laplace transform theorems and the Table of Laplace transform pairs, shown in Appendix 1. The advantage is that the complete solution is obtained simultaneously. The Laplace transform method is illustrated below with several examples. The steps in solving the linear ordinary differential equations are summarized below.

Step 1. Obtain the Laplace transform of the given differential equation term by term and incorporate the given initial conditions.

Step 2. Write the Laplace transform so obtained in the partial fraction expansion form.

Step 3. Referring to Laplace transform table, get the inverse transform of each term.

Step 4. The time function $f(t)$ is given by the sum of all the inverse transforms.

Example 2.15. *The following differential equation represents the behaviour of a particular system. Obtain its solution.*

$$\ddot{x} + 2\dot{x} + 6x = 3e^{-3t}, \ \dot{x}(0) = 0, \ x(0) = 1$$

Solution:

Applying the real differentiation theorem,

$$s^2 X(s) - sx(0) - \dot{x}(0) + 2[sX(s) - x(0)] + 6X(s) = \frac{3}{s+3}$$

Incorporating the initial conditions, we get

$$s^2 X(s) - s - 0 + 2sX(s) - 2 + 6X(s) = 3/(s + 3)$$

$$X(s)(s^2 + 2s + 6) = \frac{3}{s+3} + s + 2$$

$$= \frac{s^2 + 5s + 9}{s+3}$$

$$X(s) = \frac{s^2 + 5s + 9}{(s+3)(s^2 + 2s + 6)} = \frac{s^2 + 5s + 9}{(s+3)(s+1+j\sqrt{5})(s+1-j\sqrt{5})}$$

$$= \frac{A}{s+3} + \frac{C}{s+1+j\sqrt{5}} + \frac{C*}{s+1-\sqrt{5}}$$

$$A = (s+3)X(s)\big|_{s=-3} = \frac{s^2+5s+9}{s^2+2s+6}\bigg|_{s=-3} = \frac{1}{3}$$

$$C = (s+1+j\sqrt{5})X(s)\big|_{s=-1=j\sqrt{5}}$$

$$= \frac{s^2 + 5s + 9}{(s+3)(s+1-j\sqrt{5})}\bigg|_{s=-1-j\sqrt{5}}$$

$$= \frac{(-1-j\sqrt{5})^2 + 5(-1-j\sqrt{5}) + 9}{(2-j\sqrt{5})(-1-j\sqrt{5}+1-j\sqrt{5})}$$

$$= \frac{j3\sqrt{5}}{10 + j4\sqrt{5}} = \frac{1}{2}e^{j(\pi/2 - 41.8°)}$$

$$C* = \frac{1}{2}e^{-j(\pi/2 - 41.8°)}$$

$$\therefore \qquad X(s) = \frac{1}{3(s+3)} + \frac{\frac{1}{2}e^{j(\pi/2-41.8°)}}{s+1+j\sqrt{5}} + \frac{\frac{1}{2}e^{-j(\pi/2-41.8°)}}{s+1-j\sqrt{5}}$$

Taking inverse Laplace transform, we get

$$x(t) = \frac{1}{3}e^{-3t} + e^{-t}\sin(\sqrt{5}t + 41.8°)$$

2.11. CONVOLUTION INTEGRAL THEOREM

This theorem states that if $F_1(s)$ and $F_2(s)$ are Laplace transforms of $f_1(t)$ and $f_2(t)$ respectively, then

$$\mathcal{L}\left[\int_0^t f_1(t-\tau)f_2(\tau)d\tau\right] = F_1(s)F_2(s) \qquad \qquad ...(2.57)$$

The convolution integral is usually written as $f_1(t) * f_2(t)$ where the mathematical operation $f_1(t) * f_2(t)$ is called convolution. If we replace $t-\tau$ by T in the integral, we obtain

$$f_1(t) * f_2 = \int_0^t f_1(t-\tau)f_2(\tau)d\tau$$

$$= -\int_t^0 f_1(T)f_2(t-T)dT \left\{ \begin{array}{l} t-\tau=T \\ d\tau=-dT \\ \text{when } \tau=0, T=t \text{ lower limit} \\ \text{when } \tau=t, T=0 \text{ upper limit} \end{array} \right.$$

$$= -\int_t^0 f_1(\tau)f_2(t-\tau)d\tau$$

$$= f_2(t) * f_1(t) \qquad \qquad ...(2.58)$$

Fig. 2.8 shows $f_1(t)$, $f_1(t-\tau)$, $f_2(\tau)$ and $f_1(t-\tau) * f_2(\tau)$. The shape of the curve in Fig. 2.8 (b) depends on t.

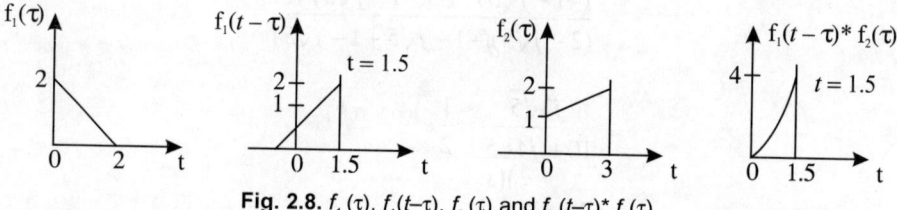

Fig. 2.8. $f_1(\tau)$, $f_1(t-\tau)$, $f_2(\tau)$ and $f_1(t-\tau)* f_2(\tau)$

The proof of Eq. (2.57) follows:

Since $f_1(t-\tau) u(t-\tau) = 0$ for $t < \tau$, we may write Eq.(2.57) as

$$f_1(t) * f_2(t) = \int_0^t f_1(t-\tau)f_2(\tau)d\tau$$

$$\mathcal{L}[f_1(t) * f_2(t)] = \mathcal{L}\left[\int_0^\infty f_1(t-\tau)u(t-\tau)f_2(\tau)d\tau\right]$$

$$= \int_0^\infty \left[\int_0^\infty f_1(t-\tau)u(t-\tau)f_2(\tau)d\tau\right]e^{-st}dt$$

Let $(t - \tau) = T$. Changing the order of integration after substituting $t - \tau = T$, we have

$$\mathcal{L}[f_1(t) * f_2(t)] = \int_0^\infty f_1(T)e^{-s(t+\tau)}dT\int_0^\infty f_2(\tau)d\tau \qquad \qquad ...(2.59)$$

$$= \int_0^\infty f_1(T)e^{-sT}dT\int_0^\infty f_2(\tau)e^{-s\tau}d\tau$$

$$= F_1(s)F_2(s)$$

Since $\mathcal{L}^{-1}[F_1(s)F_2(s)] = \int_0^\infty f_1(t-\tau)f_2(\tau)$ theorem can be gainfully used to obtain inverse Laplace transform. This method is called convolution integral method.

The procedure to find inverse Laplace transform using convolution integral theorem is as follows:

(a) Given the function $F(s)$, split it into products of two or more parts for which inverse Laplace transform can be obtained from the Laplace transform table.

(b) From the transform table find their inverse Laplace transform, $f_1(t)$ and $f_2(t)$

(c) Substitute $t - \tau$ in $f_1(t)$ and $t = \tau$ in $f_2(t)$.

(d) The time function $f(t)$ which is the inverse Laplace of $F(s)$ is obtained by evaluating the convolution integral $f_1(t) * f_2(t)$. Thus,

$$\mathcal{L}^{-1}F(s) = f(t) = \int_0^\infty f_1(t-\tau)f_2(\tau)d\tau \qquad \qquad ...(2.60)$$

If the response of a linear system for a unit step excitation is known, by applying the convolution integral theorem, its response to any arbitrary excitation can be found.

Example 2.16. *Find the inverse Laplace transform of the following function using convolution integral theorem.*

$$F(s) = \frac{1}{(s+2)(s+5)}$$

Solution:

Taking inverse Laplace transform of $F_1(s)$ and $F_2(s)$,

$$f_1(t) = e^{-2t} \text{ and } f_2(t) = e^{-5t}$$

Hence, $f(t-\tau) = e^{-2(t-\tau)} \text{ and } f_2(\tau) = e^{-5\tau}$

Therefore, applying convolution integral theorem,

$$f(t) = \int_0^t e^{-2(t-\tau)}e^{-5\tau}d\tau$$

$$= e^{-2t}\int_0^t e^{-3\tau}d\tau$$

$$= e^{-2t}\left[\frac{e^{-3\tau}}{-3}\right]_0^t$$

$$= e^{-2t}\left[\frac{-e^{-3t}+1}{3}\right]$$

$$= \frac{1}{3}e^{-2t}\left(1-e^{-3t}\right) = \frac{1}{3}\left[e^{-2t}-e^{-5t}\right]$$

2.12. SIMPLIFICATION OF PARTIAL FRACTION EXPANSION METHOD

In this section, a simple procedure to evaluate coefficients of multiple poles and complex poles is described. In the case of multiple poles, the evaluation of coefficients by repeated differentiation may be tedious when a polynomial in s occurs in the numerator of $F(s)$. The method is to determine those coefficients for real poles which can be obtained readily applying Eq.(2.38). The corresponding partial fractions are then subtracted from $F(s)$ to obtain the resultant function. By a similar method, the coefficients of the resultant function may be determined. This procedure avoids the differentiation altogether.

In case of a pair of complex poles, the quadratic factor is considered as a single fraction with a numerator polynomial in s which is one degree lower than the denominator polynomial. The coefficients of real poles are evaluated. The corresponding partial fractions are then subtracted from $F(s)$. This results in the determination of the coefficients of the numerator polynomial of the quadratic term. Using the Laplace transform pair table, the inverse Laplace transform is obtained. The above simplification procedures are illustrated with the following examples.

Example 2.17. *Find the inverse Laplace transform of the function*

$$F(s) = \frac{1}{(s+2)^3(s+3)}$$

Solution:

In partial-fraction form it is expressed as

$$F(s) = \frac{B_1}{(s+2)^3} + \frac{B_2}{(s+2)^2} + \frac{B_3}{s+2} + \frac{A}{s+3}$$

$$B_1 = (s+2)^3 F(s)\big|_{s=-2} = 1$$

$$F_1(s) = F(s) - \frac{B_1}{(s+2)^3} = \frac{1}{(s+2)^3(s+3)} - \frac{1}{(s+2)^3}$$

$$= -\frac{1}{(s+2)^2(s+3)}$$

$$= \frac{B_2}{(s+2)^2} + \frac{B_3}{s+2} + \frac{A}{s+3}$$

$$B_2 = (s+2)^2 F_1(s)\big|_{s=-2} = -1$$

$$F_2(s) = F_1(s) - \frac{B_2}{(s+2)^2} = -\frac{1}{(s+2)^2(s+3)} + \frac{1}{(s+2)^2}$$

$$= \frac{1}{(s+2)(s+3)} = \frac{B_3}{s+2} + \frac{A}{s+3}$$

$$B_3 = (s+2)F_2(s)\big|_{s=-2} = 1$$

$$A = (s+2)F_2(s)\big|_{s=-3} = -1$$

Then,

$$F(s) = \frac{1}{(s+2)^3} - \frac{1}{(s+2)^2} + \frac{1}{s+2} - \frac{1}{s+3}$$

From Appendix 1, we get $f(t)$ as

$$f(t) = \frac{1}{2}t^2 e^{-2t} - te^{-2t} + e^{-2t} - e^{-3t}$$

It may be noted that though A would have been evaluated first or along with B_1, first B_2 is evaluated and lastly A is evaluated. This procedure is followed so that $F_1(s)$ and $F_2(s)$ may be determined with ease.

Example 2.18. *Find $f(t)$ given F(s) as*

$$F(s) = \frac{20}{(s^2 + 6s + 25)(s+1)}$$

Solution:

In the modified partial fraction form,

$$F(s) = \frac{A}{s+1} + \frac{Bs+C}{s^2 + 6s + 5}$$

$$A = (s+1)F(s)\big|_{s=-1} = 1$$

$$F_1(s) = F(s) - \frac{A}{s+1} = \frac{20}{(s^2 + 6s + 25)(s+1)} - \frac{1}{s+1}$$

$$= \frac{-(s^2 + 6s + 5)}{(s^2 + 6s + 25)(s+1)}$$

$$= \frac{-(s+5)}{s^2 + 6s + 25}$$

$$= \frac{-(s+5)}{(s+3)^2 + 4^2}$$

Hence

$$B = -1, \ C = -5$$

$$F(s) = \frac{1}{s+1} - \frac{s+5}{(s+3)^2 + 4^2}$$

Taking the inverse Laplace transform,

$$f(t) = e^{-t} - 1.125 \ e^{-3t} \ \sin(4t + 63.4°)$$

2.13. ADDITIONAL EXAMPLES

Example 2.19. *The Laplace transform of x(t) is given by*

$$X(s) = \frac{s+1}{s(s^2 + 3)}$$

If $x(t)$ represents the position, find the initial values of position, velocity and acceleration.

Solution:

Applying the initial value theorem, the initial value of position is given by

$$x(0) = \lim_{t \to 0} x(t) = \lim_{s \to \infty} sX(s) = \frac{s+1}{s^2 + 3} = \frac{1 + 1/s}{s + 3/s} = 0$$

The initial value of velocity is given by

$$\dot{x}(0) = \lim_{t \to 0} \dot{x}(t)$$

$$\mathcal{L}\left[\dot{x}(t)\right] = sX(s) - x(0) = \frac{s(s+1)}{s(s^2 + 3)} = \frac{(s+1)}{(s^2 + 3)}$$

Hence, the initial velocity is given by

$$\dot{x}(0) = \lim_{s \to \infty} sX(s)$$

$$= \lim_{s \to \infty} \frac{s(s+1)}{(s^2+3)} = \frac{1+1/s}{1+3/s^2} = 1$$

Similarly,

$$\ddot{X}(s) = s^2 X(s) - s\dot{x}(0) - x(0)$$

$$= \frac{s(s+1)}{s^2+3} - 1 = \frac{s-3}{s^2+3}$$

Hence,

$$\ddot{x}(0) = \lim_{s \to \infty} s \frac{(s-3)}{(s^2+3)}$$

$$= \lim_{s \to \infty} \frac{1-3/s}{1+3/s^2} = 1$$

Thus, the initial position is zero, the initial velocity and initial acceleration are equal to unity.

Example 2.20. *If $F_1(s)$ is the Laplace transform of the first cycle of a periodic waveform, show that the Laplace transform of the periodic waveform is given by $F_1(s)/(1-e^{-Ts})$ where T is the period.*

Solution:

Since each cycle of a periodic wave occurs after a period of T starting from $t = 0$, if we denote the first cycle by $f_1(t)$, the periodic wave is then

$$f(t) = f_1(t)\, u(t) + f_1(t - T)\, u(t - T) + f_1(t - 2T)\, u(t - 2T) + \ldots$$

Taking the Laplace transform,

$$F(s) = F_1(s) + F_1(s)e^{-Ts} + F_1(s)e^{-2Ts} + F_1(s)e^{-3Ts} + \ldots$$
$$= F_1(s)[1 + e^{-Ts} + e^{-2Ts} + \ldots]$$

$$= \frac{F_1(s)}{1-e^{-Ts}}$$

since the binomial expansion of $(1 - e^{-Ts})^{-1}$ is $[1 + e^{-Ts} + e^{-2Ts} + \ldots]$

Example 2.21. *Find the Laplace transform of the pulse train of amplitude A with a pulse-width of τ and period T as shown in Fig. 2.9.*

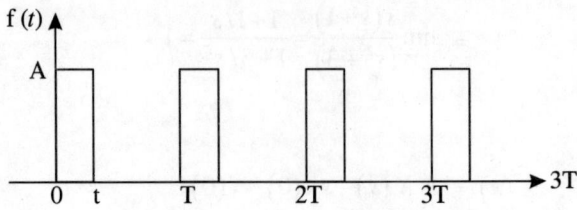

Fig. 2.9. Pulse train of amplitude *A*

Solution:

The first pulse is obtained by a step function of amplitude *A* occurring at *t* = 0 and another step function of amplitude of –*A* occurring at *t* = τ. Hence the Laplace transform of the first cycle is

$$F_1(s) = \mathcal{L}\left[Au(t) - Au(t - \tau)\right] = \frac{A}{s} - \frac{A}{s}e^{-s\tau}$$

$$= \frac{A}{s}\left(1 - e^{-s\tau}\right)$$

Hence the Laplace transform of the pulse train is

$$F(s) = \frac{F_1(s)}{1 - e^{-Ts}} = \frac{A}{s} \cdot \frac{1 - e^{-\tau s}}{1 - e^{-Ts}}$$

2.14. SUMMARY

In this chapter, we have first explained the complex variables and functions followed by definition and concept of poles and zeros. Then, we have defined Laplace transform and inverse transform. We have derived Laplace transform of simple standard functions that are used for testing the performance of control systems and their comparison. We have also discussed the important theorems and properties of Laplace transform and the advantages of using Laplace transform in the system analysis. Since it systematizes the solution of differential equations and incorporates the initial conditions, it gives the complete solution of differential equations at one stroke. We have also studied the method of finding the Laplace transform of various functions. Also discussed are the various methods of determining the time function from its Laplace transforms. Because of these properties, the Laplace transform method is extensively used in the analysis and design of feedback control systems. Finally we have discussed the convolution integral theorem and its application in obtaining the time function from the Laplace transform.

REVIEW QUESTIONS

2.1. Give four advantages of Laplace transform.

2.2. Explain poles and zeros with example.

2.3. Define Laplace transform and inverse Laplace transform.

2.4. Write down the Laplace transform of (i) unit impulse (ii) unit step (iii) unit ramp and (iv) unit parabola functions.

2.5. State the initial and final value theorem of Laplace transform.

2.6. State the convolution integral theorem.

EXERCISE

2.1. What are the advantages of using Laplace transform for the solution of system equations?

2.2. State and explain the following Laplace theorems:

(a) Linearity theorem

(b) Superposition theorem

(c) Translation in-time theorem

(d) Translation in s-domain theorem

(e) Real differentiation and real integration theorems

(f) Initial and Final value theorems.

2.3. Enumerate the steps involved in solving the linear ordinary differential equations.

2.4. State and explain the convolution integral theorem.

2.5. Show that the derivative and integral of unit step function are the unit impulse and unit ramp function respectively.

2.6. Show that the derivative and integral of a unit ramp function are unit step and unit parabolic function respectively.

2.7. Show that the first and second derivative of a unit parabolic function are unit ramp and unit step function respectively.

PROBLEMS

2.1. Draw the pole-zero diagrams of the following:

(a) $G(s) = \dfrac{K(s+4)(s+7)}{(s+1)(s+5)(s+10)}$

(b) $G(s) = \dfrac{K(s+1)(s+5)}{s(s+2)(s+3)}$

(c) $G(s) = \dfrac{K}{s^2(s+1)(s+4)}$

(d) $G(s) = \dfrac{K(s+2)}{s(s+1)^2(s+5)(s-8)}$

(e) $G(s) = \dfrac{K(s+1)}{s(s+1)(s^2+2s+2)}$

(f) $G(s) = \dfrac{K(s^2+4s+5)}{s(s+7)(s^2+2s+2)}$

2.2. Obtain the Laplace Transform of the following time functions from the first principles:

(a) $f(t) = 5$

(b) $f(t) = t + 4e^{-2t} + \sin 2t$

(c) $f(t) = 5 + \cos 5t$

(d) $f(t) = \sinh at$

(e) $f(t) = \cosh bt$

(f) $f(t) = e^{-at}\sin \omega t$

(g) $f(t) = A \sin at \cos bt$

2.3. (a) Obtain the L.T. of e^{at} from the first principles.

(b) Using the result obtained in (a), obtain the L.T. of

(i) Unit Step function

(ii) $\cos \omega t$

(iii) $\sin \omega t$

(iv) $e^{-at} \cos \omega t$

(v) $e^{-at} \sin \omega t$

(vi) e^{t}

(vii) t^{n}

2.4. Write down the L. T. of the following differential equations:

(a) $x'' + 4x' + 3x = 9$; all initial conditions are zero

(b) $x'' + 2x' + 5x = 10$; $x(0) = 2$, $x'(0) = -4$

(c) $x''' - 2x'' + 5x' = 0$; $x(0) = x'(0) = 0$, $x''(0) = 1$

(d) $x'' + 2x' + 6x = 3e^{-3t}$; $x(0) = 1$, $x'(0) = 0$

2.5. Obtain the partial–fraction expansion of

(a) $X(s) = \dfrac{2(s+2)}{s^2(s+1)(s+4)}$

(b) $X(s) = \dfrac{18(s+2)}{s(s+3)^2(s+4)}$

(c) $X(s) = \dfrac{10(s+6)}{s(s+5)(s^2+3s+1)(s^2+7s+12)}$

2.6. Obtain the inverse L.T. of the following:

(a) $F(s) = \dfrac{20}{s(s+6)(s+8)}$

(b) $F(s) = \dfrac{10(s+6)}{s(s+5)(s^2+3s+2)(s^2+7s+12)}$

(c) $F(s) = \dfrac{10(s^2+2s+2)}{s^2+9s+20}$

(d) $F(s) = \dfrac{2s^4}{(s+1)^5}$

(e) $F(s) = \dfrac{10}{(s+2)(s+1)^3}$

(f) $F(s) = \dfrac{10}{(s+3)^3(s+4)^3}$

(g) $F(s) = \dfrac{5}{s^2+6s+10}$

(h) $F(s) = \dfrac{10}{(s+2)(s^2+6s+25)}$

(i) $F(s) = \dfrac{3}{s(s^2+4)}$

(j) $F(s) = \dfrac{s+2}{s\left(s^2+9\right)}$

(k) $F(s) = \dfrac{s^2+6s+12}{(s+3)(s+4)^3}$

(l) $F(s) = \dfrac{s^2+6s+6}{(s+2)(s+3)(s+4)}$

(m) $F(s) = \dfrac{10}{(s+1)^2\left(s^2+2s+2\right)}$

(n) $F(s) = \dfrac{39}{s\left(s^2+4s+13\right)(s+3)}$

(o) $F(s) = \dfrac{0.9524\left(s^2+2s+2.1\right)}{s(s+1)\left(s^2+2s+2\right)}$

2.7. Obtain $x(t)$ of the following differential equation using L.T. method and find its steady state value

(i) $x'' + 5x' + 4x + 10 = 0,$ $x(0) = -1,$ $x'(0) = 1$

2.8. A mechanical system is governed by the differential equation

$$\dfrac{d^3x}{dt^3} + \dfrac{10d^2x}{dt^2} + 7\dfrac{dx}{dt} = 0, \qquad x(0) = 10$$

Obtain $x(t)$

2.9. The governing equation of a mechanical translational system is given by

$f(t) = Mx'' + Bx'' + Kx$ where $f(t)$ is the force applied, M is the mass, B is the coefficient of friction and K is the spring constant. (a) Obtain an expression for the displacement $x(t)$ when a sudden force of 20 Newtons is applied. Assume $x(0) = 4$, $M = 10$ kg-m, $B = 80$ N/m/sec and $K = 160$ N/m. (b) Find $x(t)$ at $t = \infty$

2.10. In the electric circuit shown in Fig. P2.9, obtain expressions for $i(t)$ and the voltage across the condenser after the switch S is closed.

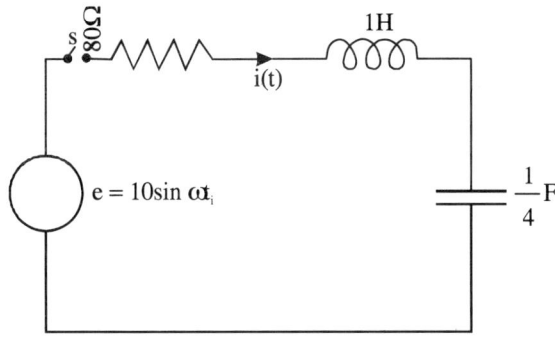

Fig. P. 2.10

2.11. In the gun control system, the relationship between the motor position θ_m and the gun position θ_g is given by

$$Js^2\theta_g + Bs\theta_g + K\theta_g = f(t) + K\frac{\theta_m}{n}$$

where

J = Moment of inertia of the gun, base etc. = 50 Kg-m²

B = Coefficient of viscous friction = 300 N/rad/sec.

K = Shaft stiffness = 450 N-m/rad

$f(t)$ = Firing torque = 0

n = Gear ratio = 10

Obtain the time response of the system if the motor is suddenly rotated by 4 radians.

2.12. Find the initial value and initial slope of the function.

$$f(s) = \frac{10(s+2)\sin\theta + 3\cos\theta}{s^2 + 4s + 13}$$

2.13. The L.T. of a time function is

$$F(s) = \frac{2s+5}{s\left(s^2 + 4s + 7\right)}$$

Find the initial value of $f(t)$, its first and second derivatives.

2.14. Find the initial and final values of $x(t)$ in Prob. 2.4

2.15. Find the initial and final values of:

(a) $F(s) = \dfrac{s+8}{(s+3)^2 + 10}$

(b) $F(s) = \dfrac{6s(s+3)}{(s+1)(s+4)(s+5)}$

(c) $F(s) = \dfrac{s+1}{s(s+2)(s^2 + s + 1)}$

2.16. Given

$$X(s) = \dfrac{K(s+1)}{s(s+2)^2 (s+3)}$$

Find the initial value of $x(t)$. Determine the value of K such that the final value is unity.

2.17. Using convolution theorem, obtain the inverse L. T. of:

(a) $X(s) = \dfrac{1}{(s+2)(s+4)}$

(b) $X(s) = \dfrac{1}{s^2 (s+9)}$

(c) $X(s) = \dfrac{6}{s^2 (s+9)}$

(d) $X(s) = \dfrac{1}{s(s+4)^2}$

Chapter 3

SYSTEM TRANSFER FUNCTIONS, BLOCK DIAGRAMS AND SIGNAL FLOW GRAPHS

3.1. INTRODUCTION

The first step in the analysis of any physical system is to obtain the description of the system in the form of an integro differential equation. This mathematical description of the system is called *mathematical model.* This model should represent the physical system as accurately as possible and at the same time it should be simple. Usually some simplifying approximations are made in deriving the models. All physical systems are inherently nonlinear. However, if the effect of the nonlinearities on the system response is small, they can be ignored. Thus the system may be considered a linear system.

A linear system is that which satisfies the homogeneity and superposition principles.

Homogeneity Principle: Consider a system with input $r(t)$ and output $c(t)$. The homogeneity principle states that if the input signal $r(t)$ is scaled by a factor α, then the output is also scaled by the same factor. Thus,

if $r_i(t) = \alpha\, r(t)$, then $c_i(t) = \alpha\, c(t)$

Superposition Principle: The superposition principle states that if a system is excited by several inputs, then the output is the sum of the individual outputs caused by the application of each input separately. Thus,

if

$$r_1(t) \rightarrow c_1(t)$$
$$r_2(t) \rightarrow c_2(t)$$
$$\dots\dots\dots\dots\dots$$
$$r_n(t) \rightarrow c_n(t)$$

Then,

$$c(t) = c_1(t) + c_2(t) + \dots c_n(t)$$

A linear system is described by linear differential equations. A differential equation is linear if its coefficients are all constants. An experimental system may be considered linear if the cause and effect are proportional.

The mathematial model of a linear system may be in the form of differential equation of transfer function which describes the cause-effect or input-output relationship of the system. It does not include any information regarding the internal structure of the system.

The behaviour of a system may also be represented by a block diagram. It represents clearly the flow of information and the functions preformed by each component of the system. Thus it shows the functional operation of the system.

While the block-diagram representation of a system simplifies the understanding of the behaviour of the system, the signal flow graph provides further simplification for large and complex systems.

In this chapter, we shall describe the methods of obtaining the transfer functions using mathematical models, block diagrams and signal flow graphs.

3.2. THE CONCEPT OF TRANSFER FUNCTION

Fig. 3.1 shows the representation of a linear system. The physical system may be mechanical, electrical, hydraulic or biological. Let $g(t)$ be the impulse response of the system, $r(t)$ be the input and $c(t)$ be the response or output. The input-output relationship of the system is generally described by an *n-th* order differential equation with constant real coefficients of the form

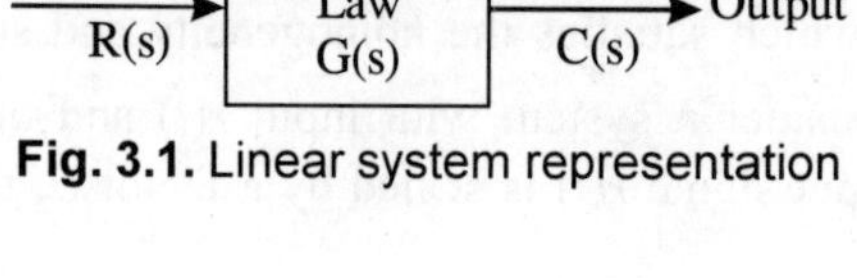

Fig. 3.1. Linear system representation

$$a_n \frac{d^n c(t)}{dt^n} + a_{n-1} \frac{d^{n-1} c(t)}{dt^{n-1}} + ... + a_1 \frac{dc(t)}{dt} + a_0 =$$

$$b_m \frac{d^m r(t)}{dt^m} + b_{m-1} \frac{d^{m-1} r(t)}{dt^{m-1}} + ... + b_1 \frac{dr(t)}{dt} + b_0 \qquad ...(3.1)$$

It is assumed that $n \geq m$. The response of the system may be determined by solving Eq. (3.1). However this is tedious. In the analysis and design of linear control systems, the Laplace Transform method is extensively used. Therefore, taking the Laplace transform of Eq. (3.1), assuming zero initial conditions, and rearranging, we get

$$G(S) = \frac{C(s)}{R(s)} = \frac{b_m s^m + b_{m-1} s^{m-1} + ... + b_1 s + b_0}{a_n s^n + a_{n-1} s^{n-1} + ... + a_1 s + a_o} \qquad ...(3.2)$$

where $G(s)$, $C(s)$ and $R(s)$ are the Laplace transforms of $g(t)$, $c(t)$ and $r(t)$ respectively.

Equation (3.2) describes the effect of the dynamic system on the input signal as it is transferred through the system to the output and hence is called transfer function. Thus *the transfer function of a system may be defined as the ratio of the Laplace transform of the output to the Laplace transform of the input, setting all initial conditions to zero*. It is defined only for a linear constant coefficient system and is independent of the system input. It is expressed only as a function of the complex variable s.

The concept of transfer function enables one to divide a complex system into a number of simpler blocks, determine the transfer function of each block and then obtain the overall system transfer function.

3.3. STEPS IN DETERMINING SYSTEM TRANSFER FUNCTIONS

The first step that must be mastered by the system engineer is the derivation of the describing equations of systems. The derivation of system equations involves application of physical laws that govern the particular system such as Newton's Laws of motion for mechanical systems. The steps in the derivation of transfer function are:

(a) First the input and output variables are identified among the system variables.

(b) The differential equation relating the input and the output is formulated.

(c) Assuming zero initial conditions, the differential equation is Laplace–transformed.

(d) The ratio of the Laplace transform of the output to the Laplace transform of the input is obtained. This is the transfer function of the system under consideration.

The method is illustrated in the following sections.

3.4. TRANSFER FUNCTIONS OF ELECTRICAL SYSTEMS

The three basic elements used extensively in electrical network systems are the resistance R, the inductance L and the capacitance C and are shown in Fig. 3.2

Electrical Element	Symbol	Representation
Resistance	R	
Inductance	L	
Capacitance	C	

Fig. 3.2. Basic electrical elements

The voltage drop across each of these elements when a current $i(t)$ is passed, is given by

$$e_R = R\,i(t)$$
$$e_L = L\,di(t)/dt$$
$$e_C = (1/C)\int i\,dt \qquad\qquad ...(3.3)$$

where e_R, e_L and e_C are the voltage drops across R, L and C respectively. Assuming zero initial conditions, the Laplace transform of Eq. (3.3) are

$$E_R(s) = R\,I(s)$$
$$E_L(s) = Ls\,I(s)$$
$$E_C(s) = (1/Cs)\,I(s) \qquad\qquad ...(3.4)$$

where R is in Ohms, L in henries and C in farads. Table 3.1 shows S.I. units of electrical components.

Electrical network equations are written by applying Kirchhoff's Laws. Kirchhoff's voltage law states that, *in any mesh or closed circuit, the algebraic sum of the voltage drops around the mesh is equal to zero.* In other words, the sum of the voltage drops equals to the applied voltage. The application of the voltage law gives rise to *mesh or loop equations.* Kirchhoff's current law states that, *at any node or junction, the algebraic sum of the currents is equal to zero.* In other words, the sum of the incoming currents is equal to the sum of the outgoing currents. The application of the current law gives rise to *nodal equations.* The following examples illustrate the derivation of transfer functions of simple electrical networks.

Table 3.1. Electrical Symbols and Units

Symbol	Quantity	SI Units
e or v	Voltage	Volts
i	Current	Amperes
L	Inductance	Henrys
C	Capacitance	Farads
R	Resistance	Ohms

Example 3.1 Mesh Analysis: *Series RLC circuit*

Consider a series RLC circuit shown in Fig. 3.3. Obtain $E_o(s)/E_i(s)$ by mesh analysis.

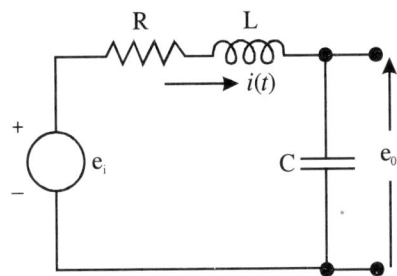

Fig. 3.3. Series RLC circuit for mesh analysis

Solution:

Applying Kirchhoff's voltage law, we obtain the differential equation

$$e_i(t) = Ri(t) + L\,di(t)/dt + (1/C)\int i(t)dt \qquad \text{...(3.5)}$$

Taking Laplace transform, we get

$$E_i(s) = RI(s) + LsI(s) - Li(0) + \frac{1}{Cs}I(s) - \frac{1}{C}i(0)$$

Assuming zero initial current, we obtain

$$E_i(s) = (R + Ls + \frac{1}{Cs})I(s)$$

or,

$$E_i(s) = \frac{1 + RCs + LCs^2}{Cs}I(s)$$

Hence,

$$I(s) = \frac{Cs}{1 + RCs + LCs^2}E_i(s) \qquad \text{...(3.6)}$$

If the current $I(s)$ is the desired output quantity, then the transfer function is given by

$$\frac{I(s)}{E_i(s)} = \frac{Cs}{1 + RCs + LCs^2} \qquad \text{...(3.7)}$$

If the voltage across the capacitor is the desired output quantity, then the transfer function $E_o(s)/E_i(s)$ is

$$\frac{E_o(s)}{E_i(s)} = \frac{1}{Cs} \cdot \frac{I(s)}{E_i(s)} = \frac{1}{1 + RCs + LCs^2} \qquad \text{...(3.8)}$$

Similarly a transfer function may be obtained for voltages across R or L as the output quantity.

For simple electrical networks, the transfer function may be defined as

$$\frac{E_o(s)}{E_i(s)} = \frac{\text{Laplace Transform of the output impedance}}{\text{Laplace Transform of the total impedance}} = \frac{Z_o(s)}{Z_T(s)} \quad ...(3.9)$$

This definition enables us to write the transfer functions of simple electrical networks directly.

Example 3.2. Nodal Analysis: *Series RLC circuit*

Obtain $E_o(s)/E_i(s)$ *of the circuit in Fig. 3.4 by Nodal analysis.*

Solution:

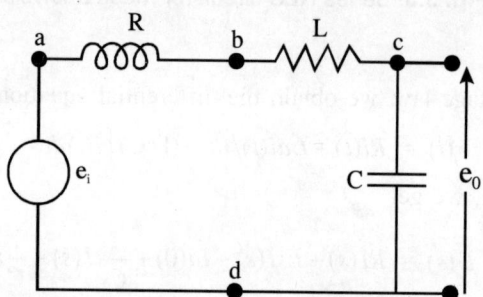

Fig. 3.4. Series RLC circuit for nodal analysis

Fig. 3.4 shows the same circuit with the nodes, *a, b, c* and *d*. The node *d* with respect to which voltages of other nodes are measured is considered as *the reference node*. Thus, the voltages at nodes *a* and *b* are written as e_a and e_b respectively. The voltage e_a is the source voltage $e_i(t)$ and is known. The unknown node voltages are e_b and e_c. Hence, two equations one at node *b* and another at node *c* are required to determine the values of e_b and e_c. Assuming zero initial conditions, the transform node equations are:

At node b: $\dfrac{E_b(s) - E_i(s)}{L_s} + \dfrac{E_b(s) - E_c(s)}{R} = 0$ ...(3.10)

At node c: $\dfrac{E_c(s) - E_b(s)}{R} + \dfrac{E_c(s)}{(1/Cs)} = 0$ ...(3.11)

Rearranging the above two equations, we get

$$-E_i(s)\left(\frac{1}{Ls}\right) + E_b(s)\left(\frac{1}{Ls} + \frac{1}{R}\right) - E_c(s)\left(\frac{1}{R}\right) = 0 \quad ...(3.12)$$

$$-E_b(s)\left(\frac{1}{R}\right) + E_c(s)\left(\frac{1}{R} + Cs\right) = 0 \quad ...(3.13)$$

Let us consider the voltage $E_c(s)$ across C as the output quantity. Then, the transfer function $E_c(s)/E_i(s)$ from Eqs. (3.12) and (3.13) is given by

$$\frac{E_c(s)}{E_i(s)} = \frac{1}{1 + RCs + LC\,s^2} \qquad ...(3.14)$$

which is same as Eq. (3.8)

Example 3.3 Mesh Analysis: *Multiloop Network*

Fig. 3.5 shows as electrical network with three meshes. Obtain $E_o(s)/E_i(s)$ by mesh analysis.

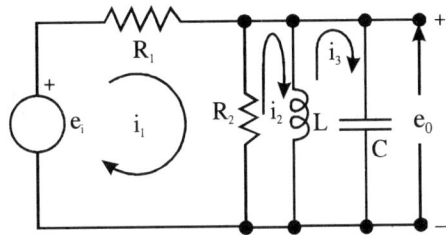

Fig. 3.5. Multiloop network for mesh analysis

Solution:

Let i_1, i_2 and i_3 represent the three mesh currents. The three mesh equations are:

$$(R_1 + R_2)\,I_1(s) - R_2 I_2(s) = E_i(s) \qquad ...(3.15)$$
$$-R_2\,I_1(s) + (R_2 + Ls)\,I_2(s) - LsI_3(s) = 0 \qquad ...(3.16)$$
$$-LsI_2(s) + (Ls + 1/Cs)\,I_3(s) = 0 \qquad ...(3.17)$$

The output voltage $E_o(s)$ is given by

$$E_o(s) = \frac{1}{\left(\dfrac{1}{R_2} + \dfrac{i}{Ls} + Cs\right)}\,I_1(s) = \frac{R_2 Ls}{R_2 + Ls + R_2 LCs^2}\,I_1(s) \qquad ...(3.18)$$

Solving the three mesh equations simultaneously, we get

$$I_1(s) = \left[\frac{R_2 + Ls + R_2 LC\,s^2}{R_1 R_2 + (R_1 + R_2)L\,s + R_1 R_2\,LC\,s^2}\right] E_i(s) \qquad ...(3.19)$$

The transfer function $I_1(s)/E_i(s)$ is given by the terms in the bracket.

By substituting Eq. (3.19) in Eq. (3.18) and rearranging, we get the required transfer function $E_o(s)/E_i(s)$ as

$$\frac{E_o(s)}{E_i(s)} = \frac{\tau_1 s}{1+\alpha\tau_1 s+LC\,s^2} \qquad\qquad ...(3.20)$$

where $\tau_1 = L/R_1$ and $\alpha = (R_1 + R_2)/R_2$.

The transfer function $E_o(s)/E_i(s)$ may be directly obtained without writing mesh equation. By definition of Eq. (3.9).

$$\frac{E_o(s)}{E_i(s)} = \frac{Z_o(s)}{Z_T(s)} \qquad\qquad ...(3.21)$$

$$Z_0(s) = \frac{R_2.Ls.(1/Cs)}{R_2 Ls + Ls(1/Cs) + (1/Cs)R_2}$$

$$= \frac{R_2 Ls}{R_2 + Ls + R_2 LC\,s^2}$$

$$Z_T(s) = R_1 + \frac{R_2 Ls}{R_2 + Ls + R_2 LC\,s^2} = \frac{R_1 R_2 + R_1 Ls + R_2 Ls + R_1 R_2 LCs^2}{R_2 + Ls + R_2\,LC\,s^2}$$

Hence,

$$\frac{E_o(s)}{E_i(s)} = \frac{R_2 Ls}{R_1 R_2 + (R_1 + R_2)Ls + R_2 LCs^2}$$

$$= \frac{\tau_1 s}{1+\alpha\tau_1 s+LC\,s^2} \qquad\qquad ...(3.22)$$

where τ_1 and α are as defined above.

Example 3.4. Nodal Analysis: *Multiloop Network*

Obtain $E_o(s)/E_i(s)$ of the circuit in Example 3.3 by nodal analysis.

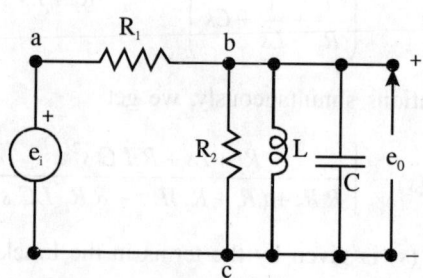

Fig. 3.6. Multiloop network for nodal analysis

Solution:

Fig. 3.6 shows the same circuit with nodes marked and node voltages. There is only one unknown node voltage e_b. Therefore only one node equation is required to obtain the transfer function $E_o(s)/E_i(s)$.

$$\frac{E_o(s) - E_i(s)}{R_1} + E_o(s)\left(\frac{1}{R_2} + \frac{1}{Ls} + Cs\right) = 0 \qquad \qquad ...(3.23)$$

or,

$$E_o(s)\left(\frac{1}{R_1} + \frac{1}{R_2} + \frac{1}{Ls} + Cs\right) - \frac{E_i(s)}{R_1} = 0 \qquad \qquad ...(3.24)$$

Hence,

$$\frac{E_o(s)}{E_i(s)} = \frac{1}{R_1} \cdot \frac{1}{\left(\dfrac{1}{R_1} + \dfrac{1}{R_2} + \dfrac{1}{Ls} + Cs\right)}$$

which after simplification yields the transfer function given in Eq.(3.20) or Eq. (3.22).

From the above two examples, it is seen that the *nodal analysis requires less number of equations and can be solved with ease.* Hence it is preferred.

3.5. RULES FOR FORMULATION OF NODE EQUATIONS

A close examination of Eq. (3.12), (3.13) and (3.24) reveal the following rules which simplify the formulation of node equations.

(a) An equation is formed for each node using Kirchhoff's current law.

(b) The equation includes the following terms:

(i) The sum of admittances of all branches connected to a node multiplied by that node voltage. This term is *positive*.

(ii) The admittance of each branch multiplied by the node voltage at the other end of the branch. Their sum is *negative*.

(iii) The sum of the terms in (i) and (ii) are equal to zero.

(c) The number of equations required is equal to the number of unknown node voltages.

With these rules, the node equations may be written directly by inspection.

3.6. MECHANICAL SYSTEMS

Practical control systems consist of mechanical and electrical components. Systems containing mechanical elements are known as mechanical systems. The mechanical systems are of two types:

   (i) Translational systems

  (ii) Rotational systems

The behaviour of mechanical systems are governed by Newton's Law of motion and D' Alembert's principle. Newton's second Law of motion states that *the force on a mass M is equal to the product of mass and its acceleration.* D' Alembert's principle states that *the algebraic sum of the forces acting along a direction is equal to zero.* In other words, the sum of the applied forces is equal to the sum of the forces resisting the motion.

3.6.1. Mechanical Translational Systems

In a translational system, the motion is along a straight line. It is a linear motion. The three basic elements in a translational system are mass, damping and elastance.

Mass M is an inertial element which stores kinetic energy. Force acting on a mass produces acceleration. Hence, the reaction force $F_m(s)$ is

$$F_m(s) = MA(s) = MsV(s) = Ms^2X(s) \qquad \qquad ...(3.25)$$

where $F_m(s)$, $A(s)$, $V(s)$ and $X(s)$ are Laplace transforms of force, acceleration, velocity and displacement respectively. *One terminal of the mass is always considered to have the motion of the reference and the other terminal the motion of the mass.*

The damping or viscous friction B is the element which absorbs energy. The damping force $F_B(s)$ is equal to the product of damping B and the relative velocity of its two ends. Thus

$$F_B(s) = B[V_1(s) - V_2(s)] = BV(s) = BsX(s) \qquad \qquad ...(3.26)$$

A dashpot is good example of viscous friction.

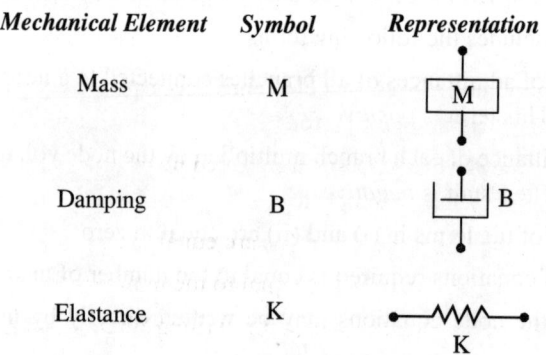

Fig. 3.7. Mechanical translational elements

The elastance or stiffness K is an element which stores potential energy. According to Hooke's Law, the force $F_k(s)$ is equal to the product of the elastance K and the relative displacement at the two ends. Thus

$$F_k(s) = K[X_1(s) - X_2(s)] = KX(s) \qquad \qquad ...(3.27)$$

A linear spring is an example of this type of element.

The network representations of mass, damping and elastance are shown in Fig. 3.7.

Table 3.2. shows symbols and SI units of translational mechanical systems components

Table 3.2. Mechanical Translational Symbols and Units

Symbol	Parameter	Unit
F	Force	Newtons
X	Displacement	Meters
V	Velocity	Meter/sec
A	Acceleration	Meter/sec^2
M	Mass	Kilograms
K	Stiffness coefficient	Newtons/ meter
B	Damping coefficient	Newtons/(meter/sec)

3.6.2. Mechanical Rotational Systems

In a rotational system, the motion is about a fixed axis and is angular. The three basic elements in a rotational system are moment of inertia J, viscous damping B and elastance K.

When a torque T is applied to a body having a moment of inertia J, it produces an angular acceleration $\alpha(s)$. The reaction torque $T_J(s)$ is equal to the product of moment of inertia and the angular acceleration. Thus,

$$T_J(s) = J\alpha(s) = Js\Omega(s) = Js^2\,\Theta(s) \qquad \qquad ...(3.28)$$

where $\alpha(s)$, $\Omega(s)$ and $\Theta(s)$ are angular acceleration angular velocity and angular displacement respectively. The element J is an energy storing element and stores rotational kinetic energy.

Viscous damping occurs when a body moves through a fluid. To produce motion of the body through fluid, a torque must be applied to overcome the reaction damping torque. The damping torque $T_B(s)$ is equal to the product of the damping B and the relative angular velocity of its two ends. Thus,

$$T_B(s) = B[\Omega_1(s) - \Omega_2(s)]$$
$$= B\Omega(s) = Bs\Theta(s)$$

A torque, applied to a spring, twists the spring by an angle $\Theta(s)$. The applied torque is transmitted through the spring and appears at the other end. The spring reaction torque $T_K(s)$ is equal to the product of elastance K and the relative angular displacement of its two ends. Thus,

$$T_K(s) = K[\Theta_1(s) - \Theta_2(s)] = K\Theta(s) \qquad \qquad ...(3.29)$$

The network representation of moment of inertia, damping and elastance is shown in Fig. 3.8.

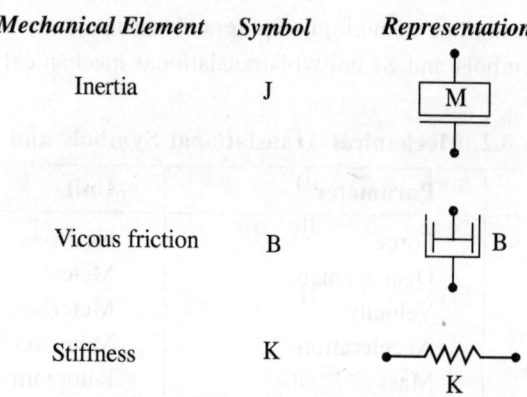

Mechanical Element	Symbol	Representation
Inertia	J	M
Vicous friction	B	B
Stiffness	K	K

Fig. 3.8. Mechanical rotational elements

Table 3.3 shows symbols and *SI* units of rotational mechanical system components.

Table 3.3. Mechanical Rotational Symbols and Units

Symbol	Parameter	Unit
T	Torque	Newton-meters
Θ	Angle	Radians
Ω	Angular Velocity	Radians/second
α	acceleration	Radians/second^2
J	Moment of inertia	Kilogram-meters^2
K	Stiffness coefficient	Newton-meter/radian
B	Damping coefficient	Newton-meter/(rad./sec.)

3.7. RULES FOR FORMULATION OF MECHANICAL SYSTEM EQUATIONS

Mechanical systems are analogous to the electrical systems. Once the transfer function of a mechanical system is derived, it may be analyzed using well-developed Laplace transform technique. The rules for formulating the mechanical system equations are similar to those for electrical system. The simplifying rules are:

(a) Draw the mechanical system and mark the resulting linear or angular displacements in the system due to the application of force or torque.

(b) With the linear or angular displacement as variable nodes, draw the mechanical network between the reference node and the variable nodes. Terminals of elements having same displacement are connected together. One end of M and J is always connected to the reference node.

(c) At each variable node, an equation is formed. This includes the following terms:

    (i) The node displacement multiplied by the elements (with appropriate Laplace operators) connected at that node. This term is *positive*.

    (ii) The displacement at the other end of each element multiplied by the element with appropriate Laplace operator. This term is *negative*.

    (iii) The sum of terms in (i) and (ii) is equal to zero.

(d) The number of equations to be written is equal to the number of unknown displacements.

3.8. TRANSFER FUNCTIONS OF MECHANICAL SYSTEMS

In this section, the method of formulating the describing equations of mechanical systems, applying the above rules, are illustrated with suitable examples. The most important step is to draw the mechanical network of the mechanical system. Once the proper network is drawn, the system equations may be simply formulated. The method of drawing the network is also explained in detail.

Example 3.5. Mechanical Translational System.

Consider a simple mechanical translational system consisting of a mass, a damper and a spring as shown in Fig. 3.9(a). Obtain X(s)/F(s).

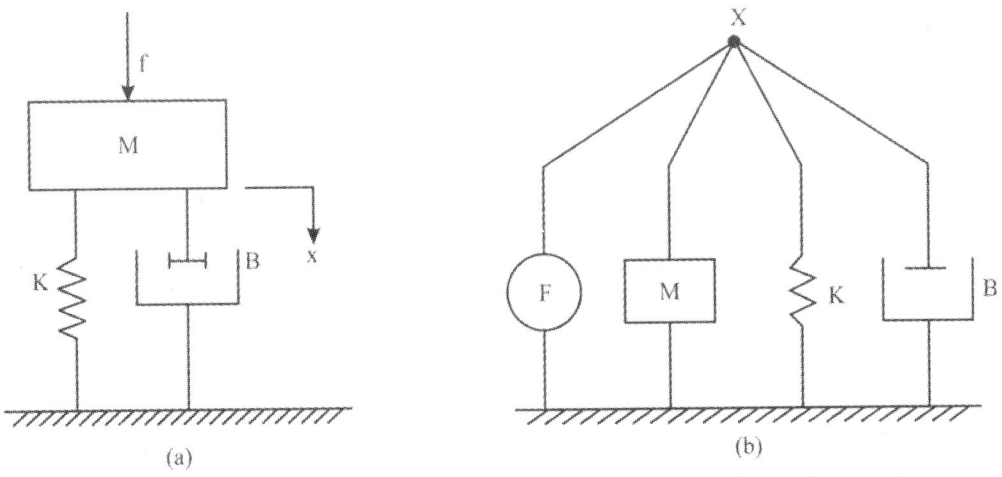

(a)   (b)

Fig. 3.9. (a) Simple mechanical translational system (b) Mechanical network

Solution:

A downward force is applied on the mass. Assuming the system is initially at rest, the mass M is displaced vertically downward on application of the force. The ends of the damper and the spring connected to the solid mass are also displaced by the same amount. The other ends of B and K are connected to the reference node. Hence, there is no displacement.

To draw the mechanical network, *first* the reference node is drawn as shown in Fig. 3.9(b) by a hatched horizontal line. The displacement node X is marked well-above the reference node. Consider now each element in the mechanical system, visualize the displacements of its two ends and draw the element between the two nodes in the mechanical network. It may be noted that *mass, inertia, applied force or torque is always drawn between the reference node and the corresponding variable node*. In this example, since there is only one displacement, all the mechanical elements are connected between the reference node and the variable node X. The resulting diagram is the mechanical network and is shown in Fig. 3.9(b). There is only one unknown node variable X and the system behaviour is described by

$$F(s) = (M s^2 + Bs + K)\, X(s) \qquad\qquad ...(3.30)$$

where $F(s)$ and $X(s)$ are Laplace transform of force and displacement respectively. The transfer function $X(s)/F(s)$ is given by

$$\frac{X(s)}{F(s)} = \frac{1}{Ms^2 + Bs + K} \qquad\qquad ...(3.31)$$

It may be noted that Eq. (3.31) is similar in form to Eq.(3.14). Since the system has only one displacement or unknown variable, it is known as *single-degree freedom system*.

Example 3.6. Mechanical Rotational system:
Consider a mechanical rotational system consisting of a disk with moment of inertia J mounted on a ball bearing having viscous friction B and attached to a shaft with elastance K as shown in Fig. 3.10(a). Obtain $\Theta_1(s)/T(s)$, $\Theta_2(s)/T(s)$ *and* $(\Theta_2(s)/\Theta_1(s)$.

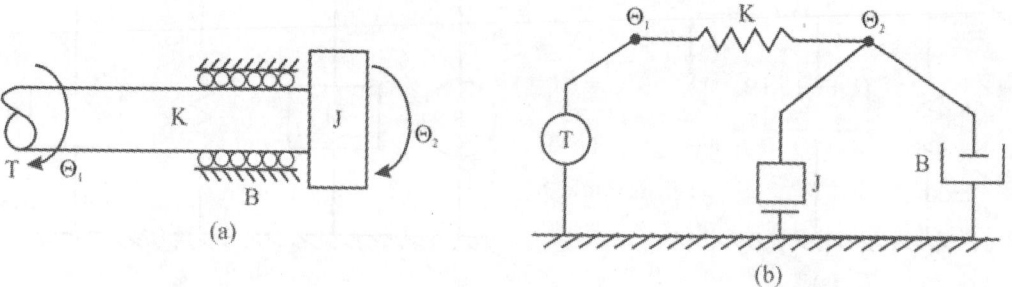

(a)

(b)

Fig. 3.10. (a) Mechanical rotational system (b) Mechanical network

Solution:

A Torque T is applied at the free end of the shaft. Because of shaft stiffness, the angular displacements at the two ends of the shafts are different. Let the displacement at the free end be θ_1 and that at the other end be θ_2. The displacement of J and B is the same as θ_2.

To draw the mechanical network, *first* the reference node is drawn as shown in Fig. 3.10(b). The two angular displacements θ_1 and θ_2 are the two variable nodes and are marked well-above the reference node as well as reasonably separated from each other, as shown in Fig. 3.10(b). Now consider each element of the mechanical rotational system, visualize the angular displacements at the two ends and draw it accordingly between the nodes in the network. The resulting diagram is the required mechanical network. Since there are two unknown variable nodes, two equations describe the system and they are:

At node θ_1: $\qquad\qquad T(s) = K\Theta_1(s) - K\Theta_2(s)$ $\qquad\qquad\qquad$ ...(3.32)

At node θ_2: $\qquad\qquad 0 = -K\Theta_1(s) + (Js^2 + Bs + K)\,\Theta_2(s)$ $\qquad\qquad$ ...(3.33)

The two equations may be solved to obtain the transfer function $\Theta_1(s)/T(s)$ or $\Theta_2(s)/T(s)$ or $\Theta_2(s)/\Theta_1(s)$.

By combining Eq. (3.32) and (3.33), we get

$$\frac{\Theta_2(s)}{T(s)} = \frac{1}{Js^2 + Bs} \qquad\qquad\qquad\qquad ...(3.34)$$

By substituting the value of $\Theta_2(s)$ obtained by Eq. (3.34) in Eq. (3.33), we get

$$\frac{\Theta_1(s)}{T(s)} = \frac{Js^2 + Bs + K}{K(Js^2 + Bs)} \qquad\qquad\qquad ...(3.35)$$

From Eq. (3.33), we get

$$\frac{\Theta_2(s)}{\Theta_1(s)} = \frac{K}{Js^2 + Bs + K} \qquad\qquad\qquad ...(3.36)$$

Since this system has two angular displacements, it is of *two-degree freedom*.

Example 3.7 Multiple Translational System

In Fig. 3.11(a), a multiple translational system consisting of M_1, M_2, B_1, B_2, K_1 and K_2 is shown. Obtain $X_2(s)/X_1(s)$.

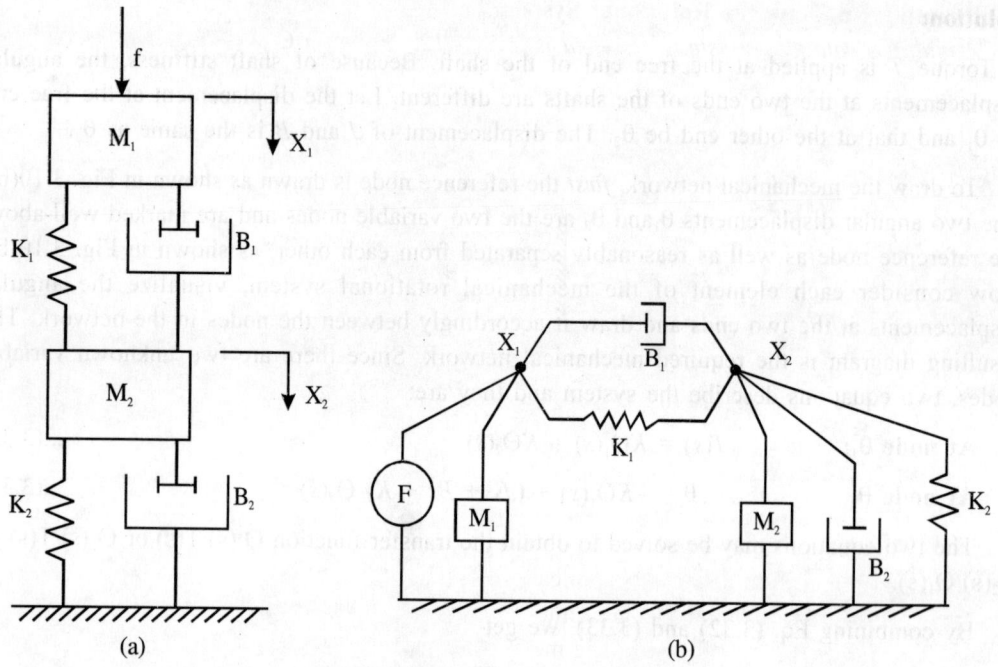

Fig. 3.11: (a) Two-degree translational system, (b) Mechanical network

Solution:

If a downward force f is applied on M_1, it is displaced by x_1. The ends of B_1 and K_1 connected to M_1 are also displaced by x_1; the other ends of B_1 and K_1 connected to M_2 are displaced by x_2. Hence M_2 and the ends of B_2 and K_2 connected to M_2 are all displaced by the same amount x_2. The other ends of B_2 and K_2, being connected to reference node, do not undergo any displacement.

To obtain the mechanical network, draw the reference line and mark the variable nodes X_1 and X_2. The force and the mass M_1 are drawn between reference node and the variable node X_1. B_1 and K_1 are drawn between X_1 and X_2 nodes. M_2, B_2 and K_2 are drawn between reference node and the node X_2. The resulting diagram is the mechanical network shown in Fig. 3.11(b).

At node X_1 : $\qquad F(s) = (M_1 s^2 + B_1 s + K_1)\, X_1(s) - (B_1 s + K_1)\, X_2(s)$ ...(3.37)

At node X_2 : $\qquad 0 = -(B_1 s + K_1)\, X_1(s) + (M_2 s^2 + B_2 s + K_2 + B_1 s + K_1)\, X_2(s)$

...(3.38)

The transfer function $X_2(s) / X_1(s)$ is given by

$$\frac{X_2(s)}{X_1(s)} = \frac{B_1 s + K_1}{M_2 s^2 + (B_1 + B_2)s + K_1 + K_2} \qquad\qquad\text{...(3.39)}$$

Example 3.8. Multiple Rotational System
A multiple rotational system is shown in Fig. 3.12(a) Obtain $\Theta_3(s)/\Theta_2(s)$.

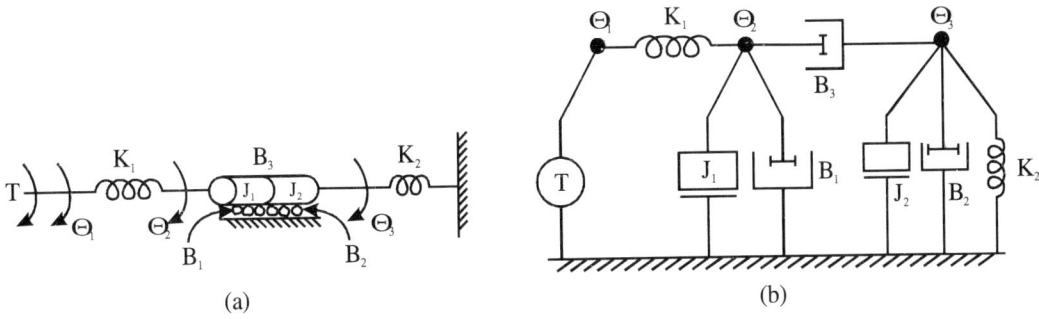

(a) (b)

Fig. 3.12: (a) Multiple rotational system; (b) Mechanical network

Solution:

It has two disks with moment of inertia J_1 and J_2. Damping exists between the disks and also between the frame and each disk. Application of a torque to the free end of the spring K_1 produces three different angular displacements: θ_1 at the free end of K_1, θ_2 at the other end of K_1 and θ_3 at the movable end of K_2. The mechanical network shown in Fig. 3.12(b) is obtained by following the steps described in the previous examples. Because there are three different displacements, this system is of *three degree freedom*. The three node equations are:

At node θ_1 : $T(s) = K_1 \Theta_1(s) - K_1 \Theta_2(s)$ ...(3.40)

At node θ_2 : $0 = -K_1 \Theta_1(s) + (J_1 s^2 + B_1 s + K_1 + B_3 s) \Theta_2(s) - B_3 s \Theta_3(s)$ ...(3.41)

At node θ_3 : $0 = -B_3 s \Theta_2(s) + (J_2 s^2 + B_2 s + B_3 s + K_2) \Theta_3(s)$ ...(3.42)

The transfer function $\Theta_3(s)/\Theta_2(s)$ is given by

$$\frac{\Theta_3(s)}{\Theta_2(s)} = \frac{B_3 s}{J_2 s^2 + (B_2 + B_1)s + K_2} \qquad ...(3.43)$$

Similarly, the other transfer functions may be derived.

3.9. GEAR TRAINS

In a mechanical system, when a load is coupled to a drive motor, the moment of inertia and damping relative to the motor are important. Gear trains are often used because

(a) It is usually economical to design servo motors which run at a much higher speed than that required by the load.

(b) It permits higher load acceleration for a given motor.

(c) It reduces the effect of load on the drive motor.

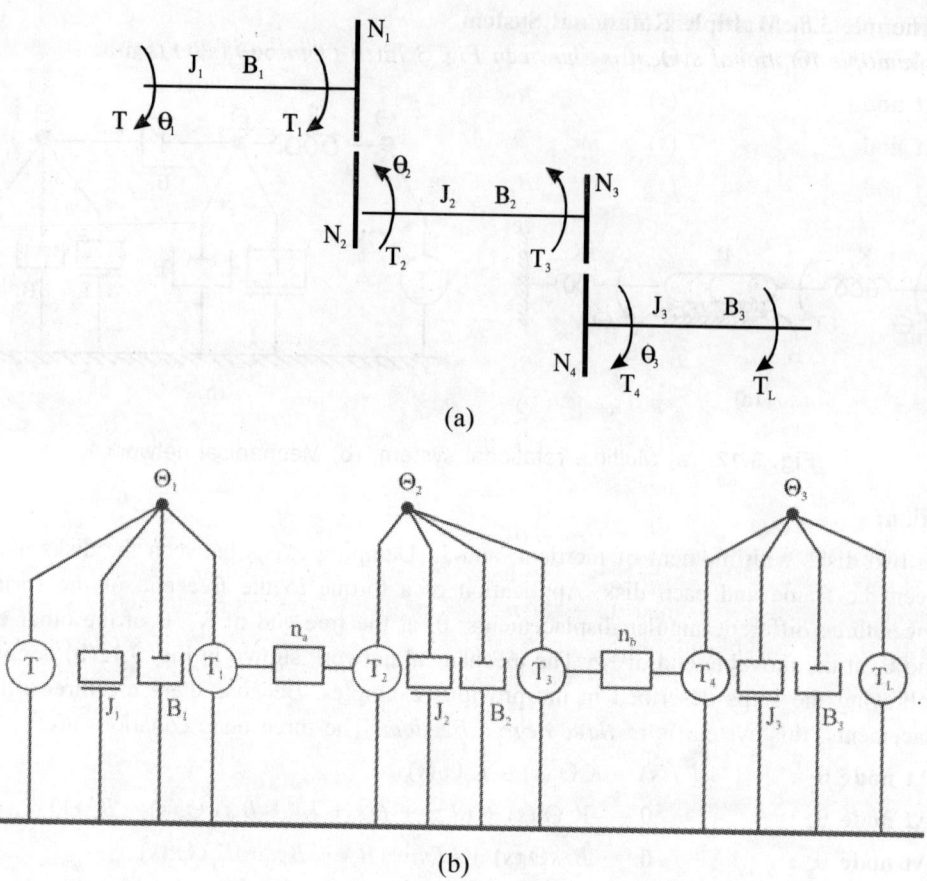

(a)

(b)

Fig. 3.13. (a) Rotational system with gears; (b) Mechanical network

Fig. 3.13 (a) shows a multiple gear rotational system. T represents the applied torque, N the number of teeth on each gear and n the gear ratio. Since the shaft length between gears is very short, the stiffness is assumed to be infinite. Application of the torque T produces three different angular displacements, θ_1, θ_2 and θ_3 as shown in Fig. 3.13(a). At each gear pair, two torques are produced, a restraining torque and a drive torque. T_1 is the load on gear 1 produced by the rest of the gear train. T_2 is the drive torque transmitted to gear 2 to drive the rest of the gear train. T_3 is the load on gear 3 and T_4 the drive load. T_L is the load torque. These torques are inversely proportional to the speed of the respective gears. The following fundamental relations are apparent:

$$T_1/T_2 = N_1/N_2 = \theta_2/\theta_1 = n_a \qquad \qquad ...(3.44)$$

$$T_3/T_4 = N_3/N_4 = \theta_3/\theta_2 = n_b \qquad \qquad ...(3.45)$$

The mechanical network for this system is shown in Fig. 3.13 (b). Since there are three variable nodes, Θ_1, Θ_2 and Θ_3, the three equations which describe the system are:

At node Θ_1 : $T(s) = (J_1s^2 + B_1s)\,\Theta_1(s) + T_1(s)$ ...(3.46)

At node Θ_2 : $T_2(s) = (J_2s^2 + B_2s)\,\Theta_2(s) + T_3(s)$ ...(3.47)

At node Θ_3 : $T_4(s) = (J_3s^2 + B_3s)\,\Theta_3(s) + T_L(s)$ ...(3.48)

Combining Eq. (3.44) to (3.48), we get

$$T(s) = (J_1s^2 + B_1s)\,\Theta_1(s) + n_a(J_2s^2 + B_2s)\,\Theta_2(s) +$$
$$n_a n_b(J_3s^2 + B_3s)\,\Theta_3(s) + n_a n_b\,T_L(s) \qquad ...(3.49)$$

In terms of $\Theta_1(s)$,

$$T(s) = (Js^2 + Bs)\,\Theta_1(s) + n_a n_b\,T_L(s) \qquad ...(3.50)$$

where J, the equivalent inertia is given by

$$J = J_1 + n_a^2 J_2 + n_a^2 n_b^2 J_3 \qquad ...(3.51)$$

and B, the equivalent damping is

$$B = B_1 + n_a^2 B_2 + n_a^2 n_b^2 B_3$$

It is apparent that, if the gear ratios, n_a and n_b are small, the load inertia and damping are negligible. If the resultant load torque $n_a n_b\,T_L$ is negligible, the transfer function $\Theta_1(s)/T(s)$ is given by

$$\frac{\theta_1(s)}{T(s)} = \frac{1}{Js^2 + Bs} \qquad ...(3.52)$$

3.10. BLOCK DIAGRAMS

System diagrams or engineering drawings of control systems are frequently congested with details and the operation of the systems cannot be readily understood from the diagrams. The diagram may also contain more details so that to make a reference to it is often tedious. *A block diagram is commonly used to represent the needed data in a compact form and to show the functions performed by each sub-system.* It is a pictorial representation of the functions performed and of the flow of signal. Each subsystem is shown by a block and labeled with the name of the component. The interconnections of the blocks are shown by directed lines drawn from one to the other. The flow of signals is always along the indicated direction. The input and the output of each block are also labeled and the transfer function is also written inside the block.

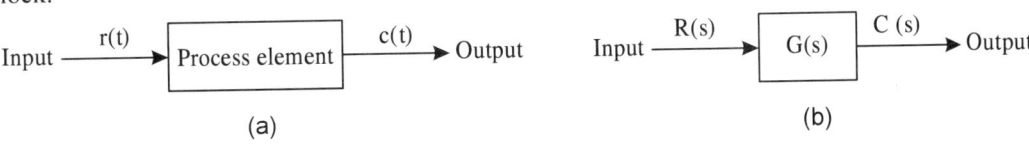

(a)   (b)

Fig. 3.14. (a) Process element; (b) block diagram representation

Fig. 3.14 shows a process element and its block diagram representation. The signal flowing into the block is the *input* and that flowing out of the block is the *output*. Thus, in a block diagram, the flow of signals is *unidirectional*.

3.10.1. Advantages

The advantages of the block diagram representation are:

(1) It is a pictorial representation of functions performed by each sub-system.

(2) It indicates more realistically the flow of signals in actual systems.

(3) The functional operation of the system can be more readily visualized by examination of the block diagram.

(4) It is possible to evaluate the contribution of each sub-system to the overall system performance.

(5) It is easy to form the overall block diagram for the entire system.

(6) It contains information concerning the dynamic behaviour of the system.

(7) It shows the similarities between various physical systems, thus enabling us to analysis them with the same technique.

3.10.2. Disadvantages

The disadvantages of the block diagram approach are:

(1) It does not contain information concerning the physical construction of the system.

(2) The main source of energy is not explicitly shown.

(3) The block diagram of a given system is not unique since a number of different block diagrams may be drawn depending upon the viewpoint of the analysis.

Any linear control system may be represented by a block diagram comprising blocks, summing points and branch points. A summing point is indicated in the block diagram by a circle with or without a cross. The plus or minus sign at each arrow-head indicates whether the signal is to be added or subtracted. The quantities being added or subtracted must have same dimensions and the same units. An error detector or a comparator is a summing point where the feedback signal is subtracted from the reference signal. A branch point is indicated by a dot from which signal fans out to other blocks or summing points undiminished.

3.11. BASIC BLOCK DIAGRAM OF A CLOSED-LOOP SYSTEM

A block diagram of a closed-loop system is shown in Fig. 3.15. Complicated systems may be reduced to this basic and standard form. The various quantities shown in the block diagram are defined below:

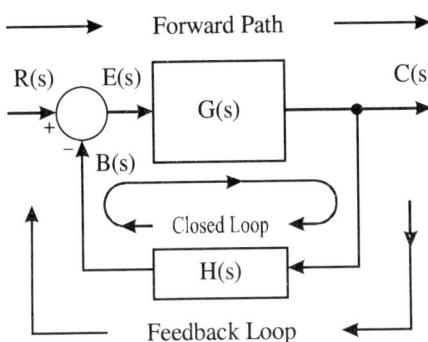

Fig. 3.15. Block diagram of a closed-loop system.

$R(s)$ is the *reference input*.

$E(s)$ is the *actuating or error signal* and is equal to $R(s) - B(s)$.

$B(s)$ is the *feedback signal*. It is equal to the product of the feedback transfer function $H(s)$ and the output $C(s)$. It has the same dimension and the same unit as those of $R(s)$.

$C(s)$ is the *output or controlled variable* and is given by the product of the transfer function $G(s)$ and the input to the block, $E(s)$.

$G(s)$ is the *forward or open-loop transfer function* and is equal to $C(s)/E(s)$.

$M(s)$ is the *closed-loop transfer function* and is equal to $C(s)/R(s)$. It relates the closed-loop system dynamics to the dynamics of forward and feedback elements.

$H(s)$ is the *feedback transfer function* and is equal to $B(s)/C(s)$. It modifies $C(s)$ to give proper output $B(s)$ so that the feedback signal $B(s)$ can be compared in the error detector.

$G(s)H(s)$ is the *loop transfer function* and is equal to $B(s)/E(s)$.

For the system shown in Fig. 3.15, the closed-loop transfer function $C(s)/R(s)$ and the error-to-input ratio $E(s)/R(s)$ are derived below.

3.11.1. Closed-loop Transfer Function M(s)

$$C(s) = G(s)\ E(s)$$
$$E(s) = R(s) - B(s) \qquad \qquad ...(3.53)$$
$$= R(s) - H(s)\ C(s) \qquad \qquad ...(3.54)$$

Substituting Eq. (3.54) in Eq. (3.53) and rearranging,

$$M(s) = \frac{C(s)}{R(s)} = \frac{G(s)}{1 + G(s)H(s)} = \frac{\text{Forward Transfer function}}{1 + \text{loop Transfer function}} \qquad ...(3.55)$$

3.11.2. Error-to-Input Ratio E(s)/R(s)

From Eq. (3.55),

$$C(s) = \frac{G(s)}{1+G(s)H(s)} \cdot R(s)$$

$$= G(s) \cdot E(s)$$

Hence,

$$\frac{E(s)}{R(s)} = \frac{1}{1+G(s)H(s)} = \frac{1}{1+\text{loop transfer function}} \qquad ...(3.56)$$

It may be noted that in Eqs. (3.55) and (3.56), the denominator is same and is given by

$$F(s) = 1 + G(s)\,H(s) \qquad ...(3.57)$$

F(s) characterizes the performance of a system and hence called *characteristic function*. If H(s) is unity, the system is a *unity feedback system*.

3.12. PROCEDURE FOR DRAWING BLOCK DIAGRAMS

The following steps help in drawing the block diagram of a system:

(a) Divide the system into a number of subsystems containing at least one functionally important component. If there is loading effect between the components, they should be combined into a single subsystem.

(b) Find the transfer function of each subsystem and represent it in a block.

(c) Assemble them into a complete block diagram according to the flow of signals.

Example 3.9. *Consider the RC circuit shown in Fig. 3.16(a) Represent it by a block diagram.*

Solution:

The given circuit may be divided into two subgroups consisting of *R* and *C* respectively. The equations in Laplace transform are:

$$I(s) = Cs[E_i(s) - E_o(s)] \qquad ...(3.58)$$

$$E_o(s) = RI(s) \qquad ...(3.59)$$

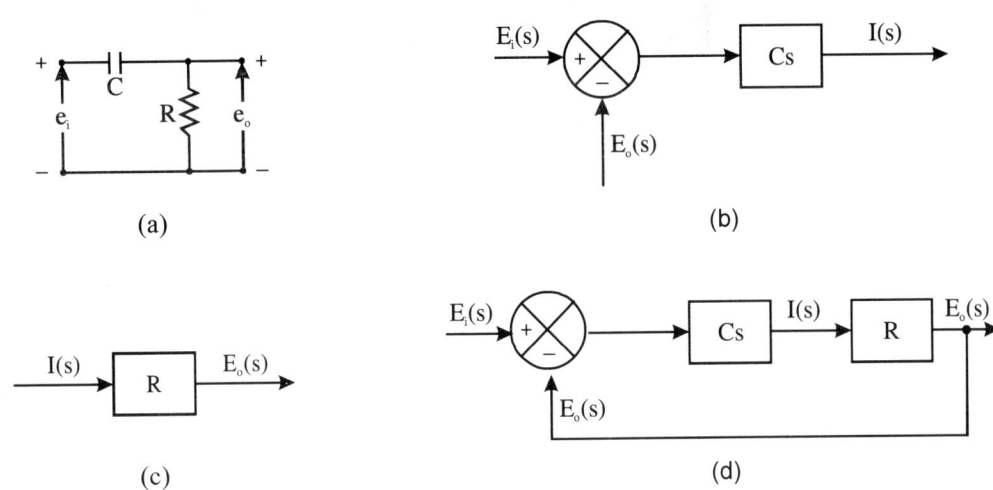

Fig. 3.16. (a) RC circuit (b) Block diagram representation of C
(c) Block diagram representation of R (d) Block diagram representation of RC network

Equation (3.58) represents a summing operation and the corresponding block diagram is shown in Fig. 3.16(b). Equation (3.59) represents the block shown in Fig. 3.16(c). The complete block diagram shown in Fig. 3.16(d) is obtained by assembling these two blocks.

3.13. BLOCK DIAGRAM REDUCTION TECHNIQUES

In general, the block diagram of practical systems is complicated involving multiple loops and several inputs. Multiple loops may arise due to inherent nature of several interconnected loops within the system or due to intentional inclusion of additional loops to obtain satisfactory performance of the system. Multiple inputs are inherent in all practical systems because of the presence of disturbances such as noise. To analyze the system, the block diagram of the system should be reduced to the basic form shown in Fig. 3.15. The reduction is carried out by means of simple transformation theorems illustrated in Table 3.4.

Table 3.4. Simple transformations to reduce a complex block diagram

S.No.	Manipulations	Original Block Diagrams	Equivalent Block Diagrams
1.	Interchanging summing points	A → (+,−) → A–B → (+,+) → A–B+C ; B↑ , C↑	A → (+,+) → A+C → (+,−) → A–B+C ; C↑ , B↑
2.	Splitting of summing point	C↓ ; A → (+,+,−) → A–B+C ; B↑	A → (+,−) → A–B → (+,+) → A–B+C ; B↑ , C↓
3.	Interchanging of blocks	A → G_1 → AG → G_2 → AG_1G_2	A → G_2 → AG_2 → G_1 → AG_1G_2
4.	Combining cascaded blocks	A → G → AG_1 → G_2 → AG_1G_2	A → G_1G_2 → AG_1G_2
5.	Combining feed forward blocks	A → G_1 → AG_1 , A → G_2 → AG_2 → (+,−) → AG_1+AG_2	A → $G_1 + G_2$ → AG_1+AG_2
6.	Shifting summing point to left of a block	A → G → AG → (+,−) → AG–B ; B↑	A → (+,−) → A–$\frac{B}{G}$ → G → AG–B ; $\frac{B}{G}$↑ ← $\frac{1}{G}$ ← B
7.	Shifting summing point to right of a block	A → (+,−) → A–B → G → AG–BG ; B↑	A → G → AG → (+,−) → AG–BG ; B ; B → G → BG
8.	Shifting branch point to left of a block	A → G → AG ; AG	A → G → AG ; A → G → AG
9.	Shifting branch point to right of a block	A → G → AG ; A	A → G → AG ; AG → $\frac{1}{G}$ → A
10.	Shifting branch point to left of summing point	A → (+,−) → A–B ; A–B ; B↑	B↓ ; A → (+,−) → A–B ; A → (+,−) → A–B ; B↑

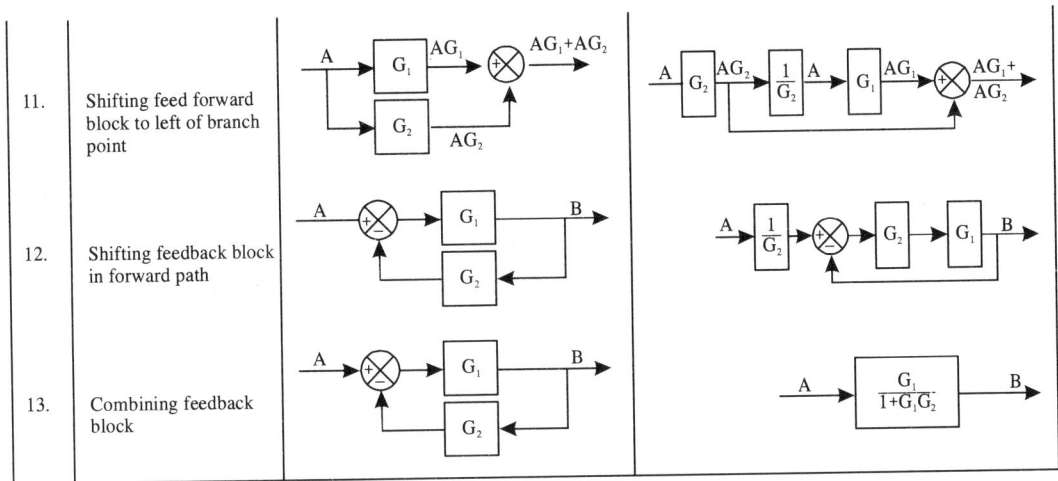

11.	Shifting feed forward block to left of branch point		
12.	Shifting feedback block in forward path		
13.	Combining feedback block		

A general rule for reducing a complicated block diagram is to:

(i) Move summing points

(ii) Interchange summing points

(iii) Rearrange branch points

(iv) Reduce the internal feedback loops.

The application of the transformation is illustrated with an example.

Example 3.10. *Consider the multiloop system shown in Fig. 3.17(a). Obtain C(s)/R(s).*

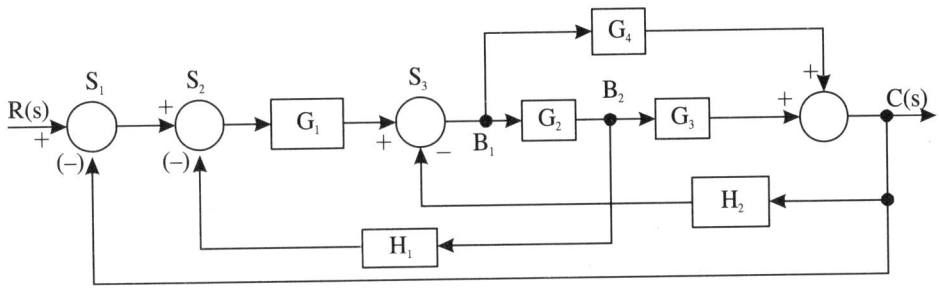

(a) Block diagram of a multiple loop system

Solution:

We have first to reduce the block diagram to the standard one as shown in Fig. 3.17(b) by moving and eliminating summing points and branch points judiciously.

It has two feedback loops and two forward paths.

(i) By moving the summing point S_3 between S_1 and S_2, we obtain Fig. 3.17(b).

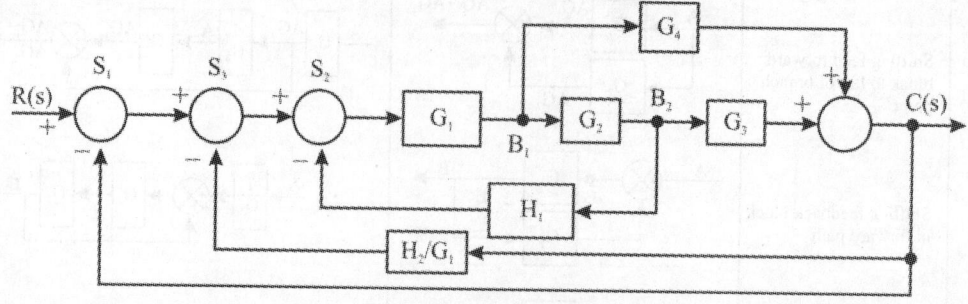

(b) Block diagram after moving S_3

(ii) Then, moving the branch point B_2 to B_1 shown in Fig. 3.17(b), we get 3.17(c).

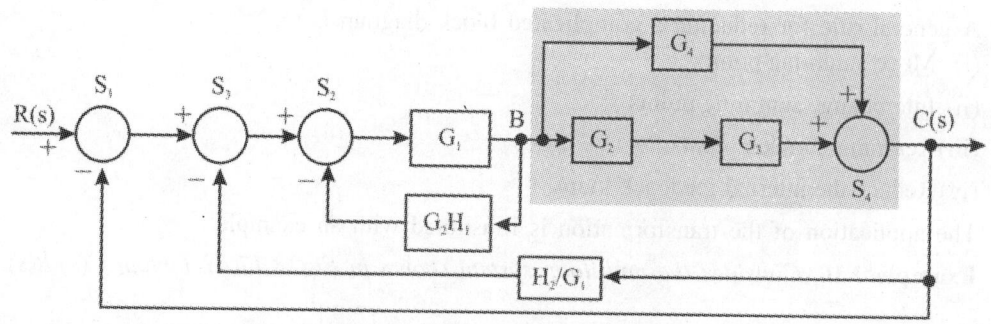

(c) Block diagram after moving B_2

(iii) Eliminating the feed-forward loop shown in Fig. 3.17(c), we obtain Fig. 3.17(d).

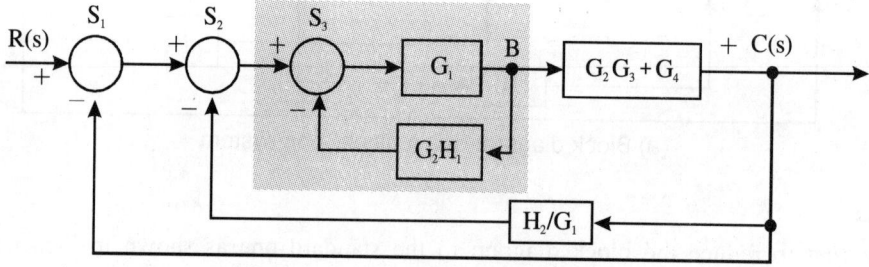

(d) Block diagram after eliminating Feed-forward loop

(iv) Eliminating the feedback loop $(G_1 G_2 H_1)$ shown in Fig. 3.17(d) will result in Fig. 3.17(e).

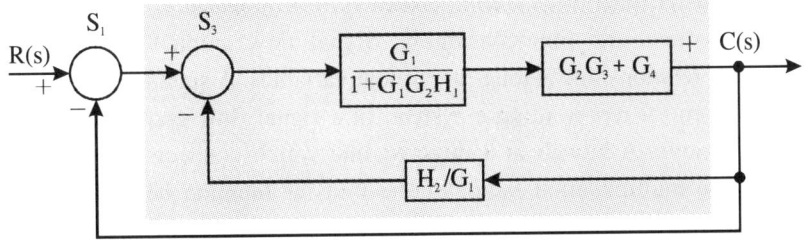

(e) Block diagram after eliminating feed-back loop $G_1 G_2 H_1$

(v) Combining feed forward gains and eliminating the feedback loop (H_2/G_1) as shown in Fig. 3.17(e) will result in Fig. 3.17(f).

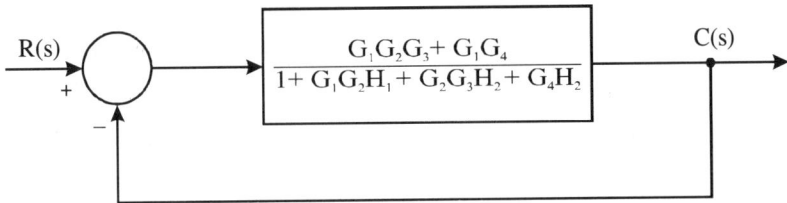

(f) Block diagram after eliminating feed-back path H_2/G_1

(vi) Elimination of unity feedback path results in Fig. 3.17(g).

$$R(s) \rightarrow \boxed{\dfrac{G_1 G_2 G_3 + G_1 G_4}{1+ G_1 G_2 H_1 + G_2 G_3 H_2 + G_4 H_2 + G_1 G_2 G_3 + G_1 G_4}} \rightarrow C(s)$$

(g) Block diagram after eliminating unity feed-back path

Fig. 3.17. Original block diagram and reduction steps

The transfer function of the closed-loop system is given by

$$M(s) = \frac{C(s)}{R(s)} = \frac{G_1 G_2 G_3 + G_1 G_4}{1+ G_1 G_2 H_1 + G_2 G_3 H_2 + G_4 H_2 + G_1 G_2 G_3 + G_1 G_4}$$

$$...(3.60)$$

3.14. SIGNAL FLOW GRAPHS

The evaluation of transfer function of a complicated system by means of block diagram reduction technique is quite tedious and time-consuming. Signal flow graph method is an alternative approach for finding relationships among the system variables. A signal flow graph is a simpler and more compact form of representing a system. In a signal flow graph, each system variable is represented by a node. A branch is a directed line which connects two nodes and acts as signal multiplier. The multiplication factor or the transfer function between the two nodes is indicated along the branch. Thus, the signal flow graph is a network of directed lines. It depicts the flow of signals from one point of a system to another and gives the relationships among the signals. Though a signal flow graph contains essentially the same information as a block diagram, the transfer function can easily be obtained from a signal flow graph by the application of a gain formula without requiring a reduction of the graph.

3.14.1. Definitions

The following terms which are frequently used in a signal flow graph are defined and indicated in Fig. 3.18.

Node : A node is a point representing a system variable.

Transmittance: The transmittance is the gain between two nodes.

Branch : A branch is a directed line segment which connects two nodes. The gain of the branch is the transmittance which is usually indicated along the branch.

Source or Input Node : A source or an input node is a node which has only outgoing branches. This node corresponds to an independent variable.

Sink or Output Node : A sink or output node is a node which has only incoming branches. This node corresponds to a dependent variable.

Mixed Node : A mixed node has both incoming and outgoing branches.

Path : A path is any continuous unidirectional succession of branches traversed in the indicated branch directions.

Forward Path : A forward path is a path from the input node to the output node along which no node is encountered more than once.

Path Gain : A path gain is the product of branch gains encountered in traversing the path.

Loop : A loop is a closed path which originates and terminates on the same node along which no other node is encountered more than once.

Loop Gain : A loop gain is the product of all branch gains of the branches forming the loop, Fig. 3.18 shows nodes and branches together with branch gains.

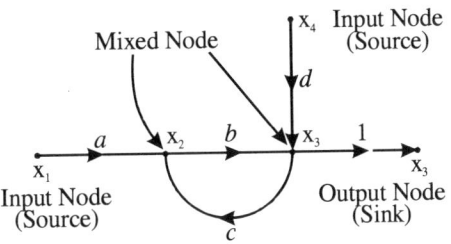

Fig. 3.18. Typical signal flow graph

3.15. BASIC PROPERTIES OF SIGNAL FLOW GRAPHS

Some of the important properties of signal flow graphs are summarized below:

(a) A node represents a variable. It adds the signals of all incoming branches and transmits this sum through all outgoing branches. The nodes are generally arranged form left to right.

(b) Signals flow along the branches only in the indicated directions.

(c) A branch from node x_j to node x_k represents the functional dependence of x_k on x_j but not the reverse.

(d) A signal from node x_j flowing along a branch to node x_k is multiplied by transmittance t_{jk} of the branch. The multiplied signal appears at the node x_k.

(e) A mixed node may be treated as an output node by adding an outgoing branch with unity gain.

(f) A signal flow graph is not unique for a given system.

3.16. SIGNAL FLOW GRAPH SIMPLIFICATION

Based upon the properties of signal flow graph, the transfer function may be determined using the following rules:

(a) Node with One Incoming Signal: Fig. 3.19 shows the node x_2 with one incoming branch with a transmittance of t_{12}.

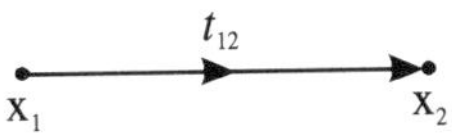

Fig. 3.19. Node with one incoming signal

The relationship is given by
$$x_2 = t_{12}\, x_1$$

(b) Node with Three Incoming Signals: Fig. 3.20 shows the node x_1 receiving three incoming signals. The relationship is given by

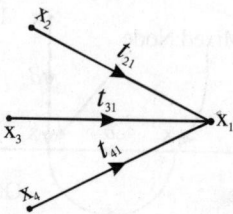

Fig. 3.20. Node with three incoming signals

$$x_1 = t_{21} x_2 + t_{31} x_3 + t_{41} x_4$$

(c) Cascaded Branches: Fig. 3.21(a) shows three branches that are connected in cascade. The simplified signal flow graph is shown in Fig. 3.21(b).

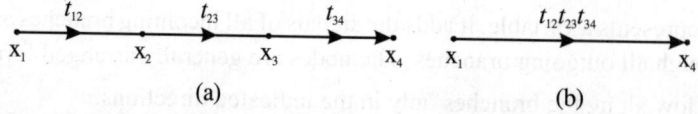

Fig. 3.21. (a) Cascaded branches; (b) Equivalent single branch representation

The relationship is given by

$$x_4 = t_{12} t_{23} t_{34} x_1$$

(d) Feed Forward Branches: In Fig. 3.22(a), x_2 has two incoming branches from the node x_1. This is replaced by a single branch as shown in Fig. 3.22(b).

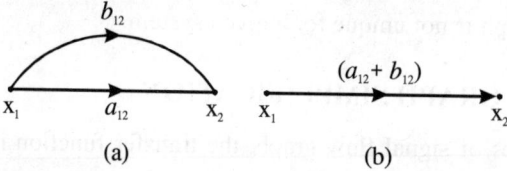

Fig. 3.22. (a) Two feed forward branches; (b) Simplified single branch

The relationship is given by

$$x_2 = (a_{12} + b_{12}) x_1$$

(e) Elimination of Mixed Node: In Fig. 3.23(a), the nodes x_1 and x_2 are mixed nodes since they have one incoming and outgoing signal. The mixed nodes can be eliminated as shown in Fig. 3.23(b).

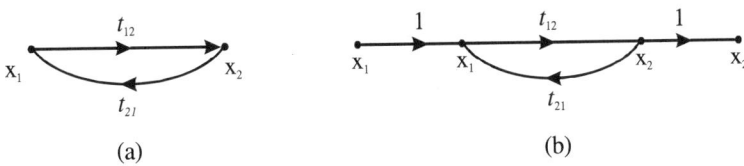

(a) (b)

Fig. 3.23. (a) Mixed nodes; (b) Input and Output node representation

(f) Eliminating Feedback Loop: Fig. 3.24(a) shows a feedback loop with forward transmittance of t_{12} and feedback transmittance of t_{21}. This can be replaced by a forward branch as shown in Fig. 3.24(b).

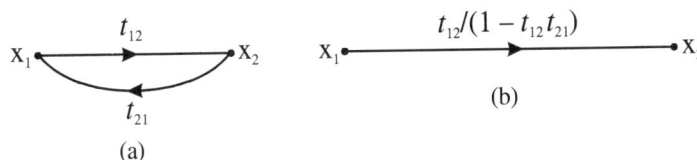

(a)

Fig. 3.24. (a) Feedback Loop; (b) Equivalent single branch representation

The relationship is given by

$$x_2 = t_{12}/(1 - t_{12}\, t_{21})$$

In applying this rule, care should be exercised because this rule cannot be applied for feedback loops touching each other. This is demonstrated in Example 3.11. In such cases, Mason's gain formula may be used.

3.17. CONSTRUCTION OF SIGNAL FLOW GRAPHS

The construction of a signal flow graph is basically a matter of following the cause and effect relations through the systems, relating each variable in terms of itself and other variables. It can be easily drawn once the equations describing the system are obtained. The steps that help construct the signal flow graphs are described below:

Step 1: Obtain the system equations. Identify the system variables. Each system variable is a node.

Step 2: Arrange the array of nodes from left to right, starting with the source node and ending with the sink node and label them.

Step 3: Interconnect the node with directed branches and mark the transmittances.

Step 4: If the sink node has outgoing branches, add a dummy node through a branch of unity transmittance.

Once all nodes are properly connected with incoming and outgoing branches, the signal flow graph is complete. The construction of signal flow graph is illustrated with examples.

Example 3.11. *Construct the signal flow graph of the electrical network shown in its Fig. 3.25(a) and obtain the transfer function* $E_0(s)/E_1(s)$.

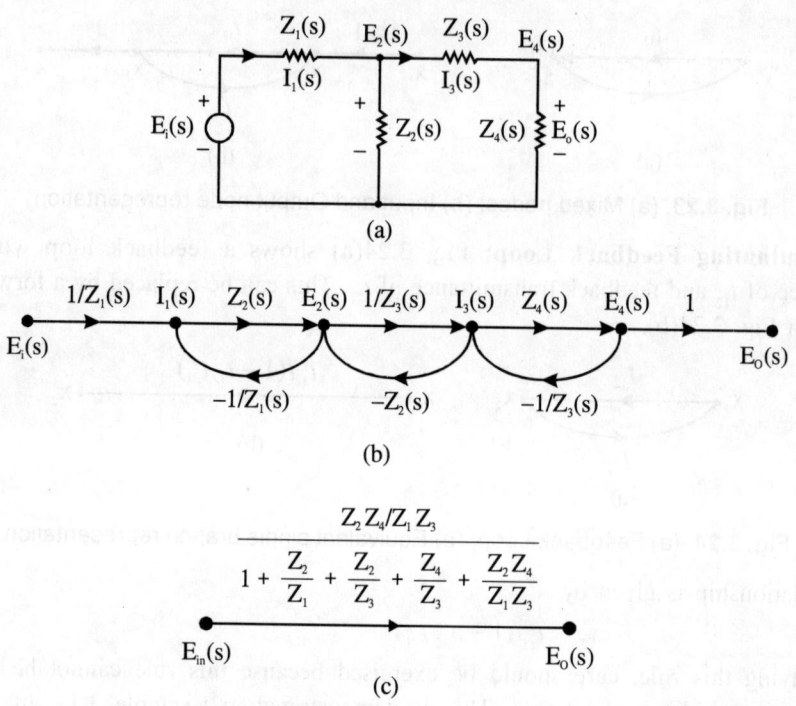

Fig. 3.25. (a) Electrical network for Example 3.11 (b) Signal flow graph (c) Final signal flow graph

Solution:

Step 1: With the indicated loop currents and node voltages, the following set of equations is written.

$$I_1(s) = [E_i(s) - E_2(s)]/Z_1(s) \qquad\qquad ...(3.61)$$

$$E_2(s) = [I_1(s) - I_3(s)]Z_2(s) \qquad\qquad ...(3.62)$$

$$I_3(s) = [E_2(s) - E_4(s)]/Z_3(s) \qquad\qquad ...(3.63)$$

$$E_4(s) = I_3(s)Z_4(s) \qquad\qquad ...(3.64)$$

$$E_0(s) = E_4(s) \qquad\qquad ...(3.65)$$

In electrical systems, currents and voltages form the system variables and hence nodes in signal flow graph. Here we have six nodes:

$$E_i(s),\ I_1(s),\ E_2(s),\ I_3(s),\ E_4(s) \text{ and } E_0(s)$$

Step 2: The six nodes are arranged in the same order from left to right as shown in Fig. 3.25(b).

Step 3: The node $I_1(s)$ has two incoming node, one from node E_i with a transmittance of $1/Z_1$ drawn as a horizontal directed branch and the other from node E_2 with a transmittance of $-1/Z_1$ drawn as a curved directed branch. This translates Eq. (3.61) into the signal flow graph.

This procedure is carried out for other nodes, thus translating Eq. (3.62) to (3.65). This results in the complete graph shown in Fig. 3.25(b).

Step 4: Applying the simplification rule, the three feedback loop may be eliminated one after another. The result is

$$\frac{E_o(s)}{E_i(s)} = \frac{Z_2 Z_4/Z_1 Z_3}{(1+\frac{Z_2}{Z_1})\,(1+\frac{Z_2}{Z_3})\,(1+\frac{Z_4}{Z_3})} \quad which\ is\ erroneous$$

The analytical solution of the simultaneous equations (3.61) to (3.65) yields the correct result as

$$\frac{E_o(s)}{E_i(s)} = \frac{Z_2 Z_4/Z_1 Z_3}{1+\frac{Z_2}{Z_1}+\frac{Z_2}{Z_3}+\frac{Z_4}{Z_3}+\frac{Z_2 Z_4}{Z_1 Z_3}} \quad\quad …(3.66)$$

Hence *elimination of feedback loop should not be applied when the feedback loops touch each other.* In Eq. (3.66), $Z_2 Z_4/Z_1 Z_3$ represents the forward path gain, $(-Z_2/Z_1)$, $(-Z_1/Z_3)$ and $(-Z_4/Z_3)$, represent the gains of first, second and third loop respectively and the last term $(Z_2 Z_4/Z_1 Z_3)$ is the product of the gains of first and third loops which are not touching each other. Therefore,

$$\frac{E_o(s)}{E_i(s)} = \frac{\text{Forward Path Gain}}{1-(\text{sum of loop gains})+\text{Product of loop gains of two non-touching loops}} \quad …(3.67)$$

Eq. (3.67) forms the basis for the Mason's gain formula.

Example 3.12. *For the signal flow graph shown in Fig. 3.26(a), obtain the transfer function* x_o/x_1 *using the simplification rule and verify by analytical solution.*

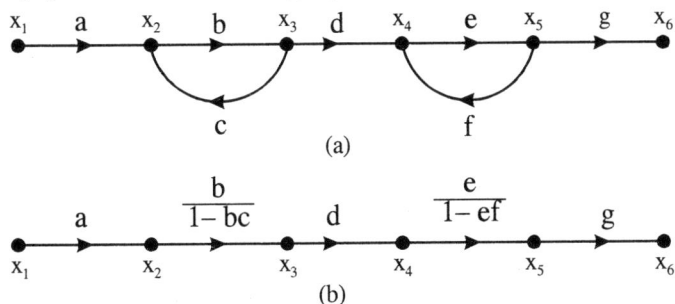

Fig. 3.26. (a) Signal flow graph for Example 3.12 (b) Simplified signal flow graph

Solution:

Eliminating the feedback loops by the simplification rule results in Fig. 3.26(b). Then the transfer function is x_o/x_1

$$\frac{x_o}{x_1} = \frac{ab\,d\,e\,g}{1-(bc+ef)+bcef} \quad\quad …(3.68)$$

This agrees with Eq. (3.67).

The equations represented by the signal flow graph are:

$$x_2 = ax_1 + cx_3 \qquad \qquad ...(3.69)$$

$$x_3 = bx_2 \qquad \qquad ...(3.70)$$

$$x_4 = dx_3 + fx_5 \qquad \qquad ...(3.71)$$

$$x_5 = ex_4 \qquad \qquad ...(3.72)$$

$$x_0 = gx_5 \qquad \qquad ...(3.73)$$

From Eq. (3.69) and (3.70)

$$x_2 = ax_1 + cx_3 = ax_1 + bcx_2$$

Therefore, x_2 is given by

$$x_2 = \frac{a}{1 - bc} x_1 \qquad \qquad ...(3.74)$$

From Eq. (3.71) and (3.72)

$$x_4 = dx_3 + fx_5 = dx_3 + efx_4$$

Therefore, x_4 is given by

$$x_4 = \frac{d}{1 - ef} x_3 \qquad \qquad ...(3.75)$$

Solving for x_0,

$$x_0 = gx_5 = gex_4 = \frac{ged}{1-ef} x_3 = \frac{bdeg}{1-ef} x_2 = \frac{abdeg}{(1-bc)(1-ef)} x_1$$

Therefore,

$$\frac{x_0}{x_1} = \frac{ab\,deg}{1-(bc+ef)+bcef}$$

This agrees with Eq. (3.68).

3.18. MASON'S GAIN FORMULA FOR SIGNAL FLOW GRAPHS

The transfer function of a system may be obtained from its block diagram by reduction technique or from the signal flow graph by using simplification rules. Since the methods involve step-by-step procedure, they are tedious and time-consuming. A simple and direct formula, developed by Mason, is preferred to obtain the transfer function from a signal flow graph by inspection. The mason's gain formula is

$$M = \frac{\sum_{k=1}^{n} P_k \Delta_k}{\Delta} \qquad \qquad ...(3.77)$$

where

M = overall gain between the input and output nodes

n = Number of forward paths.

P_k = gain of the k-th forward path

Δ = determinant of the graph or characteristic function

= 1 – (sum of all individual loop gains) + (sum of gain products of all possible combinations of two non-touching loops) – (sum of gain products of all possible combinations of three non-touching loops) + ...(3.78)

Δ_k = Determinant of the cofactor of Δ or Δ of the k-th forward path of the graph with the loops touching the k-th forward path removed. ...(3.79)

Non-touching loops are loops that do not have a common node.

The Mason formula is applicable only between the input node and the output node. Though the gain formula appears to be formidable, its application to practical system is simple. The procedure for applying the formula is outlined below:

Step 1: Identify all the forward paths present in the signal flow graph and calculate the gain of each path.

Step 2: Identify all the feedback loops present in the signal flow graph and compute the gain of each loop. The sum of all loop gains will be the second term in Eq. (3.78).

Step 3: Determine all possible combinations of two non-touching loops and obtain the product of loop gains. The sum of all such gains will be third term in Eq. (3.78).

Step 4: Repeat step 3 for three, four, ... non-touching loops and obtain the product of gains respectively. The sum of such gains for three, four, ... non-touching loops will be the fourth, fifth ...... term in Eq. (3.78).

Step 5: For each forward path indentified in Step 1, find the part of the graph not touched by the path and obtain the value of Δ_k for each path from Eq. (3.79).

Step 6: Substitute the calculated values in Eq. (3.77) and obtain the transfer function.

The application of the Mason's formula is illustrated below with suitable examples.

Example 3.13. *Apply the general gain formula and obtain the transfer function $E_o(s)/E_i(s)$ for the signal flow graph shown in Fig. 3.25(b).*

Solution:

Step 1: The signal flow graph has only one forward path. The gain is $Z_2 Z_4 / Z_1 Z_3$.

Step 2: There are three feedback loops. The gains of individual loops are: $(-Z_2/Z_1)$, $(-Z_2/Z_3)$ amd $(-Z_4/Z_3)$.

Step 3: There is only one set of two non-touching loops, the first and the third loops. Their gain product is $Z_2 Z_4 / Z_1 Z_3$.

Step 4: There are no three or more non-touching loops.

Step 5: The only forward path touches all parts of the graph. Hence $\Delta_1 = 1$.

Step 6: Substituting the above values in Eq.(3.77),

$$\frac{E_o(s)}{E_i(s)} = \frac{Z_2 Z_4 / Z_1 Z_3}{1 - \left(-\dfrac{Z_2}{Z_1} - \dfrac{Z_2}{Z_3} - \dfrac{Z_4}{Z_3} \right) + \dfrac{Z_2 Z_4}{Z_1 Z_3}}$$

$$= \frac{Z_2 Z_4 / Z_1 Z_3}{1 + \dfrac{Z_2}{Z_1} + \dfrac{Z_2}{Z_3} + \dfrac{Z_4}{Z_3} + \dfrac{Z_2 Z_4}{Z_1 Z_3}}$$

Example 3.14. *For the signal flow graph shown in Fig. 3.27, find the functional relation between source node and sink node.*

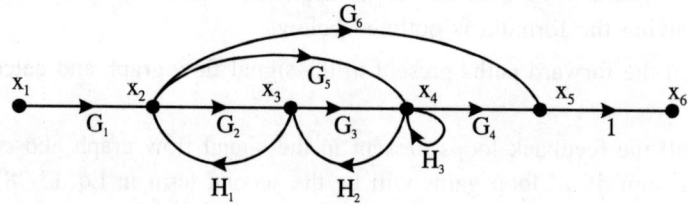

Fig. 3.27. Signal flow graph for Example 3.14

Solution:

Step 1: There are three forward paths.

Gain of path $x_1 - x_2 - x_3 - x_4 - x_5 = P_1 = G_1 G_2 G_3 G_4$
Gain of path $x_1 - x_2 - x_4 - x_5 = P_2 = G_1 G_5 G_4$
Gain of path $x_1 - x_2 - x_5 = P_3 = G_1 G_6$

Step 2: There are four individual loops. Their gains are:

Gain of loop $x_2 - x_3 - x_2 = L_1 = G_2 H_1$
Gain of loop $x_3 - x_4 - x_3 = L_2 = G_3 H_2$
Gain of loop $x_4 - x_4$ (self loop) $= L_3 = H_3$
Gain of loop $x_2 - x_4 - x_3 - x_2 = L_4 = G_5 H_1 H_2$

Step 3: There is only one set of two non-touching loops: $(x_2 - x_3 - x_2)$ and $(x_4 - x_4)$. Their gain product is $L_1 L_3 = G_2 H_1 H_3$.

Step 4: There are no three or more non-touching loops.

Step 5: The first and second forward paths are touching all the four loops, Hence $\Delta_1 = 1$, $\Delta_2 = 1$. The third forward path does not touch the second and third loops. Hence,

$$\Delta_3 = 1 - G_3 H_2 - H_3.$$

Step 6: Substituting the above values in Eq. (3.77), we get,

$$\frac{x_5}{x_1} = \frac{P_1\Delta_1 + P_2\Delta_2 + P_3\Delta_3}{1-(L_1+L_2+L_3+L_4)+L_1L_3}$$

$$= \frac{G_1G_2G_3G_4 + G_1G_4G_5 + G_1G_6(1-G_3H_2-H_3)}{1-(G_2H_1+G_3H_2+H_3+G_5H_1H_2)+G_2H_1H_3}$$

3.19. SIGNAL FLOW GRAPHS OF LINEAR SYSTEMS

The signal flow graphs are extensively used in the analysis of linear systems. A linear system is described by a set of simultaneous equations. Construction of signal flow graph for a linear system is described with an example.

Example 3.15. *Construct a signal flow graph and obtain a functional relation between* x_1 *and* x_5 *of the linear system described by the following set of equations.*

$$x_2 = t_{12}x_1 + t_{42}x_4$$
$$x_3 = t_{23}x_2 + t_{43}x_4$$
$$x_4 = t_{34}x_3 + t_{44}x_4$$
$$x_5 = t_{25}x_2 + t_{45}x_4$$

Solution:

There are five variables x_1 through x_5. Marking them as nodes and translating each of the steps as explained in Section 3.17, we get the completed signal flow graph shown in Fig. 3.28.

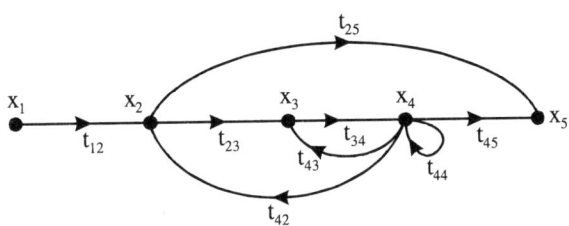

Fig. 3.28. Translation of system equations into signal flow graph

There are two forward paths with gains

$$P_1 = t_{12}t_{23}t_{34}t_{45}$$
$$P_2 = t_{12}t_{25}$$

There are three individual loops with gains

$$L_1 = t_{23}t_{34}t_{42}$$
$$L_2 = t_{34}t_{42}$$
$$L_3 = t_{44}$$

There are no two or more non-touching loops.

The first forward parth P_1 touches all loops. Hence $\Delta_1 = 1$ The second forward path P_2 does not touch the second and third loops. Hence $\Delta_2 = 1 - L_2 - L_3$. Therefore,

$$\frac{x_5}{x_1} = \frac{P_1 \Delta_1 + P_2 \Delta_2}{1 - (L_1 + L_2 + L_3)} = \frac{t_{12}t_{23}t_{34}t_{45} + t_{12}t_{25}(1 - t_{34}t_{43} - t_{44})}{1 - (t_{23}t_{34}t_{42} + t_{34}t_{43} + t_{44})}$$

3.20. CONSTRUCTION OF SIGNAL FLOW GRAPH OF CONTROL SYSTEMS

The evaluation of transfer functions of multiloop systems by block diagram reduction technique is tedious. However, it may easily be evaluated by first drawing the signal flow graph and then applying the Mason's gain formula. The procedure to draw the signal flow graph for a given block diagram is described below.

Step 1: From the block diagram, determine the nodes of signal flow graph. Adjacent summing points, adjacent branch points or a summing point preceding a branch point may be clubbed together and considered as single node. However, *a branch point preceding a summing point cannot be clubbed together.*

Step 2: Arrange the nodes in the same order as in block diagram from left to right and label them.

Step 3: Each block between the nodes in a block diagram is represented by a directed branch between the corresponding nodes in the signal flow graph with corresponding transmittances.

Step 4: Once all the blocks in the block diagram have been translated in to the signal flow graph, the signal flow graph is completed.

The above procedure is illustrated below with an example.

Example 3.16. *Construct the signal flow graph for the block diagram shown in Fig. 3.29(a) and obtain the transfer function C/R.*

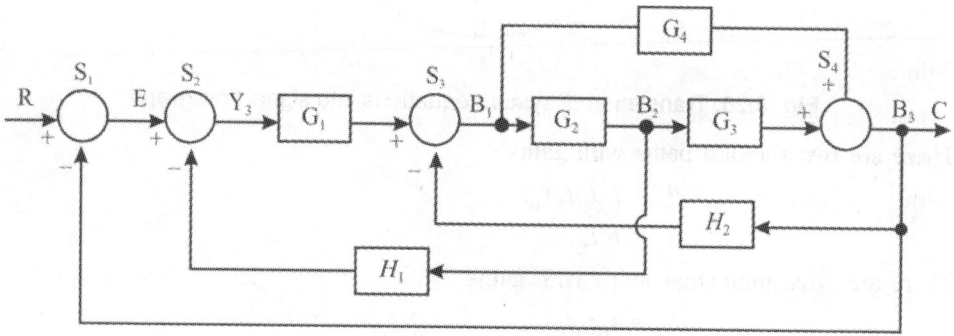

Fig. 3.29. (a) Block diagram for Example 3.16

Solution:

(S is a summing point and B is a branch point)

Step 1: There are six nodes. S_1 and S_2 will be node 1, S_3 and B_1 node 2, B_2 node 3, S_4 and B_3 node 4 and R and C nodes.

Step 2: The nodes are arranged in an array from left to right and labeled accordingly.

Step 3: All branches are now translated into directed branches between respective nodes with their transmittances marked properly.

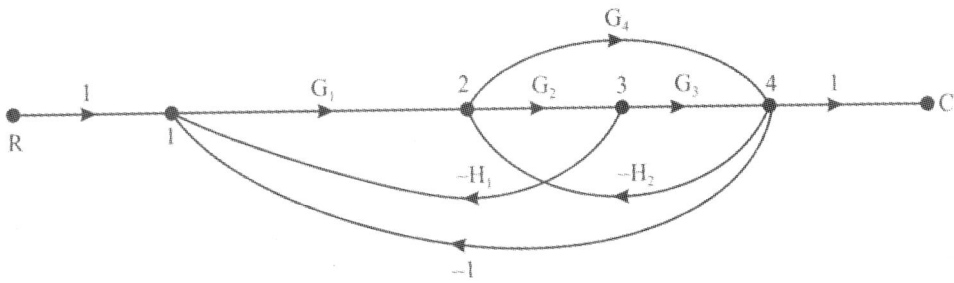

Fig. 3.29. (b) Signal flow graph of Example 3.16

Step 4: Having obtained the complete signal flow graph, apply Mason's formula to evaluate the transfer function as detailed below:

There are two forward paths with gains of P_1 and P_2.

Gain of path $R - 1 - 2 - 3 - 4 - C = P_1 = G_1G_2G_3$

Gain of path $R - 1 - 2 - 4 - C = P_2 = G_1G_4$

There are five loops with gains of L_1, L_2, L_3, L_4 and L_5

Gain of path $1 - 2 - 3 - 4 - 1 = L_1 = -G_1G_2G_3$

Gain of path $1 - 2 - 4 - 1 = L_2 = -G_1G_4$

Gain of path $2 - 3 - 4 - 2 = L_3 = -G_2G_3H_2$

Gain of path $1 - 2 - 3 - 1 = L_4 = -G_1G_2H_1$

Gain of path $2 - 4 - 2 = L_5 = -G_4H_2$

There are no two or more non-touching loops.

Both forward paths touch all the loops. Hence $\Delta_1 = 1$, $\Delta_2 = 1$. Therefore,

$$\frac{C}{R} = \frac{P_1\Delta_1 + P_2\Delta_2}{1 - (L_1 + L_2 + L_3 + L_4 + L_5)}$$

$$= \frac{G_1G_2G_3 + G_1G_4}{1 + G_1G_2G_3 + G_1G_4 + G_1G_2H_1 + G_2G_3H_2 + G_4H_2}$$

The result is same as in Eq. (3.60).

3.21. MERITS OF SIGNAL FLOW GRAPHS

The merits or advantages of signal flow graph are:

(i) It provides a graphic representation of flow of signals within the system.

(ii) In a signal flow graph it is immediately apparent where and how feedback appears in the system.

(iii) The effects of input and/or parameter changes upon signals at all points in the system are clearly illustrated.

(iv) The flow graph is easily constructed from system equations or block diagram.

(v) The input-output relation is directly obtained using Mason's gain formula.

3.22. ADDITIONAL EXAMPLES

Example 3.17. *Derive the transfer function of the T-network shown in Fig. 3.30.*

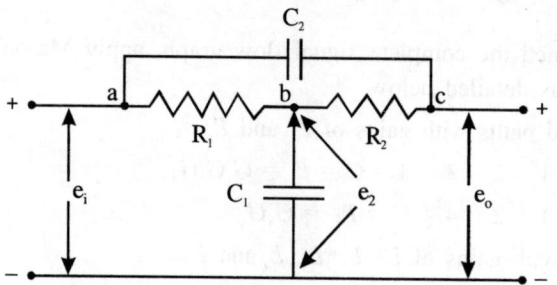

Fig. 3.30. T-network for Example 3.17

Solution:

There are two unknown nodes, *b* and *c*. The equations are:

At node *b*:
$$E_2(s)\left[\frac{1}{R_1}+\frac{1}{R_2}+C_1 s\right]-E_1(s)/R_1-E_0(s)/R_2 = 0 \qquad \text{...(3.80)}$$

At node *c*:
$$E_0(s)\left[\frac{1}{R_2}+C_2 s\right]-E_1(s)C_2 s-E_2(s)/R_2 = 0 \qquad \text{...(3.81)}$$

From Eq. (3.80), we get

$$E_2(s) = [E_1(s)/R_1+E_0(s)/R_2]\bigg/\left(\frac{1}{R_1}+\frac{1}{R_2}+C_1 s\right) \qquad \text{...(3.82)}$$

Substituting Eq. (3.82) in Eq. (3.81), and re-arranging,

$$\frac{E_o(s)}{E_i(s)} = \frac{1 + (R_1C_2 + R_2C_2)s + R_1R_2C_1C_2s^2}{1 + (R_1C_1 + R_1C_2 + R_2C_2)s + R_1R_2C_1C_2s^2} \qquad \ldots(3.83)$$

Example 3.18. *Derive the transfer function $Y_1(s)/F(s)$ for the dynamic absorber system shown in Fig. 3.31(a).*

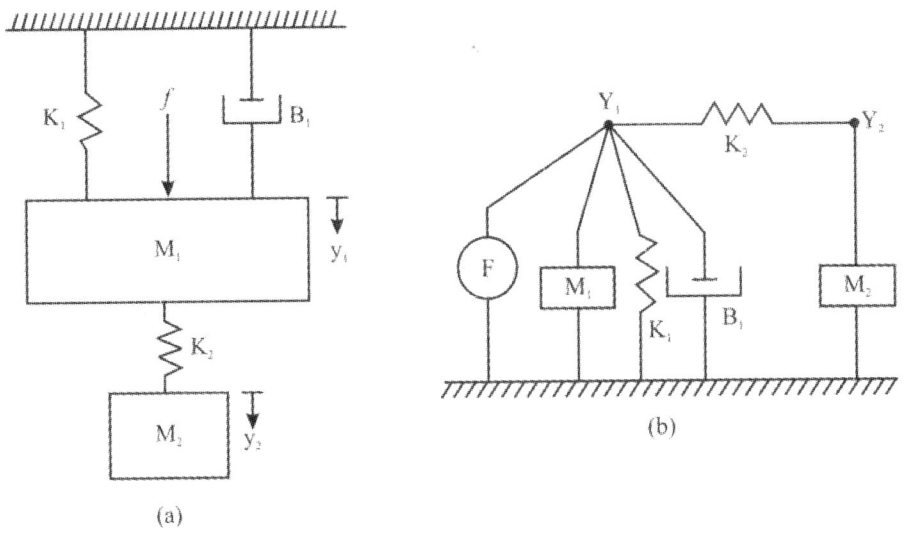

(a)

(b)

Fig. 3.31. (a) Dynamic absorber system; (b) Mechanical network

Solution:

The mechanical network is shown in Fig. 3.31(b) with the two displacements Y_1 and Y_2 as nodes. The node equations are:

At node Y_1 : $\qquad F(s) = (M_1s^2 + B_1s + K_1 + K_2)Y_1(s) - K_2Y_2(s) \qquad \ldots(3.84)$

At node Y_2 : $\qquad 0 = (M_2s^2 + K_2)\ Y_2(s) - K_2Y_1(s) \qquad \ldots(3.85)$

From Eq. (3.85), we get

$$Y_2(s) = \frac{K_2}{M_2s^2 + K_2}Y_1(s) \qquad \ldots(3.86)$$

Substituting Eq. (3.86) in Eq. (3.84) and rearranging, we get

$$\frac{Y_1(s)}{F(s)} = \frac{M_2s^2 + K_2}{M_1M_2s^4 + M_2B_1s^3 + (M_2K_1 + M_1K_2)s^2 + B_1K_2s + K_1} \qquad \ldots(3.87)$$

Example 3.19. *Fig. 3.32(a) shows an automatic voltage control system which maintains the load voltage E_o constant at 400V. The speed of the prime mover is constant. K_a is the gain of the amplifier (amp/volt), K_g is the gain of the generator (volt/amp), h is the feedback ratio, I is the load current and R_a is the armature resistance. (For the operation of the system, refer to Chapter 1) (a) Draw the block diagram and derive $E_o(s)/E_r(s)$ and $\Delta E_o(s)/\Delta I_a(s)$ under (i) Open-loop and (ii) Closed-loop conditions. (b) If $K_g = 200$, $K_a = 4$, the feedback ratio = 0.5, $R_a = 1$ and change in armature current is 20A, obtain $E_r(s)$ and $\Delta E_o(s)$ under (i) open-loop and (ii) closed-loop conditions.*

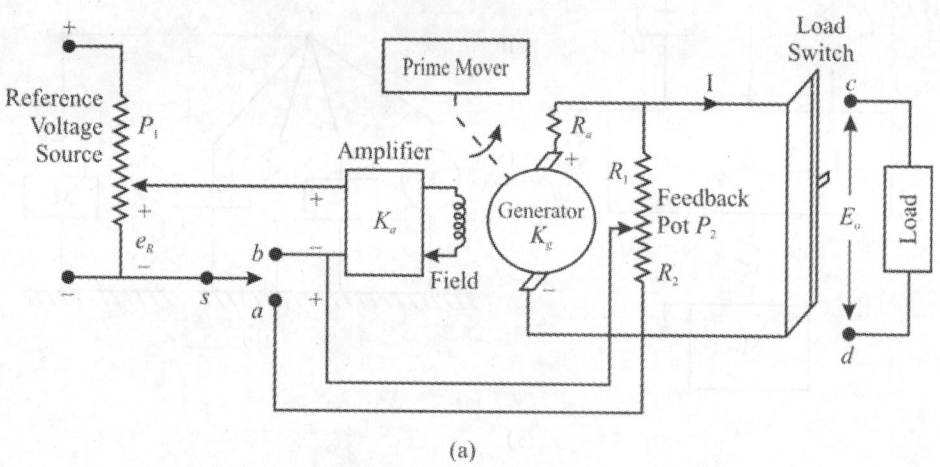

(a)

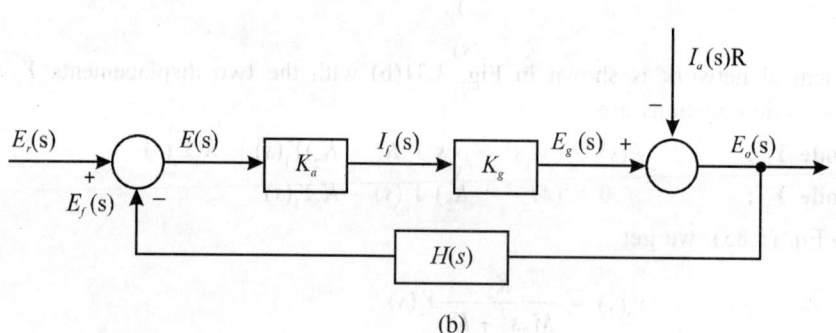

(b)

Fig. 3.32. (a) Automatic voltage regulator; (b) Block diagram

Solution:

The first block is the summing point where the reference input $E_r(s)$ is compared with feedback voltage $E_f(s)$ with appropriate polarity. The error voltage is applied to the amplifier block with gain K_a. The output of the amplifier is the field current $I_f(s)$. Therefore the next block is the generator block with K_g as its transfer function. The output voltage $E_o(s)$ is equal to $E_g(s) - R_a I_a(s)$. Hence the next block is a summer and $R_a I_a(s)$ is shown as external disturbance. Having obtained $E_o(s)$ the feedback loop is completed through a block with transfer function $H(s)$. The complete block diagram is shown in Fig. 3.32(b).

(a) (i) *Open-loop Condition:*

There is no feedback. The feedback path is open.

Hence,

$$E_o(s)/E_r(s) = K_a K_g \qquad \qquad ...(3.88)$$
$$E_o(s) = E_g(s) - I_a(s)R_a$$
$$\Delta E_o(s) = -\Delta I_a(s)R_a$$

or,

$$\Delta E_o(s)/\Delta I_a(s) = -R_a \qquad \qquad ...(3.89)$$

(ii) *Closed loop condition:*

The feedback ratio is h.

$$E_o(s)/E_r(s) = K_a K_g/(1 + K_a K_g H(s)) \qquad \qquad ...(3.90)$$
$$E_f(s) = H(s) \, E_0(s)$$
$$E_o(s) = E_g(s) - I_a(s)R_a$$
$$= K_a K_g \, E(s) - I_a(s)R_a$$
$$= K_a K_g[E_r(s) - E_f(s)] - I_a(s)R_a$$
$$= K_a K_g[E_r(s) - H(s) \, E_o(s)] - I_a(s)R_a$$

Therefore

$$E_0(s) = \frac{K_a K_g E_r(s) - I_a(s) \, R_a}{1 + K_a K_g \, H(s)} \qquad \qquad ...(3.91)$$

The ratio $\Delta E_o(s)/\Delta I_a(s)$ is given by

$$\frac{\Delta E_o(s)}{\Delta I_a(s)} = \frac{-R_a}{1 + K_a K_g \, H(s)} \qquad \qquad ...(3.92)$$

(b) (i) *open-loop condition:*

From Eq. (3.88)

$$E_r(s) = E_o(s)/K_a K_g$$
$$= 400/(4 \times 200) = 0.5 \text{ V}$$

From Eq.(3.89)

$$\Delta E_o(s) = -R_a \Delta\ I_a(s)$$
$$= 1 \times -20 = -20\ \text{V}$$

(ii) *Closed loop conditions* from Eq. (3.90),

$$E_r(s) = E_o(s)\frac{1 + K_a K_g H(s)}{K_a K_b}$$

$$= 400 \times \frac{1 + 800 \times 0.5V}{4 \times 200}$$

$$= \frac{401}{2}$$

$$= 200.5\ \text{V}$$

From Eq. (3.92)

$$\Delta E_0 = -\frac{I_a(s)R_a}{1 + K_a K_g H(s)}$$

$$= -\frac{20 \times 1}{1 + 400}$$

$$= -0.04986\ \text{V}$$

Thus, the change in the output voltage is reduced by $401 = 1 + K_a K_g H(s)$ times.

Example 3.20. *Obtain the transfer function C(s)/R(s) for the block diagram shown in Fig 3.33(a) by (i) block diagram reduction technique and (ii) signal flow graph method.*

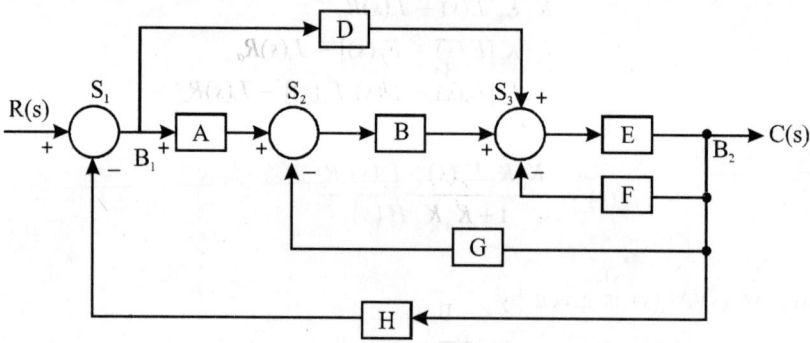

Fig. 3.33. (a) Block diagram for Example 3.20

Solution:

(i) *Block diagram reduction technique:*

Step 1: Eliminate the feedback loop F as shown in Fig. 3.33(b).

Step 2: Shift S_2 to S_3 position as in Fig. 3.33(c).

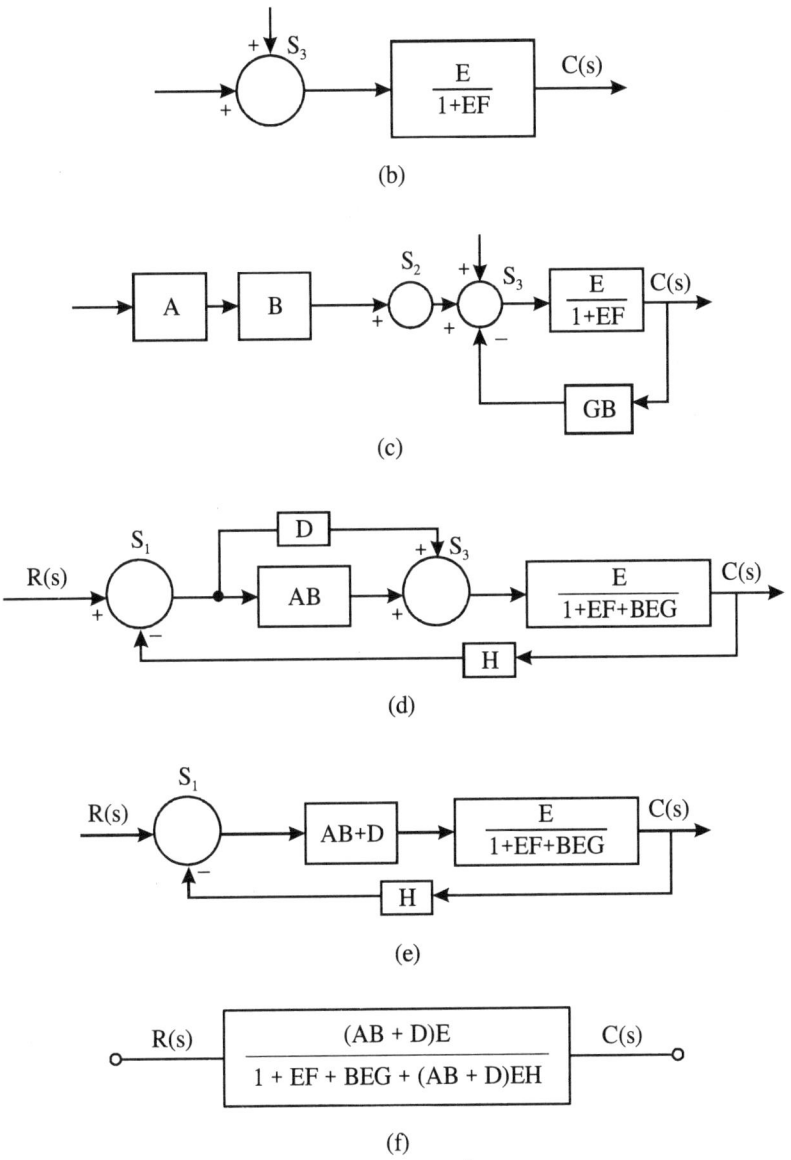

(b)

(c)

(d)

(e)

(f)

Fig. 3.33. (b) Feedback F eliminated; (c) S_2 shifted to S_3 position; (d) Blocks A and B replaced by a single block and feedback GB eliminated; (e) Block D eliminated; (f) Equivalent single block.

Step 3: Replace the cascaded blocks A and B and eliminate the feedback loop. This is shown in Fig. 3.33(d).

Step 4: Eliminate the feed-forward block D as in Fig. 3.33(e).

Step 5: Cascading and then eliminating the feedback loop results in the closed loop transfer function as shown in Fig. 33(f).

$$\frac{C(s)}{R(s)} = \frac{ABE + DE}{1 + ABEH + BEG + DEH + EF} \qquad \qquad ...(3.93)$$

(ii) *Signal flow graph method:*

Step 1: There are six nodes, $R(s)$, S_1 and B_1, S_2, S_3, B_2 and $C(s)$. Arrange them from left to right and label them accordingly.

Step 2: Transfer each block to the flow graph by a directed branch with appropriate transmittance. The complete signal flow graph is shown is Fig. 3.34.

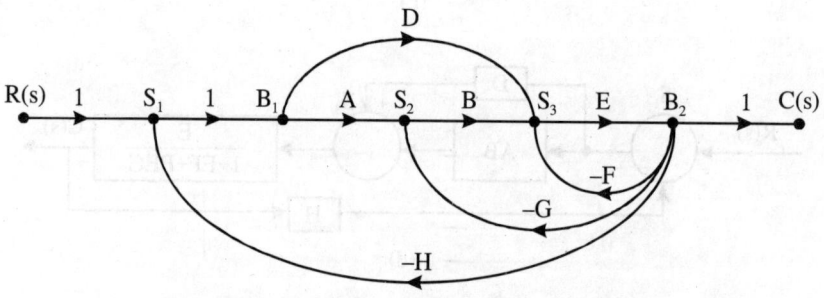

Fig. 3.34. Signal flow graph of Example 3.20

Step 3: Apply the Mason's gain formula and obtain $C(s)/R(s)$.

There are two forward paths; their gains are:

$$P_1 = ABE$$
$$P_2 = DE$$

There are four loops. Their gains are:

$$L_1 = -ABEH$$
$$L_2 = -BEG$$
$$L_3 = -DEH$$
$$L_4 = -EF$$

There are no non-touching loops. The two forward paths touch all the four loops. Hence $\Delta_1 = 1$ and $\Delta_2 = 1$. Therefore,

$$\frac{C(s)}{R(s)} = \frac{P_1 \Delta_1 + P_2 \Delta_2}{1 - \left(L_1 + L_2 + L_3 + L_4\right)}$$

$$= \frac{ABE + DE}{1 + ABEH + BEG + DEH + EF} \qquad ...(3.94)$$

Example 3.21. *A closed-loop system is represented by the following differential equation:*

$$\frac{d^3c}{dt^3} + 2\frac{d^2c}{dt^2} + 4\frac{dc}{dt} = 20e$$

where e = *r* −0.8*c. Identify the quantities c, r and e. (i) Draw a block diagram of the system and determine the overall transfer function. (ii) Construct a signal flow graph and verify the transfer function obtained.*

Solution:

    c is the output quantity

    r is the unit step input.

    e is the error voltage

(i) *Block diagram:*

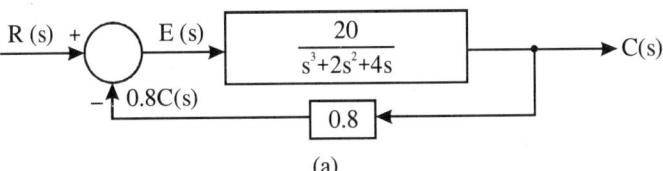

(a)

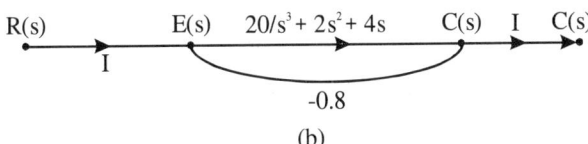

(b)

Fig. 3.35. (a) Block diagram; (b) Signal flow graph

The block diagram is shown in Fig. 3.35(a). The transfer function is

$$C(s)/R(s) = \frac{G(s)}{1 + G(s)H(s)}$$

$$= \frac{20}{s^3 + 2s^2 + 4s + 16}$$

(ii) *Signal flow graph:*

The signal flow graph is shown in Fig 3.30(b). There is only one forward path. Its gain P_1 is given by

$$P_1 = 20/(s^3 + 2s^2 + 4s)$$

Since the forward path touches the loop,

$$\Delta_1 = 1$$

There is only one loop. Its gain L_1 is given by

$$L_1 = -\frac{16}{s^3 + 2s^2 + 4s}$$

$\therefore$

$$\Delta = 1 - L_1 = 1 + \frac{16}{s^3 + 2s^2 + 4s}$$

$$\frac{C(s)}{R(s)} = \frac{P_1 \Delta_1}{\Delta} = \frac{20}{s^3 + 2s^2 + 4s + 16}$$

Example 3.22. *Find the transmittance between signal Y_1 and the output Y_5 of the signal flow graph shown in Fig. 3.36.*

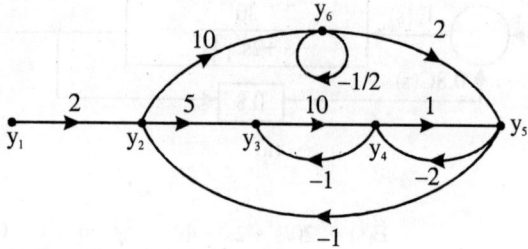

Fig. 3.36. Signal flow graph

Solution:

There are two forward paths.

Gain of path $y_1 - y_2 - y_3 - y_4 - y_5 = P_1 = 2 \times 5 \times 10 \times 1 = 100$

Gain of path $y_1 - y_2 - y_6 - y_5 = P_2 = 2 \times 10 \times 2 = 40$

There are six loops.

Gain of loop $y_3 - y_4 - y_3 = L_1 = -10$,

Gain of loop $y_4 - y_5 - y_4 = L_2 = -2$

Gain of loop $y_2 - y_3 - y_4 - y_5 - y_2 = L_3 = 5 \times 10 \times 1 \times -1 = -50$

Gain of self-loop $y_6 - y_6 = L_4 = -0.5$

Gain of $y_2 - y_6 - y_5 - y_2 = L_5 = 10 \times 2 \times - 1 = -20$

There are 3 sets of two non-touching loops. Their gains are:

$$L_1 L_4 = 5$$
$$L_2 L_4 = 1$$
$$L_3 L_4 = 25$$

$$\Delta = 1 - (L_1 + L_2 + L_3 + L_4 + L_5) + (L_1L_4 + L_2L_4 + L_3L_4)$$
$$= 1 - (-10 - 2 - 50 - 0.5 - 20) + (5 + 1 + 25) = 114.5$$

The first forward path does not touch L_4 loop. Hence

$$\Delta_1 = 1 - (L_4)$$
$$= 1 + 0.5 = 1.5$$

The second forward path does not touch L_1 loop. Hence

$$\Delta_2 = 1 - L_1$$
$$= 1 + 10 = 11$$

$$\frac{Y_5(s)}{Y_1(s)} = \frac{P_1 \Delta_1 + P_2 \Delta_2}{\Delta}$$

$$= \frac{100 \times 1.5 + 40 \times 11}{114.5}$$

$$= 5.15$$

3.23. SUMMARY

Classical methods of analysis and design of control systems are based on (i) system representation in the *s*-domain, (ii) transfer functions, (iii) block diagrams or (iv) signal flow graphs. In this chapter, we have first defined a linear system. We have discussed the basic components used in simple electrical and mechanical translational and rotational systems. We have explained the steps used to obtain the describing functions as well as transfer functions of single loop and multiloop systems. In addition, we have discussed the basic block diagram of a closed-loop control system, also explained the methods of constructing the system block diagrams and their reduction to obtain the transfer functions. We then have explained terms that are frequently used in the signal flow graph and discussed the basic properties of signal flow graphs and the methods of obtaining the transfer functions from them. The signal flow graph method is preferred since its notation is compact, more detail is conveniently shown and the transfer functions can be directly obtained by the application of Mason's gain formula.

REVIEW QUESTIONS

3.1 Define a linear system.

3.2 Define the system transfer function.

3.3 List out the various steps involved in determining system transfer function.

3.4 Distinguish between mesh and nodal analysis.

3.5 Mention the rules that simplify the formulation of node equation of electrical systems.

3.6 Distinguish between mechanical translational and rotational systems.

3.7 Mention the rules to formulate mechanical system equations.

3.8 Give four advantages of representing systems by block diagrams.

3.9 Draw the basic block diagram of a closed-loop system and explain.

3.10 Discuss the procedure for drawing a block diagram of a system.

3.11 Describe a signal flow graph.

3.12 State four properties of signal flow graph.

3.13 List out the steps involved in the construction of signal flow graph of a system.

3.14 Write down the Mason's gain formula and explain.

3.15 Mention four advantages of using signal flow graphs.

EXERCISE

3.1 Define the transfer function of an element. What are the assumptions made?

3.2 Explain the concept of transfer function with reference to a single-input single-output system.

3.3 Mention the three basic elements used in an electrical network. Write down the voltage-current relationship for each.

3.4 Enumerate the rules for formulating the node equations of an electrical network. Illustrate with an example.

3.5 What are the three basic elements used in a mechanical translational system? Write down the force-displacement and force-velocity relationship for each.

3.6 Mention the three basic elements used in a mechanical rotational system. Write down the torque-angular displacement and torque-angular velocity relationship.

3.7 Enumerate rules for formulating the mechanical system equations and illustrate with an example.

3.8 What is the advantage of block diagram reduction technique compared to writing system equations?

3.9 Compare the signal flow graph method with the block diagram.

3.10 List the rules for block diagram reduction.

3.11 Define the following terms in signal flow graph representation: (i) Source (ii) Sink (iii) path (iv) loop (v) gain

3.12 (a) Mention the advantages of signal flow graphs.

(b) State and explain clearly the Mason's formula used in signal flow graph analysis.

PROBLEMS

3.1. Derive the transfer functions $\dfrac{E_o(s)}{E_i(s)}$ for the circuit shown in Fig P 3.1

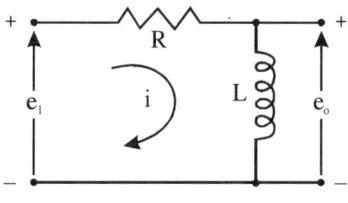

Fig. P 3.1

3.2. Determine the transfer functions of the networks shown in Fig. P 3.2 (a), (b) and (c).

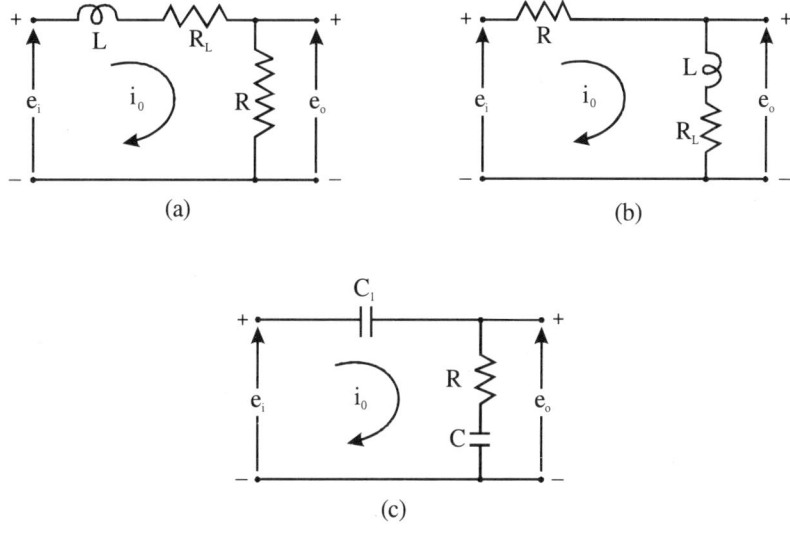

(a)

(b)

(c)

Fig. P 3.2

3.3. For the network given in Fig. P 3.3, obtain the transfer function.

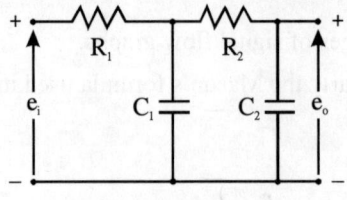

Fig. P 3.3

3.4. Determine the transfer function for the circuit shown in Fig. P 3.4

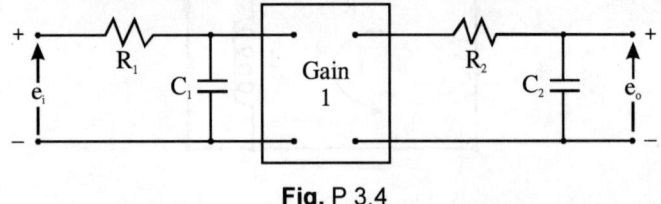

Fig. P 3.4

3.5. Find $E_o(s)/E_i(s)$ for the circuit shown in Fig. P 3.5

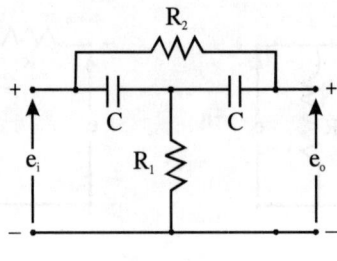

Fig. P 3.5

3.6. Obtain the transfer function $\dfrac{X(s)}{F(s)}$ for the system shown in Fig. P 3.6

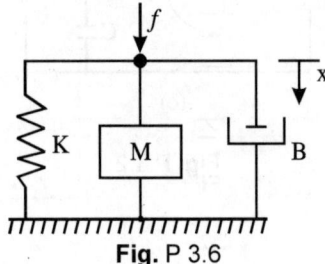

Fig. P 3.6

3.7. A force $f(t)$ is applied to the mass-dashpot-spring system shown in Fig. P3.7. Derive the transfer function $\dfrac{X(s)}{F(s)}$

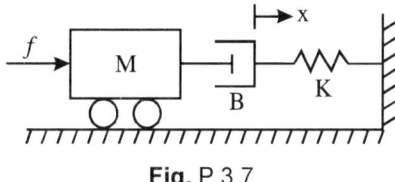

Fig. P 3.7

3.8. For each of the mechanical system shown in Fig. P 3.8, construct the mechanical network and determine $X(s)/F(s)$.

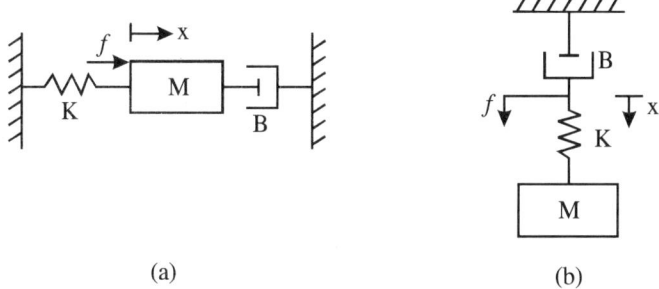

(a)   (b)

Fig. P 3.8

3.9. A schematic diagram of an accelerometer for measuring linear acceleration is shown in Fig. P 3.9. Determine the transfer function relating Y to the acceleration α of the frame.

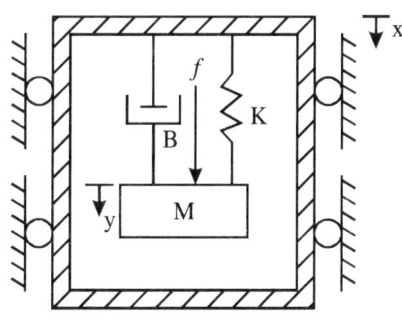

Fig. P 3.9

3.10. Two mechanical vibration absorbers are shown in Fig. P 3.10. Derive $X(s)/F(s)$.

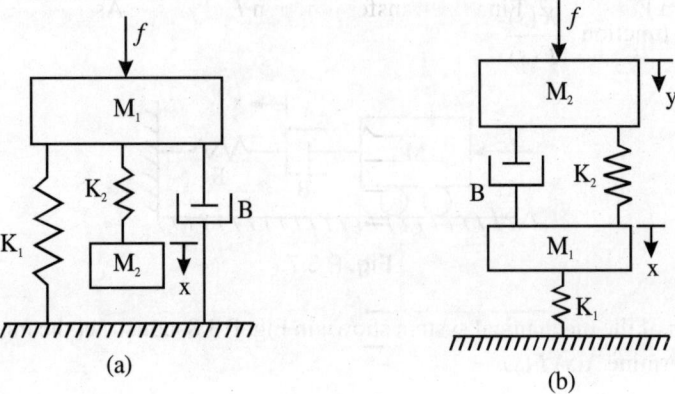

(a)   (b)

Fig. P 3.10

3.11. In Fig P 3.11, x_2 is the displacement of mass M_2 from its rest position and F is a force applied to mass M_1. The weightless cable moves the pulley without slipping. Derive a transfer function relating x_2 to f. The moment of inertia and radius of the pulley are J and r respectively.

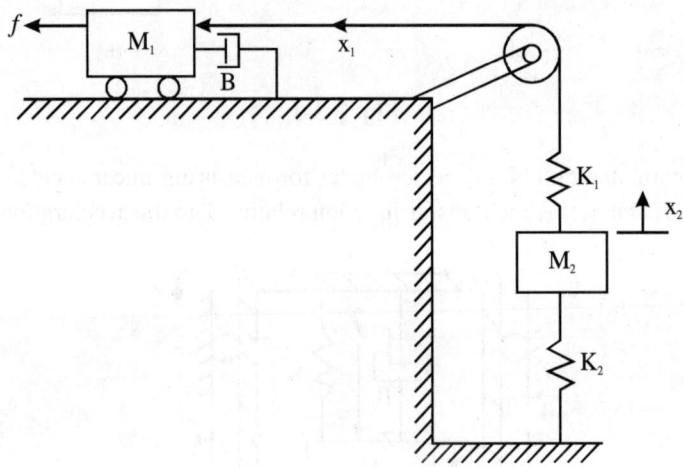

Fig. P 3.11

3.12. An operational amplifier has a gain of -1000 and the input and feedback elements are as shown in Fig. P 3.12. Find the transfer function $E_o(s)/E_i(s)$. Assume zero input current to the amplifier.

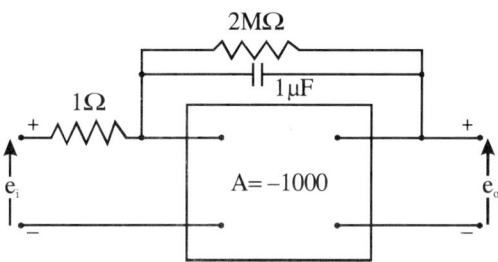

Fig. P 3.12

3.13. Determine the transfer function $\dfrac{\Theta(s)}{T(s)}$ for the Fig P 3.13.

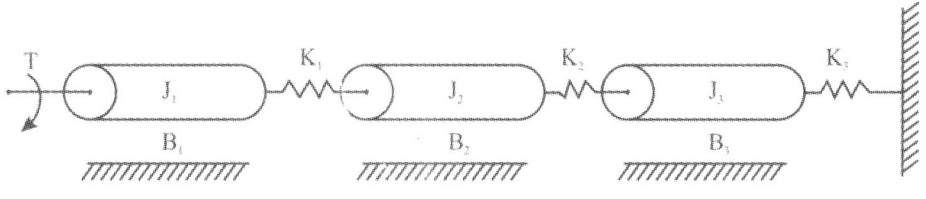

Fig. P 3.13

3.14. Draw the mechanical network for the mechanical system shown in Fig P 3.14 and derive the transfer function $X(s)/F(s)$.

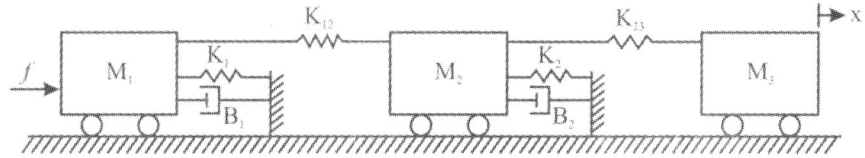

Fig. P 3.14

3.15. For the mass-spring-damper combination shown in Fig P 3.15 determine the equation relating f and x, f and y, and f and z.

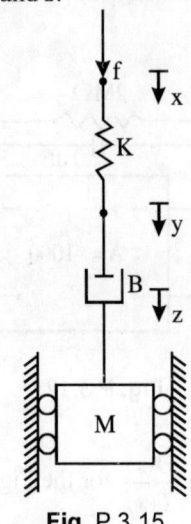

Fig. P 3.15

3.16. Derive the following transfer functions for the system shown in Fig. P 3.16.
(a) E/R, (b) C/R, (c) X/R

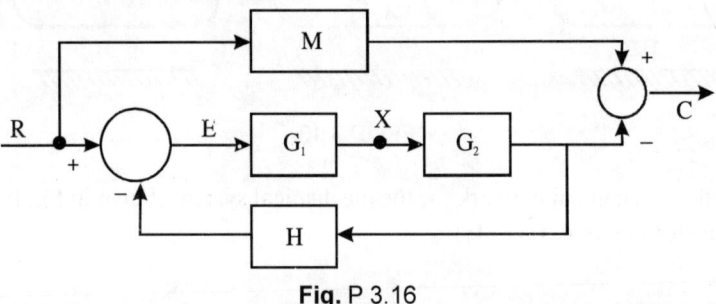

Fig. P 3.16

3.17. Referring to Fig. P.3.17, derive the following transfer functions.
(a) C/R, (b) E/R, (c) C/D, (d) E/D

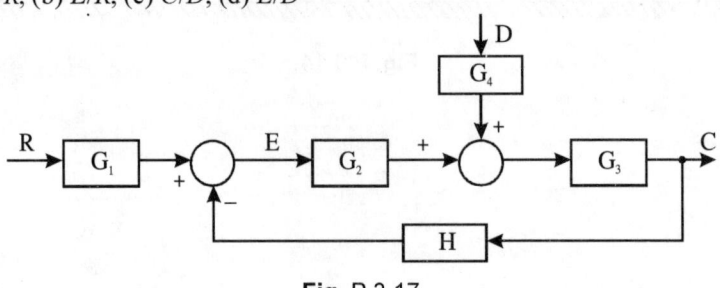

Fig. P 3.17

3.18. Find the transfer function $C(s)/R(s)$ for the system shown in Fig P.3.18.

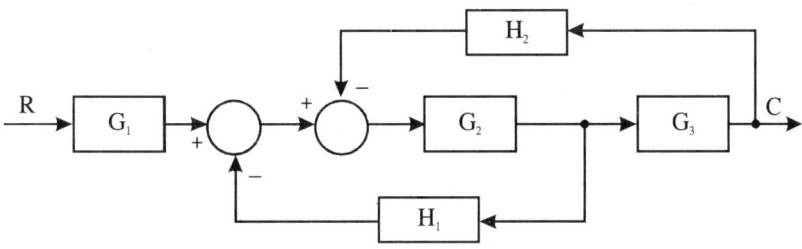

Fig. P 3.18

3.19. Find $C(s)/R(s)$ by block diagram method and signal flow graph method for the system shown in Fig. P 3.19.

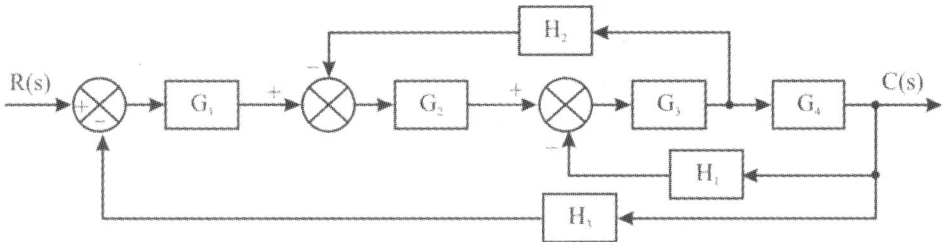

Fig. P 3.19

3.20. Find $X_1(s)$ and $X_2(s)$ for the signal flow graph shown in Fig P 3.20.

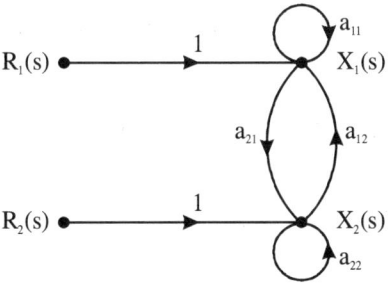

Fig. P 3.20

3.21. Find $C(s)/R(s)$ for signal flow graph shown in Fig. P. 3.21.

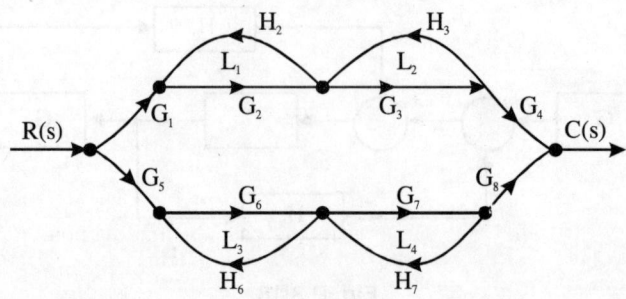

Fig. P 3.21

3.22. Find $C(s)/R(s)$ using Mason's formula for the signal flow graph shown in Fig. P 3.22.

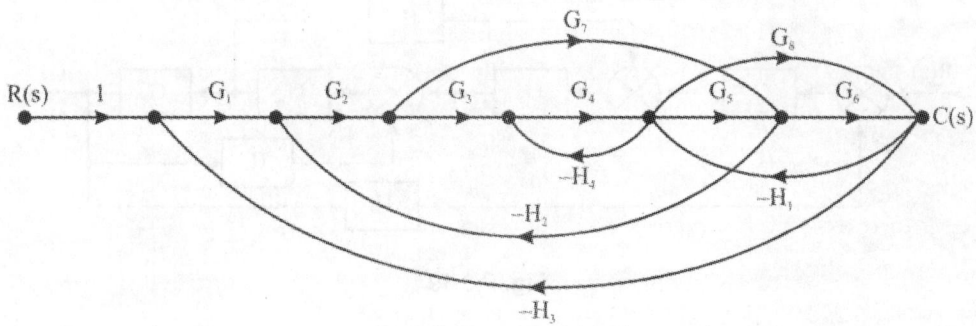

Fig. P 3.22

3.23. Obtain $E_2(s)/E_1(s)$ by (a) block diagram method and (b) Mason's formula for the system shown in Fig. P 3.23

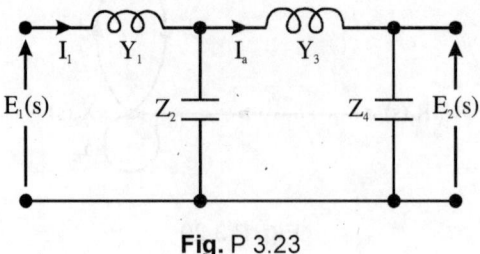

Fig. P 3.23

3.24. Draw signal flow graphs for the block diagrams shown in Figs. P 3.16 to P. 3.18 and obtain the required transfer functions.

Chapter 4

ANALOGOUS SYSTEMS AND COMPUTER SIMULATION

4.1. INTRODUCTION

Analogies are useful to compare an unfamiliar system with a familiar system. This helps to visualize the system more easily and to readily obtain the analytical solutions. *Two physical systems are said to be analogous to each other when the systems are governed by the same form of describing equations.* The equations describing electrical, mechanical and many other systems are of the same form. The existence of analogous systems and solutions makes it possible to extend the results of a familiar system to all analogous systems.

There are significant advantages to the use of electrical analogs of mechanical systems. For instance, it is not convenient to set up a mechanical mass-spring-dashpot system and test its response in the laboratory because such components are not readily available in a wide variety of sizes and are inconvenient to work with. Since electrical components are readily available and can be varied with ease, it is often convenient to study the response characteristics of an analogous electrical network system equivalent to the mechanical system of interest and adjust the component values to produce the desired results. Then the equivalent mechanical design is specified.

The differential equations of complex control systems can easily be solved on electronic computers. They are very useful in the simulation, analysis and design of complex systems.

In this chapter, we shall develop analogies between systems and discuss the simulation of systems on computers.

4.2. FORCE - CURRENT ANALOG

Consider the mechanical translational system shown in Fig. 4.1. With zero initial conditions, the governing equation of the system is

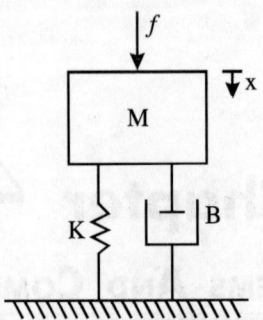

Fig. 4.1. Mass-spring-damper system

$$F(s) = (Ms^2 + Bs + K) X(s) \qquad\qquad ...(4.1)$$

Eq. (4.1) may be written as a function of velocity $V(s)$ instead of the displacement $X(s)$ as in Eq. (4.2).

$$F(s) = (Ms + B + K/s) V(s) \qquad\qquad ...(4.2)$$

Let us consider now a parallel RLC network shown in Fig. 4.2. If the node voltage is E(s), the system equation is

$$I(s) = (Cs + 1/R + 1/Ls) E(s) \qquad\qquad ...(4.3)$$

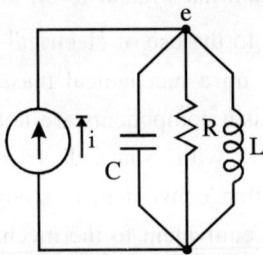

Fig. 4.2. Parallel RLC network

Examination of Eq.(4.2) and (4.3) shows that they are of the same form. Hence, the two systems are analogous to each other. The analogous quantities are shown in Table 4.1. Since the force in the mechanical system is analogous to the current in the electrical system, this analogy is known as *force-current analog*.

Table 4.1. Analogous Quantities in Force-Current Analogy

Mechanical Translational System	Electrical System
Force F	Current I
Mass M	Capacitance C
Viscous-friction Coefficient B	Reciprocal of Resistance, 1/R
Spring Constant K	Reciprocal of Inductance, 1/L
Velocity V	Voltage E

4.3. FORCE—VOLTAGE ANALOG

Let us consider the series RLC network shown in Fig 4.3. The governing equation of the system is

$$E(s) = (Ls + R + 1/Cs)\, I(s) \qquad \qquad ...(4.4)$$

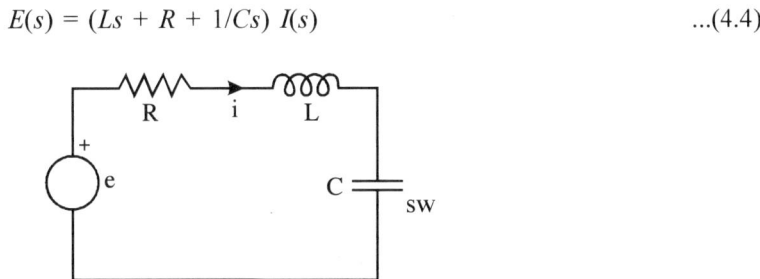

Fig. 4.3. Series RLC circuit

Equations (4.2) and (4.4) are of the same form and the two systems are analogous to each other. The analogous quantities are shown in Table 4.2. Since the force in the mechanical system is analogous to the voltage in the electrical system, this is known as *force-voltage analog*.

Table 4.2: Analogous Quantities in Force-Voltage Analogy

Mechanical Translational System	Electrical System
Force F	Voltage E
Mass M	Inductance L
Viscous-friction Coefficient B	Resistance R
Spring Constant K	Reciprocal of Capacitance, 1/C
Velocity V	Current I

4.4. TORQUE—CURRENT ANALOG

Consider a mechanical rotational system shown in Fig. 4.4(a). The system has a mass with a moment of inertia J immersed in a fluid. A torque T is applied to the mass. There is only one angular displacement. The mechanical network is shown in Fig 4.4(b).

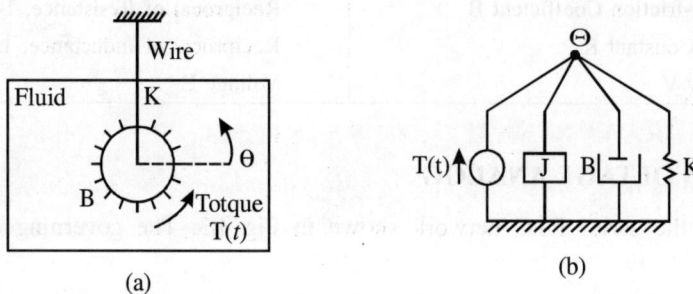

(a)   (b)

Fig. 4.4. (a) Mechanical rotational system (b) Mechanical network

The describing equation is

$$T(s) = (Js^2 + Bs + K)\ \Theta(s)$$
$$= (Js + B + K/s)\ \Omega(s) \qquad\qquad ...(4.5)$$

where Θ is the angular displacement and Ω angular velocity.

Comparing Eq. (4.5) with Eq. (4.3), they are of the same form and analogous. Since torque is analogous to current, this analogy is called *torque - current analog*. The analogous quantities are shown in Table 4.3.

Table 4.3: Analogous Quantities in Torque-Current Analogy

Mechanical Rotational System	Electrical System
Torque T	Current I
Inertia J	Capacitance C
Viscous-friction Coefficient B	Reciprocal of Resistance 1/R
Spring Constant K	Reciprocal of Inductance 1/L
Velocity V	Voltage E

4.5. TORQUE—VOLTAGE ANALOG

Compare Eq. (4.5) and (4.4). They are analogous. Since the torque is now analogous to voltage, this is known as *torque-voltage analog*. The analogous quantities are shown in Table 4.4.

Table 4.4: Analogous Quantities in Torque-Voltage Analogy

Mechanical Translational System	Electrical System
Torque T	Voltage E
Moment of inertia J	Inductance L
Viscous-friction Coefficient B	Resistance R
Spring Constant K	Reciprocal of Capacitance 1/C
Velocity V	Current I

4.6. STEPS IN DRAWING ANALOGOUS SYSTEMS

The following steps help in drawing the electrical analogs of mechanical systems.

Step 1: Draw the mechanical network of the given mechanical system and write down the equations following the rules given in Section 3.7.

Step 2: Next draw the current analog circuit. Since this analog circuit configuration is of the same form of mechanical network, this is drawn first irrespective of which analog circuit is desired. At this stage reference may be made to Table 4.1 or 4.3.

Step 3: In the analog circuit, indicate the relationship between analogous elements.

Step 4: Derive the equations of analogous circuit and check with those of the original system.

Step 5: Draw the voltage analog circuit from the above current analog circuit with the help of Table 4.2 or 4.4. Follow the rule that series connected elements in the current analog circuit will be parallel connected elements in the voltages analog circuit and *vice versa*.

4.7. ILLUSTRATIVE EXAMPLES

In this section, we shall illustrate the method of obtaining electrical analogs for mechanical systems.

Example. 4.1: *Obtain the electrical analogs of the mechanical system shown in Fig. 4.5.*

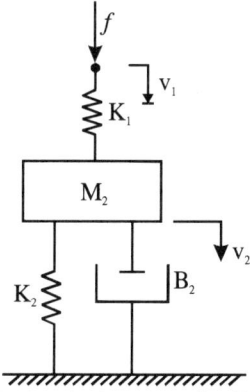

Fig. 4.5. Mechanical system for Example 4.1

Solution:

For the given system, the mechanical network is drawn as in Fig. 4.6. By referring to Table 4.1, the force–current analog is drawn as shown in Fig. 4.7.

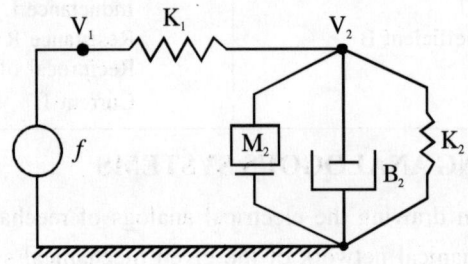

Fig. 4.6. Mechanical network

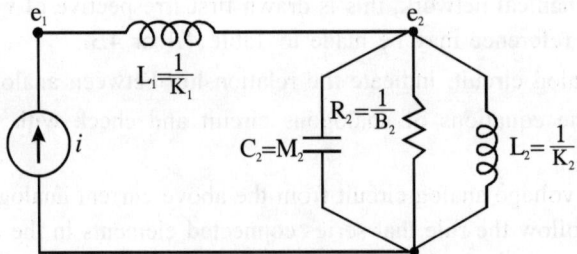

Fig. 4.7. Force-current analog

Next, the force–voltage analog is drawn from the current analog. It is shown in Fig. 4.8.

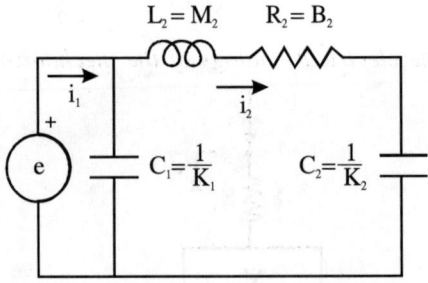

Fig. 4.8. Force-voltage analog

The mechanical network in Fig. 4.6 has two nodes. The two equations that describe the system are:

At node V_1 : $F(s) = (K_1/s) \, [V_1(s) - V_2(s)]$ ... (4.6)

At node V_2 : $\qquad 0 = (-K_1/s)V_1(s) + [M_2s + B_2 + (K_2 + K_1)/s]V_2(s)$ ... (4.7)

The describing equations of the force - current analogous system are:

At node e_1 : $\qquad I(s) = (1/L_1s)[E_1(s) - E_2(s)]$ ... (4.8)

At node e_2 : $\qquad 0 = (-1/L_1s)E_1(s) + [C_2s + 1/R_2 + 1/L_2s + 1/L_1s[E_2(s)$ ... (4.9)

From the force-voltage analog, the two mesh equations are:

For mesh I : $\qquad E(s) = (1/C_1s)[I_1(s) - I_2(s)]$ ... (4.10)

For mesh II : $\qquad 0 = (-1/C_1s)I_1(s) + [L_2s + R_2 + 1/C_2s + 1/C_1s) \cdot I_2(s)$ ... (4.11)

It is apparent that the set of equations are of the same form.

Example 4.2: *Write down the equations of motion and obtain the analogous electrical circuits for the mechanical rotational system shown in Fig. 4.9.*

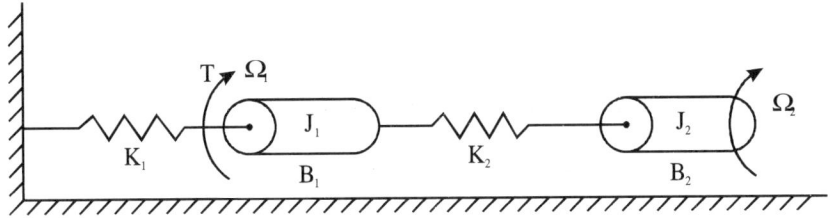

Fig. 4.9: Mechanical rotational system for Example 4.2

Solution:

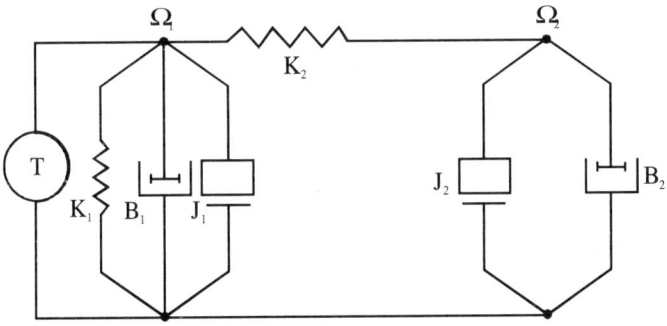

Fig. 4.10: Mechanical network

The mechanical network is drawn as in Fig 4.10. The two node equations are:

At node Ω_1 : $\qquad T(s) = [J_1s + B_1 + (K_1 + K_2)/s]\, \Omega_1(s) - (K_2/s)\, \Omega_2(s)$ ... (4.12)

At node Ω_2 : $\qquad 0 = (-K_2/s)\, \Omega_1(s) + J_2s + B_2 + K_2/s)\Omega_2(s)$ ... (4.13)

Fig. 4.11 shows the force-current analog. The node equations are:

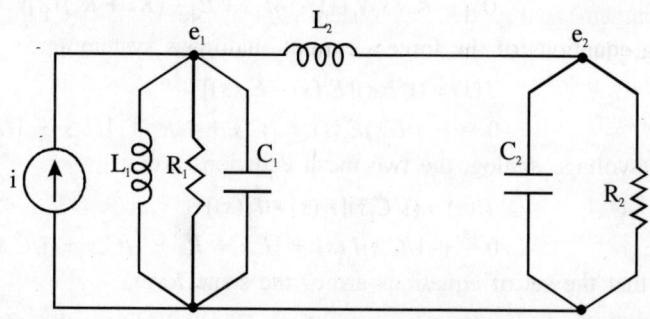

Fig. 4.11. Force-current analog

At node e_1 : $\qquad$ $I(s) = (C_1s + 1/R_1 + 1/L_1s + 1/L_2s)\,E_1(s) - (1/L_2s)\,E_2(s)$ $\quad$ ... (4.14)

At node e_2 : $\qquad$ $0 = (-1/L_2s)E_1(s) + (C_2s + 1/R_2 + 1/L_2s)\,E_2(s)$ $\qquad$ ... (4.15)

Fig 4.12 shows the force-voltage analog. The mesh equations are:

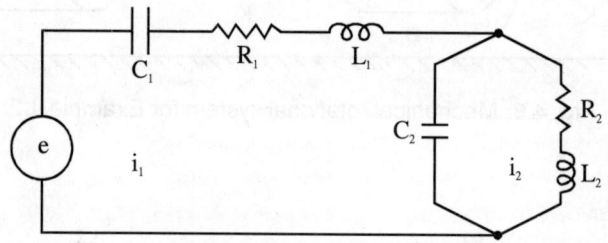

Fig. 4.12. Force-voltage analog

For mesh I : $\qquad$ $E(s) = (L_1s + R_1 + 1/C_1s + 1/C_2s)I_1(s) - (1/C_2s)I_2(s)$ $\qquad$ ... (4.16)

For mesh II : $\qquad$ $0 = (-1/C_2s)I_1(s) + (L_2s + R_2 + 1/C_2s)I_2(s)$ $\qquad$ ... (4.17)

4.8. MERITS OF FORCE-CURRENT ANALOG

The force-current analog exists for all physical systems whereas the force-voltage analog exists for certain systems only. Hence force-current analog is always preferred. The force-current analog results in an analogous electrical circuit similar to mechanical network configuration whereas the force-voltage configuration is completely different from the mechanical network configuration. Because of this fact, the force-current analog is known as *direct analog* and the force-voltage analog as *indirect analog*.

4.9. ADVANTAGES OF COMPUTER SIMULATION

A simulation study is often an important step in the evaluation and synthesis of complex control systems. When a mathematical model of a system is available, it may be utilized to investigate various designs of a planned system without actually building the system itself. The computer has become a standard tool for simulation study of continuous systems in recent years because of the following advantages.

1. The performance of a system can be observed under all conditions.
2. With a simulation model, the results of system performance can be extrapolated for prediction purposes.
3. Systems which are presently in a conceptual stage may be examined and decisions concerning future developments can be made accordingly.
4. Systems can be tested in a much reduced period of time.
5. Simulation results can be obtained at less then real experimentation cost.
6. Study of hypothetical situations can be carried out.
7. Computer simulation is often the only feasible or safe technique to analyze and evaluate a system.

A system may be simulated using analog or digital computer.

4.10. ANALOG COMPUTER SIMULATION

The analog computer is one of the most useful engineering tools available for the analysis and design of control systems. The word analog is used because the observed or measured quantities, usually voltages, are analogous to the physical variables of the system under study. Since physical variables are represented by voltages, the effect of variation in control elements and system parameters can be simply obtained by adjusting the potentiometer settings. Also, the performance of the control system can be directly observed and almost immediately on an x - y plotter or an oscilloscope.

The basic building blocks of an analog computer used to solve intergo-differential equations are integrators, summers and multipliers. The fundamental device used in developing the three basic units is an operational amplifier. In this section, we shall discuss the basic principle and elements of an analog computer and explain the methods of solution of problems and system simulation with suitable examples. Amplitude and time scaling are also discussed.

4.10.1. Operational Amplifiers

An operational amplifier is one which is used to perform various mathematical operations such as *integration, summation* and *multiplication*. It is a direct-coupled amplifier and has a very high gain of the order of 10^6 to 10^8. The internal impedance of the amplifier is very high and the input current is *essentially zero*. An amplifier with input impedance Z_i and feedback impedance Z_f is shown in Fig. 4.13.

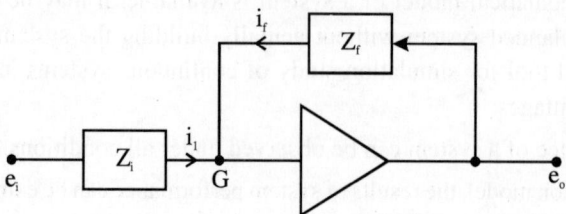

Fig. 4.13. Operational amplifier

Applying Kirchhoff's current law at the node G, we get

$$I_i(s) + I_f(s) = 0 \qquad \text{... (4.18)}$$

$$I_i(s) = [E_i(s) - E_g(s)]/Z_i(s) \qquad \text{... (4.19)}$$

$$I_f(s) = [E_0(s) - E_g(s)]/Z_f(s) \qquad \text{:.. (4.20)}$$

Substituting Eq. (4.19) and (4.20) in Eq. (4.18) and rearranging,

$$\frac{E_o(s)}{Z_f(s)} = -\frac{E_i(s)}{Z_i(s)} + E_g(s)\left(\frac{1}{Z_i(s)} + \frac{1}{Z_f(s)}\right) \qquad \text{... (4.21)}$$

Since the gain of the amplifier is very high, the voltage $E_g(s)$ will be negligible. Hence

$$\frac{E_o(s)}{E_i(s)} = -\frac{Z_f(s)}{Z_i(s)} \qquad \text{... (4.22)}$$

Eq. (4.20) is the basic equation of an operational amplifier. By properly choosing the input and feedback impedances, all mathematical operations can be performed.

4.10.2. Multiplication and Division

If we choose the input and feedback impedances to be resistances as shown in Fig. 4.14, the output voltage is obtained from Eq. (4.22) as

$$E_o(s) = -\frac{R_f}{R_i}E_i(s) \qquad \text{... (4.23)}$$

Eq. (4.23) can also be derived by the application of Kirchhoff's current law at the node G in Fig. 4.14. Thus, the output voltage is equal to the input voltage multiplied by the ratio R_f/R_i. The standard value of R_f used in analog computation is 1 meg-ohm. With proper choice of the input resistance R_i, an output voltage equal to the input voltage multiplied by a constant greater than unity can be obtained. If we choose R_i to be greater that R_f, we may obtain an output voltage which is less than the input, thus implying a division. The conventional method of obtaining a division of the input voltage is to use a potentiometer.

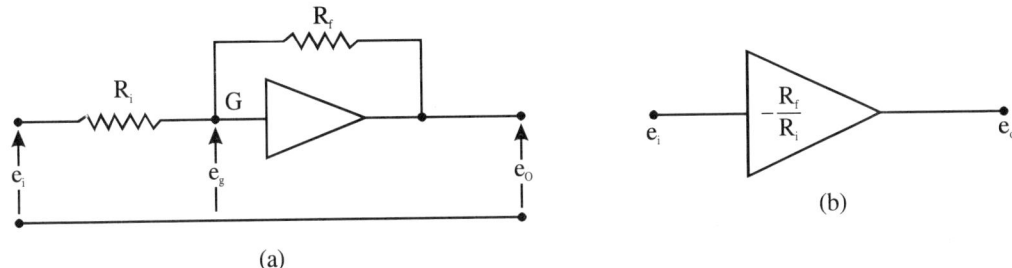

Fig. 4.14. (a) Multiplier and Devider (b) Representation

Fig. 4.15 shows a potentiometer. The output of the potentiometer is equal to a constant times the input, the constant k being adjustable between 0 and 1. A ten-turn helical potentiometer is generally used in analog computers which can be adjusted to three digits by a means of a calibrated dial. For example, to obtain an output equal to 23.4% of the input, the potentiometer constant k is adjusted to be 0.234.

The potentiometer is a useful device while programming a differentiaι equation on an analog computer. Since standard size resistors and capacitors are used, the ratio Z_f/Z_i is normally integral numbers. The coefficients of the differential equation to be solved may not be integral numbers. To obtain the exact value of coefficients, the potentiometer is used at the input in conjunction with standard ratios. While adjusting a potentiometer, it is necessary to compensate for any loading effect. This is achieved by comparing the setting of the potentiometer with the load connected to a standard reference potentiometer.

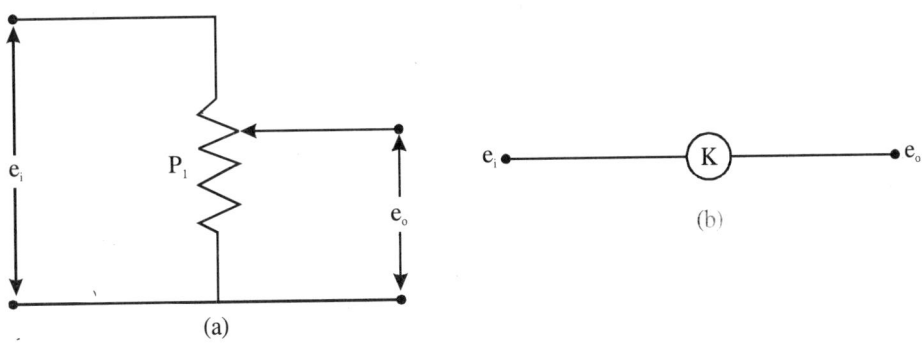

Fig. 4.15. (a) Potentiometer (b) Representation

4.10.3. Sign Inversion

Sign inverter is one whose output is, at all times, equal to the magnitude of the input but opposite in phase to the input. This can be easily implemented by choosing the input and feedback resistors of the multiplier to be equal as shown in Fig. 4.16. Thus

$$E_0(s) = -E_i(s) \qquad \qquad \text{... (4.24)}$$

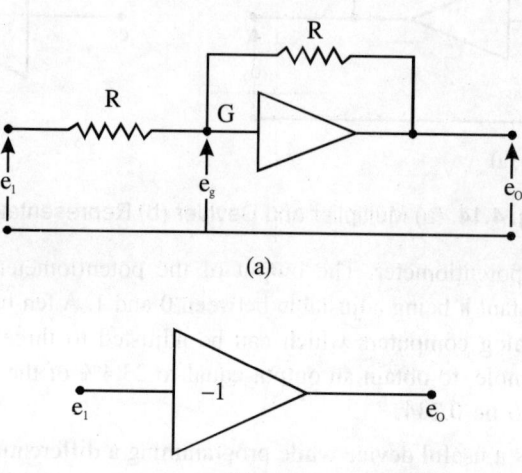

(a)

(b)

Fig. 4.16. (a) Sign changer (b) Representation

This is used, wherever necessary, to restore any sign change that is undesirable as well as in subtraction.

4.10.4. Addition and Subtraction

When two or more voltages are to be added, the setup is modified as shown in Fig 4.17. The number of input points being equal to the number of voltages to be added. A nodal equation at G can be written as

$$I_1(s) + I_2(s) + I_f(s) = 0 \qquad \qquad \text{... (4.25)}$$

Assuming $E_g = 0$, Eq. (4.25) may be written as

$$\frac{E_1(s)}{R} + \frac{E_2(s)}{R} + \frac{E_0(s)}{R} = 0 \qquad \qquad \text{... (4.26)}$$

or,

$$E_0(s) = -[E_1(s) + E_2(s)] \qquad \qquad \text{... (4.27)}$$

Thus the output voltage is the sum of the two input voltages with a change in polarity. This setup is called an *adder* or *a summing amplifier*. These voltages may be of any form and vary in time so as to be analogous to some function whose value varies with respect to some independent variable. Hence when these voltages are suitably scaled to be within the voltage range of the amplifiers, they can be used to represent physical quantities and the differential equations representing the behaviour of physical systems can be solved on an analog computer.

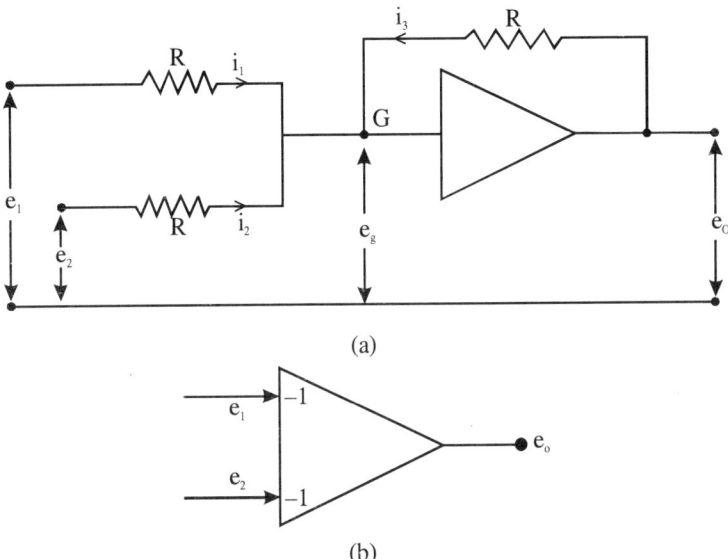

(a)

(b)

Fig. 4.17. (a) Adder (b) Representation

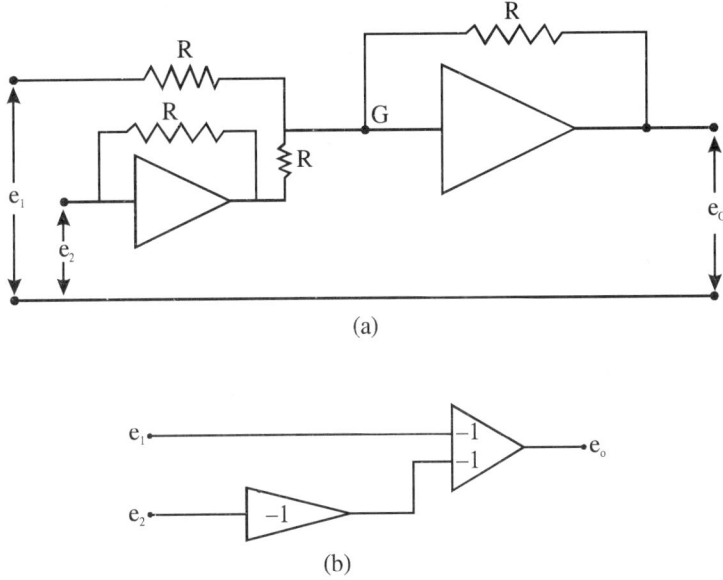

(a)

(b)

Fig. 4.18. (a) Subtractor (b) Representation

Subtraction of two or more voltages can be performed by a combination of sign inverters and adders as shown in Fig. 4.18. The voltage e_2 to be subtracted is first applied to the input of a sign inverter to obtain $(-e_2)$ and this is applied along with e_1 to the summing amplifier. The output of the setup is thus

$$E_0(s) = -[E_1(s) - E_2(s)] \qquad \qquad ...(4.28)$$

If the input resistances are chosen to be different from R_f, Eq.(4.27) becomes

$$E_0(s) = -\left[\frac{R_f}{R_1}E_1(s) + \frac{R_f}{R_2}E_2(s)\right] \qquad \qquad ... (4.29)$$

and Eq.(4.28) becomes

$$E_0(s) = -\left[\frac{R_f}{R_1}E_1(s) - \frac{R_f}{R_2}E_2(s)\right] \qquad \qquad ... (4.30)$$

Thus the input voltages can be multiplied by standard ratios or by any three-digit coefficient by employing potentiometers at the input and then added to give the output. This setup is extesively used in analog computers.

4.10.5. Integration

Integration of an input voltage can be achieved by choosing the feedback element in Fig. 4.19 to be a capacitor. Then, we have the setup shown in Fig. 4.19. The nodal equation at G is

$$\frac{E_1(s)}{R} + CsE_0(s) = 0 \qquad \qquad ... (4.31)$$

or,

$$E_0(s) = -\frac{1}{RC}\frac{E_1(s)}{s} \qquad \qquad ... (4.32)$$

Thus the output voltage is the integral of the input voltage multiplied by a constant $(1/RC)$ which may be made equal to, greater than or less than unity by choosing suitable values of R and C. The standard values of R and C are 0.1, 0.25, 0.5, 1.0 meg-ohm and 1 microfarad respectively.

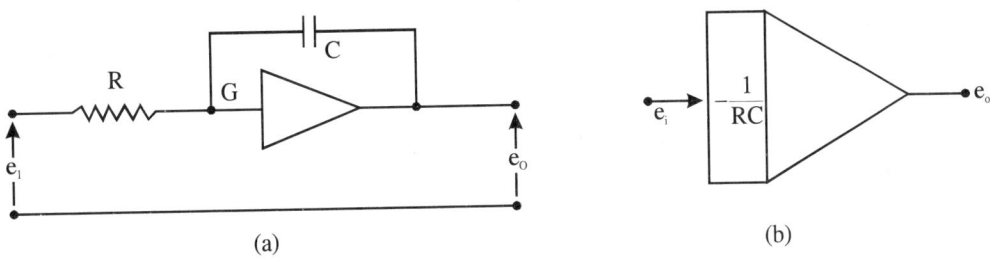

Fig. 4.19. (a) Integrator (b) Representation

When e_1 is integrated as shown in Fig. 4.19, the initial value of e_0 should also be considered. Hence whenever an integrator is used, the initial value of the output should also be incorporated. The integrator may have more than one input. In such a case, the output is equal to the integral of sum of all inputs with sign inversion.

4.10.6. Inserting Initial Conditions

If the initial values of integration are not zero, they have to be inserted before the solution of the equation starts. The output of integrators being voltages, the initial values should also be represented by voltages. These voltages are applied across the feedback capacitors of the integrators. A schematic circuit for applying initial condition voltage is shown in Fig. 4.20. The source voltage with appropriate polarity is connected across the potentiometer and the output of the potentiometer is set to correspond to the initial condition voltage. As soon as the switch is closed, the capacitor gets charged to this value. When the solution is started, the switch is opened by means of relays. The output of integrator is then given by

$$e_0(t) = -\int_o^t e_1 dt + e_0(o) \qquad \qquad \text{... (4.33)}$$

When the input voltage is zero at the start of the solution, the output voltages is equal to the initial value.

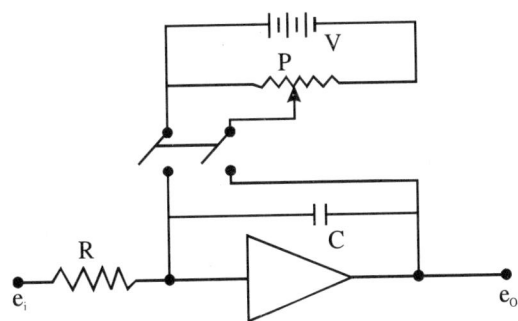

Fig. 4.20. Application of initial condition voltage

It is assumed that the initial condition voltage when applied to an integrator will appear with sign inversion at the output.

4.10.7. Differentiation

Differentiation of an input voltage can be accomplished by employing a capacitor as the input element and a resistor as the feedback element as shown in Fig. 4.21. The nodal equation is

$$CsE_1(s) + \frac{E_0(s)}{R} = 0 \qquad \qquad \text{... (4.34)}$$

or,

$$E_0(s) = -RCsE_1(s) \qquad \qquad \text{... (4.35)}$$

Thus the output voltage is the derivative of the input voltage multiplied by a negative constant. This set-up is called a differentiator and is *seldom* used in analog computation. The basic reason for avoiding differentiation is due to the fact that the differentiator accentuates noise transients leading to inaccurate and unstable operation.

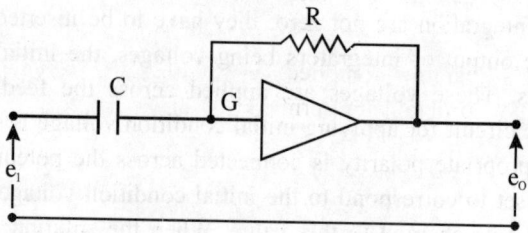

Fig. 4.21. Differentiator

Differentiation is a study of the amount of instantaneous variation or an instantaneous rate of change of the input quantity. In contrast to this, the integration is a process of averaging or smoothing out. The differentiator would not be able to distinguish between a signal and noise. Noise is any unwanted signal that occurs for a very short duration. Even when the signal is corrupted by negligibly small amount of noise occurring for a very short duration, the differentiator output will momentarily be very large since it picks up the rate of change of noise component which will be very high. Hence the output will be highly inaccurate. On the other hand, integration ignores the instantaneous change but considers only the sum of all changes. Thus, any effect produced by noise transients would be smoothed out since they will be rather small and insignificant when compared to the overall output.

The differential equations are normally solved by a process of successive integration, starting from the highest derivative, thus avoiding the use of differentiators.

4.11. PROGRAMMING A FIRST ORDER LINEAR ORDINARY DIFFERENTIAL EQUATION (LODE)

In analog computation, the differential equations are solved by successive integration of the highest order derivative in order to avoid the use of differentiators which are inherently unsuitable for computing purposes. The number of integrators required to solve a differential equation will be same as the order of the equation. In addition, a few adders and sign inverters may also be necessary as explained below.

Example 4.3. *Consider a first order differential equation given by*

$$\dot{x} + 5x = 8, \ x(0) = 2 \qquad \qquad \therefore \ (4.36)$$

Set up an analog computer simulation

Solution:

Step 1: Solve this equation for the highest order derivative.

$$\dot{x} = 8 - 5x \qquad \qquad \dots \ (4.37)$$

Step 2: The term $(-x)$ can be generated by integrating $\dot{x}$ (assumed to be available) as shown in Fig. 4.22(a).

Step 3: The output of the integrator is applied along with 8V battery to the input of an adder (as in Fig. 4.22(b)) in order to obtain the term $5x - 8$ at its output.

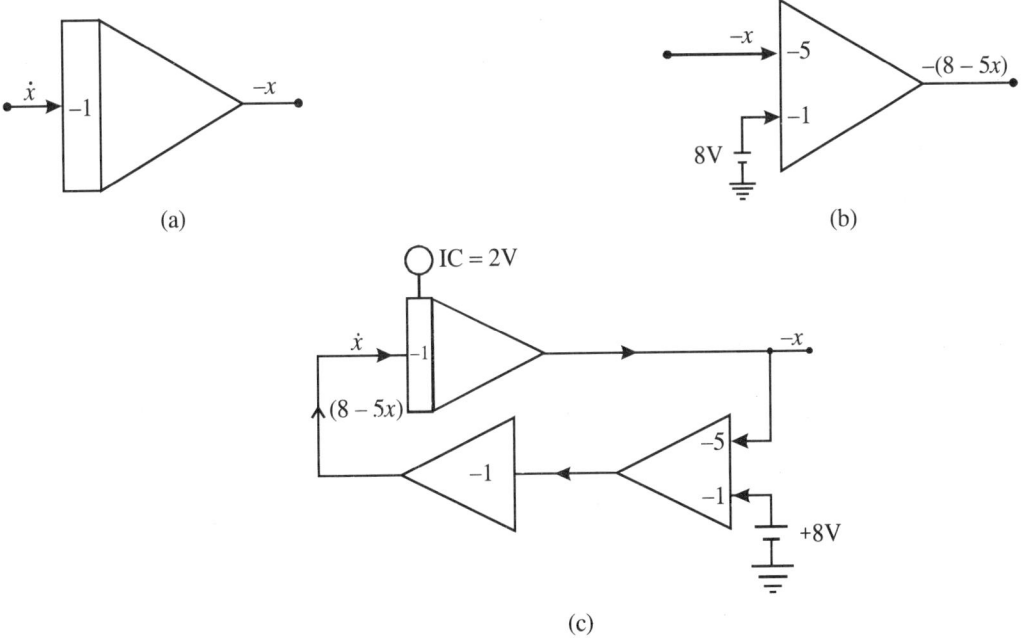

(a)

(b)

(c)

Fig. 4.22. (a) Generation of $-x$ form $\dot{x}$ (b) Obtaining $-(8-5x)$ (c) Complete setup

Step 4: Though $\dot{x}$ has been assumed to be available at Step 2, it has to be derived from some point in the setup. To obtain $\dot{x}$ form the output of the adder, an inverter is needed. The output of the inverter is $(8 - 5x)$.

Step 5: The "equal" sign of Eq. (4.37) is simulated by connecting the output of the inverter to the input of Fig. 4.22(a). *This is the most crucial step in the setup as this compels agreement between the function fed into and that given out.* The complete setup is shown in Fig. 4.22(c). The output being $(-x)$, the initial condition voltage is $-[-x(0)] = 2$V.

4.12. PROGRAMMING A SECOND ORDER LINEAR ORDINARY DIFFERENTIAL EQUATION

Example 4.4. *Consider the second order differential equation*

$$f(t) = \ddot{x} + B\dot{x} + Kx \qquad \qquad ...(4.38)$$

with $x(0) = 2$, $\dot{x}\ (0) = 1$. Set up analog computer simulation.

Solution:

Step 1: Solve the equation for the highest order derivative

$$\ddot{x} = f\left(t\right) - B\dot{x} - Kx \qquad \qquad ...(4.39)$$

Step 2: It is assumed that $\ddot{x}$ exists and is available at the starting point. This is passed through two integrators successively in order to obtain $-\dot{x}$ and x as shown in Fig. 4.23(a).

Step 3: The generated variables are multiplied by proper constants using multipliers. The signs of these terms are corrected by use of inverters wherever necessary as in Fig. 4.23(c).

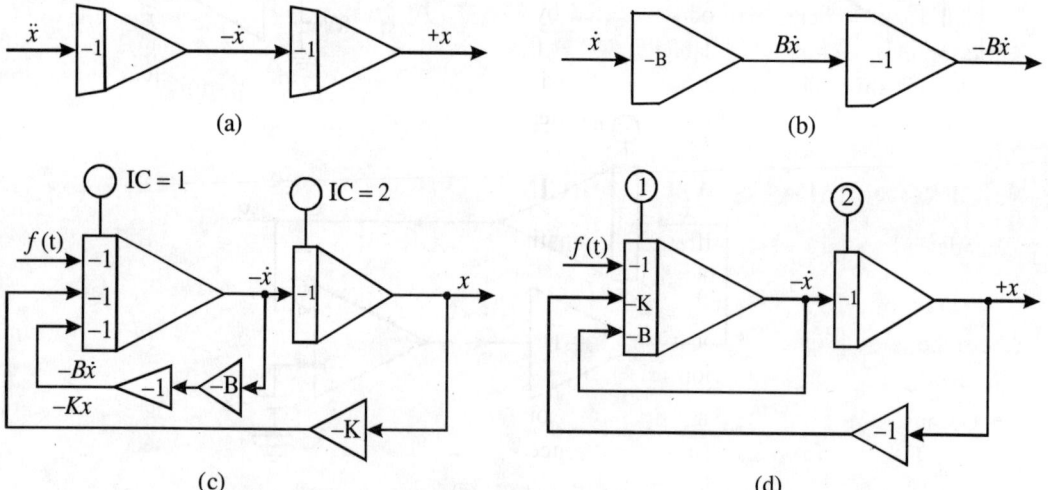

Fig. 4.23. (a) Obtaining $\dot{x}$ and x (b) Multiplication and inversion (c) Insertion of initial conditions with proper sign (d) Final analog computer setup

Step 4: The function *f(t)* is obtained from a function generator.

Step 5: Now that every term on the right-hand side of Eq. (4.39) is available with proper sign at some point in the setup, they are applied to the first integrator, thus completing the loops.

Step 6: Initial conditions with proper signs are also inserted as shown in Fig. 4.23(c).

Step 7: The final setup is obtained after eliminating unnecessary sign changers and multipliers. Fig. 4.23(d) gives the final computer setup.

4.13. PROGRAMMING *n*-TH ORDER LINEAR ORDINARY DIFFERENTIAL EQUATION

For solving the *n*-th order linear differential equation, the computer is programmed in a similar way. The steps to be followed are:

Step 1 : Set the highest order derivative equal to all the other terms in the differential equation.

Step 2: Draw, in cascade, a number of integrators equal to the order of the equation.

Step 3: Assume that the highest-order derivative is available and assume that it is the input to the first integrator. Then the outputs of the integrators will be a function of the dependent variable and its first (n – 1) derivatives.

Step 4: With these integrator outputs and the forcing function, generate the terms which are equal to the highest order derivative.

Step 5: Close the loop by feeding all these terms to the input of the first integrator.

Step 6: Setup the initial conditions on the pertinent integrators. Each initial condition appears at the output of the integrator with sign inverted.

This is the general method developed by Lord Kelvin to solve differential equations when entire forcing function is available. However this method cannot be used to program a differential equation in which the forcing function consists of the variable *x* and its derivatives but x alone is available. In such cases the canonical method is used.

4.14. PROGRAMMING WITH FORCING-FUNCTION RESTRICTION

Consider a second - order differential equation of the form

$$\ddot{y} + a_1 \dot{y} + a_o y = b_1 \dot{x} + b_o x \qquad \qquad \text{... (4.40)}$$

where the forcing function consists of the variable x and its first derivative but *only the variable x is available*. Such a situation arises often in servo-mechanical problems.

Because the computer has integrators only, we have no way of generating the derivaties directly. The general method fails and hence the *canonical method* is to be used.

In the operator notation ($p = d/dt$), Eq. (4.40) becomes

$$p^2 y + a_1 py + a_o y = b_1 px + b_o x \qquad \qquad \text{... (4.41)}$$

As in the general method, the highest-order derivative is equated to all the other terms.

$$p^2 y = b_o x - a_o y + b_1 px - a_1 py \qquad \qquad \text{... (4.42)}$$

With constant coefficients, Eq. (4.42) may be written as

$$p^2 y = (b_o x - a_o y) + p(b_1 x - a_1 y) \qquad \qquad \text{... (4.43)}$$

Next integrate twice. This is same as multiplying both sides by $(1/p^2)$. Thus,

$$y = \frac{1}{p^2}(b_o x - a_o y) + \frac{1}{p}(b_1 x - a_1 y) \qquad \qquad \text{... (4.44)}$$

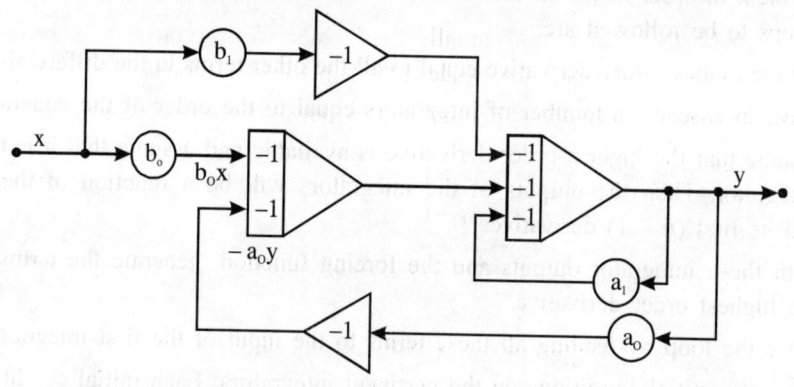

Fig. 4.24. Final analog computer setup

In contrast to the general method, in the canonical method it is assumed that y (the zero-th order derivative) is available and forms the output of the computer setup. Eq. (4.44) reveals that two integrators are required and that y can be obtained by integrating the first term $(b_o x - a_o y)$ twice, and the second term $(b_1 x - a_1 y)$ once, with proper signs. Hence the first term forms the input of the first integrator, the output of which along with the second term sign-inverted, is applied to the second integrator in order to obtain y as the output. The coefficients are obtained by using potentiometers. The complete setup is shown in Fig. 4.24.

The initial conditions of the above problem are stated in terms of the variables y and x and their derivatives. For the general method, there are no difficulties because the outputs of the integrators are also y and its derivatives. However, this is not the case in the canonical method except for the integrator whose output is y.

Fig. 4.24 shows that the output of integrator is y, the initial value of which is known. However, the output of integrator 1 is $(1/p)(a_o y - b_o x)$, the initial value of which is to be computed. This can be done by integrating Eq. (4.43) once, and rearranging terms, whereby we get

$$\frac{1}{p}(a_o y - b_o x) = (b_1 x - a_1 y) - py \qquad \qquad ... (4.45)$$

The initial values of all terms on the right side of the equation are known. Hence we can calculate the initial value of the output of integrator 1. For higher order differential equations, we could follow a similar procedure for calculating the initial values applied to the other integrator outputs.

4.15. PROGRAMMING SIMULTANEOUS DIFFERENTIAL EQUATIONS

The general method of programming is equally applicable when the problem to be programmed is in the form of a set of first or higher-order simultaneous equations.

Example 4.5: *Consider the following set of equations.*

$$\begin{aligned}
\ddot{x} - 8\dot{x} + 4y &= 12 \\
\ddot{y} - 2\dot{y} + 3x &= 0 \\
x(o) = -2, \qquad \dot{x}(o) &= 1 \\
y(o) = 1, \qquad \dot{y}(o) &= -1
\end{aligned} \qquad \qquad ... (4.46)$$

Setup an analog computer simulation to solve this problem.

Solution:

Step 1: Solve both equations for higher-order derivative.

$$\begin{aligned}
\ddot{x} &= 12 + 8\dot{x} - 4y \\
\ddot{y} &= 2\dot{y} - 3x
\end{aligned} \qquad \qquad ... (4.47)$$

Step 2: Lower order derivatives are generated by successive integration.

Step 3: The terms on the right side of Eq. (4.47) are added to obtain $\ddot{x}$ and $\ddot{y}$ as in Fig. 4.25(a) and (b)

Step 4: The loops are properly connected to complete input circuits to the adders. The final setup is shown in Fig. 4.25(c).

Step 5: All the initial conditions are inserted with proper sign at the integrator as in Fig. 4.25(c).

In preparing an analog computer solution to a differential equation, the *solution must remain* within voltage and other limits imposed by the machine. The maximum voltage that is obtainable at the output of any computing amplifier is limited to ± 100V or ± 10V depending on the

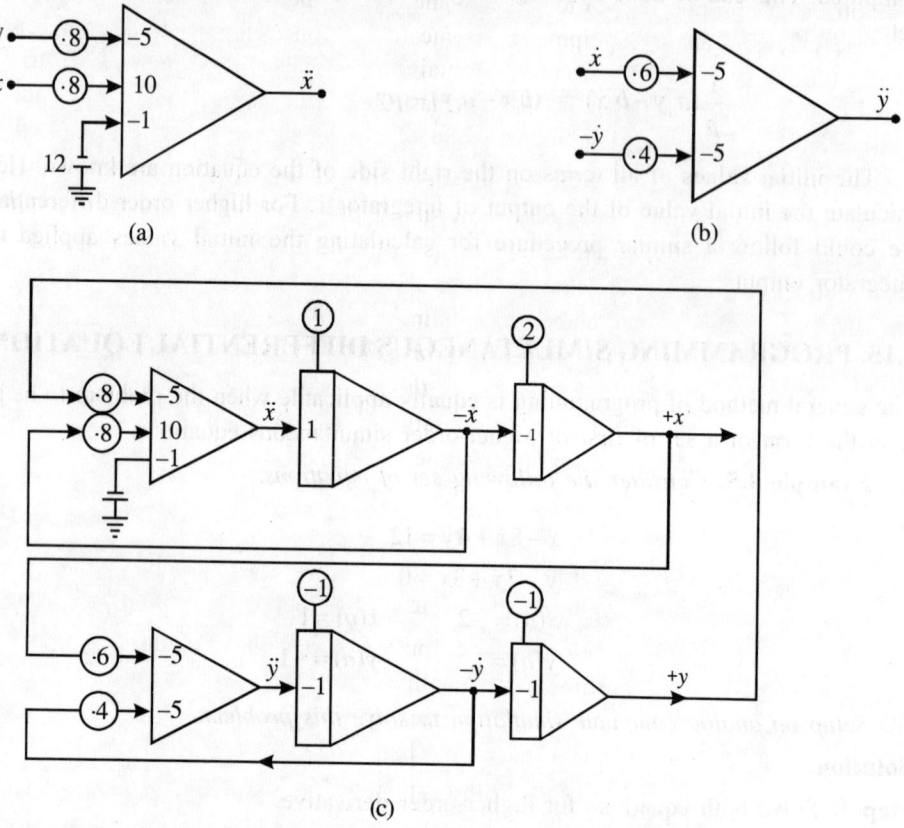

Fig. 4.25. (a) Obtaining $\ddot{x}$ (b) Obtaining $\dot{x}$ (c) Final analog computer setup

machine used. But the problem variables may have any finite magnitudes. Therefore some means must exist to relate these magnitudes, so that the problem variables may be represented in the computer by voltage (*called machine variables*) and transformed back into problem quantities. Scale factors precisely accomplish this. Thus the scale factors play a vital role in analog computer programming. We shall study these in detail in the next section.

4.16. AMPLITUDE AND TIME SCALING

The computer voltage and solution time are analogous to the dependent and independent problem variables respectively. Furthermore, when the computer is programmed directly from the problem equations, there is one-to-one correspondence between the computer and problem variables; i.e., one volt on computer is equal to one unit of dependent variable, and one unit of computer time to one unit of independent variable of the problem. However the computer has inherent limitations which bound the range of errorless operations of each of the components. In a computer solution, one wishes to take advantages of operation under conditions which lead to good accuracy. In analog equipment this means that one attempts to solve the problem in as short a time as possible, consistent with obtaining a suitable recording and one tries to keep the signal level as high as possible in order to mask the effects of noise and minor inherent machine nonlinearities, while operating in the zone of the greatest accuracy of the separate devices which comprise the computer setup.

As a rule, to ensure proper operating range and accuracy, the equations which are solved on a computer are not the original equations of the problem, but are a modified version scaled to conform to capabilities of the computer. Amplitude and time scaling are the two means at our disposal to combat computer shortcomings. *Amplitude scaling is the process of selecting factors for magnitudes of the dependent variables of the problem.* This changes the voltage level of the components. *Time scaling is the selection of scale factors for the independent variable of the problem.* This alters the duration of the time required for the computer to solve the problem. A technique of scaling, simple to implement and relatively immune to error, is described in this section.

4.16.1. Need for Amplitude Scaling

The need for amplitude scaling arises because of maximum and minimum voltage limitations.

Maximum Voltage Limitations: The output voltage of a computing amplifier is proportional to a wide rage of its input. This proportionality fails as the amplifier saturates when the problem voltage magnitudes exceed this range. In order to avoid computing errors, the magnitudes should be scaled down so as to remain within the linear range.

Minimum Voltage Limitation: Even well-designed amplifiers generate random noise and error voltages. While programming, special care should be taken to ascertain that the machine variable is never so small as comparable to noise and error voltages except for a very short time. Therefore, the magnitudes should be scaled up so as to be well above noise level.

Thus, the voltages at the output of each amplifier should not be too high or too low in order to get as accurate a solution as possible.

4.16.2. Amplitude Scaling

A convenient technique to carry out the amplitude scaling is called *normalization method.* This allows the standard values of R and C to be used and also grants the programmer a certain measure of independence of the computer to be used.

The steps to be followed in this method are:

1. An estimate is first made of the maximum value of all variables (as discussed later) that would appear in the solution of the differential equation.

2. Each variable is then normalized with respect to its maximum values so that maximum of each normalized variable is unity.

3. This maximum is made to correspond to maximum voltage of the particular computer.

Example 4.6. *Set up a computer diagram to solve the differential equation*

$$4\ddot{y} + \dot{y} + 9y = 0 \qquad \qquad \dots (4.48)$$

with the initial conditions $y(0) = 4$ *and* $\dot{y}(0) = -3/4$

Solution:

Step 1: The upper bounds of the maximum values of variables as estimated by Eq. (4.57) are,

$$y_m = |y_{max}| = 4$$
$$\dot{y}_m = |\dot{y}_{max}| = 6$$
$$\ddot{y}_m = |\ddot{y}_{max}| = 9$$

Step 2: Rewriting the equations in terms of the normalized variables, y/y_m, $\dot{y}/\dot{y}_m$ and $\ddot{y}/\ddot{y}_m$ yields

$$36(\ddot{y}/9) + 6(\dot{y}/6) + 36(y/4) = 0 \qquad \qquad \dots (4.49)$$

$$y(0)/4 = 1, \ \dot{y}(0)/6 = -1/8$$

The quantities within parentheses are normalized values.

Step 3: The higher-order derivative is equated to all the other terms. Thus,

$$(\ddot{y}/9) = -\frac{1}{6}(\dot{y}/6) - (y/4) \qquad \qquad \dots (4.50)$$

Step 4: Since the input of the first integrator is $(\ddot{y}/9)$, the output would be $-(\ddot{y}/9)$ and not $-(\dot{y}/6)$ as required by Eq. (4.50). Hence, to get $-(\dot{y}/6)$ as the output of the first integrator,

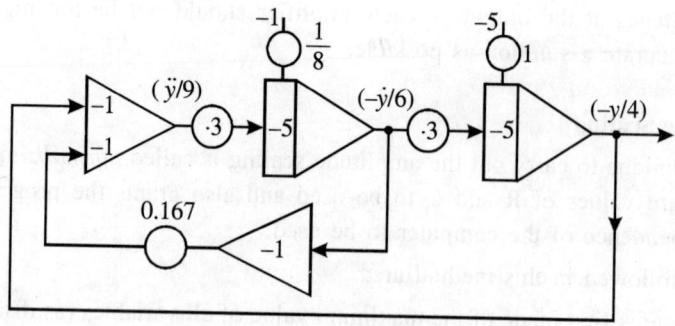

Fig. 4.26. Amplitude scaling

the input should be modified to $3/2(\ddot{y}/6)$. Similarly, the input to the second integrator is to be modified to $3/2(\dot{y}/6)$. The particular coefficients are obtained by means of potentiometers and multiplication factors of the integrators as shown in Fig. 4.26. It has been assumed that fixed voltage sources are available only for the maximum values of $+1$ unit (100V or 10V).

If $\ddot{y}$ is not needed explicitly, the summation of $(y/4)$ and $(\dot{y}/6)$ can be performed together with the integration which would yield $-(\dot{y}/6)$. An inverter is now needed for y but not for $\dot{y}$. Potentiometer settings and multiplication factors do not get altered. As a rule, an adder always precedes the first integrator in scaled computer set-up in order to enable the user to change the time-scale factor quickly with ease.

4.17. ESTIMATING MAXIMUM VALUES

The process of scaling is dependent on the judicious determination of maximum values of the variables. The conventional procedure is to make intelligent guesses of the maximum values and use them in diagramming the solution and interconnecting the amplifiers. If a guess is too small, at least one amplifier is overloaded and the solution is incorrect. In this event, an indicator on the front panel glows identifying the over-loaded amplifier but no damage is done. If a guess is too high, the full range of the corresponding amplifier will not be used, and a given absolute error in the voltage will cause a greater relative error. This condition, in which the solution is less accurate than it could be, can be detected by a voltmeter, usually incorporated in the computer.

If the first guess is poor, the computer user need merely make another guess. The only loss is that of time. When a physical system is being modeled by a differential equation, often the system will furnish clues as to the maxima. It should be noted incidentally that it is not algebraic maxima, but rather maximum magnitudes (absolute values) that are of interest for scaling.

Rules for estimating maximum values will now be discussed for three commonly encountered types of differential equations.

(a) Second-order Linear Homogeneous Equation with Constant Coefficients

Consider the differential equation

$$a\ddot{y} + b\dot{y} + cy = 0 \qquad \qquad ... (4.51)$$

The characteristic equation is

$$am^2 + bm + c = 0 \qquad \qquad ... (4.52)$$

The nature of the solution depends on the roots of the characteristic equation. The three cases are :

Case i: $b^2 - 4ac > 0$: The characteristic equation has two distinct real roots α_1 and α_2. The solution is of the form

$$y = C_1 e^{\alpha_1 t} + C_2 e^{\alpha_2 t} \qquad\qquad\qquad \text{... (4.53)}$$

where C_1 and C_2 are arbitrary constants. Maximum values of y and $\dot{y}$ when both α_1 and α_2 are negative, are

$$y_m = C_1 + C_2$$
$$\dot{y}_m = C_1 \alpha_1 + C_2 \alpha_2 \qquad\qquad\qquad (4.54)$$

If either or both of the roots are positive, maximum values can be estimated at the maximum desired value of t.

Case ii: $b^2 - 4ac = 0$: The characteristic equation has a double root and the solution is

$$y = e^{\alpha t}(C_1 + C_2 t) \qquad\qquad\qquad (4.55)$$

The maximum values can be estimated in the specified range of t.

Case iii: $b^2 - 4ac < 0$: This case is frequently encountered in practice since all practical systems are oscillatory to some extent. The characteristic equation has complex conjugate roots $\alpha \pm j\omega$, and the solution is

$$y = e^{\alpha t}(C_1 \sin \omega t + C_2 \cos \omega t) \qquad\qquad\qquad \text{... (4.56)}$$

For $\alpha \le 0$, the maximum values are estimated as

$$y_m = \left[y(o) + (\dot{y}(o)/\omega_o)^2 \right]^{1/2}$$
$$\dot{y}_m = \omega_o y_m$$
$$\ddot{y}_m = \omega_o^2 y_m$$

where $\omega_o \approx (c/a)^{1/2}$.

(b) Second-order Linear Differential Equations with Constant Co-efficients and Constant Forcing Functions

Consider the equation

$$a\ddot{y} + b\dot{y} + cy = F \qquad\qquad\qquad \text{... (4.57)}$$

Under steady-state condition, the value of y is given $y = F/c$. If y_1 satisfies Eq. (4.51), then $y_1 + (F/c)$ is a solution under dynamic condition as shown below:

$$a\ddot{y} + b\dot{y} + cy = a\left(\ddot{y}_1 + \frac{\ddot{F}}{c} \right) + b\left(\dot{y}_1 + \frac{\dot{F}}{c} \right) + c\left(y_1 + \frac{F}{c} \right)$$

$$= a\ddot{y}_1 + b\dot{y}_1 + cy_1 + F$$

$$= F$$

For $b^2 - 4ac \leq 0$ and α 0, y_l is oscillatory, and suitable estimates of the maximum values are:

$$y_m = y_1 + \left|\frac{F}{c}\right|$$

$$\dot{y}_m = \omega_o y_m$$

$$\ddot{y}_m = \omega_o^2 y_m$$

... (4.58)

where y_1 is given by

$$y_1 = \left\{\left[y(o) - (F/c)\right]^2 + \left[\dot{y}(o)/\omega_o\right]^2\right\}^{\frac{1}{2}}$$

(c) Equal Coefficient Rule

So far, rules for estimating the maximum values are given for a second-order differential equation. No definite estimate can be made for higher-order equations except when the forcing function is constant and all initial conditions are zero. Even in this case is available only an empirical rule developed by Jackson who has used it with some success. The set of estimated values can be used only when they are increasing or decreasing monotonically. When the rule is applied, the coefficients of all the derivatives in the normalized equation will be equal. Hence the name.

Let the *n*-th order differential equation be

$$a_n \frac{d^n y}{dt^n} + \ldots\ldots\ldots\ldots\ldots + a_1 \frac{dy}{dt} + a_o y = F \qquad \text{... (4.59)}$$

with all initial conditions zero.

The rule gives maxima of

$$y_m = 2F/a_o$$

$$\left(\frac{d^i y}{dt^i}\right)_m = F/a_i \qquad \text{... (4.60)}$$

4.18. NEED FOR TIME SCALING

The need for time scaling arises because

1. The undamped natural frequencies or forcing function frequencies lie outside the range of the computer. The computer can solve problems without being time-scaled up to a frequency of 10 radians per second.

2. Servo-motors used for multiplication of two variables are velocity-limited and can handle accurately only those input signals which change at a rate less than the velocity limit for the motor.

3. The output recording instruments such as $x - y$ recorders and galvanometer types have limited band-width and hence maximum recorder speed is limited.

4. In simulating a process of long duration, the solution time on the computer should be short. For example, a chemical process may take 24 hours which may have to be observed in 24 minutes on analog computers. The computer thus solves the problem at a faster rate.

5. The computer solution time should be longer in cases where the problem time is very short. For example, biological reaction may take place in one second. No worth while observation can be made if solved at that rate on the computer. This solution has to be slowed down.

4.18.1. Time Scaling

In solving differential equations on an analog computer, the problem independent variable T and the computer solution time T is related by

$$T = \alpha t \qquad \qquad \ldots (4.61)$$

If the time scale factor $\alpha = 1$, the solution takes place in real (physical world) time and the computer is then known as a *real-time computer*. When a computer is used as a part of a process control system (it is an *on-line computer*) and the solution takes place in the real time.

If $\alpha > 1$, the computer tames a longer period T to solve the problem and it is then called a *slow-time computer*. In *fast-time computer*, $\alpha < 1$, and the computer solution time T is shorter than the problem time.

Because of time scaling, the derivative of the dependent variable will get modified as:

$$\frac{dy}{dt} = \frac{dy}{d(T/\alpha)} = \alpha \frac{dy}{dT}$$

In general,

$$\frac{d^n y}{dt^n} = \alpha^n \frac{d^n y}{dT^n} \qquad \qquad \ldots (4.62)$$

Substituting $T = \alpha t$ in the integration equation

$$e_o(T) = e_o(T_o) - \frac{1}{RC} \int_{T_o}^{T} e_1(T) dT$$

we obtain

$$e_o(\alpha t) = e_o(\alpha t_o) - \frac{1}{RC} \int_{\alpha t_o}^{\alpha t} e_1(\alpha t) d(\alpha t)$$

$$= e_o(\alpha t_o) - \frac{\alpha}{RC} \int_{\alpha t_o}^{\alpha t} e_1(\alpha t) dt \qquad \qquad \ldots (4.63)$$

Therefore, after time scaling, the integrator operation can be regarded as integration with respect to problem time together with multiplication by α. If the computer solution is to represent the given problem, the gain of all the integrators must be changed by a factor $1/\alpha$. *This modification must not be made for adder inputs.*

If the input forcing function is a function of time, *f(t)*, it must also be time-scaled to *f(T/α)*. If the forcing function is generated by integration operation, a correct change of these integrator-gains will time-scale the input function automatically. However, if the input function is obtained from an external source, it must be first recorded and then rescaled. The time scaling also changes the natural frequency of the system simulated.

The time scaling has no effect on the maximum values and initial values of the parameters. (The maximum values change if time-scaling is carried out first).Thus, we make the following important statement about time-scaling on a computer diagram. *To time-scale by α, multiply the gain of each integrator by a factor $1/\alpha$.*

Let us again consider Eq. (4.48). The solution is a damped sinusoid of a period of about 4 seconds. ($\omega = 2\pi f = 3/2$; $f \approx 1/4$; $t = 1/f = 4$).

To display the solution on an oscilloscope at the rate of about one period per second, we have t = 4 and T = 1. Hence $\alpha = 0.25$. The gain of each integrator is therefore to be increased by a factor of 4. The resulting diagram is presented in Fig. 4.27. Note that the gain of each integrator is now increased by a factor of 4 from 1.5 (= 0.3 × 5) to 6 (= 0.6 × 10).

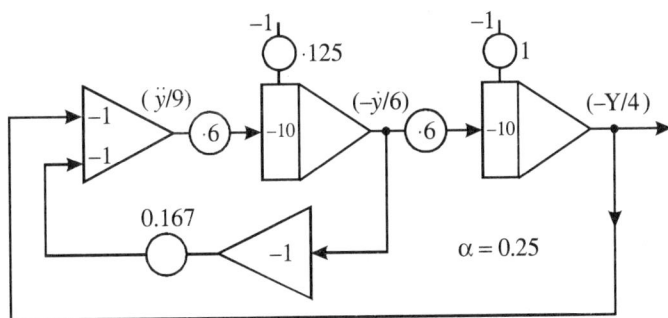

Fig. 4.27. Time-scaled computer setup

4.19. EXAMPLES OF AMPLITUDE AND TIME SCALING

(a) LODE with Constant Coefficients and Constant Forcing Function.

Example 4.7. *Consider the equation*

$$4\ddot{y} + \dot{y} + 9y = 18$$

$$y(o) = 4, \quad \dot{y}(o) = -3/4 \qquad \qquad \dots (4.64)$$

Set up amplitude and time scaled computer diagram.

Solution:

The maximum values as calculated by Eq. (4.58) are

$$y_m = 4$$
$$\dot{y}_m = 6$$
$$\ddot{y}_m = 9$$

... (4.65)

They are same as in the case of homogenous equation.

The normalized equation is

$$36(\ddot{y}/9) + 6(\dot{y}/6) + 36(y/4) = 18$$
$$(\ddot{y}/9) + 1/6(\dot{y}/9) + (y/4) = 0.5$$
$$(y(0)/4) = 1, \ (\dot{y}(0)/6) = -0.125$$

... (4.66)

If the time-scale factor is chosen to be same ($\alpha = 0.25$) as that in the case of homogeneous equation, the gain of each integrator has to be increased by a factor of 4.

The scaled computer diagram is left as an exercise. It may be noted that the only modification required when compared to Fig. 4.27 is to add one more input point to the summer with an input of 0.5 unit obtained from a constant source with a potentiometer set at 0.5.

(b) LODE with Constant Coefficients and Variable Forcing Function

Example 4.8. *Let the differential equation be*

$$4\ddot{y} + \dot{y} + 9y = 18\sin 2t$$
$$y(o) = 4, \ \dot{y}(o) = -3/4$$

... (4.67)

Set up amplitude and time scaled computer diagram.

Solution:

The same maximum values given by Eq. (4.65) may be used since, with $F = F_{max} = 18$, Eq. (4.58) yields the same values. The maximum value of $\sin 2t = 1$. The normalized equation is

$$36(\ddot{y}/9) + 6(\dot{y}/6) + 36(y/4) = 18\sin 2t$$
$$(\ddot{y}/9) + 1/6(\dot{y}/6) + (y/4) = 0.5\sin 2t$$
$$y(0/4) = 1, \ \dot{y}(0/6) = -0.125$$

... (4.68)

If the same time-scale factor is assumed ($\alpha = 0.25$), the only change we should make in Fig. 4.27 is to apply forcing function of $\dfrac{1}{2}\sin(8T)$ to one more input point of the summing amplifier.

(c) Higher-order Equation with Zero Initial Conditions

Example 4.9. *Consider the equation*

$$\dddot{y} + 3\ddot{y} + 5\dot{y} + 12y = 24$$
$$y(0) = \dot{y}(0) = \ddot{y}(0) = 0 \qquad \qquad \text{... (4.69)}$$

Set up amplitude and time scaled computer diagram

Solution:

By equal coefficient rule,

$$
\begin{aligned}
y_m &= 2 \times 24/2 = 4 \\
\dot{y}_m &= 24/5 = 4.8 \\
\ddot{y}_m &= 24/3 = 8 \\
\dddot{y}_m &= 24/1 = 24
\end{aligned}
\qquad \qquad \text{... (4.70)}
$$

Since the maximum values are monotonic, these values can be used. The equality of the coefficients can be observed in the normalized equation.

$$24(\dddot{y}/24) + 24(\ddot{y}/8) + 24(\dot{y}/48) + 48(y/4) = 24$$
$$(\dddot{y}/24) + (\ddot{y}/8) + (\dot{y}/4.8) + 2(y/4) = 1 \qquad \qquad \text{... (4.71)}$$

Let it be required to display three cycles in about four seconds. An approximate frequency may be obtained as $\omega = (12/3)^{1/2} = 2$. The time to complete three cycles is given by $wt = 6\pi$ or $t = 3\pi \approx 10$. In view of the requirement that T = 4, the time scale factor $\alpha = 0.4$. The gain of each integrator is therefore increased by a factor of 2.5. The gains required are:

For First integrator: (24/8) 2.5 = 7.5

For Second integrator: (8/4.8) 2.5 = 4.167

For Third integrator: (4.8/4) 2.5 = 3

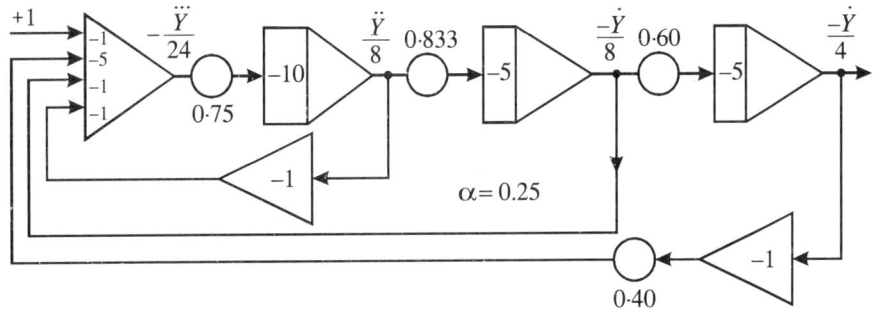

Fig. 4.28. Scaled computer setup

The scaled computer setup is shown in Fig. 4.28 with the constant forcing function being obtained from a constant source.

(d) Higher-order Equation with Initial Conditions

Example 4.10. *Let the equation be*

$$\dddot{y} + 3\ddot{y} + 5\dot{y} + 12y = 24$$

with

$$y(0) = 2, \ \dot{y}(0) = 1, \ \ddot{y}(0) = -3 \qquad \qquad ... \ (4.72)$$

Setup amplitude and time scaled computer diagram.

Solution:

As already stated, there is no defined rule by which the maximum values may be determined. A possible method is to drop the terms higher than second order in the given equation and find the maximum values applying Eq. (4.58). The maximum values of higher-order derivatives may be found by the extension of the rule $\ddot{y}_m = \omega_o^3 y_m$ and so on. Thus, ω_o being 2,

$$y_m = 2.5$$
$$\dot{y}_m = 5$$
$$\ddot{y}_m = 10$$
$$\dddot{y}_m = 20$$

It may be noted that these maximum values are to be taken only as guides. The correct values can be easily determined after a trial run of the problem on the computer. The normalized equation is

$$20(\dddot{y}/20) + 30(\ddot{y}/10) + 25(\dot{y}/5) + 30(y/2.5) = 24$$
$$(\dddot{y}/20) + 1.5(\ddot{y}/10) + 1.25(\dot{y}/5) + 1.5(y/2.5) = 1.2 \qquad ... \ (4.73)$$

If the time scale factor is taken to be 0.4, the gain of each integrator is increased by a factor of 2.5. The gains required are:

First integrator: (20/10) 2.5 = 5

Second Integrator: (10/5) 2.5 = 5

Third Integrator: (5/2.5) 2.5 = 5

The scaled computer setup is left as an exercise.

4.20. FUNCTION GENERATORS

In most problems involving analog computers, it is essential to generate certain functions which may simulate physical driving functions or non-linear phenomena which cannot be obtained with elements such as integrators, summing amplifiers, and potentiometers alone. For example, it may be required to generate a voltage of the form $e = E \sin \omega t$, i.e., to produce an oscillation with a prescribed amplitude and frequency or to generate product of variables.

In this section, we shall discuss the generation of functions involving time and space variables, and also a computer setup that would differentiate variables without using a differentiator. Non-linear function generators are not discussed here.

4.20.1. Generation of Analytic Functions

In the solution of differential equations, the forcing function is a function of time, such as step-functions, functions of t and powers of t, exponential function, sinusoidal and hyperbolic functions and functions of the form $y = a^t$. The general principle to generate the functions is to obtain them as solutions of a homogeneous differential equation of order n if:

1. The function and its first n derivatives are continuous.

2. The derivatives can be calculated.

The procedure is to differentiate a number of times until, with a proper selection of coefficients, the homogeneous equation is formed. The equation thus constructed may be linear or non-linear. In either event the solution will be the function we seek when we impose the correct initial conditions. The following examples illustrate the methods of generation of various functions of time.

(a) Step Function

A step function is a sudden and instantaneous change in the input such as an application of a D.C. source. This may be generated manually with a series connection of a D.C. source and a switch.

(b) Ramp Function

$$y = kt \qquad \qquad \text{... (4.74)}$$

Differentiating (4.74)

$$\dot{y} = k \qquad \qquad \text{... (4.75)}$$

This may be generated by applying a step function to an integrator as shown in a part of Fig. 4.29 with zero initial condition [$y(0) = 0$].

(c) Parabolic Function

$$y = Kt^2 \qquad \qquad \text{... (4.76)}$$

Differentiating twice, we form the equation

$$\ddot{y} = 2K \qquad \qquad \text{... (4.77)}$$

with $\dot{y}(0) = y(0) = 0$. This may be generated by applying a step function of magnitude $2K$ to the input of first integrator. As Eq. (4.77) is of second order, two integrators are required as shown in part of Fig. 4.29.

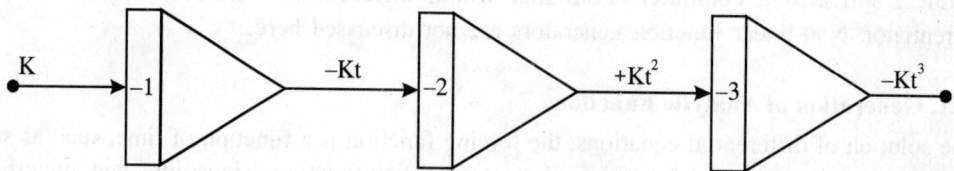

Fig. 4.29. Function generation

(d) Powers of t

When generating integral powers of t, we can use multipliers. However, a more convenient method is to use integrators in cascade. The resulting circuit is shown in Fig. 4.29. The initial condition on each integrator is zero at $t = 0$.

(e) Sine and Cosine Functions

Sine and cosine functions of dependent variables are used so frequently that a special device called a resolver is available to generate them. These trigonometric functions are also solutions of differential equations which can be implemented on a computer

Let

$$y = \sin at \qquad\qquad ... (4.78)$$

$$\dot{y} = a \cos at \qquad\qquad ... (4.79)$$

$$\ddot{y} = -a^2 \sin at = -a^2 y \qquad\qquad ... (4.80)$$

Thus, the solution of Eq. (4.80) generates sine and cosine functions as shown in Fig. 4.30. The initial value of *cosat* being unity, the initial condition is incorporated in the second integrator.

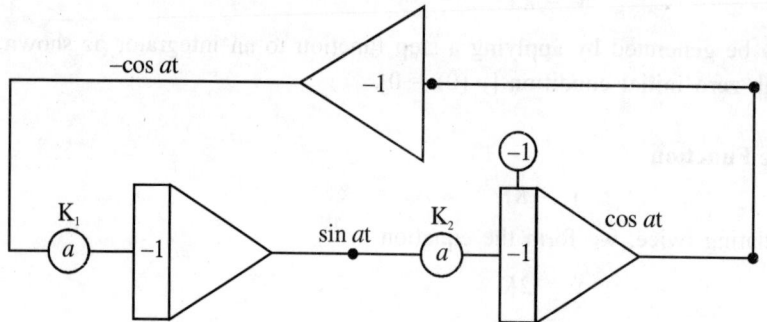

Fig. 4.30. Generation of Sine and Cosine functions

Since the potentiometers K_1 and K_2 should be set to the same value a, they are usually ganged potentiometers. By varying the value of a, the frequency of the functions are varied, and the amplitude is varied by varying the initial condition.

4.21. SIMULATION OF TRANSFER FUNCTIONS

The analog computer is sometimes used for problems other than solving a set of differential equations. One such application is simulation of system transfer functions on the computer. Computer components can be combined to form a wide range of transfer functions of system which, when excited, will produce the desired response. Because of the greater flexibility available with computer simulation and design, they are used in simulation problems in which the system is modeled, tested, and improved, in whole or in part. Computing elements have also been used to form special transfer functions for application such as oil-refinery, process control, machine-tool control, and autopilot guidance of aircraft.

4.21.1. Conventional Method

This method is best suited if the computer has plenty of amplifiers because no unusual connections of components are necessary. Each amplifier is used in the conventional form as summer or integrator with potentiometer gain settings. However more number of amplifiers is required in this method.

Example 4.11. *Generate the transfer function*

$$\frac{E_o(s)}{E_i(s)} = \frac{8\tau}{1+s\tau} \qquad \qquad ... (4.81)$$

This may be written as

$$E_o(s) = E_i(s) - \frac{E_o(s)}{s\tau} \qquad \qquad ... (4.82)$$

Assuming zero initial condition, this can be generated as shown in Fig. 4.31.

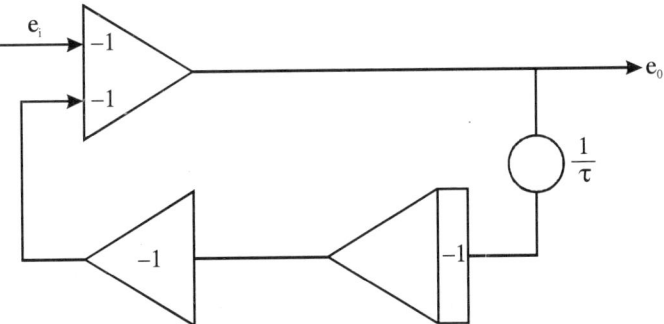

Fig. 4.31. Generation of transfer function

Example 4.12. *Generate the transfer function*

$$\frac{E_o(s)}{E_i(s)} = K\frac{s(1+s\tau_1)}{(1+s\tau_2)(1+s\tau_3)} \qquad \dots (4.83)$$

Solution:

This can be written as

$$[s^2\tau_2\tau_3 + s(\tau_2 + \tau_3) + 1]E_o(s) = (Ks^2\tau_1 + Ks)E_i(s) \qquad \dots (4.84)$$

Substituting

$$\tau_2\tau_3 = A$$
$$(\tau_2 + \tau_3) = B$$

and

$$K\tau_1 = C$$

and equating, we get

$$As^2E_o(s) = -E_o(s) - Bs\,E_o(s) + KsE_i(s) + Cs^2E_i(s)$$

Dividing by As^2,

$$E_o(s) = -\frac{1}{A}\frac{E_o(s)}{s^2} - \frac{B}{A}\frac{E_o(s)}{s} + \frac{K}{A}\frac{E_i(s)}{s} + \frac{C}{A}E_i(s) \qquad \dots (4.85)$$

The analog computer setup is shown in Fig. 4.32.

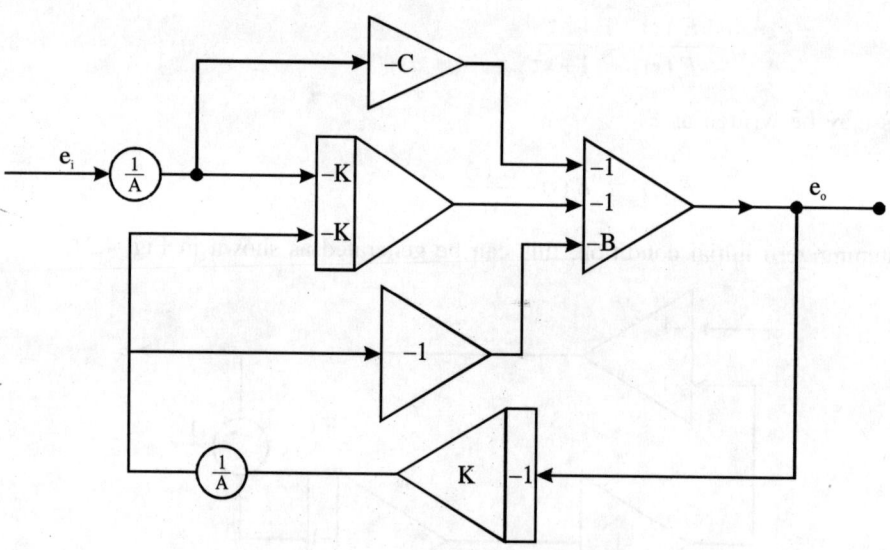

Fig. 4.32. Computer setup for Example 4.12

4.21.2. Function Generation by Transfer Function Approach

Certain functions, particularly exponential functions and their sums, can be generated by transfer function approach. The principle is to write the function to be generated as a function of Laplace operator and obtain the transfer function which, when excited by a driving function such as a unit step, or by initial conditions on the integrators, will produce the desired voltage.

Example 4.13. *Generate the function*

$$f(t) = 1 - e^{-t} \qquad \qquad \dots (4.86)$$

Solution:

The Laplace transform may be identified as

$$E_o(s) = \frac{1}{s(s+1)} = -E_i(s)\frac{Z_f(s)}{Z_i(s)} \qquad \qquad \dots (4.87)$$

Let,

$$E_i(s) = -\frac{1}{s}$$

$$Z_f(s) = \frac{1}{(s+1)}$$

$$Z_i(s) = 1$$

We can generate the function $f(t)$ with a negative unit step function as input. However, to permit freedom of choice of parameter values, we may choose

$$E_i(s) = -\frac{K_1}{s}$$

$$Z_f(s) = \frac{K_2}{(s+1)}$$

$$Z_i(s) = K_3 \qquad \qquad \dots (4.88)$$

with $K_1, K_2, K_3 = 1$. The setup in Fig. 4.33 with $R_i = K_3$, $R_f = K_2$ and $C_f = 1/K_1$ will generate the desired function $f(t)$.

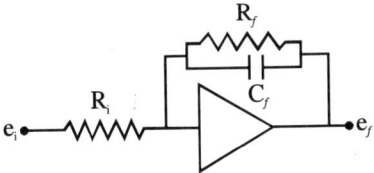

Fig. 4.33. Generation of $(1 - e^{-t})$

4.21.3. Simulation of Control System Transfer Function

A closed-loop control system is one in which the output C is constantly compared with the input R and the error E, which is the difference between the input and the output. It is used to actuate a controller which, in turn, will control the output. Thus the output is maintained in close correspondence with the input.

The relationship between C and E is termed the open-loop transfer function. Given the open-loop transfer function, the dynamic performance of the closed-loop system may be easily investigated on an analog computer.

Example 4.14. *The open-loop transfer function of a control system is*

$$\frac{C(s)}{E(s)} = G(s) = K/s(1+s)(1+s\tau) \qquad \ldots (4.89)$$

A computer diagram is to be set up in order to investigate the effect of varying the gain K and the time constant τ on the dynamic performance of the closed-loop system.

Solution:

The transfer function being in factored form can be easily simulated by means of three operational amplifiers. In Fig. 4.34 amplifier $A1$ will simulate the term $-1/(1+s)$. Amplifiers $A2$ with a potentiometer in feedback capacitor path will simulate $-1/(1+k_1s)$. As the value of potentiometer set at k_1 is varied, the time constant τ is varied. By properly choosing the value of the resistors, τ may be varied over a wide value. The amplifier $A3$ with potentiometer set at k_2 will simulate the transfer function $-k_2/sRC$. By varying k_2 and selecting a proper combination of R and C, the gain $K (= k_2/RC)$ may be widely varied, thus the open-loop transfer function is simulated.

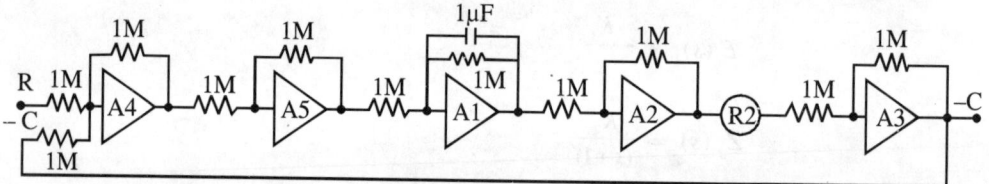

Fig. 4.34. Simulation of Transfer function for Example 3.14

Since the purpose is to simulate the closed-loop system we have to compare the output and input to obtain E. The output of Fig. 4.34 being $-C$ is added along with input R at the summing amplifier $A4$ and we obtain the output as $-E$ which is again inverted by an inverter $A5$ whose output is fed into the amplifier $A1$ thus completing the loop.

Thus the computer setup is flexible in that the gain K and the time constant τ may be independently varied and the effect may be immediately studied.

From the examples discussed in this section, it may be noted that a number of possible circuit configurations present themselves in simulating a given transfer function. Most of them are developed by trial-and-error. But, with a little practice, the pattern becomes quite clear.

4.22. DIGITAL COMPUTER SIMULATION

Simulation studies of continuous systems can also be carried out on a digital computer. The advantages of digital computers are:

(1) It is an extremely accurate and flexible tool.

(2) It can be easily programmed.

(3) It is highly reliable.

Three commonly used simulation packages are CSMP, DYNAMO and MIMIC. CSMP or Continuous System Modeling Program is available with all IBM computers. CSMP serves the same purpose as an analog computer program except that the scaling is practically eliminated. The use of CSMP is illustrated by the following example.

Example 4.15. *Write a program in CSMP to solve the second-order differential equation*

$$\ddot{c} + 5\dot{c} + 3c = r$$

with $c(0)$ and $\dot{c}(0)$ as initial conditions.

Solution:

The equation is rearranged as

$$\ddot{c} = r - 5\dot{c} - 3c$$

$$\dot{c} = \int_0^\infty \ddot{c}\,dt$$

$$c = \int_0^\infty \dot{c}\,dt$$

Let us denote the variables by

$$c = C, C(o) = C_0$$
$$\dot{c} = C_1, \dot{C}(o) = C_{10}$$
$$\ddot{c} = C_2$$
$$r = R$$

Then the CSMP main program is:

DYNAMIC

   C = INTGRL (C1, C0)

   C1 = INTGRL (C2, C10)

   C2 = R – 5. * C1 – 3. * C

| TIMER | FINTIM = 20.0, OUTDEL = 0.5 |
| PRINT | C2, C1, C |

The output for $\ddot{c}$, $\dot{c}$ and c will be printed for every 0.5 seconds up to 20 seconds.

4.23. ADDITIONAL EXAMPLES

Example 4.16. *Draw the electrical analogs of the mechanical system shown in Fig. 4.35 and obtain their governing equations.*

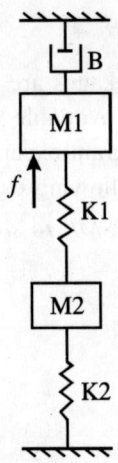

Fig. 4.35. Mechanical system for Example 4.16

Solution:

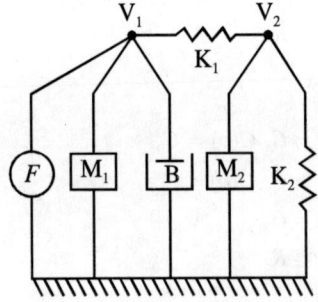

Fig. 4.36. Mechanical network

With V_1 and V_2 as nodes, the mechanical network is drawn as shown in Fig. 4.36. The force-current and the force-voltage analogs are shown in Fig. 4.37 (a) and (b) respectively. The governing equation are:

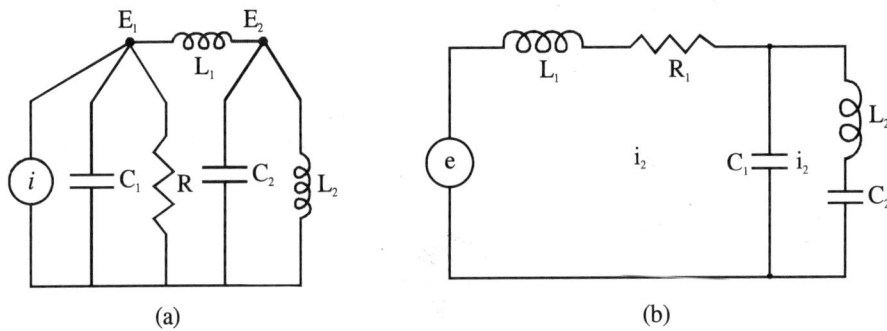

Fig. 4.37. (a) Force-current analog (b) Force-voltage analog

(a) *Mechanical Network*

At node V_1 : $F(s) = (M_1 s + B + K_1/s)V_1(s) - (K_1/s)V_2(s)$

At node V_2 : $O = [M_2 s + (K_2 + K_1)/s]V_2(s) - (K_1/s)V_1(s)$

(b) *Force-Current Analog Network*

At node E_1 : $I(s) = [C_1 s + 1/R + (1/L_1 s)]E_1(s) - (1/L_1 s)E_2(s)$

At node E_1 : $O = \left[C_2 s + \dfrac{1}{L_2 s} + \dfrac{1}{L_1 s} \right] E_2(s) = -\left(\dfrac{1}{L_1 s} \right) E_1(s)$

(c) *Force - Voltage Analog Network*

For Mesh I : $E(s) = (L_1 s + R + 1/C_1 s)\, I_1(s) - (1/C_1 s) I_2(s)$

For Mesh II : $O = (L_2 s + 1/C_2 s + 1/C_1 s) I_2(s) - (1/C_1 s)\, I_1(s)$

Example 4.17. *A position control servomechanism is represented by the differential equation*

$$58\ddot{\theta}_0 + 26.2\dot{\theta}_0 + 3.5\theta_0 = 0.7$$

Draw an analog computer diagram to enable the actual position θ_0 to be investigated. Time scaling should be introduced such that the solution is speeded up by a factor of 2. All initial condition are equal to zero.

Solution:

The maximum values are:

$$(\omega_o \sqrt{(3.5)/58} \approx 0.25)$$

$$\theta_0 \ max = 0.4$$
$$\dot{\theta}_0 \ max = 0.4 \times \omega_o = 0.4 \times .25 = 0.1$$
$$\ddot{\theta} \ max = 0.1 \times .25 = 0.025$$

The normalized equation is

$$58 \times 0.025 \left(\frac{\ddot{\theta}_0}{0.025} \right) + 26.2 \times 0.1 \left(\frac{\dot{\theta}_0}{0.1} \right) + 3.5 \times 0.4 \left(\frac{\theta_o}{0.4} \right) = 0.7$$

Or,

$$\frac{\ddot{\theta}_0}{0.025} + 1.807 \left(\frac{\dot{\theta}_0}{0.1} \right) + 0.966 \left(\frac{\theta_o}{0.4} \right) = 0.483$$

Since $\alpha = 1/2$, $(1/\alpha) = 2$. Hence

First integrator gain $= \dfrac{0.025}{0.1} \times 2 = 0.5$

Second integrator gain $= \dfrac{0.1}{0.4} \times 2 = 0.5$

The computer diagram is shown in Fig. 4.38.

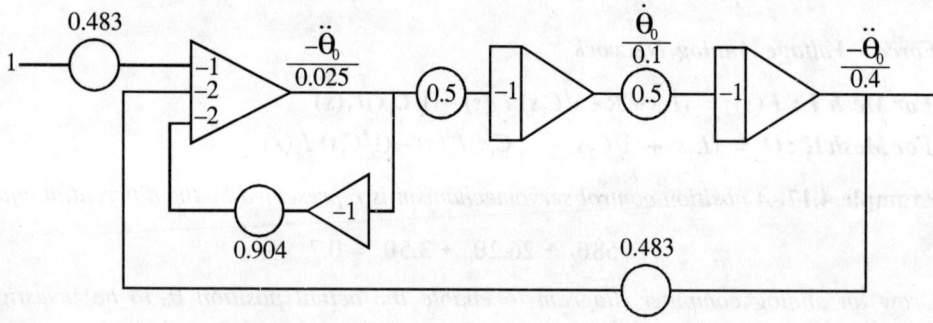

Fig. 4.38. Computer setup for Example 4.17

Example 4.18. *The open-loop transfer function of a second-order system is given by*

$$\frac{C(s)}{E(s)} = \frac{K}{s(s+p)}$$

Write a program in CSMP to simulate the system.

Solution:

Let the variables be denoted by

$$COUT = c(t)$$

$$CDOT = \dot{c}(t)$$

$$GAIN = K$$

$$ERROR = E$$

$$RIN = \text{Reference Input}$$

$$CONTL = K*E$$

The program in CSMP is

DYNAMIC

$$ERROR = RIN - COUT$$

$$CONTL = GAIN* ERROR$$

$$CDOT = REALPL (P, CONTL, 0.0)$$

$$COUT= INTGL (CDOT, 0.0)$$

PARAMETER	P = 1.5, RIN = 1.0
PARAMETER	GAIN = (2.0, 5.0, 10.0)
TIMER	FINTIM = 10.0, OUTDEL = 0.5
PRINT	COUT, CDOT

The function REALPL simulates the function with one real pole *p*.

4.24. SUMMARY

In this chapter, we have first defined an analogous system and its advantages. The system engineer always prefers to work with electrical systems since it is easier to vary the electrical parameters and determine the best combination of the electrical parameters that would give the required response and then convert them into mechanical components for final design and testing. We then have discussed the force-current and force-voltage analogous systems. We have indicated that force-current analog topology is similar to the mechanical system topology; it is easier to draw this analogy first and then obtain the force-voltage analogy. We have also discussed the method of obtaining the electrical analogous systems of mechanical systems and their simulation on analog and digital computers. The concept of analogous system exposes the fact that different physical systems can be studied with the same general mathematical model and the results of one can be extended to other analogous systems. For any physical system it is possible to obtain the force-current analog and test it safely in the laboratory and adjust the parameters to give the desired characteristics.

The recent trend in system simulation studies is to utilize analog and digital computers. Each has its own merits. For further detail on computer simulation, refer to standard books on this.

REVIEW QUESTIONS

4.1. Define analogous system.

4.2. What are the advantages of using electrical analogs?

4.3. Explain Force-Current analog.

4.4. Explain Force-Voltage analog.

4.5. Explain Torque-Current analog.

4.6. Explain Torque-Voltage analog.

4.7. What are the merits of force-current analog?

4.8. Mention four advantages of computer simulation of systems.

4.9. What is an operational or computing amplifier?

4.10. Draw a multiplier circuit using OPAMP and explain.

4.11. Draw an adder circuit using OPAMP and explain.

4.12. Draw a sign inverter circuit using OPAMP and explain.

4.13. Draw a subtractor circuit using OPAMP and explain.

4.14. Draw an integrator circuit using OPAMP and explain.

4.15. Explain how initial conditions are inserted in analog computer simulation of systems.

4.16. Explain the need for amplitude scaling in analog computer simulation of systems.

4.17. Explain the need for time scaling in analog computer simulation of systems.

4.18. Explain how the sine and cosine function are generated using OPAMP.

EXERCISE

4.1. What are analogous systems? Explain how two types of analogs are formed.

4.2. Explain force-voltage and force-current analogs. Illustrate with examples.

4.3. What do you understand by analog circuits? List the analogous quantities in force-voltage and force-current analogs.

4.4. Mention the advantages of computer simulation.

4.5. Draw the schematic diagram of an operational amplifier with input impedance Z_i and feedback impedance Z_f. Derive the transfer function $E_o(s)/E_i(s)$. Show that by proper selection of the impedances Z_i and Z_f, one can perform addition, subtraction, sign inversion, multiplication and division.

4.6. Explain how initial conditions are included in the analog computer simulation.

4.7. Differential equations are normally solved by a process of successive integration. Explain why?

4.8. Discuss the various steps involved in analog computer simulation of *n*-th order differential equation.

4.9. Explain the need for amplitude and time scaling.

4.10. Discuss the equal co-efficient rule of estimating the maximum values for amplitude time scaling.

PROBLEMS

4.1. Obtain the force-current and force-voltage analogs of the mechanical system shown in Fig. P 3.6.

4.2. Derive the analogous electrical circuits for the mechanical system shown in Fig. P 3.7.

4.3. Determine the force-voltage analog of the systems shown in Fig. P 3.8(a) and (b).

4.4. Obtain the two analogous circuits of the vibration absorbers shown in Fig. P 3.9.

4.5. Determine the force-voltage analog circuit of the mechanical rotational system shown in Fig. P 3.10(a) and (b).

4.6. Derive the analogs of the mechanical rotational system shown in Fig. P 3.13.

4.7. Derive the analogs of mechanical translational system shown in Fig. P 3.14

4.8 Derive the transfer function $V_2(s)/F(s)$ for the mechanical system shown in Fig. P 4.8 and obtain the force-current analog.

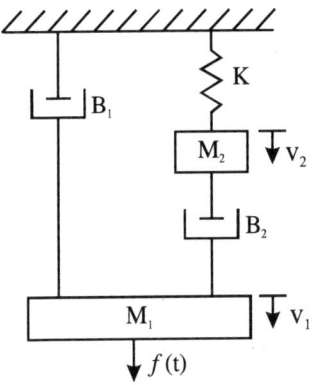

Fig. P 4.8

4.9 Express the output as a function of the inputs for each of the circuits shown in Fig. P 4.9.
Values of resistors and capacitors are in meg-ohms and microfarads respectively.

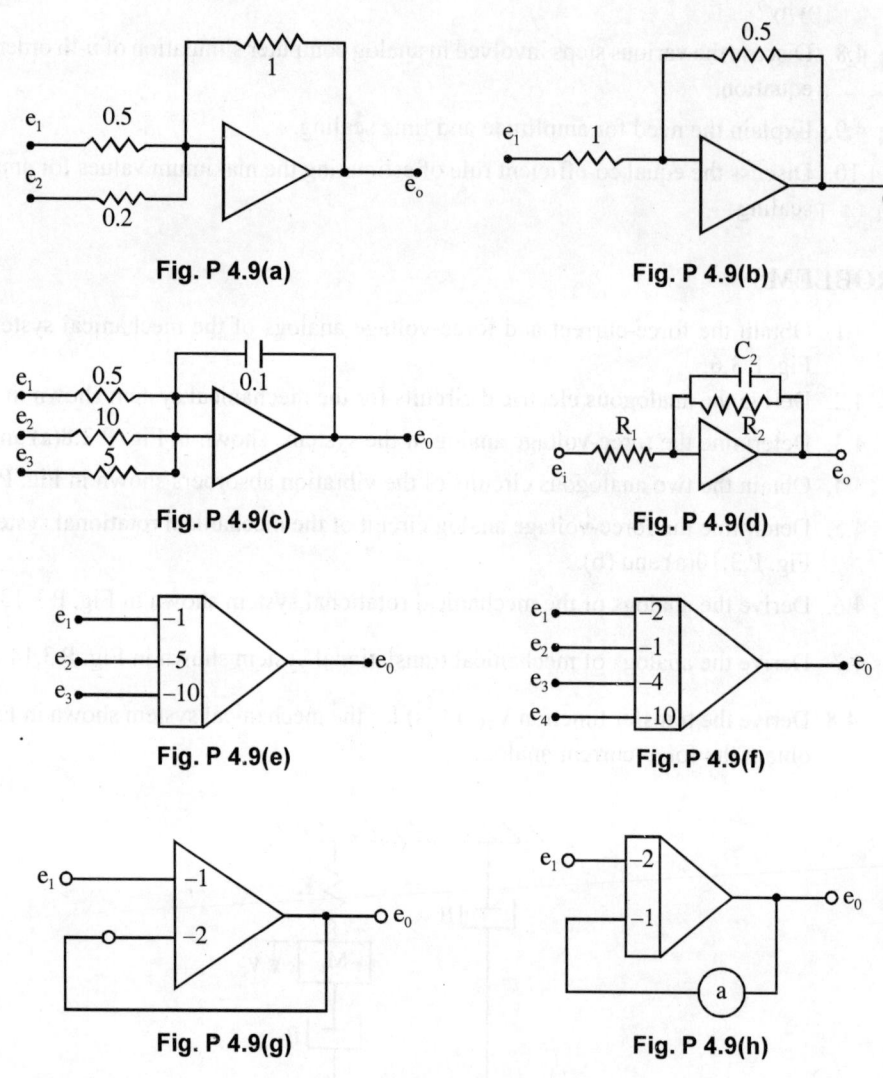

Fig. P 4.9(a)   Fig. P 4.9(b)

Fig. P 4.9(c)   Fig. P 4.9(d)

Fig. P 4.9(e)   Fig. P 4.9(f)

Fig. P 4.9(g)   Fig. P 4.9(h)

4.10 Set up analog computer diagram for the solution of the following.
   (a) $x'' + 1{\cdot}4x' + 2x = 16$, $x(0) = 2$, $x'(0) = -3$
   (b) $2x'' + 5x' - 4x = 20$, $x(0) = -1$, $x'(0) = 0$
   (c) $40x'' + 10x' + 40x = 400$, $x(0) = 1$, $x'(0) = -2$
   (d) $Mx'' + Bx' + Kx = f(t)$, $x(0) = 0$, $x'(0) = 0$

4.11 Setup magnitude scaled computer diagram for the following

(a) $x'' + 2x' + 4x = 6$, $x(0) = 3$, $x'(0) = -1$

(b) $2x'' + 10x' + 2x = 4$, $x(0) = -1$, $x'(0) = 0$

4.12 Apply time-scaling to the equations in P 4.11 with time scale factor of $\alpha = 2, 0.4, 4$ and 0.5 respectively.

4.13 Simulate the open-loop transfer function of a unity feedback control system given by $G(s) = 2/s(1 + 0.1s)(1 + 0.7s)$ on an analog computer

4.14 Refer to Fig 4.13. Select suitable values for Z_i and Z_f to obtain $E_o(s)/E_i(s) = -5/(1 + 0.2s)$.

Chapter 5

MODELING OF SYSTEMS AND COMPONENTS

5.1. INTRODUCTION

Before we attempt to analyze a complex control system, it is necessary to get familiarized with the functions and characteristics of components being used in such systems. In this chapter, the constructional details and operational characteristics of components used in Electrical, Hydraulic, Pneumatic and Inertial Guidance systems are discussed. In each case, the transfer function is derived, In deriving transfer function, simplifying assumptions are made so that the components such as error detectors, amplifiers and actuators may be represented by linear equation.

5.2. TRANSDUCERS AND ERROR DETECTORS

    (i) *Transducers:* Transducers are devices that convert one form of energy into another. The most common types of transducers develop electrical signals for input mechanical motions, hydraulic or pneumatic pressures etc. Depending upon the type of the input quantity, the transducers are called as potentiometers, tachogenerators and thermocouples. Potentiometer is an electromechanical transducer which converts mechanical shaft rotation into proportional d.c. electrical signal. Tachogenerator is one which generates electrical signals proportional to velocity of shafts. Thermocouple converts heat energy into electrical signal proportional to the temperature.

    (ii) *Error Detectors:* Error detectors are devices in which two quantities of the same type are compared. In a unity feedback system, they serve the important function of comparing the output or a function of the output with the reference input. In multiloop systems, they are used to compare two intermediate signals or the output with intermediate signals. The output of such devices is the difference between the two quantities compared and is a measure of error. The most extensively used error detectors are potentiometers, E-transformers and synchros.

5.3. POTENTIOMETERS

A potentiometer is a resistor with a movable wiper which is positioned by a mechanical control. It consists of many turns of fine wire wound very close together or conductive plastic resistance elements so that good linearity is obtained. The potentiometer should be considered as a variable voltage divider rather than a variable resistance since a voltage is applied across the resistance and a fraction of this voltage, picked off between the wiper and one end of the resistance, appears as the output. A potentiometer is shown in Fig. 5.1. The two extreme ends are connected to a constant D.C. voltage source E. The movable terminal is the wiper. The output appears between the wiper and the negative terminal and is proportional to the angle of rotation of the wiper.

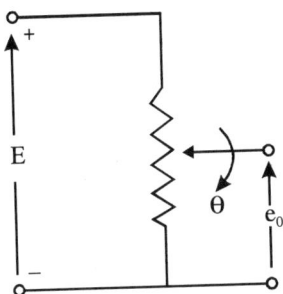

Fig. 5.1. Potentiometer

5.3.1. Resolution of a Potentiometer

The resolution of a potentiometer is an important characteristic. It is defined as the smallest change in the output voltage (resulting due to rotation the wiper) which can be detected and is expressed as a percentage of the total applied voltage. It is also expressed as a measure of fineness with which a given voltage is approached. This characteristic depends upon the pitch of the winding and the diameter of the arc upon which the wiper travels. The wiper ordinarily shorts out several turns of the wire-wound potentiometer, and the output voltage variation with wiper rotation is in the form of steps as indicated in Fig. 5.2. The smallest change in voltage that is detectable is that of one turn and thus if there are n turns,

$$\text{Resolution} = (1/n) \times 100\% \qquad \text{... (5.1)}$$

The resolution of the potentiometer limits the accuracy of a control system since the system accuracy cannot be better than that of the component. Precision wire-wound potentiometers have a resolution of about 0.001%. A carbon or the conductive plastic resistance potentiometers has zero resolution since the wiper moves along a continuous resistance path.

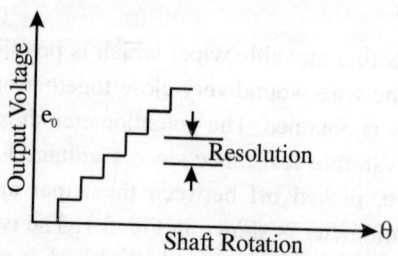

Fig. 5.2. Potentiometer output voltage

5.3.2. Potentiometer as Error Detector

As error detectors, they are always used in pairs, as shown in Fig. 5.3. The two potentiometers transform the input and output shaft positions into proportional voltages. These two voltages are compared and the difference or error voltage e appears across the two terminals a and b. The diagram shown in Fig. 5.3 represents a simple bridge circuit. When the two arms are in the same relative position, the error voltage is zero. If the relative position of the output shaft is below the input shaft position the potential at point a is higher than at b, and the polarity of the error signal is that shown in Fig. 5.3. If the position of the output shaft is above the input shaft, the potential at b is higher than that of a and the polarity of the error will be reversed. Thus *the polarity of the error voltage indicates the direction of misalignment of the two shafts.* The transfer function of the error detector is

$$K = K_s \,(\theta_r - \theta_c) = K_s \,\theta_e$$

(5.2)

θ_r = reference input shaft position in radians or degrees.

θ_c = Controlled output shaft position in radians or degrees.

K_s = is the sensitivity of the potentiometer.

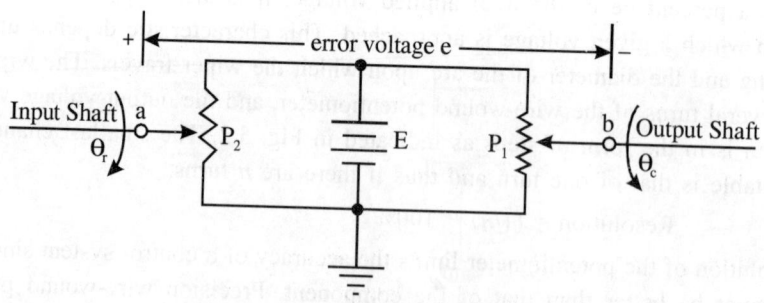

Fig. 5.3. Potentiometer error detector

which is defined as *the change in output voltage for unit radian* and is expressed in volts per radian. The block diagram representation of potentiometer is shown in Fig. 5.4.

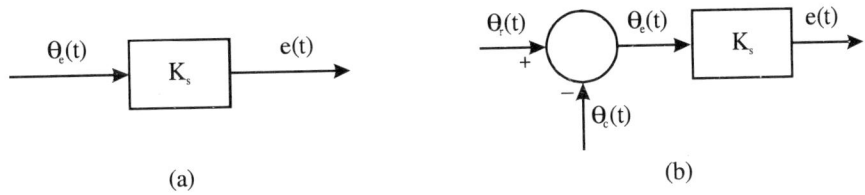

Fig. 5.4. Block diagram representation (a) potentiometer δ (b) Error detector
$θ_e(t)$ error input $θ_r$ reference input and $θ_c$ controlled output

5.3.3. Application of Potentiometers

A typical application of a pair of potentiometer as error detector in control systems is shown in Fig. 5.5. In this system, an error signal e, proportional to the misalignment between the reference input shaft and controlled output shaft, is produced at the output of the potentiometer. The amplifier amplifies this signal and supplies necessary energy for the field excitation. The motor then rotates in such a direction as to reduce the misalignment and hence, the error voltage. When the error voltage is zero, theoretically, the output shaft is in correspondence with the input shaft. The potentiometers are also used in rate-gyro transducers and pressure transducers.

5.3.4. Limitations of Potentiometer System

The application of potentiometers as error detector in control system is limited due to the following reasons.

1. Small angular displacements cannot be measured because of frictional effects.

2. The wiper-contact is the delicate part of potentiometer and may be damaged unless handled carefully.

3. The output voltage changes in discrete steps. This contributes to the servo inaccuracy.

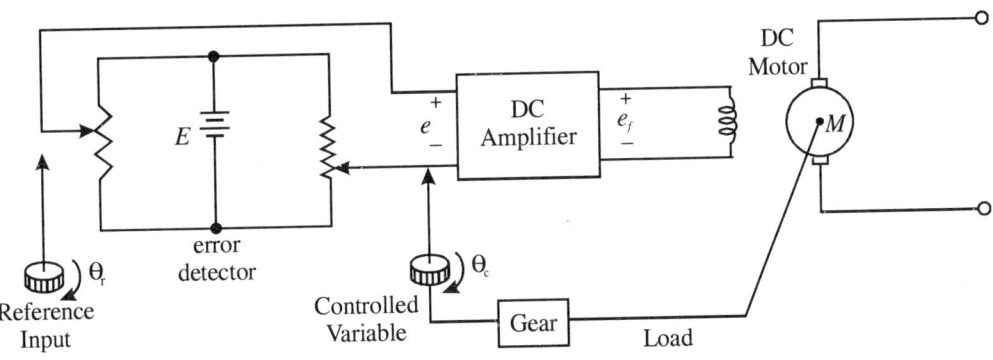

Fig. 5.5. Position control system

4. The useable angle of rotation of a potentiometer is limited to less than 360° for a single turn potentiometer.

5. The maximum sensitivity of a potentiometer is limited to 3 V/rad, because a larger applied voltage will entail a larger heat dissipation.

5.4. E–TRANSFORMER (LINEAR VARIABLE DIFFERENTIAL TRANSFORMER)

Construction: An E-Transformer or LVDT is a device used to measure small displacements. Since this device has no brushes or slip-rings, the frictional effects may be reduced to a minimum. It can be used only over a limited range of input shaft position.

Fig 5.6 shows an E-transformer used to measure small linear displacements. It consists of an E-shaped laminated-iron structure. A coil is wound on each leg. The coils on the two outer legs are identical and connected in opposition. The centre coil is normally excited by a constant a.c. voltage.

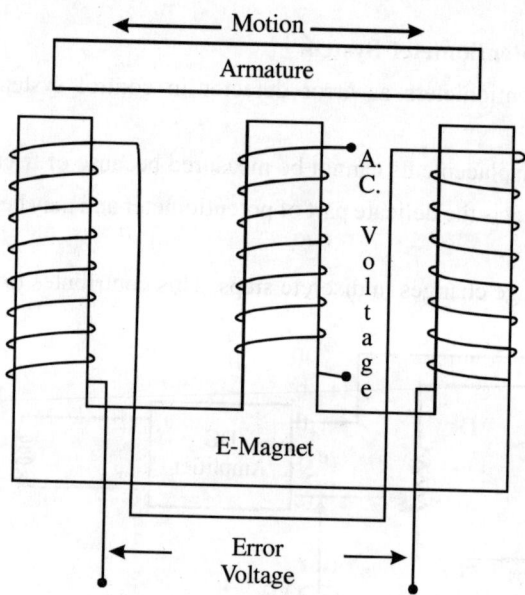

Fig. 5.6. E-transformer

Operation: When the movable armature, also made of iron laminations, is symmetrically placed with respect to the two outer legs, the magnetic field is symmetrical about the centre leg and since the outer coils are wound in opposition, the error voltage is zero. This position of the armature is referred to as the *null position*. If the mechanical displacement moves the armature to the right from the null position, the voltage induced in the right-hand coil is greater than that in the left hand coil. Hence, a net error voltage is produced which is in phase with the applied voltage. Motion of the armature to the left produces an error voltage that is 180° out of phase with that produced by motion to the right. Within limits, the amplitude of the error voltage is proportional to the displacement, whereas the phase indicates the direction of the displacement from the null position, as shown in Fig. 5.7.

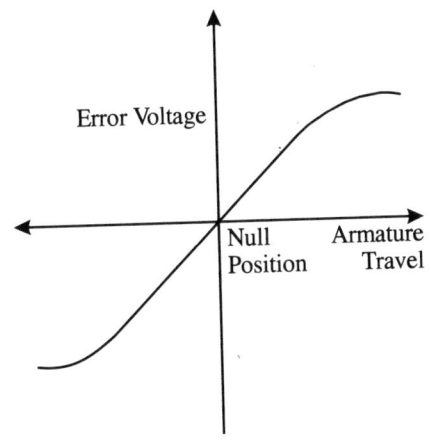

Fig. 5.7. Error voltage of the E-transformer

5.5. SYNCHROS

Synchros are the most commonly used transducers in a.c. control systems. They are also called as *Selsyn, Autosyn, Telesyn, Diehlsyn* etc., by various manufacturers but the principle of operation is the same in all cases. Depending on the particular use, synchros are given special names such as transmitter (or generator), receiver (or motor) and control transformer. The different units are discussed here.

5.5.1. Synchro Transmitter or Generator (G)

(i) **Construction:** The synchro transmitter or generator is a small three phase, two pole machine. It depends upon transformer action for its operation. It consists of a stator and a rotor. The stator is a slotted cylindrical structure and mounted inside a housing. The stator coils S_1, S_2 and S_3 are wound in the uniformly spaced slots. They are spaced 120 electrical degrees apart and are connected in *wye* similar to that of a three phase induction motor. The three stator windings constitute the secondary of the transmitter.

The rotor of the synchro transmitter has a salient pole, dumb bell-shaped structure with a single winding. The dumbbell-shaped cross section gives the effect of a concentrated flux developed by the excitation of the rotor winding. The rotor is excited by a single phase voltage applied through two slip rings, so that continuous rotation of the rotor is possible (Fig. 5.8). The rotor constitutes the primary of the transmitter.

(ii) Principle of Operation: In the normal operation, the rotor is excited with a single phase voltage. The resulting current sets up a sinusoidal flux at line frequency and this, in turn, produces alternating voltages in the stator windings by transformer action. The voltage induced in a particular winding is proportional to the angle between the axis of this winding and the axis of the rotor. Thus, the synchro generator is a single phase transformer with rotor as its primary and the three stator windings as its secondary.

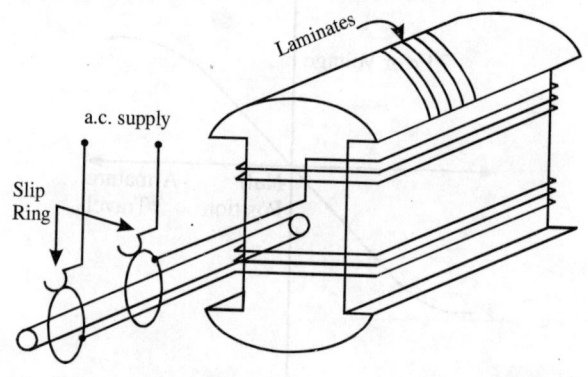

Fig. 5.8. Rotor construction

Figure 5.9 shows the schematic diagram of a synchro generator. The position of the rotor shown in Fig. 5.9 is normally defined as the *electrical zero* of the synchro which corresponds to maximum coupling with the stator coil S_2. If the rotor is excited by an a.c. voltage, then the voltages are induced across each stator winding and the neutral. They are given by

$$E_r = E \sin \omega t \qquad \qquad \text{... (5.3)}$$

$$E_{s2n} = KE \sin \omega t \qquad \qquad \text{... (5.4)}$$

$$E_{s1n} = KE \cos 120° \sin \omega t = -0.5 \, KE \sin \omega t \qquad \qquad \text{... (5.5)}$$

$$E_{s3n} = KE \cos 240° \sin \omega t = -0.5 \, KE \sin \omega t \qquad \qquad \text{... (5.6)}$$

The three terminal voltages are:

$$E_{s1s2} = E_{s1n} - E_{s2n} = -1.5 \, KE \sin \omega t \qquad \qquad \text{... (5.7)}$$

$$E_{s2s3} = E_{s2n} - E_{s3n} = 1.5 \, KE \sin \omega t \qquad \qquad \text{... (5.8)}$$

$$E_{s3s1} = E_{s3n} - E_{s1n} = 0 \qquad \qquad \text{... (5.9)}$$

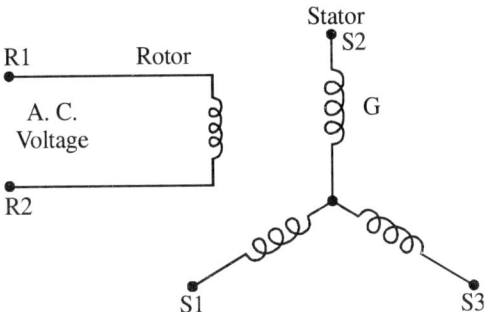

Fig. 5.9. Synchro generator

The above equations indicate that the terminal voltages are single phase voltages since they are induced by the same flux.

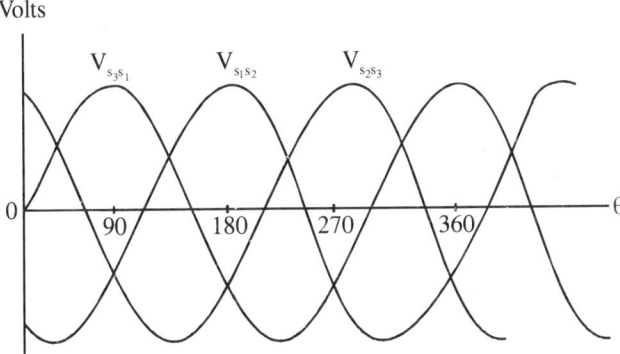

Fig. 5.10. Synchro output voltage *vs* θ

If the rotor is now allowed to rotate in a counter clockwise direction these voltage change and a plot of voltage as a function of angular position is shown in Fig. 5.10. The maxima of these voltages are displaced 120° apart. Thus any set of the three voltages describes the angular position of the rotor shaft uniquely.

5.5.2. Synchro Receiver or Motor (M)

The synchro motor or receiver is identical to the synchro generator in construction and operation except for the following:

(a) The motor is equipped with low-friction ball bearings and an inertia damper.

(b) The input to the motor is a set of three single phase voltages (usually from the generator stator coils) and the output is mechanical in the form of a shaft rotation.

5.5.3. Synchro Control Transformer (CT)

(i) Construction: The stator of a control transformer is similar to that of a synchro generator or motor but wound with larger number of turns of fine wires. It has, therefore, higher impedance per phase. This makes possible to feed several control transformers from a single generator. The rotor is cylindrical in shape with slots to accommodate the winding. The rotor winding is a single distributed one with high impedance. The air gap between the rotor and the stator is uniform owing to the rotor's cylindrical construction (Fig. 5.11(a)). Because of the uniform air gap, the magnetizing current drawn from the generator is kept to a minimum. *The control transformer rotor is not excited.*

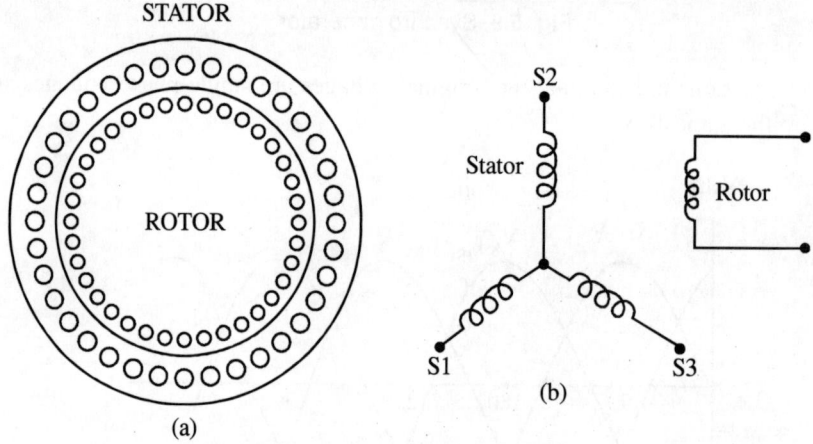

Fig. 5.11. (a) Constructional details of CT (b) Electrical zero position.

(ii) Operation: The stator of the control transformer is excited by the output of a synchrogenerator. The rotor is not excited. When the rotor axis is normal to the S_2 coil, no voltage is induced in the rotor from S_2. However, both S_1 and S_3 do induce voltages in the rotor which are equal but opposite and hence they cancel. The position of the rotor shown in Fig. 5.11(b) is called *the electrical zero position* of the control transformer. Rotated counterclockwise, the voltage induced in the rotor varies in an approximately sinusoidal manner as shown in Fig. 5.12. This voltage is called the *error voltage* and the error angle is the difference between the rotor positions of synchro generator and control transformer. Thus, the control transformer operates as a differential device which provides an electrical signal, indicative of a mechanical error. When the error angle is small, the error voltage is directly proportional to the error angle. Therefore, the transfer function of a control transformer may be written in the form

$$T(s) = K_s \ (\theta_1 - \theta_2) = K_s\theta_e \qquad\qquad \text{... (5.10)}$$

where K_s = sensitivity of the control transformer.

θ_1 = synchro generator rotor position from its null position.

θ_2 = control transformer rotor position from its electrical zero position.

θ_e = error angle

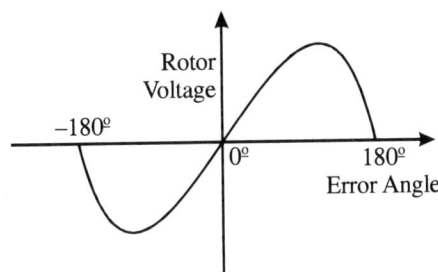

Fig. 5.12. Control transformer voltage

The output of a control transformer is usually a suppressed carrier modulated voltage with a carrier frequency equal to the frequency of the voltage exciting the stator and the modulating frequency depends on the speed of rotation of the rotor of the control transformer. In Fig. 5.12, the modulation envelope only is shown.

5.5.4. Synchro Applications

Synchros are always used in various combinations such as synchro generator-motor, generator-control transformer, generator-differential-motor etc.

(a) **Data Transmission:** A synchro generator-motor combination is commonly used for remote indication of position. In this case, the load is only a pointer or a dial. The system consists of a synchro generator whose stator is connected electrically to the stator of a remotely placed synchro motor, with the rotors excited by the same a.c. supply (Fig. 5.13). A displacement of generator rotor by an angle θ_1 causes an electrical unbalance in the system resulting in a flow of current through the stators proportional to the difference between the rotor position of the generator and motor. A torque is produced in the motor causing its rotor to rotate through an angle until the two rotors are in alignment. For small error angles, the transfer function of the system is

$$T = K_{sm} (\theta_1 - \theta_2) = K_{sm}\theta_e \qquad \qquad ... (5.11)$$

where T is the developed torque,

K_{sm} is the sensitivity of the motor,

θ_e is the error angle.

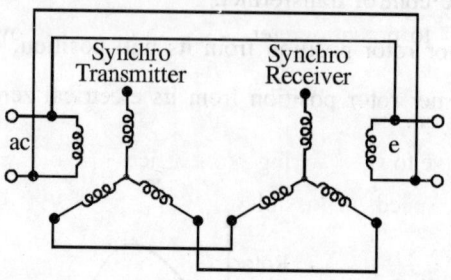

Fig. 5.13. Remote position controller

(b) **Synchro Error Detector:** A synchro generator and a control transformer can be used as an error detector (Fig. 5.14). The reference input shaft is mechanically connected to generator rotor, and the output controlled shaft to control transformer rotor. When an angular misalignment exits between the reference and controlled shafts, an error voltage of proper polarity appears at the amplifier input. The amplified error voltages is applied to the a.c. servomotor which, in turn, rotates in such a direction as to reduce the misalignment to zero. For small discrepancies, the error voltage is

$$e = K_s \ (\theta_1 - \theta_2) \qquad \qquad ... (5.12)$$

where e = Error voltages

K_s = sensitivity of synchro generator and control transformer.

θ_r = reference shaft position.

θ_c = controlled shaft position.

When the controlled shaft is aligned with reference shaft, the error voltage is zero and the motor stops.

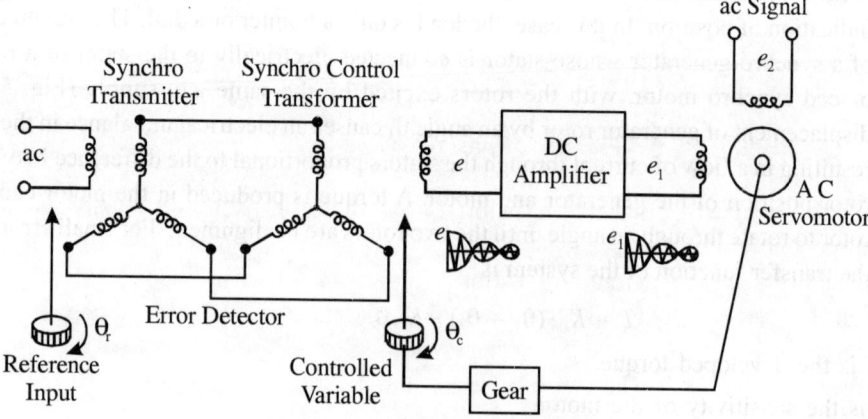

Fig. 5.14. Synchro error detector

5.5.6. Comparison between Synchros and Potentiometers

The synchros are preferred to potentiometers because of the following advantages:

1. No resolution error.
2. No wear from rotation.
3. Relatively insensitive to stray wiring capacitance.
4. Operation at higher speeds is possible.
5. Accuracy is higher.
6. Useful operating angle is 360° and is continuous.
7. Highly reliable.
8. Adaptable for multispeed operation.

5.6. PHASE SENSITIVE RECTIFIERS

Synchros can also be used to detect the errors in D.C. servos. When used the error appears as an amplitude modulated signal. This signal, before being applied to the D.C motor for producing the corrective torque, must be rectified to produce a D.C or very low frequency signal whose polarity depends on the error signal. Fig 5.15 shows a simple phase-sensitive rectifier circuit whose output is proportional to the error magnitude.

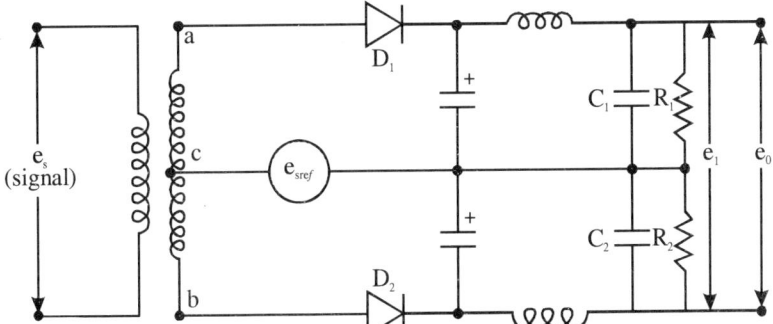

Fig. 5.15. Phase sensitive detector

The error signal is applied to a transformer with a centre-tapped secondary winding. The centre-tap is connected to a large sinusoidal signal of reference phase obtained from the same supply as that of the synchro-transmitter. Both diodes conduct during the positive half-cycles of the reference voltage e_r. The voltage across ca is $(e_r + e_s)$ *sin* ωt and the capacitor C_1 gets charged to a voltage of $(e_r + e_s)$. The voltage across cb is $(e_r - e_s)$ *sin* ωt and the capacitor C_2 gets charged upto $(e_r - e_s)$. The output voltage is given by

$$e_0 = (e_1 \sim e_2) = \pm 2e_s \qquad \qquad ...\,(5.13)$$

The circuit rectifies only the positive half-cycle and a smoothing circuit is used to filter the carrier frequency. But this should not smooth out the variations in the error signal. If a carrier frequency of t least ten times that of the highest signal frequency is used, the filter circuit design does not present any difficulty.

5.7. TACHOGENERATORS

A tachogenerator is a rotating electromagnetic device that generates an electrical output voltage proportional to shaft speed. In control systems, two types of tachogenerators are used:

1. A.C. Induction type Tachogenerator.

2. D.C. Tachogenerator

5.7.1. A.C. Tachogenerators

A.C. tachogenerator is very similar to a 2-phase induction motor. However, a tachogenerator is excited on only one phase (called *reference phase*), and a voltage of line frequency approximately proportional to the shaft speed is generated on the unexcited or *output phase*.

Fig 5.16 shows the schematic diagram of an a. c. tachogenerator. The reference phase is excited by a sinusoidal voltage of rated value. The current flowing through this phase sets up

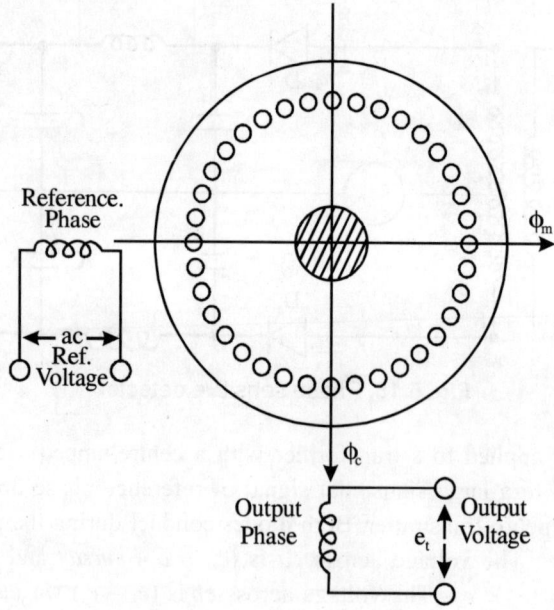

Fig. 5.16. A.C. tachogenerator

an alternating magnetic flux in the air-gap. The output phase is placed at 90 electrical degrees from the reference phase and the output is generated across this phase. When the rotor shaft is stationary, the output voltage is zero. For a well-designed tachogenerator, the rotor-bar inductance may be assumed negligible.

When the rotor shaft is rotated, a voltage is induced in the rotor by the *main flux* ϕ_m. This voltage is proportional to the speed of rotation. This voltage causes a current to flow through the cage rotor, and this rotor current, in turn, sets up an alternating flux ϕ_c called the *cross-field flux*. This cross field flux, by transformer action, induces an output voltages across the output phase.

The transfer function can be written as

$$e_t = K_t \, d\theta/dt \qquad \qquad \dots (5.14)$$

where e_t is the output voltage of the tachogenerator,

K_t is the sensitivity of the tachogenerator in volts/rad./sec. or volts/rpm,

$d\theta/dt$ is the shaft speed.

The magnitude of K_t for some available tachogenerator is in the range of 0.3 volts to 10 volts per 1000 rpm.

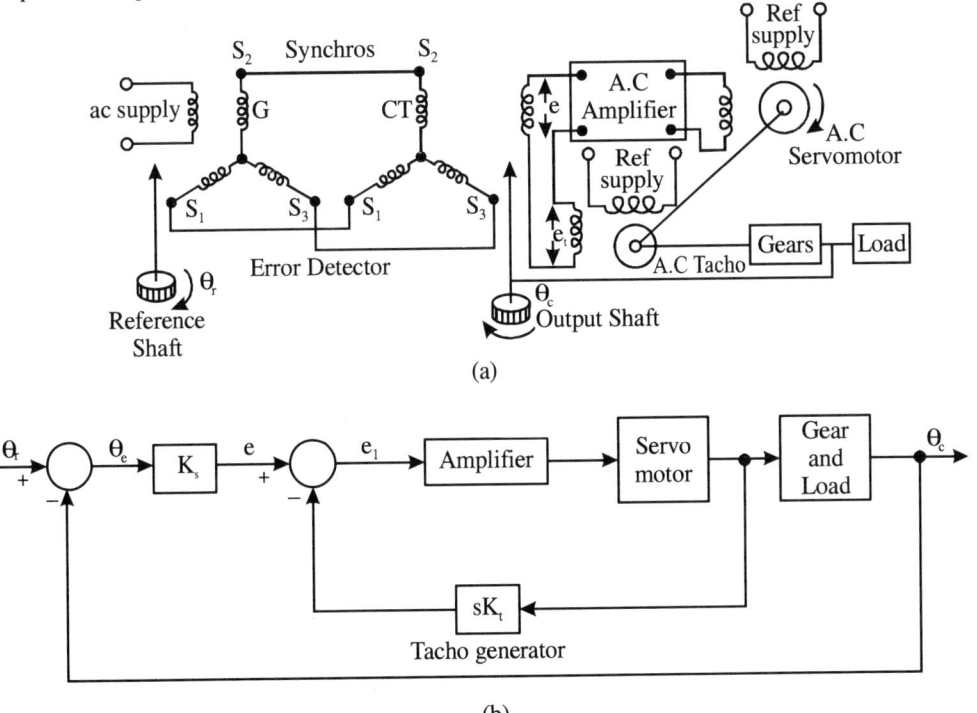

(a)

(b)

Fig. 5.17. (a) A.C. speed control system (b) Block diagram

When excited with a sinusoidal source, an ideal unit has the following characteristics:

1. The output voltage is proportional to shaft speed.
2. At any one speed, the output voltage is proportional to the input voltage.
3. The relative phase of the output voltage is either 0° or 180° depending on the direction of shaft rotation.
4. Less residual voltage is present.

The applications of A.C tachogenerators are quite extensive. A.C. tachogenerators are uniquely adaptable to A.C. control systems and when carefully designed, generate high quality output. A control system with tachogenerator feedback is known to have better stability in its response.

The disadvantages are small residual voltage at zero speed, nonlinearity and voltage gradient and phase angle errors at low speeds. A typical servo employing A.C tachogenerator is shown in Fig. 5.17.

5.7.2. D.C. Tachogenerators

D.C. tachogenerator is a small D.C generator, usually built with a stable permanent magnet to provide the field flux. When the field flux is constant, the generated voltage is directly proportional to the speed.

The commutator of a D.C. tachogenerator has relatively large number of segments so as to provide as smooth an output voltage as possible.

The important factors that differentiate precision tachogenerators from D.C. machines are:

1. Tachogenerators deliver noise-free output voltage which varies linearly with speed.
2. They run smoothly, have low static and running drag and low-inertia.
3. Cogging is prevented by skewing the armature slots and tapering the poleface contour.
   Cogging or slot-lock is the magnetic attraction between stator and rotor. This prevents the rotor from running until the control voltage reaches a certain minimum break-in value.
4. Friction losses are reduced by limiting brush pressure, using high quality anti-friction bearings, and minimizing the commutator diameter.
5. Eddy currents are reduced by laminating the armature and poles.
6. Magnetic materials with negligible hysteresis are used.
7. To improve commutation, fairly large airgap and linear magnetization range, high resistance risers, and hard brushes are used.
8. Fine wire is used to obtain high signal-to-noise ratio.

5.7.3. Comparison A.C. and D.C. Tachogenerator

The main advantages of D.C. tachogenerator are:

1. They are free from waveform and phase-shift distortions.
2. It is possible to achieve very high voltage gradients in the smallest size (10 to 20 volts/1000 rpm).
3. Temperature compensation can be achieved with ease.
4. They can be used with high-pass filters to reduce servo velocity lag.

The d.c. tachogenerators have the following limitations:

1. Arcing occurs at the brush contacts. Commutation is adversely affected by moisture, temperature variations and deposits of brush carbon on the commutator.
2. Radio noise, generated because of arcing at brushes, interferes with adjacent communication channels.
3. Output waveform has a definite inherent ripple because of finite number of commutator bars.
4. A high driving torque is required to overcome brush friction and hysteresis effects.

5.8. AMPLIFIERS

In control systems, the error signal voltage is so low that it cannot be directly used to actuate the servomotor which, in turn, applies corrective effort to the load. Before being applied to the servomotor, the error signal should therefore be amplified. For this purpose, electronic, rotating and magnetic amplifiers are used in electro-mechanical systems.

5.8.1. Electronic Amplifiers

These amplifiers are used directly to supply the servomotors in low power systems. In high power systems, they are used as pre-amplifiers. These are of two types: direct-coupled (D.C.) and A.C. coupled. In D.C servo systems, the D.C. amplifiers are used and in A.C systems A.C coupled amplifiers are used. A.C. amplifiers are preferred to D.C. amplifiers even in D.C systems because of the attendant difficulties with D.C amplifiers such as drift and noise. Drift is the tendency for the quiescent operating point to shift with time. The major causes of drift are changes in power supply voltages, changes in device characteristics due to temperature variations and aging. To overcome these difficulties and to achieve high performance, it is customary to use chopper stabilization as shown in Fig. 5.18. The modulator or chopper converts the input D.C or low-frequency signals to a relatively high frequency signal which is then amplified by a standard A.C. amplifier. The demodulator then rectifies and restores this amplified signal to its original form.

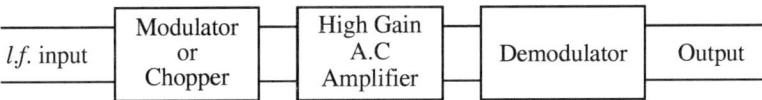

Fig. 5.18. Chopper stabilized D.C. amplifier

5.8.2. Rotating Amplifiers

(*a*) **D.C. Generator:** The simplest form of a rotating amplifier is a D.C. generator which can be used as a power amplifier. The input power to the field circuit is much lower than the output power delivered by the armature circuit. In the approximately linear portion of the magnetization curve, the voltage induced in the armature circuit at constant speed, is given by

$$E_g = K_g \, I_f \qquad \qquad ...(5.15)$$

where K_g is the generator constant in volts/amp. of field current I_f.

A schematic representation of a D.C. generator is shown in Fig. 5.19. L_f, R_f and L_a, R_a are the inductance and resistance of the field and armature circuits respectively.

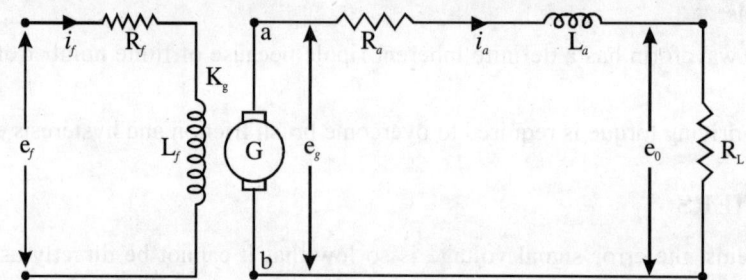

Fig. 5.19. Schematic representation of D.C. generator

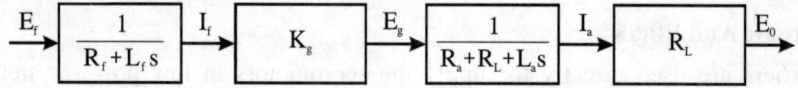

Fig. 5.20. Block diagram of D.C generator

The governing equations are:

$$I_f(s) = E_f(s)/(L_f s + R_f) \qquad \qquad ...(5.16)$$

$$E_g(s) = K_g \, I_f(s) \qquad \qquad ...(5.17)$$

$$E_g(s) = K_g \, E_f(s)/(L_f s + R_f) \qquad \qquad ...(5.18)$$

$$\therefore \qquad \frac{E_g(s)}{E_f(s)} = \frac{K_g}{R_f + L_f s} = \frac{K_g / R_f}{1 + s\tau_f} \qquad \qquad ...(5.19)$$

where $\tau_f = L_f/R_f$ is the field time constant.

The block diagram of Fig. 5.19 is shown in Fig. 5.20.

From the block diagram, we get

$$\frac{E_o(s)}{E_f(s)} = \frac{K_g R_L}{\left(R_f + L_f s\right)\left(R_a + R_L + L_a s\right)} = \frac{K_g R_L / \left(R_L + R_a\right) R_f}{\left(1 + s\tau_a\right)\left(1 + s\tau_c\right)} \qquad \dots (5.20)$$

where $\tau_f = L_f / R_f$ is the field time constant and $\tau_a = L_a/(R_a + R_L)$ is known as armature time constant.

(*b*) **Amplidyne:** An amplidyne is an electromechanical power amplifier capable of controlling large output power with only very low exciting voltage. Fig. 5.21(a) shows an elementary form of Amplidyne. The Amplidyne is fundamentally a two-pole D.C. generator with two pairs of brushes, one pair being placed along the pole axis and the other pair along an axis perpendicular to the pole axis. The pole axis is normally called the *direct axis* and the other the *quadrature axis*. The input or control voltage e_c is applied to the control field.

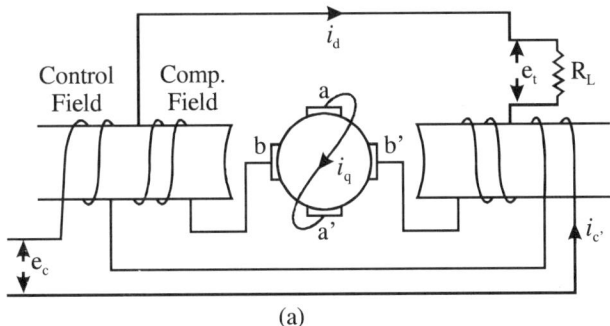

(a)

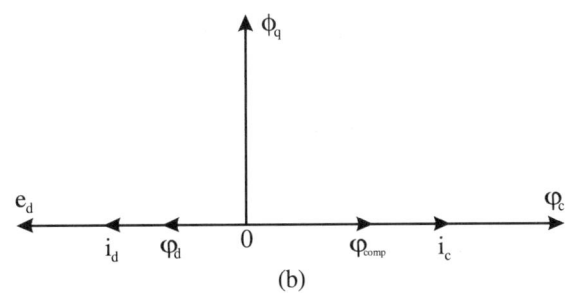

(b)

Fig. 5.21. (a) Amplidyne (b) Magnetic fluxes

The control field has a large number of turns and high impedance. Hence, the input power is small. The current i_c in the control field sets up a small flux φ_c along the direct axis. This flux φ_c induces a voltage e_q in the armature along the quadrature axis between the brushes aa'. The impedance between the brushes aa' is small and the voltage e_q is virtually shorted. Therefore a large quadrature current i_q flows. This large current i_q results from a relatively small control current i_c. The current i_q, in turn, produces a large flux φ_q which is in quadrature with the control field flux φ_c. The flux φ_q induces a voltage e_d and is applied to the load which draws a load current i_d. This current i_d sets up a flux φ_d which opposes the control flux φ_c and might cancel the action of the Amplidyne. This effect can be offset by producing a flux φ_{comp} equal in magnitude but opposite in direction to φ_d. This can be achieved by a compensating winding, located on the control field poles and connected in series with the armature so that the current i_d flows through it. Therefore, in the unsaturated condition of the core, the output voltage e_d is proportional to the input voltage. The terminal voltage e_t is reduced by the internal impedance drop.

From the above discussion, it is clear that the Amplidyne can be represented by a two-stage rotating amplifier as shown in Fig. 5.22. The transfer function may directly be obtained from the block diagram.

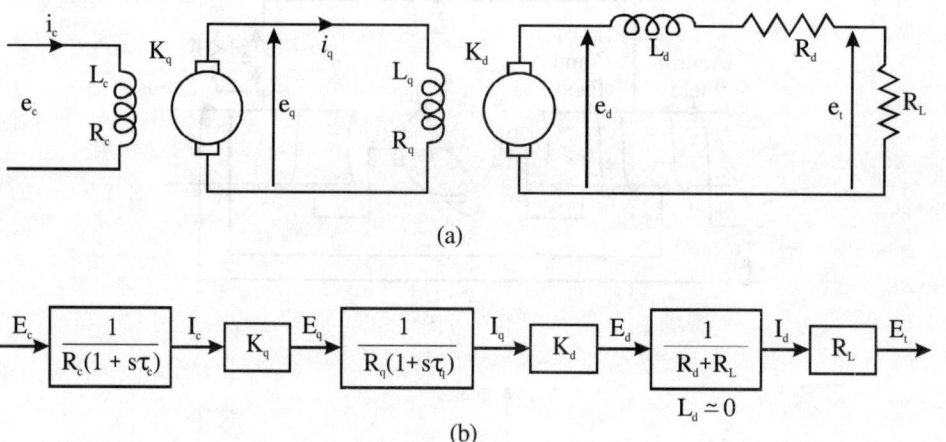

(a)

(b)

Fig. 5.22. (a) Amplidyne as two-stage amplifier; (b) Block diagram representation

The transfer function of each block is obtained from the following equations:

$$I_c(s) = \frac{E_c(s)}{(R_c + L_c s)} = \frac{E_c(s)}{R_c(1 + s\tau_c)} \qquad \qquad ...\,(5.21)$$

where $\tau_c = L_c/R_c$ is the time constant of the control winding.

$$E_q(s) = K_q I_c(s) \qquad \qquad ...\,(5.22)$$

$$I_q(s) = \frac{E_q(s)}{\left(R_q + L_q s\right)} = \frac{E_q(s)}{R_q\left(1 + s\tau_q\right)} \qquad \qquad ...\,(5.23)$$

where $\tau_q = L_q/R_q$ is the quadrature time constant.

$$E_d(s) = K_d \, I_q(s) \qquad\qquad \dots (5.24)$$

If L_d is neglected, the load is a resistive load and the current $I_d(s)$ is given by

$$I_d(s) = E_d(s)/(R_d + R_L) \qquad\qquad \dots (5.25)$$

The terminal voltage e_t is given by

$$E_t(s) = R_L \, I_d(s) \qquad\qquad \dots (5.26)$$

Hence the transfer function is given by

$$\frac{E_t(s)}{I_c(s)} = \frac{K_q K_d R_L}{R_q \left(1 + s\tau_q\right)\left(R_d + R_L\right)} \qquad\qquad \dots (5.27a)$$

Or

$$\frac{E_t(s)}{E_c(s)} = \frac{K_q K_d R_L}{R_c R_q \left(1 + s\tau_c\right)\left(1 + s\tau_q\right)\left(R_d + R_L\right)} \qquad\qquad \dots (5.27a)$$

The main advantage of the Amplidyne is its small size and compactness.

5.9. ELECTRIC ACTUATORS

Actuators are devices which convert input quantities (other than mechanical) to mechanical force or torque, thus proving the mechanical actuating forces or torques necessary for the system. Various devices that are used as actuators are electrical motors, hydraulic pumps and pneumatic controllers. In electrical control systems, all types of electrical motor are used. In this section, we shall study the characteristics of the d.c. and a.c. servomotors.

5.9.1. D.C. Servomotors

In d.c. control systems, a separately excited d.c. servomotor finds extensive use as actuators because of ease of control and of small size. The special constructional features of a d.c. servomotors are:

1. Armature inductance is so small ($L_a \approx 0$) that self-induced back emf is low.
2. The mutual coupling between the armature and the other circuits of the machine is essentially zero.
3. The total brush resistance of the motor is small ($R_b \approx 0$) compared to the overall circuit resistance.
4. The cross-magnetizing flux resulting from armature reaction is negligible.
5. The speed-torque characteristic of the motor is almost linear.

5.9.2. Methods of Exciting D.C. Servomotors

There are two ways of enerzising the d.c. servomotors. One method is to apply a constant voltage to the field circuit and a variable voltage to the armature. The field current being constant, the developed torque is proportional to the armature current. This is called *armature controlled d.c. servomotor.* In the other method, the field voltage is varied while the armature current is kept constant. In this case, the torque is proportional to the field current. This is called *field controlled d.c. servomotor.*

(a) Armature Control: In this type of control, the torque is proportional to the armature current I_a, the field flux being constant,

Hence, torque is given by

$$T(s) = K_T I_a(s) \qquad \ldots (5.28)$$

where K_T is the torque constant and is expressed as Newtons per ampere. Because the motor is rotating in a magnetic field, a back-emf E_b proportional to the speed is induced and is given by

$$E_b(s) = K_b\Omega_m(s) = K_b \, d\Theta_m(s)dt \qquad \ldots (5.29)$$

where K_b is the back emf constant and $\Omega_m(s)$ is the speed of the motor.

Fig. 5.23 shows the circuit of an armature controlled motor. By applying the Kirchhoff's voltage law to the armature circuit, we obtain

$$E_a(s) = (L_a s + R_a) I_a(s) + E_b(s) \qquad \ldots (5.30)$$

where $E_a(s)$ is the applied armature voltage, L_a is the armature inductance and R_a is the armature resistance.

Combining Eq. (5.28) and (5.30), we get

$$T(s) = K_T (E_a(s) - E_b(s))/(L_a s + R_a) \qquad \ldots (5.31)$$

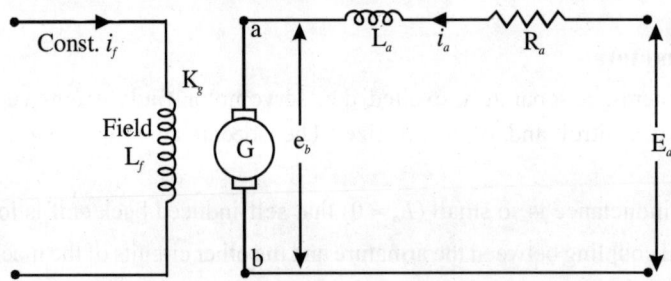

Fig. 5.23. Armature control of d.c. servomotor

This torque is used to rotate the motor consisting of moment of inertial J_m and frictional damping B_m.

Hence, an equation connecting the torque and the load may be obtained as

$$T(s) = (J_m s + B_m) \Omega_m(s)$$
$$= (J_m s + B_m)s \, \Theta_m(s) \qquad \ldots (5.32)$$

Substituting the value of $T(s)$ and $E_b(s)$ obtained in previous equations and rearranging, we obtain

$$E_a(s) = \frac{s[(L_a s + R_a)(J_m s + B_m) + K_T K_b]}{K_T} \Theta_m(s) \qquad \text{... (5.33a)}$$

$$\frac{\Theta_m(s)}{E_a(s)} = \frac{K_T}{sR_a B_m[(1 + s\tau_a)(1 + s\tau_m)] + K_T K_b s} \qquad \text{... (5.33b)}$$

where $\tau_a = L_a/R_a$ is armature time constant and $\tau_m = J_m/B_m$ is the mechanical time constant of the motor.

In the standard form, the above equation is written as

$$\frac{\Theta_m(s)}{E_a(s)} = \frac{G(s)}{1 + G(s)H(s)}$$

where $H(s) = K_b s$. Hence $G(s)$ may be shown to be equal to

$$G(s) = K_T / [sR_a B_m(1 + s\tau_a)(1 + s\,\tau_m)] \qquad \text{... (5.34)}$$

The above equations may also be obtained directly from the block diagram shown Fig. 5.24.

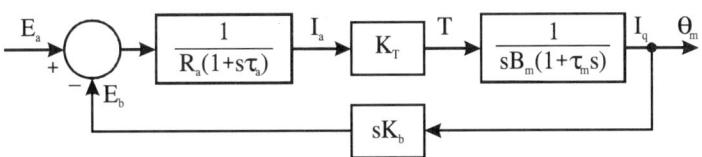

Fig. 5.24. Block diagram of armature control

If the armature inductance $L_a = 0$, Eq. (5.33b) reduces to

$$\frac{\Theta_m(s)}{E_a(s)} = \frac{K_T}{sR_a B_m(1 + s\tau_m) + K_T K_b s} \qquad \text{... (5.35)}$$

and Eq. (5.34) reduces to

$$G(s) = K_T / [s\,R_a\,B_m(1 + s\tau_m)] \qquad \text{... (5.36)}$$

(*b*) **Field Control:** Fig. 5.25 shows the schematic diagram of a field controlled motor. In this type of control, the armature current is kept constant. If the armature current is not constant, it will lead to nonlinear operation which cannot be analyzed by linear analysis methods.

Since the armature current is constant, the torque is proportional to the field flux which, in turn is proportional to field current. Therefore,

$$T(s) = K_T I_f(s) \qquad \text{... (5.37)}$$

where K_T is the torque constant.

The field current $I_f(s)$ is given by

$$I_f(s) = E_f(s)/(L_f s + R_f) \qquad (5.38)$$

where L_f is the inductance in henries and R_f is the resistance in ohms of the field circuit. Hence the torque is

$$T(s) = K_T E_f(s)/(L_f s + R_f) \qquad \text{... (5.39)}$$

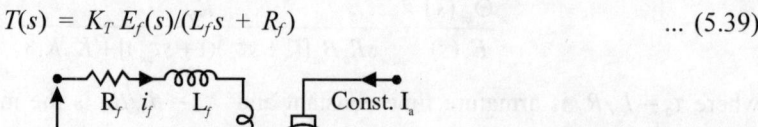

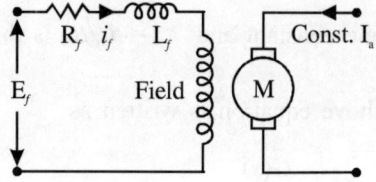

Fig. 5.25. Field control of d.c. servomotor

Also,

$$T(s) = (J_m s + B_m)s\ \Theta_m(s) \qquad \text{... (5.40)}$$

Therefore,

$$T(s) = K_T E_f(s)/(L_f s + R_f) = (J_m s + B_m)s\ \Theta_m(s) \qquad \text{... (5.41)}$$

Rearranging we have,

$$\frac{\Theta_m(s)}{E_f(s)} = \frac{K_T}{s(J_m s + B_m)(L_f s + R_f)} = \frac{K_T}{sR_f B_m (1 + s\tau_m)(1 + s\tau_f)} \qquad \text{... (5.42)}$$

where $\tau_f = L_f / R_f$ is the field time constant.

In the field control, there is no feedback path and hence the block diagram has only a forward loop as shown in Fig. 5.26.

$$E_f \longrightarrow \boxed{\dfrac{1}{R_f(1 + s\tau_f)}} \xrightarrow{I_f} \boxed{K_T} \xrightarrow{T} \boxed{\dfrac{1}{sB_m(1 + s\tau_m)}} \xrightarrow{\Theta_m}$$

Fig. 5.26. Block diagram of field control

5.9.3. Comparison between Armature and Field Control

The main differences between the armature and field controlled operation of the d.c. servomotor are:

1. The power required for field control is small while that for armature control is large.

2. The field control requires low current devices as sources such as transistors, whereas the armature control requires heavy current devices such as rotating amplifiers or silicon contolled rectifiers.

3.   The inductance in the field circuit is large when compared to the almost negligible armature inductance. Therefore the armature controlled motor is a *second order system* and the field controlled motor is a *third order system.*

4.   Because of the large inductance, the time constant of the field circuit is much larger than that of the armature circuit. Consequently, the response of a field control is not as rapid as that of the armature control.

5.   Because of the stringent requirement of constant armature current (which is very difficult to obtain in practice), the motor performance under field control is not as linear as under armature control.

6.   In the armature controlled operation, the damping is due to the presence of R_a, B_m of the motor and the induced back emf E_b, whereas in the field control, the entire damping is due to motor friction and load.

5.9.4. A.C. Servo Motors

An A.C. servo motor is basically a two-phase induction motor but differs from it in the following aspects.

1.   The servomotor is wound as two phase machines with the two stator field coils placed 90 electrical degrees apart.

2.   The voltage applied to the windings are rarely balanced. To achieve proper operation of the motor, the control winding voltage may be varied from zero to rated voltage of either polarity. While this being done, the reference field is kept at a constant value.

3.   The speed-torque curve of a servo motor should be linear and have a negative slope for stable operation. This is achieved by using high-resistance materials in rotor construction.

   (*a*)   **Principle of operation:** Fig. 5.27 shows the schematic diagram of a two phase servomotor with cage rotor. The stator coils are spaced 90 electrical degrees apart. One coil is applied with a fixed voltage E_r, called *reference voltage*. This winding is called *reference winding*. The other coil is supplied with a variable voltage E_c called *control voltage*. This winding is called the *control winding*. The frequency of the two voltages is equal but the phase relation is such that the control voltage either leads or lags by 90 degrees. When the a.c. servomotor is energized with the two voltages, the air gap magnetic flux generated by the two stator currents are also in time and space quadrature. They combine to yield a rotating magnetic field of fixed magnitude and at synchronous speed. This rotating magnetic field induces a voltage in the stationary rotor conductors. The rotor conductors, being shorted at the ends, carry heavy currents. The interaction between the current carrying rotor conductors and the rotating field develops a torque which rotates the rotor in the same direction as that of the rotating magnetic field. In the steady operation, the rotor runs at a speed slightly less than that of the synchronous rotating field. The voltage induced in the rotor conductors is proportional to the relative speed.

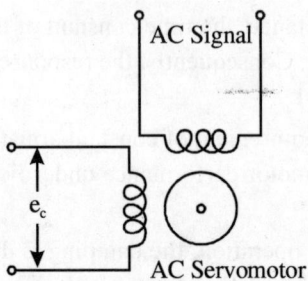

Fig. 5.27. Two-phase servomotor

(*b*)   **Transfer Function:** The control action of the A.C. servomotor is due to the application of a variable control voltage E_c. The torque-speed characteristic is a function of this voltage E_c. The characteristic is not a linear function of E_c. However, it may be considered as linear for a small portion of the characteristic. During this working range, the torque being a function of speed and E_c, may be written as

$$T(E_c, \omega) = K_v E_v + K_\omega \omega \qquad \qquad ... (5.43)$$

where K_v and K_ω are constants and equal to $dT/dE_c\big|_{\omega = constant}$ and $dT/d\omega\big|_{E_c = constant}$ respectively. For a motor with inertia J_m and damping B_m, the torque equation is

$$T(s) = (J_m s + B_m)s\ \Theta_m(s) \qquad \qquad ... (5.44)$$

Equating the torque equations, we get

$$(J_m s + B_m)s\ \Theta_m(s) = K_v E_c + K_\omega s\ \Theta_m(s) \qquad \qquad ... (5.45)$$

From the above equation, the transfer function is given by

$$\frac{\Theta_m(s)}{E_c(s)} = \frac{K_v}{s[J_m s + (B_m - K_\omega)]} \qquad \qquad ... (5.46)$$

For the operation of the motor to be stable, the term $(B_m - K_\omega)$ must be always positive which implies that the slope of the torque-speed characteristic K_ω for a fixed E_c must be negative. The negative slope of the torque-speed characteristic ensures stable operation.

The block diagram representation of an A.C. servomotor is shown in Fig. 5.28. The transfer function can be easily derived from this block diagram.

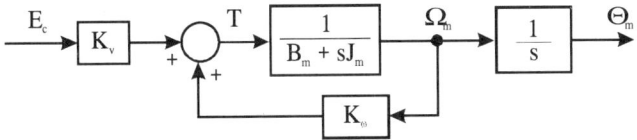

Fig. 5.28. Block diagram of a.c. servomotor

5.9.5. Comparison between D.C. and A.C. Servomotors

D.C. servomotors are extensively used wherever it is possible and preferred to A.C servomotors because of the following advantages:

1. D.C. servomotors can deliver higher output from a small frame size.
2. They can be driven from a low-power amplifier.
3. Simple stabilization techniques can be adopted.
4. They have negligible residual voltage.
5. They are more efficient.
6. They can be adapted for variable speed systems.
7. The can be used as d.c. tachogenerators.

However, the D.C. servomotors do have certain drawbacks. They are:

1. Maintenance is difficult.
2. They are sources of radio interference because of the commutator segments and brushes.
3. Friction is more due the presence of brushes and commutator.
4. Commutator and brushes occupy a large percentage of the available motor volume.
5. Sensitivity is poor for weak signals.
6. In the field-controlled motors, problems of stabilization are present due to hysteresis, and highly inductive input circuit.

5.10. HYDRAULIC SERVOMOTOR

A hydraulic servomotor is shown in Fig. 5.29. It is essentially a valve controlled hydraulic power amplifier and actuator. Little power is required for positioning the valve which controls large power, thus providing large power amplification. The oil is supplied at a constant high pressure. The oil draining to the sump is at low pressure. The motion of the value regulates the flow of oil to the either side of the main piston. A small input motion $x(t)$ results in a large change of oil flow resulting in a large displacement $y(t)$ of the output shaft.

Let Q be the rate of flow of oil to the main cylinder and $p = P_1 - P_2$ be the pressure difference across the main piston. In general, Q is a function of x and p.

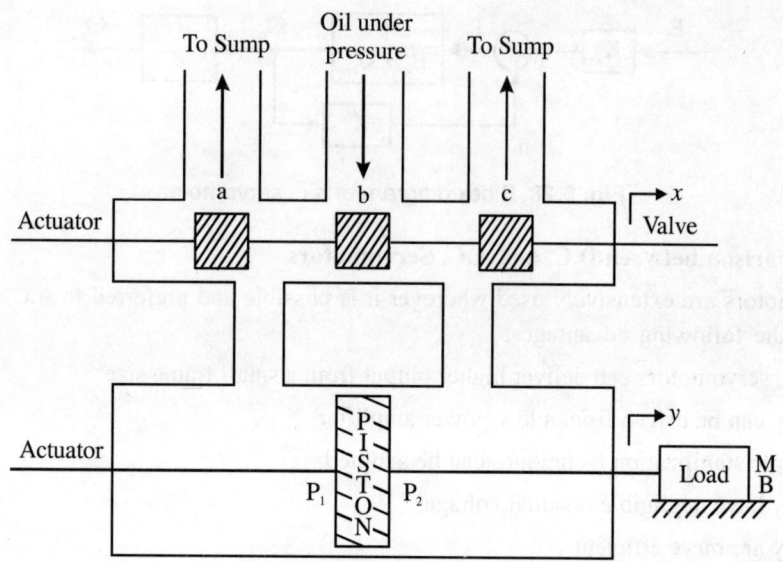

Fig. 5.29. Hydraulic servomotor

Thus

$$Q = f(x, p) \qquad\qquad\qquad \text{... (5.47)}$$

Considering the partial derivatives,

$$dQ = \frac{\delta Q}{\delta x}\,dx + \frac{\delta Q}{\delta p}\,dp \qquad\qquad\qquad \text{... (5.48)}$$

Integrating Eq. (5.48) we get

$$Q = K_x\, x - K_p\, p$$

Or

$$p = (K_x\, x - Q)/K_p \qquad\qquad\qquad \text{(5.49)}$$

where $K_x = \partial Q/\partial x$ and $K_p = -\partial Q/\partial p$. The force developed by the main piston is given by

$$F = Ap \qquad\qquad\qquad \text{... (5.50)}$$

where A is the area of the piston. As this force is applied to the load, we have

$$F = Ap = M\, y'' + B y' \qquad\qquad\qquad \text{... (5.51)}$$

Substituting the value of p from Eq. (5.49), we get

$$A(K_x\, x - Q)/K_p = M y'' + B y' \qquad\qquad\qquad \text{... (5.52)}$$

The flow rate Q is related to the main piston movement y by

$$Q = Ay' \qquad\qquad\qquad \text{... (5.53)}$$

Substituting Eq. (5.53) in Eq. (5.52) and rearranging, we have

$$AK_x\, x/K_p = My'' + (B + A^2/K_p)y' \qquad \text{... (5.54)}$$

Assuming zero initial conditions and taking the Laplace transform of Eq. (5.54), the transfer function $Y(s)/X(s)$ is given by

$$\frac{Y(s)}{X(s)} = \frac{AK_x\,/\,K_p}{s(Ms + B + A^2\,/\,K_p)} \qquad \text{... (5.55)}$$

5.11. PNEUMATIC ACTUATOR

Fig. 5.30 shows a common type of pneumatic actuator. The operating fluid is the compressed air. When the compressed air at a pressure P_i is injected through the input manifold, it pushes the diaphragm and the plunger attached to the diaphragm. This displaces the plunger by an amount y. The plunger system has a load of mass M, viscous friction B and spring constant K. The pressure force acting on the diaphragm is AP_i where A is the area of the diaphragm. This force is used to position the load. Hence,

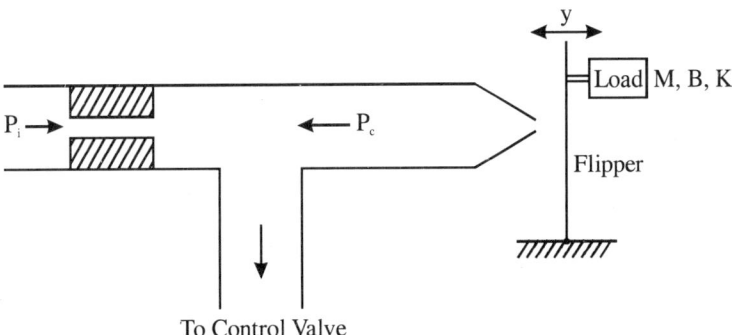

Fig. 5.30. Pneumatic actuator

$$AP_i(s) = (Ms^2 + Bs + K)Y(s) \qquad \text{... (5.56)}$$

Therefore the transfer function $Y(s)/P_i(s)$ is given by

$$Y(s)/P_i(s) = A/(Ms^2 + Bs + K) \qquad \text{... (5.57)}$$

5.12. GYROSCOPE

Gyroscope is a mechanical device used to sense angular motions. It is commonly used in inertial guidance and auto-pilot systems. Fig. 5.31 shows a single degree – freedom gyroscope. Basically, it consists of a disk or wheel. The spinning wheel or rotor is mounted on an axle and supported by a frame work, called *gimbals* so that the disk may turn in any direction. It has three axes, the *input axis*, the *output axis* and the *spin axis*. Relative to the case, it is free to move about the output axis. The input signal which is the turning rate of angle about the input axis is obtained from the resulting motion of the gimbal about the output axis relative to the case. The inner gimbal has a moment of inertia J about the output axis.

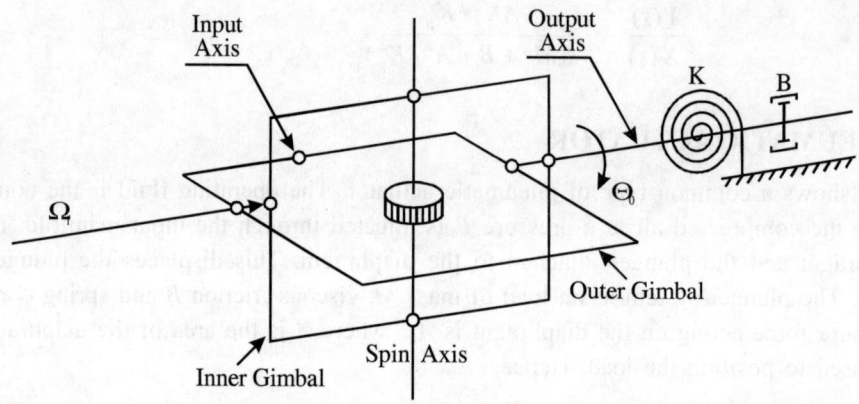

Fig. 5.31. Single-degree freedom gyroscope

The equation of motion of the gyroscope can be obtained by equating the rate of change of angular momentum to the sum of the external torques. The change in angular momentum about the output axis is the sum of the changes in momentum about the output axis $J\theta_0''$ and in the rotor momentum $-H\omega_i$, where H is a constant. The external torques are the sum of the damping torque $-B\theta_0'$ and the spring torque $-K\theta_0$. Hence, the equation of motion of the gyro system assuming zero initial conditions is given by

$$J\theta_0'' - H\omega_i = -B\theta_0' - K\theta_0 \qquad \qquad \text{... (5.58)}$$

or,

$$(Js^2 + Bs + K)\,\Theta_0(s) = H\Omega_i(s) \qquad \qquad \text{... (5.59)}$$

Then the transfer function is given by

$$\frac{\Theta_o(s)}{\Omega_i(s)} = \frac{H}{Js^2 + Bs + K} \qquad \qquad \text{... (5.60)}$$

For a constant input, the steady-state value of $\Theta_0(s)$ is given by

$$\lim_{s \to 0} \frac{sH}{(Js^2 + Bs + K)} \frac{\Omega_i}{s} = \frac{H\Omega_i}{K} \qquad \qquad \text{... (5.61)}$$

In the absence of the spring, the transfer function is

$$\frac{\Theta_0(s)}{\Omega_i(s)} = \frac{H}{s(Js+B)} \qquad \dots (5.62)$$

This gyro is known as an *integration gyro.*

5.13. D.C. POSITION CONTROL SYSTEM

A schematic diagram of a D.C. position control system is shown in Fig. 5.32(a). The pair of potentiometers functions as transducers and error detector. The sensitivity of both potentiometers is K_s. The D.C. amplifier has a gain of K_a volt/volt. The torque constant of the D.C. motor is K_T in N-m/amp. The equivalent moment of inertia and the equivalent viscous damping referred to the motor side are J_e and B_e respectively. (For principle of operation, refer section 1.8.).

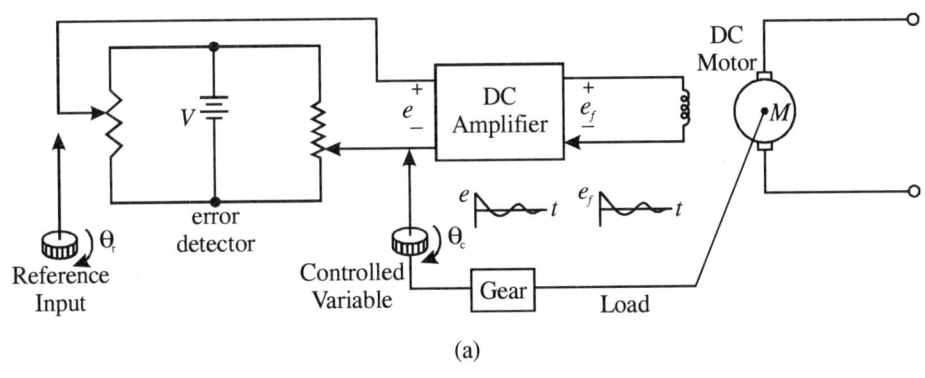

(a)

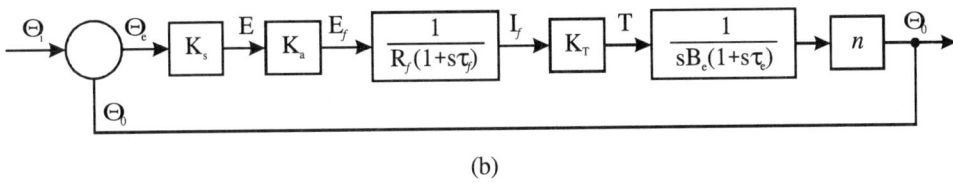

(b)

Fig. 5.32. (a) D.C. position control system (b) Block diagram

The block diagram of the system is shown in Fig. 5.32(b). The transfer function is given by

$$\frac{\Theta_0(s)}{\Theta_i(s)} = \frac{K_s K_a K_T n}{sR_f B_e(1+s\tau_f)(1+s\tau_e) + K_s K_a K_T n} \qquad \dots (5.63)$$

where $n = N_1/N_2$ is the gear ratio, τ_f is field time constant and τ_e is the effective armature time constant.

5.14. A.C. POSITION CONTROL SYSTEM

Fig. 5.33(a) shows the schematic diagram of an A.C. servo system used to position the load so that the output shaft follows the input shaft as closely as possible. Here the reference input is the input shaft position θ_i and the controlled output is the output shaft position θ_o . The synchro transmitter and a control transformer are used as transducers and error detector. The rotor of the synchro transmitter is connected mechanically to the input shaft and the rotor of the control transformer to the output shaft. When the two shafts are aligned, the error voltage is zero and the motor does not turn. When the input shaft is turned by an angle θ_i, there exists misalignment and an error voltage of relative polarity corresponding to the misalignment appears at the input of the amplifier. This error voltage e is amplified and the output of the amplifier drives the motor in such a direction as to reduce the error.

The sensitivity of the synchro pair is K_s. The gain of the A.C. amplifier is K_a volts/volt. The torque of the a.c. servomotor is a function of the control voltage and the speed. K_v and K_ω are the two torque constants. K_v is the rate of change of torque with respect to the control voltage at a constant motor speed and K_ω is with respect to the speed for a constant control voltage. The block diagram of the A.C. position control system is shown in Fig. 5.33(b). The transfer function is given by

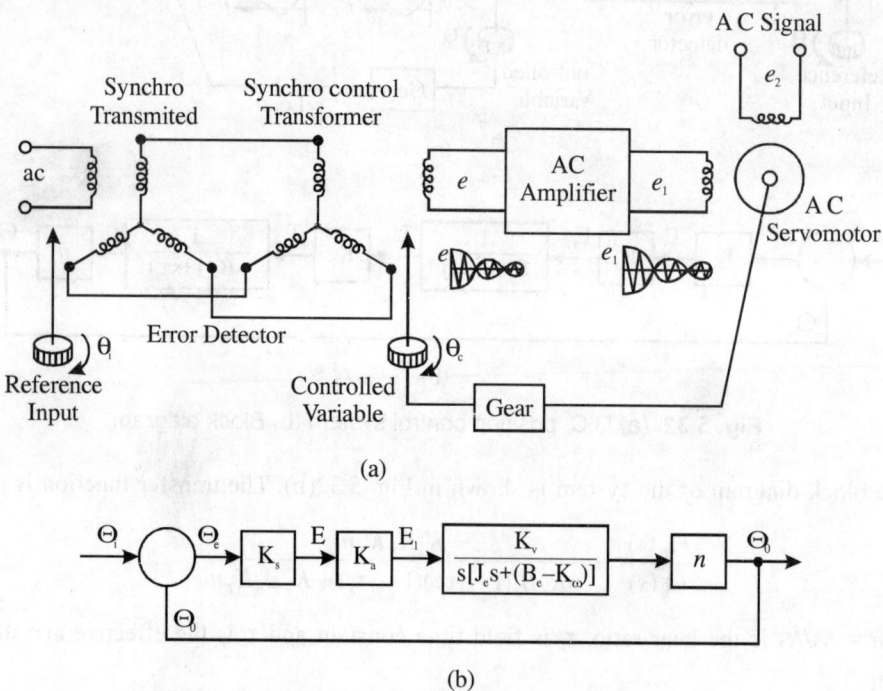

(a)

(b)

Fig. 5.33. (a) A.C. position control system, (b) Block diagram

$$\frac{\Theta_0(s)}{\Theta_i(s)} = \frac{K_s K_a K_T n}{s(J_e s + B_e - K_\omega) + K_s K_a K_T n} \qquad \dots (5.64)$$

where J_e and B_e are the equivalent moment of inertia and viscous friction referred to the motor side and are given by

$$J_e = J_m + n^2 J_L \qquad \dots (5.65)$$
$$B_e = B_m + n^2 B_L \qquad \dots (5.66)$$

where the suffix m refers to motor and the suffix L to the load.

5.15. D.C. MOTOR SPEED CONTROL SYSTEM

In many control systems one of the requirements is to maintain the speed constant under varying load conditions. A commonly used speed control system is shown in Fig. 5.34. It consists of a potentiometer P which sets the desired speed, an amplifier (A) to amplify the error voltage and supply the field current, a field-controlled motor (M) to drive the load and a tachogenerator TG to sense the speed and generate the feedback voltage. The setting of P provides the required input voltage for the desired speed. The tachogenerator voltage which is proportional to the speed is compared with the input voltage. Under steady-state conditions when the motor runs at the desired speed, the feedback voltage is equal to the input voltage and hence the error voltage is zero.

If due to load fluctuations the speed of the motor changes, a corrective action is initiated. For instance, if the speed increases, the feedback voltage from the tachometer increases and the input to the amplifier is negative. This reduces the field current of the motor. The motor torque, being a function of the field current, reduces and this reduction in torque restores the motor speed to the desired value set by the potentiometer P. If the speed decreases for any reason, a reverse action is set up and the speed of the motor is again restored.

The block diagram of the speed control system is shown in Fig. 5.34(b). K_a is the amplifier gain in amp/volt, K_T is the torque constant in N-m/amp, K_T is the tachogenerator constant. J_e is the equivalent inertia and B_e is the equivalent viscous damping referred to motor side and $\tau_e = J_e/B_e$ is the effective motor time constant. The transfer function of the speed control system is

$$\frac{\Omega_0(s)}{E_i(s)} = \frac{K_a K_T}{B_e(1 + s\tau_e) + K_a K_T K_s} \qquad \dots (5.67)$$

5.16. TEMPERATURE CONTROL SYSTEM

The temperature control system is an example of industrial process control. The process to be controlled is the temperature of a tank which is to be kept constant. Fig. 5.35(a) shows a simple temperature controller. The potentiometer P is calibrated in terms of the desired temperature and its setting provides the reference input e_r. The actual temperature of the tank θ is the output variable and is measured by a thermocouple immersed in the liquid. The thermocouple produces a voltage e_{th} which is proportional to θ. This voltage is amplified to produce a voltage e_f the feedback voltage. The error voltage is the difference between e_r and e_f and is the actuating signal. It is amplified to produce the solenoied voltage e_s. The solenoid current i_s produces a proportional force f which acts on the solenoid armature and valve to control the valve position y. The solenoid armature and the valve together have a mass M, viscous friction, B and elastance K. The valve position controls the hot steam flow q into the heating coil in the tank. The resulting tank temperature θ is directly proportional to the steam flow.

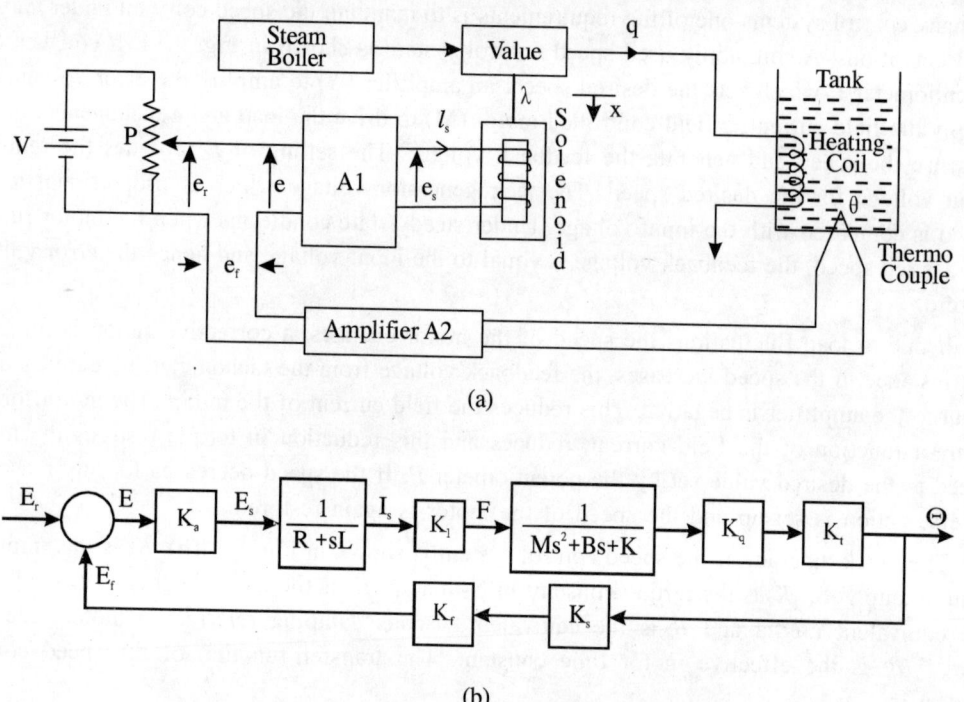

(a)

(b)

Fig. 5.35. (a) Temperature control system; (b) Block diagram

The operation of the system is a follows:

If, for any reason, the tank temperature decreases, this causes an increase in the actuating signal e. This increases the solenoid current and hence the force. This increased force acting on the armature and valve causes an increase in the hot steam flow through the heating coil in the tank. Therefore, the temperature θ rises to the original set value. If the tank temperature increases, a reverse action is set in and θ is again restored to its set value. Under steady-state operation, the feedback voltage e_f equals the input e_r and hence the error is zero. The hot steam flow is stabilized at the steady-state value that maintains the temperature θ at the set value.

The block diagram of the system is shown in Fig. 5.35(b). K_q is the proportionality constant between steam flow and valve opening and K_t is between tank temperature and steam flow. The transfer function is given by

$$\frac{\Theta(s)}{E_r(s)} = \frac{K_a K_1 K_q K_t}{(R + sL)(Ms^2 + Bs + K) + K_a K_1 K_q K_t K_s K_f} \qquad \dots 5.68$$

5.17. SYSTEM WITH TRANSPORTATION LAG

Thus for we have considered systems whose transfer functions are ratios of finite polynomials in s. However, pure time delays occur in various types of system such as hydraulic, pneumatic and mechanical translational system. The effect of this time lag is to delay the signal $x(t)$ at its input by T at its output where T is the time delay. Consider the thickness control of sheet metal rolling out of a steel mill. Fig. 5.36(a) shows a simplified diagram of thickness control system.

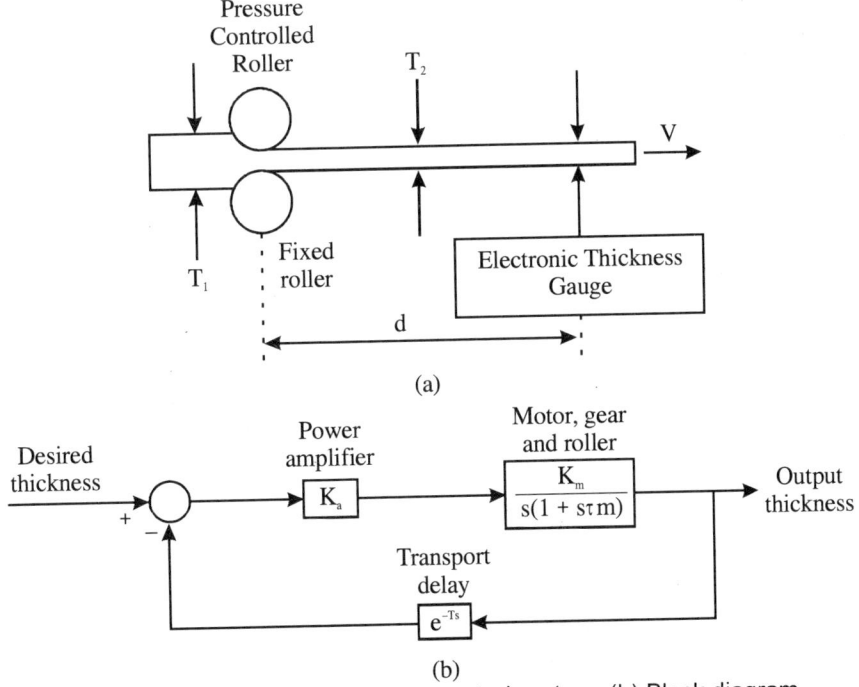

(a)

(b)

Fig. 5.36. (a) Metal thickness control system; (b) Block diagram

The thickness is sensed at a distance d downstream from the rollers controlling the thickness. If the velocity of the output sheet is v m/sec. then the delay $T = d/v$ sec. There are two rollers which control the thickness of the sheet. One roller is fixed and the position of the second is controlled by a motor and gear arrangement. The feedback signal $B(s) = e^{-Ts} C(s)$. The block diagram of this system is shown in Fig. 5.36(b). The transfer function is

$$\frac{C(s)}{R(s)} = \frac{K_a K_m}{s(1+s\tau_m) + K_a K_m e^{-Ts}} \qquad \text{... (5.69)}$$

where τ_m is the motor time constant.

5.18. SOLVED PROBLEMS

Example 5.1. *A linear 100 kΩ potentiometer with an arc length of 344° is connected to the 12V stabilized d.c. supply. Calculate (a) the potentiometer constant K_s in volts/rad. and (b) the output voltage when a 100 kΩ load resistor is connected to the wiper which is set at one half of the arc length.*

Solution:

(a) Arc length $= 344° = 344 \times 2\pi/360$ rad.

Potentiometer constant K_s = voltage/arc length $= 12 \times 360/(344 \times 2\pi) = 2.0$V/rad.

(b) At the midpoint of the potentiometer, the no-load voltage is 6V. However, when a load resistor is connected, the output voltage is decreased because of the loading effect. Fig. 5.37 shows the 100kΩ load connected.

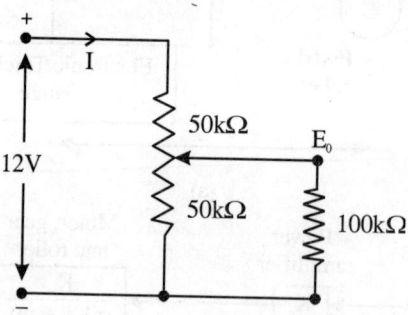

Fig. 5.37. Potentiometer with load

The equivalent resistance is given by

$$R_{eq} = (50 \times 100)/(50 + 100)$$
$$= 100/3 \ k\Omega$$

The total resistance is given by
$$R_{Total} = 50 + (100/3) = 250/3 \ k\Omega$$
The current I is given by
$$I = (12/250/3) = 144 \ mA$$
The output voltage is given by
$$E_o = I \ x \ R_{eq} = 144 \ x \ 100/3 = 4.8 \ volts$$

Example 5.2. *Show that, in an armature controlled D.C. motor, the torque constant K_T and the back emf constant K_b are equal.*

Solution:

The mechanical power of the motor is $T\omega$ and the electrical power is $E_b I_a$.
$$T = K_T \ I_a$$
$$E_b = K_b \ \omega$$
By equating the two powers, we get
$$K_T \ I_a \ \omega = K_b \ \omega \ I_a$$
or,
$$K_T = K_b$$

Example 5.3. *Determine the transfer function $\Theta(s)/E_a(s)$ of an armature controlled d.c. motor with the following parameters: $E_a = 26V$, $La = 0$, $j = 0.183$.*

Solution:

Given: $L_a \approx 0$, $J = 0.183 \times 10^{-5} \ kg - m^2$, $B = 0.21 \times 10^{-4} \ N - m/rad./sec.$ $E_a = 26V$,

The stalling torque $T_s = 0.0706$ N–m and the no-load speed is 520 rad./sec. $E_a = 26V$,

The back *emf* constant K_b is given by
$$K_b = E_a/\omega = 26/520 = 0.05 \ V$$
The torque constant K_T is given by
$$K_T = 0.05 = T_s \ R_s/E_a \ N - m$$
The armature resistance R_a is given by
$$\therefore \qquad\qquad R_a = 0.05 \times 26/0.0706 = 18.4\Omega$$
The transfer function of an armature controlled D.C motor is
$$\frac{\Theta(s)}{E_a(s)} = \frac{K_T}{s(R_a Js + BR_a + K_T K_b)}$$

$$= \frac{0 \cdot 05}{s \left(18 \cdot 4 \times 0 \cdot 183 \times 10^{-5} s + 0 \cdot 21 \times 10^{-4} \times 18 \cdot 4 + 0 \cdot 05 \times 0 \cdot 05\right)}$$

$$= \frac{1485}{s(s + 189)}$$

Example 5.4. *Determine the transfer function of a separately excited d.c. generator with the following parameters: $L_f = 20$ H, $R_f = 100$ Ω, $L_a = 0.5$ H, $R_a = 1$Ω. The generator supplies power to an inductive load of 10H and 200 ohms. The generator constant $K_g = 200$ V/amp.*

Solution:

Given: $L_f = 20$ H, $R_f = 100$ Ω, $L_a = 0.5$ H, $R_a = 1$Ω, $K_g = 200$ V/A, $R_L = 200$Ω, $L_L = 10$ H

The load impedance is given by

$$Z_L = 200 + 10s = 200(1 + 0.05s)$$

The field impedance is given by

$$Z_f = 100 + 20s = 100(1 + 0.2s)$$

The armature impedance is given by

$$Z_a = 1 + 0.5s$$

The block diagram of the system is shown in Fig. 5.38. The transfer function is given by

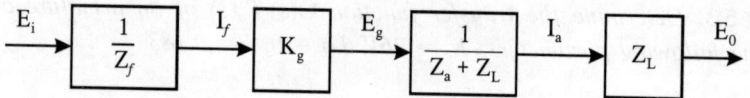

Fig. 5.38. Block diagram for Example 5.4

$$E_o(s)/E_i(s) = K_g Z_L/[Z_f(Z_a + Z_L)]$$

$$Z_a + Z_L = (1 + 0.5s) + (200 + 10s)$$

$$= (201 + 10.5s) = 201(1 + 0.52s)$$

$$E_o(s)/E_i(s) = 200 \times 200(1 + 0.05)/[100 (1 + 0.2s)((201 + 10.5s)]$$

$$= 1.99/(1 + 0.2s)(1 + 0.52s)$$

Example 5.5. *A separately excited 460V, 52.3 rad./s, 74.6kW D.C. servo motor is directly coupled to a load having an inertia of 106kg – m². The motor details are: the effective inductance of the armature circuit = 0.015H, mean value of the armature resistance = 0.1 ohm, inertia of the armature = 9.7kg – m². Derive the transfer function relating the output angular velocity to the applied voltage.*

Solution:

Given: $R_a = 0.1 \ \Omega$, $L_a = 0.015$ H, $J = 106 + 9.7 = 115.7$ kg $- m^2$, $\omega = 52.3$

The armature time constant is given by

$$\tau_a = L_a/R_a = 0.015/0.1 = 0.15$$

Since

$$E_a I_a = K_b I_a \theta',$$

$$K_b = E_a/\theta' = 460/52.3 = 8.8 \ \text{V/rad./s}$$

$$P = E_a I_a = K_b I_a \omega' = T \omega' = K_T I_a \omega'$$

Hence,

$$K_b = K_T \ \text{when} \ K_b \ \text{is in V/rad./s.}$$

$$\Omega(s)/E_a(s) = \frac{K_T}{JR_a s (1 + \tau_a s) + K_T K_b}$$

$$= 8.8/[115.7 \times 0.1s \ (1 + 0.15s) + 77.44]$$

$$= 0.76/[s(1 + 0.15s) + 6.69]$$

Example 5.6. *An a.c. servo motor has the following parameters at no load:* $K_v = 2.27$ *N − m/V,* $E_c = 12$ *V,* $K_\omega = -0.136$ *N − m/rad./s,* $J_m = 0.227$ *N-m/rad./s². Find (a) the initial acceleration, (b) Steady state speed, (c) Acceleration when the speed of the motor is 50 rad./s. (d) If the motor is having a constant load orque of 9 N–m, find (i) the initial acceleration, (ii) steady-state speed and (iii) acceleration when the speed is 50 rad./s.*

Solution:

Given:
$$K_v = 2.27 \ N - m/V, \ E_c = 12 \ V, \ K_\omega = - \ 0.136 \ N - m/rad./s,$$
$$J_m = 0.227 \ N - m/rad./s^2 \ T_L = 9 \ N - m,$$

(a) The initial torque is given by

$$T = K_v E_c + K_\omega \omega$$

$$= 2.27 \times 12 - 0.136 \times 0 = 27.24 \ N–m$$

The initial acceleration α is given by

$$\alpha = T/J_m = 27.24/0.227 = 120 \ \text{rad./s}$$

(b) In the steady-state, the speed being constant, there is no acceleration. Hence, the accelerating torque is zero. Therefore,

$$K_\omega \omega = K_v E_c = 27.24$$

$$\omega = 27.24/0.136 = 120 \ \text{rad./s}$$

(c) When $\omega = 50$ rad./s,

$$T = K_v E_c + K_\omega \omega$$

$$= 2.27 \times 12 - 0.136 \times 50 = 6.8 \ \text{N–m}$$

The acceleration α is given by

$$\alpha = T/J_m = 6.8/0.227 = 90 \ \text{rad./s}^2$$

(d) The load torque $T_L = 9\ N - m$. Hence, the accelerating torque is given by

$$T = K_v E_c + K_\omega\ \omega$$
$$T = 27.24 - 9 = 18.24$$

(i) The initial acceleration α is given by

$$\therefore \alpha = T/J_m = 18.24/0.227 = 80.4\ \text{rad./s}^2$$

(ii) In the steady-state, $T = 0$. Hence,

$$0 = 27.24 - 0.136\ \omega - 9 = 18.24 - 0.136\ \omega$$
$$\therefore \omega = 18.24/0/136 = 134.1\ \text{rad./s}$$

(iii) When $\omega = 50\ \text{rad./s}$,

$$T = 27.24 - 6.8 - 9 = 11.44\ N - m$$
$$\therefore \alpha = T/J_m = 11.44/0.227 = 50.4\ \text{rad./s}$$

5.19. SUMMARY

Analysis of control systems demands a thorough knowledge of the components of the system. In this chapter, we have first discussed the transducers and error detectors, the important system components that are extensively used in control systems. The system components that are discussed are potentiometers, LVDT, synchros, A.C. and D.C. tachogenerators, electronic amplifiers, Amplidyne, rotating amplifiers, D.C. and A.C. servomotors, Hydraulic servomotor, pneumatic actuators and gyroscope. We have studied the constructional features and operating characteristics of the potentiometer, defined its resolution, its application as an error detector and its limitations. Then, the LVDT have been explained in detail and its applications have been described. We then have described the constructional features and operating characteristics of synchro motor, synchro generator and synchro control transformer. The applications of synchros as data transmitters and error detectors have been presented. We have then carried out a comparative study of synchros and potentiometers as system components. The details of D.C. and A.C. tachogenerators have been described and their performance has been compared. We have then presented electronic and electric actuators and discussed their construction and operation. The armature and field control of D.C. generator are discussed and compared their performance. The D.C. and A.C. servomotors have been explained in detail and their characteristics are compared. Finally, we have also discussed the construction and operation of gyroscopes. The application of each of these components are also illustrated with suitable examples. Many practical systems have been described and their transfer functions obtained from the combinations of the component transfer functions.

REVIEW QUESTIONS

5.1. Define a transducer and an error detector.

5.2. What is meant by the resolution of a potentiometer?

5.3. Mention the applications of potentiometers.

5.4. What are the limitations of potentiometer?

5.5. What is LVDT?

5.6. What is a synchro generator?

5.7. What is a synchro motor?

5.8. What is a synchro control transformer?

5.9. Mention the applications of synchros.

5.10. Compare synchros and potentiometers.

5.11. What is a tachogenerator? Mention its types.

5.12. What are the advantages of d.c. tachogenerator?

5.13. What are the disadvantages of d.c. tachogenerator?

5.14. What is a rotating amplifier? Give an example.

5.15. What is an amplidyne?

5.16. What are the methods of exciting a d.c. servo motor?

5.17. What are the advantages of d.c. servo motor?

5.18. What are the disadvantages of d.c. servo motor?

5.19. What is a gyroscope?

EXERCISE

5.1. (a) Define the terms transducers and error detectors.

    (b) Explain how a pair of potentiometers can be used as an error detector.

    (c) What are the limitations of potentiometers?

5.2. Describe the constructional features of synchro-generator, synchro-motor and control transformer.

5.3. Show that the terminal voltages of a synchro-generator are single phase voltages.

5.4. (a) Explain the principle of operation of a synchro error detector.

    (b) What are the merits of synchro error detector compared to potentiometer error detector?

5.5. (a) Describe the constructional features of an A.C. tachogenerator.

    (b) Draw the schematic diagram of an A.C servomechanism using tachogenerator feedback.

5.6. (a) Draw the block diagram of the servomechanism in 5.5(b).

    (b) Obtain the transfer function of the above control system.

5.7. What are the important factors that differentiate precision D.C. tachogenerator from D.C. machines?

5.8. Bring out the merits and limitations of D.C. and A.C tachogenerators.

5.9. Derive the transfer function of a separately excited D.C. generator.

5.10. Describe the constructional features of an amplidyne and explain its operation.

5.11. Show that the amplidyne can be represented by a two-stage rotating amplifier and obtain its transfer function.

5.12. (a) Mention the special constructional features of D.C. servo motor.

      (b) Draw the schematic diagram of a D.C. servomechanism using D.C. servo motor and obtain its transfer function.

5.13. (a) Explain the two methods of controlling the speed of a D.C. servo motor.

      (b) Derive the transfer function in each case.

      (c) Compare the characteristics of the above two types of control.

5.14. Describe the constructional features and principle of operation of an A.C. servo motor.

5.15. (a) With a schematic diagram, describe the application of an A.C. servo motor.

      (b) Draw its block diagram and obtain its transfer function.

5.16. (a) Schematically represent a D.C. generator to study its control action and explain ow the action of amplification is taking place.

      (b) Derive the transfer function of the above system when delivering power to a resistance load.

PROBLEMS

5.1. A single turn potentiometer has a resistance of 1K ohm.

    (a) If the measured resistance at its midpoint is 515 ohm, what is its linearity?

    (b) If the measured resistance at its quarter point is 230 ohm, what is its linearity?

5.2. A two-turn, 200 windin g, 10K ohm potentiometer has a reference voltage of 20V applied across its terminals. Assume that the rotation of the potentiometer is full 720°. Determine the potentiometer constant or gain in

    (a) Volts/degree

    (b) resolution or volts/winding

    (c) Volts/rad

    (d) Volts/turn.

5.3. A ten-turn potentiometer having a resistance of 50K ohm with 1% linearity uses 40V supply.

    (a) Determine the potentiometer constant in volts/turn.

(b) Find the range of voltage at the mid-point setting. Assume unloaded condition.

(c) Assume that the potentiometer is perfectly linear. Find the voltage at the midpoint when it is loaded with (i) 500K ohm and (ii) 25K ohm.

5.4. A synchro control transformer rotor has a peak output of 60V.

(a) Assuming its characteristic is linear, it gives an output of 75% of its peak value when the rotor angle is 45°. Find its sensitivity in volts/deg.

(b) Determing the output when the rotor angle is 10° using the sensitivity obtained in (a).

(c) What is actual output for a rotor angle of 10° using the sine wave?

5.5. A synchro control transformer has an output voltage equal to 75 sin θ where θ is the rotor angle. What is the approximate sensitivity in the linear range?

5.6. In synchro generator (SG) and control transformer (CT) combination, the CT rotor output is given by 90 sin $(\theta_T - \theta_g)$ where θ_T and θ_g are the angle of CT rotor and that of SG rotor with respect to their zero positions respectively.

(a) With zero positions of both at 0°, find the CT rotor output when $\theta_g = 50°$ and $\theta_T = 120°$.

(b) In the above case, if the zero position of CT rotor is 0° and that of SG rotor is 30°, find the CT rotor output.

5.7. A separately excited d.c. motor has an armature inductance L_a and resistance R_a. The back emf E_b is proportional to motor-shaft speed. The torque T generated by the motor is proportional to the armature current I_a. J is the combined inertia of the motor armature and the load. B is the total viscous friction acting on the output shaft. Determine the transfer function between the input armature voltage E_a and the angular position of the output shaft θ. Assume K_b is the back-emf constant and K_T the torque constant.

5.8. A d.c. tachogenerator has an armature inductance L_a and resistance R_a. The output voltage is E_0. The back emf is $E_b = K_b \, (d\theta/dt)$. If it is connected to a load resistance R_L, determine its transfer function. $E_0(s)/\Theta(s)$.

5.9. The stall torque of a servo motor is 6 N-m at an armature voltage of 20V.

(a) Find its motor constant K_V.

(b) If the slope of the torque-speed curve is 0.5, Find the speed at no load.

5.10. An a.c. servometer has a moment of inertia of 1.36×10^{-4} kg-m². Its stalling torque is 0.136 N-m when the applied voltage is 10V. The no-load speed is 1593 rpm.

(a) Find the motor time constant.

(b) Find the transfer function $\Omega(s)/E_a(s)$ and $\Theta(s)/E_a(s)$.

5.11. A d.c. motor has a stalling torque of 400 N-m and a no-load velocity of 800 rad/sec when the applied voltage is 10 Volts. If the moment of inertia is. 0.8 kg-m², find the transfer function.

5.12. If the motor in Prob. 5.11 is connected to a load of 100 N-m through a gear with a ratio of 0.5,

(a) Find the steady state motor speed and load speed.

(b) Find the value of the load torque that would stall the motor.

5.13. The gyroscope often used in autopilots, automatic gun sights etc. is shown in Fig. 5.31. Assume that the speed of the rotor is constant and that the total developed torque about the output axis is

$$T_0 = H \, d\theta_i/dt$$

where H is a constant. The inner gimbal exerts a moment of inertia J about the output axis. Find the transfer function $\Theta_0(s)/\Theta_1(s)$.

5.14. The differential equation describing the dynamic operation of one-degree-freedom gyroscope shown in Fig. 5.31 is

$$J d^2\theta_0/dt^2 + B d\theta_0/dt + K = H\omega_i$$

where ω_i is the angular velocity of the gyroscope about the input axis., θ_0, the angular position of the spin-axis is the measured output of the gyroscope. H is the angular momentum stored in the spinning wheel, J is the moment of inertia of the wheel about the output axis, B is the viscous friction coefficient about the output axis and K is the spring constant of the restraining spring attached to the spin axis.

(a) Determine the transfer function $\Theta_0(s)/\Omega_i(s)$ and show that the steady state output is proportional to the magnitude of a constant rate input. This type of gyroscope is called a *rate gyro*.

(b) Determine the transfer function between $\Theta_0(s)/\Omega_i(s)$ with the restraing spring removed. Show that the output is proportional to the intergral of the input rate. This type of gyroscope is called an *intergrating gyro*.

5.15. A magnetic amplifier with a low output impedance is in cascade with a low-pass filter and a preamplifier. The amplifier has a high input impedance and a gain of 1 and is used as error detector.

The transfer function of the magnetic amplifier is $20/(s + 1)$ and the value of the resistor in the low pass filter is 50Ω.

(a) Find the value of capacitor in the low pass filter so that the damping ratio of the closed loop transfer function is 0.7.

(b) Calculate the settling time of the resulting system.

5.16. The transfer function of a large microwave antenna with drive motor and amplidyne is approximated by

$$G(s) = 100/(s^2 + 14s + 100)$$

This is preceded by a magnetic amplifier with a transfer function $G_1 = K_a/(1 + 0.2s)$.

(a) Determine the sensitivity of the system to a change in K_a.

(b) If a disturbance $D(s) = 15/s$ occurs after the magnetic amplifier, determine K_a so that steady-state error is $0.2°$ when $R(s) = 0$.

(c) Find the error if there is no feedback.

Chapter **6**

TRANSIENT RESPONSE ANALYSIS

6.1. INTRODUCTION

The first step in analyzing a control system is to derive a mathematical model of the system. The methods of obtaining the model of a system have been discussed in detail in Chapters 3 and 5. Once such a model is obtained, we may analyze the performance of the system by various methods. One such method is the time-domain analysis. Since control systems are inherently time-dependent systems, the time domain response is the prime interest in the analysis and design of control systems. The input signals to many practical control systems are not known ahead of time, and in many cases they are random in nature and hence cannot be expressed analytically. In some special cases only, the input signal is known in advance and may be represented by analytical expressions or as a specific curve as in the case of the automatic cutting tools. For the purposes of analysis and design, we must have a basis for comparing the performance of various control systems. One such way of doing this is to specify standard test signals and compare the responses of various systems to these input signals. The use of test signals is justified because there exists a correlation between the characteristics of a system to a typical test input signal and the capability of the system to cope with actual input signals.

6.2. TYPICAL TEST SIGNALS

The commonly used test input signals are:

  (a) Unit step function
  (b) Unit ramp function
  (c) Unit parabolic function
  (d) Unit impulse function.

They are defined below.

(a) *Unit step input function:* The unit step function represents an instantaneous change of unit magnitude in the reference input. For instance, in a position control system, this signal represents a sudden rotation of the input shaft by one radian or degree. The step input is also termed position or displacement input. The unit step function is analytically defined as

$$r(t) = 1 \qquad t > 0$$
$$= 0 \qquad t < 0 \qquad \qquad \text{... (6.1)}$$

It is *not defined at t = 0*. It is usually represented by

$$r(t) = u(t) \qquad \qquad \text{... (6.2)}$$

Taking the Laplace transform,

$$R(s) = 1/s \qquad \qquad \text{... (6.3)}$$

If the step function is of magnitude R. then

$$R(s) = R/s \qquad \qquad \text{... (6.4)}$$

The unit step function is shown in Fig. 6.1(a).

(b) *Unit Ramp Input Function*: The unit ramp function is one which increases linearly with respect to time with a slope of unity. It is shown in Fig. 6.1(b) and is defined as

$$r(t) = t \qquad t \geq 0 \qquad \qquad \text{... (6.5)}$$
$$= 0 \qquad t < 0$$

It is usually represented by

$$r(t) = tu(t) \qquad \qquad \text{... (6.6)}$$

Its Laplace transform is

$$R(s) = 1/s^2 \qquad \qquad \text{... (6.7)}$$

If the slope of ramp function is R, then

$$R(s) = R/s^2 \qquad \qquad \text{... (6.8)}$$

The ramp input is also termed as *velocity input*.

(c) *Unit parabolic Input Function*: The graphical representation of the parabolic function is shown in Fig. 6.1(c). It is defined as

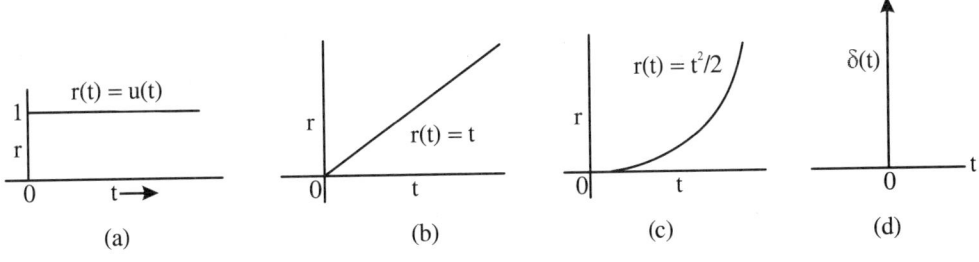

Fig. 6.1. (a) Unit step function (b) Unit ramp function
(c) Unit parabolic function (d) Unit impulse function

$$r(t) = t^2/2 \qquad t \geq 0$$
$$= 0 \qquad t < 0 \qquad \qquad \text{... (6.9)}$$

It is represented by

$$r(t) = \frac{1}{2} t^2 u\ (t) \qquad \qquad \text{... (6.10)}$$

Its Laplace transform is

$$R(s) = 1/s^3 \qquad \qquad \text{... (6.11)}$$

The parabolic function represents *acceleration*. If the acceleration is R, then

$$R(s) = R/s^3 \qquad \qquad \text{... (6.12)}$$

(d) *Unit Impulse Function*: The unit impulse function is also known as Dirac-delta funtion. It is shown in Fig 6.1(d) and is denoted by $\delta(t)$. It is defined as

$$\delta(t) = \infty \qquad t = 0$$
$$\delta(t) = 0 \qquad t \neq 0$$

$$\int_{-\infty}^{\infty} \delta(t)dt = 1 \qquad \qquad \text{... (6.13)}$$

The Laplace transform of the unit impulse function is

$$\mathcal{L}(\delta(t)) = 1. \qquad \qquad \text{... (6.14)}$$

Eq. (6.14) represents the area of the unit impulse function. If the impulse function has an area of A, then

$$\mathcal{L}(A\delta(t)) = A. \qquad \qquad \text{... (6.15)}$$

The above test signals are known as *singularity functions*. They can be obtained from one another by successive integration or differentiation. For example, the integral of the impulse function is the step function, the integral of the step function is the ramp function and the integral of the ramp function is the parabolic function.

For the purpose of analyzing a system, the test signals are determined according to the form of the input to which the system is subjected to most frequently under normal operation. If the inputs to a system are gradually changing functions of time, then a ramp function is a good test signal. If a system is subjected to sudden disturbances, a step function may be used as the test signal. For a system is subjected to shock inputs, the impulse function is a good test signal. The control systems designed on the basis of test signals generally perform satisfactorily to actual inputs. By the use of these test signals, the performance of various systems may be compared on the same basis.

6.3. TRANSIENT AND STEADY -STATE RESPONSE

The time response of a control system consists of two parts:

(i) The transient response

(ii) The steady-state response.

The transient response is the response of the system from the initial state to the steady state. It disappears as time *t* approaches infinity. All physical systems do exhibit transient phenomenon because of inherently present energy storing elements such as inertia, mass and inductance. Transient response reveals the nature of the system response during the transition period and also gives an indication of the speed of response. The transient response is significant for the dynamic behaviour of the system and hence its control is very important.

The steady-state response is the manner in which the system output behaves as t approaches infinity and is that part of the response which remains after the transient has subsided. The steady-state response is also important since it gives an indication of the final accuracy of the system. If the output of the system at the steady-state does not exactly agree with the input, the system is said to have a steady-state error. *The time-domain analysis of a system involves the evaluation of the transient and steady-state responses as well as the time required to reach the steady-state and the steady-state error.* The system design in time-domain involves the design of controllers that meet the time-domain specifications which are usually in terms of the transient and steady-state performances.

6.4. FIRST-ORDER SYSTEMS

Consider the *RC* lag network shown in Fig 6.2. The transfer function is given by

$$E_0(s)/E_i(s) \;=\; C(s)/R(s) = \frac{1/Cs}{R+1/Cs} = \frac{1}{1+RCs} = \frac{1}{1+\tau s} \qquad \dots (6.16)$$

The highest power of *s* in the denominator polynomial of the transfer function is known as the *order of the system.*

We shall analyze the response of this system to unit impulse, unit step, unit ramp and unit parabolic functions, *assuming all initial conditions to be zero.* All systems having the same transfer function will have the same output response to the same input. Only the physical interpretation is different.

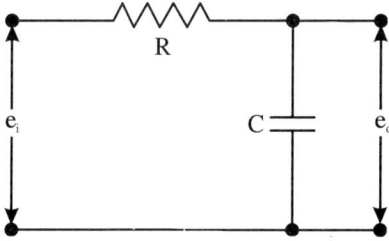

Fig. 6.2. RC lag network

6.4.1. Unit Impulse Response

For unit impulse input, $R(s) = 1$. The output $C(s)$ is given by

$$C(s) = R(s) [1/(1 + \tau s)] \qquad \qquad \dots (6.17)$$
$$= 1/(1 + \tau s) \qquad \qquad \dots (6.18)$$

Taking the inverse Laplace transform, we get

$$c(t) = 1/\tau \cdot e^{-t/\tau} \quad t \geq 0 \qquad \qquad \dots (6.19)$$

Here $\tau = RC$ is the **time constant** of the system.

The response is shown in Fig. 6.3. The important characteristics are:

(a) The initial value is $1/\tau$

(b) The initial slope is $-1/\tau^2$

(c) The transient part is $(1/\tau) \cdot e^{-t/\tau}$

(d) There is no steady-state part. The response approaches zero as t approaches infinity.

(e) The slope decreases monotonically from $-1/\tau^2$ at $t = 0$ to zero at $t = \infty$.

(f) The response curve decreases to 36.8, 13.5, 5, 1.8 and 0.7% of its initial value at $t = \tau$, 2τ, 3τ, 4τ, and 5τ respectively. The length of the time taken to decrease to 2% of its initial value is called the *setting time* t_s and equal to 4τ approximately.

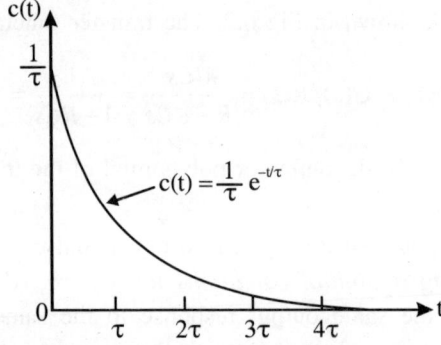

Fig. 6.3. Unit impulse response of a first-order system

6.4.2. Unit Step Response

The Laplace transform of $r(t) = u(t)$ is $R(s) = 1/s$. Hence

$$C(s) = \frac{1}{s(1 + s\tau)} \qquad \qquad \dots (6.20)$$

Expanding $C(s)$ into partial fractions, we get

$$C(s) = 1/s - \tau/(1+s\tau) = \frac{1}{s} - \frac{1}{s+1/\tau} \qquad \text{... (6.21)}$$

Taking the inverse Laplace transform, the response $c(t)$ is given by

$$c(t) = 1 - e^{-t/\tau} \quad t \geq 0 \qquad \text{... (6.22)}$$

The response curve $c(t)$ is shown in Fig. 6.4

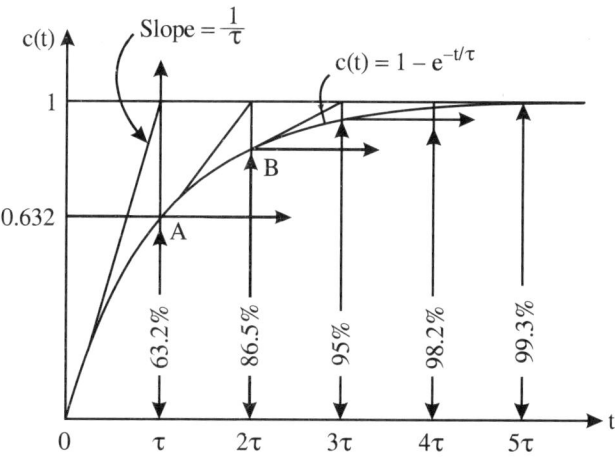

Fig. 6.4. Unit step response of first-order system

The response $c(t)$ is initially zero, increases exponentially and finally becomes unity. The important characteristics of such exponential response curves are:

(a) The transient part of the response is $e^{-t/\tau}$ which disappears as t approaches infinity.

(b) The steady-state error $e_{ss} = r(t) - c(\infty) = 0$

(c) When $t = \tau$, $c(t) = 1 - e^{-1} = 0.632$. The response of the system has reached 63.2% of the final value at $t = \tau$. The smaller the time constant τ, the faster is the system response.

(d) The slope of the response curve at $t = 0$ is given by

$$dc(t)/dt = (1/\tau) e^{-t/\tau} \qquad \text{... (6.23)}$$

The initial slope is $1/\tau$. The output will reach the final value at $t = \tau$ if it maintained its initial speed of response.

(f) The slope of the response curve decreases monotonically from $1/\tau$ at $t = 0$ to zero at $t = \infty$. This is apparent from Eq. (6.23).

(g) The response curve reaches 63.2, 86.5, 95, 98.2 and 99.3% of its final value at $t = \tau$, 2τ, 3τ, 4τ and 5τ respectively. Thus, the response curve remains within 2% of its final value for $t \geq 4\tau$. For all practical purposes, the system is considered to have reached the steady-state when its response remains within 2% of its final value. *The time to reach the steady-state is called the setting time t_s and is given by*

$$t_s = 4\tau \qquad \qquad \text{... (6.24)}$$

(*h*) The transient part of the response curve is the graph from $t = 0$ to $t = 4\tau$.

(*i*) The steady-state part of the curve is the graph from $t = 4\tau$ to $t = \infty$.

The first-order system follows a unit step function with no error.

6.4.3. Unit Ramp Response

The Laplace transform of the unit ramp function $r(t) = t\, u(t)$ is $R(s) = 1/s^2$. Therefore, the output of the system is

$$C(s) = \frac{1}{s^2(1+s\tau)} \qquad \qquad \text{... (6.25)}$$

$$= \frac{1}{s^2} - \frac{\tau}{s} + \frac{\tau^2}{1+s\tau} = \frac{1}{s^2} - \frac{\tau}{s} + \frac{\tau}{s+1/\tau} \qquad \qquad \text{... (6.26)}$$

The time response $c(t)$ is given by

$$c(t) = t - \tau + \tau\, e^{-t/\tau} \qquad t \geq 0 \qquad \qquad \text{... (6.27)}$$

The response of the system is shown in Fig 6.5. The important characteristics are:

(a) The transient part of the response is $\tau\, e^{-t/\tau}$

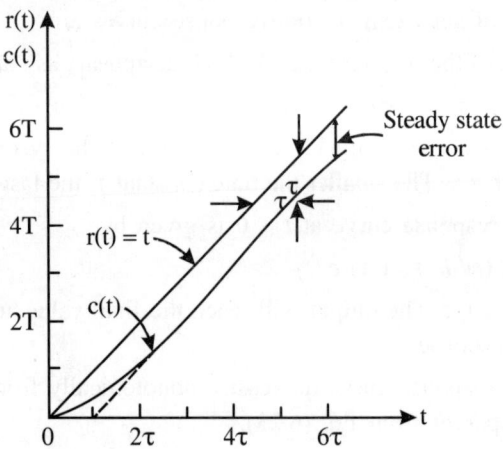

Fig. 6.5. Unit ramp response of first-order system

(b) The steady-state part is $(t - \tau)$

(c) The steady-state error is τ.

(d) At $t = \tau$, the error is 0.368τ. The smaller the time constant, the smaller is the steady-state error.

(e) The slope of the response curve increases monotonically from zero at $t = 0$ to unity at $t = \infty$.

(f) The settling time t_s is equal to 4τ.

The first order system follows the ramp input with a steady-state error equal to τ.

6.4.4. Unit Parabolic Response

The Laplace transform of unit parabolic function $r(t) = \frac{1}{2} t^2 u(t)$ is $R(s) = 1/s^3$. Therefore, the output response is

$$C(s) = \frac{1}{s^3(1+s\tau)} \qquad \qquad \text{... (6.28)}$$

$$= \frac{1}{s^3} - \frac{\tau}{s^2} + \frac{\tau^2}{s} - \frac{\tau^3}{1+s\tau} = \frac{1}{s^3} - \frac{\tau}{s^2} + \frac{\tau^2}{s} - \frac{\tau^2}{s+1/\tau}$$

Taking the inverse Laplace transform

$$c(t) = (1/2)t^2 - \tau t + \tau^2 - \tau^2 e^{t/\tau} \qquad t \geq 0 \qquad \text{... (6.30)}$$

The steady-state error increases to a large value as t approaches infinity. In other words, *the first-order system will not be able to follow the parabolic input.*

6.4.5. Relationship among Responses

An important property of a linear time-invariant system is that its responses to the test signals are related to each other. If the system response for any one test signal is known, the response to other test signals can be easily derived. For example, for the unit parabolic input, the response $c(t)$ is given by Eq. (6.30). The unit ramp input is the derivative of the unit parabolic input. Hence the response of the system to unit ramp input can be obtained by differentiating Eq. (6.30). Thus, the unit step response can be obtained by differentiating Eq. (6.27) and the unit impulse response is the time-derivative of unit step response. Similarly, the step response of a system can be obtained by integrating its unit impulse response and by determining the integration constant from zero output initial conditions

6.5. SECOND-ORDER SYSTEMS

In this section, we shall first consider a position control system, obtain its transfer function, generalize the transfer function and obtain the response of the generalized transfer function so that the analysis may be extended to other second-order systems.

Consider the position control system shown in Fig. 6.6(a). The objective of this control system is to position the mechanical load in accordance with the reference input. A pair of potentiometers functions as as a transducer and error-sensing device. The angular displacement θ is the reference input to the system. The output signal of the input potentiometer is proportional to the angular displacement. The output shaft position determines the angular position c of the output potentiometer. The error signal is the difference between e_r and e_c. This error signal is amplified by the amplifier whose gain in K_a. The amplifier output is applied to the armature circuit of the D.C. motor. The field excitation is kept constant.

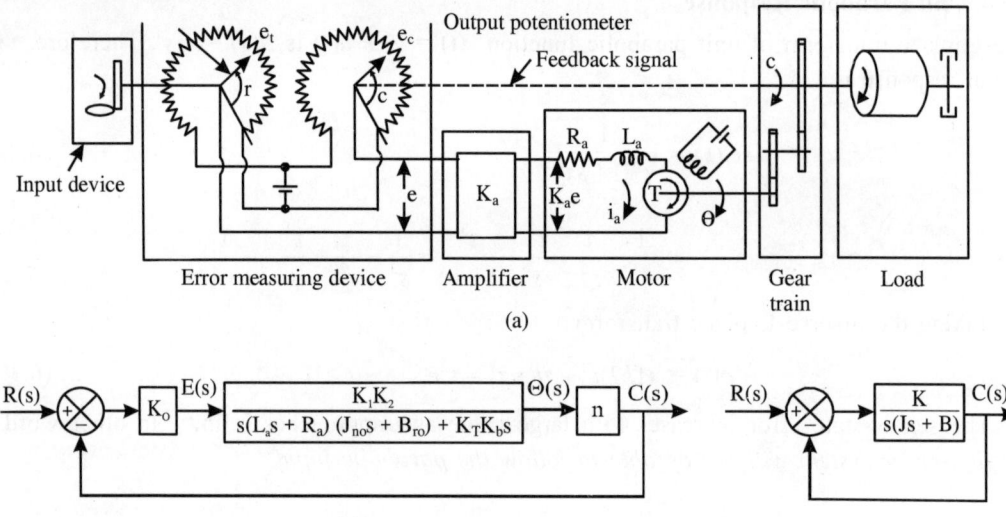

(a)

$$\frac{C(s)}{R(s)} = \frac{K}{Js^2 + Bs + K} = \frac{K/J}{s^2 + (B/J)s + (K/J)} \qquad \text{... (6.31)}$$

Fig. 6.6. (a) Position control system (b) Block diagram (c) Reduced block diagram

If an error exists between the two shaft positions, the motor develops a torque to rotate the mechanical load in such a way as to reduce the error to zero. The block diagram of the system is shown in Fig. 6.6(b) and the reduced block diagram in Fig. 6.6 (c). The closed-loop transfer function assuming $La \simeq 0$ is given by

$$\frac{C(s)}{R(s)} = \frac{K}{Js^2 + Bs + K} = \frac{K/J}{s^2 + (B/J)s + (K/J)} \qquad \text{... (6.31)}$$

where

$K = K_s\, K_a\, K_t n/R_a$

$J\ (= J_m + n^2 J_L)$ is the moment of inertia referred to the motor shaft, J_m is the inertia of the motor, J_L is the inertia of load and gear train.

$B [=B_m + n^2B_L + (K_t K_b/R_a)]$ is the viscous friction referred to the motor shaft.

B_m is the viscous friction of the motor, B_L is that of load and the gear train.

n is the gear ratio.

K_a is the gain of the amplifier.

K_T is the torque constant of the motor.

K_b is the back emf constant of the motor.

K_s is the sensitivity of the potentiometers.

Equation (6.31) may be written in the factored form as

$$\frac{C(s)}{R(s)} = \frac{K/J}{\left(s + B/2J + \sqrt{(B/2J)^2 - K/J}\right)\left(s + B/2J - \sqrt{(B/2J)^2 - K/J}\right)} \qquad ...(6.32)$$

The closed-loop poles are real if $(B^2 - 4JK) \geq 0$ and complex if $(B^2 - 4JK) < 0$. The condition $(B^2 - 4JK) = 0$ is called *critically damped*, $(B^2 - 4JK) > 0$ is *over-damped*, $(B^2 - 4JK) < 0$ is *under-damped*. If the viscous friction $B = 0$, the system is *undamped*. Substitution of the following notations result in the standard form of a general second-order closed-loop transfer function

$$K/J = \omega_n^2, B/J = 2\zeta\omega_n = 2\alpha \qquad ...(6.33)$$

$$\frac{C(s)}{R(s)} = \frac{\omega_n^2}{s^2 + 2\zeta\omega_n s + \omega_n^2} \qquad ... (6.34)$$

where ω_n is the undamped natural frequency, ζ is the damping ratio of the system and α is the damping factor since it controls the damping of the system and the rate of rise and fall of the time response of the system. The damping ratio ζ is the ratio of the actual damping B to the critical damping $B_c = 2\sqrt{JK}$

$$\zeta = B/B_c = B/2\sqrt{JK} \qquad ... (6.35)$$

6.5.1. Unit Step Response

The generalized closed-loop transfer function of a second-order system may be represented in the block diagram form as shown in Fig. 6.7. The dynamic behaviour of second-order systems can thus be described in terms of two variables ζ and ω_n. If ζ lies between 0 and 1, the closed-loop poles are complex conjugates and lie in the left-half of the s-plane. The system is then under-damped and the transient response is decaying oscillations. If $\zeta = 1$, the system is critically damped. If $\zeta > 1$, the system is over-damped, If $\zeta = 0$, the transient response does not die out and the system response continues to oscillate.

We shall now obtain the response of the second-order system shown in Fig. 6.7 for (i) under-damped, (ii) undamped, (iii) critically damped and (iv) over-damped cases when the input function is a unit step.

(*a*) **Under-damped case** (0 < ζ < 1): In this case, the closed-loop transfer function $C(s)/R(s)$ is written as

$$\frac{C(s)}{R(s)} = \frac{\omega_n^2}{(s+\zeta\omega_n + j\omega_d)(s+\zeta\omega_n - j\omega_d)} \qquad \ldots (6.36)$$

where $\omega_d = \omega_n\sqrt{1-\zeta^2}$ is the damped natural frequency of the system. For unit step input, $C(s)$ is given by

$$C(s) = \frac{\omega_n^2}{s(s^2 + 2\zeta\omega_n s + \omega_n^2)} \qquad \ldots (6.37)$$

Taking the inverse transform of $C(s)$ (see Example 2.13)

$$e(t) = 1 - \frac{e^{-\zeta\omega_n t}}{\sqrt{1-\zeta^2}} \sin(\omega_d t + \theta) \qquad t \geq 0 \qquad \ldots (6.38)$$

where $\theta = \cos^{-1}\zeta = \tan^{-1}(\sqrt{1-\zeta^2}/\zeta)$. The above response shows that the frequency of the transient response is ω_d which varies with ζ. The error signal is given by

$$e(t) = r(t) - c(t)$$

$$= \frac{e^{-\zeta\omega_n t}}{\sqrt{1-\zeta^2}} \sin(\omega_d t + \theta) \qquad \ldots (6.39)$$

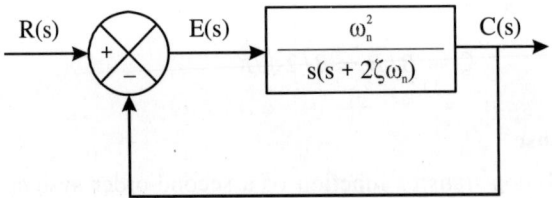

$$R(s) \qquad E(s) \qquad \frac{\omega_n^2}{s(s + 2\zeta\omega_n)} \qquad C(s)$$

Fig. 6.7. Block diagram of a second-order system

The error signal also exhibits a damped sinusoidal oscillation of frequency ω_d. At steady state, as *t* approaches ∞, the error is zero. Fig. 6.8 shows the variation of $c(t)$ as a function of time *t*.

(*b*) **Undamped case** (ζ = 0): The response of the second-order system for unit step input when the system is undamped may be obtained by substituting ζ = 0 in Eq. (6.38). Hence

$$c(t) = 1 - \sin(\omega_n t + \pi/2)$$

$$= 1 - \cos \omega_n t$$

The output response thus continues to oscillate at the undamped natural frequency of the system. The response is shown in Fig. 6.8.

(*c*) **Critically Damped case:** ($\zeta = 1$) In this case, the two poles of the system are negative, real and equal. For unit step input, $C(s)$ is

$$C(s) = \frac{\omega_n^2}{s(s + \omega_n)^2} \qquad\qquad \text{... (6.41)}$$

Therefore,

$$c(t) = 1 - e^{-\omega_n t}(1 + \omega_n t) \qquad\qquad t \geq 0 \qquad\qquad \text{... (6.42)}$$

The time response is shown in Fig 6.8.

(*d*) **Over-damped case** ($\zeta > 1$): In this case the two closed-loop system poles are negative real and unequal. For unit step input, the response $C(s)$ is given by

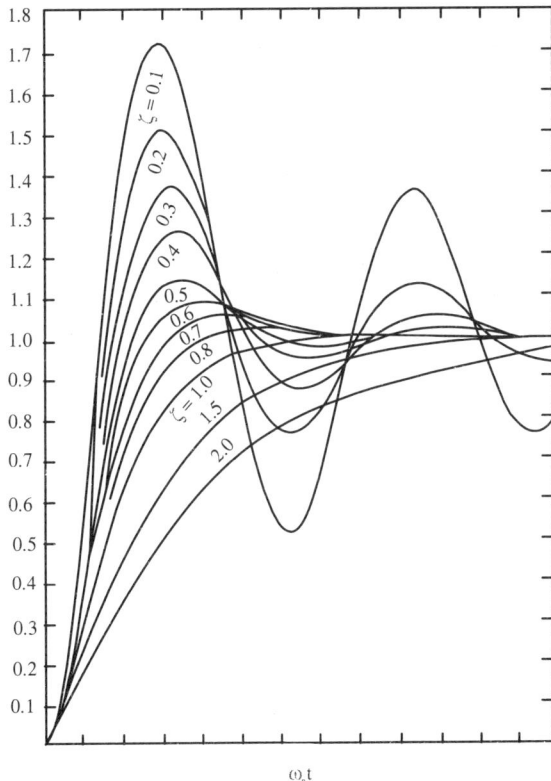

Fig. 6.8. Unit step response of a second-order system

$$C(s) = \frac{\omega_n^2}{s(s^2 + 2\zeta\omega_n s + \omega_n^2)} \qquad \qquad \text{... (6.43)}$$

$$= \frac{\omega_n^2}{s(s + p_1)(s + p_2)} \qquad \qquad \text{... (6.44a)}$$

where $p_1 = \omega_n(\zeta + \sqrt{\zeta^2 - 1})$ and $p_2 = \omega_n(\zeta - \sqrt{\zeta^2 - 1})$ (6.44b)

In partial fraction form,

$$C(s) = 1/s + \frac{\omega_n}{2p_1\sqrt{\zeta^2 - 1}\,(s + p_1)} - \frac{\omega_n}{2p_2\sqrt{\zeta^2 - 1}\,(s + p_2)} \qquad \text{... (6.45)}$$

Taking the inverse Laplace transform and rearranging,

$$c(t) = 1 + \frac{\omega_n}{2\sqrt{\zeta^2 - 1}}\left(\frac{e^{-p_1 t}}{p_1} - \frac{e^{-p_2 t}}{p_2}\right) \qquad \text{... (6.46)}$$

Thus the response has two transient terms with two time constants τ_1 and τ_2 given by

$$\tau_1 = 1 / \left[\omega_n(\zeta - \sqrt{\zeta^2 - 1})\right]$$

$$\tau_2 = 1 / \left[\omega_n(\zeta + \sqrt{\zeta^2 - 1})\right] \qquad \text{... (6.47)}$$

For values of $\zeta \gg 1$, $\tau_1 \ll \tau_2$ and hence τ_1 may be neglected. This implies that the pole $-p_2$ is located very much closer to the $j\omega$-axis than $-p_1$. This approximation reduces the second-order system to a first-order one with an approximate transfer function.

$$\frac{C(s)}{R(s)} = \frac{p_2}{s + p_2} \qquad \qquad \text{... (6.48)}$$

In deriving the above transfer function, it is assumed that the initial and final values of Eqs. (6.44) and (6.48) are the same. The unit step response of Eq. (6.48) is given by

$$c(t) = 1 - e^{-p_2 t} \qquad \qquad \text{... (6.49)}$$

The response of a second-order systems for values of ζ varying from zero to 2 is shown in Fig. 6.8. These curves correspond to $c(t)$ given by Eq. (6.39), (6.40), (6.42) and (6.46). The time response of a second-order system with ζ between 0.5 and 0.8 is faster than the critically damped or over-damped response. An over-damped response is always sluggish in responding to any input.

6.5.2. Roots of Characteristic Equation and Time Response

The general form of the closed-loop transfer function of a second-order system is given by

$$\frac{C(s)}{R(s)} = \frac{\omega_n^2}{s^2 + 2\zeta\omega_n s + \omega_n^2} \qquad \text{... (6.50)}$$

The denominator polynomial of $C(s)/R(s)$ when equated to zero is called the *characteristic equation*. Thus the general expression for the characteristic equation of a second-order system is

$$s^2 + 2\zeta\omega_n s + \omega_n^2 = 0 \qquad \text{... (6.51)}$$

This equation characterizes the behaviour of a second-order system in the time-domain and hence called characteristic equation. The roots of the characteristic equation are the poles of the closed-loop transfer function $C(s)/R(s)$. The two roots of the characteristic equation are

$$s_1, s_2 = -\zeta\omega_n \pm j\omega_n \sqrt{1-\zeta^2} \qquad 0 < \zeta < 1$$

$$= -\alpha \pm j\omega_d \qquad \text{... (6.52)}$$

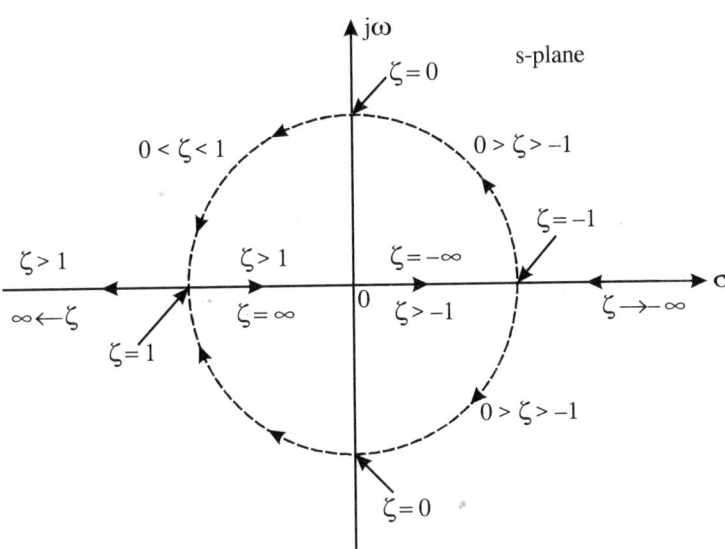

Fig. 6.9. Relationship between poles, and α, ζ, ω_n, and ω_d

The location of the roots and the relationship between the roots and a, ζ, ω_n and ω_d are shown in Fig. 6.9. The effect of the roots of the characteristic equation on the damping of the second-order system is illustrated in Fig. 6.10. When $\zeta = 0$, the two roots are complex conjugate pair and are located on the $j\omega$- axis. This represents undamped case and the system response is sustained oscillations. The system in this condition is said to be *marginally or critically stable*. For $0 < \zeta < 1$, the roots are complex conjugate pairs with negative real parts. The transient response is oscillatory and finally dies out. When $\zeta = 1$, the roots are negative real and equal. The response of such a system exhibits little oscillation and the system is said to be *critically damped*. For $\zeta > 1$, the roots are negative and unequal. The response is sluggish with no oscillations whatsoever. Such a system is called *over-damped system*. For $\zeta < 0$, the real part of the roots is positive and the system response increases indefinitely. Such a system is known as *unstable system*. The location of the roots of the characteristic equation in the s-plane and the corresponding time responses are shown in Fig. 6.10. Thus, *for a system to be stable, the roots of the characteristic equation must lie in the left half of the s-plane.*

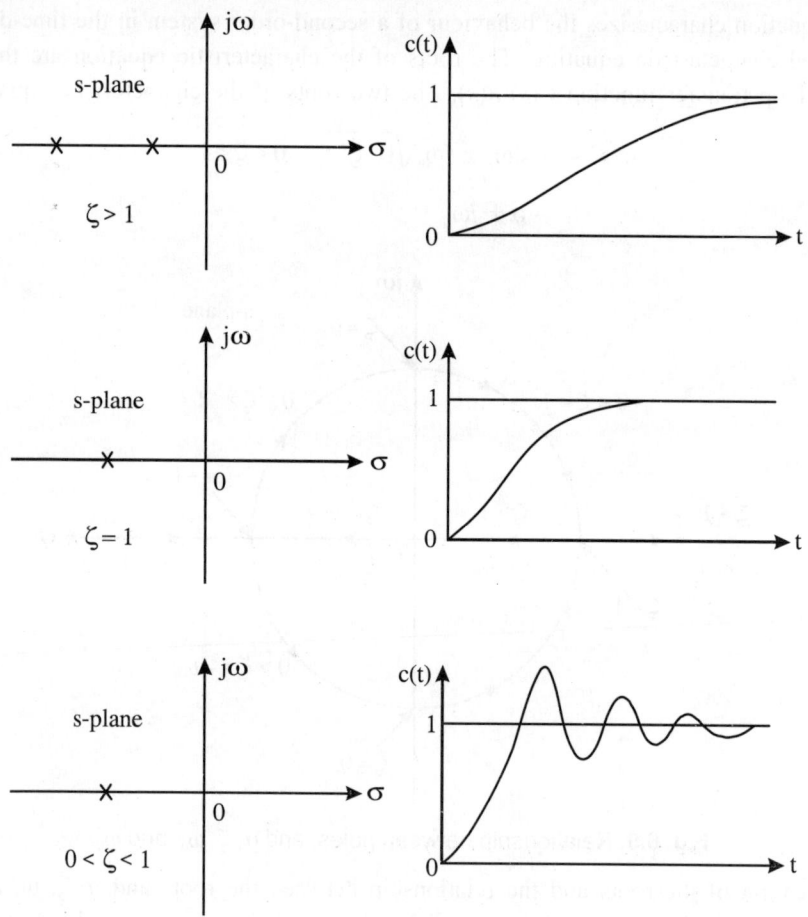

Fig. 6.10: Pole location and system response

6.5.3. Definitions of Transient Response Specifications

Since control systems are inherently dynamic systems, the desired performance characteristics are usually specified in terms of time-domain quantities for specific input signal. Frequently, they are specified in terms of the transient response to a unit step input as it is easy to generate. The time response of a system to a unit step input depends on the initial conditions. For convenience in comparing the transient responses of various systems, all initial conditions are assumed to be zero. The important transient response specifications are:

1. Delay time t_d
2. Rise time t_r
3. Peak time t_p
4. Maximum overshoot M_o
5. Setting time t_s

These specifications are defined below and illustrated in Fig. 6.11.

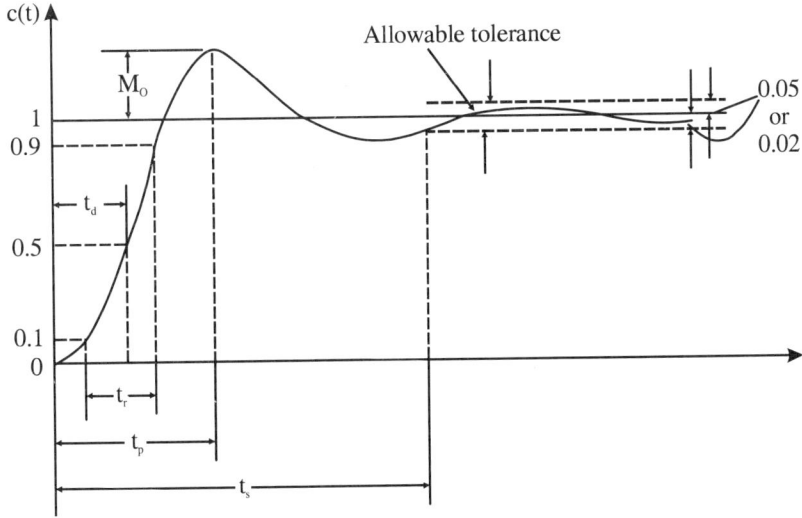

Fig. 6.11. Transient Response Specifications

(*i*) **Delay time t_d:** The delay time is the time required for the time response to reach 50% of the final value for the first time.

(*ii*) **Rise Time t_r:** The rise time is the time required for the time response to rise from 10% to 90% or 0% to 100% of its final value for the first time. The 10% to 90% rise time is normally used for over-damped systems and the 0% to 100% rise time for under-damped systems.

(*iii*) **Peak time t_p:** The peak time is the time required for the response to reach the first peak of the overshoot.

(iv) Maximum Over-shoot M_o: The maximum overshoot is the maximum peak value of the response curve measured from unity. The maximum percentage overshoot is given by

$$\text{Maximum percentage overshoot} = \frac{c(t_p) - c(\infty)}{c(\infty)} \times 100\% \qquad \ldots (6.53)$$

This is indicative of the relative stability of the system.

(v) Setting time t_s: The settling time is the time required for the response to reach and stay within ± 2% or ±5% of the final value. The settling time depends on the largest time constant of the system.

6.5.4. Expressions for Transient Response Specifications

We shall now obtain the transient response specifications of the second-order system described by Eq. (6.34) in terms of ζ and ω_n. The system is assumed to be under-damped.

(i) Rise time t_r: The response $c(t)$ for an under-damped system at $t = t_r$ is $c(t_r) = 1$. From Eq. (6.38)

$$c(t_r) = 1 = 1 - \frac{e^{-\zeta\omega_n t_r}}{\sqrt{1-\zeta^2}} \sin(\omega_d t_r + \theta)$$

or,
$$\qquad \ldots (6.54)$$

$$\frac{e^{-\zeta\omega_n t_r}}{\sqrt{1-\zeta^2}} \sin(\omega_d t_r + \theta) = 0$$

which implies that

$$(\sin \omega_d t_r + \theta) = 0$$

i.e.,

$$\sin \omega_d t_r \, \cos\theta + \cos \omega_d t_r. \, \sin\theta = 0$$

or,

$$\tan \omega_d t_r = -\tan\theta = \tan(\pi - \theta)$$

Therefore

$$t_r = \frac{\pi - \theta}{\omega_d} \qquad \ldots (6.55)$$

where

$$\tan \theta = \sqrt{1-\zeta^2} / \zeta$$

(ii) Peak time t_p : At $t = t_p$, the slope of the response curve is zero. Therefore, by equating the time-derivative of $c(t)$ in Eq. (6.38) to zero, we obtain the peak time.

$$\frac{dc(t)}{dt}\bigg|_{t=t_p} = -\frac{e^{-\zeta\omega_n t_p}}{\sqrt{1-\zeta^2}} \cdot \omega_n\sqrt{1-\zeta^2} \cdot \cos(\omega_d t_p + \theta) + \frac{\zeta\omega_n}{\sqrt{1-\zeta^2}} \cdot e^{\zeta\omega_n} t_p \sin(\omega_d t_p + \theta) = 0$$

On simplification, we get

$$\tan(\omega_d t_p + \theta) = \frac{\sqrt{1-\zeta^2}}{\zeta} = \tan\theta = \tan(n\pi + \theta), \ n = 0, 1, 2, \ \dots$$

or,

$$\omega_d t_p = n\pi$$

Since the peak time corresponds to the first peak overshoot, $n = 1$ and

$$\omega_d t_p = \pi$$

$$\therefore \quad t_p = \frac{\pi}{\omega_d} = \frac{\pi}{\omega_n\sqrt{1-\zeta^2}} \qquad \dots (6.56)$$

(iii) Maximum overshoot M_o: The maximum overshoot occurs at $t = t_p = \pi/\omega_d$. Therefore,

$$M_0 = c(t_p) - 1$$

$$= -\frac{e^{-\zeta\omega_n \pi/\omega_d}}{\sqrt{1-\zeta^2}} \sin\left(\omega_d . \pi/\omega_d + \theta\right)$$

$$= \frac{e^{-\pi\zeta/\sqrt{-\zeta^2}}}{\sqrt{1-\zeta^2}} \sin\theta \qquad \dots(6.57)$$

$$= e^{-\pi\zeta/\sqrt{1-\zeta^2}}$$

since $\sin\theta = \sqrt{1-\zeta^2}$. The maximum percentage overshoot is

$$\text{Maximum percentage overshoot} = e^{-\pi\zeta/\sqrt{1-\zeta^2}} \times 100\% \qquad \dots(6.58)$$

The maximum overshoot depends on ζ only.

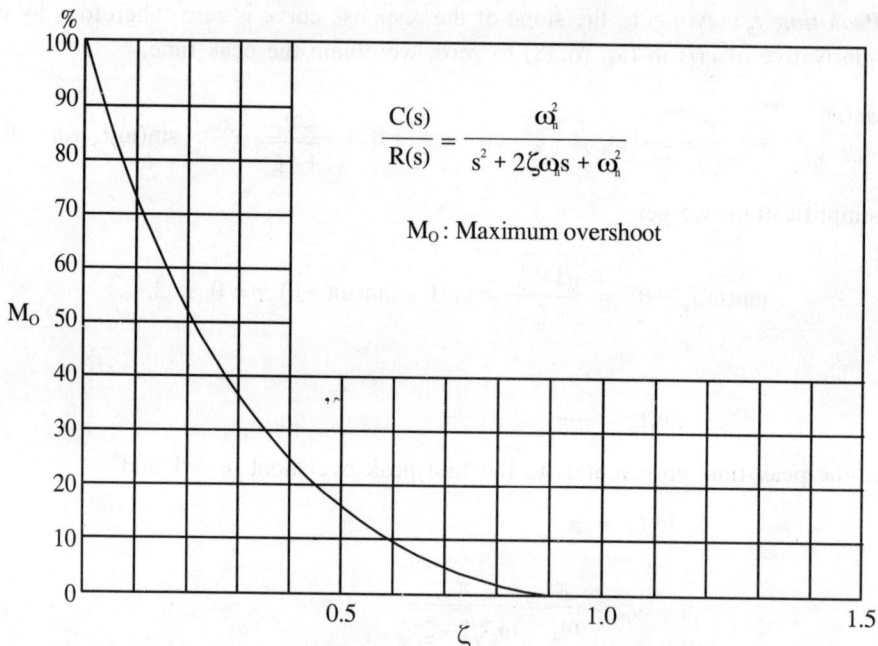

Fig. 6.12. Relationship between ζ and M_p

Fig. 6.12 shows the relationship between the damping ratio ζ and the maximum percentage overshoot. In order to limit the maximum percentage overshoot, the damping ratio ζ should lie between 0.4 and 0.8 so as the maximum percentage overshoot is between 25% and 2.5%.

(iv) Setting time t_s : The output response of a second-order system for unit step input decays exponentially with a time constant $\tau = 1/\zeta\omega_n$. For $0 < \zeta < 0.9$, the settling time t_s for ± 2% tolerance band is 4τ and for ± 5% tolerance band it is 3τ.

$$t_s = 4\tau = 4/\zeta\omega_n \qquad (\pm\ 2\%\ \text{tolerance band}) \qquad \ldots (6.59)$$

$$= 3\tau = 3/\zeta\omega_n \qquad (\pm\ 5\%\ \text{tolerance band}) \qquad \ldots (6.60)$$

Example 6.1. *The characteristic equation of a second-order system is given by* $s^2 + 6s + 25 = 0$. *Determine t_r, t_p, M_o, t_s, the time at which the first undershoot occurs, the frequency of the damped oscillations and the number of oscillations before reaching the steady-state when excited by a unit step input.*

Solution:

The characteristic equation is

$$s^2 + 2\zeta\omega_n s + \omega_n^{\,2} = s^2 + 6s + 25 = 0$$

Hence,

$$\omega_n = \sqrt{25} = 5 \text{ rad/sec}$$

$$\zeta = 6/2\omega_n = 0.6$$

$$\omega_d = \omega_n\sqrt{1-\zeta^2} = 5\times0.8 = 4 \text{ rad/sec}$$

$$t_r = \frac{\pi-\theta}{\omega_d}$$

$$= \frac{\pi-\tan^{-1}\sqrt{1-\zeta^2}/\zeta}{4}$$

$$= \frac{\pi-\tan^{-1}(4/3)}{4}$$

$$= (3.14-0.93)/4 = 0.553 \text{ sec}$$

$$t_p = \pi/\omega_d = 3.14/4 = 0.785 \text{ sec}$$

$$M_o = e^{-0.6 \times 3.14/0.8} = 0.095$$

The percentage overshoot is 9.5%.

$$t_s = 4/\zeta\omega_n = 4/3 = 1.33 \text{ sec} \quad (\text{for} \pm 2\% \text{ tolerance band})$$

$$t_s = 3/\zeta\omega_n = 3/3 = 1.0 \text{ sec} \quad (\text{for} \pm 5\% \text{ tolerance band})$$

The first undershoot occurs at time t_u given by

$$t_u = 2\pi/\omega_d = 1.57 \text{ sec}$$

The frequency of damped oscillation is $f_d = \omega_d/2\pi = 0.637 \ Hz$

The number of oscillations before reaching steady-state is

$$N_s = t_s.f_d = 0.85 \text{ cycles for } \pm 2\% \text{ tolerance band}$$

$$= 0.64 \text{ cycles for } \pm 5\% \text{ tolerance band}$$

6.5.5. Unit Ramp Response

The response of an under-damped second-order system to a unit ramp input is given by

$$C(s) = \frac{\omega_n^2}{s^2(s^2 + 2\zeta\omega_n s + \omega_n^2)} \quad\quad ... (6.61)$$

since $R(s) = 1/s^2$. From Laplace transform table $c(t)$ is

$$c(t) = t - \frac{2\zeta}{\omega_n} + \frac{e^{-\zeta\omega_n t}}{\omega_n \sqrt{1 - \zeta^2}} \sin(\omega_d t + \theta) \qquad \text{... (6.62)}$$

where $\phi = 2 \tan^{-1}\left(\sqrt{1 - \zeta^2} / \zeta\right)$. The response is shown in Fig. 6.13. The output angular velocity equals the input angular velocity or ramp after the transient oscillations have died out. In other words, the output shaft position lags behind the input shaft position by a small amount known as velocity error or velocity lag. However this error must exist for proper operation of the system since the motor requires an input from the amplifier, to provide the necessary torque.

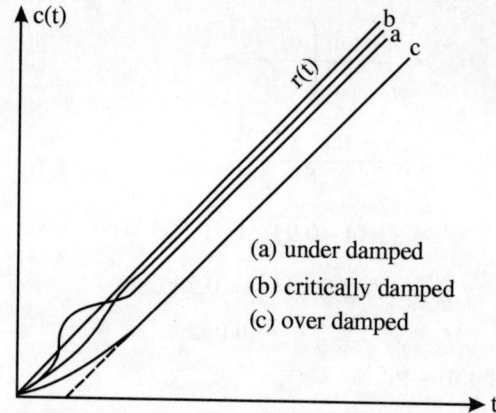

(a) under damped
(b) critically damped
(c) over damped

Fig. 6.13. Unit ramp response of second-order system

This velocity error is given by

$$\text{Velocity error} = r(t) - c(\infty)$$

$$= t - \left(t - \frac{2\zeta}{\omega_n}\right)$$

$$= 2\zeta/\omega_n \qquad \text{... (6.63)}$$

This steady-state error can be decreased either by decreasing ζ or by increasing ω_n. Since ζ is usually chosen on the basis of maximum overshoot and settling time, it is decreased in practice by increasing ω_n which depends on system gain K.

6.5.6. Unit Impulse Response

The Laplace transform of a unit impulse input is $R(s) = 1$. Therefore the response $C(s)$ of a second – order system is

$$C(s) = \frac{\omega_n^2}{s^2 + 2\zeta\omega_n s + \omega_n^2} \qquad \text{... (6.64)}$$

(a) *Under-damped case: For $0 < \zeta < 1$:*

$$C(s) = \frac{A}{s + \zeta\omega_n - j\omega_d} + \frac{B}{s + \zeta\omega_n + j\omega_d}$$

$$= \frac{\omega_n / \sqrt{1 - \zeta^2}}{2j(s + \zeta\omega_n - j\omega_d)} - \frac{\omega_n / \sqrt{1 - \zeta^2}}{2j(s + \zeta\omega_n + j\omega_d)}$$

Therefore,

$$c(t) = \frac{\omega_n}{\sqrt{1 - \zeta^2}} \left(\frac{e^{-(\zeta\omega_n t - j\omega_d t)} - e^{-(\zeta\omega_n t + j\omega_d t)}}{2j} \right)$$

$$= \frac{\omega_n}{\sqrt{1 - \zeta^2}} \, e^{-\zeta\omega_n t} . \sin\omega_d t \qquad t \geq 0 \qquad\qquad \dots (6.65)$$

(b) *Undamped case:* $\zeta = 0$

$$c(t) = \omega_n \sin \omega_n t$$

(c) *Critically damped case: For* $\zeta = 1$:

$$C(s) = \frac{\omega_n^2}{(s + \omega_n)^2}$$

and

$$c(t) = \omega_n^2 . t . e^{-\omega_n t} \qquad\qquad (t \geq 0) \qquad\qquad \dots (6.66)$$

(d) *Over-damped case: For* $\zeta > 1$,

$$C(s) = \frac{\omega_n^2}{(s + \zeta\omega_n - \omega_d)(s + \zeta\omega_n + \omega_d)}$$

and

$$c(t) = \frac{\omega_n}{2\sqrt{\zeta^2 - 1}} \left[e^{-(\zeta - \sqrt{\zeta^2 - 1})\omega_n t} - e^{-(\zeta + \sqrt{\zeta^2 - 1})\omega_n t} \right] \quad t \geq 0 \qquad\qquad \dots (6.67)$$

Fig. 6.14 shows a family of unit impulse response curves for various values of ζ for the under-damped, critically damped and over-damped case. They are functions of ζ only since $c(t)/\omega_n$ is plotted against the dimensionless variable $\omega_n t$. For the under-damped case, the unit impulse response $c(t)$ oscillates about zero whereas for critically damped or over-damped cases there is no oscillation and the response is always positive. Hence, if $c(t)$ does not change sign, the system is either over-damped or critically damped. Therefore, the step response of such systems does not overshoot but increases monotonically and approaches a constant value.

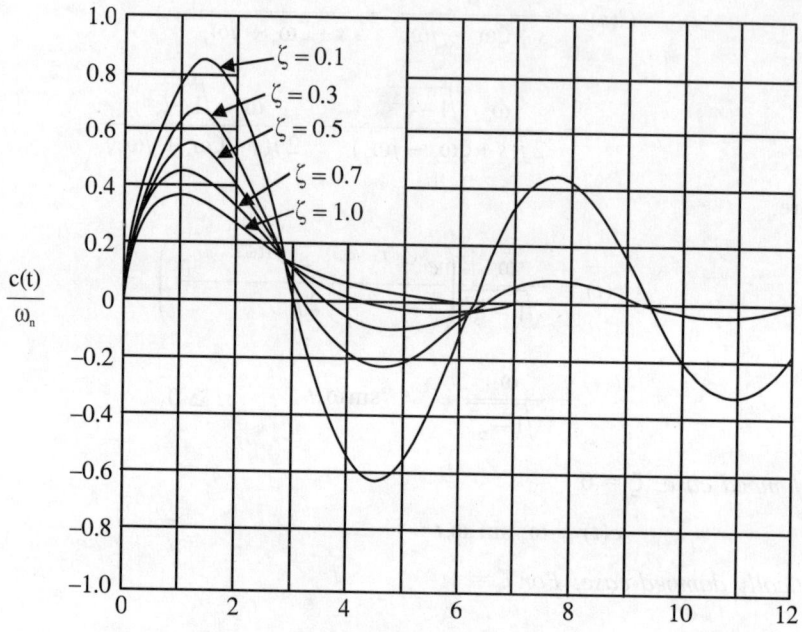

Fig. 6.14. Unit impulse response for various values of ζ

The time at which the maximum overshoot for the unit impulse response of the under-damped system occurs can be determined by equating the time-derivative of $c(t)$ to zero

$$\frac{dc(t)}{dt} = -\zeta\omega_n \, e^{-\zeta\omega_n t} \sin\omega_d t + e^{-\zeta\omega_n t} . \, \omega_d . \cos\omega_d t = 0$$

or,

$$\tan \omega_d t = \omega_d/\zeta\omega_n = \sqrt{1-\zeta^2}/\zeta = \tan \theta$$

Hence

$$t_p = \frac{\theta}{\omega_n\sqrt{1-\zeta^2}} \qquad \text{for } 0 < \zeta < 1. \qquad \qquad ...(6.68)$$

The maximum response is

$$c(t)_{max} = \omega_n \exp\left(-\zeta\theta/\sqrt{1-\zeta^2}\right) \qquad \qquad ...(6.69)$$

6.5.7. Effect of Adding a Pole

The closed-loop transfer function of a second-order system with the addition of a pole is of a third-order and has three poles of which one must be real and the other two may be real or complex depending on the value of system gain K. Assuming that it has a complex pair of poles, and a real pole at $s = -1/\tau$, the transfer function may be written as

$$\frac{C(s)}{R(s)} = \frac{\omega_n^2}{(1+\tau s)(s^2 + 2\zeta\omega_n s + \omega_n^2)} \qquad \text{... (6.70)}$$

If the system gain is such that the roots of the quadratic function is negative real and unequal, $\zeta > 1$ and the system is over-damped. The transient response of the system for unit step response is given by

$$C(s) = \frac{p\omega_n^2}{s(s+p)(s+p_1)(s+p_2)} \qquad (\zeta - 1)$$

$$= \frac{A_1}{s} + \frac{A_2}{(s+p)} + \frac{A_3}{(s+p_1)} + \frac{A_4}{(s+p_2)} \qquad \text{... (6.70a)}$$

where $p = 1/\tau$.

Therefore

$$c(t) = A_1 + A_2 e^{-pt} + A_3 e^{-p_1 t} + A_4 e^{-p_2 t} \qquad \text{... (6.71)}$$

If the system gain is such that the quadratic roots are negative and equal, $\zeta = 1$ and the system is critically damped. Then the response of the system is given by

$$C(s) = \frac{\omega_n^2 p}{s(s+p)(s+p_1)^2} \qquad (\zeta = 1) \qquad \text{... (6.72)}$$

and

$$c(t) = A_1 + A_2 te^{-pt} + A_3 te^{-p_1 t} + A_4 e^{-p_1 t} \qquad \text{... (6.73)}$$

If the system gain is further increased, the roots are complex pairs and $c(t)$ will be of the form

$$c(t) = A_1 + A_2 e^{-pt} + A_3 e^{-p_1 t} + A_4 e^{-p_2 t} \qquad \text{... (6.74)}$$

When $\zeta = 0$, then

$$c(t) = A_1 + A_2 e^{-pt} + A_3 \sin \omega_n t \qquad \zeta = 0 \qquad \text{... (6.75)}$$

Fig. 6.15 shows the unit step response of the system with $\omega_n = 1$, $\zeta = 0.5$ and $\tau = 0, 0.5, 1, 2$ and 4. From the curves it can be seen that the rise time increases and the maximum overshoot decreases as the pole at $s = -p$ is moved towards the origin. Thus, *the effect of adding a pole to a second-order system is to reduce the over-shoot and increase the setting time.*

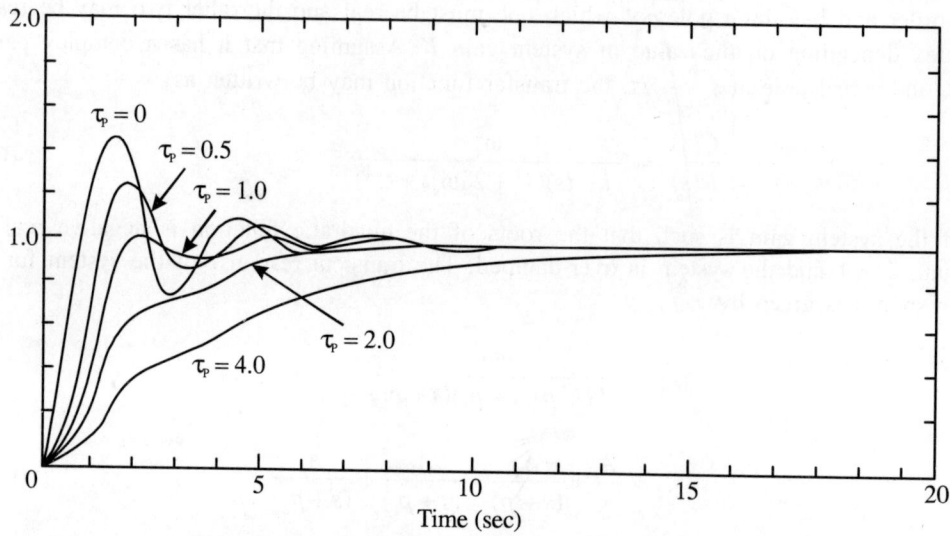

Fig. 6.15. Effect of additional pole on unit step response

6.5.8. Effect of Adding a Zero

The closed-loop transfer function with the addition of a zero at $s = -1/\tau$, is

$$\frac{C(s)}{R(s)} = \frac{\omega_n^2(1+s\tau)}{s^2+2\zeta\omega_n s+\omega_n^2} \qquad \text{... (6.76)}$$

The response of the above system for unit step input is shown in Fig 6.16 with $\omega_n = 1$, $\zeta = 0.5$ and $\tau = 0, 1, 3, 6$ and 10. It can be seen that *the addition of a zero at $s = -1/\tau$ decreases the rise time and increases the maximum overshoot.*

6.5.9. Performance Characteristics and System Gain

We shall now discuss how the time response performance characteristics of a second-order system vary when the system gain is varied from zero to infinity.

Consider the closed-loop transfer function of a position control system with system gain K.

$$\frac{C(s)}{R(s)} = \frac{400K}{s^2+48.5s+400K} \qquad \text{... (7.77)}$$

The unit step response is given by Eq. (6.38) where $\omega_n = 20\sqrt{K}$ and $\zeta = 1.2125/\sqrt{K}$. Thus, the natural undamped frequency of the system is proportional to $\sqrt{K}$ and the damping ratio is inversely proportional to $\sqrt{K}$. The product of ω_n and ζ is a constant and equal to 24.25. The time responses of the system for three typical values of *K* are shown in Fig. 6.17. Table 6.1 gives the performance characteristics of the unit step responses for the three value of *K* used.

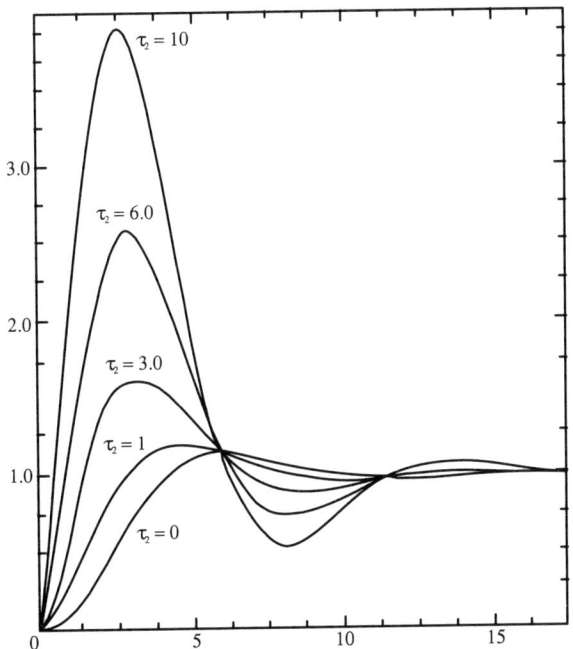

Fig. 6.16. Effect of additional zero on unit step response

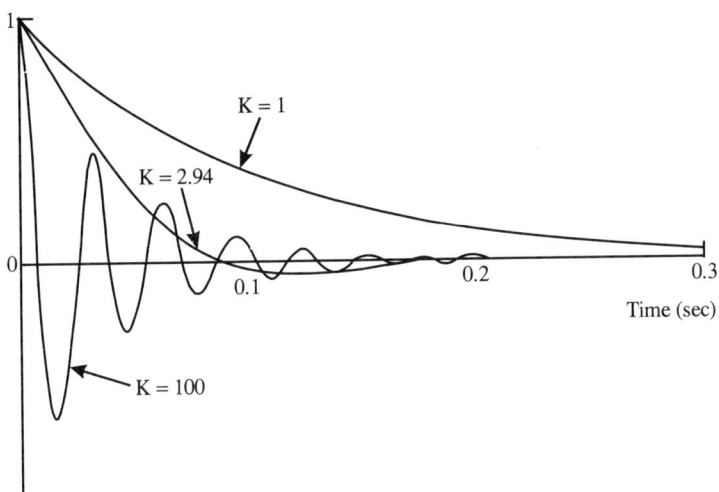

Fig. 6.17. Time response *vs* system gain *K*

Table 6.1. Comparison of Performance

Gain K	ζ	ω_n	Maximum Overshoot	t_d	t_r	t_s	t_p
1.0	1.2125	20	0	0.1	0.215	0.115	—
2.94	0.707	34.3	0.043	0.044	0.065	0.086	0.13
100.	0.12125	200	0.681	0.005	0.006	0.31	0.016

When the value of K is set to a very low value ($K = 1$), the damping ratio is 1.2125 since the system is over-damped, the step response approaches the final value without any oscillations and overshoot. The rise time, the delay time and the setting time are all very long. When $K = 2.94$, the damping ration is 0.707 and the overshoot is only 4.3 percent. The rise time is moderate but the setting time is 0.086 sec. When K is set to 100, the damping ration is 0.12125. The system is under-damped and the overshoot is 68.1 percent which is very high. However the rise time and delay time are very short. The settling time is 0.31sec. It can be seen that the best possible time response is obtained with $K = 2.94$.

6.5.10. Poles and System Gain

In practice, the evaluation of unit step response for each change in the system gain K is time-consuming and laborious. However the performance of a system can be predicted by investigating the roots of the characteristic equation as K is varied. The characteristic equation of the system described by Eq. (6.77) is given by

$$s^2 + 48.5s + 400K = 0 \qquad \qquad \text{... (6.78)}$$

The roots which are also the poles of the closed-loop transfer function are

$$s_1 = -24.25 + \sqrt{588 - 400K} \qquad \qquad \text{... (6.79)}$$

$$s_2 = -24.25 - \sqrt{588 - 400K} \qquad \qquad \text{... (6.80)}$$

The roots of the characteristic equation for $K = 1$, 2.94, 100 and ∞ are tabulated below.

Table 6.2. System poles for various values of system gain

K	s_1	s_2
1	-10.54	-37.96
2.94	$-24.25 + j\ 24.25$	$-24.25 - j\ 24.25$
100	$-24.25 + j\ 198.5$	$-24.25 - j\ 198.5$
∞	$-24.25 + j\ \infty$	$-24.25 - j\ \infty$

Thus the roots of the characteristic equation or the poles of the closed-loop transfer function of a second-order system lie in the left half of the s-plane as K is varied from 0 to ∞. Hence *a second-order system is always or absolutely stable. The trajectories of the roots as K is varied from 0 to ∞ is called root locus* and are extensively used for the analysis and design of linear control systems.

6.6. UNIT STEP RESPONSE OF A THIRD-ORDER SYSTEM

Consider a third-order closed-loop transfer function given by

$$\frac{C(s)}{R(s)} = \frac{400,000K}{s^3 + 1040s^2 + 48,500s + 400,000K} \qquad \ldots (6.81)$$

The characteristic equation is

$$s^3 + 1040s^2 + 48,500s + 400,000K = 0 \qquad \ldots (6.82)$$

For $K = 1$,

$$\frac{C(s)}{R(s)} = \frac{400,000}{(s+10.66)(s+37.85)(s+991.5)} \qquad \ldots (6.83)$$

The unit step response is

$$c(t) = 1 - 1.407\,e^{-10.66t} + 0.408\,e^{-37.85t} - 0.00043e^{-991.5t} \qquad \ldots (6.84)$$

It can be seen that the response is dominated by the two roots at $s = -10.66$ and -37.85. The last term in Eq. (6.84) due to the third root at -991.5 decays very rapidly and its magnitude at $t = 0$ is also small. Thus, we can conclude that the roots closer to the imaginary axis dominate the response and those relatively far away to the left of the $j\omega$ –axis contribute very little. The roots which dominate the response are known as *dominant roots*.

When $K = 2.94$, Eq. (6.81) becomes

$$\frac{C(s)}{R(s)} = \frac{1,176,000}{(s+992.32)(s^2 + 47.68s + 1185.24)} \qquad \ldots (6.85)$$

The roots are -992.32, $-23.84 \pm j\,24.83$. The complex roots dominate the transient response since their real part is closer to the $j\omega$-axis compared to the real root at -992.32. Therefore the third-order system may be approximated to a second-order system by letting $s = 0$ in the less dominant pole-factors. Thus Eq. (6.85) becomes

$$\frac{C(s)}{R(s)} = \frac{1185.2}{s^2 + 47.68s + 1185.2} \qquad \ldots (6.86)$$

The unit step response is given by Eq. (6.38) with $\omega_n = 34.43$ and $\zeta = 0.6925$. Since these parameter values are not the true values of the third-order system they are called *relative natural undamped frequency* and *relative damping ratio* and are determined from the dominant roots.

For $K = 126$, the roots of the characteristic equation are $\pm j$ 220.2 and -1040. The system has a pair of complex roots on the $j\omega$-axis and the response is oscillatory and is given by

$$c(t) = 1 - 0.043 e^{1040t} - 0.9783 \sin(220.2t + 78°) \qquad \dots (6.87)$$

The system is said to be *marginally stable* or *on the verge of instability.* If K is increased to 126.1, the complex roots lie in the right half of the s-plane, the oscillatory term in Eq. (6.87) will increase exponentially and the system becomes unstable. Thus *the third-order system is liable to become unstable. The value of K for which the system is marginally stable is the maximum gain of the system for which the system is stable.* It was shown earlier that second-order systems are absolutely stable as K varies from 0 to ∞. Fig. 6.18 shows the unit step responses of the third-order system for $K = 1$, 2.94, 126 and $K = 200$.

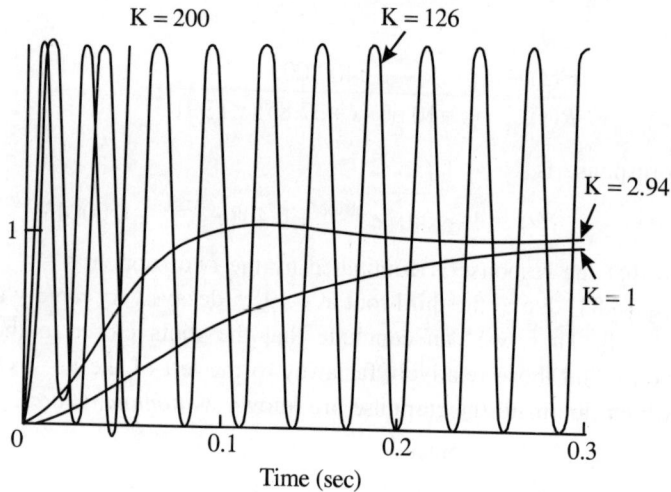

Fig. 6.18. Unit step response of a third-order system

6.7. TRANSIENT RESPONSE OF HIGHER-ORDER SYSTEMS

Let us consider the system shown in Fig. 6.19. The closed-loop transfer function is

$$\frac{C(s)}{R(s)} = \frac{G(s)}{1 + G(s)H(s)} \qquad \dots (6.88)$$

In general, $G(s)$ and $H(s)$ are ratios of polynomials in s. Thus

$$G(s) = N_1(s)/D_1(s) \text{ and } H(s) = N_2(s)/D_2(s)$$

where $N_i(s)$ and $D_i(s)$ are polynomials in s. Eq.(6.88) may be written as

$$\frac{C(s)}{R(s)} = \frac{N_1(s)D_2(s)}{D_1(s)D_2(s) + N_1(s)N_2(s)}$$

$$= \frac{b_m s^m + b_{m-1}s^{m-1} + \cdots\cdots + b_0}{a_n s^n + a_{n-1}s^{n-1} + \cdots\cdots + a_0} \qquad n \geq m$$

and in the factored form,

$$\frac{C(s)}{R(s)} = \frac{K(s+z_1)(s+z_2)\cdots\cdots(s+z_m)}{(s+p_1)(s+p_2)\cdots\cdots(s+p_n)} \qquad \text{... (6.89)}$$

Assuming all the closed-loop poles are distinct, the unit step response of Eq. (6.89) is

$$C(s) = \frac{A}{s} + \sum_{i=1}^{n}\frac{A_i}{s+p_i} \qquad \text{... (6.90)}$$

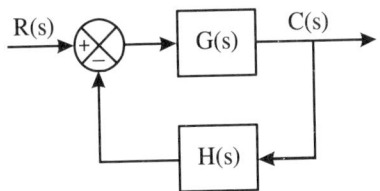

Fig. 6.19. Higher-order system

where A_i is the residue of the pole at $s = -p_i$ and A is the residue at $s = 0$.

For a stable system, all the closed-loop poles lie in the left half of the s-plane and the relative magnitudes of the residues determine the relative importance of the poles. If a closed-loop zero is close to a closed-loop pole, then the residue of this pole is small and the corresponding transient response term is also small. Thus, *a pair of closely located pole and zero will effectively cancel each other. If a pole is located very far from the origin, its effect on the transient response is small and may be neglected.*

If the closed-loop transfer function has real and complex conjugate poles, then the characteristic equation consists of first-and second-order terms and Eq. (6.89) my be written as

$$\frac{C(s)}{R(s)} = \frac{K\prod\limits_{i=1}^{m}(s+z_i)}{\prod\limits_{j=1}^{q}(s+p_j)\prod\limits_{k=1}^{r}(s^2 + 2\zeta_k\omega_k s + \omega_k^2)} \qquad \text{... (6.91)}$$

where $q + 2r = n$. Assuming the real poles to be distinct, $C(s)$ may be written in partial fraction expansion form

$$C(s) = \frac{A}{s} + \sum_{j=1}^{q} \frac{A_j}{s+p_j} + \sum_{k=1}^{r} \frac{B_k(s+\zeta_k\omega_k)+C_k\omega_k\sqrt{1-\zeta_k^2}}{s^2+2\zeta_k\omega_k s+\omega^2_K} \qquad \dots \text{(6.92)}$$

Thus the response of a higher-order system is composed of a number of terms involving simple functions found in the responses of first and second-order systems. The unit step response is

$$c(t) = A + \sum_{j=1}^{q} A_j e^{-p_j t} + \sum_{k=1}^{r} B_k e^{-\zeta_k \omega_k t} + \cos \omega_k \sqrt{1-\zeta^2} \ t +$$

$$\sum_{k=1}^{r} C_k e^{-\zeta_k \omega_k t} \sin \omega_k \sqrt{1-\zeta^2} \ t \qquad \dots \text{(6.93)}$$

The steady-state output $c(\infty) = A$.

From the above discussion, it is apparent that the poles of the input $R(s)$ yield the steady-state response terms while the poles of $C(s)/R(s)$ appear as exponential response terms. The zeros of $C(s)/R(s)$ affect the magnitudes and signs of the residues.

(i) *Dominant Closed-loop Poles:* The relative dominance of closed-loop poles is determined by the real parts of the closed-loop poles and by the relative magnitudes of the residues which depend on both closed-loop poles and zeros. If the ratios of the real parts exceed five and there are no zeros closely, then closed-loop poles nearest to the $j\omega$-axis will dominate in the transient response because their corresponding terms in the transient response decay slowly. Such poles are known as *dominant poles*. The dominant closed-loop poles frequently occur in the form of a complex conjugate pair. The gain of higher-order systems is often adjusted so that there will be a pair of dominant complex conjugate closed-loop poles because the presence of these poles in a stable system reduces the effect of nonlinearities.

Though all closed-loop poles lie in the left half of the s-plane, the transient response characteristic may not be satisfactory. If a dominant complex conjugate pole pair lies close to the $j\omega$-axis the transient response may be excessively oscillatory or very slow. In order to guarantee fast yet well-damped transient response characteristics, it is therefore necessary that the *closed-loop poles of the system lie within a particular region in the complex plane, shown in Fig. 6.20.*

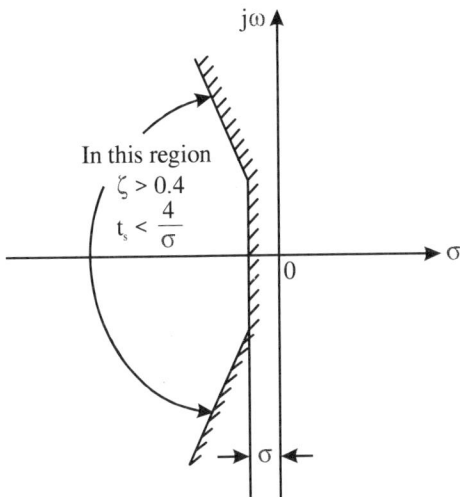

Fig. 6.20. Region of closed loop poles for stable system

6.8. SOLVED PROBLEMS

Example 6.2. *The response of a second-order system for unit ramp is*

$$C(s) = \frac{\omega_n^2}{s^2(s^2 + 2\zeta\omega_n s + \omega_n^2)}$$

Find its time response for (a) $0 < \zeta < 1$, (b) $\zeta = 1$, (c) $\zeta > 1$ and (d) $\zeta = 0$.

Solution:

(a) For $0 < \zeta < 1$:

$$C(s) = \frac{A}{s^2} + C_1(s)$$

$$A = \frac{\omega_n^2}{s^2 + 2\zeta\omega_n s + \omega_n^2}\bigg|_{s=0} = 1$$

$$C(s) = \frac{1}{s^2} + C_1(s)$$

$$C_1(s) = C(s) - \frac{1}{s^2} = -\frac{s + 2\zeta\omega_n}{s(s^2 + 2\zeta\omega_n s + \omega_n^2)}$$

$$\frac{s+2\zeta\omega_n}{s\left(s^2+2\zeta\omega_n s+\omega_n^2\right)} = \frac{B}{s}+C_2(s)$$

$$B = \left.\frac{s+2\zeta\omega_n}{s^2+2\zeta\omega_n s+\omega_n^2}\right|_{s=0} = \frac{2\zeta}{\omega_n}$$

$$C_2(s) = \frac{s+2\zeta\omega_n}{s(s^2+2\zeta\omega_n s+\omega_n^2)} - \frac{2\zeta}{\omega_n s}$$

$$= \frac{1}{s^2+2\zeta\omega_n s+\omega_n^2} + \frac{2\zeta\omega_n}{s(s^2+2\zeta\omega_n s+\omega_n^2)} - \frac{2\zeta}{\omega_n s}$$

$$= \frac{1}{s^2+2\zeta\omega_n s+\omega_n^2} - \frac{2\zeta}{\omega_n s}\left(\frac{s+2\zeta\omega_n}{s^2+2\zeta\omega_n s+\omega_n^2}\right)$$

$$= \frac{2\zeta^2-1}{s^2+2\zeta\omega_n s+\omega_n^2} - \frac{2\zeta}{\omega_n}\frac{s+\zeta\omega_n}{s^2+2\zeta\omega_n s+\omega_n^2}$$

Therefore,

$$C(s) = \frac{1}{s^2}-\frac{2\zeta}{\omega_n s}+\frac{2\zeta^2-1}{s^2+2\zeta\omega_n s+\omega_n^2}+\frac{2\zeta}{\omega_n}\frac{s+\zeta\omega_n}{s^2+2\zeta\omega_n s+\omega_n^2} \qquad \dots (6.94)$$

Taking the inverse Laplace transform, we obtain

$$c(t) = t - \frac{2\zeta}{\omega_n}+\frac{2\zeta^2-1}{\omega_n} - e^{-\zeta\omega_n t}\sin\omega_d t + \frac{2\zeta}{\omega_n}e^{-\zeta\omega_n t}\cos\omega_d t$$

$$= t - \frac{2\zeta}{\omega_n}+\frac{e^{-\zeta\omega_n t}}{\omega_n\sqrt{1-\zeta^2}}\cdot\sin\left(\omega_n\sqrt{1-\zeta^2}t+\phi\right) \qquad (0<\zeta<1)$$

when $\phi = \tan^{-1}2\zeta\sqrt{1-\zeta^2}\big/\left(2\zeta^2-1\right)=2\tan^{-1}\sqrt{1-\zeta^2}\big/\zeta$.

(b) When $\zeta = 1$: $C(s)$ becomes

$$C(s) = \frac{\omega_n^2}{s^2(s+\omega_n)^2} = \frac{A}{s^2}+\frac{B}{s}+\frac{C}{(s+\omega_n)^2}+\frac{D}{s+\omega_n}$$

$$= \frac{1}{s^2}-\frac{2\zeta}{\omega_n s}+\frac{1}{(s+\omega_n)^2}+\frac{2}{\omega_n}\frac{1}{s+\omega_n}$$

Therefore,

$$c(t) = t - \frac{2\zeta}{\omega_n} + te^{-\omega_n t} + \frac{2}{\omega_n}e^{-\omega_n t}$$

$$= t - \frac{2\zeta}{\omega_n} + \frac{2}{\omega_n}e^{-(1+\omega_n t/2)} \qquad \zeta = 1 \qquad \qquad \dots (6.96)$$

(c) For $\zeta > 1$

$$C(s) = \frac{\omega_n^2}{s^2(s + p_1)(s + p_2)}$$

where p_1 and p_2 are given in Eq. (6.44(b)).

In partial fraction form

$$C(s) = \frac{A}{s^2} + \frac{B}{s} + \frac{C}{s + p_1} + \frac{D}{s + p_2}$$

where, using partial fraction theorem,

$$A = 1$$
$$B = 2\zeta/\omega_n$$
$$C = \omega_n^2/p_1^2(p_2 - p_1)$$
$$D = \omega_n^2/p_2^2(p_1 - p_2)$$

Taking the inverse Laplace transform,

$$c(t) = t - \frac{2\zeta}{\omega_n} - \frac{\omega_n^2}{p_1 - p_2}\left(\frac{e^{-p_1 t}}{p_1^2} - \frac{e^{-p_2 t}}{p_2^2}\right) \qquad \qquad \dots (6.97)$$

(d) For when $\zeta = 0$

Substituting $\zeta = 0$ in Eq. (6.95), we get

$$c(t) = t - \frac{1}{\omega_n}\sin \omega_n t \qquad \zeta = 0 \qquad \qquad \dots (6.98)$$

Example 6.3. *Obtain the steady-state error for a unit ramp input of the unity feedback systems with the following closed-loop transfer functions*

(a) $C(s)/R(s) = \omega_n^2/(s^2 + 2\zeta\omega_n s + \omega_n^2)$

(b) $C(s)/R(s) = (1 + \tau s)\omega_n^2/(s^2 + 2\zeta\omega_n s + \omega_n^2)$

Examine the possibilities of reducing the error to zero.

Solution:

(a)
$$E(s) = R(s) - C(s)$$

$$= \frac{s(s + 2\zeta\omega_n)}{s^2(s^2 + 2\zeta\omega_n s + \omega_n^2)}$$

$$e_{ss} = \lim_{s \to 0} s.E(s) = 2\zeta/\omega_n$$

The steady-state error in this case cannot be reduced to zero.

(b)
$$E(s) = \frac{s(s + 2\zeta\omega_n - \tau\omega_n^2)}{s^2(s^2 + 2\zeta\omega_n s + \omega_n^2)}$$

$$e_{ss} = \lim_{s \to 0} s.E(s) = (2\zeta - \tau\omega_n)/\omega_n$$

The steady state error can be reduced to zero if $\tau = 2\zeta/\omega_n$.

$C(s)$ may be written as

$$C(s) = \frac{\omega_n^2}{s^2 + 2\zeta\omega_n s + \omega_n^2}(1 + \tau s) \, R(s)$$

Thus, the input to the system is the reference input $R(s)$ and τ times its derivative. This type of control is known as *proportional and derivative input control* and τ is the derivative time constant.

Example 6.4. *The forward transfer function of a control system is given by* $G(s) = 4/s(s + 1)$. *(a) For a unity feedback, find the percentage overshoot. (b) If the feedback transfer function is* $(1 + \tau s)$, *find* τ *and maximum overshoot for* $\zeta = 0.6$.

Solution:

(a) *Unity feedback*:

$$G(s) = 4/s(s + 1)$$
$$C(s)/R(s) = 4/s(s^2 + s + 4)$$

$$\omega_n = \sqrt{4} = 2 \, \text{rad./sec}$$

$$\zeta = 1/2\omega_n = 0.25$$

Maximum Overshoot $M_0 = e^{-\pi\zeta/\sqrt{1-\zeta^2}}$

$$= e^{-0.25\pi/\sqrt{1-(0.25)^2}}$$

$$= 0.444$$

Percentage maximum overshoot = 44.4%

(b) Non-unity feedback:

$$H(s) = (1 + \tau s)$$

$$C(s)/R(s) = G(s)/1 + G(s)H(s)$$

$$= 4/(s^2 + s + 4\tau s + 4)$$

$$\omega_n = \sqrt{4} = 2$$

$$\zeta = 0.6 = \frac{1 + 4\tau}{2\omega_n} = (1 + 4\tau)/4$$

$$\tau = 0.35 \text{ sec}$$

Percentage maximum overshoot

$$= e^{-0.6\pi/\sqrt{1-(0.6)^2}} \times 100\%$$

$$= 9.48\%.$$

Example 6.5. *The closed-loop transfer function of a system is given by*

$$\frac{C(s)}{R(s)} = \frac{6s + 1}{s^5 + 12s^4 + 55s^3 + 120s^2 + 124s + 48}$$

Determine its time response when a step input of magnitude 6 is applied. Obtain its steady-state output.

Solution:

$$C(s) = \frac{6(6s + 1)}{s(s^2 + 12s^4 + 55s^3 + 120s^2 + 124s + 48)}$$

$$= \frac{6(6s + 1)}{s(s + 1)(s + 3)(s + 4)(s + 2)^2}$$

$$= \frac{A}{s} + \frac{B}{s + 1} + \frac{C}{s + 3} + \frac{D}{s + 4} + \frac{E}{(s + 2)^2} + \frac{F}{(s + 2)}$$

Using partial fraction theorem,

$$A = s\, C(s)|_{s=0} = 0.125$$

$$B = (s + 1)\, C(s)|_{s=-1} = 5$$

$$C = (s+3) C(s)|_{s=-3} = -17$$

$$D = (s+4) C(s)|_{s=-4} = 2.875$$

$$E = (s+2)^2 C(s)|_{s=-2} = -16.5$$

$$F = d/ds\left[(s+2)^2 C(s)\right]_{s=-2} = 9$$

Substituting the values and taking inverse Laplace transform

$$c(t) = 0.125 + 5e^{-t} - 17e^{-3t} + 2.875e^{-4t} - 16.5te^{-2t} + 9e^{-2t}$$

The steady-state output is

$$c(\infty) = 0.125$$

Example 6.6. *The forward transfer function of a unity feedback system is $K/s(1 + \tau s)$. The maximum overshoot for a unit step input is to be reduced from 80% to 40% Find the factor by which the system gain K is to be reduced.*

Solution:

Maximum overshoot $M_0 = e^{-\pi\zeta/\sqrt{1-\zeta^2}}$ ... (6.99)

Let

$$m = In\ M_0 = -\pi\zeta/\sqrt{1-\zeta^2}$$

$$\zeta = \sqrt{m^2/(m^2 + \pi^2)}$$ ... (6.100)

(a) For $M_0 = 80\%$

$$m = In\ 0.8 = -0.223$$

$$\zeta_1 = \sqrt{(-0.223)^2/\left[(-0.223)^2 + \pi^2\right]}$$

$$= 0.07$$

(b) For $M_0 = 40\%$

$$m = \ln\ 0.4 = -\ 0.96$$

$$\zeta_2 = \sqrt{0.916^2/(.916^2 + \pi^2)}$$

$$= 0.28$$

The forward transfer function is

$$G(s) = K/s(1 + \tau s)$$

$\therefore$ $\qquad$ $C(s)/R(s) = (K/\tau)/(s^2 + s/\tau + K/\tau)$

$$\omega_n^2 = K/\tau$$

$$\zeta = \frac{1}{\tau} \cdot \frac{1}{2\omega_n} = \frac{1}{2\tau} \cdot \frac{1}{\sqrt{K/\tau}} = \frac{1}{2\sqrt{K\tau}}$$

or

$$K \propto \frac{1}{\zeta^2}$$

$$K_2/K_1 = (\zeta_1/\zeta_2)^2 = 1/16$$

The system gain should be reduced by a factor of 16.

Example 6.7: *The transfer function of a positional control system is given by*

$$\Theta_0(s)/\Theta_1(s) = 1/(Js^2 + Bs + K)$$

For a unit step input, a maximum overshoot of 9% was observed at $t = 1$ sec. The steady-state error was 0.1 radian.. Find the values of J, B and K.

Solution:

$$\theta_{0ss} = \lim_{s \to 0} s \frac{1}{(Js^2 + Bs + K)} \frac{1}{s} = 1/K = 0.1$$

Hence,

$$K = 10$$
$$M_0 = 0.09$$

Using Eqs. (6.99) and (6.100)

$$\zeta = 10$$

$$t_p = \pi/\omega_n\sqrt{1-\zeta^2} = 1$$

$$\omega_n = \pi/\sqrt{1-0.6^2} = 3.93 \, \text{rad./sec.}$$

$$\frac{\Theta_0(s)}{\Theta_i(s)} = \frac{1/J}{s^2 + (B/J)s + K/J}$$

$$K/J = \omega_n^2 = 15.4$$

$$J = K/\omega_n^2 = 10/15.4 = 0.65 \ Kg - m^2$$

$$\zeta = 0.6 = \frac{B}{2\sqrt{KJ}}$$

$$B = 0.6 \times 2\sqrt{10 \times 0.65}$$

$$= 3.06 \ N - m/(\text{rad/sec})$$

Example 6.8. *A closed loop transfer function has a forward and feedback transfer function* $G(s) = K_1/s^2$ *and* $H(s) = 1 + K_2s$. *Find the values of* K_1 *and* K_2, *so that the overshoot of unit step response is 25% and the peak time is 2 seconds.*

Solution:

$$\frac{C(s)}{R(s)} = \frac{G(s)}{1 + G(s)H(s)} = \frac{K_1/s^2}{1 + \frac{K_1}{s^2}(1 + K_2s)}$$

$$= \frac{K_1}{s^2 + K_1 K_2 s + K_1}$$

From the above equation

$$\omega_n = \sqrt{K_1}, \ 2\zeta\omega_n = K_1 K_2$$

Hence $\zeta = K_1 K_2 / 2\omega_n = K_2 \sqrt{K_1}/2$

$$e^{-\pi\zeta/\sqrt{1-\zeta^2}} = 0.25$$

$$e^{\pi\zeta/\sqrt{1-\zeta^2}} = 4$$

Taking natural logarithm

$$\frac{\pi\zeta}{\sqrt{1-\zeta^2}} = In \ 4 = 1.386$$

From this, we get

$$\zeta = 0.4$$

$$t_p = \frac{\pi}{\omega_n\sqrt{1-\zeta^2}} = \frac{\pi}{\sqrt{K_1(1-\zeta^2)}} = 2$$

$$\sqrt{K_1} = \pi/2\sqrt{1-\zeta^2}$$

$$K_1 = 2.93$$

$$\zeta = 0.4 = \frac{K_2\sqrt{K_1}}{2} = \frac{K_2\sqrt{2.93}}{2}$$

$$K_2 = 0.47$$

Example 6.9. *A second – order servo consists of an inertia J = 20 kg-m², viscous damping coefficient B = 100 N-m per rad. per sec. and a proportional controller which supplies a torque of 1000 N-m/rad. error. The system is initially at rest. Determine the percentage overshoot for a unit step input and the settling time for 5% oscillations.*

Solution:

$$\frac{C(s)}{R(s)} = \frac{K}{Js^2 + Bs + K} = \frac{1000}{20s^2 + 100s + 1000}$$

$$= \frac{50}{s^2 + 5s + 50}$$

From the above equation,

$$\omega_n = \sqrt{50} = 7.07$$

$$\zeta = 5/2\omega_n = 0.3536$$

$$M_0 = e^{-\pi\zeta/\sqrt{1-\zeta^2}} = e^{-1.187} = 0.305 \text{ or } 30.5\%$$

The settling time for 5% oscillations is

$$t_s \text{ (for 5\% oscilations)} = 3/\zeta\omega_n = 1.2 \text{ sec.}$$

Example 6.10. *Measurements conducted on a servomechanism gave the system response to be*

$$c(t) = 1 + 0.2\, e^{-60t} - 1.2\, e^{-10t}$$

(a) Obtain the expression for closed-loop transfer function.

(b) Determine its undamped natural frequency and damping ratio.

Solution:
The Laplace transform of c(t) is

$$C(s) = \frac{1}{s} + \frac{0.2}{s+60} - \frac{1.2}{s+10}$$

$$= \frac{600}{s(s+60)(s+10)}$$

$$\omega_n^2 = 600.$$

Hence

$$\omega_n = \sqrt{600} = 24.5$$

$$2\zeta\omega_n = 70$$

$$\zeta = 70/2 \times 24.5 = 1.43 \text{ (Overdamped)}$$

Example 6.11. *A unity feedback control system is characterized by an open-loop transfer function*

$$G(s) = K/s(s + 10)$$

Determine the gain K so that the system will have a damping ratio of 0.5. For this value of K, determine settling time, the peak overshoot and the time at peak overshoot for a unit step input.

Solution:

(a)

$$\frac{C(s)}{R(s)} = \frac{K}{s^2 + 10s + K}$$

$$\zeta = \frac{10}{2\omega_a} = 0.5$$

$$\omega_n = 10 \text{ rad/sec.}$$

$$K = \omega_n^2 = 100$$

(b)

$$t_s = \frac{4}{\zeta\omega_n} = \frac{4}{0.5 \times 10} = 0.8 \text{ sec.}$$

(c)

$$t_p = \pi/(\omega_n\sqrt{1 - \zeta^2}) = \frac{3.14}{10 \times \sqrt{3}/2} = 0.363$$

$$M_o = e^{-\pi\zeta/\sqrt{1-\zeta^2}} = e^{-1.813} = 0.164 \text{ or } 16.4\%$$

Example 6.12. *A control system has a forward and feedback transfer function given by*

$$G(s) = 1/(s + 2) \text{ and } H(s) = 3/(s + 6).$$

(*a*) *Determine its transient response for unit step input.*

(*b*) *Calculate the percentage maximum overshoot.*

Solution:

(a) The closed-loop transfer function is given by

$$\frac{C(s)}{R(s)} = \frac{G(s)}{1 + G(s)H(s)}$$

$$= \frac{1/(s+2)}{1 + \left[1/(s+2)\right]\left[3/(s+6)\right]}$$

$$= \frac{s+6}{s^2 + 8s + 15}$$

For unit step input $R(s) = 1/s$. Therefore

$$C(s) = \frac{s+6}{s(s+3)(s+5)}$$

$$= \frac{0.4}{s} - \frac{0.5}{s+3} + \frac{0.1}{s+5}$$

$$c(t) = 0.4 - 0.5\,e^{-3t} + 0.1e^{-5t}$$

(b)
$$\omega_n = \sqrt{15} = 3.873$$

$$\zeta = \frac{8}{2\omega_n} = \frac{4}{\sqrt{15}} = 1.033$$

Since the damping ratio is greater than 1, the system is over-damped. Hence there is no overshoot. $M_o = 0\%$

6.9. SUMMARY

In this chapter, we have described the typical test signals such as unit impulse, unit step, unit ramp functions commonly employed to assess the performance of control systems. We have represented these signals in Laplace transform so as to obtain the response easily. This helps to compare the characteristics of various systems. We have then obtained the response of first-order system for unit impulse, unit step, unit ramp functions and the important characteristics in each case have been brought out. Similarly, the responses of second-order, third-order and higher-order systems for these typical input signals have been obtained and discussed in detail. Especially, for a second-order systems, we have elaborately studied its performance under various conditions such as undamped, under-damped, critically damped and over-damped cases. The performance specifications in time domain such as M_o, t_p, t_r, t_s have been defined and discussed.

We have also explained the effect of adding a pole or zero and its effects on the system response. We have also discussed the effect of varying system gain on the time-domain performance specifications.

REVIEW QUESTIONS

6.1. What are the standard test signals used in control systems?

6.2. Define a unit impulse function and explain.

6.3. Define a unit step function and explain.

6.4. Define a unit ramp function and explain.

6.5. Define a unit parabolic function and explain.

6.6. Write down the transfer function of a first order system and its response to standard test-signals.

6.7. What is meant by time-constant? What is its importance?

6.8. Write down the transfer function of a second-order system and its response to unit impulse and unit step signals.

6.9. What is meant by undamped, under-damped, critically damped and over-damped system?

6.10. Define delay time, rise time and setting time.

6.11. Define peak time and maximum overshoot.

6.12. Explain the effect of adding a pole to a system.

6.13. Explain the effect of adding a zero to a system.

6.14. What is meant by dominant poles? Explain.

EXERCISE

6.1. Explain why under-damping is preferred over critical damping.

6.2. (a) What are the advantages of time response analysis over frequency response analysis?

(b) Sketch the unit step response of a second-order system and explain the time domain specifications of the system.

6.3. Derive expression for the percentage maximum overshoot of an under damped second-order system for an unit step input.

6.4 What do you understand by the terms steady-state response and transient response of a system?

6.5 (a) Obtain the under-damped time response of a second-order system to a unit step input.

(b) Sketch the response of the system and explain the terms 'overshoot' and 'settling time'.

6.6. For a second-order system, the maximum overshoot of a step response is only a function of the damping ratio. Prove this statement. Sketch the variation of percent overshoot as a function of damping ratio.

6.7. Write a note on the use of PID controllers.

6.8. Analyze a second-order control system with unity feedback for a unit step acceleration (i.e. unit parabolic) input and derive an expression for its time response.

6.9. Derive the response of a first-order system subjected to a ramp input.

6.10. Obtain the response of a second-order system to a ramp input.

PROBLEMS

6.1 Find the impulse response of a first-order system at $t = 1$, 2 and 3 when the time constant is 1, 2 and 3. Also find the initial slope and initial value.

6.2 Find the unit step response of the first-order system for $t = 1$, 2 and 3 when $t = 1, 2, 3$. Also the initial slope and settling time.

6.3. The transfer function of a second-order system is given by

$$\frac{C(s)}{R(s)} = \frac{K/J}{s^2 + (B/J)s + K/J}$$

(a) If $K = 6, J = 1$ and $B = 5$, find its unit step response. What is the steady-state value? Which is the transient part of the response and what is its value at $t = \infty$. Find the values of the resonant frequency and damping ratio.

(b) If the constants are $K = 2, J = 1$ and $B = 2$, repeat (a).

(c) If $K = 9, J = 1$ and $B = 0$ repeat (a).

6.4. The open-loop transfer function of a unity feedback control system is given by

$$\frac{C(s)}{E(s)} = G(s) = \frac{9}{s(s+3)}$$

Find the natural frequency of resonance, damping ratio, damped frequency and time constant.

6.5. Determine $\omega_n, \omega_d, t_r, t_p, t_s, M_o, t_u$ (the time at which first undershoot occurs) and the number of oscillations before reaching the steady state value for the system in Problem 6.4.

6.6. An antenna-positioning servomechanism has the closed-loop transfer function given by

$$C(s)/R(s) = 25/(s^2 + 4s + 25)$$

(a) If the system is given a sudden rotation of 0.1 radian, find its response. Also find ω_n, ζ, M_o, t_p and t_s.

(b)   After an hour, the wind creates a disturbance torque of 1/–n which opposes the input
signals, find the new shaft position.

6.7.   The impulse response of a system is $\sin 2t$. Determine the transfer function of the system.

6.8.   (*a*) The unit step response of a given system is

$$y(t) = 1 - 7/3 e^{-t} + 3/2 e^{-2t} - 1/6 e^{-4t}$$

Find the transfer function.

(b)   Find its impulse response and check its transfer function.

6.9.   (a)   Determine the transfer function of a system having a gain of 3, zeros at $-2 \pm j1$ and
poles at -3 and at $-1 \pm j1$.

(b)   Obtain its unit step response.

6.10.   Determine the unit step response of a system which has the transfer function

$$T(s) = 4/(s^2 - 1)\,(s^2 + 1)$$

6.11.   The transfer function of a unity feedback position control system is given by

$$\Theta_m(s)/\Theta_r(s) = 6/(s^2 + 5s + 6)$$

(a)   Find the response Θ_m for a step input of 1 rad./sec.

(b)   How long does the system take to reach steady-state?

(c)   Is the system over-damped or under-damped?

6.12.   Repeat Prob. 6.11 if the transfer function is

$$T(s) = 2/(s^2 + 2s + 2)$$

6.13.   The characteristic equation of a second-order system is given by

$$F(s) = s^2 + 0.6s + 9 = 0$$

Find (*a*) the natural resonant frequency ω_n, (*b*) the damping ratio ζ, (*c*) the damped resonant
frequency ω_d, (*d*) the time constant τ and (*e*) the maximum or peak percentage overshoot.

6.14.   A step input of 0.1 rad. is applied to a second-order positional control system having a
transfer function

$$T(s) = 25/(s^2 + 5s + 25)$$

(a)   Find the steady-state motor shaft position θ_m.

(b)   Determine ω_n and ζ.

(c)   Is the system over-damped or under-damped?

(d)   If the system gain is doubled, find the new value of ζ.

6.15.   The characteristic equation of a system is $(s^2 + 2s + 4)$. Find ζ and ω_n

6.16. A positional servomechanism has a forward gain of 0.2, a moment of inertia of 1.25 kg –m^2 and a viscous friction of 5N-m/rad/sec. If a disturbance torque of 10N-m is applied to the system, find (a) steady-state shaft position θ_m in radians (b) steady-state error in radians.

6.17. An input θ_R of 0.1 rad. is applied to a servo having a forward gain of 100 and feedback gain of unity. If a disturbance torque of 1N-m occurs at the shaft, find (a) θ_{mss} and (b) e_{ss}.

6.18. The forward transfer function is $G(s) = 50/s(s + 4)$ and the feedback is 2. Find (a) ω_n, (b) ζ, (c) ω_d and (d) percentage overshoot.

6.19. In a system, $G(s) = 2/(s^2 + 2s + 25)$ and $H(s) = 2s$. Find (a) ω_n, (b) ζ, (c) ω_d and (d) M_o.

6.20. The closed loop transfer function of a system is given by $10K/(s^2 + 50s + 100K)$. It is excited by a unit step.

(a) If $K = 25$, find percentage overshoot. (b) Find K for critical damping. (c) Find the steady state error for each value of K above.

6.21. A closed-loop servo has J = 1.30×10^{-6} kg – m², B = 6.6×10^{-5}N-m/rad/sec, $\zeta = 0.25$ and develops a torque of 2.72×10^{-4} N-m/volt. The error detector has a constant of 1V/deg. The motor-load gear ratio is 10:1 (a) Find ω_n. (b) If the system gain K = $K_s K_a K_t$ where K_s is the error detector constant, K_a = amplifier gain and K_t is the motor-torque constant, find K_a.

6.22. The transfer function of a third-order system is given by

$$C(s)/R(s) = \frac{30}{s^3 + 10s^2 + 31s + 30}$$

Find its unit step response.

6.23. A unity feedback system with a forward transfer function

$$G(s) = 80/s(s + 2)$$

has a rate feedback transfer function $H(s) = as$.

(a) Determine the damping ratio and natural frequency of oscillations in the absence of the derivative feedback ($a = 0$).

(b) Determine the value of 'a' such that the damping ratio is 0.7.

6.24. The open-loop transfer function of a unity feedback system is

$$G(s) = A/s(1 + sT)$$

By what factor should the gain A be multiplied so that the damping ratio be increased from 0.2 to 0.6?

6.25. The forward and feedback transfer functions of a system are

$$G(s) = 10/s^2 , \quad H(s) = as + b$$

Determine the value of a and b to give a maximum overshoot of 16% and a time constant of 1 sec. for a unit step input. Time constant is defined as the inverse of the damping factor.

6.26. The forward transfer function of a unity feedback system has two cascaded blocks with transfer function

$$G_1(s) = 3/2s \text{ and } G_2(s) = 2/(s+4)$$

At $t = 0$, a sudden disturbance of unit magnitude occurs between the two blocks. Determine the response $c(t)$ for $t > 0$ for the disturbance.

6.27. The open-loop transfer function of a unity feedback system is

$$G(s) = 1/s(s+1)$$

Determine the rise time, peak time, settling time and peak overshoot.

6.28. An under-damped second-order system exhibited 17% overshoot for a step input and the frequency of oscillation was 0.5 Hz. Calculate the resonant peak and the resonant frequency of the system.

6.29. The forward and feedback transfer functions of a system is

$$G(s) = 1/(s+0.8) \text{ and } H(s) = 3/(s+2.5)$$

Determine the time response of the system for a unit step input.

Chapter 7

STEADY-STATE RESPONSE ANALYSIS

7.1. INTRODUCTION

In the previous chapter, we have defined the transient response and derived the necessary expressions. In addition, another important time-domain specification is the *steady-state error*. Imperfections in the system components such as *dead-zone, static friction, amplifier drift* and *deterioration due to ageing* will cause steady-state error. The steady-state error is a measure of the accuracy of a system when a specific type of input is applied. The steady-state performance of a stable control system is often judged by the steady-state error due to step, ramp or parabolic inputs. A system may not have steady-state error to a step input but may exhibit nonzero steady-state error to ramp input. The steady-state error of a control system depends on the type of the input as well as the type of the open-loop transfer function. In this chapter, we shall investigate the steady-state performance of a system to the three typical inputs.

7.2. SYSTEM TYPES

Control systems may be classified according to their ability to follow step, ramp, parabolic etc. inputs, since actual inputs to a system may be considered as the combinations of such inputs.

Generally, the open-loop transfer function of a control system may be represented by

$$G(s)H(s) = \frac{K(1+T_1 s)(1+T_2 s)...(1+T_m s)}{s^\ell (1+T_a s)(1+t_b s)...(1+t_p s)} \qquad \text{... (7.1)}$$

where K and T's are constants. *The type of a control system refers to the number of poles of $G(s)H(s)$ at the origin, $s = 0$.* Hence, the system described by Eq. (7.1) is of type ℓ since there are ℓ poles at the origin. If $\ell = 0$, there are zero poles at the origin and hence it is Type 0 system. If $\ell = 1$, one pole is located at the origin and hence it is Type 1 system. If $\ell = 2$, it is Type 2 system and so on. *The order of the system* described by Eq. (7.1) *is equal to $n = \ell + p$, the highest order of the denominator polynomial.*

The properties of system types are:

Type 0: A constant actuating signal results in a constant value of the controlled variable.

Type 1: A constant actuating signal results in a constant rate of change of the controlled variable.

Type 2: A constant actuating signal results in a constant second derivative of the controlled variable.

7.3. STEADY-STATE ERROR

Generally, the steady-state errors of linear control systems depend on the input and system type. If the reference input $r(t)$ and the controlled output $c(t)$ are both of the same type such as voltage, position, etc, and they are at the same signal level, then the steady-state error is

$$e(\infty) = r(t) - c(\infty) \qquad \qquad \cdots (7.2)$$

However, in practice, it is sometimes impossible to have both signals either of the same type or at the same level, Therefore, a non-unity element $H(s)$, is usually inserted in the feedback path as shown in Fig. 7.1 in order that the feedback signal is of the same type and at the same level. From Fig. 7.1, the error is given by

$$E(s) = R(s) - B(s)$$
$$= R(s) - H(s)C(s)$$
$$= R(s) - G(s)H(s)E(s)$$

or

$$\frac{E(s)}{R(s)} = \frac{1}{1+G(s)H(s)} \qquad \qquad \cdots (7.3)$$

The steady-state error e_{ss} is

$$e_{ss} = \lim_{t \to \infty} e(t) = \lim_{s \to 0} sE(s) = \lim_{s \to 0} s\frac{1}{1+G(s)H(s)}R(s) \qquad \cdots (7.4)$$

We shall now consider the effects of the step, ramp and parabolic inputs on Type 0, 1 and 2 systems.

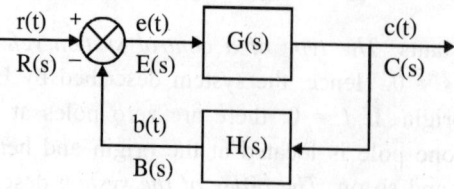

Fig. 7.1. Non-unity feedback system

7.3.1. Static Error Coefficients

The three static (steady-state) error coefficients, K_p, K_v and K_a for step, ramp and parabolic inputs respectively are defined below.

(*i*) *Step Input:* The steady-state error of a control system for step input of magnitude R is given by

$$e_{ss} = \lim_{s \to 0} sE(s) = \lim_{t \to 0} s \frac{1}{1 + G(s)H(s)} R(s)$$

$$= \lim_{s \to 0} \frac{s(R/s)}{1 + G(s)H(s)} = \lim_{s \to 0} \frac{R}{1 + G(s)H(s)}$$

$$= \frac{R}{1 + \lim_{s \to 0} G(s)H(s)} = \frac{R}{1 + K_p} \qquad \qquad \text{... (7.5)}$$

Thus, the static step error coefficient K_p is defined by

$$K_p = \lim_{s \to 0} G(s)H(s) \qquad \qquad \text{... (7.6)}$$

(*ii*) *Ramp Input:* The steady-state error for a ramp input $R(s) = R/s^2$ is

$$e_{ss} = \lim_{s \to 0} \frac{s \cdot R(s)}{1 + G(s)H(s)} = \lim_{s \to 0} \frac{sR/s^2}{1 + G(s)H(s)}$$

$$= \lim_{s \to 0} \frac{R}{s + sG(s)H(s)} = \frac{R}{\lim_{s \to 0} sG(s)H(s)} = \frac{R}{K_v} \qquad \qquad \text{... (7.7)}$$

The static ramp error coefficient K_v is defined by

$$K_v = \lim_{s \to 0} sG(s)H(s) \qquad \qquad \text{... (7.8)}$$

(*iii*) *Parabolic Input:* The steady-state error for a parabolic input $R(s) = R/s^3$ is

$$e_{ss} = \lim_{s \to 0} \frac{s \cdot R(s)}{1 + G(s)H(s)} = \lim_{s \to 0} \frac{s \cdot R/s^3}{1 + G(s)H(s)}$$

$$= \lim_{s \to 0} \frac{R}{s^2 + s^2 G(s)H(s)} = \lim_{s \to 0} \frac{R}{s^2 G(s)H(s)} = \frac{R}{K_a} \qquad \qquad \text{... (7.9)}$$

The static parabolic error coefficient K_a is defined by

$$K_a = \lim_{s \to 0} s^2 G(s)H(s) \qquad \qquad \text{... (7.10)}$$

We shall now analyze the static (steady-state) error of various system types for step, ramp and parabolic inputs.

7.3.2. Type 0 System

The general transfer function of a Type 0 system is

$$G(s)H(s) = \frac{K(1+sT_1)(1+sT_2)}{(1+sT_a)(1+sT_b)} \qquad \text{... (7.11)}$$

(*a*) *Step Input:* The static error coefficient for step input is

$$K_p = \lim_{s \to 0} G(s)H(s)$$

$$= \lim_{s \to 0} \frac{K(1+sT_1)(1+sT_2)}{(1+sT_a)(1+sT_b)} = K \qquad \text{... (7.12)}$$

Thus, the steady-state error e_{ss} for step input is

$$e_{ss} = \frac{R}{1+K_p} = \frac{R}{1+K} = \text{Constant} \qquad \text{... (7.13)}$$

(*b*) *Ramp Input:* The static ramp error coefficient K_v is

$$K_v = \lim_{s \to 0} sG(s)H(s) = 0 \qquad \text{... (7.14)}$$

and the steady-state error e_{ss} for ramp input is

$$e_{ss} = R/K_v = \infty \qquad \text{... (7.15)}$$

(*c*) *Parabolic Input:* The static parabolic error co-efficient K_a is

$$K_a = \lim_{s \to 0} s^2 G(s)H(s) = 0 \qquad \text{... (7.16)}$$

and the steady-state error for parabolic input is

$$e_{ss} = R/K_a = \infty \qquad \text{... (7.17)}$$

From the foregoing analysis, it is clear that a Type 0 system will produce a constant steady-state error for constant (step) input signals. As the error for ramp and parabolic inputs is infinity, it implies that *Type 0 system cannot follow ramp and parabolic inputs.*

7.3.3. Type 1 System

The general form of transfer function of Type 1 system is

$$G(s)H(s) = \frac{K(1+sT_1)(1+sT_2)}{s(1+sT_a)(1+sT_b)} \qquad \text{... (7.18)}$$

(*a*) **Step Input:** The static step error coefficient K_p is

$$K_p = \lim_{s \to 0} G(s)H(s) = \infty \qquad \qquad ... (7.19)$$

Therefore, the steady-state error e_{ss} for step input is

$$e_{ss} = R/(1 + K_p) = 0 \qquad \qquad ... (7.20)$$

(*b*) **Ramp Input:** The static ramp error coefficient K_v is

$$K_v = \lim_{s \to 0} sG(s)H(s) = K \qquad \qquad ... (7.21)$$

and the steady-state error for ramp input is

$$e_{ss} = R/K_v = R/K = \text{Constant} \qquad \qquad ... (7.22)$$

(*c*) **Parabolic Input:** The static parabolic error coefficient K_a is

$$K_a = \lim_{s \to 0} s^2 G(s)H(s) = 0 \qquad \qquad ... (7.23)$$

The steady-state error e_{ss} for parabolic input is

$$e_{ss} = R/K_a = \infty \qquad \qquad ... (7.24)$$

From the above analysis, it may be noted that *Type* 1 *system will follow a step input with zero steady-state error and a ramp input with constant steady-state error. It cannot follow the parabolic input.*

7.3.4. Type 2 System

The general form of transfer function of a Type 2 system is

$$G(s)H(s) = \frac{K(1 + sT_1)(1 + sT_2)}{s^2(1 + sT_a)(1 + sT_b)} \qquad \qquad ... (7.25)$$

(*a*) **Step Input:** The static step error constant K_p is

$$K_p = \lim_{s \to 0} G(s)H(s) = \infty \qquad \qquad ... (7.26)$$

The steady-state error e_{ss} for a step input is

$$e_{ss} = R/(1 + K_p) = 0 \qquad \qquad ... (7.27)$$

(*b*) **Ramp Input:** The static ramp error constant K_v is

$$K_v = \lim_{s \to 0} sG(s)H(s) = \infty \qquad \qquad ... (7.28)$$

The steady-state error e_{ss} for a ramp input is

$$e_{ss} = R/K_v = 0 \qquad \qquad ... (7.29)$$

(*c*) **Parabolic Input:** The static parabolic error constant K_a is

$$K_a = \lim_{s \to 0} s^2 G(s)H(s) = K \qquad \qquad ... (7.30)$$

The steady-state error e_{ss} for a parabolic input is

$$e_{ss} = R/K_a = R/K = \text{Constant} \qquad \ldots (7.31)$$

From the above discussion, it may be seen that *Type 2 system will follow step and ramp inputs with zero steady-state error and a parabolic input with a constant steady-state error.*

The results obtained in this section are summarized in Table 7.1

Table 7.1. Steady-state Error Characteristics

System Type	K_p	K_v	K_a	Step Error	Ramp Error	Parabolic Error
0	K	0	0	$e_{ss} = R/(1 + K)$	$e_{ss} = \infty$	$e_{ss} = \infty$
1	∞	K	0	$e_{ss} = 0$	$e_{ss} = R/K_v$	$e_{ss} = \infty$
2	∞	∞	K	$e_{ss} = 0$	$e_{ss} = 0$	$e_{ss} = R/K_a$

The above error constants are significant in error analysis only when the input is step, ramp or parabolic function and is applicable only to stable systems. It may be noted that the steady-state performance can be improved by increasing the type of the system by adding an integrator. However, this increases the order of the system resulting into a less stable system.

The main disadvantage of the static error constants is that they are not useful when the input is other than the three basic types. The steady-state error e_{ss} obtained through the static error coefficients is either zero, finite or infinite. Thus they do not provide any information on how the error varies with time. A more general representation of the steady state error is by error series discussed in the next section.

7.3.5. Dynamic Error Series

In this section, the concept of error constants is generalized to include inputs of any arbitrary function of time. The error $E(s)$ is given by

$$E(s) = \frac{R(s)}{1 + G(s)H(s)} = R(s) \cdot W_e(s) \qquad \ldots (7.32)$$

where $W_e(s)$ is known as *error transfer function* and is given by

$$W_e(s) = 1/[1 + G(s)H(s)] \qquad \ldots (7.33)$$

The error signal $e(t)$ may be written, using convolution integral, as

$$e(t) = \int_{-\infty}^{t} w_e(\tau) r(t - \tau) \qquad \ldots (7.34)$$

where $w_e(\tau)$ is the inverse Laplace transform of $W_e(s)$.

Assuming that the first n derivatives of $r(t)$ exist for all values of t, the function $r(t - \tau)$ can be expand into a Taylor series.

$$r(t - \tau) = r(t) - \tau\dot{r}(t) + \frac{\tau^2}{2!}\ddot{r}(t) - \frac{\tau^3}{3!}\dddot{r}(t) + \ldots\ldots\ldots \qquad \ldots (7.35)$$

where the dot represents the derivative with respect to time.

The limit of Eq. (7.34) may be taken from 0 to t, since $r(t) = 0$ for $t < 0$. Substituting Eq. (7.35) into Eq. (7.34), we get

$$e(t) = \int_0^t \omega_e(\tau)\left[r(t) - \tau\dot{r}(t) + \frac{\tau^2}{2!}\ddot{r}(t) - \frac{\tau^3}{3!}\dddot{r}(t) + \ldots\ldots\ldots\right]d\tau \qquad \ldots (7.36)$$

$$= r(t)\int_0^t \omega_e(\tau)d\tau - \dot{r}(t)\int_0^t \tau w_e(\tau)d\tau + \ddot{r}(t)\int_0^t \frac{\tau^2}{2!}w_e(\tau)d\tau - \ldots\ldots$$

$$\ldots (7.37)$$

The steady-state error is obtained by

$$e_{ss} = \lim_{t\to\infty}e(t) = \lim_{t\to\infty}e_s(t) \qquad \ldots (7.38)$$

where e_s the steady-state part of $e(t)$, is given by

$$e_s = r_s(t)\int_0^\infty w_e(\tau)d\tau - \dot{r}_s(t)\int_0^\infty \tau w_e(\tau) + \ddot{r}_s(t)\int_0^\infty \frac{\tau^2}{2!}w_e(\tau)d\tau \quad \ldots (7.39)$$

where $r_s(t)$ is the steady-state part of $r(t)$. By defining

$$C_i = (-1)^i\int_0^\infty \tau^i w_e(\tau)d\tau, \qquad\qquad i = 0, 1, 2,\ldots\ldots\ldots \qquad \ldots (7.40)$$

Eq. (7.39) becomes

$$e_s(t) = C_0 r_s(t) + \frac{C_1\dot{r}_s(t)}{1!} + \frac{C_2\ddot{r}_s(t)}{2!} + \ldots\ldots\ldots + \frac{C_n r_s(n)(t)}{n!} \qquad \ldots (7.41)$$

Eq. (7.41) is called the *error series* and the coefficients C_i are known as *generalized error coefficients* or *dynamic error coefficients*.

Evaluating the generalized error constants from Eq. (7.40) is not simple. However they may be evaluated directly from the error transfer function $W_e(s)$.

$$W_s(s) = \mathcal{L}[w_e(\tau)] = \int_0^\infty w_e(\tau)e^{-st}d\tau \qquad \ldots (7.42)$$

Taking the limit on both sides of Eq.(7.42) as s approaches zero,

$$\lim_{s \to 0} W_e(s) = \lim_{s \to 0} \int_0^\infty w_e(\tau) e^{-s\tau} d\tau = C_0$$

The derivative of $W_e(s)$ with respect to s is given by

$$\lim_{s \to 0} \frac{dW_e(s)}{ds} = \lim_{s \to 0} - \int_0^\infty \tau w_e(\tau) e^{-s\tau} d\tau = C_1 \qquad \qquad \dots (7.43)$$

In general,

$$C_i = \lim_{s \to 0} \left[d^i W_e(s) / ds^i \right] \qquad \qquad \dots (7.44)$$

An advantage of error series is that the steady-state error can be obtained in terms of the input function and its derivatives.

Example 7.1. *The open-loop transfer function of a unity feedback system is given by* $G(s) = K/(s + 1)$. *Find the steady state error for unit step, ramp and parabolic inputs by using* (a) *static error constants and* (b) *error series.*

Solution:

(a) Static Error Constants are:

$$K_p = \lim_{s \to 0} G(s) = K$$

$$K_v = \lim_{s \to 0} sG(s) = 0$$

$$K_a = \lim_{s \to 0} s^2 G(s) = 0$$

Therefore, the steady-state error is

$$e_{ss} = 1/(1 + K) \qquad \qquad \text{for unit step input}$$

$$e_{ss} = \infty \qquad \qquad \text{for unit ramp input}$$

$$e_{ss} = \infty \qquad \qquad \text{for unit parabolic input}$$

(b) The error transfer function $W_e(s)$ is

$$W_e(s) = \frac{1}{1 + G(s)} = \frac{s+1}{s+K+1}$$

The generalised error constants are:

$$C_0 = \lim_{s \to 0} W_e(s) = 1/(K+1)$$

$$C_1 = \lim_{s \to 0} \frac{dW_e(s)}{ds} = K/(K+1)^2$$

$$C_2 = \lim_{s \to 0} \frac{d^2 W_e(s)}{ds} = -2K/(K+1)^3$$

Therefore the error series e_s is

$$e_s = \frac{1}{K+1} r_s(t) + \frac{K}{(K+1)^2} \dot{r}_s(t) + \frac{-K}{(K+1)^2} \ddot{r}_s(t) + \dots$$

(i) For unit step input:

$$r_s(t) = 1, \ \dot{r}_s(t) = \ddot{r}_s(t) = 0$$

Hence

$$e_s(t) = \frac{1}{K+1}$$

(ii) For unit ramp input:

$$r_s(t) = t, \ \dot{r}_s(\tau) = 1, \ \ddot{r}_s(t) = 0$$

Hence

$$e_s(t) = \frac{1}{K+1} t + \frac{K}{(K+1)^2}$$

(iii) For unit parabolic input:

$$r_s(t) = t^2/2, \ \dot{r}_1(t) = t, \ \ddot{r}_1(t) = 1$$

$$e_s(t) = \frac{1}{K+1} \frac{t^2}{2} + \frac{Kt}{(K+1)^2} - \frac{K}{(K+1)^3}$$

From the above, it is apparent that the static error constants fail to indicate the exact manner the steady-state error varies with time whereas the error series provides the necessary information.

Example 7.2. *The open-loop transfer function of a unity feedback system is*

$$G(s) = K/(s + 1)$$

Find the error series when r(t) = sin ωt.

Solution:

$$r_s(t) = \sin \omega t$$

$$\dot{r}_s(t) = \omega \cos \omega t$$

$$\ddot{r}_s(t) = -\omega^2 \sin \omega t$$

The error series is given by

$$e_s(t) = \left(C_0 - \frac{C_2}{2!}\omega^2 + \frac{C_4}{4!}\omega^4 - \ldots\ldots\ldots \right)\sin\omega t + \left(C_1\omega - \frac{C_3}{3!}\omega^2 + \ldots\ldots \right)\cos\omega t$$

Thus the error series is an infinite series. The convergence of the series depends on K. Assuming $K = 100$ and $\omega = 2$,

$$C_0 = 1/(K + 1) = 0.0099$$
$$C_1 = K/(K + 1)^2 = 0.0098$$
$$C_2 = -2K/(K + 1)^3 = -0.000194$$
$$C_3 = 6K/(K + 1)^4 = 0.096 \times 10^{-6}$$

Therefore $e_s(t)$ is

$$e_s(t) = \left(0.0099 + \frac{0.000194}{2} \times 4 \right)\sin 2t + 0.0196\cos 2t$$

$$= 0.01029 \sin 2t + 0.0196 \cos 2t$$

$$= 0.02215 \sin (2t + 62.3°)$$

7.4. ERROR CRITERIA

In the design of control systems, the performance specifications may be given in terms of the transient-response behaviour to specific inputs or in terms of a performance index. *A performance index is a number which indicates the goodness of the system performance.* If a control system is designed so that the values of system parameters result in a minimum or maximum value of the selected performance index, then the system is considered optimal. The optimal values of the parameters directly depend upon the selected performance index.

Many different performance indices may be used in the design of control systems. In general, the requirements of a performance index are:

(*a*) *Selectivity:* An optimal adjustment of the system parameters must be clearly distinguished from non-optimal adjustments.

(*b*) *Un-ambiguity:* A performance index must yield a single value without ambiguity.

(*c*) *Dependence on Parameters:* A performance index must be a function of system parameters and exhibit a minimum or maximum.

(*d*) *Ease of Computation:* A performance index must be easily computed, analytically or experimentally.

7.5. ERROR PERFORMANCE INDICES

In this section, we shall discuss several error criteria in which the performance indices are integrals of some function or weighted function of the error. The values of these integrals can be obtained as functions of system parameters. Therefore, once the performance index is specified, the optimal system can be designed by adjusting the system parameters to yield the smallest value of the integral.

Among the various performance indices proposed in the literature, we shall discuss the following four.

(*i*) ISE criterion: $\int_0^\infty e^2(t)\,dt$

(*ii*) ITSE criterion: $\int_0^\infty te^2(t)\,dt$

(*iii*) IAE criterion: $\int_0^\infty |e(t)|\,dt$

(*iv*) ITAE criterion: $\int_0^\infty t|e(t)|\,dt$

Consider a control system whose desired output is $x(t)$ and the actual output is $y(t)$. Then the error $e(t)$ is

$$e(t) = x(t) - y(t) \qquad\qquad \text{... (7.49)}$$

If $e(t)$ is zero as t approaches infinity, the performance index is zero. If $e(t)$ does not approach zero as t tends to infinity, then we can define the error $e(t)$ as

$$e(t) = y(\infty) - y(t) \qquad\qquad \text{... (7.50)}$$

and the performance index will yield a finite value.

7.5.1. Integral Square-Error (ISE) Criterion

The performance index as defined by ISE criterion is

$$J = \int_0^\infty e^2(t)\,dt \qquad\qquad \text{... (7.51)}$$

The upper limit may be replaced by T which is chosen to be large such that $e(t)$ for $t > T$ is negligible. *The optimum system is that which minimizes this integral.* This performance index is extensively used in the design of control systems because of the case of computing the integral analytically and experimentally. Fig.7.2 shows $x(t)$, $y(t)$, $e(t)$ and $\int e^2(t)dt$ where $x(t)$, the desired output is a unit step. The integral of $e^2(t)$ from 0 to T is the total area under the curve $e^2(t)$.

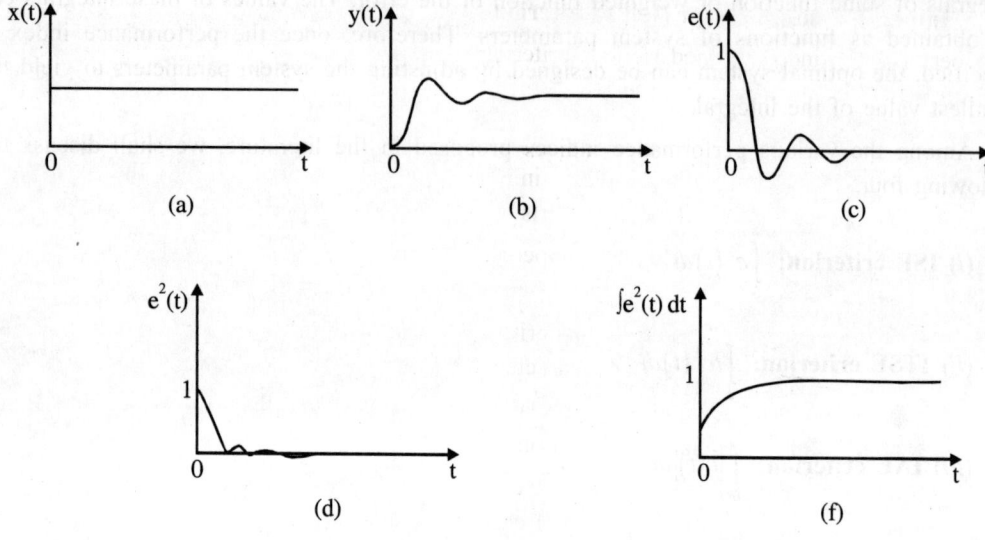

Fig. 7.2. ISE criterion

The characteristics of this index are:

(a) It weighs large errors heavily and small errors lightly.

(b) It is not very selective as seen from Fig.7.2; a change in ζ from 0.5 to 0.7 does not yield much change in the value of the integral.

(c) A system designed by this criterion is fast and oscillatory and therefore has poor relative stability.

For some system such as space craft systems, the minimization of the performance index results in the minimization of power consumption.

7.5.2. Integral-of-Time-Multiplied Square-Error (ITSE) Criterion

The performance index based on ITSE criterion is

$$J = \int_0^\infty te^2(t)dt \qquad\qquad\qquad ... (7.52)$$

The optimum system is that which minimizes this integral.

The characteristics of this criterion are:

(a) In unit-step response, a large initial error is lightly weighed.

(b) The error occurring later in the transient response is penalized heavily.

(c) This has better selectivity compared to ISE criterion.

7.5.3. Integral Absolute-Error (IAE) Criterion.

The performance index defined by IAE criterion is

$$J = \int_0^\infty |e| dt \qquad\qquad \text{... (7.53)}$$

The optimum system is that which minimizes the error. This is one of the most easily applied performance indices. The characteristics of this error criterion are:

(a) By this criterion, highly under-damped and highly over-damped systems cannot be made optimum.

(b) An optimum system based on this criterion is that which has reasonable damping and a satisfactory transient-response characteristic.

(c) The selectivity is poor and cannot be easily evaluated analytically. However it is particularly useful for analog computer simulation.

Fig. 7.3 shows $e(t)$ and $|e(t)|$ versus t curves. The minimization of the IAE criterion is directly related to the minimization of fuel consumption of spacecraft systems.

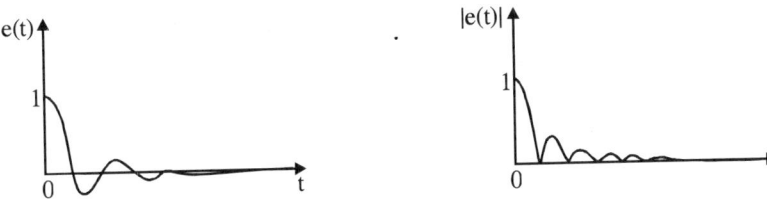

Fig. 7.3. IAE criterion

7.5.4. Integral-of-Time -multiplied Absolute-Error (ITAE) Criterion

According to the above criterion, the optimum system is that which minimizes the performance index

$$J = \int_0^\infty t|e(t)| dt \qquad\qquad \text{... (7.54)}$$

The characteristics of this performance criterion are:

(a) A large initial error in unit-step response is lightly weighed and errors occurring late in the transient response are penalized heavily.

(b) A system designed based on this criterion has small overshoot in the transient response and the oscillations are well damped.

(c) The criterion possesses a good selectivity.

(d) It can easily be measured experimentally.

(e) However, it is difficult to evaluate analytically.

7.5.5. Comparison of Error Criteria

The performance curves of the above criteria are shown in Fig. 7.4.

The system considered for comparison is

$$C(s)/R(s) = 1/\left(s^2 + 2\zeta s + 1\right)$$

... (7.55)

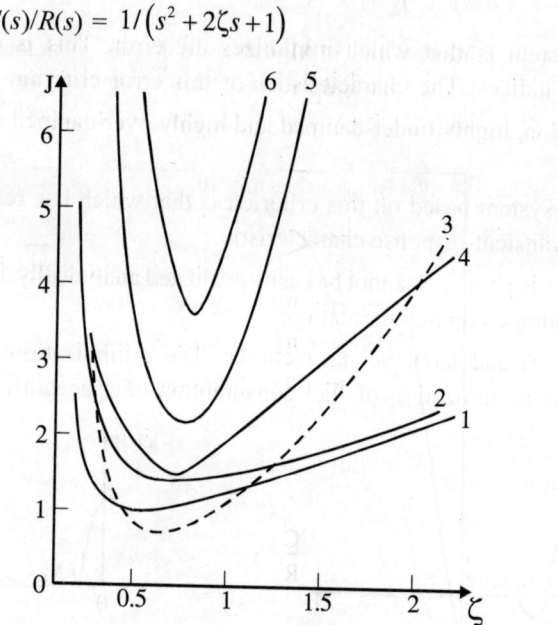

Fig. 7.4. Error criteria performance curves

1. $J = \int_0^\infty e^2(t)\, dt$

2. $J = \int_0^\infty \left[e^2(t) + \dot{e}^2(t)\right] dt$

3. $J = \int_0^\infty t\, e^2(t)\, dt$

4. $J = \int_0^\infty \left|e(t)\right| dt$

5. $J = \int_0^\infty t\left|e(t)\right| dt$

6. $J = \int_0^\infty t\left[\left|e(t)\right| + \left|\dot{e}(t)\right|\right] dt$

Examination of these curves reveals the following:

(1) The ISE criterion does not possess good selectivity since the curve is rather flat near the point where the performance index is the minimum.

(2) Other performance indices have good selectivity.

(3) For the second-order system considered, $\zeta = 0.7$ corresponds to the optimal or near optimal value for all the performance indices.

When $\zeta = 0.7$, the second-order system exhibits a swift response to a step input with approximately 5% overshoot.

7.5.6. Application of ITAE Criterion to *n-th* Order System

Consider the closed-loop transfer function of the form

$$\frac{C(s)}{R(s)} = \frac{a_0}{s^n + a_{n-1}s^{n-1} + \dots\dots a_s + a_0} \qquad \dots (7.56)$$

The coefficients that minimize the ITAE criterion has been determined. Table 7.2 shows these coefficients. Fig. 7.5 shows the step response curves of the optimal systems. Table 7.3 shows the coefficients for ramp input.

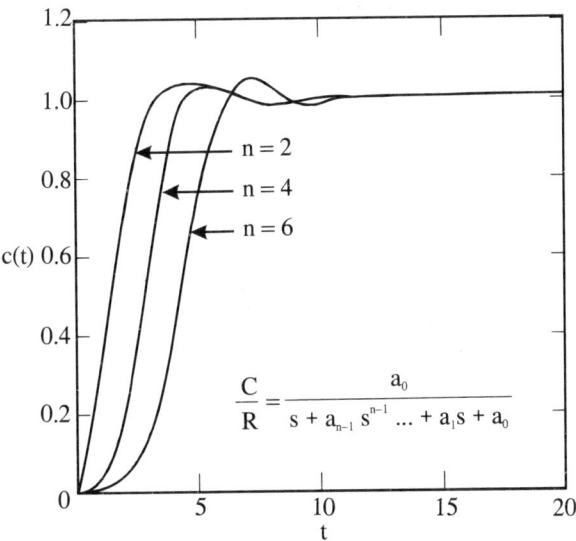

Fig. 7.5. Optimal system step response

Table 7.2. Optimal value of coefficients based on ITAE criterion for step input

$$\frac{C(s)}{R(s)} = \frac{a_0}{s^n + a_{n-1}s^{n-1} + \dots + a_1 s + a_0}$$

$$a_0 = \omega_n^n$$

$n = 1$	$a_o = \omega_n$				
$n = 2$	$a_1 = 1.4\omega_n$				
$n = 3$	$a_1 = 2.15\,\omega_n^2$	$a_2 = 1.75\omega_n$			
$n = 4$	$a_1 = 2.7\,\omega_n^3$	$a_2 = 3.4\,\omega_n^2$	$a_3 = 2.1\omega_n$		
$n = 5$	$a_1 = 3.4\,\omega_n^4$	$a_2 = 5.5\,\omega_n^3$	$a_3 = 5\,\omega_n^2$	$a_4 = 2.8\omega_n$	
$n = 6$	$a_1 = 3.95\,\omega_n^5$	$a_2 = 7.45\,\omega_n^4$	$a_3 = 8.6\,\omega_n^3$	$a_4 = 6.6\,\omega_n^2$	$a_5 = 3.25\omega_n$

Table 7.3. Optimal values of coefficients based on IATE criterion for ramp input

[Transfer function $T(s) = (b_1s + b_0)/(s^n + a_{n-1}s^{n-1} + \ldots + a_2s^2 + a_1s + a_0)$]

$n = 2$	$s^2 + 3.2\omega_0 s + \omega_0^2$
$n = 3$	$s^3 + 1.75\ \omega_0 s^2 + 3.25\ \omega_0^2 s + \omega_0^3$
$n = 4$	$s^4 + 2.41\ \omega_0 s^3 + 4.93\ \omega_0^2 s^2 + 5.14\omega_0 s + \omega_0^4$
$n = 5$	$s^5 + 2.19\ \omega_0 s^4 + 6.5\ \omega_0^2 s^3 + 6.3\omega_0^3 s^2 + 5.24\omega_0^4 s + \omega_0^5$

7.6. INTRODUCTION TO SYSTEM OPTIMIZATION

In this section, we shall solve simple optimization problem using certain error criteria by Laplace transform method.

Example 7.3. *Consider a unity feedback system with the forward transfer function* $G(s) = 1/s(s + 2\zeta)$. *Assuming the system is initially at rest and at $t = 0$, a unit step input is applied to the system, find the optimal value of $\zeta > 0$ such that ISE criterion is minimum.*

Solution:

If $F(s)$ is the Laplace transform of $f(t)$, then the Laplace transform of $\int_0^\infty f(t)dt = F(s)/s$. Applying the final value theorem,

$$\lim_{t \to \infty} \int_0^t f(t)dt = \lim_{s \to 0} s \cdot \frac{F(s)}{s} = \lim_{s \to 0} F(s) \qquad \ldots (7.57)$$

The ISE criterion is

$$J = \int_0^\infty e^2(t)dt = \lim_{s \to 0} F(s)$$

where $F(s)$ is the Laplace transform of $e^2(t)$. For the given system

$$\frac{E(S)}{R(s)} = \left(s^2 + 2\zeta s\right)\Big/\left(s^2 + 2\zeta s + 1\right)$$

For a unit step input, $R(s) = 1/s$. Hence,

$$E(s) = \frac{s^2 + 2\zeta s}{s\left(s^2 + 2\zeta s + 1\right)} = \frac{s + 2\zeta}{s^2 + 2\zeta s + 1}$$

$$= \frac{s + \zeta}{s^2 + 2\zeta s + 1} + \frac{\zeta}{s^2 + 2\zeta s + 1}$$

Taking the inverse Laplace,

$$e(t) = e^{-\zeta t}\left[\cos\sqrt{1-\zeta^2}\, t + \frac{\zeta}{\sqrt{1-\zeta^2}}\sin\sqrt{1-\zeta^2}\, t\right] \qquad t \geq 0$$

Hence

$$e^2(t) = e^{-2\zeta t}\left[\frac{1}{2\zeta(1-\zeta^2)} + \frac{1}{2}\frac{1-2\zeta^2}{1-\zeta^2}\cos 2\sqrt{1-\zeta^2}\, t + \frac{\zeta}{\sqrt{1-\zeta^2}}\sin 2\sqrt{1-\zeta^2}\, t\right]$$

Therefore,

$$F(s) = \mathcal{L}\left[e^2(t)\right] = \frac{1}{2\sqrt{(1-\zeta^2)}}\frac{1}{s+2\zeta} + \frac{1-2\zeta^2}{2(1-\zeta^2)}\cdot\frac{s+2\zeta}{(s+2\zeta)^2+4(1-\zeta^2)}$$

$$+ \frac{\zeta}{\sqrt{1-\zeta^2}}\frac{2\sqrt{1-\zeta^2}}{(s+2\zeta)^2+4(1-\zeta^2)} \qquad\qquad \text{... (7.58)}$$

$$J = \lim_{s\to 0} F(s)$$

$$= \frac{1}{2\sqrt{1-\zeta^2}}\cdot\frac{1}{2\zeta} + \frac{1-2\zeta^2}{2(1-\zeta^2)}\cdot\frac{2\zeta}{4} + \frac{2\zeta}{\sqrt{1-\zeta^2}}\cdot\frac{2\sqrt{1-\zeta^2}}{4}$$

$$= \frac{1}{4\zeta\sqrt{1-\zeta^2}} + \frac{(1-2\zeta^2)\zeta}{4(1-\zeta^2)} + \frac{\zeta}{2}$$

Neglecting higher order terms in ζ, we get

$$J = \frac{3}{4}\zeta + \frac{1}{4\zeta} \qquad\qquad \text{... (7.59)}$$

The optimal value of J is obtained by equating $dJ/d\zeta$ to zero, and solving for ζ.

$$dJ/d\zeta = \frac{3}{4} - \frac{1}{4\zeta^2} = 0$$

or,

$$\zeta = \frac{1}{\sqrt{3}} = 0.577$$

Hence,

$$\min J = \frac{3\zeta}{4} + \frac{1}{4\zeta}$$

$$= 0.433 + 0.433$$

$$= 0.866$$

Example 7.4. *For the system in Example 7.3, find the optimal value of ζ for ITSE criterion to be minimum.*

Solution:

The ITSE criterion is

$$J = \int_0^\infty t\, e^2(t)\, dt$$

The Laplace transform of $t e^2(t)$ is

$$\mathcal{L}\left[t e^2(t) \right] = -\frac{d}{ds} F(s) \tag{7.60}$$

where $F(s)$ is the Laplace transform of $e^2(t)$ and is given by Eq.(7.58)

$$-\frac{d}{ds} F(s) = \frac{1}{2(1-\zeta^2)} \frac{1}{(s+2\zeta)^2} + \frac{1-2\zeta^2}{2(1-\zeta^2)} \cdot \frac{(s+2\zeta)^2 - 4(1-\zeta^2)}{\left[(s+2\zeta)^2 + 4(1-\zeta^2) \right]^2}$$

$$+ \frac{4\zeta(s+2\zeta)}{\left[(s+2\zeta)^2 + 4(1-\zeta^2) \right]^2}$$

Applying final value theorem,

$$J = \lim_{s \to 0} \left[-\frac{d}{ds} F(s) \right]$$

$$= \frac{1}{2(1-\zeta^2)} \frac{1}{4\zeta^2} + \frac{1-2\zeta^2}{2(1-\zeta^2)} \cdot \frac{2\zeta^2 - 1}{4} + \frac{\zeta^2}{2}$$

$$= \zeta^2 + \frac{1}{8\zeta^2}$$

Differentiating J with respect to ζ and equating it to zero

$$dJ/d\zeta = 2\zeta - \frac{1}{4\zeta^3} = 0$$

or,

$$\zeta = 0.595$$

and,

$$\min J = 0.595^2 + \frac{1}{8(0.595)^2} = 0.707$$

7.7. ADDITIONAL EXAMPLES

Example 7.5. *The open-loop transfer function G(s)H(s) is*

$$G(s)H(s) = K/s^2(s+5)(s+2)$$

Find (a) e_{ss} for the three standard inputs and (b) the error series.

(a) Steady state error:

$$K_p = \lim_{s \to 0} G(s)H(s) = \lim_{s \to 0} \frac{K}{s^2(s+2)(s+5)} = \infty$$

$$K_v = \lim_{s \to 0} sG(s)H(s) = \infty$$

$$K_s = \lim_{s \to 0} s^2 G(s)H(s) = \lim_{s \to 0} \frac{K}{(s+2)(s+5)} = \frac{K}{10}$$

The static-state error for:

(i) *Unit step input is*

$$e_{ss} = \frac{1}{1+K_p} = 0$$

(ii) *Unit ramp input is*

$$e_{ss} = 1/K_V = 0$$

(iii) *Unit parabolic input is*

$$e_{ss} = 1/K_a = 10/K$$

(b) The unit ramp input is $r(t) = t^2/2$. Hence

$$\dot{r}(t) = t \text{ and } \ddot{r}(t) = 1$$

$$W_e(s) = 1/[1+G(s)H(s)]$$

$$= \frac{s^2(s+2)(s+5)}{s^2(s+2)(s+5)+K}$$

$$C_0 = W_e(s)\,|_{s=0} = 0$$
$$C_1 = dW_e(s)/ds\,|_{s=0} = 0$$
$$C_2 = d^2\,W_e(s)/ds^2\,|_{s=0} = 10/K$$

Error series is given by

$$C_0 r(t) + C_1 \dot{r}(t) + C_2 \ddot{r}(t) = 10/K$$

Example 7.6. *Compute IAE and ITAE criteria for the system*

$$C(s)/R(s) = 1/(s^2 + 2\zeta s + 1) \qquad (\zeta \geq 1)$$

Assume the system is initially at rest and is subjected to a unit-step input.

Solution:

For a unit step input, $R(s) = 1/s$. The error $E(s)$ is

$$E(s) = \frac{s^2 + 2\zeta s}{s(s^2 + 2\zeta s + 1)} = \frac{s + 2\zeta}{s^2 + 2\zeta s + 1}$$

For $\zeta \geq 1$, there is no overshoot. Hence $|e(t)| = e(t)$ for $t \geq 0$.

(a) *The IAE criterion is*

$$J = \int_0^\infty |e(t)|\,dt = \int_0^\infty e(t)\,dt = \underset{s \to 0}{lt}\, E(s) = 2\zeta$$

(b) *The ITAE criterion is*

$$J = \int_0^\infty t\,|e(t)| = \int_0^\infty te(t)\,dt$$

$$= \lim_{s \to 0}\left[-\frac{d}{ds} E(s)\right]$$

$$= \lim_{s \to 0}\left[-\frac{(s^2 + 2\zeta s + 1) - (s + 2\zeta)(2s + 2\zeta)}{(s^2 + 2\zeta + 1)^2}\right]$$

$$= 4\zeta^2 - 1.$$

Example 7.7. *The closed loop transfer function of a system is*

$$\frac{C(s)}{R(s)} = \frac{1}{(s^2 + 2\zeta s + 1)}$$

Compute $\int_0^\infty c^2(t)dt$ for a unit impulse input.

Solution:

For a unit impulse input $R(s) = 1$.

Hence,

$$C(s) = \frac{1}{s^2 + 2\zeta s + 1}$$

$$c(t) = \frac{1}{\sqrt{1-\zeta^2}} e^{-\zeta t} \sin\sqrt{1-\zeta^2}\, t$$

$$c^2(t) = \frac{1}{\sqrt{1-\zeta^2}} e^{-2\zeta t} \sin\sqrt{1-\zeta^2}\, t$$

$$= \frac{1}{1-\zeta^2} e^{-2\zeta t} \frac{1}{2}\left(1 - \cos 2\sqrt{1-\zeta^2}\, t\right)$$

$$\mathcal{L}\left[c^2(t)\right] = \frac{1}{2(1-\zeta^2)}\left[\frac{1}{s+2\zeta} - \frac{s+2\zeta}{(s+2\zeta)^2 + 4(1-\zeta^2)}\right]$$

$$\int_0^\infty c^2(t)dt = \lim_{s\to 0}\left[c^2(t)\right] = \frac{1}{4\zeta}$$

Example 7.8. *Given* $x' + 2x' + x = 0$ *for* $0 < \zeta \le 0.9$. *Find the value of* ζ *which minimizes the performance index*

$$J = \int_0^\infty \left[x^2(t) + \dot{x}^2(t)\right]dt$$

The initial conditions are $x(0) = 1$, $\dot{x}(0) = 0$.

Solution:

The solution of the given equation is

$$x(t) = K_1 e^{\lambda_1 t} + K_2 e^{\lambda_2 t}$$

$$\lambda_1 = -\zeta + \sqrt{\zeta^2 - 1}$$

$$\lambda_2 = -\zeta - \sqrt{\zeta^2 - 1}$$

$$\dot{x}(t) = K_1\lambda_1 e^{\lambda_1 t} + K_2\lambda_2 e^{\lambda_2 t}$$

$$x^2(t) = K_1^2 e^{2\lambda_1 t} + K_2^2 e^{2\lambda_2 t} + 2K_1K_2 e^{(\lambda_1+\lambda_2)t}$$

$$\dot{x}^2(t) = K_1^2\lambda_1^2 e^{2\lambda_1 t} + K_2^2\lambda_2^2 e^{2\lambda_2 t} + 2K_1K_2\lambda_1\lambda_2\, e^{(\lambda_1+\lambda_2)t}$$

$$J = \int_0^\infty [x^2(t) + \dot{x}^2(t)]dt$$

$$J = \int_0^\infty [K_1^2(1+\lambda_1^2)e^{2\lambda_1 t} + 2K_1K_2(1+\lambda_1\lambda_2)e^{(\lambda_1+\lambda_2)t}$$

$$+ K_2^2(1+\lambda_2^2)e^{2\lambda_2 t}]dt$$

Integrating and substituting the upper limit ($t = \infty$) and low limit ($t = 0$) and since the real parts of λ_1 and λ_2 are negative, we get

$$J = -\left[\frac{K_1^2(1+\lambda_1^2)}{2\lambda_1} + \frac{2K_1K_2(1+\lambda_1\lambda_2)}{\lambda_1+\lambda_2} + \frac{K_2^2(1+\lambda_2^2)}{2\lambda_2}\right]$$

since

$$\lambda_1 + \lambda_2 = -2\zeta,\ \lambda_1\lambda_2 = 1$$

$$J = \zeta(K_1^2 + K_2^2) + \frac{2K_1K_2}{\zeta}$$

$$x(0) = 1 = K_1 + K_2$$

$$\dot{x}(0) = 0 = K_1\lambda_1 + K_2\lambda_2$$

$$K_1 = \lambda_2/(\lambda_2 - \lambda_1) \text{ and } K_2 = -\lambda_1/(\lambda_2 - \lambda_1)$$

$$J = \zeta + 1/2\zeta$$

$$dJ/d\zeta = 1 - \frac{1}{2\zeta^2} = 0$$

$$\zeta = 1/\sqrt{2}$$

$$J = 1/\sqrt{2} + \sqrt{2}/2$$

$$= 1.414$$

Example 7.9. *For the system considered in Ex.(7.3), compute the performance index*

$$J = \int_0^\infty t[|e(t) + \dot{e}(t)|]dt$$

Assume the system is initially at rest and is subjected to a unit step input.

Solution:

$$E(s) = \frac{s^2 + 2\zeta s}{s^2 + 2\zeta s + 1} \cdot \frac{1}{s} = \frac{s + 2\zeta}{s^2 + 2\zeta s + 1}$$

For $\zeta \geq 1$, there is no overshoot. Hence $|e(t)| = e(t)$

Since $e'(t) < 0$, $|e'(t)| = -e'(t)$ for $t \geq 0$.

$$J = \int_0^\infty t\left[|e(t) + e'(t)|\right]dt = \int_0^\infty \left[te(t) - t\dot{e}(t)\right]$$

$$= \lim_{s \to 0}\left[-\frac{d}{ds}E(s) + \frac{d}{ds}sE(s)\right]$$

$$= \lim_{s \to 0}\left[(-1 + s)\frac{dE(s)}{ds} + E(s)\right]$$

$$= 4\zeta^2 + 2\zeta - 1$$

7.8. SUMMARY

In this chapter, we have first classified the systems as Type 0, Type 1 and Type 2 systems. We have then defined the steady-state or static error and error coefficients. We have derived expressions for position, velocity and acceleration error coefficients K_p, K_v and K_a respectively for Type 0, Type 1 and Type 2 systems. The steady-state performance of control systems has been analysed in terms of system type and the type of inputs. It has been shown that Type 0 system follows the step input with constant error and does not follow the ramp or parabolic inputs. Similarly, it has been shown that Type 1 system follows the step input with no error, the ramp input with constant error and does not follow the parabolic input; and Type 2 system follows the step and ramp inputs with zero error and the parabolic input with constant error. The main disadvantage of static error coefficients is that they are not useful when the input signal is other than the three basis signals. Hence, to overcome this difficulty, we have then defined the error series that are applicable for any type of input.. The advantages of error series have been brought out clearly. The various commonly used performance indices such as ISE, ITSE, IAE and ITAE criteria have been defined. The requirements of performance indices are selectivity, un-ambiguity, dependence on parameters and ease of computation. We have also compared the error criteria. Then, we have introduced system optimization and solved some system optimization problems using error criteria. Illustrative examples have been worked out to make the concept clear.

REVIEW QUESTIONS

7.1. What is meant by steady-state response?

7.2. What is meant by type of a system?

7.3. Define the steady-state error.

7.4. Define the static error coefficients.

7.5. What is an error series? What is its advantage?

7.6. What are the important requirements of a performance index?

7.7. Mention the various performance indices.

7.8. What are the characteristics of ISE criterion?

7.9. What are the characteristics of ITSE criterion?

7.10. What are the characteristics of IAE criterion?

7.11. What are the characteristics of ITAE criterion?

EXERCISE

7.1. Distinguish between type and order of a system.

7.2. Explain the essential features of Type 0, 1 and 2 systems.

7.3. Find the steady-state errors for Type 0, 1 and 2 systems when unit step, ramp and parabolic inputs are applied to these systems

7.4. What is meant by dynamic or generalised error coefficients?

7.5. Derive the static error coefficients K_p, K_v and K_a

7.6. Distinguish between static and generalised (dynamic) error coefficients.

7.7. Explain the meaning of the terms: Position, velocity and acceleration error constant. Tabulate their values for Type 0, 1 and 2 systems.

7.8. Derive the error constants for step, ramp and parabolic inputs to the system. Also deduce the effects of these constants on Type 0, 1 and 2 system.

7.9. Explain the various error performance indices and compare them.

PROBLEMS

7.1. Determine the static error coefficients for unity feedback systems whose forward transfer functions are

(a) $26/s^2(s-1)(s^2+6s+13)$.

(b) $100(s-1)/s(s+2)(1+0.25s)$.

7.2. The open-loop transfer function of a servo system with unity feedback is

$G(s) = 10/s(1+0.1s)$.

Evaluate the static error constants.

7.3. Find the steady-state error of the system in problem 7.2 when subjected to an input of

$r(t) = a_0 + a_1 t + a_2 t^2/2$

7.4. The open-loop transfer function of a unity feedback system is

$G(s) = 4(s+1)/s^2(s+2)$.

(i) Find the position, velocity and acceleration error coefficients.

(ii) State the Type and order of the system.

(iii) Calculate the steady-state error when the input is $3 - 2t + 3t^2$. Sate the assumptions made in calculating the steady-state error.

7.5. A unity feedback control system has the loop transfer function

$G(s) = 10/s(s+1)(s+2)$.

Calculate the static error constants.

7.6. A unity feedback system has a loop transfer function

$G(s) = 20/s(s+1)$.

Obtain the steady-state error for the system when it is subjected to an input $r(t) = 1 + t + t^2$.

7.7. The forward and feedback transfer function of a system is

$G(s) = 1/(s+0.8)$ and $H(s) = 3/(s+2.5)$.

Determine the type of the system and evaluate the static error coefficients.

7.8. A unity feed back system has

$G(s) = K/s(s+1)(0.5s+1)$ and $r(t) = 5t$. It is desired that for the ramp input the steady state error is less then or equal to 0.1. What minimum value K must have for this condition to be satisfied? For this value of K, is the system stable or unstable?

7.9. The open-loop transfer function of a unity feedback control system is

$G(s) = K/s(s+1)(s+2)$.

It is desired that the steady state error must be limited to 0.1 for a unit ramp input. Find the minimum value of K for this condition.

7.10. A unity feedback system has a loop transfer function

$G(s) = K/s(s+1)(s+2)$.

The steady-state error for a ramp input of $r(t) = 10t$ is required to be less than 1.0. Find the value of K.

7.11. The open-loop transfer function of a system is given by

$G(s) H(s) = K/s(s + 2)(s + 10)$.

Determine the value of K for which the steady state error to a step velocity input of $r(t) = t$ is 0.1.

7.12. A servo system is being designed to keep the antenna of tracking radar pointed at a flying target. The servo system must be able to follow a target traveling a straight line course with a speed of 960 kmph with a maximum permissible error of 0.1 degree. The shortest distance from antenna to target is 360 m. Determine the value of the velocity error coefficient K_v in order to satisfy these specifications.

7.13. The open-loop transfer function of a unity feedback control system is

$G(s) = 500/s(1 + 0.1s)$.

Evaluate the error series for the system Determine the steady state error for an input $r(t) = 1 + 2t + t^2$.

7.14. A unity feedback control system has the open-loop transfer function

$G(s) = 1/s(s + 1)$.

Find the generalized error coefficients and steady-state error for the input

$r(t) = 4 + 6t + 2t^2$.

7.15. The forward transfer function of a unity feedback system is

$G(s) = 1.06/s(s + 1)(s + 2)$.

Find the final value of integral of the error signal for a unit step input.

7.16. A very accurate and rapidly responding control system is required for a system which allows live actors seemingly to perform inside of a complex miniature sets. A two-camera system is employed where one camera is trained on the actor and the other on the miniature set. The challenge is to obtain rapid and accurate coordination of the two cameras by using sensor information from the foreground camera to control the movement of the background camera. The closed-loop transfer function is

$$\frac{C(s)}{R(s)} = K_a K_m \omega_0^2 \big/ \left(s^3 + 2\zeta\omega_0 s^2 + \omega_0^2 s + K_a K_m \omega_0^2 \right)$$

where K_a and K_m are the gain of amplifier and motor respectively. Obtain the optimum coefficient for the above transfer function based on ITAE criterion.

7.17. A space telescope is to be launched to carry out astronomical experiments. The pointing control system is desired to achieve 0.01 minute of arc and track solar objects with apparent motion up to 0.21 minute of arc per second. The forward and feedback transfer functions are $G(s) = K_1/s^2$ and $H(s) = K_2(1 + s)$. (a) Determine the loop gain $K = K_1K_2$ so that the response to a step command is as rapid as reasonable with an overshoot of less than 5% (b) Determine the steady state error for a unit step and ramp input. (c) Obtain the value of K based on ITAE criterion for step input.

7.18. The Japanese National Railway has a train called Bullet express that travels between Tokyo and Osaka on the Tokaido line. This train travels 500 km in three hours and 10 minutes at an average speed of 150 kmph. In order to maintain a desired speed, a speed control system is used which yields a zero steady state error to ramp input. Determine a optimum third order system for an ITAE performance criterion and $\omega_n = 4$.

Chapter 8

TIME-DOMAIN DESIGN OF CONTROL SYSTEMS

8.1. INTRODUCTION

Control systems are designed to perform specific tasks. Compensation is the technique used to improve the system performance such as stability, transient and steady-state response by using additional circuits or equipment. The time-domain design refers to the utilization of the time-domain properties of the system to be designed.

We discussed, in the previous chapters, the time-domain characteristics of control systems represented by the transient and the steady-state responses when certain test signals are applied. These test signals are usually either a step function or a ramp function, depending on the design objective. For a step function input, the performance specifications are maximum overshoot, rise time and settling time. The design objective is to have the controlled variable behave in the desired ways. In this chapter, we shall discuss the time-domain methods for the design and compensation of control systems.

8.2. CONTROLLER CONFIGURATIONS

A controller is a sub-systems used in series or in feedback loop of a system to improve certain performance characteristics of the system. The control system designer at the outset decides the basic composition of the overall system and the place where the controller is to be placed relative to the components of the controlled process. This is called *fixed-configuration design*. The problem then is to design the elements of the controller.

Several commonly used system configurations with controller compensation are shown in Fig 8.1. The most commonly used system configuration is *series* or *cascade compensation* shown in Fig. 8.1(a). In this case, the controller is placed *in series with* the controlled process. Fig. 8.1(b) shows the controller placed in the minor feedback loop and hence it is called *feedback* or *minor loop compensation*. In the above two schemes there is only one controller in each system and hence they are *one-degree-of-freedom* controllers. The main disadvantage of this type is that the performance criteria that can be realized are limited. For example, if a system is designed to achieve a certain amount of relative stability, it may have poor sensitivity to parameter variations. Or, if the characteristic roots are selected to provide a certain amount of relative damping, the maximum overshoot of the step response may be still excessive, owing to the zero of the closed-loop transfer function.

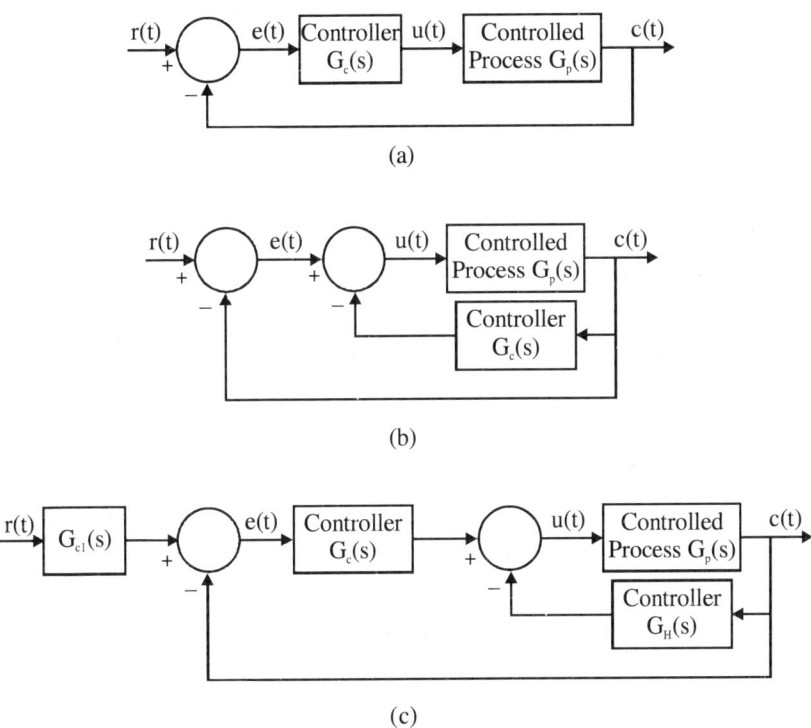

Fig. 8.1. (a) Series or cascade compensation (b) Feedback or minor-loop compensation (c) Series feedback compensation

To have more flexibility in the design, it is usual practice to use compensation schemes with *two degrees of freedom*. Fig 8.1(c) shows such a scheme with a controller in series and another in the feedback path. This is known as *series-feedback compensation*.

8.3. CHOICE BETWEEN CASCADE AND FEEDBACK COMPENSATORS

For improving the system performance, we can use either a cascade or feedback compensation. The choice between the two is influenced by the following factors:

1. ***Design Procedure:*** The compensator design should be simple and direct. Cascade compensator design is simpler and direct than feedback compensators.

2. *Availability:* Compensators of a type may not exist or may not be practical.

3. *Type of Input signal:* If the system utilizes 400Hz carrier, cascade compensator design is simpler.

4. *Size, weight and cost:* Small size, light weight and less costly compensator should be selected.

5. *Environmental conditions:* When the system is subjected to sudden and rapid changes in altitude and temperature, feedback compensation is more desirable.

6. *Noise:* In an electrically noisy situation, feedback compensators are preferred since they reduce noise.

7. *Response Time:* Faster response time is usally achieved by feedback compensation.

8. *Isolation:* When one part of a control system is to be isolated from the other, feedback compensation will achieve the desired isolation.

9. *Dominant Poles*: When the transfer function of the process $G_p(s)$ has a pair of dominant complex poles, feedback compensators are desirable.

10. *Other Factors:* Availability of components and the designer's experience and preferences influence the choice of the compensators.

8.4. FIGURE OF MERIT

The time response of an under-damped system subjected to a unit step input, is shown in Fig. 8.2. The following figures of merit or performance specifications in time domain are used to judge the system performance.

(a) M_p : *Peak Overshoot* is the maximum amplitude of the system response at the first overshoot over and above the unit step input and is given by $e^{-\pi\zeta/\sqrt{(1-\zeta^2)}}$.

(b) t_p : *Peak Time* is the time at which the peak overshoot occurs and is equal to $\pi/\omega_n\sqrt{1-\zeta^2}$.

(c) t_s : *Settling time* is the time for the response to first reach and thereafter remain within ± 2% of the final value and is equal to $4/\zeta\omega_n$.

(d) *N: Number of Oscillations* in the response is the number of overshoots and undershoots up to the settling time.

$$N = \text{Settling Time/period} = \frac{4/\zeta\omega_n}{2\pi/\omega_d}$$

$$= \frac{2}{\pi}\frac{\sqrt{1-\zeta^2}}{\zeta}$$

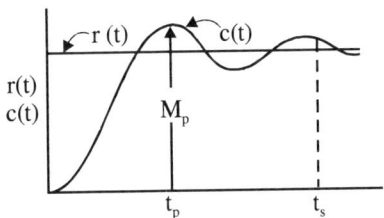

Fig. 8.2. Unit step response of an under-damped system

8.5. TYPES OF CONTROLLERS

In this section, we shall discuss the basic types of controllers:

(i) Proportional (P) controller.

(ii) Proportional and Derivative (PD) controller

(iii) Proportional and Integral (PI) controller

(iv) Proportional, Integral and Derivative (PID) controller.

8.5.1. Proportional Controller

For a proportional controller, the relationship between the output $m(t)$ and the actuating error signal $e(t)$ is

$$m(t) = K_p\, e(t) \qquad\qquad ...(8.1)$$

Thus, the transfer function $G_c(s)$ of the controller is

$$G_c(s) = \frac{M(s)}{E(s)} = K_p \qquad\qquad ...(8.2)$$

where K_p is the proportional gain. Whatever may be the actual mechanism and the form of the operating power, *the proportional controller is essentially an amplifier with adjustable gain.* A block diagram of such a controller is shown in Fig 8.3. In all examples we have discussed so far, the controller is the proportional controller or an amplifier with constant gain K.

Since under-damped systems have quick response to input signals, all practical systems are under-damped systems. Such a system exhibits exponentially decaying oscillations in the transient response.

An over-damped system though sluggish in its response, can be made to respond faster by increasing the forward path gain. This also reduces the steady-state error. However, the maximum overshoot is increased. It is therefore necessary to introduce some form of control so as to reduce the maximum overshoot as much as possible without sacrificing the steady-state accuracy. This can be achieved by PD, PI and PID controllers.

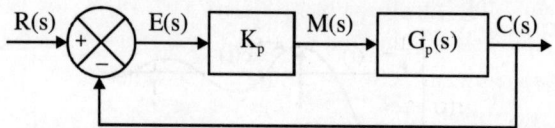

Fig.8.3. Block diagram of a proportional controller

8.5.2. PD Controller

The control action of a PD controller is defined by
$$m(t) = K_p\, e(t) + K_D\, de(t)/dt \qquad\qquad ...(8.3)$$

The transfer function is given by
$$G_c(s) = M(s)/E(s) = K_P + K_D s = K_P(1 + T_D s) \qquad ...(8.4)$$
where K_D is the derivative constant and T_D is the derivative time constant. Both K_P and T_D are adjustable. In derivative controllers, the output is proportional to the rate of change of the actuating error signal. Therefore, the derivative control is also called *rate control*. The derivative time T_D is the time interval by which the rate action advances the effect of a proportional control action.

The block diagram of a control system with PD controller is shown in Fig 8.4. The transfer function of the process is
$$G_p(s) = \omega_n^2 / s(s + 2\zeta\omega_n) \qquad\qquad ...(8.5)$$
The transfer function of the PD controller is given by
$$G_c(s) = K_P + K_D s \qquad\qquad ...(8.6)$$
The open-loop transfer function of the overall system is
$$G(s) = \frac{C(s)}{E(s)} = G_c(s)G_P(s) = \frac{\omega_n^2}{s}\frac{(K_P + K_D s)}{(s + 2\zeta\omega_n)} \qquad ...(8.7)$$

From Eq.(8.7) it is apparent that the derivative control introduces a zero at $s = -K_p/K_D = -1/T_D$ to the open-loop transfer function.

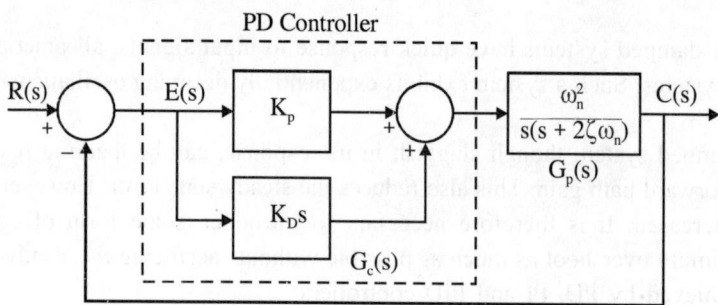

Fig. 8.4. Proportional and derivative controller

Let us now investigate the effect of the derivative control on the time-domain response of a control system. Fig. 8.5(a) shows the unit step response of a system with proportional control only. Fig. 8.5(b) and (c) show the corresponding error signal $e(t)$ and its derivative, $de(t)/dt$ respectively. The system response $c(t)$ shown in Fig. 8.5(a) has a high maximum overshoot, and the system is somewhat oscillatory and takes long time to settle down to the final value within the prescribed limits. In practice, this type of response may be objectionable for many control purpose.

Let us examine the causes for such a response. Assume that a motor is used to drive the system. During the time interval $0 < t < t_1$, the error signal is positive and the motor develops a positive torque. The amount of this positive torque is more than required to reach the desired response and hence the response overshoots. During the interval $t_1 < t < t_2$ the error is negative and the corresponding motor torque is also negative. The negative torque tends to reduce the output acceleration and thus limits the overshoot. However, the torque developed in the interval $t_2 < t < t_3$ during which the error is still negative causes the response to reverse. But, the negative torque developed during this interval is so large that the response undershoots. The positive motor torque developed during the time interval $t_3 < t < t_4$, limits the under-shoot and that during the interval $t_4 < t < t_5$ causes to the response to reverse and to overshoot. Since the system is assumed to be stable, the amplitude of the error signal reduces with each oscillation and the response eventually settles down to its final desired value.

Thus, the factors that contribute to a high overshoot are:

1. The positive correcting torque in the interval $0 < t < t_1$ is too large.

2. The negative retarding torque in the interval $t_1 < t < t_2$ is inadequate.

Therefore the logical approach to reduce the overshoot in the step response of the system is to decrease the amount of positive correcting torque and to increase the retarding torque. Similarly the negative corrective torque in the interval $t_2 < t < t_3$ should be reduced, and the positive retarding torque in the interval $t_3 < t < t_4$ should be increased to reduce the under-shoot.

As shown in Fig 8.5 (c), the time derivative of $e(t)$ is negative during the time interval $0 < t < t_1$. When this signal is added to the original error signal, the net actuating signal to the motor is reduced and hence the motor torque developed is reduced. During the interval $t_1 < t < t_2$, both $e(t)$ and $de(t)/dt$ are negative and hence the negative retarding torque developed is greater than that of the proportional case. These will therefore result in a smaller overshoot. In this time interval, $e(t)$ and $de(t)/dt$ have opposite signs. Therefore the negative torque is reduced and hence the under-shoot is also reduced. Thus, the PD controller gives precisely the desired compensation effect.

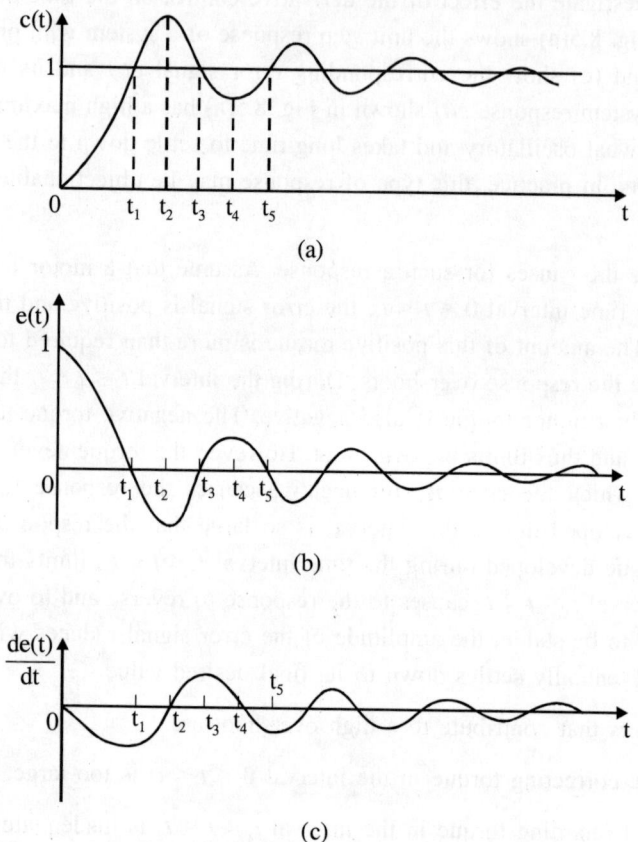

Fig. 8.5. (a) Unit step response with proportional controller (b) Error signal (c) Derivative of the error signal

The derivative control is essentially an *anticipatory type of control* since the time derivative of the error signal represents the slope of the error signal. Usually in a linear system, if *e(t)* or the slope of *e(t)* due to a step input is large, the overshoot is also high.

The derivative control measures the instantaneous slope of e(t), predicts the large overshoot ahead of time and takes proper correcting effort before the overshoot actually occurs.

The derivative control has no effect on the steady-state error of a system *if the steady-state error is constant with respect to time* since the time derivative of this error is zero. However, if the steady- state error varies with time, a torque is again developed proportional to *de(t)/dt* which reduces the magnitude of the error.

While the PD controller improves the transient response of a system, it amplifies the noise signals and may cause saturation in the actuator.

The closed-loop transfer function of the PD system is

$$\frac{C(s)}{R(s)} = \frac{\omega_n^2(K_P + K_D s)}{s^2 + (2\zeta\omega_n + \omega_n^2 K_D)s + \omega_n^2 K_P} \qquad ...(8.8)$$

Thus, from the denominator polynomial of Eq. (8.8), it is seen that the damping factor is increased which results in a reduced maximum overshoot and settling time. In spite of increased damping, the rise time is faster with derivative control because of the presence of a zero-factor in the transfer function. The design of the PD controller is illustrated below by an example.

Example 8.1. *The open-loop transfer function of a system is given by* $G(s) = 400K_p /$ *s(s+48.5). Design a PD controller such that the system is critically damped. Assume* $K_p = 100$

Solution:
The characteristic equation of the given system without PD control is

$$F(s) = s^2 + 48.5s + 40,000 = 0 \qquad ...(8.9)$$

Hence,

$$\omega_n = \sqrt{40,000} = 200$$
$$\zeta = 48.5/2 \times 200 = 0.1215$$

The maximum overshoot is

$$e^{-\pi\varsigma/\sqrt{1-\varsigma^2}} \times 100 = 68\%$$

If now the derivative control is included, the characteristic equation is

$$F(s) = s^2 + (48.5 + 400K_D)s + 400K_P = 0 \qquad ...(8.10)$$

Hence, $\omega_n = \sqrt{400K_P} = 200$. For critical damping

$$\zeta = (48.5 + 400K_D)/2 \times 200 = 1 \qquad ...(8.11)$$

Therefore,

$$K_D = 0.87875 \qquad ...(8.12)$$

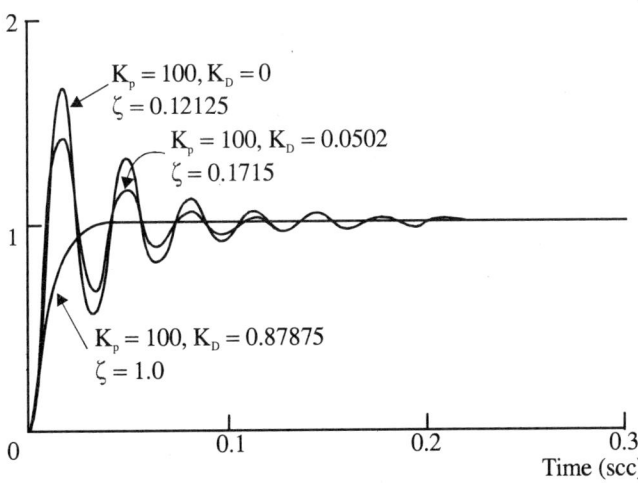

Fig. 8.6. Unit step response with and without PD control

The unit step responses of the system with and without derivative control are shown in Fig. 8.6. Without derivative control and with $K_p = 100$, the response has an overshoot of 68% and is somewhat oscillatory. With $K_D = 0.87875$, the damping becomes critical and hence there is no overshoot and the system settles down rapidly.

Thus the PD Controller, when designed properly, could yield a system with fast rise time, a little or no overshoot and small settling time.

8.5.3. PI Controller

The control action of a PI controller is defined by

$$m(t) = K_p e(t) + K_I \int_0^t e(t) dt \qquad ...(8.13)$$

The transfer function of the controller is

$$G_c(s) = \frac{M(s)}{E(s)} = \frac{K_P + K_I}{s} = K_P \left(1 + \frac{1}{T_I s} \right) \qquad ...(8.14)$$

where K_I is the integral constant and T_I is the integral time constant, both K_p and T_I are adjustable. The integral time constant adjusts the integral action while K_p affects the proportional and integral parts of the control action. The reciprocal of T_I is called *reset rate. The reset rate is the number of times per minute that the proportional part of the control action is repeated.* Reset rate is measured in terms of repeats per minute.

Fig. 8.7 shows the block diagram of a control system with PI controller with a second-order process. The transfer function of the PI controller is

$$G_c(s) = K_P + \frac{K_I}{s} \qquad ...(8.15)$$

Hence, the open-loop transfer function of the system is

$$G(s) = G_c(s).G_P(s) = \frac{\omega_n^2 (K_p + K_I/s)}{s(s + 2\zeta\omega_n)} = \frac{\omega_n^2 (K_p s + K_I)}{s^2(s + 2\zeta\omega_n)} \qquad ...(8.16)$$

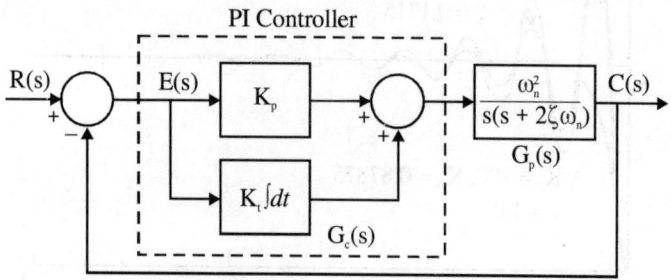

Fig. 8.7. PI controller

Thus, the PI controller introduces a zero at $s = -K_I/K_p$ and a pole at $s = 0$ to the open-loop transfer function. The order and type of the system is therefore increased by one. This reduces the steady-state error by one order, that is, if the steady-state error is constant to a given input, the PI controller reduces it to zero. For the system represented by Eq. (8.16), the steady-state error will be zero for a ramp input and constant for a parabolic input since the system is Type 2. However, because the system becomes third-order, it may be less stable than the original second-order system or may become unstable if the parameters K_p and K_I are not properly chosen.

In the case of a system with PD control, the value of K_p is important for a Type 1 system since it affects the magnitude of the steady-state error for a ramp input. When a Type 1 system is converted to a Type 2 system by a PI controller, K_p no longer affects the steady-state error. The steady-state error is always zero for a ramp input. Then the problem is to choose the proper combination of K_p and K_I so that the transient response is satisfactory. The design of PI controller is illustrated with an example below.

Example 8.2. *Design a PI controller for the system in Example 8.1 such that $\zeta = 0.85$ and the system response with controller is very close to the response without controller.*

Solution:

The characteristic equation without the controller is

$$F(s) = s^2 + 48.5s + 400K_p = 0 \qquad \qquad ...(8.17)$$
$$\zeta = 0.85, \ \omega_n = 48.5/(2 \times 0.85) = 28.5$$

Since,

$$\omega_n^2 = 28.5^2 = 400K_p \qquad \qquad ...(8.18)$$

Hence $\qquad \qquad K_p = 2$

The closed-loop transfer function with PI controller is

$$\frac{C(s)}{R(s)} = \frac{400(K_p s + K_I)}{s^3 + 48.5s^2 + 400K_p s + 400K_I} \qquad \qquad ...(8.19)$$

With $K_p = 2$

$$\frac{C(s)}{R(s)} = \frac{400 \times 2(s + 0.5K_I)}{s^3 + 48.5 \ s^2 + 800s + 400K_I} \qquad \qquad ...(8.20)$$

Since the responses have to be very close to the original response, Eq. (8.20) may be written as, with the same characteristic equation as in Eq. (8.17).

$$\frac{C(s)}{R(s)} = \frac{800}{(s + 0.5K_I)} \frac{(s + 0.5K_I)}{(s^2 + bs + 800)} \qquad \qquad ...(8.21)$$

Equating the coefficients of s^2 and s terms of the denominator polynomials of Eq. (8.20) and (8.21), we get

$$b + 0.5K_I = 48.5 \qquad \qquad ...(8.22)$$
$$48.5K_I + 800 = 800 \qquad \qquad ...(8.23)$$

This implies that the value of K_I should be small. Choosing $b = 48.4$ so that the two responses be closer, we obtain

$$0.5K_I = 0.1 \hspace{4cm} ...(8.24)$$

or,

$$K_I = 0.2$$

Substituting the value of K_I in Eq. (8.21), we have

$$\frac{C(s)}{R(s)} = \frac{800(s+0.1)}{(s+0.1)(s^2+48.4s+800)} = \frac{800}{s^2+48.4s+800} \hspace{2cm} ...(8.25)$$

The response for various values of K_p and K_I is shown in Fig 8.8.

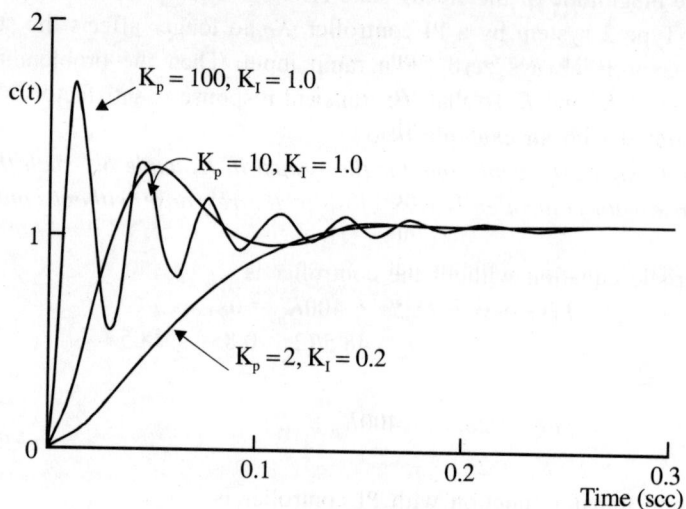

Fig. 8.8. Response of PI control

8.5.4. PID Controller

The PD controller improves the transient response of the system but has no effect on the steady-state error when it is constant. The PI controller improves the steady-state error by one order, simultaneously allowing a transient response with a little or no overshoot but the rise time may be quite slow. This leads to the use of a PID controller which utilizes the best properties of PI and PD controllers.

The PID controller is the one which is commonly used in practical control systems. The control action of the PID controller is defined by

$$c(t) = K_p e(t) + K_D \frac{de(t)}{dt} + K_I \int_0^t e(t)\,dt \hspace{2cm} ...(8.26)$$

The transfer function of the PID controller is

$$G_c(s) = \frac{M(s)}{E(s)} = K_p + K_D s + \frac{K_I}{s} \qquad \qquad ...(8.27)$$

The design problem is to determine the values of K_p, K_D and K_I so that the performance of the system is as prescribed. Fig. 8.9(a) shows the block diagram of PID controller and Fig. 8.9(b) the unit step response of a second-order system given by Eq. (8.5) with the PID controller for $K_p = 100$, $K_D = 0.87875$ and $K_I = 10$. In this case, the system being of Type 2, the steady-state error to a ramp input is zero and the transient response is reasonably good with a maximum overshoot of 7% and a fast rise time.

8.6. DESIGN OF CONTROLLERS

The design of controllers for control systems may be generally be regarded as a filter design problem. The *PD controller is a high-pass filter* and is often referred to as *phase-lead controller* since it introduces positive phase angle to the system over some specified frequency range. The *PI controller is a low-pass filter* and is also known as *phase-lag controller* since the corresponding phase angle introduced is negative. The PID controller is the simplest form of controllers that utilize the derivative and integral control actions in the compensation of control systems. This is a band-pass or band attenuating filter depending upon the controller parameters. The concepts are utilized in the frequency-domain design of controllers.

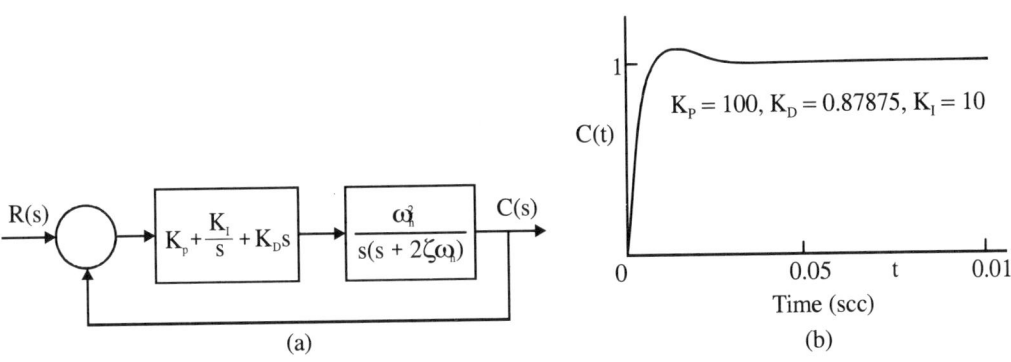

Fig. 8.9. (a) Block diagram of a PID controller (b) Unit step response of a second-order system with PID control

There are obvious advantages in designing the controllers using passive network elements only. The transfer function of a simple controller that can be realized by passive resistor-capacitor network is

$$G_c(s) = \frac{s + z_1}{s + p_1} \qquad \qquad ...(8.28)$$

If $p_1 > z_1$, the controller is a high-pass or phase-lead controller and if $p_1 < z_1$, it is a low-pass or phase-lag controller.

8.6.1. Phase-Lead (PD) Controller

Fig. 8.10 shows the network realization of the phase-lead (PD) controller with $p_1 > z_1$ in Eq. (8.28). Neglecting loading effect, the transfer function of the network is

$$\frac{E_o(s)}{E_i(s)} = \frac{\text{Output impedance}}{\text{Total circuit impedance}} \qquad ...(8.29)$$

$$= \frac{R_2}{\left[\dfrac{R_1 \cdot \dfrac{1}{C_1 s}}{R_1 + \dfrac{1}{C_1 s}}\right] + R_2} \qquad ...(8.30)$$

$$= \frac{R_2(1 + R_1 C_1 s)}{R_1 + R_2(1 + R_1 C_1 s)} \qquad ...(8.31)$$

$$= \frac{R_2}{R_1 + R_2} \cdot \frac{1 + R_1 C_1 s}{1 + \dfrac{R_1 R_2}{R_1 + R_2} C_1 s} \qquad ...(8.32)$$

$$= \frac{1}{a} \frac{1 + aTs}{1 + Ts}$$

$$= (s + 1/aT)/(s + 1/T) \qquad ...(8.34)$$

where

$$a = (R_1 + R_2)/R_2, \qquad a > 1. \qquad ...(8.35)$$

$$T = R_1 R_2 C_1/(R_1 + R_2) \qquad ...(8.36)$$

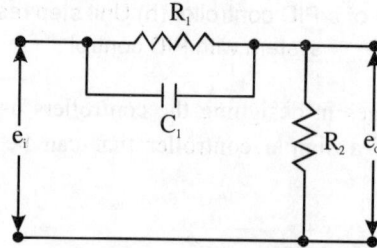

Fig. 8.10. PD controller network

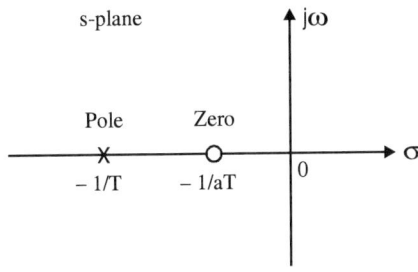

Fig. 8.11. Pole-zero location of PD controller

From Eq.(8.34), it is seen that the transfer function of a phase-lead network has a real zero at $s = -1/aT$ and a real pole at $s = -1/T$. The pole-zero configuration of the phase-lead network is shown in Fig.8.11. The pole and zero can be located at any point on the negative real axis by varying the values of a and T. Since $a > 1$, *the zero is always located to the right of the pole*, and the distance between the pole and zero is determined by the constant a. The phase-lead controller improves the relative and absolute stability of a closed-loop system because of the zero to the right of the pole.

The design of a phase-lead controller is illustrated with the following example.

Example 8.3. *Fig. 8.12 shows the block diagram of a sun-seeker control system mounted on a space vehicle so that it will track the sun with high accuracy.* θ_r *represents the reference angle of the solar ray and* θ_o *denotes the vehicle axis. The objective of the control system is to maintain the error e(t) between* θ_r *and* θ_o *near zero. The forward transfer function*

$$G(s) = \frac{2500\,K}{s(s + 25)}.$$

1.   *The steady-state error* e_{ss} *due to a ramp input should be less than or equal to 1% of the final value.*

2.   *The maximum overshoot should be less than 10%.*

  Design a phase-lead controller to meet the above specifications.

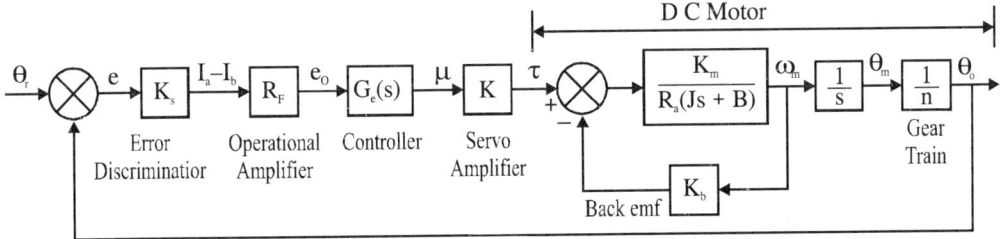

Fig. 8.12. Block diagram of a sun-seeker system

Solutions:

The open-loop transfer function of the uncompensated system is

$$G(s) = \frac{\Theta_0(s)}{E(s)} = \frac{K_s R_F K_m K / n}{sR_a(Js + B) + K_m K_b s} \qquad ...(8.37)$$

Substituting the numerical values, we get

$$G(s) = \frac{\Theta_0(s)}{E(s)} = \frac{2500K}{s(s+25)} \qquad ...(8.38)$$

The value of K is determined from the steady-state error requirement. Applying the final value theorem, we have

$$e_{ss} = \lim_{s \to 0} \frac{s.\Theta_r(s)}{1 + G(s)} \qquad ...(8.39)$$

For unit ramp input, $\Theta_r(s) = 1/s^2$. Substituting for $\Theta_r(s)$ in Eq, (8.39), we get

$$e_{ss} = \lim_{s \to 0} \frac{s \cdot (1/s^2)}{1 + \dfrac{2500K}{s(s+25)}} = \lim_{s \to 0} \frac{1}{s + \dfrac{2500K}{s+25}} = \frac{0.01}{K} \leq 0.01 \qquad ...(8.40)$$

Therefore K must be greater than or equal to 1. For $K = 1$, the characteristic equation of the uncompensated system is

$$F(s) = s^2 + 25s + 2500 = 0 \qquad ...(8.41)$$

From Eq.(8.41), $\omega_n = \sqrt{2500} = 50$ and $\varsigma = 25/(2 \times 50) = 0.25$

This corresponds to a maximum overshoot of 44.4%. The unit step response of the system with $K = 1$ is shown in Fig. 8.13.

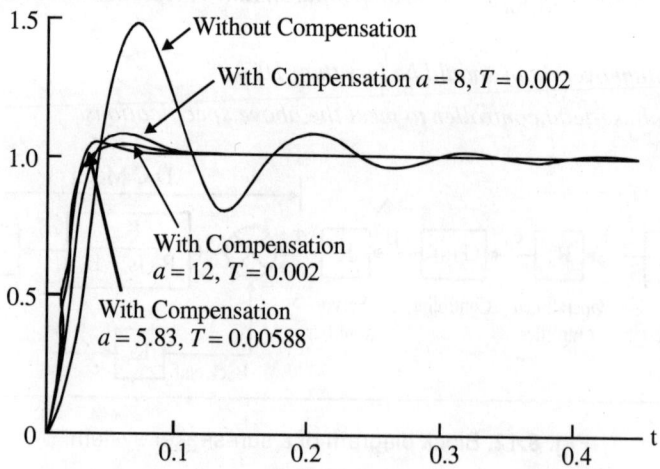

Fig. 8.13. System response with and without PD control

Assuming that the phase-lead controller is placed in series as shown in Fig 8.12, the open-loop transfer function of the overall system is given by

$$G(s) = \frac{C(s)}{E(s)} = \frac{2500(1 + aTs)}{s(s + 25)(1 + Ts)} \qquad ...(8.42)$$

The attenuation $1/a$ due to the phase-lead controller is assumed to be absorbed in the system gain.

The characteristic equation of the compensated system is

$$F(s) = s(s + 25)(1 + Ts) + 2500(1 + aTs) = 0 \qquad ...(8.43)$$

Since $Ts \ll 1$, we have

$$F(s) = s(s + 25) + 2500(1 + aTs) = 0 \qquad ...(8.44)$$
$$= s^2 + 25(1 + 100aT)s + 2500 = 0$$

As the maximum overshoot should be less than 10%, it is safe to assume $\zeta = 0.707$. Then, from Eq. (8.44), we have

$$\zeta = 0.707 = 25aT + 0.25 \qquad ...(8.45)$$

or,

$$aT = 0.0183 \qquad ...(8.46)$$

Selecting $T = 0.002$, we get $a = 9.15$. Since we have ignored the effect of pole at $-1/T$, it is safer to select a larger value for a than that calculated. Therefore choose $a = 12$. Hence the transfer function of the phase-lead controller is

$$G_c(s) = \frac{1 + 0.024s}{1 + 0.002s} = \frac{s + 41.67}{s + 500} \qquad ...(8.47)$$

The response of the system with and without phase lead controller is shown in Fig. 8.13. It is assumed that the attenuation of $1/a$ is taken care of by amplifiers in the forward path of the system.

8.6.2. Phase-Lag (PI) Controller

We may use a phase-lag controller to improve the performance of the control system. The PI controller in its simplest form is shown in Fig. 8.14. If we replace a by b in Eq. (8.44) and set $b < 1$, we have the transfer function of the phase-lag network. Thus,

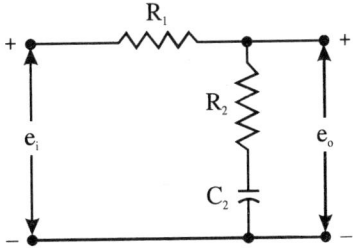

Fig. 8.14. PI controller network

$$G_c(s) = \frac{E_0(s)}{E_i(s)} = \frac{1+bTs}{1+Ts} \qquad b < 1 \qquad\qquad ...(8.48)$$

The transfer function derived from Fig. 8.12 with infinite source and output load impedances is

$$G_c(s) = \frac{E_o(s)}{E_i(s)} = \frac{1+R_2C_2s}{1+(R_1+R_2)C_2s} \qquad\qquad ...(8.49)$$

Comparing Eq. (8.48) and (8.49), we get

$$bT = R_2C_2 \qquad\qquad ...(8.50)$$

and

$$T = (R_1 + R_2)C_2$$

∴

$$b = \frac{R_1}{R_1+R_2}, \qquad\qquad b < 1 \qquad\qquad ...(8.51)$$

From Eq.(8.48), it can be seen that the transfer function of the phase-lag controller has a real zero at $s = -1/bT$ and a real pole at $s = -1/T$. Since $b < 1$, *the pole is always located to the right of the zero,* and distance between the pole and zero is determined by b. The pole-zero configurations is shown in Fig. 8.15.

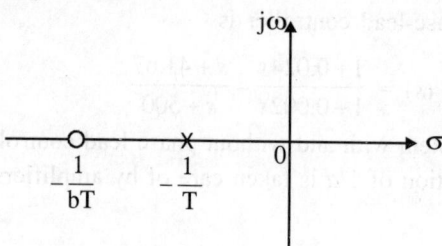

Fig. 8.15. Pole-zero locations of a PI controller

The design philosophy of the phase-lag compensator is that, for effective control, the pole and zero of the controller transfer function should be placed close together, and then the combination should be located relatively close to the origin in the s-plane. The design of a phase-lag compensator is illustrated with the following example.

Example 8.4. *Design a phase-lag controller for the system described in Example 8.3.*
Solution:
The open-loop transfer function of the system is

$$G(s) = \frac{\Theta_o(s)}{E(s)} = \frac{2500K}{s(s+25)}$$

To satisfy the steady-state error condition, the value of K should be unity. The open-loop transfer function of the compensated system is

$$G_c(s)G(s) = \frac{2500(1 + bTs)}{s(s + 25)(1 + Ts)}$$

$$= \frac{2500b(s + 1/bT)}{s(s + 25)(s + 1/T)} \qquad \text{...(8.52)}$$

Since the pole and zero should be very close to the origin in the s-plane, the values of T and bT are chosen to be large. Hence Eq. (8.52) is approximated to

$$G_c(s) \, G(s) \approx \frac{2500b}{s(s + 25)} \qquad \text{...(8.53)}$$

The characteristic equations is

$$F(s) = s^2 + 25s + 2500b = 0 \qquad \text{...(8.54)}$$

Hence, to have over-shoot less than 10%, we choose ζ to be 0.707. Therefore, from Eq. (8.54),

$$\zeta = 0.707 = 25/2 \times \sqrt{2500b} \qquad \text{...(8.55)}$$

or,

$$b = 1/8 = 0.125 \qquad \text{..(8.56)}$$

The pole of $G_c(s)$ should be placed as close to the origin as possible. Let the pole be arbitrarily placed at –0.05. Hence $T = 20$ and $bT = 2.5$. Thus, the transfer function $G_c(s)$ is

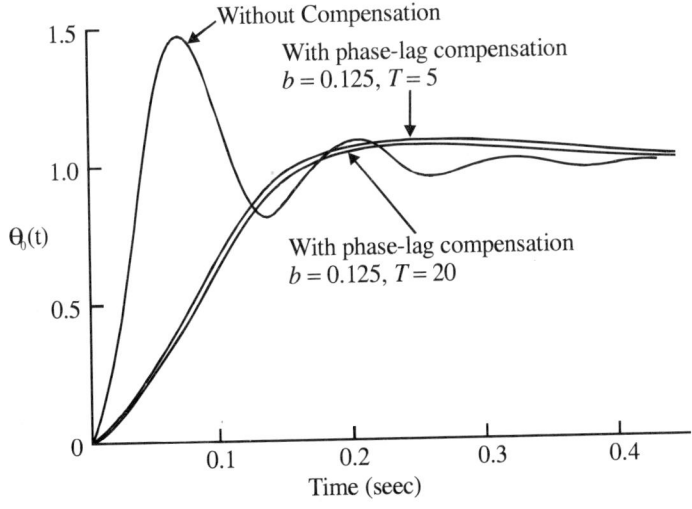

Fig. 8.16. System response with and without PI controller

$$G_c(s) = \frac{1+2.5s}{1+20s} \qquad \qquad ...(8.57)$$

The response of the system with and without compensator is shown in Fig. 8.16.

8.6.3. Lag-Lead (PID) Controller

The phase-lead controller generally improves the transient response of the system but increases the natural resonance frequency ω_n of the closed-loop system. On the other hand, phase-lag control reduces the overshoot and improves the steady-state response but usually results in a longer rise time. Thus, each has its own merits and limitations. However there are systems that cannot be satisfactorily improved by phase-lead or phase-lag controller alone. In such situations, it is but natural to use a combination of phase lead and lag controllers so that the advantages of both are utilized.

The network realization of the phase lag-lead controller is shown in Fig. 8.17. Neglecting the loading effect, the transfer function of the network is

$$G_c(s) = \frac{E_o(s)}{E_i(s)} = \frac{(1+R_1C_1s)(1+R_2C_2s)}{1+(R_1C_1+R_2C_2+R_1C_2)s+s^2} \qquad ...(8.58)$$

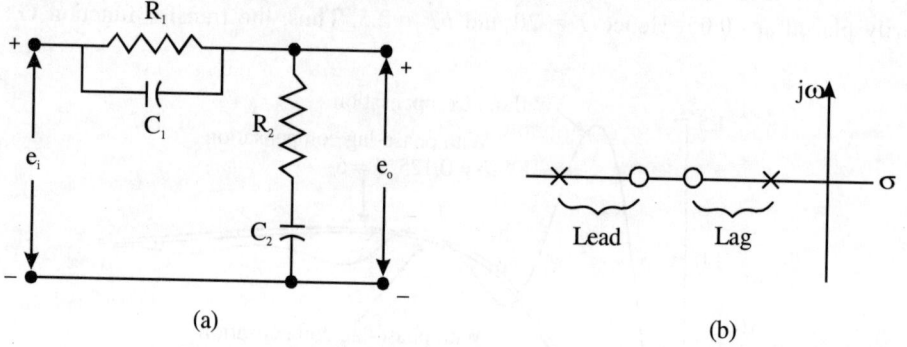

(a)   (b)

Fig. 8.17. (a) PID controller network; (b) Pole-zero locations

This can be written as

$$G_c(s) = \left(\frac{1+aT_1s}{1+T_1s}\right)\left(\frac{1+bT_2s}{1+T_2s}\right) \qquad ...(8.59)$$

where $a > 1$ and $b < 1$. Comparing Eq. (8.58) and (8.59) we have

$$aT_1 = R_1C_1 \qquad\qquad ...(8.60)$$
$$bT_2 = R_2C_2 \qquad\qquad ...(8.61)$$
$$T_1T_2 = R_1R_2C_1C_2 \qquad\qquad ...(8.62)$$
$$abT_1T_2 = R_1R_2C_1C_2 \qquad\qquad ...(8.63)$$

Therefore,

$$ab = 1 \qquad \qquad ...(8.64)$$

Eq. (8.64) implies that a and b cannot be independently specified.

Example 8.5. *Design a PID Controller for the system described in Example 8.3.*

Solutions:

The open-loop transfer function of the compensated system is

$$G_c(s)G_p(S) = \frac{2500}{s(s+25)} \frac{(1+aT_1s)}{(1+T_1s)} \frac{(1+bT_2s)}{(1+T_2s)} \qquad ..(8.65)$$

with $K = 1$ to meet the steady-state error condition. The phase-lead section is first designed as in Example 8.3, with $a = 10$. Then $T = 0.00183 \approx 0.002$. Since $ab = 1$ and $a = 10$, $b = 0.1$. Again selecting $T_2 = 20$ as in Example 8.4, $bT_2 = 2$. The compensated transfer function is

$$G_c(s)G_c(s) = \frac{2500(1+0.02s)(1+2s)}{s(s+25)(1+0.002s)(1+20s)}$$

$$= \frac{2500(s+50)(s+0.5)}{s(s+25)(s+500)(s+0.05)} \qquad ...(8.66)$$

The unit step response of the system, with PID controller has an improved rise time and reduced overshoot.

8.7. BRIDGED–T CONTROLLER

The system considered so far in this chapter contains only real poles. However, many industrial controlled processes have transfer functions with one or more complex-conjugate poles. If the complex-conjugate poles are very close to the imaginary axis in the s-plane, it may be very difficult to control such processes. It is therefore necessary to cancel the undesired complex-conjugate poles by properly designed controllers. Hence the controllers should have a transfer function with the zeros identical with the poles of the process transfer function. For exampled, if the process transfer function is

$$G_p(s) = \frac{K}{s(s^2+2s+20)}$$

the complex-conjugate poles cause instability for K greater than 40. Therefore, the transfer function of the series controller must be of the form

$$G_c(s) = \frac{s^2+2s+20}{s^2+2\zeta\omega_n s+\omega_n^2} \qquad ...(8.67)$$

The constants ζ and ω_n are determined to satisfy the performance specifications of the closed-loop system. Since the poles of the process transfer function are cancelled with the zeros of the controller transfer functions, this method of compensation is known as *pole-zero cancellation control*.

Transfer functions with complex poles and zeros can be realized with the bridged-T networks shown in Fig. 8.18 consisting of RC elements only. Since there are two configurations, we shall denote them as Type 1 and Type 2 networks. Assuming zero input source impedance and infinite output impedance, the transfer functions of the bridged-T networks and the ω_n and ζ parameters of the two networks are:

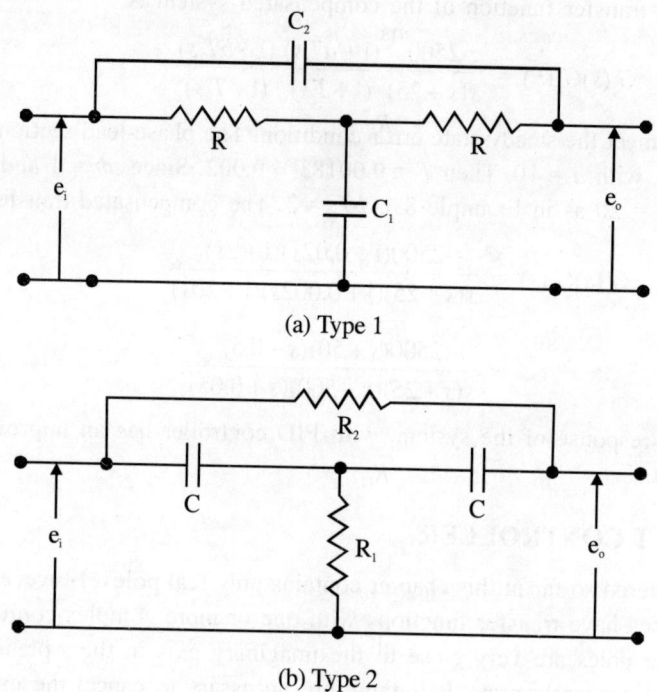

(a) Type 1

(b) Type 2

Fig. 8.18. Bridged-T networks

Bridged-T Type-1 Networks: The transfer function of Type-1 bridged-T network is given by

$$\frac{E_o(s)}{E_i(s)} = \frac{1+2RC_2 s + R^2 C_1 C_2 s^2}{1+R(C_1+2C_2)s+R^2 C_1 C_2 s^2}$$

$$= \frac{s^2+(2/RC_1)s+(1/R^2 C_1 C_2)}{s^2+(C_1+2C_2)s/RC_1 C_2+(1/R^2 C_1 C_2)} \qquad \qquad ...(8.68)$$

$$\omega_{nz} = \pm 1/R\sqrt{C_1 C_2} \qquad \qquad ...(8.69)$$

$$\zeta_z = \sqrt{C_1 C_2} \qquad \qquad ...(8.70)$$

$$\omega_{np} = \pm 1/(R\sqrt{C_1 C_2}) = \omega_{nz} \qquad ...(8.71)$$

$$\zeta_p = \frac{C_1 + 2C_2}{2\sqrt{C_1 C_2}} = \frac{1 + 2C_2/C_1}{2\sqrt{C_2/C_1}} = \frac{1 + 2\zeta_z^2}{2\zeta_z} \qquad ...(8.72)$$

Bridged-T Type-2 Networks: The transfer function of Type-2 bridged-T network is given by

$$\frac{E_o(s)}{E_i(s)} = \frac{1 + 2R_1 Cs + C^2 R_1 R_2 s^2}{1 + C(R_2 + 2R_1)s + C^2 R_1 R_2 s^2}$$

$$= \frac{s^2 + (2/CR_2)s + (1/C^2 R_1 R_2)}{s^2 + (R_2 + 2R_1)s/CR_1 R_2 + (1/C^2 R_1 R_2)} \qquad ...(8.73)$$

$$\omega_{nz} = \pm 1/\left(C\sqrt{R_1 R_2}\right) \qquad ...(8.74)$$

$$\zeta_z = \sqrt{R_1/R_2} \qquad ...(8.75)$$

$$\omega_{np} = \pm 1/\left(C\sqrt{R_1 R_2}\right) = \omega_{nz} \qquad ...(8.76)$$

$$\zeta_p = \frac{R_2 + 2R_1}{2\sqrt{R_1 R_2}} = \frac{1 + 2\zeta_z^2}{2\zeta_z} \qquad ...(8.77)$$

The limitations of the pole-zero cancellation design are:

1. In practical system the cancellation of poles and zeros of the process by the zeros and poles of the controller is rarely exact.

2. Due to ageing of the system components and changes in the operating environment, the poles and zeroes of the transfer functions may vary.

Example 8.6. *Consider a controlled process with the transfer function*

$$G_p(s) = -\frac{K(1 + 10s)}{s(1 + 0.2s + 0.25s^2)} \qquad ...(8.78)$$

Design a bridged-T controller to improve stability.

Solution:

Although the closed-loop system is always stable, ζ is very low and hence the step response of the system is oscillatory for any positive K. The unit step response of the system for $K = 1$ is shown in Fig. 8.19.

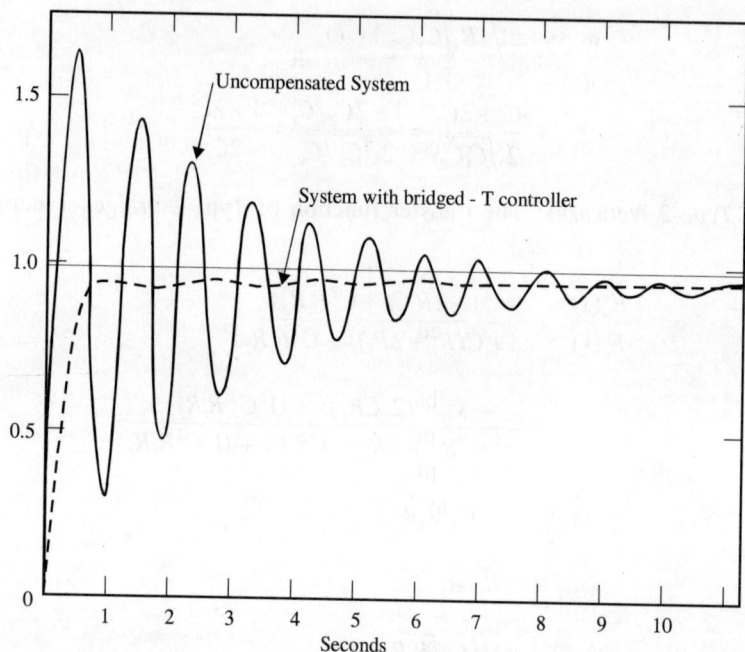

Fig. 8.19. Unit step response of uncompensated and compensated system

To improve the relative stability of the system, let a bridged-T network be used in cascade. The complex-conjugate zeros of the network are identical with the complex-conjugate poles of the process. Hence, the problem is to determine the location of the poles of the network. Therefore, the transfer function of the network should be of the form.

$$G_c(s) = \frac{s^2 + 0.8s + 4}{s^2 + 2\zeta_p \omega_{np} s + \omega_{np}^2} \qquad \ldots(8.79)$$

From Eq. (8.79) we have

$$\omega_{nz} = \sqrt{4} = 2 \qquad \ldots(8.80)$$
$$\zeta_z = 0.8/(2 \times 2) = 0.2 \qquad .(8.81)$$

Hence,

$$\omega_{np} = \omega_{nz} = 2 \qquad \ldots(8.82)$$

$$\zeta_p = \frac{1 + 2\zeta_z^2}{2\zeta_z} = 2.7 \qquad \ldots(8.83)$$

Therefore, the transfer function of the bridged-T controller is

$$G_c(s) = \frac{s^2 + 0.8s + 4}{s^2 + 10.8s + 4} = \frac{s^2 + 0.8s + 4}{(s + 0.384)(s + 10.42)} \qquad \ldots(8.84)$$

The open-loop transfer function of the compensated system is

$$G(s) = G_c(s)G_p(s) = \frac{40(s+0.1)}{s(s+0.384)(s+10.42)} \qquad \qquad ...(8.85)$$

The unit step response of the compensated system is shown in Fig. 8.19. In this case, the oscillation is less in magnitude but takes a long time to settle down to the final steady-state value of unity. The relative stability of the system is thus greatly improved.

8.8. MINOR-LOOP FEEDBACK CONTROLLERS

The series or cascade controllers discussed in previous sections are the most commonly used compensators. However, depending on the nature of the system, it is sometimes advantageous to locate the controller in a minor feedback loop as shown in Fig. 8.1(b). For instance, a tachometer in the minor feedback loop improves the stability of the system by feeding back the derivative of the output signal. In principle, any of the three controllers can be used in the minor feedback loop with varying degrees of effectiveness. We shall discuss details of these control schemes.

8.8.1. Rate-Feedback Controller

Fig. 8.20 shows a control system with a rate feedback control. In this scheme, the derivative of the output signal is feedback to actuate the plant. If the output variable is mechanical displacement, a tachometer may be used to convert the mechanical displacement into an electrical signal which is proportional to the derivative of the displacement or velocity. In Fig 8.21, the transfer function of the tachometer is denoted by $K_t s$ where K_t is the tachometer constant expressed in volts per unit velocity.

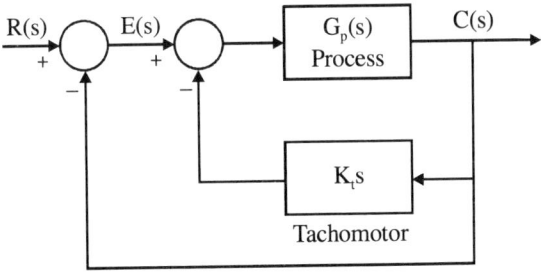

Fig. 8.20. System with rate-feedback

Let the transfer function of the process is given by

$$G_p(s) = \frac{\omega_n^2}{s(s+2\zeta\omega_n)} \qquad \qquad ...(8.86)$$

The closed-loop transfer function with rate or tachometer feedback is

$$\frac{C(s)}{R(s)} = \frac{\omega_n^2}{s^2 + (2\zeta\omega_n + K_t\omega_n^2)s + \omega_n^2} \qquad ...(8.87)$$

Thus, it can be seen from Eq. (8.87), that the effect of the tachometer feedback is to increase the damping of the closed-loop system since the coefficient of the s-term is increased by a factor $K_t\omega_n^2$. In this respect, the rate feedback has exactly the same effect as the PD control. However the PD control introduces a zero at $s = -K_p/K_D$. Hence the response of the system with rate feedback will not be identical with that obtained with PD control because the presence of the zero affects the closed-loop system response.

The open-loop transfer function with rate feedback is given by

$$\frac{C(s)}{E(s)} = G(s) = \frac{\omega_n^2}{s(s + 2\zeta\omega_n + K_t\omega_n^2)} \qquad ...(8.88)$$

The system is still Type 1 and therefore the steady-state error is not altered. When the input is a step function, the steady-state error is zero. However, when the input is a unit ramp function, the steady-state error is increased by an amount K_t and is equal to $(2\zeta + K_t\omega_n)/\omega_n$, whereas with PD control, the steady-state error is $2\zeta/\omega_n$ only.

8.8.2. Phase-Lead Controller

Let us consider the sun-seeker system of Example 8.1. Assume that the controller is located in the minor feedback loop as in Fig 8.1(b). The open-loop transfer function of the compensated system is

$$\frac{\Theta_o(s)}{E(s)} = \frac{G_p(s)}{1 + G_p(s)G_H(s)} \qquad ...(8.89)$$

where the transfer function of the process or plant is

$$G_p(s) = \frac{2500}{s(s + 25)} \qquad ...(8.90)$$

and the transfer function of the feedback controller is

$$G_H(s) = \frac{1 + aTs}{1 + Ts} \qquad ...(8.91)$$

If $a > 1$, $G_H(s)$ represents a phase-lead controller and if $a < 1$, it is a phase-lag controller. Substitution of Eq. (8.90) and (8.91) in Eq. (8.89) yields

$$\frac{\Theta_o(s)}{E(s)} = \frac{2500(1 + Ts)}{s(s + 25)(1 + Ts) + 2500(1 + aTs)} \qquad ...(8.92)$$

The closed-loop transfer function of the compensated system is

$$\frac{\Theta_o(s)}{\Theta_r(s)} = \frac{2500(1+Ts)}{s(s+25)(1+Ts)+2500(1+aTs)+2500(1+Ts)} \qquad ...(8.93)$$

The characteristic equation of the compensated system is

$$F(s) = s(s + 25)(1 + Ts)+2500(1 + aTs)+2500(1 + Ts) = 0 \quad ...(8.94)$$

Since Ts << 1 for phase-lead controller, Eq. (8.94) reduces to

$$F(s) = s(s + 25) + 2500(1 + aTs) + 2500 = 0$$
$$= s^2 + 25(1 + 100\ aT)s + 5000 = 0 \qquad ...(8.95)$$

Thus, the natural resonance frequency ω_n of the feedback compensated system is increased by $\sqrt{2}$ times and ζ reduces to 0.5 compared to cascade compensation.

Example 8.7. *For the sun-seeker system of Example 8.1, design a feedback phase-lead controller to meet the same specifications.*

Solution:

The characteristic equation is given in Eq. (8.95) and the value of ζ is again selected to be 0.707.

Hence,

$$F(s) = s(s + 25) + 2500(1 + aTs) + 2500 = 0$$
$$= s^2 + 25(1 + aT)s + 5000 = 0 \qquad ...(8.96)$$

Therefore,

$$\zeta = 0.707 = 25(1 + 100\ aT)/(2 \times 50 \times \sqrt{2})$$
$$= 0.707 \times 0.25\ (1 + 100\ aT)$$
$$aT = 0.03 \qquad ...(8.97)$$

Choosing $a = 12$ as in Examples 8.3, we get $T = 0.025$. The transfer function of the compensator is

$$G_H(s) = \frac{1+0.03s}{1+0.025s} = \frac{s+33.3}{s+400} \qquad ...(8.98)$$

It is assumed that the attenuation of $(1/a)$ is taken care of by the amplifiers in the forward path of the system.

As shown in Section 6.5.8, as a zero of the closed-loop transfer function is moved closer to the origin the maximum overshoot may dramatically increase. The unit step response of a second-order system is shown in Fig. 8.21 as the zero of the closed-loop transfer function moves closer to the origin. From the response, it is clear that the maximum overshoot can increase several times. Since the zero of the phase-lag controller is intentionally placed very close to the origin, it would give rise to a large increase in the overshoot even though the roots of the characteristic equations are chosen properly. In fact, if we use the phase-lag controller designed in Eq. (8.57) in the minor feedback loop for the sun-seeker system, the roots of the characteristic equation would still be at −0.412, −12.32 ± j12.32, but the maximum overshoot would be in excess of 600 percent. Because of this reason, *the phase-lag controller is not used in the feedback loop.*

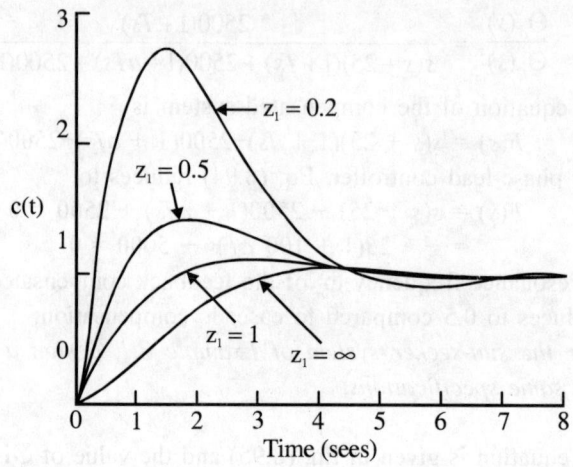

Fig. 8.21. Unit step response of a second-order system as the zero moves towards the origin.

8.9. ADDITIONAL EXAMPLES

Example 8.8. *A control system process with unity feed-back is described by*

$$G(s) = \frac{K}{s(1+0.1s)(1+0.2s)} \qquad ...(8.99)$$

With K = 100, the system is unstable, since the maximum value of K = 15 for marginal stability. Design a phase-lag controller such that compensated system has K = 100 and ζ = 0.707.

Solution:

The transfer function of the process is given by

$$G_p(s) = \frac{100}{s(1+0.1s)(1+0.2s)} = \frac{5000}{s(s+10)(s+5)} \qquad ...(8.100)$$

The compensator transfer function is

$$G_c(s) = \frac{1+aTs}{1+Ts} \qquad ...(8.101)$$

Hence,

$$G_p(s)\, G_c(s) = 5000(1 + aTs)/[s(s + 5)(s + 10)(1 + Ts)] \qquad ...(8.102)$$

Since aT and T are chosen to be large, Eq. (8.102) is approximated to

$$G_c(s)G_p(s) = \frac{5000a}{s(s+5)(s+10)} \qquad ...(8.103)$$

The characteristic equation of the approximated closed-loop system is

$$F(s) = s(s + 5)(s + 10) + 5000 \; a = 0$$

$$= s^3 + 15s^2 + 50s + 5000 \; a = 0 \qquad ...(8.104)$$

Since the compensated system should have $\zeta = 0.707$, the characteristic equation should contain a second-order term with $\zeta = 0.707$.

$$F(s) = (s+b)(s^2 + 1.414\omega_n s + \omega_n^2) = 0 \qquad ...(8.105)$$

Equating coefficients of similar terms, we have

$$b + 1.414\omega_n = 15 \qquad ...(8.106)$$

$$1.414 \; b\omega_n + \omega_n^2 = 50 \qquad ...(8.107)$$

$$b\omega_n^2 = 5000a \qquad ...(8.108)$$

From Eq. (8.106), $b = (15 - 1.414 \; \omega_n)$. Substituting this value of b in Eq. (8.107) and simplifying we get

$$\omega_n^2 - 21.2\omega_n + 50 = 0 \qquad ...(8.109)$$

Hence,

$$\omega_n = 18.5 \text{ or } 2.7 \qquad ...(8.110)$$

Since $\omega_n = 18.5$ will not satisfy Eq.(8.106),

$$\omega_n = 2.7 \qquad ....(8.111)$$

From Eq. (8.106), we get

$$b = 11.2 \qquad ...(8.112)$$

Hence a from Eq. (8.108) is

$$a = b\omega_n^2/5000 = 0.0163 \qquad ...(8.113)$$

Since the pole of the phase-lag compensator has to be very close to the origin, choose $1/T = 0.001$. Hence $T = 100$, $aT = 1.63$ and $1/aT = 0.6135$. The transfer function of the phase lag compensator is

$$G_c(s) = \frac{1+1.63s}{1+100s} = 0.0163 \left(\frac{s+0.6135}{s+0.01} \right) \qquad ...(8.114)$$

The compensated system transfer function is

$$G_c(s) \; G_p(s) = \frac{81 \cdot 5(s+0.6135)}{s(s+5)(s+10)(s+0.01)} \qquad ...(8.115)$$

Example 8.9. *The open-loop uncompensated system transfer function is $G(s)H(s)$ = K/s^2. The specifications are: $t_s \leq 4$ sec, percentage overshoot $\leq 20\%$. Design a suitable controller.*

Solution:

$$G(s) = K/s^2$$

For 20% overshoot, $\zeta \geq 0.45$

$$t_s = 4 = 4/\zeta\omega_n$$

$$\omega_n = 2.22, \; K = \omega_n^2 = 5$$

To provide suitable margin for settling time, choose $K = 10$

$$G_c(s) = \frac{1+aTs}{1+Ts}$$

$$G_pG_c(s) = 10(1 + aTs)/s^2(1 + Ts)$$

$$F(s) = s^2(1 + Ts) + 10(1 + aTs) = 0$$

$Ts \ll 1$; hence

$$\zeta = 0.45 = \frac{10aT}{2\sqrt{10}}$$

$$aT = 0.285$$

Select $a = 10$, then $T = 0.0285$

The controller transfer function is

$$G_c(s) = \frac{1+0.285s}{1+0.0285s} = \frac{s+3.51}{s+35.1}$$

Example 8.10. *The forward transfer function of a unity feedback system is given by $G(s) = K/s(s + 5)$. Design a PD controller so that the steady-state error to a ramp input is less than 3.0 percent and to obtain a rapid response to a unit step input with 25 percent or less overshoot.*

Solution:

$$K_v = \lim_{s \to 0} sG(s) = K/5$$

$$e_{ss} = 1/K_v = 0.03$$

Hence, $K_v = 30$ and $K = 150$

$$F(s) = s^2 + 5s + 150 = 0$$

$$\omega_n = \sqrt{150} = 12.25$$

$$\zeta = 5/(2 \times 12.25) = 0.2$$

Max. over shoot $= e^{-\pi\zeta/\sqrt{1-\zeta^2}} = 52.68\%$

If the derivative control is included,

$$F(s) = s^2 + (5 + K_D)s + 150 = 0$$

$$\zeta = 0.45 \text{ for 25 percent overshoot}$$

$$0.45 = (5 + K_D)/(2 \times 12.25)$$

$$K_D = 6.03$$

$$G_pG_c = \frac{150 + 6.03s}{s^2 + 11.03s + 150}$$

Example 8.11. *The uncompensated loop transfer function of a unity feedback temperature control system is $G(s)H(s) = K_1/(2s + 1) (0.5s + 1)$. Design a PI compensator of the form $G_c(s) = K_2 + K_3/s$ to have an overshoot less than or equal to 10 percent and zero steady-state error for a step input.*

Solution:

$$\frac{C(s)}{R(s)} = \frac{K_1 K_2 \left(1 + K_3 / K_2 s\right)}{(s + 0.5)(s + 2) + K_1 \, K_2 \left(1 + K_3 / K_2 s\right)}$$

$$= \frac{K_1 K_2 \left(s + K_3 / K_2\right)}{s^2 + 2.5s^2 + \left(1 + K_1 K_2\right)s + K_1 \, K_2 \left(K_3 / K_2\right)}$$

The compensator zero is selected so that the real part of the complex root is to be $-\zeta\omega_n$ -0.75 so that $t_s = 4/\zeta\omega_n = 5.33$ sec.

To have an overshoot of 10 percent or less, ζ is selected to be 0.6

Hence $\omega_n = 0.75/0.6 = 1.25$ and $K_3/K_2 = \zeta\omega_n = 0.75$.

$$F(s) = s^3 + 2.5s^2 + (1 + K_1 K_2)s + 0.75 K_1 K_2$$
$$= (s + a)(s^2 + 2\zeta\omega_n s + \omega_n^2)$$
$$= (s + a)(s^2 + 1.5s + 1.56)$$

Equating the similar coefficients,

$$a + 1.5 = 2.5; \therefore a = 1$$

$$\frac{3}{4} K_1 K_2 = 1.56a$$

$$K_1 K_2 = 2.08$$

K_2 is selected as 1. Hence $K_1 = 2.08$, $K_3 = 0.75$.

Example 8.12. *The design of a remotely operated vehicle (ROV) for undersea exploration requires the control of heading or direction of the vehicle. The system uses an electric motor and propeller to drive the ROV, and an electrically controlled rudder provides the steering. The forward transfer function of the system with unity feedback is*

$$G(s) = \frac{K}{s(s + 5)^2}$$

Design a phase-lag controller to obtain a rapid response to a unit step input with 25 percent overshoot or less and a steady-state error of less than 3.3 percent of the unit ramp input

Solution:

$$K_V = \lim_{s \to 0} sG(s) = K/25$$

$$e_{ss} = 1/K_V = 0.0333$$

Hence $K_V = 30$

Therefore $K = 25K_V = 750$

For 25 percent overshoot, $\zeta = 0.45$.

$$G_c(s) = \frac{1 + aTs}{1 + Ts}$$

$$G_p G_c(s) = 750(1 + aTs)/s(s + 5)^2 (1 + Ts)$$

Since the pole and zero are to be very close to origin, T and aT are chosen to be large.

$$G_p G_c(s) = 750a/s(s + 5)^2$$

$$F(s) = s^2 + 10s^2 + 25s + 750a = 0$$

For 25 percent overshoot, $\zeta = 0.45$.

$$s^3 + 10s^2 + 25s + 750a = (s + b)(s^2 + 2\zeta\omega_n s + \zeta_n^2)$$

$$= (s + b)(s^2 + 0.9\omega_n s + \omega_n^2)$$

Equating the similar coefficients,

$$b + 0.9\omega_n = 10$$

$$0.9\,\omega_n b + \omega_n^2 = 25$$

$$b\omega_n^2 = 750a$$

Solving, we get

$$\omega_n = 2.63, \; b = 7.63 \text{ and } a = 0.07$$

With $T = 20$,

$$G_c(s) = \frac{1 + 0.14s}{1 + 20s}$$

With the attenuation adjusted,

$$G_c = \frac{s + 7.1}{s + 0.05}$$

8.10. SUMMARY

In this chapter, we have discussed the need for compensators and their various configurations such as fixed, series or cascade, feedback and minor loop configurations. We have also discussed the factors such as design procedure, availability, type of input signal, size, weight and cost, environmental conditions, noise, response time, isolation, dominant poles etc. to be considered in selection of cascade and feedback compensator for improving the system performance. The time-domain performance specifications used to judge the system performance such as M_p, t_p, t_s and N have been defined. The design of PD, PI, PID and feedback controllers have been explained to meet the given performance specifications in the time-domain. We have also discussed the bridged-T controllers, derived their transfer functions and their limitations. In most practical cases, the design method used is a *trial-and-error approach*. The designer seeks to satisfy all performance specifications by means of trial-and-error method. After a system is designed he checks to ascertain whether the designed system satisfies all the performance specifications. If it does not, then the design process is repeated by adjusting parameters or changing the system configuration until the given specifications are met. Although the design is based on a trial-and-error procedure, an experienced designer may be able to design an acceptable system without using many trials.

REVIEWS QUESTIONS

8.1. What is a controller? Mention the types.

8.2. Mention the figures of merit of performance specifications of a system in time domain.

8.3. What is a proportional, PD, PI and PID controller?

8.4. Explain a phase-lag, phase-lead and phase lag-lead controller

8.5. Draw the bridged-T controller.

EXERCISE

8.1. Discuss the various controller configurations.

8.2. Explain clearly with suitable block diagrams PI, PD and PID controllers.

8.3. Write a note of PID controller.

8.4. Bring out merits and demerits of PI, PD and PID controllers.

8.5. Enumerate the factors that determine the choice between cascade and feedback compensators.

8.6. Explain the time domain performance criteria.

8.7. Discuss the factors that contribute for high overshoot in an underdamped system and explain how the derivative control is able to reduce the overshoot.

8.8. Explain the rate feedback controller and discuss the effect on system performance.

8.9. What is meant by pole-zero cancellation control? Derive the transfer function of Type 1 and Type 2 bridged-T networks.

8.10. Show that in the bridged-T network $\zeta_p = \dfrac{1 + 2\zeta_z^2}{2\zeta_z}$ where ζ_p and ζ_z are the damping ratio due to pole and zero factors.

PROBLEMS

8.1. Design a PD controller for Example 8.9 with $\zeta = 0.707$.

8.2. Design a PI controller for Example 8.10 with $\zeta = 0.85$.

8.3. Design a phase-lead compensator with $\zeta = 0.707$ for Prob. 8.1.

8.4. Design a phase-lag compensator for Prob. 8.1 with $\zeta = 0.707$

8.5. Design a phase lag-lead compensator for Prob. 8.1 with $\zeta = 0.707$

8.6. Design a rate feedback controller for Prob. 8.1 with $\zeta = 0.707$

8.7. Design a cascade bridged-T compensator for the system $G_p(s) = \dfrac{150(1 + 2s)}{s(s^2 + s + 4)}$ to improve its stability.

Chapter 9

SYSTEM STABILITY IN S-DOMAIN

9.1. INTRODUCTION

The transfer functions for various types of systems have been developed previously. These transfer functions have some basic characteristics that permit transient and steady-state response analyses of control systems. A very important characteristic of a control system is its stability. The stability characteristic of a linear system is determined from its characteristic equation. The simplest stability criterion is the Routh's stability criterion. It provides a means of determining the system stability without evaluating the roots of the characteristic equation. In this chapter, we shall discuss the Routh's stability criteria.

9.2. THE CONCEPT OF STABILITY

A stable system is defined as a system with a bounded output. In other words, if a system is subjected to a bounded input or disturbance and the output response is bounded in magnitude, then the system is said to be stable. From this definition of stability, it follows that a linear system is stable *if and only if* the absolute value of its impulse response integrated over an infinite range is finite. This can be proved as outlined below.

The input-output relation of a linear system can be expressed by the convolution integral

$$c(t) = \left| \int_0^\infty r(t-\tau) g(\tau) d\tau \right| \qquad \qquad ...(9.1)$$

where $g(\tau)$ is the impulse response of the system. The absolute value of $c(t)$ is given by

$$|c(t)| = \left| \int_0^\infty r(t-\tau) g(\tau) d\tau \right| \qquad \qquad ...(9.2)$$

$$\leq \int_0^\infty \left| r(t-r) \right| \left\| g(\tau) \right| d\tau \qquad \qquad ...(9.3)$$

Since the absolute value of integral is not greater than the integral of the absolute value of the integrand.

If $r(t)$ is bounded, then

$$= |r(t)| \leq N < \infty \qquad ...(9.4)$$

Hence,

$$= |c(t)| \leq \int_0^\infty N|g(\tau)|d\tau = N\int_0^\infty |g(\tau)|d\tau \qquad ...(9.5)$$

If $c(t)$ is to be bounded output, then

$$N\int_0^\infty |g(\tau)|d\tau \leq M < \infty \qquad ...(9.6)$$

or,

$$= \int_0^\infty |g(\tau)|d\tau \leq P < \infty \qquad ...(9.7)$$

where N and M are finite Eq. (9.7) implies that the area under the absolute-value curve of the impulse response $g(t)$ evaluated from $t = 0$ to $t = \infty$ must be finite. We shall show now that the above impulse response requirement is linked to the restrictions on the roots of the characteristic equation of a system.

9.3. CHARACTERISTIC EQUATION AND STABILITY

By definition, the transfer function $G(s)$ of a system and its impulse response $g(t)$ are related by

$$G(s) = \mathcal{L}[g(t)] = \int_0^\infty g(t) e^{-st} dt \qquad ...(9.8)$$

The absolute value of $G(s)$ is given by

$$|G(s)| \leq \int_0^\infty |g(t)||e^{-st}|dt \qquad ...(9.9)$$

The roots of the characteristic equation of a system $F(s) = 1 + G(s)H(s) = 0$ are the poles of $G(s)$. When s takes on these values, $|G(s)| = \infty$. Since $s = \sigma + j\omega$, the absolute value of $e^{-st} = |e^{-\sigma t}|$. Hence Eq. (9.9) becomes

$$\infty \leq \int_0^\infty |g(t)||e^{-\sigma t}|dt \qquad ...(9.10)$$

If one or more roots of the characteristic equation lie on the $j\omega$ axis or in the right half of the s-plane, then $\sigma \geq 0$ and $|e^{-st}| \leq 1$. Therefore Eq. (9.10) is written as

$$\infty \leq \int_0^\infty |g(t)|dt \qquad ...(9.11)$$

for $Re(s) = \sigma > 0$.

Eq. (9.11) contradicts the stability criterion given in Eq. (9.7). We therefore conclude that for a linear system to be stable, *the roots of the characteristic equation must all lie in the left half of the s-plane.*

Thus, the stability of linear systems can be determined by checking whether any roots of the characteristic equation are in the right half of the *s*-plane or on the *j*ω-axis. Fig. 9.1 shows the regions of stability and instability in the *s*-plane. Unstable region includes the imaginary axis excluding the origin.

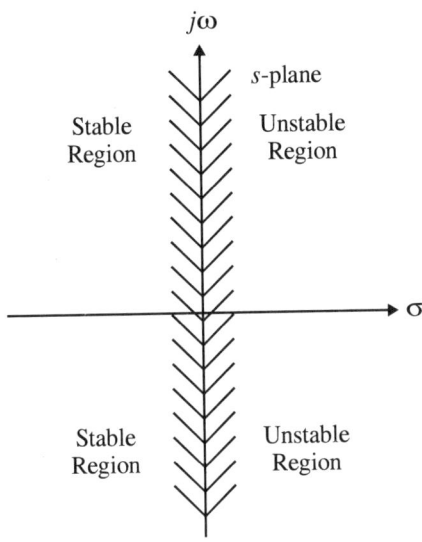

Fig. 9.1. Regions of stability

9.4. METHODS OF DETERMINING STABILITY

The stability of linear systems may be checked by investigating impulse response or by finding the roots of the characteristic equation. However, these criteria are difficult to implement. For example, the impulse response is obtained by taking the inverse Laplace transform of the transfer function which is not always simple. The determination of the roots of the characteristic equation of order greater than four can only be carried out by a digital computer. To determine the stability of a system, it is sufficient to find the existence of the roots of the characteristic equation with positive real parts. Therefore the algorithms should be simple and straight-forward to apply. The following methods are frequently used.

1. Routh-Hurwitz Criterion
2. Nyquist Criterion
3. Root Locus Plot
4. Bode Plot and Nichols Plot

9.4.1. Routh-Hurwitz Criterion

This is an algebraic method. It provides information of the absolute stability of systems. This criterion tests whether any root of the characteristics equation lies in the right half of the s-plane. The number of roots that lies on the $j\omega$-axis and in the right half of the s-plane are also indicated.

9.4.2. Nyquist Criterion

This is a semi-graphical method. It provides information on the number of poles of the closed-loop transfer function by observing the behaviour of the Nyquist plot of the open-loop transfer function, provided the relative location of the zeros of the closed-loop transfer function is known. The poles of $C(s)/R(s)$ are the roots of the characteristic equation $F(s)$.

9.4.3. Root Locus Plot

This is a plot of loci of roots of the characteristic equation when a system parameter, usually the system gain K, is varied. If the root locus lies in the right half of the s-plane, the closed-loop system is unstable.

9.4.4. Bode Plot and Nichols Plot

These are plots of $G(s)H(s)$. They are used to determine the stability of the closed-loop systems. However, the methods are applicable only if $G(s)\,H(s)$ has no poles and zeros in the right half of the s-plane.

In the following sections, we shall discuss the Routh-Hurwitz stability criterion in detail. The other methods are discussed in subsequent chapters.

9.5. ROUTH-HURWITZ CRITERION

This criterion is used to determine the location of roots (zeros) of a polynomial with constant real coefficients with respect to the left half and the right half of the s-plane, without actually solving for the roots.

The closed-loop transfer function, in general, is the ratio of two polynomials and it is given by

$$\frac{C(s)}{R(s)} = \frac{G(s)}{1 + G(s)H(s)} = \frac{b_0 s^m + b_1 s^{m-1} + \ldots\ldots + b_{m-1}s_1 + b_m}{a_0 s^n + a_1 s^{n-1} + \ldots\ldots + a_{n-1}s + a_n} \qquad \ldots(9.12)$$

The characteristic equation $F(s) = 1 + G(s)\,H(s) = 0$ is given by

$$F(s) = 1 + G(s)\,H(s) = a_0 s^n + a_1 s^{n-1} + \ldots + a_{n-1}s + a_n = 0 \qquad \ldots(9.13)$$

where all a_i are real numbers. If $a_n = 0$, divide by s to obtain the equation in the form of Eq. (9.13). All powers of s from s^n to s^0 must be present in the characteristic equation. If any coefficient other than a_n are zero, or, if all the coefficients do not have the same sign, then there are roots on the $j\omega$-axis or roots with positive real parts, and the system is unstable.

9.5.1 Necessary Conditions

The necessary conditions of the system to be stable are:

1. All the coefficients of the characteristic equation must be present.

2. All the coefficients must have the same sign.

These two necessary conditions can be easily checked by inspection. However, they are *not sufficient conditions*. It is quite possible that a polynomial satisfying the above two conditions may still have zeros in the right half of s-plane.

The *necessary and sufficient condition* that all the roots of Eq. (9.13) lie in the left half of the *s*-plane is that the polynomial's Hurwitz determinants, D_k, k = 1, 2, 3, ...*n* must all be positive.

The Hurwitz determinants of Eq.(9.13) are given by

$$D_1 = a_1, \quad D_2 = \begin{vmatrix} a_1 & a_3 \\ a_0 & a_2 \end{vmatrix}, \quad D_3 = \begin{vmatrix} a_1 & a_3 & a_5 \\ a_0 & a_2 & a_4 \\ 0 & a_1 & a_3 \end{vmatrix} \quad \cdots$$

$$D_n = \begin{vmatrix} a_1 & a_3 & a_5 & a_7 & \cdots & a_{2n-1} \\ a_0 & a_2 & a_4 & a_6 & \cdots & a_{2n-2} \\ 0 & a_1 & a_3 & a_5 & \cdots & a_{2n-3} \\ 0 & a_0 & a_2 & a_4 & \cdots & a_{2n-4} \\ \cdots & \cdots & \cdots & \cdots & \cdots & \cdots \\ 0 & 0 & \cdots & \cdots & \cdots & a_n \end{vmatrix} \quad \text{...(9.14)}$$

In Eq. (9.14), the coefficients with indices larger than n or with negative indices are taken to be equal to zero.

The application of the Hurwitz determinants is formidable for high-order systems because of the labour involved in evaluating the determinants. However, Routh has simplified this into a tabular form. The first step is to arrange the coefficients of the characteristic polynomial in the first two rows of the *Routh table*. The first row consists of the first, third, fifth, ... coefficients, and the second row consists of the second, fourth, sixth, ... coefficients. These coefficients are then used to evaluate the rest of the rows of the Routh table.

Routh table

s^n	a_0	a_2	a_4	a_6	...
s^{n-1}	a_1	a_3	a_5	a_7	...
s^{n-2}	c_1	c_2	c_3	...	
s^{n-3}	d_1	d_2	d_3	...	
...	...	...	...		
s^1	j_1				
s^0	k_1				

The constant c_1, c_2, c_3, ... in the third row are evaluated by

$$c_1 = (a_1a_2 - a_0a_3)/a_1$$
$$c_2 = (a_1a_4 - a_0a_5)/a_1$$
$$c_{3...} = (a_1a_6 - a_0a_7)/a_1 \qquad\qquad ...(9.15)$$

Then, the fourth row is formed by using s^{n-1} and s^{n-2} rows. The coefficients d_1, d_2, d_3, ... are evaluated by

$$d_1 = (c_1a_3 - a_1c_2)/c_1$$
$$d_2 = (c_1a_5 - a_1c_3)/c_1$$
$$d_{3...} = (c_1a_7 - a_1c_4)/c_1 \qquad\qquad ...(9.16)$$

The rest of rows are formed in the same way up to the s^0 row. The last two rows contain only one term each. The previous two rows contain only two terms each and so on. Thus the table is triangular.

Routh's stability criterion states that *the number of roots of the characteristic equation with positive real parts is equal to the number of changes of sign of the coefficients in the first column.* Therefore, for a system to be stable, the necessary and sufficient condition is stated below.

9.5.2. Necessary and Sufficient Conditions

The necessary and sufficient conditions for the system to be stable are:
1. All coefficients of the characteristic equation must be present.
2. All coefficients must have the same sign.
3. All terms in the first column of the Routh table should have the same sign.

All terms in the first column of the Routh table are usually positive. If there is a negative term in the first column of the Routh table, the system is unstable.

It is shown in [56] that a system is unstable if there is a negative element in any position in any row. Thus, it is not necessary to compute the Routh table further if any negative number is encountered.

The relationship between the elements in the first column of the Routh table and the Hurwitz determinants are:

$$D_1 = a_1, \; D_2 = D_1 c_1, \; D_3 = D_2 d_1, \; ..., \; D_n = a_n D_{n-1}$$

Therefore, if all the Hurwitz determinants are positive, then the terms in the first column of the Routh table would also be of the same sign.

The following examples illustrate the application of the Routh-Hurwitz stability criterion.

Example 9.1. *By using Routh stability criterion, find the stability of a system with the following characteristic equation.*

$$F(s) = s^4 + 3.5s^3 + 3.75s^3 + 1.75s + 0.5 = 0$$

Solution:

The above function satisfies the necessary conditions since $F(s)$ has no missing terms and all coefficients are of the same sign. To check the stability, we have to check whether the sufficient condition is satisfied. Therefore, the Routh table is prepared.

Routh table

s^4	1	3.75	0.5
s^3	3.5	1.75	
s^2	3.25	0.5	
s^1	1.21		
s^0	0.5		

Since all terms in the first column are of the same sign, the given system is *stable*.

Example 9.2. *By applying Routh-Hurwitz criterion, find the stability of the system with the characteristic equation* $F(s) = s^5 + s^4 + 10s^3 + 72s^2 + 152s + 240 = 0.$

Solution:

Since $F(s)$ has no missing terms and all coefficients are of the same sign, the necessary conditions are satisfied. However, the sufficient condition must still be checked. The Routh table formed by using the procedure described above is:

Routh table

Sign	s^5	1	10	152
changes	s^4	1	72	240
1	s^3	−62	−88	
2	s^2	70.6	240	
	s^1	122.6		
	s^0	240		

Inspection of the first column elements reveals that there are *two sign* changes, (i) from +1 to −62 and (ii) from −62 to +70.6. Therefore $F(s)$ has *two roots* in the right half of the s-plane. Thus, the system is *unstable*.

The criterion gives the number of roots with positive real parts but *does not give the exact values of the roots*. The roots of the characteristic equation are: $-3, -1 \pm j\sqrt{3}, +2 \pm j4$. This confirms that there are two roots with positive real parts.

9.5.3. Routh Array with too large or small coefficients

In evaluating a Routh array, the numbers often become too large or too small which is inconvenient to work with. In such cases, any complete row in the array may be divided or multiplied by an arbitrary positive constant before proceeding, without changing the conclusions as to root locations. This is illustrated in Example 9.3.

Example 9.3. *A particular control system has characteristic equation given by* $F(s) = 400s^5 + 600s^4 + 800s^2 + 300s^2 + 800s + 600 = 0$. *Is this system stable? Discuss the root locations.*

Solution:

Routh table

s^5	400	800	800	
	1	2	2	dividing by 400
s^4	600	300	600	
	2	1	2	dividing by 300
s^3	1.5	1		
s^2	−0.33	2		
s^1	10			
s^0	2			

Sign changes: 1, 2 (for s^3, s^2, s^1 rows)

The s^5 and s^4 rows have been divided by 400 and 300 respectively, thus simplifying the further calculations. There are *two sign* changes, (i) from 1.5 to −0.33 and (ii) from −0.33 to +10. Thus, there are two poles on the right half plane. Hence the system is *unstable*.

9.6. SPECIAL CASES

While evaluating the coefficients of Routh array, it may so happen that:

1. The first element in any one row of the Routh table is *zero* but the other elements are not zero.

2. All the elements in a row of the Routh table are *zero*.

9.6.1. Zero Elements in the First Column

When a zero appears as the first element of a row, evaluation of elements of the next row involves division by zero and hence the Routh test breaks down. This implies that *at least one root of the characteristic equation lies in the right half plane.* This difficulty can be overcome by any one of the following methods.

(a) Substitute a small positive number ε for the zero term and then evaluate the remaining elements as usual. The stability conclusion is made from the resulting sign changes as $\varepsilon \to 0$.

(b) Substitute $s = 1/s$ in the original polynomial. The resulting polynomial in s has the same number of roots located in each half plane as the original polynomial in s.

(c) Multiply the given polynomial by $(s + a)$ where a is a positive number and then carry on as usual.

These methods are illustrated with following examples.

Example 9.4. *Consider the characteristic equation* $F(s) = s^5 + 2s^4 + 3s^3 + 6s^2 + 2s + 1 = 0$. *Find the number of roots in each half plane.*

Solution:

(*a*) *Zero in the first column is replaced by* ε

The Routh table is

Routh table

s^5	1	3	2	
s^4	2	6	1	
s^3	$(0)\varepsilon$	1.5		zero replaced by ε
s^2	$(6\varepsilon - 3)/\varepsilon$	1		
s^1	$1.5 - \varepsilon^2/(6\varepsilon - 3)$			
s^0	1			

As $\varepsilon \to 0$ from the positive side, $(6\varepsilon - 3)/\varepsilon$ approaches $-\infty$ and $1.5 - \varepsilon^2/(6\varepsilon - 3)$ approaches $+1.5$. Therefore there are *two sign* changes and hence *two roots are in the right half plane* and three roots in the left half plane. The system is *unstable.*

(*b*) *Substitution of s=1/s in the characteristic equation and simplification*

Substituting $s = 1/s$, we get

$$F(1/s) = F_1(s) = s^5 + 2s^4 + 6s^3 + 3s^2 + 2s + 1 = 0$$

The new Routh table is

Routh table

s^5	1	6	2	
s^4	2	3	1	

		s^3	4.5	1.5	Multiplying by 2/3
Sign changes			3	1	
1	s^2	7/3	1		
	s^1	–2/7			
2	s^0	1			

As before there are *two sign* changes in the first column elements leading to the same conclusion.

(*c*) *Multiplying by* (*s* + *a*)

Multiplying by (*s* + 1), we get.

$$(s + 1)\, F(s) = s^6 + 3s^5 + 5s^4 + 9s^3 + 8s^2 + 3s + 1 = 0$$

The Routh table is

Routh table

		1	5	8	1
Sign changes	s^6	1	5	8	1
	s^5	3	9	3	
1	s^4	2	7	1	
	s^3	–1.5	3		
2	s^2	11	1		
	s^1	69/22			
	s^0	1			

As before there are *two sign* changes. Two roots are in the right half plane and three roots in the left half plane. The system is *unstable*.

9.6.2. A Zero Row

In the case of *all elements in a row* of the Routh table *are zero*, it implies one of the following conditions:

1. Pairs of roots on the $j\omega$-axis
2. Pairs of complex conjugate roots forming symmetry about the origin of the *s*-plane.
3. Pairs of real roots with opposite sign.

Generally, the elements in an odd row turn out to be zeros. This difficulty may be overcome by forming an equation, called *auxiliary equation A(s)* using the elements of the row just above the row of zeros. This auxiliary polynomial *A(s)* consists only of even powers of *s* and its highest-order term has the power of *s* indicated in the reference column to the left of this row.

The roots of auxiliary equation are the roots of original equation. The auxiliary equation is then differentiated with respect to *s* and the resulting coefficients are substituted into the zero-row. The table is then completed and the results are interpreted in the usual manner.

Example 9.5. *Find the locations of the roots of the characteristic equation* $F(s) = s^4 + 2s^3 + 11s^2 + 18s + 18 = 0$.

Solution:

The Routh table is

Routh table

s^4	1	11	18
s^3	2	18	
s^2	2	18	
	1	9	divided by 2
s^1	(0)		Results in a zero s^1-row. Form $A(s)$ with
	2		elements in the s^2-row. Thus $A(s) = s^2 + 9$.
s^0	9		$dA(s)\ ds = 2s$. Replace 0 by 2 in the s^1 row.

Since there are no sign changes, no root is lying in the right half plane. However, the roots of the $A(s) = 0$ are $\pm j3$. Thus, there is a *pair of roots on the jω-axis*.

9.7. SYSTEMS WITH TRANSPORTATION LAG

All the feedback systems discussed so far are linear, lumped parameter models whose transfer functions are the ratios of algebraic polynomials. In such cases, the system output appears immediately on application of the input signal without any *time delay* or *lag*. In certain systems such as *pipe-lines* and *belt conveyors*, the system output does not appear immediately on the application of the input but only after a finite time delay. This type of pure time delay is known as *transportation lag* or *dead time*. In systems with transportation lag, the output *c(t)* is the same as the input *r(t)* but delayed by τ sec. This can be expressed mathematically as

$$c(t) = r(t - \tau) \qquad\qquad ...(9.17)$$

Taking the Laplace transform of Eq. (9.17), we get

$$C(s) = R(s)e^{-\tau s} \qquad\qquad ...(9.18)$$

Hence, the transfer function of a constant dead time is given by

$$\frac{\acute{G}(s)}{R(s)} = e^{-\tau s} \qquad\qquad ...(9.19)$$

The transfer function $e^{-\tau s}$ may be expressed into a series

$$e^{-\tau s} = 1 - \tau s + \frac{\tau^2 s^2}{2!} - \frac{\tau^3 s^3}{3!} + \ldots\ldots\ldots \qquad \ldots(9.20)$$

If the transportation lag is very small, the second and higher-order terms in Eq. (9.20) may be neglected. The application of the Routh stability criteria to systems with transportation lag is illustrated with an example.

Example 9.6. *The forward transfer function of a system with transportation lag and unity feedback is given by*

$$G(s) = \frac{Ke^{-\tau s}}{s(s+a)}$$

Find the maximum value of K for which the system is just stable.

Solution:

The characteristic equation is given by

$$F(s) = 1 + G(s) = 1 + \frac{Ke^{-\tau s}}{s(s+a)} = 0$$

Neglecting second and higher-order terms in the series expansion of $e^{-\tau s}$, the characteristic equation becomes

$$F(s) = s^2 + as + K(1 - \tau s) = s^2 + (a - K\tau)s + K = 0$$

The Routh stability table is given by

Routh table

s^2	1	K
s^1	$a - K\tau$	
s^0	K	

Equating the s^1-row test function to zero, we obtain the value of K_{max} as

$$K_{max} = a/\tau$$

Without transportation lag, the given second-order system is stable for all positive values of K. However, if a transportation lag is introduced into the system, the maximum value of K is reduced. Thus *the system becomes less stable.*

9.8. SYSTEMS WITH COMPLEX COEFFICIENTS

We have considered so far system with constant coefficients. Consider now that the coefficients of the characteristic equation are complex. We shall apply the Routh stability criteria to system with complex coefficients.

Let the characteristic equation of a system be

$$F(s) = \sum_{i=0}^{n} (a_i + jb_i)s^i + K = 0 \qquad ...(9.21)$$

Eq.(9.21) may be expressed as

$$F(s) = F_1(s) + jF_2(s) = 0 \qquad ...(9.22)$$

where

$$F_1(s) = Re\ F(s) \qquad ...(9.23)$$

$$F_2(s) = Im\ F(s) \qquad ...(9.24)$$

Eq. (9.22) will be satisfied *if and only if*

$$F_1(s) = 0 \qquad ...(9.25)$$

and

$$F_2(s) = 0 \qquad ...(9.26)$$

The system will be stable if $F_1(s) = 0$ and $F_2(s) = 0$ satisfy the Routh stability criteria.

9.9. APPLICATION OF ROUTH STABILITY CRITERION TO CONTROL SYSTEM ANALYSIS

In linear control system analysis, the Routh stability criterion has limited use mainly because it does not suggest how to improve the relative stability or how to stabilize an unstable system. However, it is often used to determine the range of gain K or some other parameter for which a closed-loop system is stable. This information is readily obtained by setting up the Routh table as a general function of the parameter of interest and determining the condition of that parameter required to force all the terms in the first column elements to have the same sign. The system gain K is usually the variable parameter since it determines the location of the poles and therefore the stability of the system.

Example 9.7. *Find the range of K of a system with the following closed-loop transfer function to be stable.*

$$C(s)/R(s) = K(s + 2)/s(s + 5)(s^2 + 2s + 5) + K(s + 2)$$

Solution:

The characteristic equation is

$$F(s) = s^4 + 7s^3 + 15s^2 + (25 + K)s + 2K = 0.$$

The Routh table is

Routh table

s^4	1		15	2K
s^3	7		25 + K	

s^2	80 – K		2K	
s^1	(80 – K)(25 + K) –98K/(80 – K)			
s^0	2K			

The term $(80 - K)$ from the s^2 row imposes the restriction $K < 80$. The s^0 row requires $K > 0$. The numerator of the first term in the s^1 row is $-K^2 - 43\ K + 200$ and should be greater than or equal to zero.

Equating to zero and solving for K, we get $K \leq 28.1$ or -71.1. K cannot be negative because of s^0 row requirement. Hence $0 < K < 28.1$ satisfies the s^2 row condition as well. For $K = 28.1$, the s^1 row is zero and the characteristic equation has a pair of complex roots on the $j\omega$-axis. The roots are obtained from the auxiliary equation with $K = 28.1$

$$A(s) = 51.9\ s^2 + 393.4 = 0$$

or

$$s = \pm\ j2.75$$

9.10. ADDITIONAL EXAMPLES

Example 9.8. *Discuss the root locations of the characteristic equation*

$$F(s) = s^7 + 4s^6 + 5s^5 + 2s^4 + 4s^3 + 16s^2 + 20s + 8 = 0.$$

Solution:

The Rough table is

Routh table

s^7	1	5	4	20	
s^6	4	2	16	8	
	2	1	8	4	after dividing by 2
s^5	9/2	0	18		
	1	0	4		after dividing by 9/2
s^4	1	0	4		
s^3	0	0			$A(s) = s^4 + 4$
					$dA(s)/ds = 4s^3$
s^3	4	0			Coefficients of $dA(s)/ds = 4$.
s^2	(0) ε	4			
s^1	–16/ε				
s^0	4				

There are *two sign* changes. Hence two roots are in the right half plane. The two roots are the roots of $A(s)$. The roots of $A(s) = s^4 + 4 = 0$ are $s = -1 \pm j1, 1 \pm j1$.

Thus the roots $1 \pm j1$ are in the right half plane. The four roots of $A(s)$ form symmetry about the origin of the s-plane.

Example 9.9. *Find the root location of the characteristic equation*

$$F(s) = s^4 + 5s^3 + 5s^2 - 5s - 6 = 0.$$

Solution:

The Routh table is

Routh table

s^4	1	5	-6
s^3	5	-5	
	1	-1		after dividing by 5
s^2	6	-6	
	1	-1		after dividing by 6
s^1	(0)			$A(s) = s^2 - 1$
	2			$dA(s)/ds = 2$
s^0	-1		

Sign changes 1

There is only *one sign* change in the first column elements. Hence one root lies in the right half plane. The root is found from the auxiliary equation $A(s) = s^2 - 1 = 0$. They are $s = \pm 1$. Three roots lie in the left half plane.

9.11. SUMMARY

In this chapter, we have explained the concept of stability and characteristic equation and stability. We have then defined the various methods of determining the stability of the systems such as Routh-Hurwitz, Nyquist, root locus, Bode and Nichol's stability criterion. We have discussed in this chapter the method of finding stability in *s*-domain by the application of Routh-Hurwitz algorithm. The necessary and sufficient conditions have been stated and illustrated with examples. Special cases when all elements in a row are zero or when the first element in a row is zero have been considered. The method of determining the stability of a system with transportation lag or with complex coefficients has also been discussed. The above concept has been fully explained with adequate illustrative examples.

REVIEW QUESTIONS

9.1. Define a stable system.

9.2. How do you check the stability of a system?

9.3. Mention the various methods of determining the system stability.

9.4. State the necessary and sufficient conditions of Routh stability criterion.

9.5. Mention the difficulties that may arise in the application of Routh stability criterion.

9.6. What is meant by transportation lag?

9.7. How is the stability of systems with transportation lag determined?

EXERCISE

9.1. Explain the concept of stability of a system.

9.2. Describe the various methods of determining the system stability.

9.3. State and explain the (i) necessary and (ii) necessary and sufficient conditions of Routh-Hurwitz stability criterion.

9.4. What are the difficulties that may arise while applying the Routh stability criterion and how can they be overcome?

9.5. How are the systems with delay analysed?

9.6. Discuss the advantages and limitations of Routh-Hurwitz stability criterion.

9.7. Explain how stability analysis is carried out for a given linear feedback control system.

9.8. What are the conditions required for a system to be stable? How Routh-Hurwitz criterion helps in deciding a stable system?

PROBLEMS

9.1. Determine whether the following functions are stable. If unstable, find the number of roots in RHP.

(a) $s^4 + 5s^3 + 18s^2 + 34s + 20 = 0$

(b) $s^4 + 8s^3 + 32s^2 + 80s + 100 = 0$

(c) $3s^4 + 5s^3 + 3s^2 + 5s + 8 = 0$

(d) $s^5 + 10s^4 + 30s^3 + 80s^2 + 344 + 480 = 0$

(e) $s^4 + 6s^3 + 26s^2 + 56s + 80 = 0$

(f) $s^5 + 3s^4 + 28s^3 + 226s^2 + 600s + 400 = 0$

(g) $s^6 + 3s^5 + 2s^4 + 9s^3 + 5s^2 + 12s + 20 = 0$

9.2. Determine the range of values of K for which each of the following function is stable.

(a) $s^4 + s^3 + s^2 + s + K = 0$

(b) $s^4 + Ks^3 + (K + 4)s^2 + (K + 3)s + 4 = 0$

(c) $s^4 + 9s^3 + 23s^2 + (15 + K)s + 2K = 0$

(d) $s^3 + 2s^2 + (2 + K)s + 3K = 0$

(e) $s^4 + 3s^3 + 3s^2 + 2s + K = 0$

9.3. Determine the range of values of K for which the unit step response function $X(s)$ is stable.

(a) $X(s) = K/s[s(s + 2) (s^2 + s + 10 + K)]$

(b) $X(s) = K/s[s(0.02s + 1) (0.01s + 1) + K]$

(c) $X(s) = K(s + 5)/s[(s + 5) (s^2 + 8s + 20) + K(s + 5)]$

(d) $X(s) = K/s [(s + 1) (s + 2) (s + 5) + K]$

(e) $X(s) = K/s[s^3 + 6s^2 + 11s + 6 + K]$

9.4. A certain feedback control system with unity feedback has a transfer function. $G(s) = K/s(1 + 0.02 \ s) (1 + 0.01s)$. Determine the value of K that just makes the system unstable.

9.5. The closed-loop transfer function of system is given by $C(s)/R(s) = 50/[s(1 + Ts) (1 + 0.5s) + 50]$. Determine the value of T such that the system just becomes unstable and the resulting frequency of oscillation.

9.6. A unity feedback control system has a forward transfer function given by $G(s) = K/s(s^2 + 6s + 12)$. For what value of K does the system become unstable?

9.7. A unity feedback system has an open loop gain $G(s) = K/s(s + 2)(s + 4)$. Find the range of K for stable operation.

9.8. Investigate the characteristic equation given below for the distribution of roots in the RHP.

$F(s) = 2s^7 + s^6 + 2s^5 + 6s^4 + 7s^3 + 10s^2 + 6s + 4 = 0.$

9.9. The characteristic equation of a system is $F(s) = s^3 + 3Ks^2 + (K + 2)s + 4 = 0$. Determine the range of K for the stability of the system.

9.10. Apply Routh criterion to obtain the number of roots in the positive half of the complex plane for the polynomial $s^5 + 2s^4 + 24s^3 + 48s^2 - 25s - 50$.

9.11. The open loop transfer function of a unity feedback system is $G(s) = K/(s + 2) (s + 4)(s^2 + 6s + 25)$. By applying Routh's criterion, discuss the stability of closed-loop system as a function of K. Determine the values of K which will cause sustained oscillations.

9.12. Find the value of K for which the system is stable if $G(s) = K/s(s + 1) (s + 2)$.

9.13. The loop transfer function of a control system is given by $G(s)H(s) = K(s + 3)/s(s + 1)(s + 6)$. Determine the range of K for which the system is stable.

9.14. By means of Routh-Hurwitz stability criterion, determine the stability of the system with the characteristic equation $s^5 + s^4 + 2s^3 + 2s^2 + 3s + 15 = 0$. Determine the number of roots of the equation which are in the right half of s-plane.

9.15. The open loop transfer function of a system with unity feed back is $G(s)H(s) = K/2$ $(s + 1)(s + 0.6)(3s + 1)$. Determine the value of K for stability using Routh-Hurwitz criterion.

9.16. For the unity feedback system with $G(s) = \dfrac{K(1+0.2s)(1+0.25s)}{s(1+0.001s)(1+0.0005s)}$, find the range of

K for which the system is stable.

9.17. Investigate the stability of the system with the characteristic equation $s^4 + 2s^3 + 7s^2 + 10s + 10 = 0$.

9.18. The open loop transfer function of a unity feedback system is $G(s) = K(s + 2)/s(s + 1)$ $(s + 3)(s + 5)$. Determine the value of K for which the system is just stable.

9.19. Use the Routh criterion and determine the number of roots of the following polynomial with real part between 0 and -2. $F(s) = s^4 + 10s^3 + 36s^2 + 70s + 75$.

9.20. The data head of a large disk-storage device is moved to different positions on the spinning disk and rapid accurate response is required. The transfer function of head

dynamics and controller are $G_h(s) = 1/[s(s + 2)(s + 3)]$ and $G_c(s) = \dfrac{K(s+a)}{s+1}$. Determine

the range of K and a for which the system is stable.

9.21. A system has $G(s) = K/s(1 + sT)$, $H(s) = 1$. Determine the values of K and T so that the roots of the characteristic equation will lie to the left of the line $s = -4$.

Chapter 10

ROOT LOCUS TECHNIQUE

10.1. INTRODUCTION

In the analysis and design of control systems, it is necessary to investigate the performance of a system when one or more of its parameters vary over a wide range. The purpose may be to select the appropriate value of a system parameter such as gain or to study parameter variations due to ageing of the system components or environmental changes.

In Chapter 9, the importance of the characteristic equation in the study of the dynamic performance of linear control systems was clearly brought out. Finding the root of the characteristic equation is, in general, laborious for higher-order systems. The classical method of factoring polynomials is not convenient because, as the gain of the system varies, the computations must be repeated. The Routh-Hurwitz criterion indicates whether the system is stable or not, or, it gives the absolute stability of the system. It does not give the degree of stability of the system in terms of overshoot and setting time.

A simple method for finding the roots of the characteristics equation has been developed by W.R. Evans in 1948 and used extensively in control system engineering. This method is called the *root locus method* which not only indicates the *absolute stability but also the degree of stability* for a stable system. The root locus is a plot of the roots of the characteristic equation of the closed-loop system as a function of the gain. The value of the gain can then be obtained from the root locus. The roots corresponding to a specific value of gain can then be located on the root locus. The root locus method is a graphical approach which yields a clear indication of the effect of gain adjustment with relatively small effort compared to other methods.

The root locus method can, in general, be applied to study the behaviour of root of any high-order polynomial.

In this chapter, we shall study the basic properties of the root locus and show how to construct the root locus with the aid of some simple rules.

10.2. ROOT LOCUS OF SECOND-ORDER SYSTEM

To get a better insight into the root locus plots, consider the second-order closed-loop transfer function

$$\frac{C(s)}{R(s)} = \frac{K}{s^2 + 2s + K} \qquad \text{... (10.1)}$$

When the transfer function is expressed with the coefficients of the highest powers of s in both the numerator and denominator equal to unity, the value of K is known as *loop sensitivity*. The problem is to determine the roots of the characteristic equation as K varies from 0 to ∞ and to plot these roots in the s-plane.

The roots of the characteristic equation

$$F(s) = s^2 + 2s + K = 0 \qquad \text{... (10.2)}$$

are given by

$$s_{1,2} = -1 \pm \sqrt{1 - K} \qquad \text{... (10.3)}$$

For $K = 0$, the roots are $s_1 = 0$ and $s_2 = -2$ which are also the poles of the open-loop transfer function

$$G(s)H(s) = \frac{K}{s(s+2)} \qquad \text{... (10.4)}$$

When $K = 1$, then $s_1 = s_2 = -1$. Thus, when K lies between 0 and 1, both the roots are real and lie on the negative real axis of the s-plane. One root lies between 0 and -1, and the other between -2 and -1. When $K > 1$, the roots are complex and are given by

$$s_{1,2} = -1 \pm j\sqrt{K - 1} \qquad \text{... (10.5)}$$

For $K > 1$, the real part of the roots is constant and equal to -1.

The roots of the characteristic equation are determined for a number of values of K varying from 0 to ∞ and are plotted in Fig. 10.1. Curve drawn through these points is called the *root locus*. The root locus has two branches and all possible values of roots of the characteristic equation lie on these two branches for all values of K from 0 to ∞. Each branch is calibrated with K as a parameter. The arrows show the direction of increasing values of K. From the root locus, the roots that best fit the system performance specifications can be selected. Corresponding to these selected roots, the required value of K can be determined from the plot. Once the roots have been selected, the time response can be obtained.

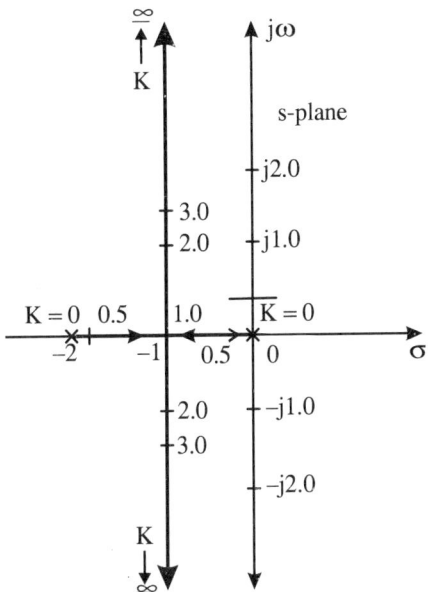

Fig. 10.1. Root locus of a second-order system

From the root locus, it is possible to determine the variation in system performance with respect to variation in loop sensitivity K. For example, a root with a damping ratio ζ is obtained with the angle $\theta = \cos^{-1} \zeta$ as indicated in Fig. 10.2.

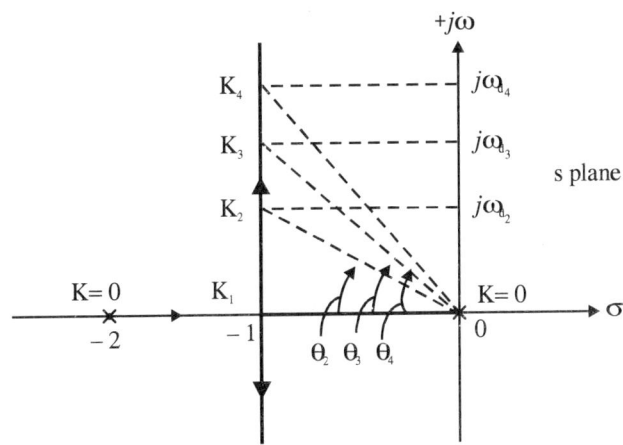

Fig. 10.2. Loop sensitivity *vs* ζ

Analysis of the root locus reveals the following characteristics when the gain of the system is increased.

(a) The damping ratio ζ is decreased. This increases the overshoot in the time-response of the system.

(b) The undamped natural frequency ω_n is increased. The value of ω_n is the distance from origin to the complex root.

(c) The damped natural frequency ω_d also increases. The value of ω_d is the imaginary part of the complex root and is equal to the frequency of transient response of the system.

(d) The rate of decay α remains constant for all values of K equal to or greater than K_1. This is true for second order system only.

(e) The root locus is parallel to $j\omega$-axis for $K > K_1$ and $\alpha = -\zeta\,\omega_n$ is constant. The root locus, thus, does not cross the $j\omega$-axis and enter the right half of the s-plane. This implies that *a second-order system can never become unstable as K is varied from 0 to* ∞.

The root locus of any control system can thus be analyzed to obtain an idea of the variation in its time response when its loop sensitivity K varies.

10.3. BASIC PROPERTIES OF ROOT LOCUS

Consider the system shown in Fig. 10.3. The closed-loop transfer function is

$$\frac{C(s)}{R(s)} = \frac{KG(s)}{1 + KG(s)H(s)} \qquad \ldots (10.6)$$

where $G(s)H(s)$ is expressed as

$$G(s)H(s) = \frac{s^m + b_{m-1}\,s^{m-1} + \ldots + b_0}{s^n + a_{n-1}\,s^n + \ldots + a_0} \qquad \ldots (10.7)$$

$G(s)H(s)$ *does not contain the variable parameter K.*

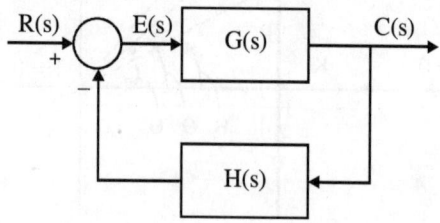

Fig. 10.3. Block diagram of a closed-loop system

The characteristic equation of this closed-loop system is

$$F(s) = 1 + KG(s)H(s) = 0 \qquad \qquad \text{... (10.8)}$$

Then Eq. (10.8) may be expressed as

$$G(s)H(s) = -\frac{1}{K} \qquad \qquad \text{... (10.9)}$$

Since $G(s)H(s)$ is a complex quantity, Eq. (10.9) may be split into two equations by equating the magnitudes and angles of both sides. We therefore obtain

(i) Angle condition:

$$\underline{|G(s)H(s)} = (2k+1)180° \qquad (K \geq 0) \qquad \text{... (10.10)}$$

where $k = 0, 1, 2, \dots\dots$

(ii) Magnitude Condition:

$$|G(s)H(s)| = 1/|K| \qquad (0 < K < \infty) \qquad \text{... (10.11)}$$

The values of s which fulfill the angle and magnitude conditions are the roots of the characteristic equation or the closed-loop poles. The locus is the plot of points on the complex plane which satisfy the angle condition alone. The roots of the characteristic equation corresponding to a given value of the gain can be determined from the magnitude condition, or, using this condition, the values of K on the loci can be determined.

10.4. OPEN LOOP TRANSFER FUNCTION

The open-loop transfer function $G(s)H(s)$ is, in general, of the from

$$G(s)H(s) = \frac{K(s+b_1)}{s^{\ell}(s+a_1)(s+a_2)} \qquad \text{... (10.12)}$$

Inspection of Eq. (10.12) reveals that the open-loop transfer function has a zero at $s = -b_1$ multiple poles at $s = 0$, and simple poles $s = -a_1$, and $s = -a_2$. Denoting the zeros by z and poles by p, the generalized open-loop transfer function may be written as

$$G(s)H(s) = \frac{K(s+z_1)(s+z_2)\dots\dots(s+z_m)}{s^{\ell}(s+p_1)(s+p_2)\dots\dots(s+p_k)} \qquad \text{... (10.13)}$$

where $n = \ell + k$ is the degree of the denominator polynomial and m is the degree of the numerator polynomial.

In cases where the variable parameter K does not appear as a multiplying factor as in Eq. (10.13), we can always condition it in the form of Eq. (10.13).

Consider an open-loop transfer function

$$G(s)H(s) = \frac{s^2 + (b_1 + c_1K)s + b_0}{s(s+a_1)(s+a_2)} \qquad \text{... (10.14)}$$

The characteristic equation is

$$F(s) = 1 + G(s)H(s) = s(s+a_1)(s+a_2) + s^2 + (b_1 + c_1K)s + b_0 = 0$$
$$\text{... (10.15)}$$

Dividing both sides of Eq. (10.15) by the factor that does not contain K, we obtain,

$$F(s) = 1 + \frac{c_1Ks}{s(s+a_1)(s+a_2) + s^2 + b_1s + b_0} \qquad \text{... (10.17)}$$

$$= 1 + G(s)H(s) \qquad \text{... (10.17)}$$

where the new open-loop transfer function is

$$G(s)H(s) = \frac{Kc_1s}{s(s+a_1)(s+a_2) + s^2 + b_1s + b_0} \qquad \text{... (10.18)}$$

which is of the standard form given in Eq. (10.13)

10.5. APPLICATION OF MAGNITUDE AND ANGLE CONDITIONS

Once the open-loop transfer function $G(s)H(s)$ is obtained in the proper form, the poles and zeros are plotted in the s-plane. Then the root locus is constructed using the angle condition. This is illustrated with an example.

Example 10.1. *Consider the open-loop transfer function*

$$G(s)H(s) = \frac{K(s+z_1)}{s(s+p_1)(s+p_2)} \qquad \text{... (10.19)}$$

Explain the construction of root-locus using angle condition.

Solution:

The locations of poles and zeros are arbitrarily assumed as shown in Fig. 10.4. Next, let us select an arbitrary point s_1 in the s-plane and draw vectors directing from the poles and zeros of $G(s)H(s)$ to the point s_1. If the point s_1 is to be on the root locus, it must satisfy the angle condition

$$\angle(s_1 + z_1) - \angle s_1 - \angle(s_1 + p_1) - \angle(s_1 + p_2) = (2k + 1)\pi, \ k = 0, 1, 2, \ldots. \qquad \text{... (10.20)}$$

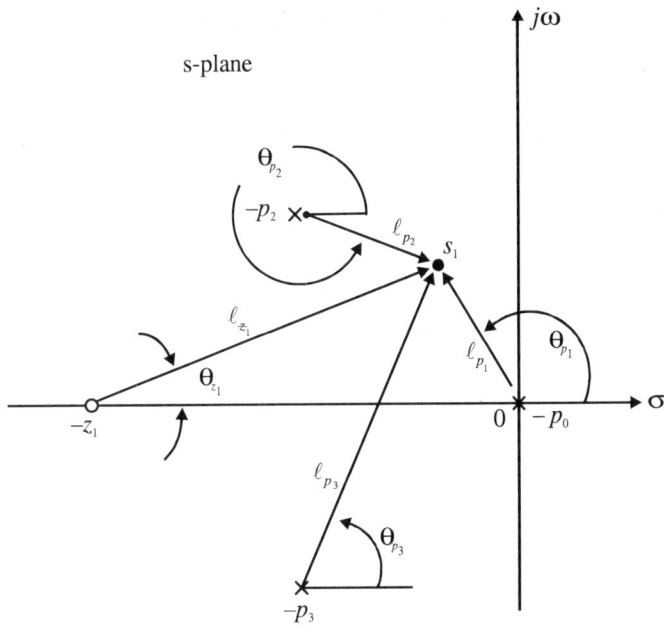

Fig. 10.4. Angle condition

As shown in Fig. 10.4, the angles θ_{z1}, θ_{p0}, θ_{p1} and θ_{p2} are the angles of the vectors measured with the positive real axis as zero reference in the counter-clockwise direction.

Therefore

$$\theta_{z1} - \left(\theta_{p0} + \theta_{p1} + \theta_{p2}\right) = (2k+1)\pi \qquad 0 \le k \le \infty \qquad \qquad \text{... (10.21)}$$

If Eq. (10.21) is not satisfied, select another search point until it is satisfied. A sufficient number of points in the s-plane that satisfy the angle condition are located and the root locus is drawn as curves passing through these points.

From the magnitude condition, we may obtain the value of loop sensitivity K. Let ℓ_{z1}, ℓ_{p0}, ℓ_{p1} and ℓ_{p2} be the lengths of the vectors drawn from zeros and poles of the open-loop transfer function $G(s)H(s)$. Then K is given by

$$K = \frac{\ell_{p0} \cdot \ell_{p1} \cdot \ell_{p2}}{\ell_{z1}}$$

$$= \frac{\text{Product of vector-lengths drawn from poles of } G(s)H(s)}{\text{Product of vector-lengths drawn from zeros of } G(s)H(s)}$$

$$\text{... (10.22)}$$

10.6. GENERAL RULES FOR CONSTRUCTING ROOT LOCUS

To facilitate the application of root locus method, the following rules for constructing the root loci are established for $K > 0$. These rules are based upon the interpretation of the angle condition and analysis of characteristic equation. They should be regarded as an aid to the construction of the root loci since they do not give exact plot. The reader should be able to extend the rules for $K < 0$.

Rule 1. Number of Branches of Root locus: The number of branches of the root loci is equal to the number of zeros or poles of the open-loop transfer function $G(s)H(s)$, whichever is greater. The remaining zeros or poles are assumed to lie at infinity.

A branch of the root loci is the locus of one root as K varies from 0 to ∞. Hence the number of branches of the root loci must be equal to the number of roots of the characteristic equation.

Rule 2. Starting Points of the Root Loci: The starting points of the root loci when $K = 0$ are the poles of $G(s)H(s)$.

From the magnitude condition,

$$|G(s)H(s)| = \frac{(s+z_1)(s+z_2)..........}{(s+p_1)(s+p_2)..........} = 1/K \qquad \text{... (10.23)}$$

As K approaches zero, $|G(s)H(s)|$ approaches infinity. To satisfy this condition, s approaches $-p_i$ ($i = 1$ to n).

Rule 3. Ending Points of Root Loci: The ending points of the root loci when $K = \infty$ are the zeros of $G(s)H(s)$.

From Eq. (10.23), as K approaches infinitely, $|G(s)H(s)|$ approaches zero. This condition is satisfied as s approaches $-z_j$, ($j = 1$ to m).

Rule 4. Real Axis Locus: If the total number of real poles and zeros of $G(s)H(s)$ to the right of the search point s_1 on the real axis is odd, then this lies on the root locus.

The above rule is based on the following observations:

(a) The sum of the angles of the vectors drawn from the complex-conjugate poles and zeros to any point s_1 on the real axis is zero.

(b) Real poles and zeros lying to the right of the point s_1 on the real axis contribute to the angle condition. The real poles and zeros lying to the left of s_1 contribute zero degrees.

(c) Each real zero of G(s)H(s) lying to the right of the point s_1 contributes $+180°$ and each pole of $G(s)H(s)$ lying to the right of s_1 contributes $-180°$.

Thus the last observation show that, for s_1 to be a point on the root locus, there must be an odd number of real poles and zeros of $G(s)H(s)$ to the right of s_1.

Rule 5. Asymptotes of Root Locus as s approaches Infinity: For large values of s, the root locus approaches the asymptotes with angles given by

$$\theta_{k+1} = \frac{(2k+1)\pi}{n-m}, k = 0,1,.................(n-m-1) \qquad \text{... (10.24)}$$

and there are $(n - m)$ asymptotes where n is number of poles and m is the number of zeros of $G(s)H(s)$.

This can be proved as indicated below.

$$\lim_{s \to \infty} G(s)H(s) = \lim_{s \to \infty} \left| K \frac{\prod_{j=1}^{m}(s+z_j)}{\prod_{i=1}^{n}(s+p_i)} \right| \frac{K}{s^{n-m}} = -1 \qquad \text{... (10.25)}$$

Therefore,

$$\underline{|K} = \underline{|s^{n-m}} = (2k+1)\pi$$

or,

$$(n-m)\underline{|s} = (2k+1)\pi$$

Hence

$$\theta_{k+1} = \frac{(2k+1)\pi}{n-m} \qquad as \ s \to \infty$$

Rule 6. Real Axis Intercept of the Asymptotes or Centroid: The intersection of the asymptotes or centroid of the root locus on the real axis is given by

$$\sigma_o = \frac{\sum_{1}^{n} \text{Re}(p_i) - \sum_{1}^{m} \text{Re}(z_j)}{n-m} \qquad \text{... (10.26)}$$

The proof of Eq. (10.26) follows:

Let the open-loop transfer function be given by

$$G(s)H(s) = K \frac{s^m + b_1 s^{m-1} + ...}{s^n + a_1 s^{n-1} + ...} \qquad \text{... (10.27)}$$

The characteristic equation is

$$F(s) = 1 + G(s)H(s) = 1 + K \frac{s^m + b_1 s^{m-1} + ...}{s^n + a_1 s^{n-1} + ...} = 0$$

Rearranging we get,

$$\frac{s^n + a_1 s^{n-1} + ...}{s^m + b_1 s^{m-1} + ...} = -K \qquad \text{... (10.28)}$$

Carrying out the fraction on the left side by the process of long division and neglecting all but the first two terms as $s \to \infty$, we get

$$s^{n-m} + (a_1 - b_1)s^{n-m-1} = -K \qquad \qquad ...(10.29)$$

or,

$$s\left(1 + \frac{a_1 - b_1}{s}\right)^{(n-m)^{-1}} = -K^{(n-m)^{-1}} \qquad \qquad ...(10.30)$$

Expanding the left side of Eq. (10.30) by Binomial theorem, we obtain

$$s\left[1 + \frac{a_1 - b_1}{(n-m)s} + ...\right] = -K^{(n-m)^{-1}} \qquad \qquad ...(10.31)$$

Considering the first two terms only, we get

$$s + \frac{a_1 - b_1}{n - m} = -K^{(n-m)^{-1}} \qquad \qquad ...(10.32)$$

or,

$$\sigma_0 + j\omega + \frac{a_1 - b_1}{n - m} = -K^{(n-m)^{-1}} \qquad \qquad ...(10.32)$$

At the real-axis intercept, $K = 0$ and $j\omega = 0$. Hence,

$$\sigma_0 = -\frac{a_1 - b_1}{n - m} = \frac{-a_1 - (-b_1)}{n - m} \qquad \qquad ...(10.33)$$

where

$-a_1$ = sum of the roots of $s^n + a_1 s^{n-1} + ...$

      = sum of the real part of poles of $G(s)H(s)$

$-b_1$ = sum of the roots of $s^m + b_1 s^{m-1} + ...$

      = sum of the real part of zeros of $G(s)H(s)$

Rule 7. Angle of Departure from Complex poles and Angle of Arrival at Complex zeros: The angle of departure from complex poles or arrival at complex zeros of $G(s)H(s)$ denotes the behaviour of the roots locus near that pole or zero. The angles may be found by the application of angle condition.

Consider the pole-zero locations of $G(s)H(s)$ as shown in Fig. 10.5. To find the angle of departure of root locus from p_2, assume the search point is at p_2 itself and draw vector-lines from all the other poles and zeros to p_2. Applying the angle condition,

$$\phi_0 + \phi_1 + \phi_2 + \phi_3 - \psi_1 = (2k + 1)\,180° \qquad \qquad ...(10.34\ (a))$$

The departure angle ϕ_2 is given by (with $\phi_3 = 90°$)

$$\phi_2 = (2k + 1)\ 180° - (\phi_0 + \phi_1 + 90° - \psi_1)$$

Similarly, the angle of approach of the root locus to a complex zero can be determined.

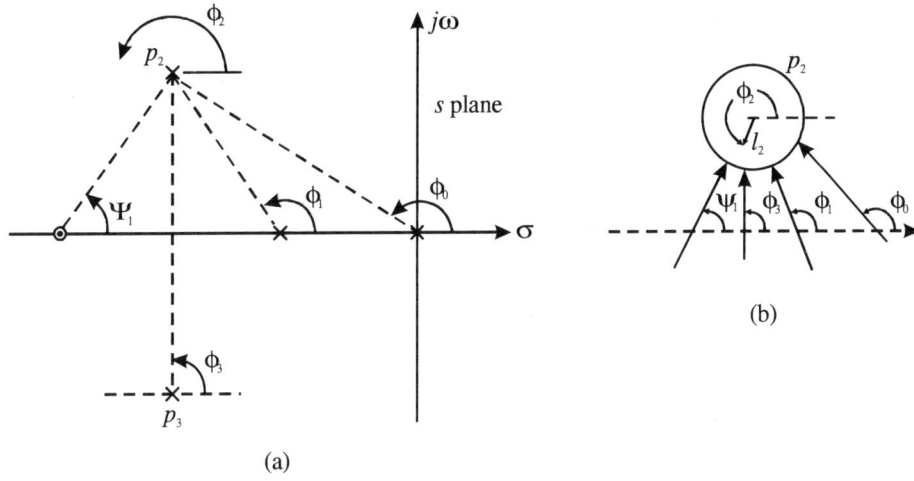

Fig. 10.5. Pole-zero location for calculating *angle* of approach

Consider the pole-zero locations of $G(s)H(s)$ as shown in Fig. 10.6. The angle of approach ψ_1 is given by

$$\psi_1 = (\phi_0 + \phi_1 + \phi_2 - 90°) - (2k + 1)\ 180° \qquad \qquad ...\ (10.34\ b)$$

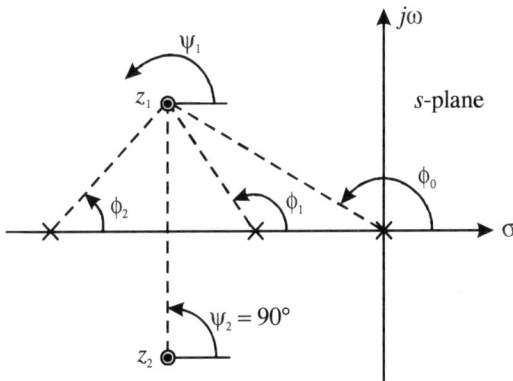

Fig. 10.6. Pole-zero location for calculating angle of departure

Rule 8. Breakaway points of Root locus: The breakaway points of the root locus $1 + KG(s)H(s) = 0$ must satisfy

$$dG(s)H(s)/ds = 0 \qquad \qquad \text{... (10.35)}$$

This may be proved as outlined below:

The branches of the rot locus start at the open-loop poles where $K = 0$ and end at the open-loop zeros where $K = \infty$.

Consider the case where the root locus has branches on the real axis between two poles. There must be a point at which the two branches meet and breakaway from the real axis and enter the complex region of the s-plane in order to approach zeros. The point at which the two branches breakaway from the real axis is called *breakaway point*. Similarly for two zeros on the real axis, the branches are coming from the complex region and enter the real axis is called *break-in point*. At these points, the characteristic equation

$$F(s) = 1 + KG(s)H(s) = 1 + KQ(s)/P(s) = 0$$

This may be written as

$$F(s) = P(s) + KQ(s) = 0 \qquad \qquad \text{... (10.36)}$$

For a small increment in K, we have

$$P(s) + (K + \Delta K)Q(s) = 1 + \frac{\Delta K Q(s)}{P(s) + KQ(s)} = 0 \qquad \qquad \text{... (10.37)}$$

Since the denominator is the original characteristic equation and at the breakaway point multiple roots exist, we may write

$$\frac{Q(s)}{P(s) + KQ(s)} = \frac{C_i}{(s - s_i)^n} = \frac{C_i}{(\Delta s)^n} \qquad \qquad \text{... (10.38)}$$

Substituting Eq. (10.38) in Eq. (10.37), we obtain

$$1 + \frac{\Delta K C_i}{(\Delta s)^n} = 0 \qquad \qquad \text{... (10.39)}$$

or,

$$\frac{\Delta K}{\Delta s} = -\frac{(\Delta s)^{n-1}}{C_i} \qquad \qquad \text{... (10.40)}$$

As Δs approaches zero,

$$\frac{dK}{ds} = 0 \qquad \qquad \text{... (10.41)}$$

at the breakaway or break-in points. From the characteristic equation,

$$K = -1/G(s)H(s)$$

Hence $dK/ds = 0$ is equivalent to

$$dG(s)H(s)/ds = 0$$

Rule 9. Intersection of the Root Locus with Imaginary Axis: The point where the root locus crosses the imaginary axis of the s-plane and the corresponding values of K may be determined by Routh stability criterion.

The frequency at the cross-over point is the undamped natural frequency of oscillation and the value of K is the maximum value for which the closed-loop system is just or *marginally stable*.

Rule 10. Symmetry of the Root-Locus: The root locus is symmetrical with respect to the real axis of the s-plane.

This is self-evident since for real coefficients of the characteristic equation, the roots must be real or in complex conjugate pairs.

Rule 11. Calculation of K on the Root Locus: Once the root locus has been constructed, the value of K at any point s_1 on the root locus can be determined by the use of Eq. (10.22).

These construction rules are summarized in Table 10.1.

Table 10.1. Rules for construction of root locus

1. Number of separate branches.	Number of poles or zeros of $G(s)H(s)$ whichever is greater.
2. Starting points.	Poles of $G(s)H(s)$. (Poles at infinity are also included)
3. Ending points.	Zeros of $G(s)H(s)$. (zeros at infinity are also included)
4. Real axis root Locus.	Section to the right of which the number of real poles and zeros are odd.
5. Asymptotes of root locus.	Lines making angles given by $\theta_{k+1} = (2k + 1)\pi/(n - m)$
6. Intersections of asymptotes.	$\sigma = (\Sigma \, Re \text{ poles} - \Sigma \, Re \text{ zeros of } G(s)H(s))/(n - m)$
7. Angles of departure/arrival.	$\angle G(s_1)H(s_1) = (2k + 1)\pi$. s_1 is the test point.
8. Breakaway point.	Roots of $dK/ds = 0$.
9. Intersection of root locus with $j\omega$-axis.	Obtained from Routh stability criteria.
10. Symmetry about real axis.	Symmetrical about real axis.
11. Value of K at any point s_1 on root locus.	$|K| = 1/|G(s_1)H(s_1)|$

Table 10.2 shows typical pole-zero configurations with the corresponding root loci.

Table 10.2. Pole-zero locations and roo-locus

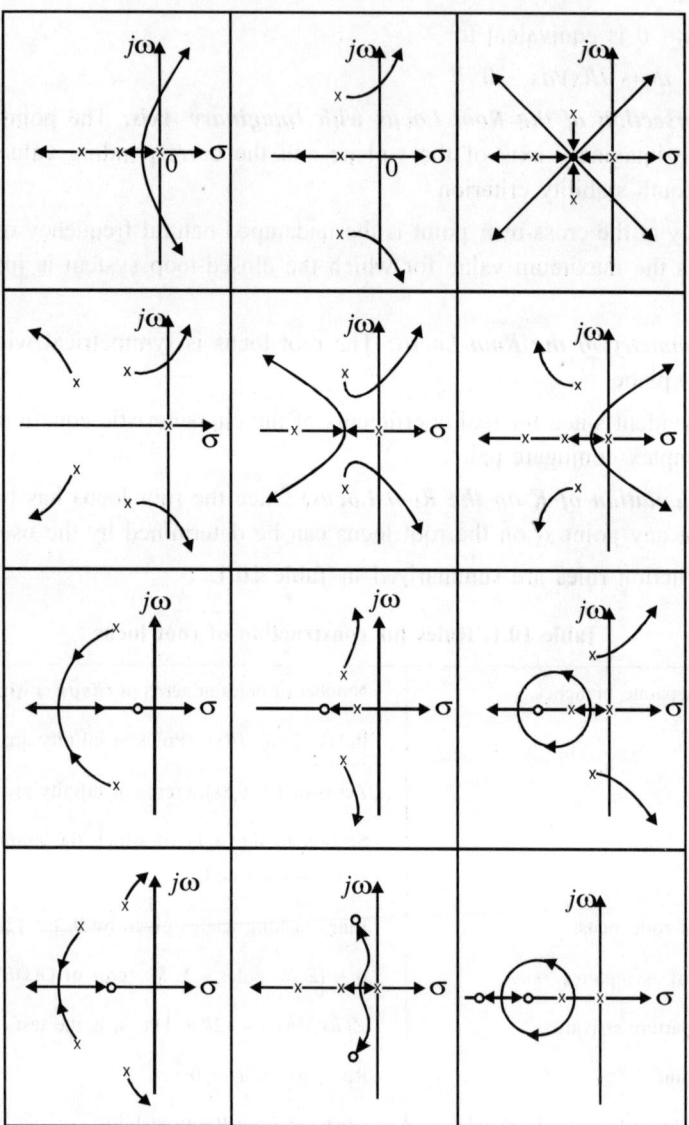

10.7. CONSTRUCTION OF ROOT-LOCUS

The application of the root locus construction rules and drawing of the root-locus are illustrated with examples.

Example 10.2. *Sketch the root locus of the system whose open-loop transfer function is given by*

$$G(s)H(s) = \frac{K}{s(s+2)(s+3)}$$

Solution:

The open loop transfer function is

$$G(s)H(s) = \frac{K}{s(s+2)(s+3)}$$

1. There are there branches since there are three poles.

2. The starting points are at the open-loop poles. They are $s = 0$, $s = -2$, and $s = -3$.

3. Since there are no finite zeros, all the three branches end *at infinity*.

4. The sections of real axis that lie on the root locus are: 0 to -2 and -3 to $-\infty$. Mark them with thick lines.

5. There are 3 asymptotes since there are three finite poles ($n = 3$) and no finite zeros ($m = 0$). The angles of the asymptotes are given by

$$\theta_{k+1} = \frac{(2k+1)180^o}{n-m} = \frac{(2k+1)180^o}{3}$$

for $k = 0$	$\theta_1 = 60°$
for $k = 1$	$\theta_2 = 180°$
for $k = 2$	$\theta_3 = 300°$

6. The real-axis intercept of asymptotes or centroid is

$$\sigma_o = \frac{0-2-3}{3} = -1.67$$

Draw the asymptotes at −1.67 with angles of 60°, 180° and 300° with the real axis.

7. There are no complex poles or zeros.

8. The breakaway point on the real axis lies between 0 and −2 and is found from $dK/ds = 0$.

From the $F(s)$, K is given by

$$-K = s^3 + 5s^2 + 6s$$

$$dK/ds = 3s^2 + 10s + 6 = 0$$

or,

$$s_{1,2} = \frac{-10 \pm \sqrt{100-72}}{6} = -2.55 \text{ or } -0.79$$

Since the breakaway point must be between 0 and –2, $s_1 = -0.79$ is the breakaway point. The value of K at the breakaway point is

$$K = 0.79 \times 1.21 \times 2.21 = 2.1$$

9. The intersection point with the imaginary axis is found from the Routh stability criterion. The characteristic equation is

$$F(s) = s^3 + 5s^2 + 6s + K = 0$$

The Routh table is

Routh table

s^3	1	6
s^2	5	K
s^1	$\dfrac{30-K}{5}$	
s^0	K	

For the system to be stable, the condition to be satisfied is: $K > 0$ and $K < 30$. Therefore. the value of K at the crossover point is 30. The auxiliary equation is

$$5s^2 + 30 = 0$$

or

$$s = \pm j\sqrt{6} = \pm j2.45$$

10. Sketch the root locus and check for the real axis symmetry. The root locus is shown in Fig. 10.7.

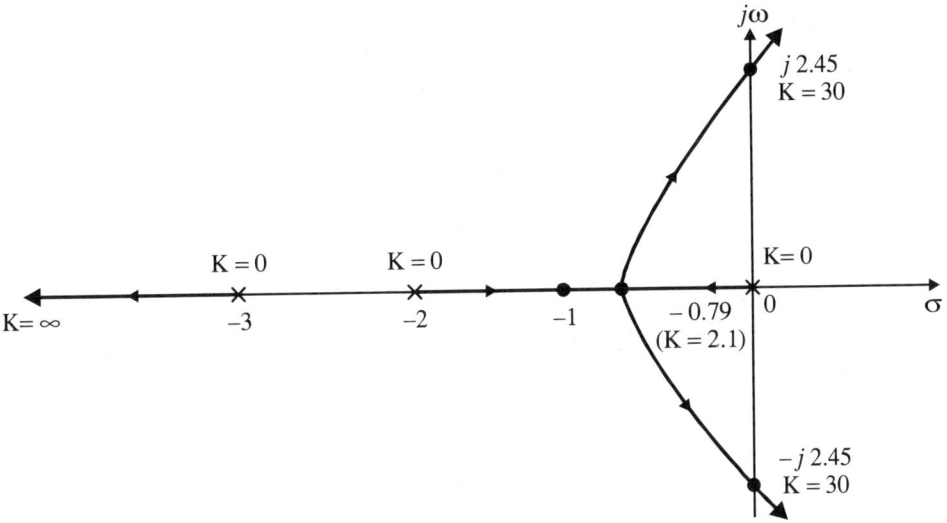

Fig. 10.7. Root-locus of $G(s)H(s) = K/s(s + 2)(s + 3)$

Example 10.3. *The open-loop transfer function of a system is*

$$G(s)H(s) = \frac{K(s+1)}{s(s+2)(s+3)}$$

Draw the root-locus.

Solution:

The openloop transfer function is

$$G(s)H(s) = \frac{K(s+1)}{s(s+2)(s+3)}$$

1. The open-loop transfer function has three poles and one zero. Hence the number of branches is three.

2. The starting point are poles of $G(s)H(s)$. They are $s = 0$, $s = -2$ and $s = -3$ ($n = 3$).

3. The ending points are the zeros of $G(s)H(s)$. One zero is at $s = -1$; other two zeros are at infinity since the branches must end at zeros. Locate the poles and finite zero in the s-plane.

4. The sections of the real axis that lie on the root locus are 0 to -1 and -2 to -3. Mark them with thick lines.

5. There are 2 asymptotes since $n = 3$ and $m = 1$. The angles are:

$$\text{for } k = 0 \qquad\qquad \theta_1 = 180/2 = 90°$$
$$\text{for } k = 1 \qquad\qquad \theta_2 = 540/2 = 270°$$

6. The real-axis intercept or centroid is

$$\sigma_o = \frac{(0-2-3)-(-1)}{2} = -2$$

Draw the asymptotes at –2 with angles of 90° and 270° with the real axis.

7. There are no complex poles or zeros.

8. The breakaway point on the real-axis lies between –2 and –3 and is found from $dK/ds = 0$. From the $F(s)$, K is given by

$$-K = s(s+2)(s+3)/s+1$$
$$= \frac{s^3 + s^2 + 6s}{s+1}$$

Hence

$$\frac{dK}{ds} = \frac{(s+1)(3s^2+10s+6)-(s^3+5s^2+6s)\cdot 1}{(s+1)^2} = 0$$
$$= 2s^3 + 8s^2 + 10s + 6 = 0$$

Because the real axis symmetry, the breakaway point will be near –2.5. By trial and error method, the roots of the above equation may be found to be –2.47, –0.765 ± j0.8. Since the break away point must be on the real axis, it is at –2.47.

The value of K at $s_1 = -2.47$ is

$$K = \frac{2.47 \times .47 \times .53}{1.47} = 0.42$$

9. The intersection point with the imaginary axis is found by the application of Routh stability criterion to $F(s)$. The characteristic equation is

$$F(s) = 1 + G(s)H(s) = s(s+2)(s+3) + K(s+1)$$
$$= s^3 + 5s^2 + (6+K)s + K = 0$$

The Routh Table is

Routh Table

s^3	1	$6+K$
s^2	5	K
s^1	$(30+5K-K)/5$	
s^0	K	

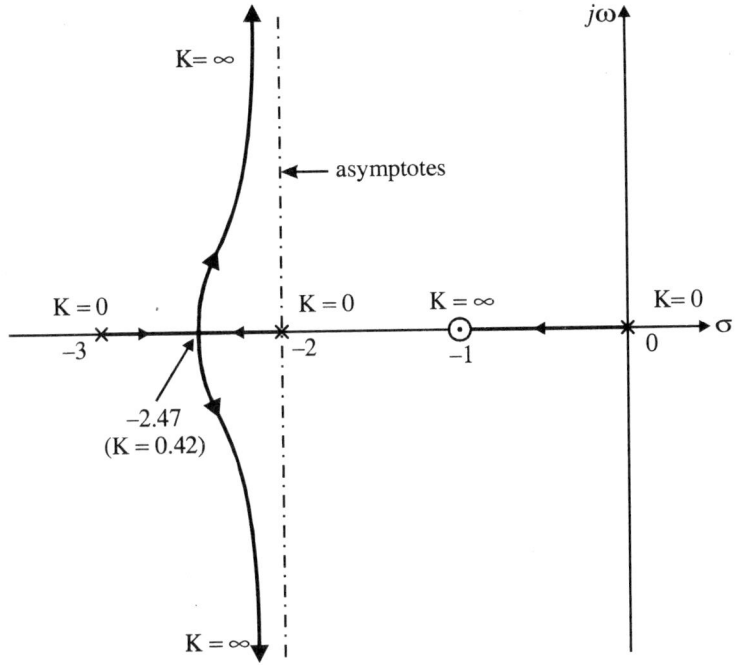

Fig. 10.8. Root locus of $G(s)H(s) = K (s + 1)(s + 2)(s + 3)$

From s^1 and s^0 row $K > 0$ or $K > -75$. The system is stable for $K > 0$ as the root locus does not cross the $j\omega$-axis.

10. Sketch the root locus and check for real axis symmetry. The root locus is show in Fig 10.8.

Example 10.4. *Sketch the root locus of the system whose forward transfer function G(s) and feedback transfer function H(s) are*

$$G(s) = \frac{2600K}{s(s^2 + 100s + 2600)}, H(s) = \frac{25}{s + 25}$$

Solution:

The open-loop transfer function $G(s)H(s)$ is

$$G(s)H(s) = \frac{65000K}{s(s + 25)(s^2 + 100s + 2600)}$$

and the characteristic equation is

$$F(s) = s(s + 25)(s^2 + 100s + 2600) + 65,000K = 0$$

1. The open-loop transfer function has four poles. They are $s = 0$, $s = -25$, $s = -50 + j10$ and $s = -50 - j10$. Mark them on the s-plane. There are thus *four* branches.

2. The starting points are the poles of $G(s)H(s)$.

3. The ending points are the zeros of $G(s)H(s)$. Since it has no finite zeros, we have to assume that the four zeros are *at infinity* at which the four branches end.

4. The section of the real axis that lies on the root locus is between 0 and -25. Mark it with thick line.

5. There are 4 asymptotes since $n = 4$ and $m = 0$. The angles of the asymptotes are :

$$\theta_{K+1} = \frac{(2k+1)180^\circ}{n-m} = \frac{(2k+1)180^\circ}{4}$$

for $k = 0$ $\theta_1 = 45^\circ$

for $k = 1$ $\theta_2 = 135^\circ$

for $k = 2$ $\theta_3 = 225^\circ$

for $k = 3$ $\theta_4 = 315^\circ$

6. The real-axis intercept of the asymptotes or centroid is

$$\sigma_o = \frac{(0 - 25 - 50 - 50)}{4} = -31.25$$

Draw the four asymptotes at -31.25 making angles of 45°, 135°, 225° and 315° with the real axis.

7. The angle of departure ϕ_3 from the pole $(-50 + j10)$ is obtained from the angle condition. From Fig. 10.9,

$$\phi_0 + \phi_1 + \phi_2 + \phi_3 = (2k+1)180^\circ$$

$$\phi_0 = 180 - \tan^{-1} 10/50 = 168.7^\circ$$

$$\phi_1 = 180 - \tan^{-1} 10/25 = 158.2^\circ$$

$$168.7^\circ + 158.2^\circ + 90 + \phi_3 = (2k+1)180^\circ$$

$$\phi_3 = 123.1^\circ$$

Because of real-axis symmetry, the angle of departure of root locus from the pole $(-50 - j10)$ is -123.1°.

8. The breakaway point on the real axis is between 0 and -25. It is found from $dK/ds = 0$. From $F(s)$, K is given by

$$-K = s(s+25)\,(s^2+100s+2600)\,/65{,}000$$
$$= (s^4+125s^3+5100s^2+65000s)/65000$$

$$\frac{dK}{ds} = 4s^3+375s^2+10200s+65000/65000 = 0$$

$$s_1 = -9.2$$

The value of K can be easily obtained by substituting $s = -9.2$.

$$-K = \frac{-9.2\times-15.8\times1764.6}{65000} = -3.95$$

Therefore $K = 3.95$ at the breakaway point.

9. The intersection point of the imaginary axis is found from the Routh criterion.

   The characteristic equtions is

$$F(s) = s^4+125s^3+5100s^2+65000s+65{,}000K = 0.$$

   The Routh Table is

Routh table

s^4	1	5100	65,000K
s^3	1	520	(after dividing by 125)
s^2	1	14.2K	(after dividing by 4580)
s^1	520 $-$ 14.2K		
s^0	14.2K		

Pure imaginary roots exist when s^1 row is zero. This occurs when $K = 520/14.2 = 36.6$. The auxiliary equation is $s^2+14.2K = 0$ and the imaginary roots are:

$$s = \pm\sqrt{14.2\times35.6} = \pm j22.8$$

10. Sketch the root locus and check for the real-axis symmetry. The root locus is shown in Fig. 10.10.

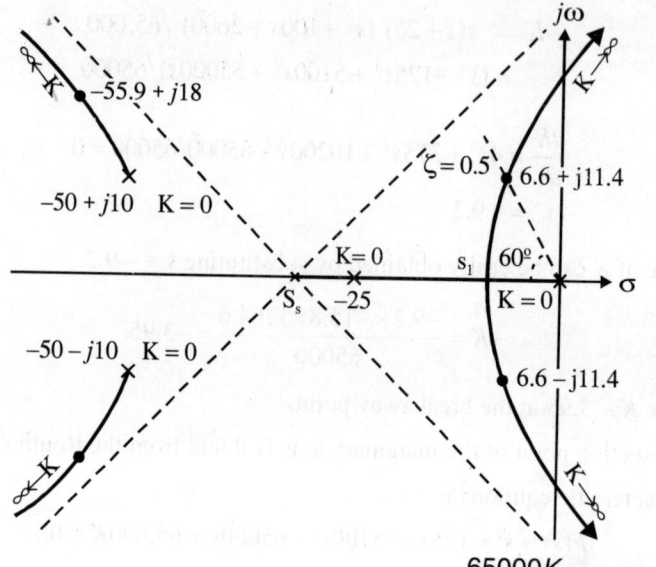

Fig. 10.10. Root locus of $G(s)H(s) = \dfrac{65000K}{s(s+25)\left(s^2+100s+2600\right)}$

10.8. OBTAINING TIME AND FREQUENCY RESPONSE FROM ROOT LOCUS

The method of obtaining the time and frequency response of a system from its root locus plot is illustrated with suitable example.

Example 10.5. *For the system considered in Example 10.4, obtain the time response from its root locus for $\zeta = 0.5$.*

Solution:

The root locus of the system is shown in Fig. 10.10

The radial line for $\zeta = 0.5$ is drawn on the graph as shown at an angle θ given by

$$\theta = \cos^{-1}\zeta = \cos^{-1}0.5 = 60°$$

The dominant roots read from the graph are

$$s_{1,2} = -6.6 \pm j11.4$$

The gain K is obtained from the expression

$$K = \frac{|s\,||\,s+25\,||\,s+50-j10\,||\,s+50+j10|}{65000}\Big|s=-6.6+j11.4$$

$$= 9.23$$

The other roots of the characteristic equation are evaluated as explained below. With $K = 9.23$, the characteristic equation is

$$s^4 + 125s^3 + 5100s^2 + 65{,}000s + 598{,}800 = 0$$

The quadratic factor representing the dominant roots is

$$(s + 6.6 + j11.4)\,(s + 6.6 - j11.4) = s^2 + 13.2s + 173.5$$

Dividing the characteristic equation by this factor and ignoring the remainder, we get

$$s^2 + 112s + 3450 = 0$$

The roots of the above quadratic factor are

$$s_{3,4} = -56 \pm j18$$

Alternative method of finding the other roots is by using Grant's rule which gives the real part of the roots. Using this value, we may find the exact root. According to Grants rule,

$$\sum_{j=1}^{n} p_j = \sum_{i=1}^{n} r_j$$

where p_j are the open-loop poles and r_j are the roots described by the root locus. For our example,

$$(0 - 25 - 50 - 50) = (-6.6 + j11.4 - 6.6 - j11.4 + \sigma + j\omega_d + \sigma - j\omega_d)$$

This yields the intersection point to be

$$\sigma = -55.9$$

By drawing a perpendicular line at $\sigma = -55.9$, the other roots are read at the intersection points of this line with the root locus. The roots thus obtained are

$$s_{3,4} = -55.9 \pm j18.$$

The closed-loop transfer function is given by

$$\frac{C(s)}{R(s)} = \frac{N_1(s)D_2(s)}{\text{Factor determined from root locus}} \tag{10.43}$$

where $N_1(s)$ is the numerator polynomial of $G(s)$ and $D_2(s)$ is the denominator polynomial of $H(s)$. Substituting these values we get

$$\frac{C(s)}{R(s)} = \frac{24{,}000(s + 25)}{(s^2 + 13.2s + 173.5)(s + 55.9 + j18)(s + 55.9 - j18)}$$

The response $c(t)$ for a unit step input is given by the inverse Laplace transform of $C(s)$.

$$C(s) = \frac{A_0}{s} + \frac{A_1}{s + 6.6 - j11.4} + \frac{A_2}{s + 6.6 + j11.4} + \frac{A_3}{s + 55.9 - j18}$$

$$+ \frac{A_4}{s + 55.9 + j18}$$

The constants are

$$A_0 = 1, A_1 = 0.604\underline{|-201.7°}, A_3 = 0.14\underline{|-63.9°}$$

Then the time response $c(t)$ is

$$c(t) = 1 + 1.21e^{-6.6t}\sin(11.4t - 111.7°) + 0.28e^{-55.9t}\sin(18t + 26.1°)$$

The time response is show in Fig. 10.11.

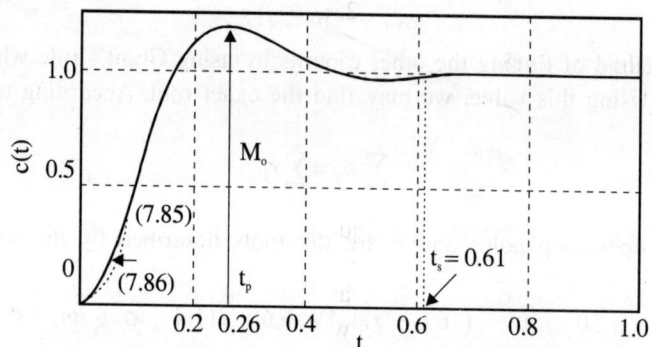

Fig. 10.11. Time response obtained from root locus

Example 10.6. *For the above example, obtain the frequency response from its root locus plot.*

Solution:

The frequency response can easily be obtained from the root locus once the system gain is determined for the desired performance.

The closed-loop transfer function is

$$\frac{C(s)}{R(s)} = \frac{24,000(s+25)}{(s+6.6-j11.4)(s+6.6+j11.4)(s+55.9-j18)(s+55.9+j18)} \quad \cdots (10.44)$$

To obtain the frequency response, $C(j\omega)/R(j\omega)$ is given by

$$\frac{C(j\omega)}{R(j\omega)} = \frac{24,000(j\omega+25)}{(j\omega+6.6-j11.4)(j\omega+6.6+j11.4)(j\omega+55.9-j18)(j\omega+55.9+j18)}$$

$$\cdots (10.45)$$

For any frequency ω_i, lines are drawn from each of the factors of Eq. (10.45) as shown in Fig. 10.12.

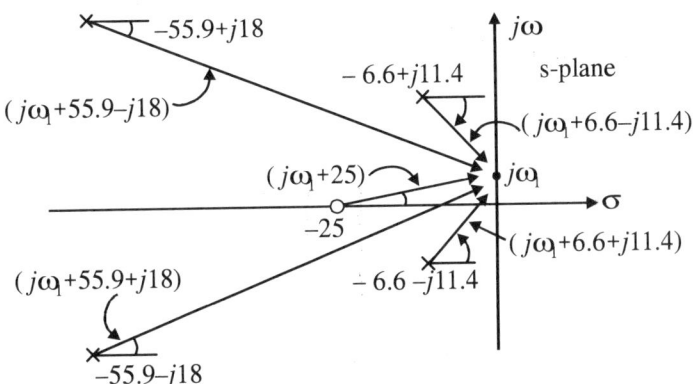

Fig. 10.12. Obtaining Frequency response from root locus

The magnitudes of these directed lines and the angles they make with the real axis are measured. Angles measured counter-clockwise are positive and clockwise are negative. Substituting these values in Eq. (10.45), the magnitude and phase angle of $C(j\omega)/R(j\omega)$ are obtained for the range of frequency of interest and plotted. The frequency response is shown in Fig. 10.13.

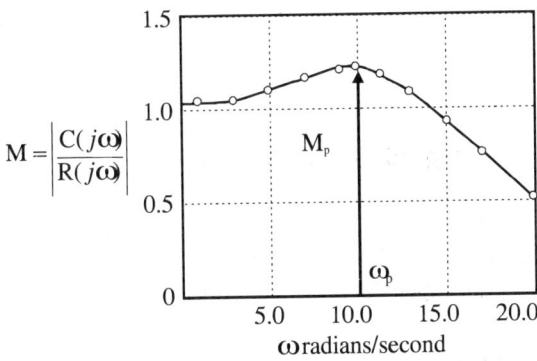

Fig. 10.13. Frequency response obtained from root locus

10.9. EFFECT OF ADDING POLES AND ZEROS

Let us investigate the effects on the root locus when poles and zeros are added to $G(s)H(s)$.

(*i*) *Addition of Poles:* In general, adding a pole to the function $G(s)H(s)$ in the left half of s-planes has the effect of pushing the original root-locus toward the right-half plane. This is illustrated by several examples.

Consider the open-loop transfer function

$$G(s)H(s) = \frac{K}{s(s+a)} \qquad a > 0 \qquad\qquad \dots (10.47)$$

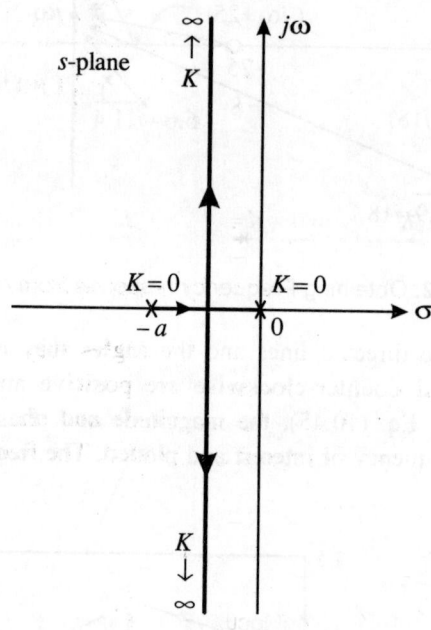

Fig. 10.14. Root locus of $K/s(s + a)$

The root-locus of the characteristic equation $1 + G(s)H(s) = 0$ is shown in Fig. 10.14. The starting point of the root-locus are the poles of $G(s)H(s)$ at $s = 0$ and $s = -a$. If we now introduce a pole at $s = -b$, the open-loop transfer function is

$$G(s)H(s) = \frac{K}{s(s+a)(s+b)} \qquad b > a \qquad\qquad \dots (10.47)$$

The introduction of the additional pole at $s = -b$ causes the complex part of the root-locus to bend toward the right half of the s-plane. The angles of asymptotes are changed from $\pm 90°$ to $\pm 60°$ and $180°$. The breakaway point is also moved to the right. For $a = 1$ and $b = 2$, the breakaway point is moved from -0.5 to -0.422 on the real axis. The root-locus is shown in Fig. 10.15. A comparison between the root-locus shown in Fig. 10.14 and 10.15 reveals that the system with root locus shown in Fig. 10.15 becomes unstable if the value of K exceeds the critical value whereas the system with the root-locus given in Fig. 10.14 does not cross the imaginary axis and hence is always stable for $0 \le K \le \infty$.

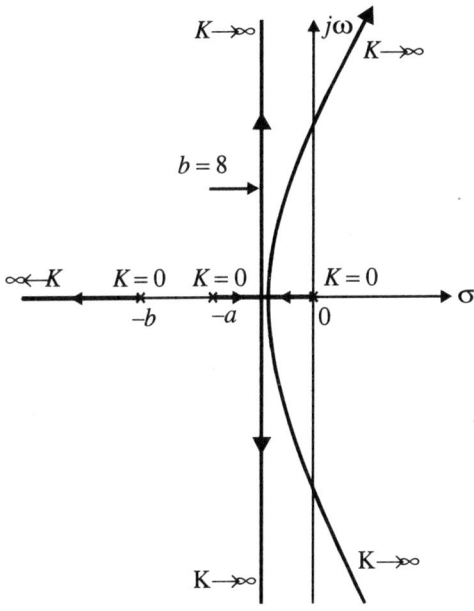

Fig. 10.15. Root locus of third-order system

If we add another pole to $G(s)H(s)$ at $s = -c$, the system becomes a *fourth order system*. The root locus for $c > b$ is shown in Fig. 10.16. It is seen that the complex part of the root-locus is further moved to the right. The angles of the asymptotes are $\pm 45°$ and $\pm 135°$. In this case, the stability condition becomes even more restricted.

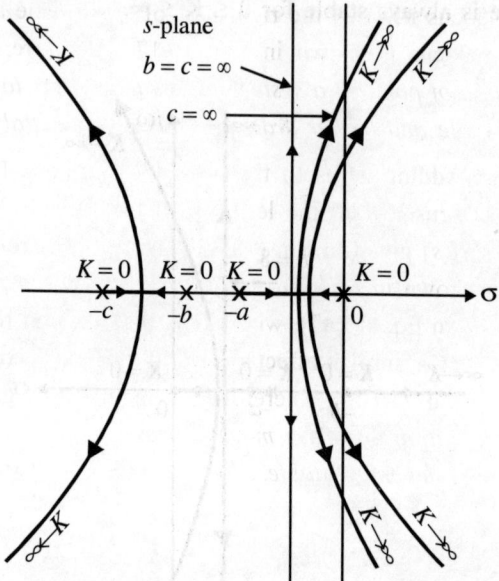

Fig. 10.16: Root locus $G(s)H(s) = K/s(s + a)(s + b)(s + c)$

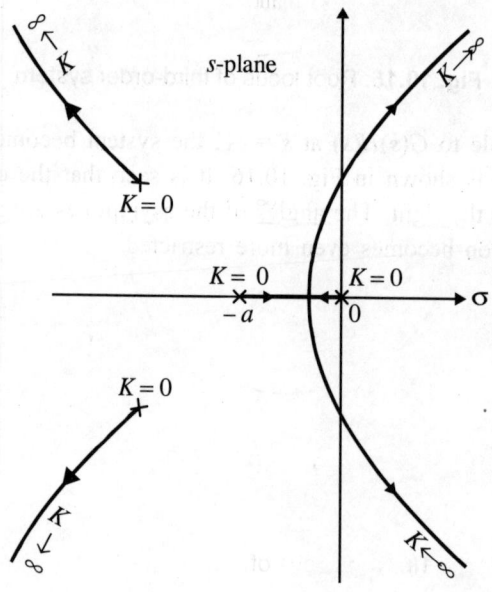

Fig. 10.17. Root locus $G(s)H(s) = K/s(s + a)(s^2 + 2\zeta\omega_n s + \omega_n^2)$

An addition of a pair of complex conjugate poles to the original two pole system shifts the complex part of the root locus towards the right half of the *s*-plane as in the case of addition of simple poles. The root-locus is shown in Fig. 10.17. Therefore, we may draw a general conclusion that *the addition of poles to a system causes the system to be less stable. A second-order system is always stable and higher-order systems are less stable.*

(ii) Addition of Zero: Adding zeros to the open loop transfer function G(s)H(s) has the effect of moving the root locus toward the left half of the *s*-plane. When a zero at $s = -b$ is added to the function G(s)H(s) given in Eq. (10.46), the resultant root-locus bends toward the left and forms a circle as shown in Fig. 10.18 with $b > a$. When a zero at $s = -c$ is added to the function G(s)H(s) given in Eq. (10.47), with $c > b$, the resultant root-locus is shown in Fig. 10.19. Fig. 10.20 shows that a similar effect results if a pair of complex conjugate zeros is added to G(s)H(s) given in Eq. (10.46). Therefore, we may draw a general conclusion that the *addition of zeros to the function G(s)H(s) moves the root locus towards the left half of the s-plane and the system becomes more stable.* Thus, *an unstable system becomes stable by the addition of zeros.*

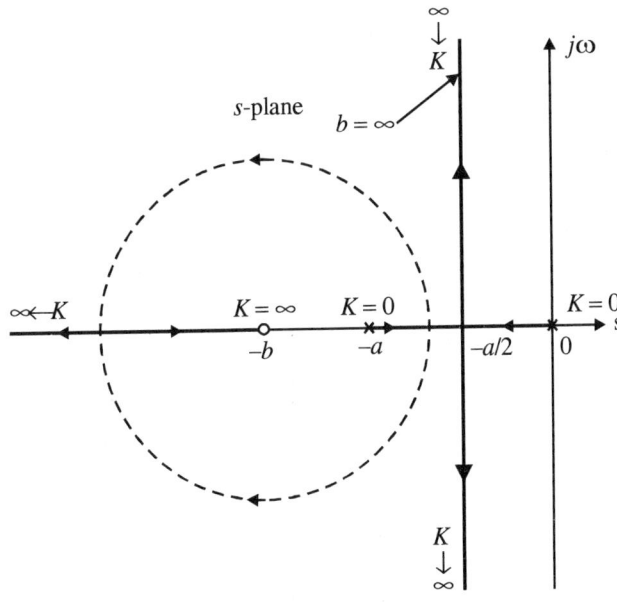

Fig. 10.18. Root locus of G(s)H(s) = K(s + b)/s(s + a)

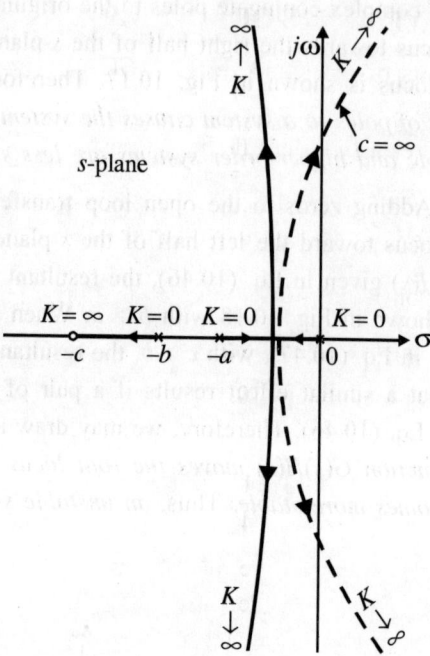

Fig. 10.19. Root locus of $G(s)H(s) = K(s + b) (s + c)/s(s + a)$

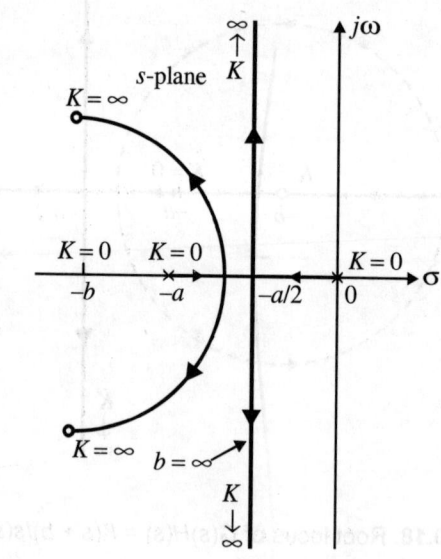

Fig. 10.20. Root locus of $G(s)H(s) = K(s^2 + 2\zeta\omega_n s + \omega_n^2)/s(s + a)$

10.10. EFFECTS OF MOVEMENTS OF POLES AND ZEROS

The effects of the movements of poles and zeros of $G(s)H(s)$ on the root-locus is illustrated with suitable example.

(*i*) *Movement of Pole:* Let us consider the open-loop transfer function

$$G(s)H(s) = \frac{K(s+1)}{s^2(s+a)} \qquad \qquad \dots (10.48)$$

The characteristic equation is given by

$$F(s) = s^2(s+a) + K(s+1) = 0 \qquad \qquad \dots (10.49)$$

Let us investigate the root-loci of Eq. (10.49) for several values of a. For an arbitrary value of a, the break away point is given by

$$s = -\frac{a+3}{4} \pm \frac{1}{4}\sqrt{a^2 - 10a + 9} \qquad \qquad \dots (10.50)$$

When $a = 10$, the breakaway points are at $s = -2.5$ and $s = -4$. The corresponding root-locus is shown in Fig. 10.21. For $a > 10$, the breakaway points are real and negative and hence the root-locus will be similar to that shown in Fig. 10.21.

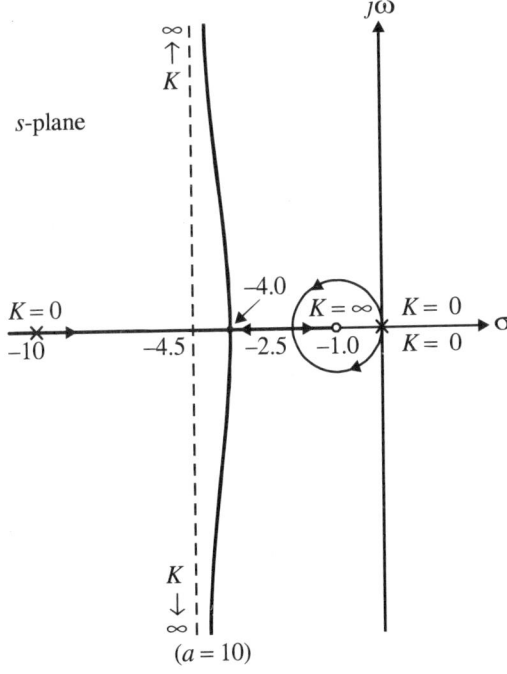

$(a = 10)$

Fig. 10.21. Root locus of $G(s)H(s) = K(s+1)/s^2(s+10)$

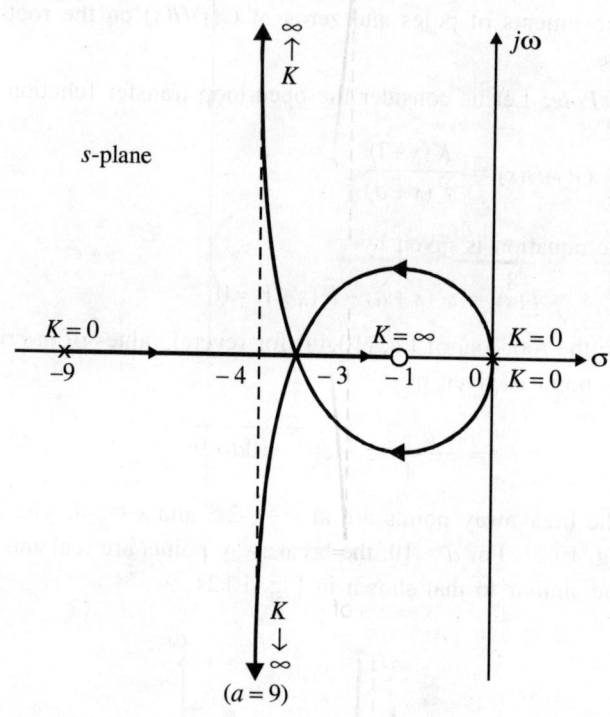

Fig. 10.22. Root locus of $G(s)H(s) = K(s + 1)/s^2(s + 9)$

When $a = 9$, the breakaway points are $s = -3$. Thus the breakaway points merge into a single point and the resultant root-locus is shown in Fig. 10.22. It may be noted that a change of pole from -10 to -9 causes a considerable change to the root-locus.

For $a < 9$, the values of s given by Eq. (10.50) do not satisfy the characteristic equation given in Eq. (10.49). This means that there are no finite, nonzero breakaway points for $a < 9$. Fig. 10.23 shows the root-locus for $a = 8$.

As the pole at $s = -a$ is further moved to the right, the complex portion of the root locus is further pushed to the right half of s-plane. Fig. 10.24 shows the root locus for $a = 3$.

When $a = b$, the pole at $s = -a$ and the zero at $s = -b$ cancel each other, the system degenerates into a second-order system and the root locus lies on the imaginary axis. This is shown in Fig. 10.25. Thus, as the pole is moved towards right, the system becomes less and less stable.

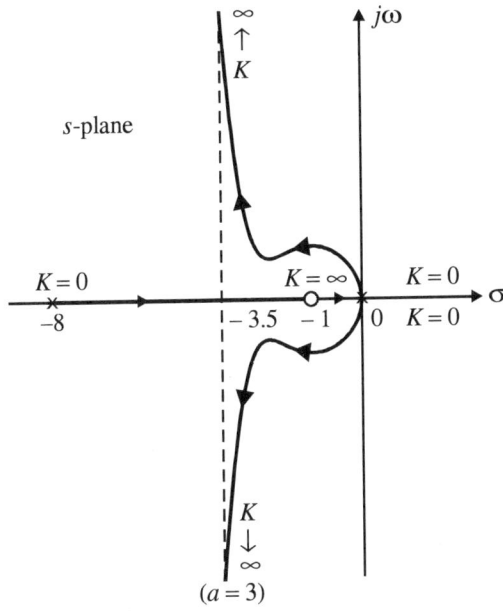

Fig. 10.23. Root locus of $G(s)H(s) = K(s + 1)/s^2(s + 8)$

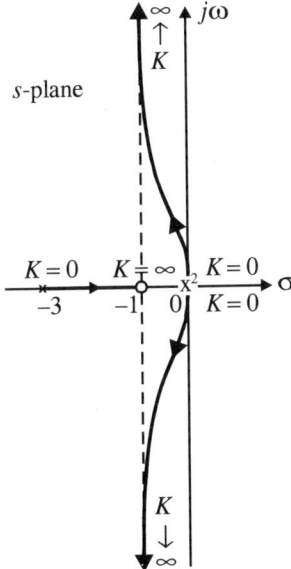

Fig. 10.24. Root locus of $G(s)H(s) = K(s + 1)/s^2(s + 3)$

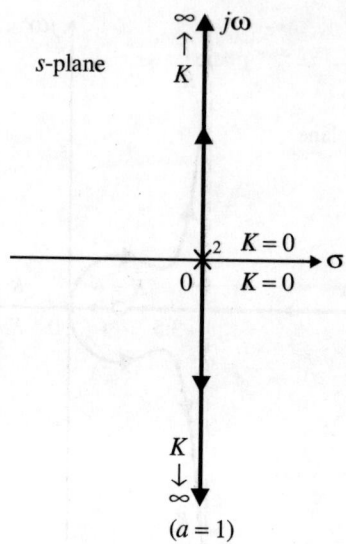

$(a = 1)$

Fig. 10.25. Root locus of $G(s)H(s) = K(s + 1)/s^2(s + 1)$

(*ii*) *Movement of Zeros:* Consider the open-loop transfer function

$$G(s)H(s) = \frac{K(s+b)}{s^2(s+9)} \qquad \dots (10.51)$$

The characteristic equation is

$$F(s) = s^2(s+9) + K(s+b) = 0 \qquad \dots (10.52)$$

For any arbitrary value of b, the break-in point is given by

$$s = -\frac{3b+9}{4} \pm \frac{1}{4}\sqrt{9b^2 - 90b + 81} \qquad \dots (10.53)$$

When $b = 10$, the break-in point given by Eq. (10.53) does not satisfy Eq. (10.52) and the system is unstable. The two branches of the locus from the origin enters the right half of s-plane and ultimately asymptotic to the straight line drawn at $\sigma_o = 0.5$ making an angle of $\pm 90°$ with the real axis as shown in Fig. 10.26. For all values of $b > 9$, the root-locus will be similar.

When $b = a$, the pole at $s = -9$ and the zero at $b = -9$ cancel each other and the system degenerates into a second-order one. The root-locus lies entirely on the $j\omega$-axis as shown in Fig. 10.27.

As the zero at $s = -b$ is moved further to the right, the complex portion of the root-locus moves towards the left-half plane resulting into a stable system. The root locus is similar to Fig. 10.28 for $b = 8$, $\sigma_o = -0.5$.

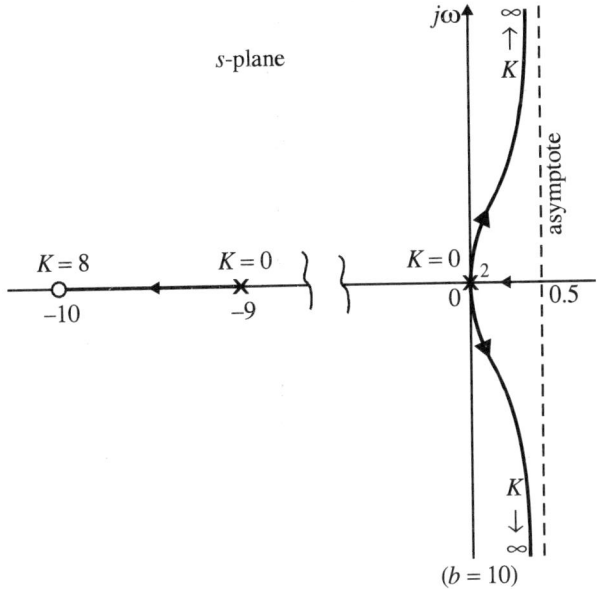

Fig. 10.26. Root locus of $G(s)H(s) = K(s + 10)/s^2(s + 9)$

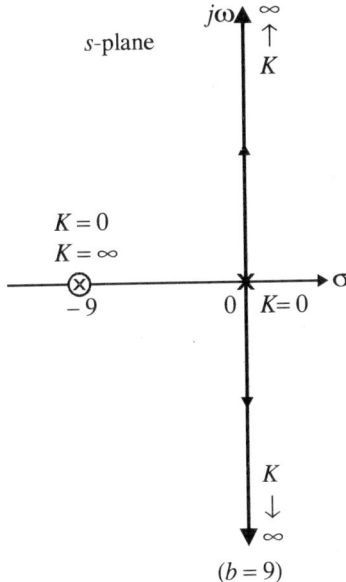

Fig. 10.27. Root locus of $G(s)H(s) = K(s + 9)/s^2(s + 9)$

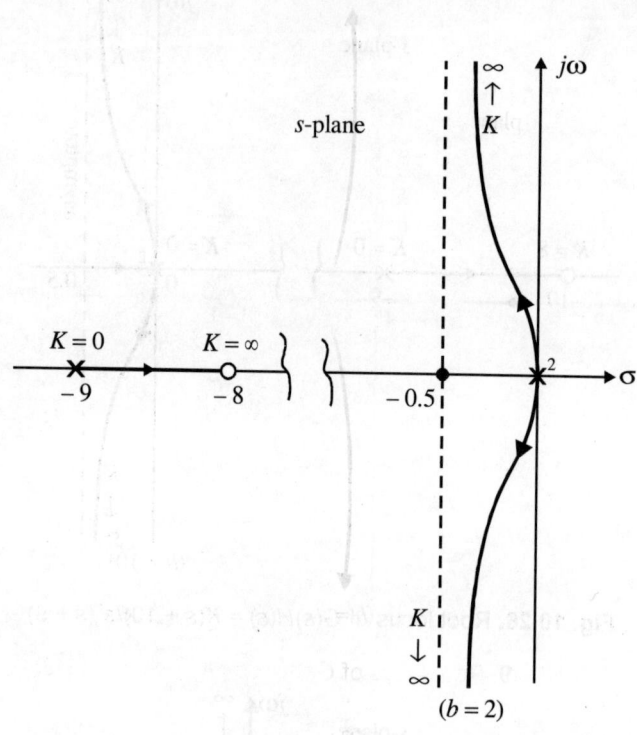

Fig. 10.28. Root locus of $G(s)H(s) = K(s + 2)/s^2(s + 9)$

When $b = 2$, there is no breakaway point. $\sigma_0 = -3.5$. The root locus is similar to Fig. 10.28.

When $b = 1$, the breakaway point occur at $s = -3$ and $\sigma_0 = -4$. The root locus is similar to Fig. 10.29.

When b lies between 0 and 1, the root locus is similar to Fig. 10.30 with two breakaway points. For example, when $b = 0.5$, the breakaway point are -1.80 and -4.17.

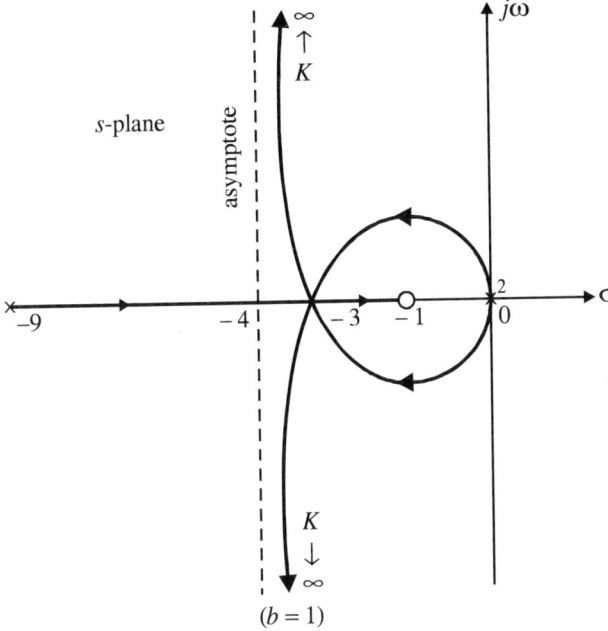

Fig. 10.29. Root locus of $G(s)H(s) = K(s + 1)/s^2(s + 9)$

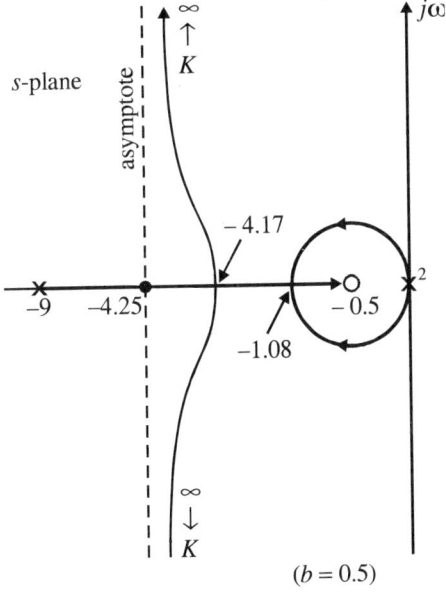

Fig. 10.30. Root locus of $G(s)H(s) = K(s + 0.5)/s^2(s + 9)$

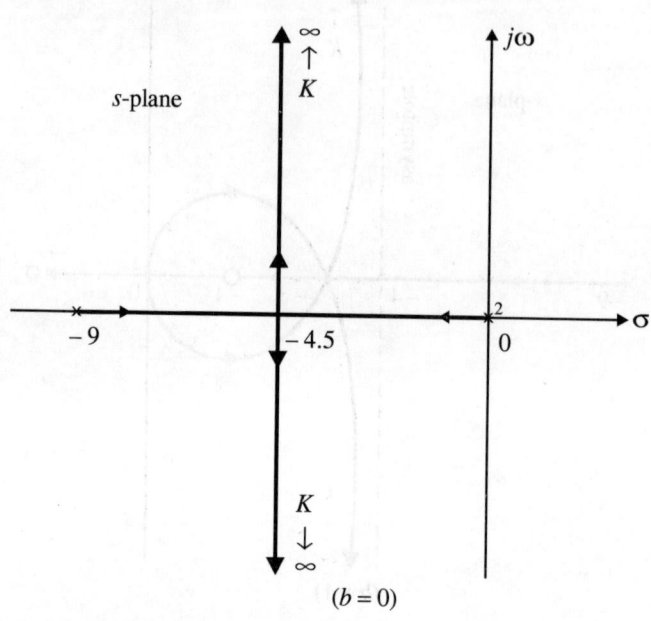

Fig. 10.31. Root locus of $G(s)H(s) = K(s + 0)/s^2(s + 9)$

When $b = 0$, one of the two poles at $s = 0$ cancels with the zero at $s = 0$ and system again degenerates into a second-order one with a breakaway point at $s = -4.5$ as shown in Fig. 10.31.

Thus, as a zero is moved towards right, the system become more stable. Thus *a zero tends to stabilize a given system whereas a pole tends to destabilize a given system.*

10.11. DERIVATIVE AND RATE FEEDBACK CONTROL

The block diagram of a positional servomechanism with proportional control, proportional-plus-derivative control and proportional-plus-rate (velocity or tachometer) feedback is shown in Fig. 32(*a*), (*b*) and (*c*) respectively.

The root locus for the proportional system is shown in Fig. 10.33(a). The closed-loop poles for $K = 5$ are located at $s_{1,2} = -0.1 \pm j\,2.234$.

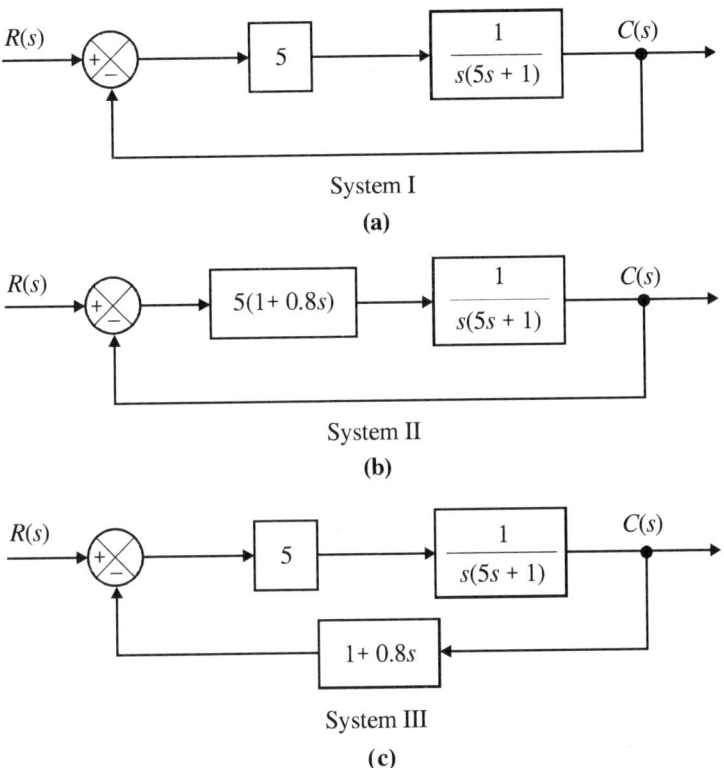

Fig. 10.32. Block diagram of (*a*) proportional control (*b*) proportional and derivative control (*c*) proportional and rate feedback

The open-loop transfer function for both proportional-plus-derivative (PD) and proportional-plus-rate (PR) feedback is the same as given below.

$$G(s)H(s) = \frac{5(1+0.8s)}{s(5s+1)} \qquad \qquad \text{... (10.54)}$$

Therefore, the root-locus for these two systems are identical and is shown is Fig. 10.33(b). However, the closed-loop transfer functions differ because the $G(s)$ and $H(s)$ for each are different. The closed-loop poles for these two systems for $K = 5$ are:

$$s_{1,2} = -0.5 \pm j0.866$$

The closed-loop transfer function for the proportional-plus derivative control system is

$$\frac{C_2(s)}{R(s)} = \frac{1+0.8s}{(s+0.5+j0.866)(s+0.5-j0.866)} \qquad (10.55)$$

The unit impulse response is given by

$$C_2(s) = \frac{0.4 + j0.346}{s + 0.5 + j0.866} + \frac{0.4 - j0.346}{s + 0.5 - j0.866}$$

Taking the inverse Laplace transform, the unit impulse response is

$$c_2(t)_{im} = e^{-0.5t}(0.8 \cos 0.866\,t + 0.693 \sin 0.866\,t)$$

The unit step response is given by

$$c_2(t)_{st} = 1 + 1.155e^{-0.5t}(0.8 \sin 0.866t + 0.755 \cos 0.866t) \dots (10.57)$$

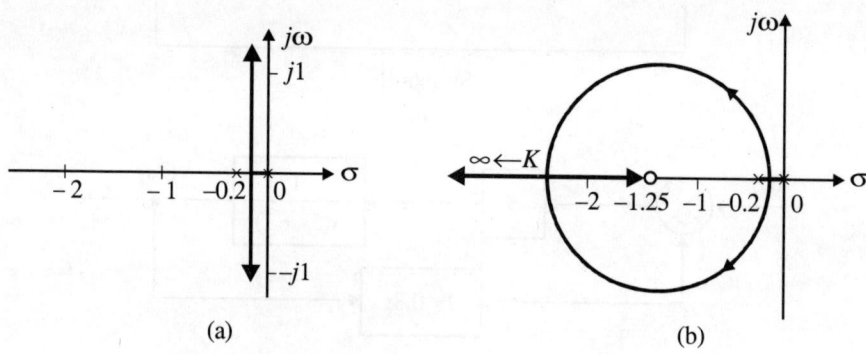

(a)   (b)

Fig. 10.33. (a) Root locus for the proportional system
(b) Root locus for PD and PR feedback system

The closed-loop transfer function for the proportional-plus-rate feedback control system is

$$\frac{C_3(s)}{R(s)} = \frac{1}{(s + 0.5 + j0.866)(s + 0.5 - j0.866)} \qquad \dots (10.58)$$

The unit impulse response is given by

$$C_3(s) = j\,0.577/(s + 0.5 + j0.866) - j\,0.577/(s + 0.5 - j\,0.866).$$

Taking the inverse Laplace transform, the unit impulse response is

$$c_3(t)_{im} = 1.155e^{-0.5t} \sin 0.866t \qquad \dots (10.59)$$

The unit step response is given by

$$c_3(t)_{st} = 1 - e^{-0.5t}(\cos 0.866t + 0.577 \sin 0.866t) \qquad \dots (10.60)$$

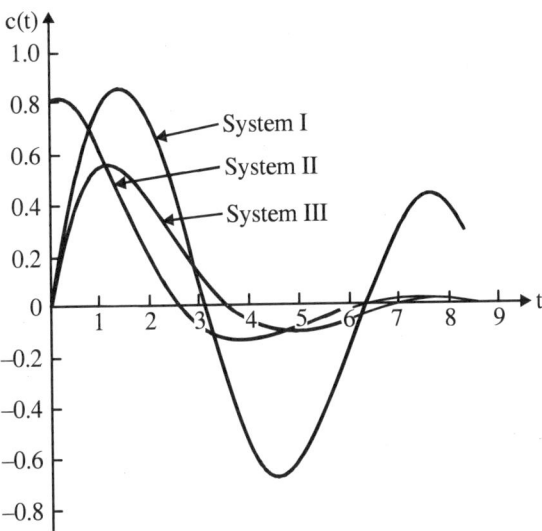

Fig. 10.34. Unit impulse response

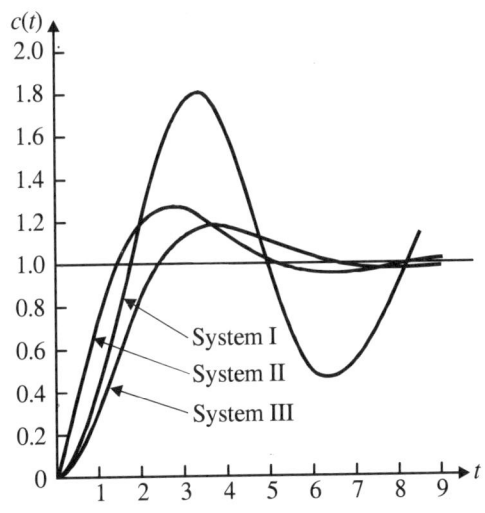

Fig. 10.35. Unit step response

The unit impulse responses and the unit step responses for the three systems are shown in Fig. 10.34 and Fig. 10.35 respectively.

It can be seen from Fig. 10.35, that the system with proportional-plus-derivative control exhibits (system II) the shortest rise time whereas the rate feedback system (system III) exhibits the longest rise time. However, the rate feedback system has the least maximum overshoot and hence has the best relative stability of all the three systems.

The unit ramp response of the system with derivative control exhibits rapid response and less steady-state error to a ramp input. This is because of the fact that the derivative control responds to the rate of change of error signal and can therefore produce early corrective action before the magnitude of the error becomes large. Since the rate feedback responds to the change of the output signal, its action is delayed and hence its steady-state error is high.

10.12. CONDITIONALLY STABLE SYSTEMS

The root locus of the system in Fig. 10.36(a) is shown in Fig. 10.36(b). This plot is constructed by applying the construction rules for root locus. From the root locus, it is seen that the given system is stable only for limited ranges of values of K: $0 < K < 14$ and $64 < K < 195$. The system becomes unstable for $14 < K < 64$ and $K > 195$. Such a system is called a *conditionally stable system*. Conditionally stable systems are not desirable. However they do occur in practice, particularly in systems which have unstable feed forward paths. Such a feed forward path may occur if the system has a minor feedback loop. Conditionally stable systems should be avoided as far as possible since the system will become unstable when the gain reduces below a critical value for some reason or other. Introduction of proper compensation may eliminate the conditional stability.

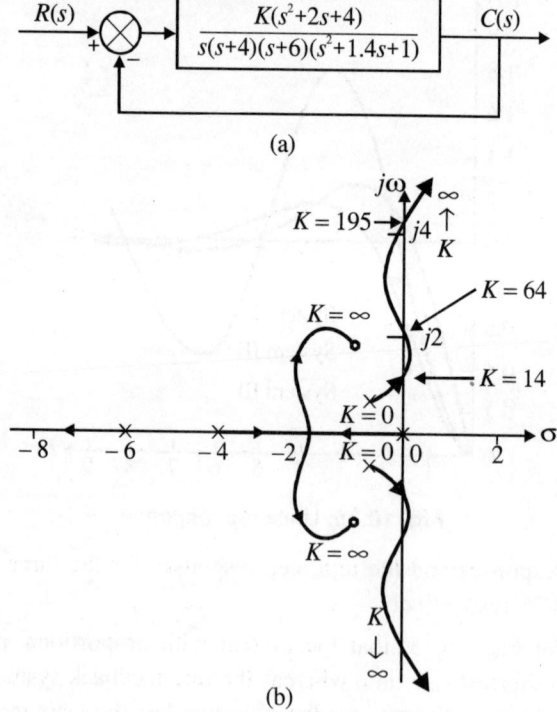

Fig. 10.36. (a) Conditionally stable system (b) Root locus of conditionally stable system

10.13. NON-MINIMUM PHASE SYSTEMS

A minimum phase system is one which has all its poles and zeros in the left half of the s-plane. A non-minimum phase system is one which has at least one pole or zero in the right half of the s-plane.

Consider the system shown in Fig. 10.37(a)) with unity feedback and the forward transfer function.

$$G(s) = \frac{K(1 - T_a s)}{s(1 + Ts)} \qquad (T_a > 0) \qquad \qquad \ldots (10.61)$$

This has a zero in the right half of the *s*-plane and hence a non-minimum phase system. For this system, the angle condition becomes

$$G(s)H(s) = \left| -\frac{K(sT_a - 1)}{s(sT + 1)} \right.$$

$$= \left| \frac{K(sT_a - 1)}{s(sT + 1)} + 180° \right.$$

$$= (2k + 2)180° \qquad \qquad k = 0,1,2,\ldots\ldots$$

or,

$$\left| \frac{K(sTa - 1)}{s(sT + 1)} = (2k)180°, \qquad k = 0,1,2,\ldots\ldots \qquad \qquad \ldots (10.62) \right.$$

The root locus can be obtained from the angle condition given in Eq. (10.62). The root locus for the system of Eq. (10.61) is shown in Fig. 10.37(b). It is seen from the root locus that the system is stable if the system gain K is less than T_a.

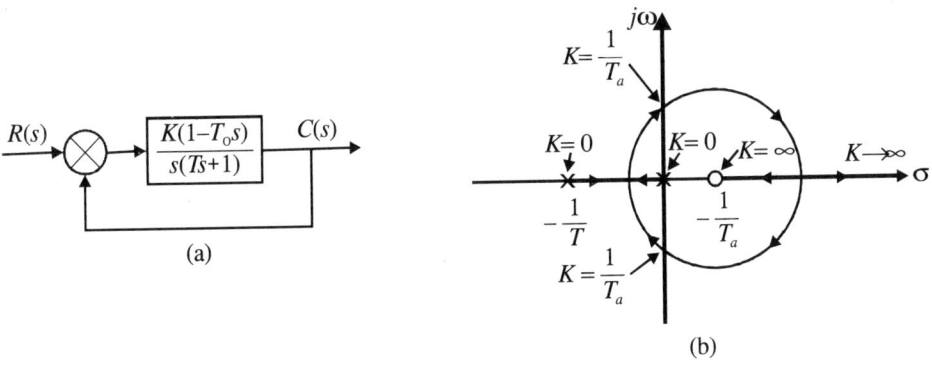

Fig. 10.37. (a) Non-minimum phase system (b) Root locus of non-minimum phase system

10.14. SYSTEMS WITH TRANSPORTATION LAG

Consider the system shown in Fig. 10.38 with transportation lag. The characteristic equation is given by

$$F(s) = 1 + G(s) = 1 + \frac{Ke^{-Ts}}{s(s+p)} = 0 \qquad \text{... (10.63)}$$

Therefore,

$$\frac{Ke^{-Ts}}{s(s+p)} = -1$$

Thus, the angle condition is

$$\left| \frac{Ke^{-Ts}}{s(s+p)} \right| = \underline{|e^{-Ts}} - \underline{|s} - \underline{|s+p} = (2k+1)180° \qquad k = 0,1,2,......$$

To find the angle of e^{-Ts}, we substitute $s = \sigma + j\omega$.

$$\underline{|e^{-Ts}} = \underline{|e^{-\sigma T - j\omega T}} = \underline{|e^{-j\omega T}} = -\omega T \text{ rads.} = -57.3\omega T \text{ degrees}$$

Angle of $e^{-\sigma T} = 0$ since it is a real quantity. Hence the above equation becomes

$$-57.3\omega T - \underline{|s} - \underline{|s+p} = (2k+1)180° \qquad k = 0, 1, 2, ...$$

Since T is the transportation lag or dead time, the angle of e^{-Ts} depends on ω only. Thus the angle and magnitude conditions are:

(i) Angle Condition:

$$57.3\omega T + \underline{|s} + \underline{|s+p} = (2k+1)180° \qquad k = 1, 2, ... \qquad \text{... (10.64)}$$

(ii) Magnitude Condition:

$$K = |s||s+p|e^{\sigma T} \qquad \text{... (10.65)}$$

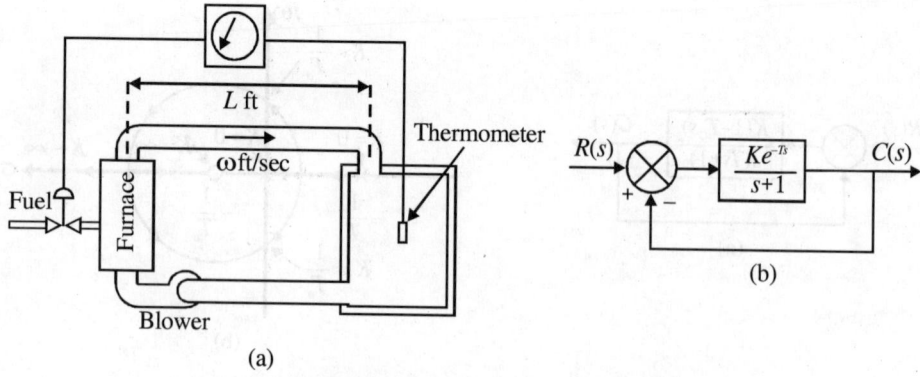

(a)

Fig. 10.38. (a) System with delay (b) Block diagram

The same system without dead time has only two branches and is stable for all values of $K > 0$. With transportation lag, the system has an infinite number of branches. The values of K where the branches cross the imaginary axis are

$$K = \omega\sqrt{p^2 + \omega^2} \qquad \qquad \text{... (10.66)}$$

where ω is the cross-over frequency. The two branches which are closest to the origin have the largest influence on the stability of the system and they are therefore called the *primary branches*. The maximum value of K is therefore is given by,

$$K_{max} = \omega_1\sqrt{p^2 + \omega_1^2} \qquad \qquad \text{... (10.67)}$$

where ω_1 is the cross-over frequency of the primary branch. The root locus is shown in Fig.10.39.

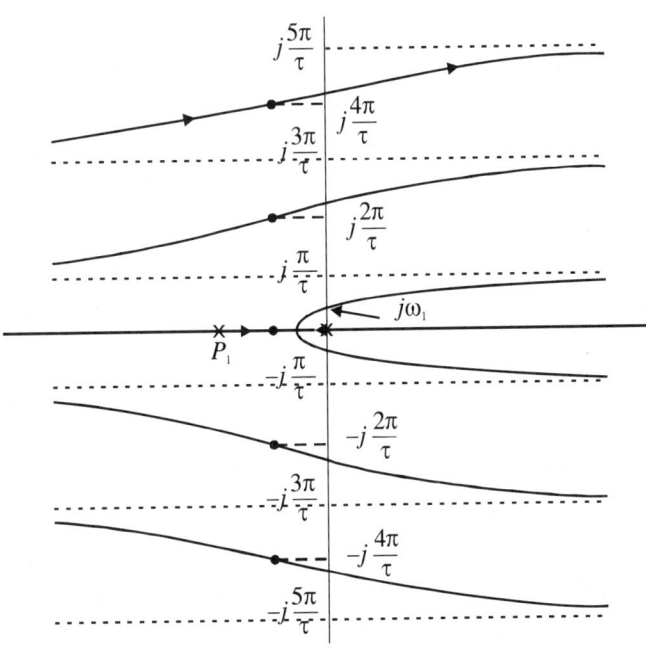

Fig. 10.39. Root locus of time-delay system

If the transportation lag is very small and the second and higher-order derivatives of the input function are small, then e^{-Ts} may be approximated by

$$e^{-Ts} \simeq 1 - Ts$$

or,

$$e^{-Ts} \simeq \frac{1}{1+Ts}$$

10.15. ROOT CONTOURS

The root locus technique thus far discussed is restricted to only one variable parameter K. However, in many design problems, the effects of varying parameters other than K need to be investigated. The effects of these parameters can be easily studied by the same root locus method. *When two or more parameters are varied, the corresponding root loci are known as root contours.*

Consider the characteristic equation

$$Q(s) + K_1 P_1(s) + K_2 P_2(s) = 0 \qquad \qquad \text{... (10.68)}$$

where K_1 and K_2 are the variable parameters and $Q(s)$, $P_1(s)$ and $P_2(s)$ are polynomials of s. To draw the root contours, first, one of the parameters, say K_2, is set to zero and the root locus is drawn for the equation

$$1 + \frac{K_1 P_1(s)}{Q(s)} = 0 \qquad \qquad \text{... (10.69)}$$

as K_1 varies from 0 to ∞. Then, a suitable value of K_1 is selected that results in satisfactory root locations. With this value of K_1 which is kept constant, the root locus is obtained for the equation

$$1 + \frac{K_2 P_2(s)}{Q(s) + K_1 P_1(s)} = 0 \qquad \qquad \text{... (10.70)}$$

as K_2 is varied from 0 to ∞. Thus, for two variables, this two-step method of evaluating the effect of K_1 and K_2 is carried out as a two-root locus procedure. For multi-parameter problem, the same procedure is followed as many times as the number of parameters.

One important feature is that the root contours of the Eq. (10.70) must all start at the points that lie on the root locus of the Eq. (10.69). This is illustrated with suitable examples.

Example 10.7. *Consider the third-order characteristic equation with two variable parameters K_1 and K_2 given by*

$$s^3 + s^2 + K_2 s + K_1 = 0 \qquad \qquad \text{... (10.71)}$$

Obtain the root contour.

Solution:

The characteristic equation $F(s)$ is

$$F(s) = s^3 + s^2 + K_2 s + K_1 = 0$$

Assuming $K_2 = 0$, we obtain

$$s^3 + s^2 + K_1 = 0$$

or,

$$1 + \frac{K_1}{s^3 + s^2} = 1 + \frac{K_1}{s^2(s+1)} = 0 \qquad \text{... (10.72)}$$

The root locus of Eq. (10.72) is drawn as K_1 is varied from 0 to ∞ as shown in Fig. 10.40(a).

Next, select a suitable value of K_1 say a_1. By dividing Eq. (10.71) by terms that do not contain K_2 we get.

$$1 + \frac{K_2 s}{s^3 + s^2 + a_1} = 0 \qquad \text{... (10.73)}$$

Then the root locus of Eq. (10.73) is drawn as K_2 is varied from 0 to ∞ for the selected value of $K_1 = a_1$. This gives the root contour of Eq. (10.71). The root contour always starts on a point on the root locus of Eq. (10.72) as shown in Fig. 10.40(b).

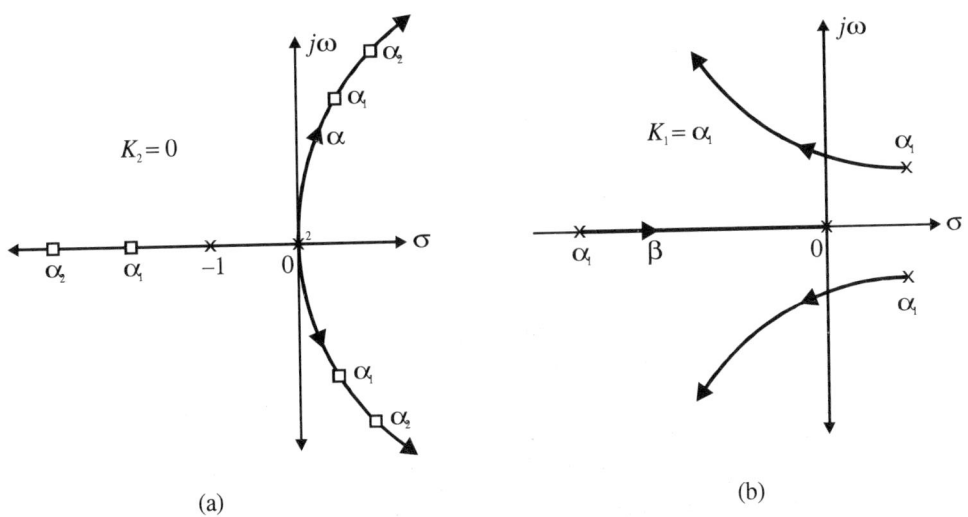

(a)

(b)

Fig. 10.40. (a) Root locus of Eq. (10.72) as K_1 is varied
(b) Root locus of Eq.(10.73) as K_2 is varied $(d_1$ and d_2 are values of $K_1)$

Example 10.8. *Consider the open-loop transfer function with gain K and a zero as variable parameter.*

$$G(s)H(s) = \frac{K(1+Ts)}{s(s+1)(s+2)} \qquad \qquad \dots (10.74)$$

Obtain the root contour.

Solution:

The characteristic equation of the system is

$$F(s) = s(s+1)(s+2) + K(1+Ts) = 0 \qquad \qquad \dots (10.75)$$

First setting $T = 0$, we obtain,

$$s(s+1)(s+2) + K = 0 \qquad \qquad \dots (10.76)$$

or,

$$1 + \frac{K}{s(s+1)(s+2)} = 0 \qquad \qquad \dots (10.77)$$

The root locus of Eq. (10.77) is sketched in Fig. 10.41 (*dashed line*) as K is varied from 0 to ∞. For a specified value of $K = a$, Eq. (10.75) may be written as

$$1 + \frac{Tas}{s(s+1)(s+2) + a} = 0 \qquad \qquad \dots (10.78)$$

The points that correspond to $T = 0$ on the root contour are at the roots of $s(s + 1)(s + 2) + a = 0$. If a is selected as 20, the poles are at -3.85, $0.425 \pm j\, 2.235$ and a zero at the origin. The root contour of Eq. (10.78) as T varies from 0 to ∞ is also sketched in Fig. 10.41 (*solid lines*). The intersection of the asymptotes for root contours is

$$\sigma_0 = \frac{-3.85 + 0.425 + 0.425}{3-1} = -1.5$$

Therefore the intersection of asymptotes is always at $s = -1.5$ regardless of the value of K. Fig. 10.41 shows root contours for three values of K. This contour also shows that addition of a zero to the open-loop transfer function moves the characteristic equation roots towards the left in the s-plane thus improving the relative stability of the closed-loop system. For $K = 20$, the system which is originally unstable is stabilized for all values of T greater than 0.233. However, ζ cannot be greater than 0.3.

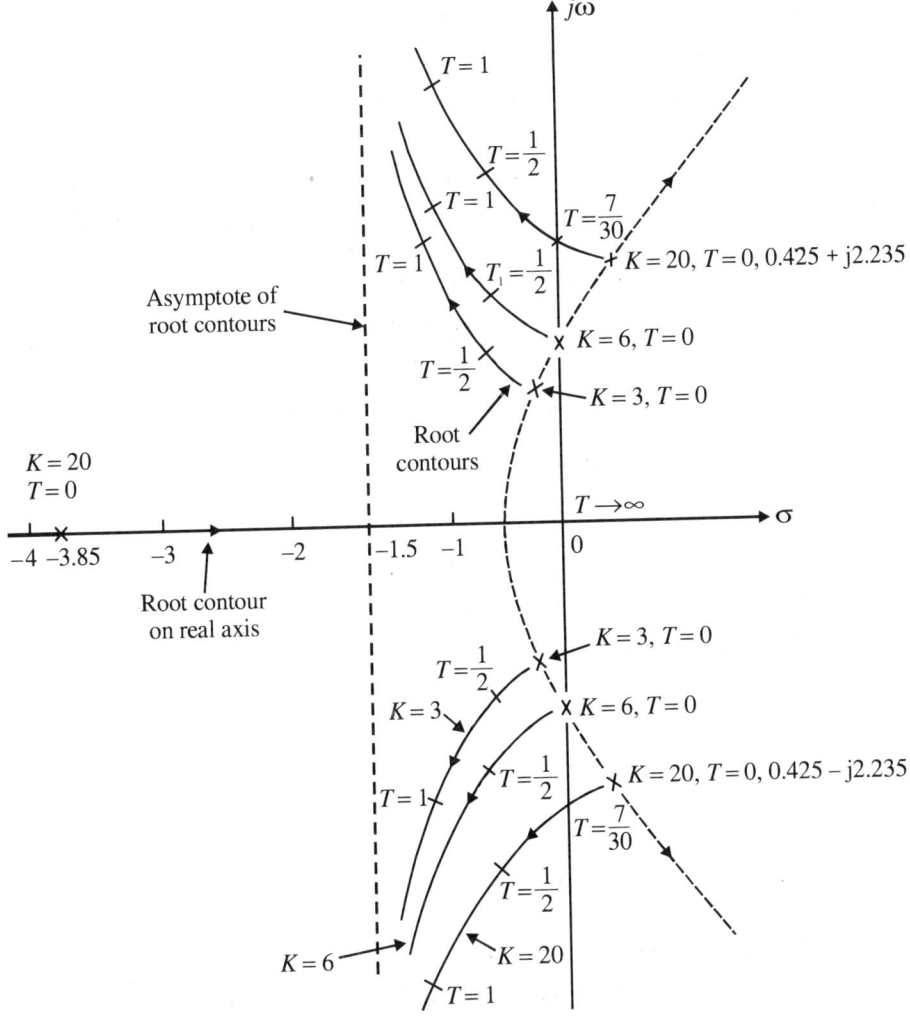

Fig. 10.41. Root contours

10.16. ADDITIONAL EXAMPLES

Example 10.9. *Sketch the root locus of the function* $G(s)H(s) = K/s(s + 2)(s^2 + 2s + 2)$.

Solution:

The openloop transfer functions is $G(s)H(s) = \dfrac{K}{s(s+2)(s^2+2s+2)}$.

1. There are four poles. Hence, there are 4 branches.

2. The starting points are poles: $0, -2, -1+j1, -1-j1$. Mark them on the s-plane.

3. The ending points are zeros. There are no finite zeros. Hence all branches move towards infinity.

4. The section of real axis that lies on the root locus is: 0 to -2. Mark this with thick line.

5. There are 4 asymptotes. The angles of asymptotes are given by $\theta_{k+1} = (2k + 1) \, 180°$.

$$\text{for } k = 0, \qquad \theta_1 = 180/4 = 45°$$
$$\text{for } k = 1, \qquad \theta_2 = 540/4 = 135°$$
$$\text{for } k = 2, \qquad \theta_3 = 90/4 = 225°$$
$$\text{for } k = 3, \qquad \theta_4 = 1260/4 = 315°$$

6. The real axis intercept of asymptotes is

$$\sigma_0 = \frac{0-2-1-1}{4} = -1$$

Draw the 4 asymptotes at -1 making the above angles with the real axis.

7. The angle of departure from the complex pole $-1+j1$ is

$$\phi_3 = 180° - \phi_0 - \phi_1 - \phi_2$$
$$\phi_0 = 180 - \tan^{-1}1 = 135°;$$
$$\phi_1 = \tan^{-1}1 = 45°;$$
$$\phi_2 = 90°$$

$\therefore$
$$\phi_3 = 180° - 135° - 45° - 90° = -90°$$

The angle of departure form $-1-j1$ is $+90°$.

8. The breakaway point on the real axis should lie between 0 and -2.

$$|K| = s(s+2)(s^2 + 2s + 2)$$
$$= s^4 + 4s^3 + 6s^2 + 4s$$
$$\frac{dK}{ds} = 4s^3 + 12s^2 + 12s + 4 = 0$$

or,

$$s^3 + 3s^2 + 3s + 1 = 0$$

or,

$$(s+1)^3 = 0$$

Hence the breakaway point is -1. The value of K at this point is 1.

9. The intersection point with the $j\omega$-axis is obtained from Routh table.

    The characteristic equation is

    $$F(s) = s^4 + 4s^3 + 6s^2 + 4s + K = 0$$

    The Routh table is:

    Routh table

s^4	1	6	K
s^3	4	4	
s^2	5	K	
s^1	$(20-4K)/5$		
s^0	K		

    From s^1-row $K = 5$. The auxiliary equation is $A(s) = 5s^2 + K = 5s^2 + 5 = 0$. Hence $s = \pm j1$.

10. The root locus is shown in Fig. 10.42 and all the above details are marked. It is symmetrical about the real axis.

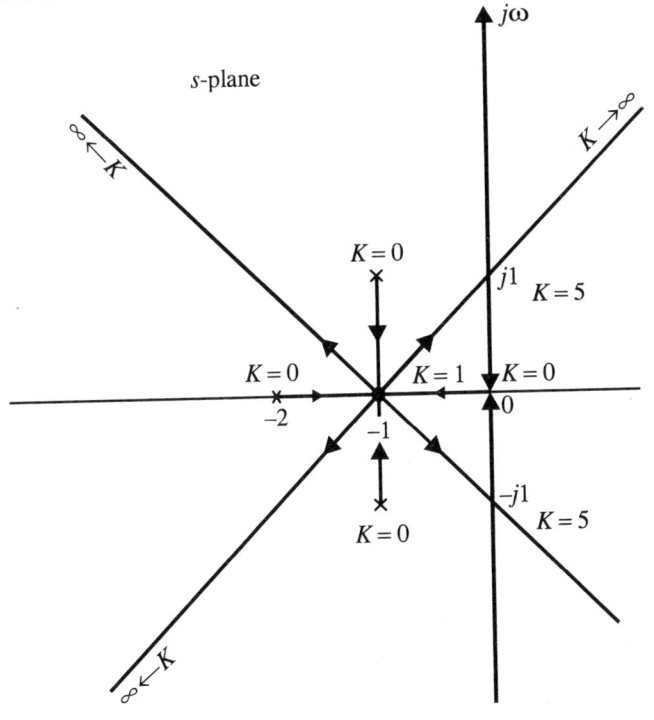

Fig. 10.42. Root locus of $G(s)H(s) = K/s(s + 2)(s^2 + 2s + 2)$

This is a special case since the poles are placed symmetrically about $(-1, j0)$ point. Hence the complex part of the root locus merges with the asymptotes. The breakaway point as well as the intersection of asymptotes is $(-1, j0)$.

Example 10.10. *Sketch the root locus given the characteristic equation* $F(s) = s(s + 4)$ $(s^2 + 4s + 20) + K = 0.$

Solution:

The $G(s)H(s)$ function is obtained by dividing both sides of the above equation by factors which are independent of K. Thus,

$$F(s) = 1 + G(s)H(s) = 1 + \frac{K}{s(s+4)(s^2 + 4s + 20)}$$

Hence,

$$G(s)H(s) = \frac{K}{s(s+4)(s^2 + 4s + 20)}$$

1. There are 4 poles only. Hence the number of branches is 4.
2. The starting points are the poles: $0, -4, -2 + j4, -2 - j4$.
3. The ending points are the zeros. Since there are no finite zeros, the 4 branches move towards infinity.
4. The sections of the real axis lying on the root locus is 0 to -4.
5. There are four asymptotes. The angles of asymptotes are:

$$\theta_{k+1} = (2k + 1)\ 180°$$

$$\text{For } k = 0,\ \theta_1 = 180/4 = 45°$$
$$\text{For } k = 1,\ \theta_2 = 540/4 = 135°$$
$$\text{For } k = 2,\ \theta_3 = 900/4 = 225°$$
$$\text{For } k = 3,\ \theta_4 = 1260/4 = 315°$$

6. The real axis intercept or the origin of asymptotes is

$$\sigma_0 = \frac{0 - 4 - 2 - 2}{4} = -2$$

Four asymptotes are drawn making the above angles with real axis at $(-2, j0)$ points.

7. The angle of departure from the pole $(-2 + j4)$ is

$$\phi_3 = 180° - 135° - 45° - 90° = -90°$$

The angle of departure from the pole $(-2 - j4)$ is $+90°$.

8. The breakaway point should lie between 0 and – 4 on the real axis.

$$|K| = s(s+4)(s^2 +4s+20)$$
$$= s^4 +8s^3 +36s^2 +80s$$
$$\frac{dK}{ds} = 4s^3 +24s^2 +72s+80 = 0$$

or,

$$s^3 +6s^2 +18s+20 = 0$$

Because of symmetry about $(-1, j0)$ point, try $s = -2$.

$$(-2)^3 +6(-2)^2 +18(-2)+20 = 0$$

Hence of breakaway point is –2 on the real axis. The value of $K = 64$. The other breakaway points are obtained as detailed below.

$$s^3 + 6s^2 + 18s + 20 = (s + 2)(s^2 + 4s + 10) = 0$$
$$s^2 + 4s + 10 = s + 2 \pm j\, 2.45$$

Hence, the other two breakaway points are $-2 \pm j\, 2.45$. These being complex, we have to check whether they satisfy the given characteristic equation and yield any real value for K.

$$|K| = |(-2 +j\, 2.45)|\,|(2 +j\, 2.45)|\,|1.55|\,|6.45| = 100.$$

Therefore, they are also breakaway points.

9. The intersection point with the $j\omega$-axis is obtained from Routh table.

The characteristic equation is

$$F(s) = s^4 + 8s^3 + 36s^2 + 80s + K = 0$$

The Routh table is

Routh table

s^4	1	36	K
s^3	8	80	
s^2	26	K	
s^1	$(2080-8K)/26$		
s^0	K		

From the s^1-row, $K = 2080/8 = 260$. The auxiliary equation is

$$A(s) = 26s^2 + 260 = 0, \text{ hence } s = \pm\, j\sqrt{10}$$

10. The root locus with all the details is shown in Fig. 10.43. It is symmetrical about the real axis.

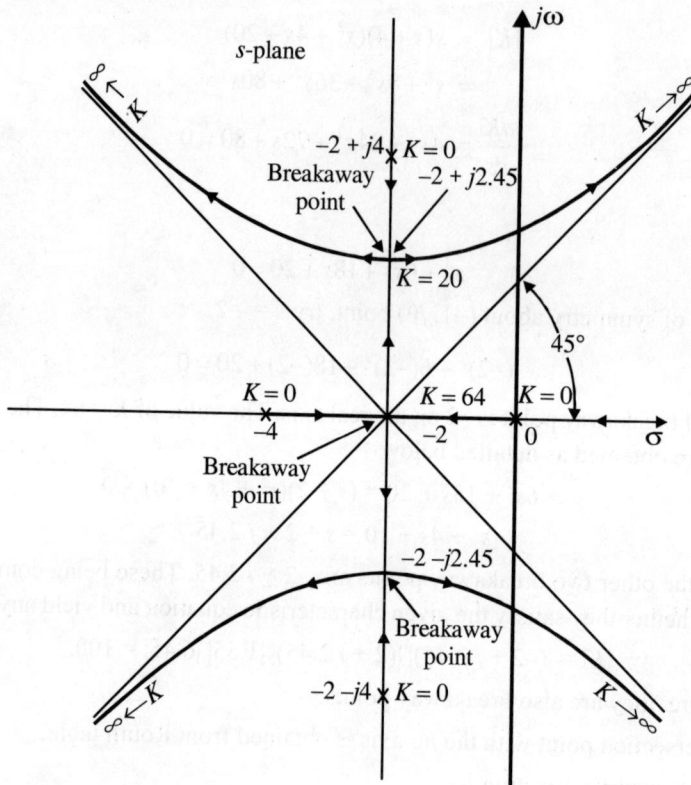

Fig. 10.43. Root locus of $G(s)H(s) = \dfrac{K}{s(s+4)\left(s^2 + 4s + 20\right)}$

Example 10.11. *The open-loop transfer function of an airplane with an autopilot in the longitudinal mode is approximated by*

$$G(s)H(s) = \frac{K(s+1)}{s(s-1)(s^2 + 4s + 16)}$$

Sketch the root locus.

Solution:

$$G(s)H(s) = \frac{K(s+1)}{s(s-1)\left(s^2 + 4s + 16\right)}$$

1. There are 4 poles; hence 4 branches.
2. Starting points are the poles: $0, +1, -2+j2\sqrt{3}, -2-j2\sqrt{3}$
3. Ending points are the zeros: -1 and infinity.
4. Real axis branches: Between 0 and 1 and between -1 and ∞.
5. The angle of asymptotes are:

   $\theta_1 = 60°$

   $\theta_2 = 180°$

   $\theta_3 = 300°$
6. The real axis intercept $\sigma_0 = -2/3$.

7. Angle of departure at $s = -2+j2\sqrt{3}$ is.

$$\underline{|-2+j2\sqrt{3}} + \underline{|-2+j2\sqrt{3}-1} + 90° + \theta_D - \underline{|-2+j2\sqrt{3}+1} = (2k+1)180°$$

$$\tan^{-1}\frac{2\sqrt{3}}{-2} + \tan^{-1}\frac{2\sqrt{3}}{-3} + 90° + \theta_D - \tan^{-1}\frac{2\sqrt{3}}{-1} = (2k+1)180°$$

$$120° + 130.9° + 90° + \theta_D - 106.1° = 180°$$

$$\theta_D = -54.8°$$

8. Breakaway points:

$$|K| = \frac{s(s-1)(s^2+4s+16)}{s+1}$$

$$\frac{d|K|}{ds} = \frac{3s^4+10s^2+21s^2+24s-16}{(s+1)^2} = 0$$

$s = 0.46, -2.22$. Real roots only required. Breakaway point: 0.46, Breakin point: -2.22.

9. Crossing point of $j\omega$-axis: The characteristic equation is given by

$$F(s) = s^4 + 3s^3 + 12s^2 + (K-16)s + K = 0.$$

Routh Array

s^4	1	12	K
s^3	3	$K-16$	
s^2	$\dfrac{52-K}{3}$	K	
s^1	$\dfrac{-K^2+59K-832}{3(52-K)}$		
s^0	K		

(i) From s°-row, $K > 0$

(ii) From s'-row, $K < 52$

(iii) From s^2 row, $K = 23.32$ or 35.68; i.e., $23.32 \leq K \leq 35.68$.

$$\text{For } K = 23.32, \ s = \pm j1.56.$$

$$\text{For } K = 35.68, \ s = \pm j2.56.$$

The system is a *conditionally stable system* as seen in Fig.10.44. The system is stable when K values are between 23.32 and 35.68.

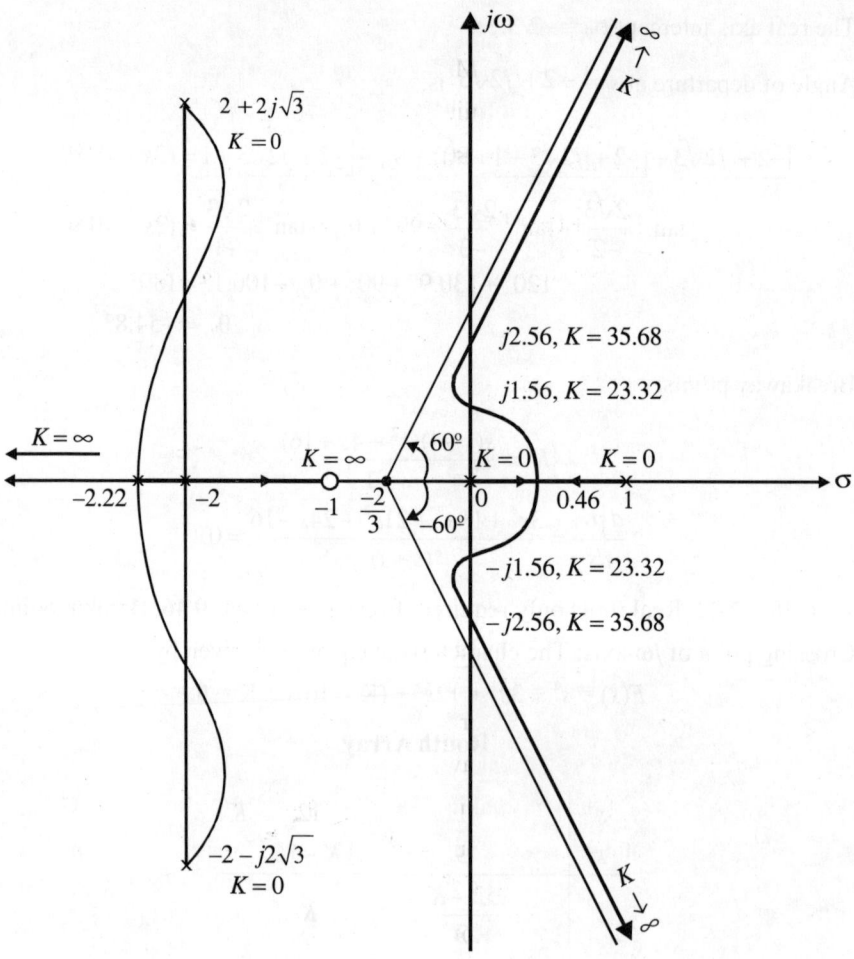

Fig. 10.44. Root locus of $G(s)H(s) = \dfrac{K(s+1)}{s(s-1)(s^2+4s+16)}$

Example 10.12. *(a) Sketch the root locus of unstable open-loop transfer function*

$G(s)H(s) = \dfrac{10(s+1)}{s(s-3)}$. *(b) Find its unit step response and comment.*

Solution:

(a) Root-locus:

The openloop transfer function is

$$G(s)H(s) = K(s + 1)/s(s - 3)$$

1. There are 2 poles; hence 2 branches.
2. Starting points are poles: 0 and +3. Mark them in the s-plane.
3. Ending points are zeros: −1 and infinity.
4. Real axis branch: 0 to 3 and −1 to −∞.
5. The angle of asymptotes is

   $\theta_1 = 180°$.
6. The real axis intercept is

   $$\sigma_0 = \frac{0 + 3 - 1}{1} = +2$$

7. Breakaway points:

$$|K| = (s^2 - 3s)/(s+1)$$

$$\frac{d|K|}{ds} = \frac{(s+1)(2s-3) - (s^2 - 3s)}{(s+1)^2} = 0$$

$$s^2 + 2s - 3 = 0$$

$$s = \frac{-2 \pm \sqrt{4 + 12}}{2}$$

$$= 1 \text{ or} - 3.$$

$$\text{Breakaway point} = 1$$

$$\text{Breakin point} = - 3$$

8. $j\omega$-axis crossing is obtained from Routh table. The characteristic equation is:

$$F(s) = s^2 - 3s + K(s + 1) = 0.$$

Routh table

s^2	1	K
s^1	$K - 3$	
s^0	K	

(i) From s'-row, $K > 3$

(ii) The auxiliary equation is

$$A(s) = s^2 + 3 = 0$$

$$s = \pm j\sqrt{3}$$

The root locus is shown in Fig. 10.45.

(b) The unit step response:

$$\frac{C(s)}{R(s)} = \frac{10(s+1)}{[s(s-3)+10(s+1)]}$$

$$= \frac{10(s+1)}{(s+2)(s+5)}$$

For $R(s) = 1/s$,

$$C(s) = \frac{10(s+1)}{s(s+2)(s+5)} = \frac{1}{s} + \frac{1.67}{s+2} - \frac{2.67}{s+5}$$

$$c(t) = 1 + 1.67\,e^{-2t} - 2.67\,e^{-5t}$$

Though the open-loop system is unstable, the closed loop system is stable since $K > 3$ ($K = 10$) and the closed loop transfer function poles lie in the LHP.

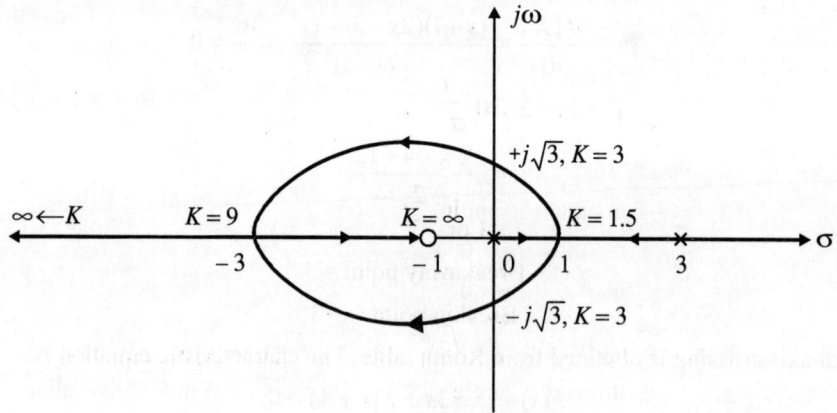

Fig. 10.45. Root locus of Example 10.12

Example 10.13. *Show that a part of the root locus of* $G(s)H(s) = \dfrac{(s+3)}{s(s+2)}$ *is circular.*

Solution:

We use angle condition to show that a part of root locus is circular

$$\underline{|G(s)H(s)} = \underline{|s+3} - \underline{|s} - \underline{|s+2} = 180°$$

At $s = \sigma + j\omega$

$$\underline{|\sigma + j\omega + 3} - \underline{|\sigma + j\omega} - \underline{|\sigma + j\omega + 2} = 180°$$

$$\tan^{-1}\frac{\omega}{\sigma+3} - \tan^{-1}\frac{\omega}{\sigma} = 180° + \tan^{-1}\frac{\omega}{\sigma+2}$$

Taking tangent of both sides, we get

$$\tan\left(\tan^{-1}\frac{\omega}{\sigma+3} - \tan^{-1}\frac{\omega}{\sigma}\right) = \frac{\dfrac{\omega}{\sigma+3} - \dfrac{\omega}{\sigma}}{1 + \left(\dfrac{\omega}{\sigma+3}\right)\cdot\left(\dfrac{\omega}{\sigma}\right)}$$

$$= \frac{-3\omega}{\sigma(\sigma+3) + \omega^2}$$

$$\tan\left(180° + \tan^{-1}\frac{\omega}{\sigma+2}\right) = \frac{0 + \dfrac{\omega}{\sigma+2}}{1 - 0\left(\dfrac{\omega}{\sigma+2}\right)} = \frac{\omega}{\sigma+2}$$

$$\therefore \quad \frac{-3\omega}{\sigma(\sigma+3) + \omega^2} = \frac{\omega}{\sigma+2}$$

$$\omega\sigma^2 + 3\omega\sigma + \omega^3 = -3\omega\sigma - 6\omega$$

Dividing both sides by ω and rearranging, we get

$$\sigma^2 + 6\sigma + \omega^2 + 6 = 0$$

$$(\sigma^2 + 6\sigma + 9) + \omega^2 + 6 - 9 = 0$$

$$(\sigma+3)^2 + \omega^2 = \sqrt{3}^2$$

This represents the equation of a circle with centre at $(-3, 0)$ and with radius $\sqrt{3}$.

10.17. SUMMARY

The relative stability and time-domain performance of a closed-loop system is directly related to the location of closed-loop system poles, or, the roots of the characteristic equation. In this chapter, the root locus method is developed to obtain graphically the roots of the characteristic equation. By utilizing the root locus method, we have investigated the movement of the characteristic roots on the s-plane as the system parameters are varied. We have also obtained the magnitude and angle conditions in order to draw the root locus. From the root locus, we have shown that it is possible to determine the variation in system performance as the system gain K is varied over a wide range. It has been shown that a second-order system never becomes unstable even though the system sensitivity K is varied from zero to infinity.

We have explained the general rules of obtaining the root locus of systems and also the method of determining the system gain K for any set of roots of the characteristic equation. We have studied the effect of addition of poles and zeros on the root locus and hence on the system response. We have also explained the stable, conditionally or marginally stable and unstable systems and their root locus. We have described the method of obtaining the time and frequency response from the root locus. We have also discussed the effect of derivative and feedback control. We have then defined and differentiated between the minimum and non-minimum phase systems. We have derived the magnitude and angle conditions for non-minimum phase systems and described their root locus. Finally, we have dealt with root contours that are root loci obtained by the variation of parameter other than the system gain K.

The root locus method can be extended to solve the roots of any polynomial provided it is arranged as a ratio of factored polynomials which is equal to ±1. Thus,

$$N(s)/D(s) = \pm 1.$$

Once this form is obtained, the roots of the polynomial can be obtained by using the procedure described in this chapter.

The root locus method provides a valuable means of displaying the closed-loop performance characteristics of a control system in the complex plane. It permits the analysis of the feedback systems and provides a basis for the selection of the system gain in order to best meet the given performance specifications. Since the closed-loop poles are obtained directly from the root locus, the form of the time response is immediately known. If the performance specifications cannot be met, the root locus can be utilized to determine the appropriate compensation to yield the desired results. This is discussed in the next chapter.

REVIEW QUESTIONS

10.1. Define a root-locus.

10.2. Sketch the root-locus of a two-pole system.

10.3. Explain the angle and magnitude conditions.

10.4. Why must the number of branches of root loci equal the number of open-loop poles?

10.5. State the rule to find real axis root locus.

10.6. How do you find the number of asymptotes of root locus and the angles between them?

10.7. Explain angle of departure and angle of arrival.

10.8. What is the effect of adding a pole or zero?

10.9. What is meant by conditionally stable systems?

10.10. Distinguish between minimum and non-minimum phase systems.

10.11. What are root contours?

EXERCISE

10.1. (a) Explain the basis of root locus plot.

(b) Explain the different rules with proof wherever possible for the construction of root locus plot.

10.2. Explain why the number of branches of the root locus is equal to the number of open-loop poles.

10.3. Enumerate the rules used in drawing the root locus of a given function.

10.4. Explain the method of obtaining loop sensitivity K at any point on the root locus.

10.5. How does root locus change due to (i) addition of zeros and (ii) addition of poles? How does the stability of the system affected in each case?

10.6. What is the effect of moving a negative pole towards the imaginary axis on the root locus and the system stability?

10.7. What is the effect of moving a negative zero towards the imaginary axis on the root locus and the system stability?

10.8. What is meant by conditionally stable system? Sketch the root locus of such a system.

10.9. What is a non-minimum phase system? What is the angle condition for such systems?

10.10. Obtain the magnitude and angle condition for systems with time delay.

PROBLEMS

10.1. Determine the breakaway points on the root locus of the following functions:

(a) $\dfrac{K}{s(s+3)^2}$

(b) $\dfrac{K(s+2)}{(s^2+2s+4)}$

(c) $\dfrac{K}{s(s+1)(s+3)(s+4)}$

(d) $\dfrac{K(s+1)}{s^2(s+9)}$

10.2. Determine the angle and magnitude of $G(s)H(s) = \dfrac{16(s+1)}{s(s+2)(s+4)}$ at

(a) $s = -j2$ (b) $s = -2 + j2$

(c) $s = -4 + j2$

10.3. Find the departure angle of the root locus from the pole at $s = -10 + j10$ for

$$G(s)H(s) = \frac{K(s+8)}{(s+14)(s+10+j10)(s+10-j10)}$$

10.4. Find the departure angle of the root locus from the pole $s = -15 + j9$ for

$$G(s)H(s) = \frac{K}{(s+5)(s+10)(s+15+j9)(s+15-j9)}$$

10.5. Determine the departure angle of the root locus from the pole $s = j2$ for

$$G(s)H(s) = \frac{K(s+1+j)(s+1-j)}{s(s+2j)(s-2j)}$$

10.6. Find the arrival angle of the root locus to $s = 1 + j1$ for GH in Prob. 10.5.

10.7. Determine the arrival angle of the root locus to s $= -7 + j5$ for

$$G(s)H(s) = \frac{K(s+7+j5)(s+7-j5)}{(s+3)(s+5)(s+10)}$$

10.8. Which part of the real axis is on the root-locus for

(a) $G(s)H(s) = \dfrac{K(s+2)}{(s+1)(s^2+6s+10)}$ (b) $G(s)H(s) = \dfrac{K}{s(s+1)^2(s+2)}$

10.9. For the function $G(s)H(s) = \dfrac{K(s+3)}{(s+1)(s+2)}$ prove that a part of the root locus is circular.

Find the breakaway points.

10.10. The open-loop transfer function of a unity feedback system is

$$G(s) = \frac{K(s+2)}{(s+1)^2}$$

Prove that a part of the root locus is circular. Find $C(s)/R(s)$ for $K = 2$.

10.11. The forward and feedback transfer functions of a system are

$$G(s) = \frac{K(s+2)}{(s+1)}, H(s) = \frac{1}{(s+1)}$$

Show that the root locus plot for this problem is the same as in Prob. 10.10. Obtain $C(s)/R(s)$ for $K = 2$.

10.12. For the unity feedback open-loop function

$$G(s) = \frac{K}{s(s^2 + 8s + 17)}$$

(a) Sketch the root-locus.

(b) Find the values of K for which the system just oscillates

(c) Find the value of K for $\zeta = 0.5$.

10.13. A certain feedback control system with unity feedback has a transfer function

$$G(s) = \frac{K}{s(1 + 0.02s)(1 + 0.01s)}$$

(a) Plot the locus of the roots of $1 + G(s) = 0$ as the static loop sensitivity is varied from zero to infinity

(b) From the root locus plot determine the value of K for a damping ratio of 0.383.

10.14. (a) For positive values of gain, sketch the root-locus for unity feedback control system having the open-loop transfer function

$$G(s) = \frac{K}{s(s^2 + 6s + 12)}$$

(b) For what values of K the system becomes unstable?

10.15. (a) Sketch the root locus plot of a unity feedback system with open loop gain

$$G(s) = \frac{K}{s(s + 2)(s + 4)}$$

(b) Find the range of K for stable operation.

10.16. A unity feedback control system has the open-loop transfer function

$$G(s) = \frac{K(s + 1)}{s(s - 1)}$$

(a) Show that the root loci of the complex roots are parts of a circle with centre $(-1, 0)$ and radius 2.

(b) Determine the range of K for stable operation.

(c) Find the marginal value of K and the frequency of oscillation.

10.17. Sketch the root locus for $G(s)H(s) = K(s+4)/(s+1)(s+2)(s+5)$ and comment on the system stability.

10.18. Sketch the root locus for $G(s)H(s) = K(s+1)/(s^2 + 3s + 3.25)$.

10.19. Sketch the root locus for $G(s)H(s) = \dfrac{K(s+2)(s+3)}{s(s+1)}$.

10.20. The forward transfer function of a unity feedback transportation lag system is

$$G(s) = \frac{2e^{-0.3s}}{(s+1)}$$. Sketch the root locus.

Chapter 11

FREQUENCY RESPONSE ANALYSIS

11.1. INTRODUCTION

A practical and important alternative approach to the analysis and design of a control system is the frequency response method. The frequency response of the system is defined as the steady-state response of the system to a sinusoidal input whose frequency is varied over the range of interest. The sinusoid is a unique signal since, when applied to a linear system, the resulting output and the signals throughout the system are sinusoidal in the steady-state. They differ from the input waveform only in amplitude and phase angle.

The advantages of the frequency response approach are:

1. The stability of the system may be investigated without determining the roots of the characteristic equation.

2. The sinusoidal test signals for various ranges of frequency and amplitude are readily available.

3. The experimental determination of the frequency response of a system is simple and most reliable.

4. The transfer function of a system can be obtained from the experimentally determined frequency response of a system.

5. A system may be designed in the frequency domain such that the efects of undesirable noise and disturbances are negligible.

6. The transfer function of a system in the frequency domain can simply be obtained from the transfer function $G(s)$ by replacing s by $j\omega$.

The disadvantages are:

1. There is no direct correlation between transient and frequency responses.

2. The frequency response approach is a graphical method with inherent approximations.

In this chapter, we shall discuss three graphical plots of the system transfer functions: Bode plots, polar plots and magnitude-versus-phase angle plot.

11.2. SINUSOIDAL TRANSFER FUNCTION

Let us consider the linear time-invariant system shown in Fig.11.1. The transfer function $C(s)/R(s)$ is given by

$$\frac{C(s)}{R(s)} = G(s) \qquad\qquad ...(11.1)$$

In general, $G(s)$ is the ratio of two polynomials

$$G(s) = \frac{N(s)}{D(s)} \qquad\qquad ...(11.2)$$

$$= \frac{N(s)}{(s + p_1)(s + p_2)...(s + p_n)}$$

The input $r(t) = R \sin \omega t$ and its Laplace transform is $R\omega/(s^2 + \omega^2)$. Hence

$$C(s) = G(s)R(s) = \frac{N(s)R(s)}{D(s)}$$

$$= N(s)R\omega/D(s)(s^2 + \omega^2)$$

$$= \frac{A_1}{s + j\omega} + \frac{A_1}{s - j\omega} + \frac{B_1}{s + p_1} + \frac{B_2}{s + p_2} + ... \frac{B_n}{s + p_n} \qquad ...(11.3)$$

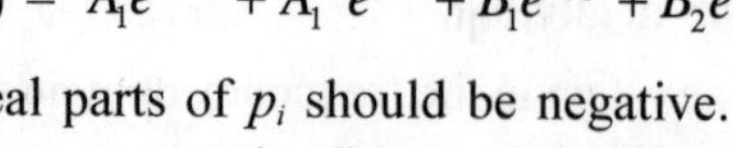

R(s) ———→ | G(s) | ———→ C(s)

Fig. 11.1. Linear time-invariant system

Taking the inverse Laplace transform, we obtain

$$c(t) = A_1 e^{-j\omega t} + A_1^* e^{j\omega t} + B_1 e^{-p_1 t} + B_2 e^{-p_2 1} + ... B_n e^{-p_n t} \quad (t \geq 0) \quad ...(11.4)$$

For a stable system, the real parts of p_i should be negative. Hence, in the steady-state as t approaches infinity, all the terms except the first two disappear. Thus,

$$c(t) = A_1 e^{-j\omega t} + A_1^* e^{j\omega t}$$

where

$$A_1 = \left. \frac{G(s)R\omega(s+j\omega)}{(s^2+\omega^2)} \right|_{s=-j\omega} = \frac{-RG(-j\omega)}{2j} \qquad ...(11.5)$$

$$A_1^* = \left. \frac{G(s)R\omega(s-j\omega)}{s^2+\omega^2} \right|_{s=j\omega} = \frac{RG(j\omega)}{2j} \qquad ...(11.6)$$

$G(j\omega)$ being a complex quantity, it may be written as

$$G(j\omega) = |G(j\omega)|e^{j\phi} \qquad ...(11.7)$$

where

$$|G(j\omega)| = \text{Magnitude or absolute value}$$

$$\phi = \underline{|G(j\omega)|} = \tan^{-1}\left(\frac{\text{Im. Part of } G(j\omega)}{\text{Re. Part of } G(j\omega)}\right)$$

Similarly,

$$G(-j\omega) = |G(j\omega)|e^{-j\phi}$$

Therefore, from Eq. (11.4)

$$c(t) = R|G(j\omega)|\frac{e^{j(\omega t+\phi)} - e^{-j(\omega t+\phi)}}{2j}$$

$$= R|G(j\omega)|\sin(\omega t + \phi)$$

$$= A \sin(\omega t + \phi) \qquad ...(11.8)$$

where

$$A = R|G(j\omega)|$$

$$\underline{|G(j\omega)|} = \underline{|C(j\omega)|} - \underline{|R(j\omega)|}$$

Therefore, the frequency response characteristics of a system to a sinusoidal input can be directly obtained from

$$\frac{C(j\omega)}{R(j\omega)} = G(j\omega)$$

Example11.1. Find the steady-state frequency response of a system with $G(s) = K/(1 + Ts)$ for $r(t) = R \sin\omega t$.

Solution:

$$G(s) = K/(1 + Ts)$$

$$G(j\omega) = K/(1 + j\omega T)$$

$$|G(j\omega)| = K/\sqrt{1+(\omega T)^2}$$

$$\phi = \underline{|G(j\omega)} = -\tan^{-1}\omega T$$

From Eq. (11.8),

$$c(t) = \frac{KR}{\sqrt{1+(\omega T)^2}}\sin(\omega t + \phi) \qquad\qquad ...(11.9)$$

The output varies from KR at low frequency to zero as ω approaches infinity. The phase angle varies from zero at low frequency to $-90°$ as ω approaches infinity.

11.3. FREQUENCY-DOMAIN SPECIFICATIONS

A control system should be stable and its performance should also satisfy a set of specifications. In the design of systems, the following frequency domain specifications are often used in practice.

(i) Resonance peak M_p

(ii) Resonance frequency ω_p

(iii) Bandwidth BW

(iv) Cut-off rate

(*i*) *Resonance Peak M_p:* The resonance peak M_p is the maximum value of the response $C(j\omega)/R(j\omega)$. In general, *the magnitude of M_p gives an indication of the relative stability of a system.* The larger the magnitude of M_p, the larger is the overshoot in the step response. For practical design purposes, the optimum value of M_p lies between 1.1 and 1.5.

(*ii*) *Resonant Frequency ω_p:* The resonant frequency ω_p is the frequency at which the resonance peak M_p occurs.

(*iii*) *Bandwidth BW:* The bandwidth BW is the frequency range from 0 to the frequency ω_0 at which the magnitude drops to 70.7% of its zero-frequency value or 3 dB down from the zero-frequency gain. *Bandwidth gives a measure of the transient response properties.* A larger bandwidth corresponds to a faster rise time because higher-frequency signals are passed on to the output. If the bandwidth is small, the transient response will generally be slow and sluggish since only signals of relatively low-frequencies are passed. The larger the bandwidth, the larger the noise present in the system.

(*iv*) *Cut-off Rate:* In general, bandwidth alone is not adequate to limit the noise. It is also necessary to specify the cut-off rate at high frequencies. However, a steep cut-off characteristic demands a large value of M_p which corresponds to a relatively less stable system.

The above performance criteria are illustrated in Fig. 11.2.

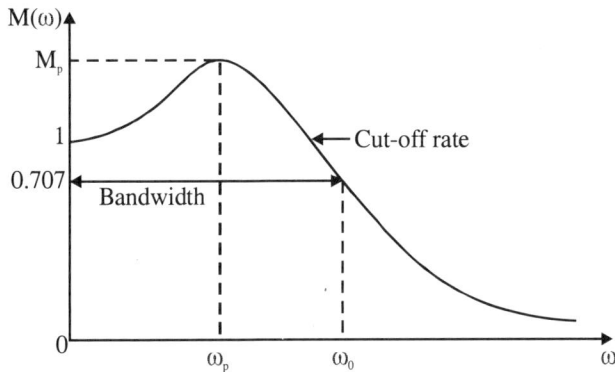

Fig. 11.2. Performance criteria

11.4. FREQUENCY-DOMAIN CHARACTERISTICS OF SECOND-ORDER SYSTEMS

Only for a second-order system, a direct and simple relationship can be obtained for M_p and ω_p in terms of system parameters. Consider the sinusoidal steady-state closed-loop transfer function of a second-order system.

$$M(j\omega) = \frac{C(j\omega)}{R(j\omega)} = \frac{\omega_n^2}{(j\omega)^2 + 2\zeta\omega_n(j\omega) + \omega_n^2} \qquad ...(11.10)$$

$$= \frac{1}{1 - (\omega/\omega_n)^2 + j2(\omega/\omega_n)\zeta}$$

$$= \frac{1}{1 - u^2 + j2u\zeta} \qquad ...(11.11)$$

where $u = \omega/\omega_n$. The magnitude and phase of $M(ju)$ are:

$$|M(ju)| = M(u) = \left[\left(1 - u^2\right)^2 + \left(2u\zeta\right)^2\right]^{-1/2} \qquad ...(11.12)$$

$$\underline{M(ju)} = \phi(u) = -\tan^{-1}\frac{2u\zeta}{1 - u^2} \qquad ...(11.13)$$

The resonant frequency ω_p is determined by setting the derivative of $M(u)$ with respect to u to zero. Thus,

$$\frac{dM(u)}{du} = -\frac{1}{2}\left[\left(1 - u^2\right)^2 + \left(2\zeta u\right)^2\right]^{-1/2}\left(4u^3 - 4u + 8u\zeta^2\right) = 0 \qquad ...(11.14)$$

Therefore,

$$4u^2 - 4u + 8u\zeta^2 = 0 \qquad\qquad ...(11.15)$$

or,

$$u_p = \sqrt{1 - 2\zeta^2} \qquad\qquad ...(11.16)$$

Hence,

$$\omega_p = \omega_n \sqrt{1 - 2\zeta^2} \qquad\qquad ...(11.17)$$

Eq. (11.17) is valid only for $\zeta \le 0.707$ since frequency is a real quantity. For all values of $\zeta > 0.707$, $\omega_p = 0$ and $M_P = 1$.

Substituting Eq. (11.16) in Eq. (11.11) and simplifying, we get

$$M_p = 1/2\zeta\sqrt{1 - \zeta^2} \qquad (\zeta \le 0.707) \qquad\qquad ...(11.18)$$

Thus M_P is a function of ζ only whereas ω_p is a function of ω_n and ζ.

The frequency ω_b at which $M(u)$ drops to 70.7% of its zero-frequency gain is the bandwidth of the system. Therefore,

$$M(u_b) = 0.707 = \left[\left(1 - u_b^2\right) + \left(2\zeta u_b\right)^2 \right]^{-1/2} \qquad\qquad ...(11.19)$$

or,

$$2 = \left(1 - u_b^2\right)^2 + \left(2\zeta u_b\right)^2 \qquad\qquad ...(11.20)$$

Solution of Eq. (11.20) gives

$$\omega_b^2 = \omega_p^2 \pm \omega_n^2 \sqrt{\left(1 - 2\zeta^2\right)^2 + 1} \qquad\qquad ...(11.21)$$

or,

$$\omega_b^2 = \omega_p^2 \pm \sqrt{\omega_p^4 + \omega_n^4} \qquad\qquad ...(11.22)$$

For any value of ζ, ω_b must be a positive real quantity. Hence the plus sign should be chosen in Eq. (11.22). Therefore,

$$\omega_b = BW = \left(\omega_p^2 + \sqrt{\omega_p^4 + \omega_n^4} \right)^{1/2} \qquad\qquad ...(11.23)$$

For a fixed ω_n, as the damping ratio ζ is increased from zero to 0.707, ω_p and BW decrease and M_P increases. These three parameters are also known as *figure-of-merit* of a system.

11.5. CORRELATION BETWEEN STEP TRANSIENT RESPONSE AND FREQUENCY RESPONSE

The transient response and the frequency response of a second-order system can be easily correlated from the simple relationships established in the previous section. The correlations are:

1. The maximum response for a unit step input in the time-domain depends upon ζ only.

2. The resonance peak M_P also depends on ζ only.

3. The rise time t_r increases with ζ.

4. The bandwidth decreases with ζ for a fixed ω_n.

5. The rise time and bandwidth are inversely proportional to each other.

6. The larger the bandwidth, the larger is M_P.

The above relationships also hold good in case of higher-order systems provided they can be approximated to a second-order system.

11.6. FREQUENCY RESPONSE

In general, the closed-loop transfer functions

$$M(j\omega) = \frac{C(j\omega)}{R(j\omega)} = \frac{G(j\omega)}{1 + G(j\omega)H(j\omega)} \qquad \qquad ...(11.24)$$

$$= ReM(j\omega) + jIm\ M(j\omega) \qquad \qquad ...(11.25)$$

$$= M(\omega)\big|\phi(\omega) \qquad \qquad ...(11.26)$$

where $M(\omega)$ is the amplitude of $M(j\omega)$ and is given by

$$M(\omega) = \left| \frac{G(\omega)}{1 + G(j\omega)H(j\omega)} \right| \qquad \qquad ...(11.27)$$

$$\phi(\omega) = \left\lfloor \frac{G(\omega)}{1 + G(j\omega)H(j\omega)} \right. \qquad \qquad ...(11.28)$$

$$= \big\lfloor G(j\omega) - \big\lfloor 1 + G(j\omega)H(j\omega)$$

Thus, it can be seen from Eq. (11.27) and (11.28), the magnitude and phase angle of a system when excited by a sinusoidal input are function of ω. The typical gain and phase characteristics are shown in Fig. 11.3.

The frequency response shown in Fig. 11.3 is similar to that of a low-pass filter. Hence, the majority of control systems have the characteristics of low-pass filter.

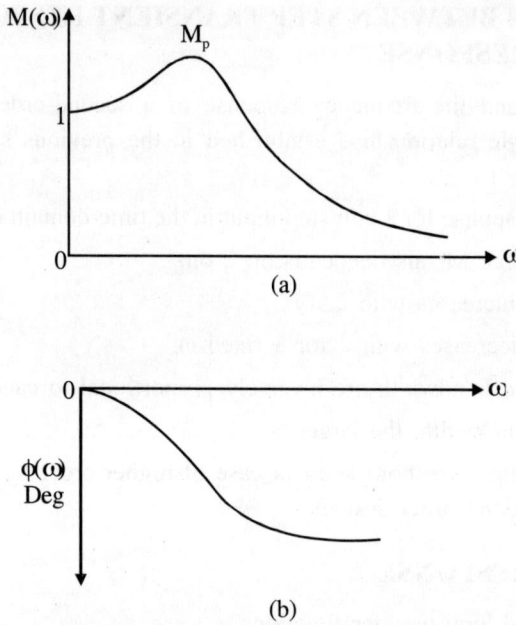

Fig. 11.3. Frequency response (a) Magnitude response (b) Phase response

11.7. FREQUENCY-DOMAIN PLOTS

For analyzing the performance of the closed-loop control systems, in the frequency domain, the open-loop transfer function is generally investigated. In general the function $G(j\omega)$ may be written as

$$G(j\omega) = \left|G(j\omega)\right|\underline{\phi} \qquad\qquad ...(11.29)$$

where $\left|G(j\omega)\right|$ is the magnitude and ϕ is the phase angle of $G(j\omega)$. Thus, in the frequency-domain plots, three variables, the magnitude of $G(j\omega)$ [or $G(j\omega)H(j\omega)$], the phase angle and the frequency are involved. These may be plotted in the following three-forms:

    (i) Bode plot

    (ii) Polar plot

    (iii) Lm - ϕ diagram

11.7.1. Bode Plot

This is the plot of the magnitude in decibels and phase angle ϕ in degrees *versus* ω. Thus, this plot consists of two graphs, *amplitude plot* and *phase plot*.

11.7.2. Polar Plot

This is a plot of the magnitude *versus* phase angle ϕ in degrees in the polar coordinates with ω as a parameter on the curve.

11.7.3. Log-Magnitude versus-Phase Plot (Lm-ϕ diagram)

This is a plot of the magnitude in decibels *versus* the phase angle ϕ in degrees in rectangular coordinates with ω as a parameter on the curve.

These frequency plots are described in the following sections.

11.8. BODE PLOT

The Bode plot is basically an approximation method since the magnitude of $G(j\omega)$ in decibels as a function ω is approximated by straight-line segments.

The advantages of Bode plots are:

1. Bode plot is relatively easy to construct.

2. The use of logarithmic scale for frequency expands the low-frequency range which is of primary importance and compresses the high-frequency range.

3. The mathematical operations of multiplication and division are transformed in to addition and subtraction.

4. The transfer function can be obtained from the plot easily.

5. This plot may be used to generate data necessary for polar plot and log-magnitude-versus-phase plot.

6. The Bode plot can easily be modified as compensation is added.

We shall now define terms used with Bode plots.

(*i*) *Decibel:* This is the unit used for the logarithm of the magnitude. Transfer functions are ratios of output to input to the system or block. When logarithm is used with transfer functions, the output and the input need not necessarily be in the same unit. For example, output many be displacement and the input voltage.

(*ii*) *Log Magnitude:* The logarithm (to base 10) of the transfer function $G(j\omega)$ expressed in decibels is

$$G_{dB} = 20\log|G(j\omega)|\ dB \qquad\qquad ...(11.30)$$

$G(j\omega)$ being a function of frequency, G_{dB} is also a function of frequency.

(*iii*) *Octave:* These are the units to express frequency ratios or bands. An *octave* is a frequency band from f_1 to f_2 where $f_2/f_1 = 2$. The frequency band from 2.4 kHz to 4.8 kHz is one octave in width. The number of octaves in the frequency range from f_1 to f_2 is given by

$$\log(f_1/f_2)/\log 2 = 3.32\ \log(f_2/f_1)\ \text{octaves} \qquad\qquad ...(11.31)$$

(*iv*) **Decade:** A decade is a frequency band from f_1 to f_2 where $f_2/f_1 = 10$. The frequency band from 2.4 KHz to 24 KHz is one *decade* in width. The number of decades from f_1 to f_2 is

$$\log(f_2/f_1) = \text{decades} \qquad \qquad \qquad ...(11.32)$$

11.9. GENERAL TRANSFER FUNCTION IN FREQUENCY DOMAIN

The generalized form of transfer function in frequency domain is

$$G(j\omega) = \frac{K(1 + j\omega T_1)...}{(j\omega)^l (1 + j\omega T_a)(1 + j2\zeta u - u^2)} \qquad \qquad ...(11.33)$$

where K is the system gain. The logarithm of magnitude of $G(j\omega)$ is

$$G_{dB} = 20[\log|K| + \log|1 + j\omega T_1| + ... -l \log|j\omega|$$

$$-\log|1 + j\omega T_a| - \log|(1 + j2\zeta u - u^2)|] \qquad \qquad ...(11.34)$$

The phase angle ϕ is given by

$$\phi = \underline{|K} + \underline{|1 + j\omega T_1} + ... - l \times \underline{|j\omega} - \underline{|1 + j\omega T_a}$$

$$-\underline{|1 + j2\zeta u - u^2} \qquad \qquad ...(11.35)$$

$$= \underline{|K} + \tan^{-1}\omega T_1 + ... - l \times 90° - \tan^{-1}\omega T_a$$

$$-\tan^{-1}(2\zeta u)/(1 - u^2) \qquad \qquad ...(11.36)$$

If K is positive, $\angle K = 0$ and if negative, $\angle K = -180°$. The G_{dB} and ϕ are plotted as functions of the log of frequency and the resulting curves are the Bode plots of the log magnitude diagram and the phase angle diagram.

11.10. DRAWING OF BODE PLOTS

The generalized form of a transfer function given by Eq. (11.33) has four basic types of factors.

1. K ...(11.37)
2. $(j\omega)^{\pm l}$ ...(11.38)
3. $(1 + j\omega T)^{\pm r}$ ...(11.39)
4. $1/(1 + 2\zeta u - u^2)$ ...(11.40)

For each of the above factors, the Bode plot can easily be drawn. Then all these curves can be added together graphically to obtain the complete Bode plot. The procedure can further be simplified by using straight-line or asymptotic approximations.

11.10.1. Constant Term K

Since K is a constant, independent of frequency,

$$G_{dB} = 20 \log K = \text{constant} \qquad \qquad ...(11.41)$$

$$\phi = \angle K = 0° \text{ since } K \text{ is positive} \qquad \qquad ...(11.42)$$

The Bode plot for the constant term K is a horizontal line and the phase angle is zero as shown in Fig. 11.4. The constant term K raises or lowers the log magnitude diagram of the complete transfer function by an amount 20 log K.

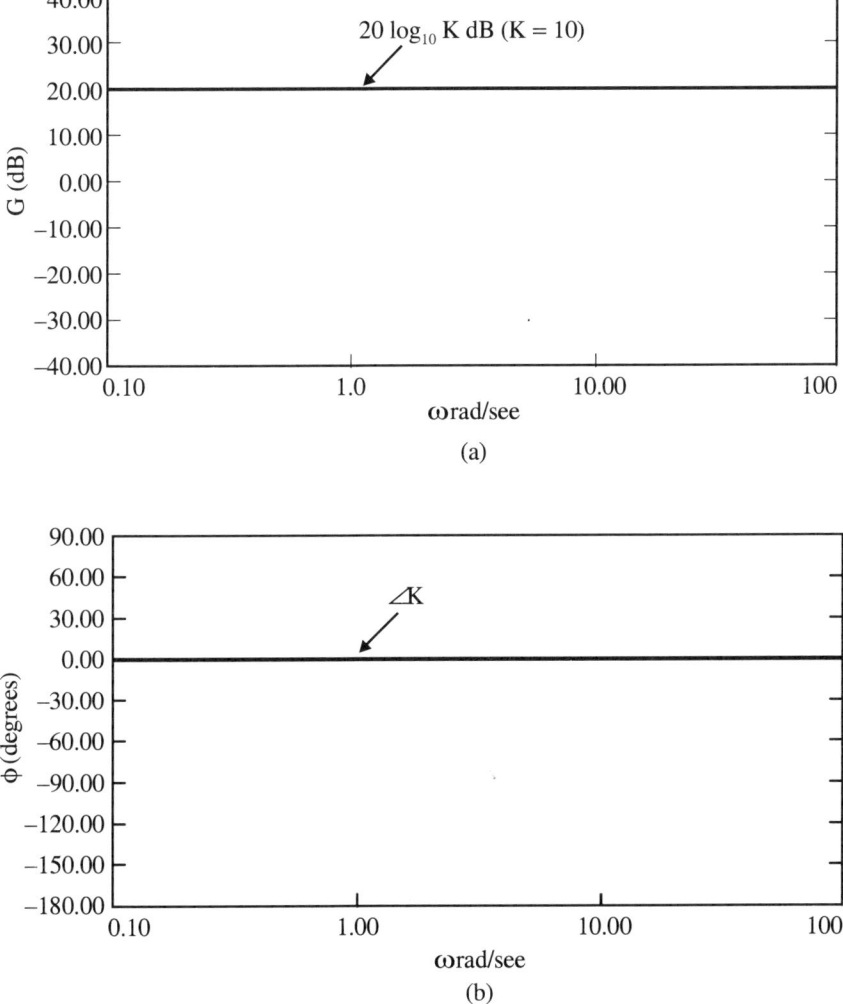

Fig. 11.4. Bode plot for constant K (a) Magnitude plot (b) Phase angle plot.

11.10.2. Poles and Zeros at the origin $(j\omega)^{\pm l}$

The magnitude of $(j\omega)^{\pm l}$ is given by

$$|G(j\omega)| = \omega^{\pm l} \qquad\qquad ...(11.43)$$

The log magnitude G_{dB} is given by

$$G_{dB} = 20\log |G(j\omega)| = \pm 20l \log \omega \qquad\qquad ...(11.44)$$

Eq. (11.44) represents the equation of a straight line with a slope given by Eq. (11.45).

$$dG_{db}/d(\log \omega) = \pm 20l \ dB/\text{decade} \qquad\qquad ...(11.45)$$

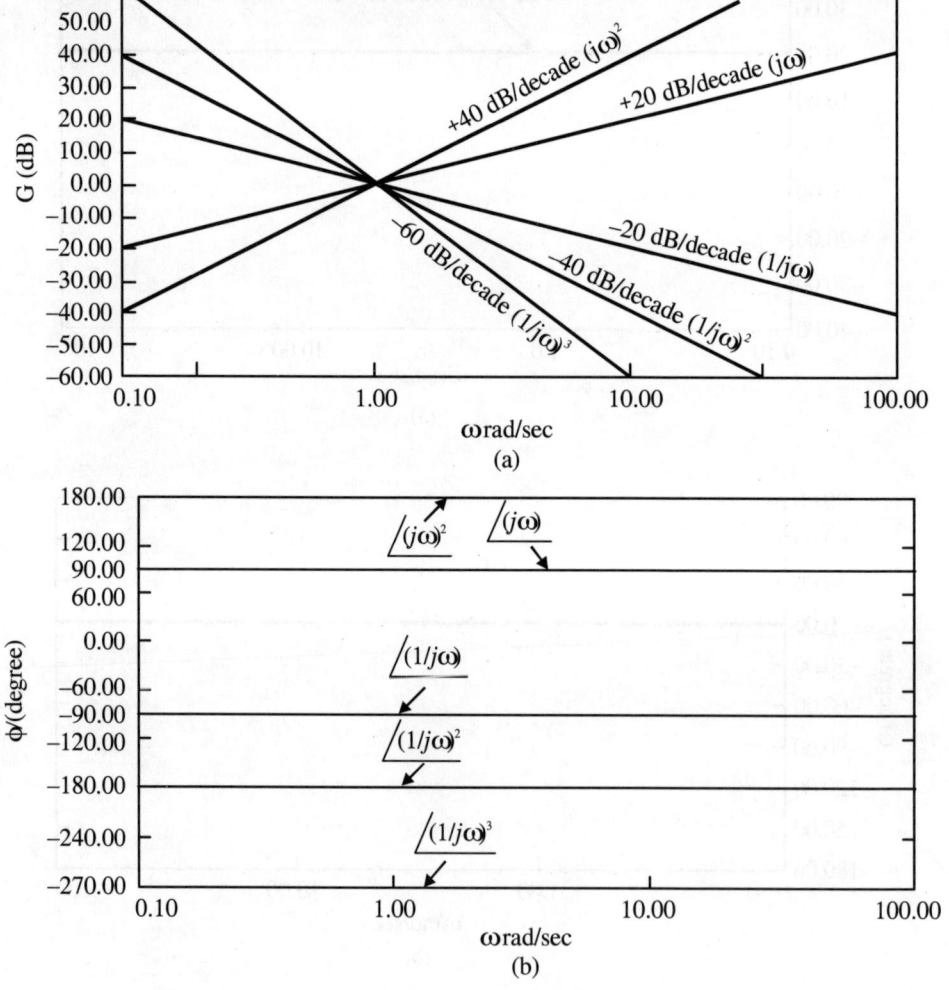

Fig. 11.5. Bode plot for $(j\omega)^{\pm l}$ (a) Magnitude plot (b) Phase angle plot

$$= \pm 20l/3.32 = \pm 6l \ dB/\text{octave} \qquad \qquad ...(11.46)$$

When $\omega = 1$, $G_{dB} = 0$. Thus the log magnitude plot of $(j\omega)^{\pm l}$ is a straight line passing through the 0 dB point at $\omega = 1$ and having a slope of $\pm 6l \ dB/\text{octave}$ or $\pm 20l \ dB/\text{decade}$.

The phase of $(j\omega)^{\pm l}$ is given by

$$\phi = \pm l \times 90° \qquad \qquad ...(11.47)$$

The log magnitude and phase angle diagrams of $(j\omega)^{\pm l}$ are shown in Fig. 11.5.

11.10.3. Pole and Zero Factors $(1 + j\omega T)^{\pm r}$

Consider

$$G(j\omega) = (1 + j\omega T)^{\pm r} \qquad \qquad ...(11.48)$$

where T is a real constant. The magnitude is given by

$$|G(j\omega)| = (1 + \omega^2 T^2)^{\pm r/2} \qquad \qquad ...(11.49)$$

Therefore,

$$G_{dB} = 20\log(1 + \omega^2 T^2)^{\pm r/2} \qquad \qquad ...(11.50)$$

To obtain straight line approximations, we approximate G_{dB} at low and high frequencies. At low frequencies, $\omega T \ll 1$ and hence can be neglected. Then

$$G_{dB} = 20\log 1 = 0dB \quad (\omega T \ll 1) \qquad \qquad ...(11.51)$$

Thus, in the low frequency range, the log magnitude diagram is same as 0 dB line. At high frequencies, $\omega T \gg 1$ and hence

$$G_{db} = 20 \log (\omega^2 T^2)^{\pm r/2}$$

$$= 20 \ r \log \omega T \qquad \qquad ...(11.52)$$

Eq. (11.52) represents a straight line with a slope of $\pm 20 \ r \ dB/\text{decade}$ and passes through the 0-dB point when $\omega T = 1$.

The frequency

$$\omega = 1/T \qquad \qquad ...(11.53)$$

is known as *corner frequency* since the low-frequency approximate plot and the high-frequency approximate plot intersect at this frequency and form the shape of a corner at this frequency as shown in Fig, 11.6. The exact magnitude plot is a smooth curve as shown in figure and deviates only slightly from the straight line approximation. The exact and approximate values of $(1 + j\omega T)$ are compared at some significant frequencies in Table 11.1.

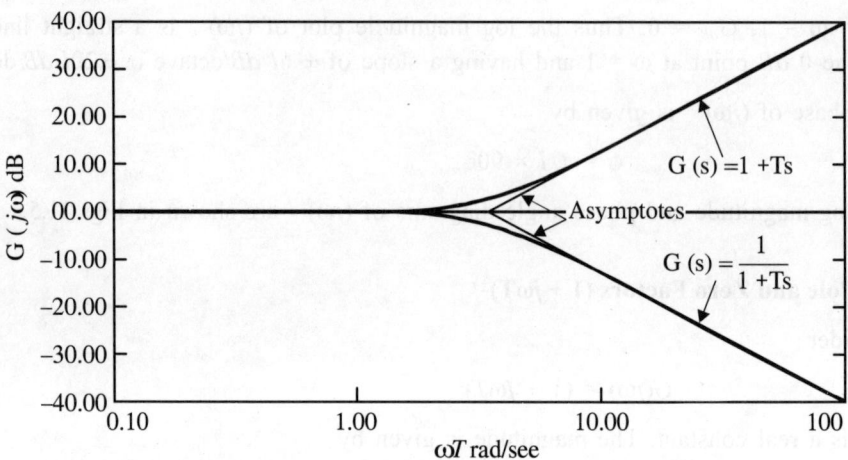

Fig. 11.6. Magnitude plot of $(1 + Ts)^{\pm 1}$

Table 11.1. Values of G_{dB} of $(1 + j\omega T)$ at significant frequencies

Frequency ω (rad./sec.)	Exact value(dB)	Approx. value(dB)	Error (dB)	Exact ϕ(deg.)
1/10T (One decade below)	0.0431	0	0.043	5.7
1/2T (One octave below)	1.0	0	1.0	26.6
1/T (Corner frequency)	3.0	0	3.0	45.0
2/T (One octave above)	7.0	6	1.0	63.4
10/T (One decade above)	20.043	20	0.043	84.3

(Note: For $(1 + j\omega T)^{\pm r}$, the values and the error are to be multiplied by $\pm r$.)

The error between the exact and the approximate values is symmetrical with respect to the corner frequency. The error is 3 dB at the corner frequency, 1 dB at one octave above $(2/T)$ and one octave below $(1/2T)$ the corner frequency and 0.043 dB at one decade above $(10T)$ and one decade below $(1/10T)$ the corner frequency. These facts simplify the drawing of log magnitude diagram for $(1 + j\omega T)$.

The procedure for drawing Bode plot is outlined below:

1. Locate the corner frequency $\omega = 1/T$ on the plot.

2. Draw the 0-dB horizontal line and 20 dB/decade line passing through $\omega = 1/T$.

3. If necessary, obtain the exact curve by using Table 11.1.

The phase angle of $G(j\omega)$ is given by

$$\phi = \tan^{-1} \omega T \qquad\qquad ...(11.54)$$

The exact values of ϕ at some significant frequencies are given in Table 11.1. A smooth curve may be drawn through these points as shown in Fig. 11.7.

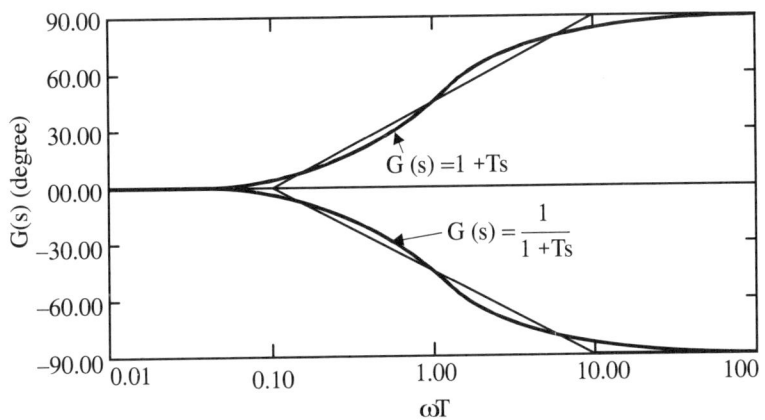

Fig. 11.7. Phase angle plot of $(1 + Ts)^{\pm 1}$

11.10.4. Quadratic Poles

The quadratic term is

$$G(j\omega) = (1 - u^2 + j2\zeta u)^{-1} \quad (\zeta \leq 0.707)$$

Substituting $u = \omega/\omega_n$, we get

$$G(j\omega) = [1-(\omega/\omega_n)^2 + j2\zeta\omega/\omega_n]^{-1} \qquad ...(11.55)$$

$$|G(j\omega)| = \left[\left(1-\omega^2/\omega_n^2\right)^2 + 4\zeta^2\omega^2/\omega_n^2\right]^{-1/2} \qquad ...(11.56)$$

At low frequencies, $\omega/\omega_n \ll 1$. Then

$$|G(j\omega)| = 1 \qquad ...(11.57)$$

and

$$G_{dB} = 20 \log 1 = 0 \ dB \qquad ...(11.58)$$

Thus, at low frequencies, the approximate log magnitude plot is a straight line that lies on 0-dB axis of the Bode plot.

At high frequencies, $\omega/\omega_n \gg 1$ and hence

$$G(j\omega) = -\omega^2/\omega_n^2 \qquad ...(11.59)$$

Therefore

$$G_{dB} = -40\log(\omega/\omega_n) \ dB \qquad ...(11.60)$$

Eq. (11.60) represents a straight line with a slope of $-40 \ dB$/decade and passes through the 0-dB point at $\omega = \omega_n$. Hence ω_n is the corner frequency with $\zeta < 0.707$.

When $\omega = \omega_n$, $|G(j\omega)| = 1/2\zeta$ from Eq. (11.56) and the log magnitude is

$$G_{dB} = -20\log(2\zeta)dB \qquad \qquad \qquad \text{...(11.61)}$$

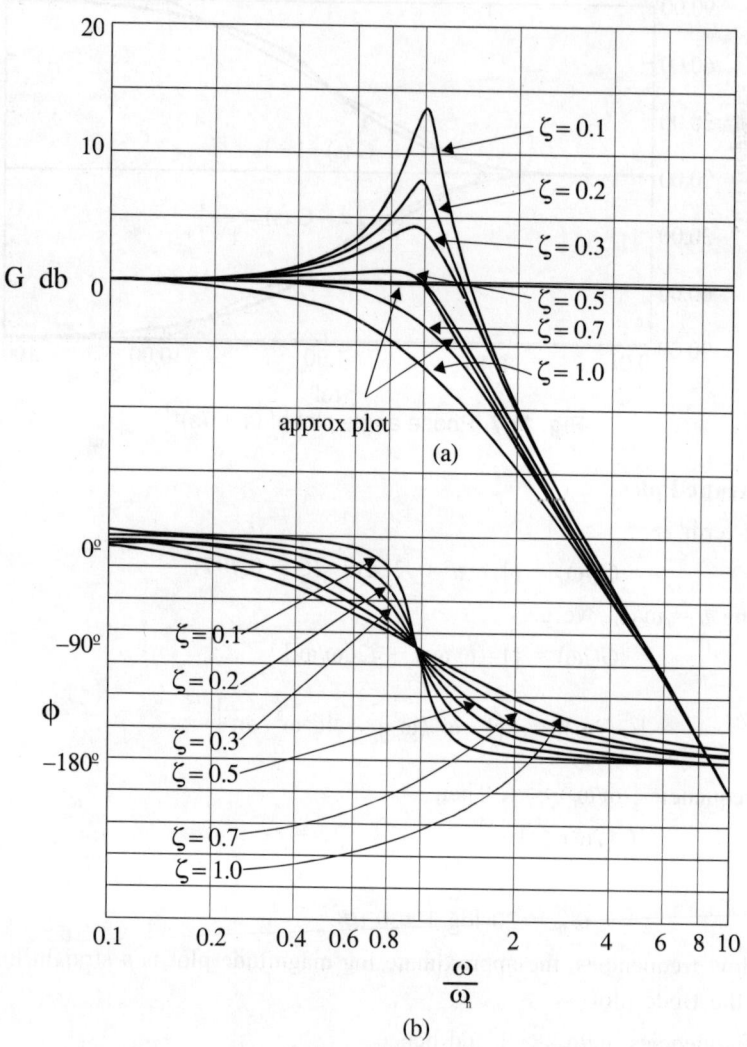

Fig. 11.8. (a) Magnitude plot (b) Phase plot

Thus, around the frequency $\omega = \omega_n$, the log magnitude value depends on the value ζ and hence may differ strikingly from the straight line approximation. The exact and approximate magnitude plots are shown in Fig 11.8(a). The procedure to draw the log magnitude diagram of a quadratic term are :

1. Locate the corner frequency $\omega = \omega_n$.

2. Draw the 0-*dB* line and –40 *dB/decade* line so as to pass through $\omega = \omega_n$.

3. The actual magnitude curve is obtained by locating the points from Fig 11.7.

The phase angle of $G(j\omega)$ is given by

$$\phi = -\tan^{-1}\left(2\zeta\omega/\omega_n\right)/\left(1-\omega^2/\omega_n^2\right) \qquad \qquad ...(11.62)$$

and is plotted as shown in Fig. 11.8(b) for various values of ζ. For all practical purpose, the straight line approximation is adequate.

11.11. EXAMPLE FOR DRAWING BODE PLOTS

Consider the open-loop transfer function of a system

$$G(j\omega)H(j\omega) = \frac{4(1+j0.5\omega)}{j\omega(1+j2\omega)\left[1+j\,0.05\omega-(0.125\omega)^2\right]} \qquad \qquad ...(11.63)$$

This function has totally *five factors* and for each factor the straight line approximation and exact curves are shown in Fig.11.9. The phase curve for each factor is shown in Fig. 11.10. They are added algebraically to obtain the composite characteristic. The pertinent details for each factor are given in Table 11.2.

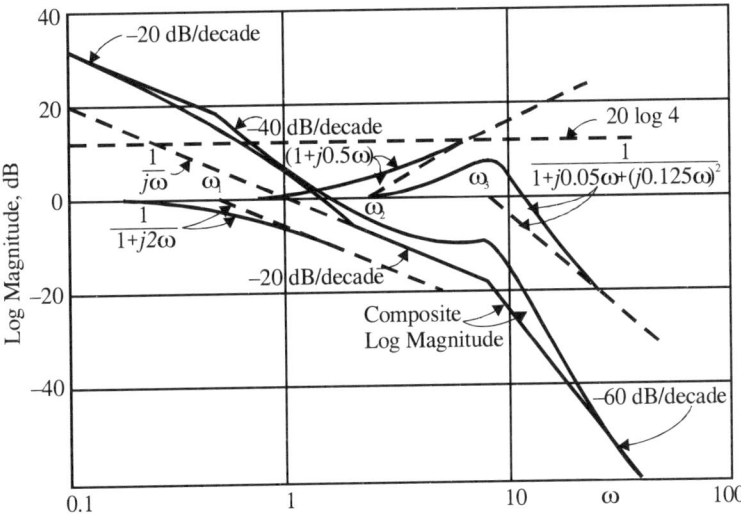

Fig. 11.9. Magnitude plot for each factor (Approximate and exact and composite plot)

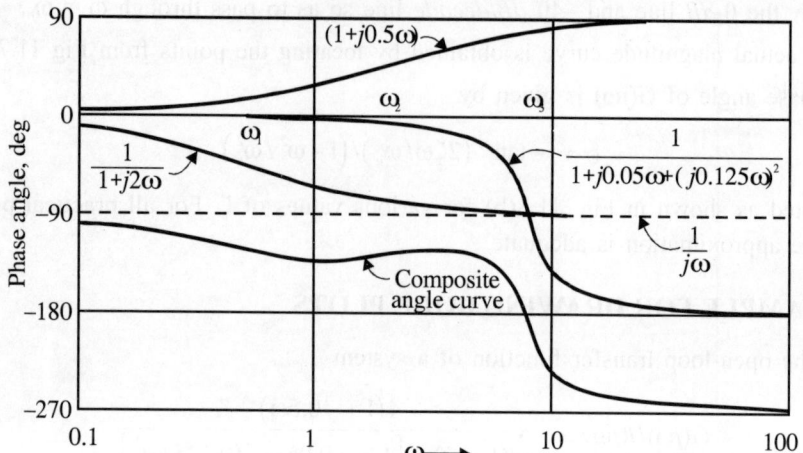

Fig. 11.10. Phase angle plot for each factor and composite plot

Table 11.2. Log magnitude-angle diagram for various factors in Eq. (11.63)

Factor	Corner frequency	Log magnitude G_{dB}	Phase angle ϕ
4	None	+12 *dB*.	0°
$(j\omega)^{-1}$	None	Slope of –20*dB*/decade passing through ω =1.	–90°
$(1 + j2\omega)^{-1}$	$\omega_1 = 0.5$	Zero slope $\omega < \omega_1$; –20*dB*/decade $\omega > \omega_1$.	Varies from 0° to – 90°
$(1 + j0.5\omega)$	$\omega_2 = 2.0$	Zero slope $\omega < \omega_2$; + 20*dB*/decade $\omega > \omega_2$.	Varies from 0° to +90°
$[1+ j0.05\omega–(0.125\omega)^2]^{-1}$ ($\omega_n = 8$; $\zeta = 0.2$)	$\omega_3 = 8.0$	Zero slope $\omega < \omega_3$; –40*dB*/decade $\omega > \omega_3$.	Varies from 0 to –180°

11.12. SYSTEM TYPE AND LOG-MAGNITUDE PLOT

The drawing of the Bode plot can be made easier if the system Type and gain are known. This is described for Type 0, 1 and 2 systems.

11.12.1. Type 0 system: The open loop transfer function of a Type 0 system is of the form

$$G(j\omega)H(j\omega) = K_0/(1 + j\omega T_a)(1 + j\omega T_b) \qquad T_a > T_b \qquad \qquad ...(11.64)$$

At low frequencies, $\omega T_a \ll 1$ and $\omega T_b \ll 1$, $G_{dB} = 20 \log K_0$ which is a constant. The slope of the log-magnitude diagram is zero below the corner frequency $\omega_1 = 1/T_a$, –20 *dB*/decade above the corner frequency ω_1 and –40*dB*/decade above $\omega_2 = 1/T_b$. The plot is shown in Fig. 11.11.

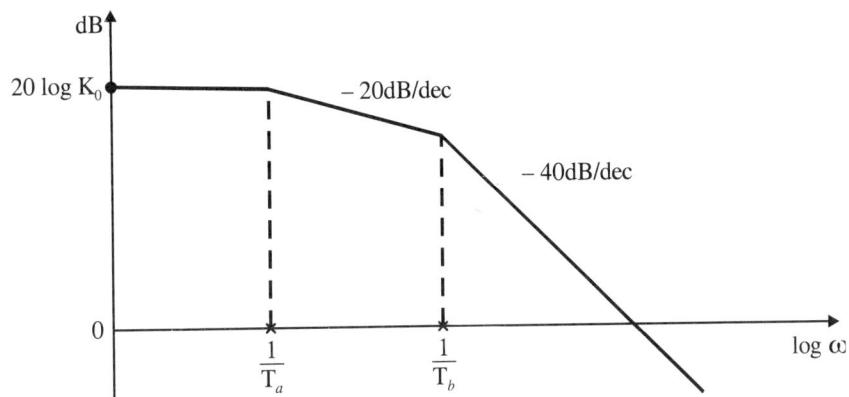

Fig. 11.11. Magnitude plot of Type 0 system

1. The initial or starting value is $20 \log K_0$.

2. The initial slope at low frequencies is $0\ dB$.

3. The gain K_0 equals the static positional error coefficient K_p, since $\lim\limits_{\omega \to 0} G(j\omega)H(j\omega) = K_p$.

4. The final slope is $20\ (Z - P)$ dB/decade where Z is the number of zeros and P is the number of poles.

11.12.2. Type 1 System: The open-loop transfer function of a Type 1 system is

$$G(j\omega)H(j\omega) = K_1/j\omega(1 + j\omega T_a) \qquad\qquad ...(11.65)$$

At low frequencies, $\omega \ll 1/T_a$ and hence

$$G_{dB} = 20\log K_1/\omega \qquad\qquad ...(11.66)$$

and the slope is $-20 dB$/decade. At $\omega = K_1$, $G_{dB} = 0$. If the corner frequency $\omega_1 = 1/T_a$ is greater than K_1, the low-frequency plot crosses the 0-dB axis at a value of $\omega_x = K_1$ as shown in Fig. 11.12(a). If $\omega_1 < K_1$, the low frequency plot may be extended to cross the 0-dB axis and the frequency at the intersection point is $\omega_x = K_1$ as shown in Fig 11.12(b).

The characteristic of Type 1 systems are:

1. The starting value of log magnitude plot is $20\log (K_1/\omega)$.

2. The initial slope is $-20 dB$/decade.

3. The interception of the low-frequency slope of $-20 dB$/decade or its extension with the 0-dB axis occurs at ω_x where $\omega_x = K_1$.

4. The gain K_1 is the static velocity error coefficient K_V since $\lim\limits_{\omega \to 0} j\omega \cdot G(j\omega)H(j\omega) = K_V$

5. The final slope is $20\ (Z–P)\ dB$/decade.

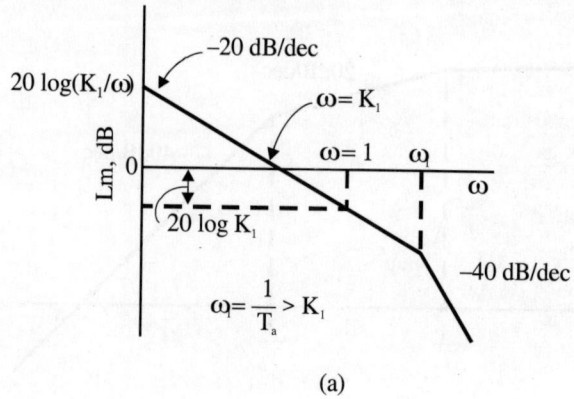

(a)

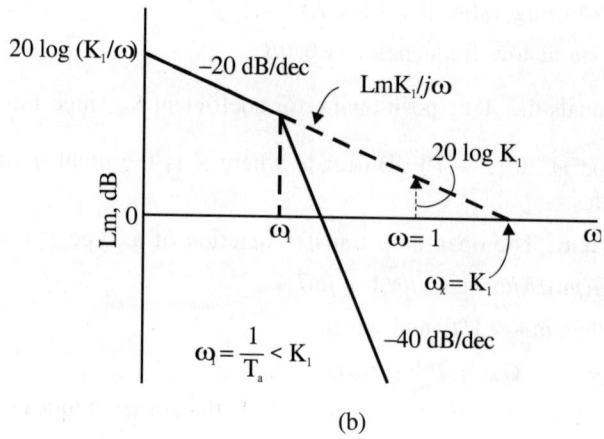

(b)

Fig.11.12. Magnitude plot of Type 1 system

11.12.3. Type 2 System: The open-loop transfer function of a Type 2 system is of the form

$$G(j\omega) = K_2/(j\omega)^2(1 + j\omega T_a)$$

At low frequencies, $\omega \ll 1/T_a$ and hence

$$G_{dB} = 20\log K_2/\omega^2 \qquad\qquad ...(11.68)$$

and the slope is -40 dB/decade. At $\omega^2 = K_2$, $G_{dB} = 0$ dB. Therefore, the intercept of the initial slope of -40 dB/decade or its extension with the 0-dB axis occurs at a frequency ω_y where $\omega_y^2 = K_2$. Fig. 11.13 shows the log magnitude diagram of Type 2 system.

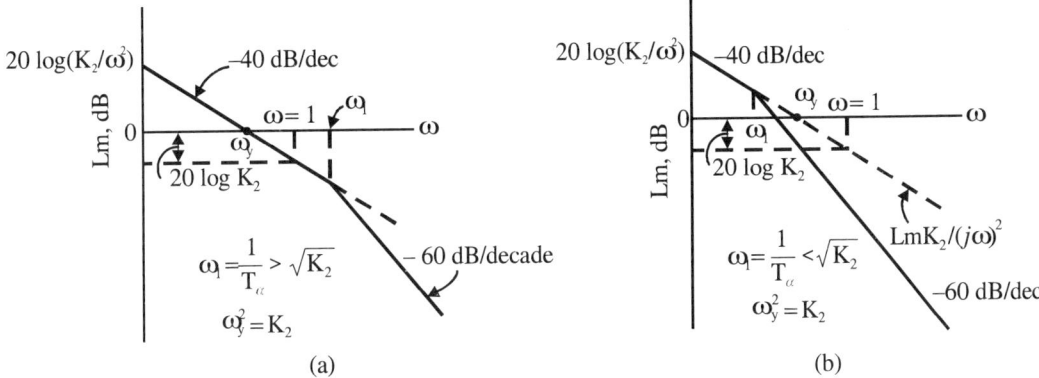

Fig. 11.13. Magnitude plot of Type 2 system

The characteristics of Type 2 system are:

1. The initial value of the log magnitude plot is $20 \log (K_2/\omega^2)$.

2. The initial slope is $-40 dB$/decade.

3. The interception of the initial slope or its extension with 0-dB axis occurs at a frequency ω_y where $\omega_y^2 = K_2$.

4. The gain K_2 is the static acceleration error coefficient since $K_a = \lim_{\omega \to 0} (j\omega)^2 G(j\omega) H(j\omega)$.

5. The final slope is given by $20 (Z–P)$ dB/decade.

11.13. DIRECT METHOD OF DRAWING BODE PLOTS

We shall now describe a direct method of drawing the Bode plot without drawing the plot for each factor of the given $G(j\omega)H(j\omega)$. From the above discussions and from Fig.11.9, the following characteristics of log magnitude plot are summarized.

1. The initial value of log magnitude plot depends on the Type of the system. It is equal to $20\log K$ for Type 0 system, $20 \log (K/\omega)$ for Type 1 system and $20\log (K/\omega^2)$ for Type 2 system. In general, the initial value is given by $20\log (K/\omega^l)$ where l is the Type of the system .

2. The initial slope also depends on the Type. For Type 0 system the initial slope is 0 dB/decade, For Type 1 system, -20 dB/decade and for Type 2 system -40 dB/decade. In general, the initial slope is given by

   Initial slope = $-20l$ dB/decade.

3. The slope of the log magnitude plot changes at each corner frequency. For poles, it

changes by –20 *r dB*/decade for poles and +20 *r dB*/decade for zeros where *r* is the multiplicity of first-order poles or zeros. For a second-order poles, the slope changes by –40*dB*/decade and for second-order zeros by +40*dB*/decade.

4. The final slope is 20 (*Z* – *P*) *dB*/decade.

5. The gain *K* can easily be obtained from the Bode plot.

The direct method of drawing the log-magnitude plot is illustrated with an example using the above characteristics.

Example 11.2. *Draw Bode plot for the open-loop transfer function given in Eq. (11.63).*

$$G(j\omega)H(j\omega) = \frac{4(1+ j0.5\omega)}{j\omega(1+ j2\omega)\left[1+ j0.05\omega- (0.125\omega)^2\right]}$$

Solution:

Step 1: The system is Type 1. Hence the initial value is 20log (4/ω). At the initial value of ω = 0.1, G_{dB} = 20log (4/0.1)= 20 × 1.6 = 32 *dB*.

Step 2: The initial slope is –20 *dB*/decade.

Step 3: The corner frequencies are: poles at 0, 0.5, 8 and zero at 2.

Step 4: Draw a straight line starting from 32 *dB* with a slope of –20 *dB*/decade up to the first corner frequency ω_1 = 0.5.

Step 5: The corner frequency ω_1 = 0.5 is due to a pole. Therefore the slope above ω_1 is the slope below ω_1 + the slope due to the pole at ω_1. Hence the slope above ω_1 is (–20) +(– 20) = –40 *dB*/decade.

Step 6: Continue the plot with a straight line of slope of –40 *dB*/decade till ω_2 = 2.

Step 7: The corner frequency ω_2 is due to a zero. Therefore, this will add a slope of + 20*dB*/ decade. Hence the slope above ω_2 is (–40 + 20) = –20*dB*/decade.

Step 8: Continue the plot by drawing a straight line with a slope of –20*dB*/decade up to ω_3 = 8.

Step 9: The corner frequency ω_3 = 8 is due to a second-order pole. Hence this will contribute a slope of –40*dB*/decade. Therefore, the slope above ω_3 = 8 is –60/decade.

Step 10: Check the final slope which is given by 20 (*Z*–*P*) *dB*/decade. In this case, *Z* = 1, *P* = 4. Hence the final slope is –60*dB*/decade which is the same as the final slope of plot drawn. The log magnitude plot is shown in Fig. 11.14.

Step 11: The phase angle plot is obtained by drawing a smooth curve joining the points computed from Eq. (11.69) at significant frequencies given in Table 11.1 shown in Fig. 11.14(b).

$$\phi = \tan^{-1}0.5\omega–90°–\tan^{-1}2\omega–\tan^{-1}0.05\omega/[1–0.125\omega^2] \qquad ...(11.69)$$

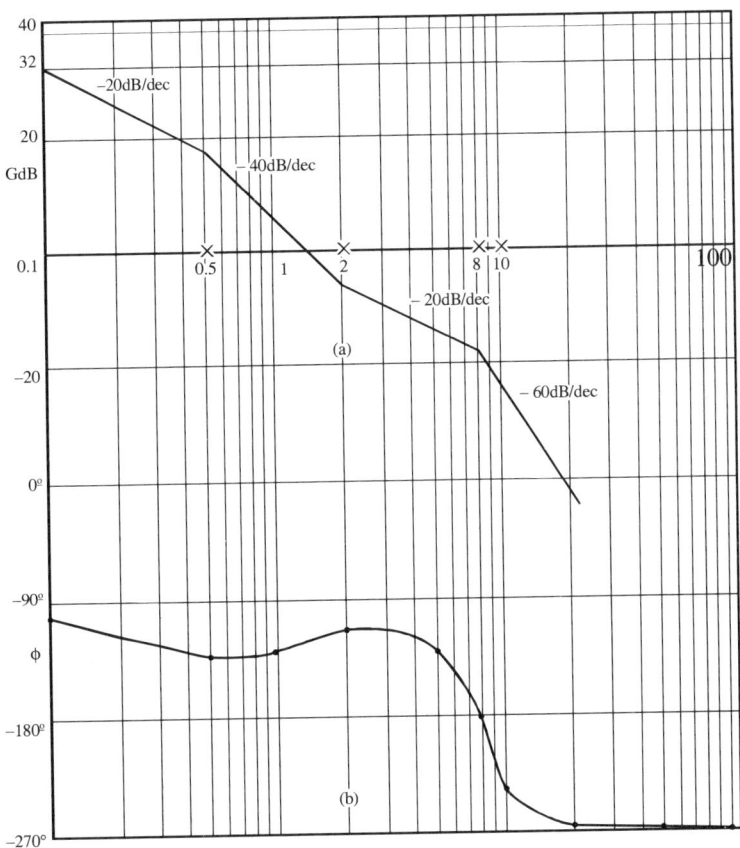

Fig. 11.14. (a) Magnitude plot (b) Phase angle plot of $\dfrac{4(1+j0.5\omega)}{j\omega(1+j2\omega)\left[1+j0.05\omega-(0.125\omega)^2\right]}$

11.14. DETERMINATION OF TRANSFER FUNCTIONS FROM BODE PLOT

When the mathematical expression for the transfer function of a system is not known, the Bode plot is used to determine the approximate transfer function. The magnitude and angle of the ratio of the output to the input can be obtained experimentally for a sinusoidal input of constant magnitude and variable frequency. These data are used to draw the Bode plot Then, asymptotes are drawn on the plotted curve bearing in mind that the slopes must be multiples of ±20 *dB/* decade. The system type and the approximate time constant can be obtained from these asymptotes. Thus the transfer function can be determined. We shall illustrate the method by an example.

Example 11.3. *Determine the open-loop transfer function for the log magnitude plot show in Fig 11.15.*

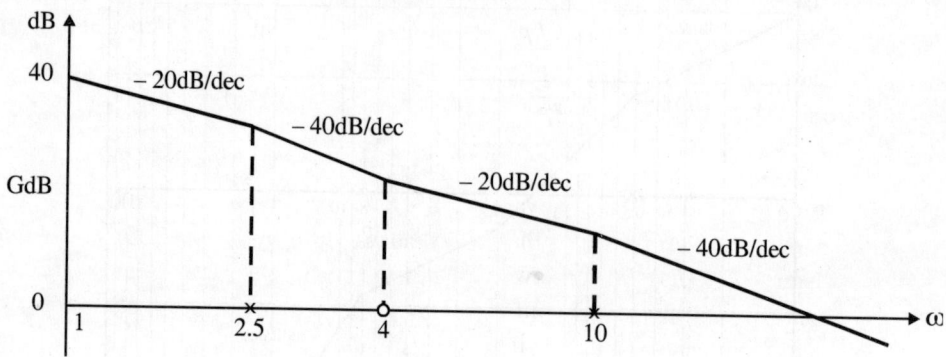

Fig. 11.15. Magnitude plot of determine G(s) H(s)

Solution:

From Fig. 11.15, we can infer the following:

1. Since the initial slope is –20 *dB*/decade, system Type is 1.

2. Since the system Type is 1, the initial value is 20log K/ω. At $\omega = 1$, $G_{dB} = 40 = 20\log$ K; hence $K = 100$.

3. The corner frequencies are poles at $\omega = 2.5$ and 10 since the slope decreases by 20 *dB*/decade after these frequencies; and zero at $\omega = 4$ since the slope increases by 20 *dB*/decade after $\omega = 4$.

4. Therefore,

$$G(j\omega)H(j\omega) = \frac{100(1+j\omega/4)}{j\omega(1+j\omega/2.5)(1+j\omega/10)}$$

$$= \frac{100(1+j0.25\omega)}{j\omega(1+j0.04\omega)(1+j0.1\omega)}$$

11.15. CHARACTERISTICS OF MINIMUM AND NON-MINIMUM PHASE SYSTEM

Transfer functions with poles and zeros in the left half of *s*-plane only are called *minimum phase* transfer functions whereas those with poles and/or zeros in the right half of the *s*-plane also are known as *non-minimum phase* transfer functions. System with minimum phase transfer function is called *minimum phase systems* and that with non-minimum phase transfer function is called *non-minimum phase system*. The range of phase angle of the system with same magnitude characteristic is minimum for minimum phase system, and greater than this minimum for non-minimum phase system. This is illustrated with an example.

Example 11.4. *Determine the range of phase angle of the two system with transfer functions*

$$G_1(j\omega) = \frac{1+j\omega T}{1+j\omega T_a}, \quad G_2(j\omega) = \frac{1-j\omega T}{1+j\omega T_a}$$

Solution:

The magnitude of $G_1(j\omega)$ and $G_2(j\omega)$ is always same. However, the phase angle ϕ_1 of $G_1(j\omega)$ is given by $\phi_1 = \tan^{-1}\omega T - \tan^{-1}\omega T_a$ whereas that of $G_2(j\omega)$ is given by $\phi_2 = -(\tan^{-1}\omega T + \tan^{-1}\omega T_a)$. Thus, the phase angle of $G_1(j\omega)$ varies between 0 and $-90°$ ending at 0; whereas that of $G_2(j\omega)$ varies from 0 to $-180°$. The pole-zero configurations and the phase angle plots are shown in Fig. 11.16. For minimum phase systems, if the magnitude curve is specified over the entire frequency range from zero to infinity, then the phase angle curve is determined uniquely and *vice verse*. However, this is not true for non-minimum phase systems.

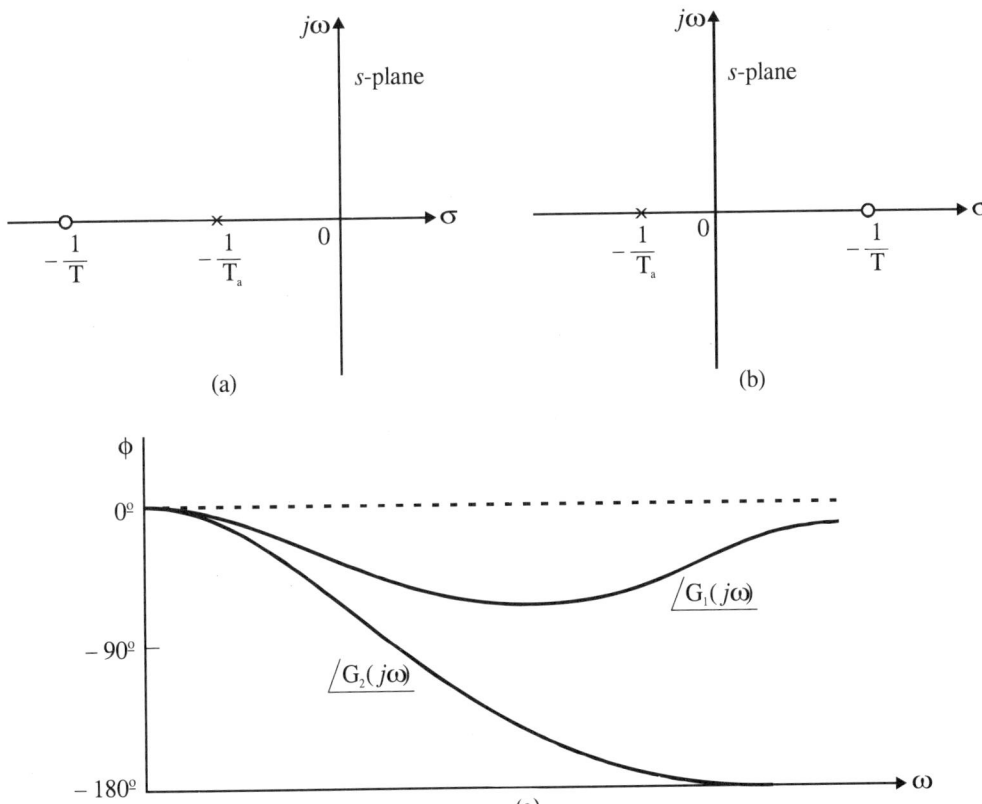

Fig. 11.16. (a) Pole-zero configuration of $G_1(j\omega)$ (b) Pole-zero configuration of $G_2(j\omega)$
(c) Phase angle plots of $G_1(j\omega)$ and $G_2(j\omega)$

Non-minimum phase situations may be due to the presence of a non-minimum phase element in the system or due to an unstable minor loop.

For a minimum phase system, the phase angle at $\omega = \infty$ is 90 $(Z–P)$ degrees where Z and P are number of zeros and poles of the transfer function. For a non-minimum phase system, it differs from this value. However, in both systems, the slope of the log-magnitude plot at $\omega = \infty$ is 20 $(Z–P)$ dB/decade. Therefore, it is possible to distinguish a minimum phase system from a non-minimum phase system by examining both the slope of the high-frequency asymptote of the log magnitude curve and the phase angle at $\omega = \infty$. If the slope at $\omega = \infty$ is 20 $(Z–P)$ dB/decade and the phase angle is 90 $(Z – P)$ degrees, than the system is minimum phase system.

The response of non-minimum phase systems is sluggish because of excessive phase lag. In most practical systems, excessive phase lag is to be avoided. For fast response, minimum phase system should be used.

11.16. BODE PLOT OF SYSTEM WITH TRANSPORTATION LAG

Transportation lag is a non-minimum phase element. It has excessive phase lag at high frequencies. The gain is unity and hence no attenuation. The sinusoidal transfer function of the transportation lag is

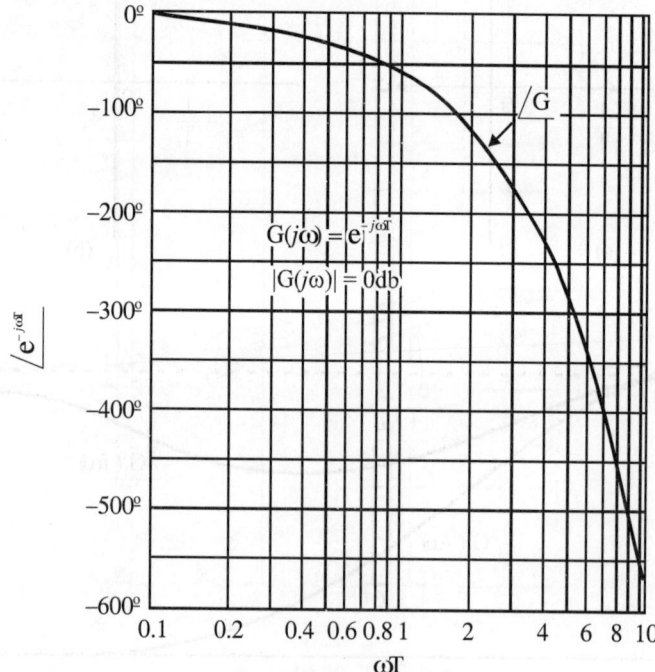

Fig. 11.17. Phase angle variation of transportation lag

$$G(j\omega) = e^{-j\omega T} \qquad \qquad ...(11.70)$$

The magnitude of $G(j\omega)$ is

$$|G(j\omega)| = 1 \qquad \qquad ...(11.71)$$

Hence the log magnitude of the transportation lag is equal to 0 *dB*.

The phase angle is

$$\phi = \underline{|G(j\omega)} = -\omega T \qquad \qquad ...(11.72)$$

Thus, the phase angle varies linearly with ω. The phase angle characteristic of transportation lag is shown in Fig. 11.17.

Example 11.5. *Draw the Bode diagram of the following transfer function.*

$$G(j\omega) = e^{-j0.5\omega}/(1 + j\omega)$$

Solution:

The log magnitude is

$$20 \ \log |G(j\omega)| = 20\log|1| - 20\log|1 + j\omega|$$
$$= -20 \ \log |1 + j\omega|$$

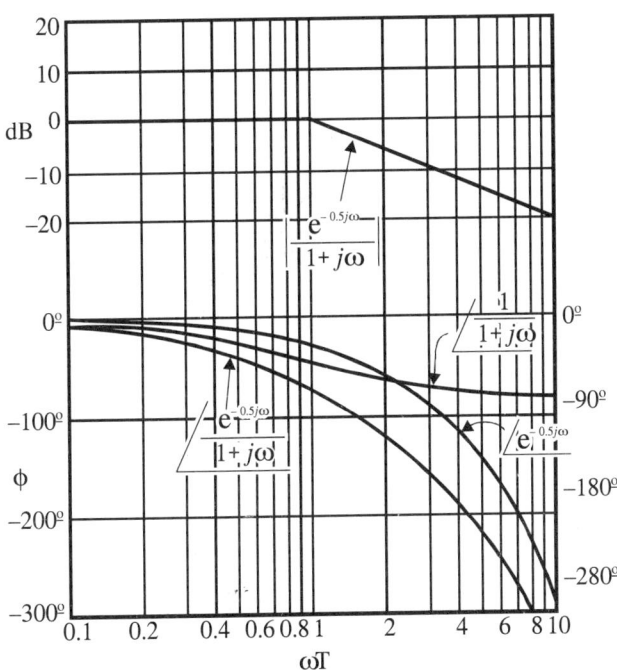

Fig. 11.18. G_{dB} and ϕ curves for $G(j\omega) = e^{-j0.5\omega}/(1 + j\omega)$

The phase angle is

$$\phi = \underline{|e^{-j0.5\omega}} + \underline{|1/(1+j\omega)}$$

The lag magnitude and phase angle curves are shown in Fig. 11.18.

11.17. POLAR PLOTS

The polar plot of a sinusoidal transfer function $G(j\omega)H(j\omega)$ is a plot of the magnitude of $G(j\omega)H(j\omega)$ verses the phase angle on the polar coordinates as ω is varied from 0 to ∞. For any frequency $\omega = \omega_i$, the magnitude and phase of $G(j\omega_i)H(j\omega_i)$ are represented by a phasor with corresponding magnitude and phase angle. In polar plots, phase angles measured clockwise are negative and counterclockwise are positive. A typical polar plot is shown in Fig. 11.19. Each point on the polar plot represents the phasor drawn from the origin to that point at a specific value of ω_i. It is important to show the frequency graduation on the plot. The projection of $G(j\omega)H(j\omega)$ on the real and imaginary axes are its real and imaginary part respectively. To construct the plots, both the magnitude and phase angle must be calculated for each frequency ω_i. Since the Bode plot is easy to construct, the data for drawing the polar plots may be obtained directly from the Bode plot if it is drawn first.

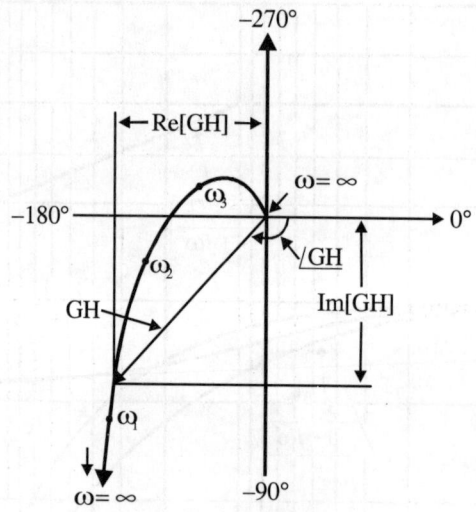

Fig. 11.19: Typical Polar Plot

In general, if a polar plot of the product of two different transfer functions is desired, it is convenient to draw the Bode plot first and then obtain the data to draw polar plots. An advantage in using polar plots is that the frequency response characteristic of the system is shown over the entire frequency range on a single plot. However, the plot does not clearly indicate the contribution of each of the individual factors of the open-loop transfer function.

11.17.1. Constant Term K

Since K is a constant and independent of frequency and since its phase angle is $0°$, it is represented by a point on the positive real axis of the polar plot.

11.17.2. Poles and Zeros at the origin $(j\omega)^{\pm 1}$

The polar plot of $j\omega$ is the positive imaginary axis and that of $(1/j\omega)$ is the negative imaginary axis since, in the first case, the magnitude is ω and the phase angle is $+90°$ and, in the second case, the magnitude is $(1/\omega)$ and the phase is $-90°$. For $1/(j\omega)^2$, the magnitude is $1/\omega^2$ and the phase angle is $-180°$. The plots are shown in Fig. 11.20.

11.17.3. Pole and Zero Factors $(1+j\omega T)^{\pm r}$

For the sinusoidal transfer function

$$G(j\omega) = 1 + j\omega T$$

$$|G(j\omega)| = \sqrt{1+(\omega T)^2} \qquad ...(11.73)$$

$$\phi = \tan^{-1} \omega T \qquad ...(11.74)$$

The magnitude of $G(j\omega)$ at $\omega = 0$ and $\omega = 1/T$ are 1 and $\sqrt{2}$ and the phase angle is 0 and $+45°$ respectively. As ω approaches infinity, the magnitude approaches infinity and the phase angle $+90°$. Therefore, the polar plot of $(1 + j\omega T)$ is a straight line from the point $(1, 0°)$ extending to $+\infty$ and parallel to the $j\omega$-axis as shown in Fig. 11.20(d).

If the sinusoidal transfer function is

$$G(j\omega) = 1/(1 + j\omega T) \qquad ...(11.75)$$

then the magnitude at $\omega = 0$ and $\omega = 1/T$ is 1 and $1\sqrt{2}$ and the phase angle is $0°$ and $-45°$. As ω approaches infinity, the magnitude approaches zero and the phase angle approaches $-90°$. Therefore, the polar plot is a semicircle as ω is varied from 0 to ∞ as shown in Fig. 11.20(e). The centre is located at 0.5 on the real axis and the radius is 0.5.

To prove that the polar plot is a semicircle, consider any point A on the polar plot and draw phasor 0A and A1. The phasor $0A = 1/(1 + j\omega T)$ and A1 is $1 - 1/(1 + j\omega T) = j\omega T/(1 + j\omega T)$. Then

$$|0A|^2 + |A1|^2 = \frac{1}{1+\omega^2 T^2} + \frac{\omega^2 T^2}{1+\omega^2 T^2} = 1$$

Therefore, the angle 0A1 is always $90°$. Hence the polar plot is a semicircle.

11.17.4. Quadratic poles

The sinusoidal transfer function is

$$G(j\omega) = 1/\left(1 - \omega^2/\omega_n^2 + j2\zeta\omega/\omega_n\right) \ (\zeta > 0) \qquad ...(11.76)$$

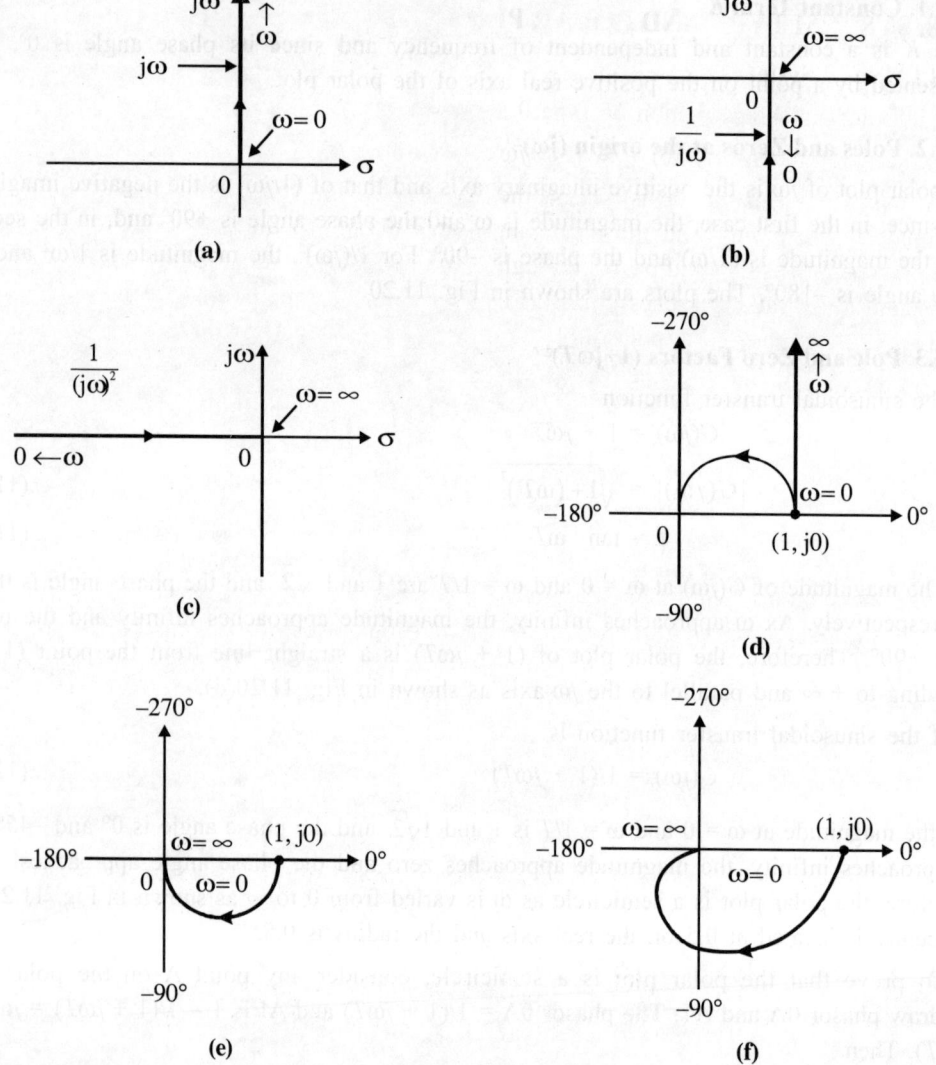

Fig. 11.20. (a) Polar plot for jω (b) Polar plot for (1/jω) (c) Polar plot for 1/(jω)² (d) Polar plot for 1

$$+ j\omega T \text{ (e) Polar plot for } 1/(1 + j\omega T) \text{ (f) Polar plot for } 1 \Big/ \Big(1 - \frac{\omega^2}{\omega_n^2} + j2\zeta\frac{\omega}{\omega_n}\Big).$$

At ω = 0, the magnitude is 1 and the phase angle is 0° and at ω = ∞, the magnitude is 0 and the phase angle is –180°. Thus, the polar plot starts at (1, 0°) and ends at (0, –180°) as ω varies from 0 to ∞.

The polar plot is shown in Fig 11.20.(f).

11.18. SYSTEM TYPE AND POLAR PLOT

11.18.1. Type 0 System

The sinusoidal transfer function of Type 0 system is of the form

$$G(j\omega)H(j\omega) = K_0/(1 + j\omega T_a)(1 + j\omega T_b) \qquad \qquad ...(11.77)$$

At $\omega = 0$, the magnitude is K_0 and the phase angle is $0°$. As ω approaches infinity, the magnitude tends to zero and the phase angle to $-180°$. Thus the polar plot of this transfer function starts at $(K_0, 0°)$ and passes through fourth and third quadrants and ends at $-180°$ with zero magnitude as ω approaches infinity. Fig. 11.21(a) shows the pole locations of Eq. (11.77) and Fig. 11.21(b) shows the polar plot.

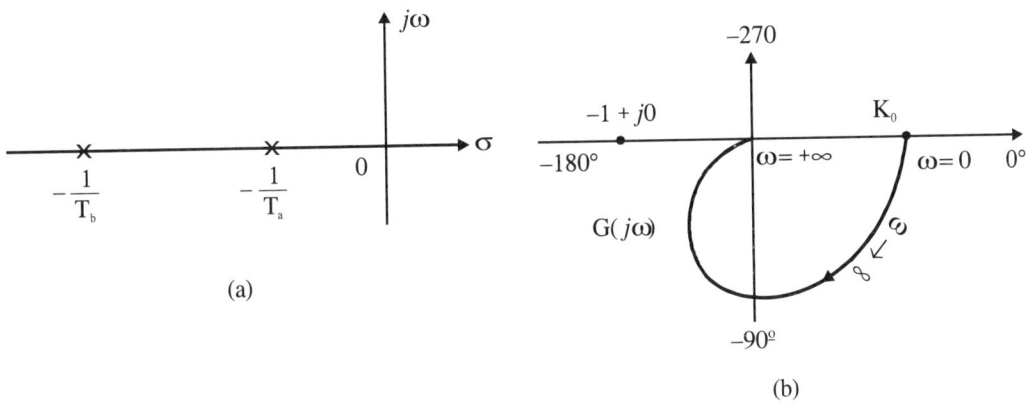

Fig. 11.21. (a) Pole-zero location of Type 0 second-order system (b) Polar plot.

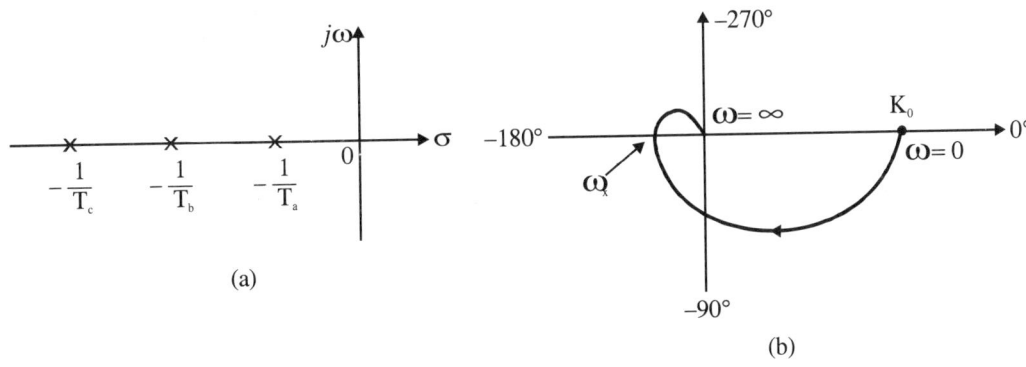

Fig. 11.22. (a) Pole-zero locations of Type 0 third-order system (b) Polar plot

If now a pole is added, then

$$G(j\omega)H(j\omega) = K_0/(1 + j\omega T_a)(1 + j\omega T_b)(1 + j\omega T_c) \qquad \qquad ...(11.78)$$

The pole locations are shown in Fig. 11.22(a) and the polar plot starts at $(K_0, 0°)$ at $\omega = 0$ and passes through fourth, third and second quadrants in that order and ends at $-270°$ with zero magnitude as ω approaches infinity as shown in Fig. 11.22(b). In this case, *the plot crosses the real axis at ω_x at which the imaginary part of the transfer function is zero.*

If now a zero is added, then

$$G(j\omega)H(j\omega) = K_0(1 + j\omega T_1)/(1 + j\omega T_a)(1 + j\omega T_b)(1 + j\omega T_c) \quad ...(11.79)$$

The polar plot for T_a and $T_b > T_1$ and $T_1 > T_c$ is shown in Fig. 11.23. *Because of the presence of zero, there is a dent in the polar plot.*

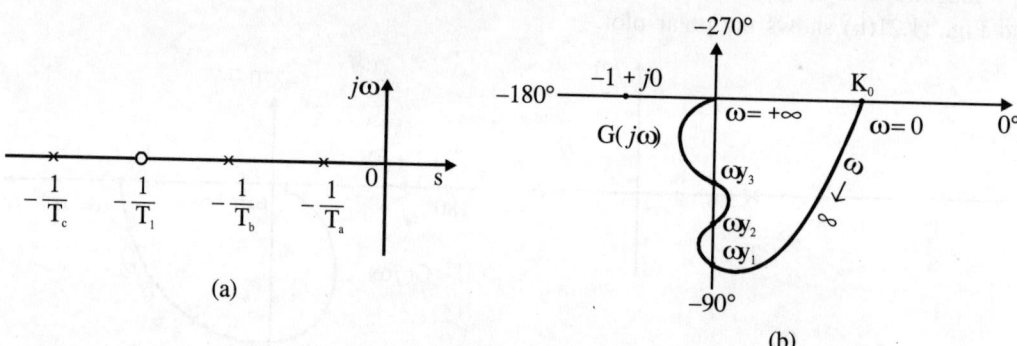

(a)

(b)

Fig. 11.23. (a) Pole-zero locations of type 0 third-order system with a zero (b) Polar plot

The characteristics of the polar plots of a Type 0 system are:

1. The polar plot always starts at a value K_0, the positional error coefficient on the positive real axis for $\omega = 0$ and ends at zero magnitude and tangent to one of the major axes at $\omega = \infty$.

2. The final phase angle is $90°$ $(Z - P)$.

3. The magnitude varies monotonically if all corner frequencies are poles.

4. If a zero is present, the polar plot will have a dent.

11.18.2. Type 1 System

The transfer function in the frequency domain of a Type 1 system is given by

$$G(j\omega)H(j\omega) = \frac{K_\infty}{j\omega(1 + j\omega T_a)(1 + j\omega T_b)(1 + j\omega T_c)} \quad ...(11.80)$$

At $\omega = 0$, the magnitude is ∞ and the phase angle is $-90°$. As ω is increased, the magnitude goes on decreasing and phase increases in the clockwise direction. Finally, when $\omega = \infty$, the magnitude is zero and the phase angle is $-360°$. Fig. 11.24 shows the polar plot. As ω approaches zero, the magnitude of Eq. (11.80) approaches infinity asymptotically to a line parallel to the

– 90° axis but displaced to the left of origin. The true asymptote is determined by finding the real part as ω approaches zero.

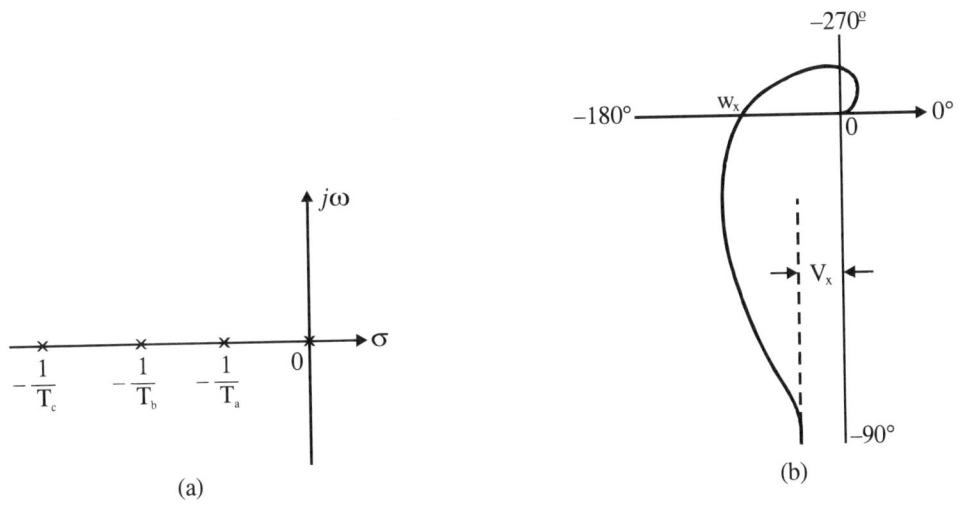

Fig. 11.24. (a) Pole-zero locations of Type 1 fourth-order system (b) Polar plot

This is given by

$$V_x = \lim_{\omega \to 0} \mathrm{Re}\left[G(j\omega)H(j\omega)\right] \qquad \qquad ...(11.81)$$

In this case,

$$V_x = -K_1(T_a + T_b + T_c) \qquad \qquad ...(11.82)$$

The polar plot crosses the negative real axis at a frequency ω_x. The value of ω_x is determined by equating the imaginary part of $G(j\omega)H(j\omega)$ to zero. Thus,

$$Im[G(j\omega)H(j\omega)] = 0 \qquad \qquad ...(11.83)$$

In this case,

$$\omega_x = (T_a T_b + T_b T_c + T_c T_a)^{-1/2} \qquad \qquad ...(11.84)$$

Eq. (11.84) is useful in the analysis of system stability.

If a zero is added, the polar plot will have a dent and the final phase angle will be –270°.

The characteristics of the polar plots of Type 1 system are:

1. The polar plot always starts at infinity and tangent to a line parallel to –90° axis.

2. The initial phase angle is –90°.

3. The final phase angle is 90° (Z–P).

4. If a zero is present, the polar plot will have a dent.

11.18.3. Type 2 system

The open-loop transfer function of Type 2 system is of the form

$$G(j\omega)H(j\omega) = \frac{K_2}{(j\omega)^2 (1+j\omega T_a)(1+j\omega T_b)} \qquad ...(11.85)$$

For $\omega = 0$, the magnitude is ∞ and the phase angle is $-180°$. As ω approaches ∞, the magnitude reduces to zero and the phase angle is $-360°$.

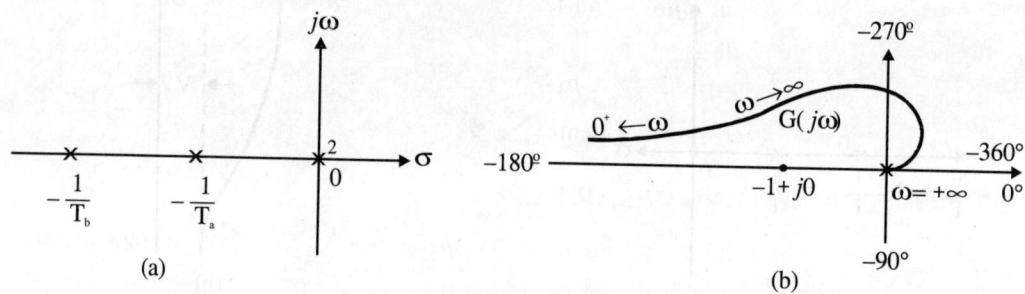

Fig. 11.25. (a) Pole-zero locations of Type 2 foruth-order system

For the transfer function of Eq. (11.85), the polar plot shown in Fig. 11.25 is a smooth curve. The introduction of a zero will form a dent in the plot and the final phase angle will be $-270°$. Eq (11.85) represents the transfer function of a system which is unstable. To make the system stable, a zero nearer to $j\omega$-axis is usually added. The new transfer function with a zero and a pole added is

$$G(j\omega)H(j\omega) = \frac{K_2 (1+j\omega T_1)}{(j\omega)^2 (1+\omega T_a)(1+\omega T_b)(1+\omega T_c)} \qquad ...(11.86)$$

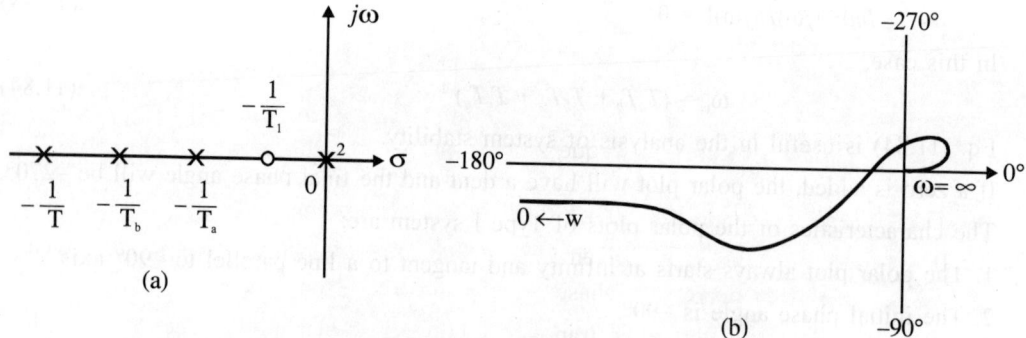

Fig. 11.26. (a) Pole-zero locations of Type 2 fifth-order system with a zero

The polar plot is shown in Fig. 11.26. At low frequencies, the plot is below negative the real axis since the first corner frequency $\omega = 1/T_1$ is a zero which contributes $+90°$. Therefore, the polar plot starts near $-180°$ and increases in the positive direction as ω is varied from 0. Then, the other corner frequencies being poles the polar plot crosses the negative real axis and increases in the negative direction. The plot crosses the negative real axis at $\omega = \omega_x$ and the final the phase angle is $-360°$.

The characteristic of the polar plots of Type 2 system are:

1. The polar plot starts at infinity and tangent to a line parallel to $-180°$ axis.

2. The initial phase angle is $-180°$

3. The final phase angle is $90° \, (Z - P)$.

4. The presence of a zero forms a dent in the plot.

11.19. STEPS IN DRAWING POLAR PLOTS

The following steps are useful in drawing the polar plot of a given sinusoidal transfer function.

1. From the given transfer function, determine the system type. The low-frequency polar plots of the different system types are shown in Fig. 11.27.

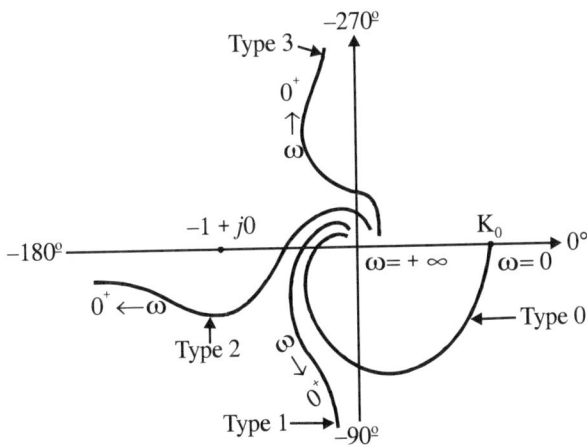

Fig. 11.27. Low-frequency polar plots of system types

2. The initial phase angle is $(-90°) \times l$ where l is the system type.

3. If there are no zeros in the given transfer function, the polar plot is a smooth curve with continuously decreasing phase angle as ω varies from 0 to ∞.

4. If there are zeros in the given transfer function, the polar plot has dents since phase angle does not decrease continuously.

5. The final magnitude is zero at $\omega = \infty$.

6. The final phase angle is $(Z - P)$ 90°.

Example 11.6. *Sketch the polar plot of the transfer function,* $G(s) = 1/s(1 + Ts)$.

Solution:

The sinusoidal transfer function is

$$G(j\omega) = 1/j\omega(1+j\omega T)$$

This is Type 1 system. Hence, the initial magnitude is infinity and initial phase angle is −90°. The initial portion of the plot is tangential to the line $V_x = -T$. The final phase angle is −180° since there are two poles and no zeros. Since there are no zeros, the plot is a smooth curve as shown in Fig. 11.28.

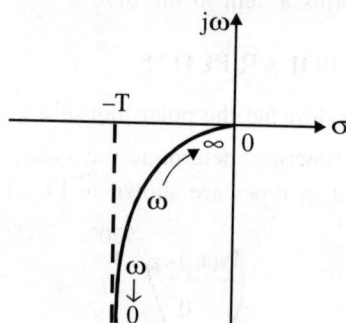

Fig. 11.28. Polar plot of Type 1 second-order system

11.20. POLAR PLOT OF TRANSPORTATION LAG

The transfer function of the transportation lag is

$$G(j\omega) = e^{-j\omega T} \qquad\qquad\qquad ...(11.87)$$

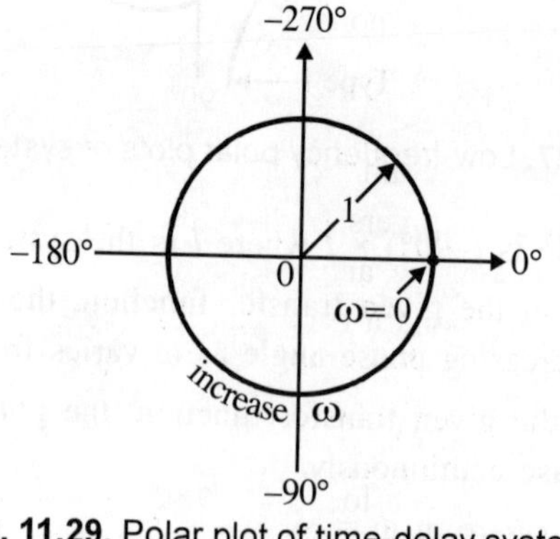

Fig. 11.29. Polar plot of time-delay system

The magnitude is always unity and the phase angle varies linearly with w. Hence the polar plot is a unit circle with centre at origin as shown in Fig. 11.29.

Example11.7 *Sketch the Polar plot of*

$$G(j\omega) = e^{-j\omega L}/(1 + j\omega T)$$

Solution:

This is a Type 0 system. Hence the initial value is $1\angle 0°$. As ω varies from 0 to ∞, the magnitude goes on decreasing as the magnitude is given by

$$|G(j\omega)| = (1 + \omega^2 T^2)^{-1/2}$$

The phase angle ϕ is given by

$$\phi = -\omega L - \tan^{-1}\omega T$$

The phase angle also decreases monotonically and indefinitely. Hence the polar plot is a spiral as shown in Fig. 11.30.

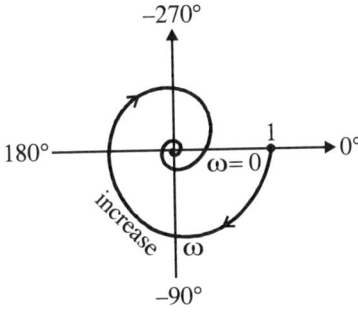

Fig. 11.30. Polar plot of Example 11.7

11.21. LOG. MAGNITUDE VERSUS PHASE ANGLE PLOT (Lm-ϕ DIAGRAM)

An alternative approach to graphically portraying the frequency response characteristic is to plot the logarithmic magnitude in *dB* versus the phase angle ϕ for a range of frequencies. The curve is graduated in terms of ω. These plots are also known as *Nichols plot or chart.* In this plot, the two graph of the Bode diagram are combined into one. Since the Bode diagram is easier to construct, the log-magnitude versus phase diagram is easily constructed by obtaining the values of log-magnitude and phase angle from the Bode plot. In the *Lm–ϕ* diagram, an increase in gain shifts the curve up and a decrease in the gain shifts the curve down *while the shape of the curve remains same.*

The advantages of Nichols plot are:

(a) The relative stability of a closed-loop system can be quickly determined from the plot.

(b) Compensation networks can be easily designed.

Example 11.8. *Draw the log-magnitude versus phase plot for the transfer function*

$$G(j\omega)H(j\omega) = \frac{5}{j\omega(1+0.5\,j\omega)(1+j\omega/6)}$$

Solution:

The magnitude of the above transfer function is

$$|G(j\omega)H(j\omega)| = \frac{5}{\omega\left[\left(1+0.25\omega^2\right)\left(1+\omega^2/36\right)\right]^{1/2}}$$

The phase angle ϕ is given by

$$\phi = -90° - \tan^{-1}0.5\omega - \tan^{-1}(\omega/6)$$

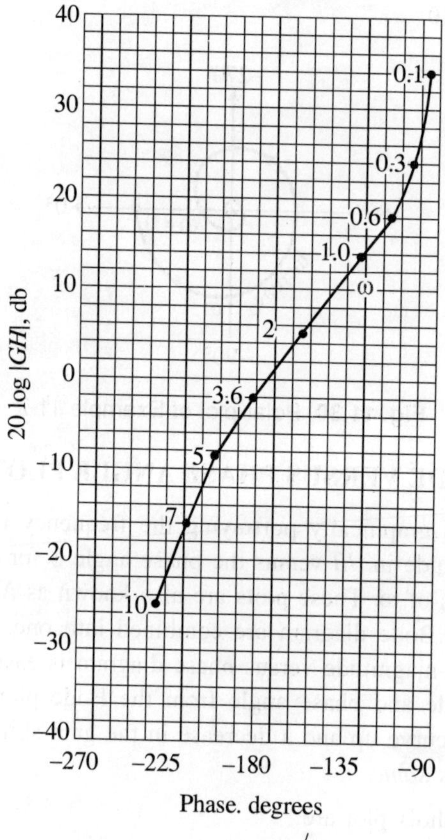

Fig. 11.31. Lm-ϕ diagram of $GH = 5\Big/ j\omega(1+0.5\,j\omega)\left(1+j\dfrac{\omega}{6}\right)$

The magnitude and phase angles are computed for a frequency range of interest, usually from $\omega = 0$ to ω_h at which the phase angle ϕ is about $-225°$ since the plot around 0 *dB* and $-180°$ is of primary interest from the system stability considerations. The plot is shown in Fig.11.31.

11.22. ADDITIONAL EXAMPLES

Example 11.9. *Draw the Bode Plot of the transfer function*

$$G(s) = \frac{10(s+10)}{s(s+2)(s+5)}$$

Solution:

The sinusoidal transfer function is

$$G(j\omega) = \frac{10(1+j0.1\omega)}{j\omega(1+j0.5\omega)(1+j0.2\omega)}$$

This is a Type 1 system. Therefore,

1. The initial slope is -20 *dB*/decade.

2. The corner frequencies are: Zero at $\omega_1 = 10$ and poles at 0, 2, and 5.

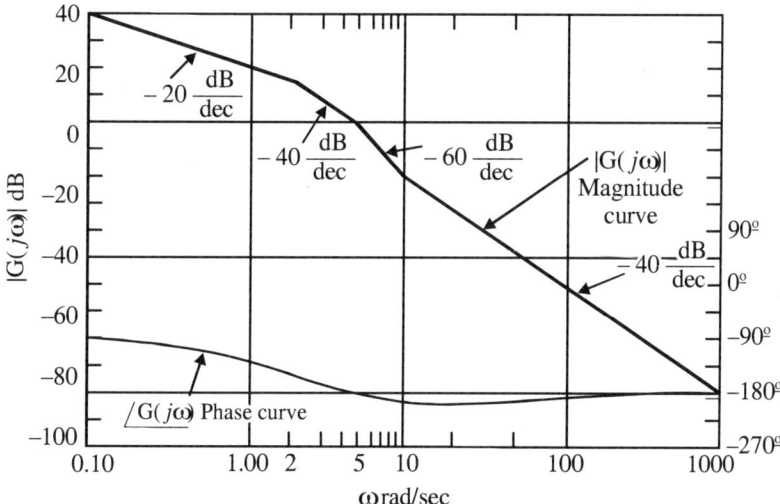

Fig. 11.32. Bode plot of $G(j\omega) = \dfrac{10(1+j0.1\omega)}{j\omega(1+j0.5\omega)(1+j0.2\omega)}$

3. Selecting the frequency at the origin as 0.1, the initial value is

$G_{dB} = 20\log K/\omega = 20\log 10/0.1 = 40\ dB$.

4. The slope up to $\omega = 2$ is $-20\ dB$/dec, from $\omega = 2$ to $\omega = 5$, it $-40\ dB$/dec; from $\omega = 5$ to $\omega = 10$, it is $-60\ dB$/dec; for $\omega > 10$, the slope is $-40\ dB$/dec.

5. The initial phase angle is $-90°$. The final phase angle is $-180°$.

6. Since the zero is the last corner frequency, the phase angle will increase beyond $-180°$ and then tend towards $-180°$.

The Bode plot is shown in Fig. 11.32.

Example 11.10. *Sketch the polar plot the transfer function*

$$G(s) = 10(s + 10)/s(s + 2)(s + 5).$$

Solution:

The sinusoidal transfer function is:

$$G(j\omega) = 10(1 + j0.1\omega)/j\omega(1 + j0.5\omega)(1 + j0.2\omega)$$

Poles are at 0, -2 and -5 and zero at -10. This being a Type 1 system, the polar plot starts at $-90°$ with infinite magnitude. As the zero at $\omega = 10$ is the last corner frequency, the phase angle decreases smoothly, crosses the $-180°$ and finally tends to $-180°$ with zero magnitude. The location of poles and zeros is shown in Fig. 11.33(a) and the polar plot in Fig 11.33(b)

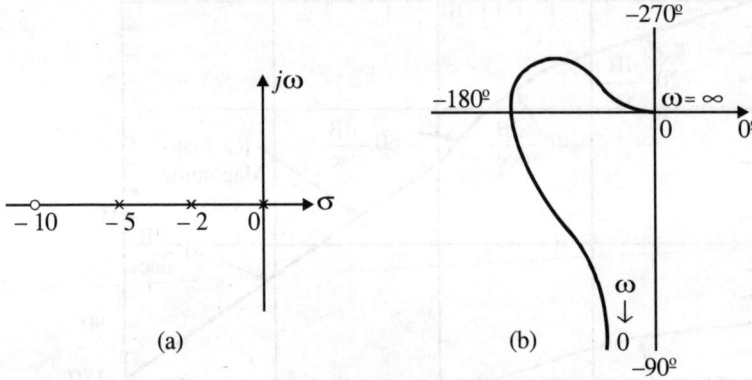

Fig. 11.33. (a) Pole-zero locations (b) Polar plot

Example 11.11. *Draw the log-magnitude versus phase plot for the transfer function*

$$G(j\omega) = 10(1 + j0.1\omega)/j\omega(1 + j0.5\omega)(1 + j0.2\omega)$$

Solution:

$$|G(j\omega)| = 10(1 + 0.01\omega^2)^{1/2}/\{\omega[(1 + 0.25\omega^2)(1 + 0.04\omega^2)]^{1/2}\}$$

Hence

$$G_{dB} = 20\log |G(j\omega)|$$

The phase angle is given by

$$\phi = \tan^{-1}0.1\omega - 90° - \tan^{-1}0.5\omega - \tan^{-1}0.2\omega$$

G_{dB} and ϕ are calculated at significant frequencies, say 0.1, 0.2, 0.5, 2, 5, 10, 20 and 100. The log-magnitude verses phase plot is shown in Fig. 11.34.

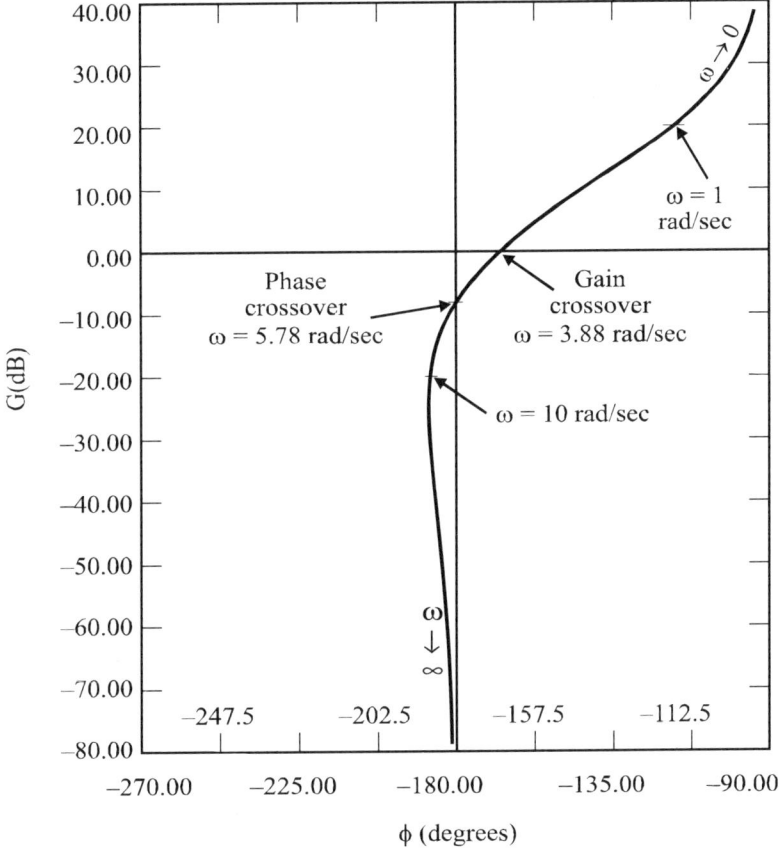

Fig. 11.34. Lm-ϕ diagram of $G(j\omega) = \dfrac{10(1+j0.1\omega)}{j\omega(1+j0.5\omega)(1+j0.2\omega)}$

Example 11.12. *Evaluate the transfer function for the plot shown in Fig 11.35.*

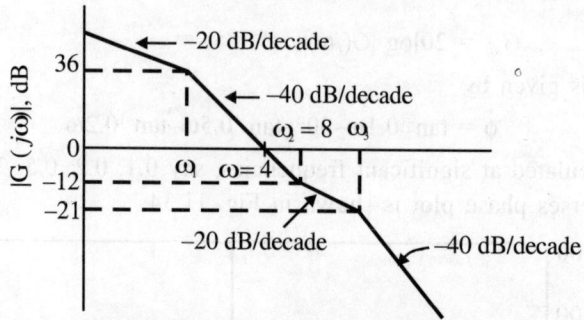

Fig. 11.35. G_{dB} plot for Example 11.12

Solution:

(1) The initial slop is $-20dB/$dec. Hence Type 1 system.

(2) The change in slope at ω_1 and ω_3 is $-20dB/$dec. Hence they are poles.

(3) At $\omega_2 = 8$, the change in slope is $+20dB/$dec. Hence it is a zero.

The transfer function is therefore written as:

$$G(j\omega) = K(1 + j\omega/8)/[j\omega(1 + j\omega/\omega_1)(1 + j\omega/\omega_3)]$$

(4) To find ω_1, we make use of $G(j\omega)$ at $\omega = 4$. The slope in this region is $-40dB/$dec. Therefore,

$$36 = -40\log \omega_1$$
$$0 = -40\log 4$$

Subtraction of the above two equations yields

$$36 = 40\log 4 - 40\log \omega_1$$
$$= 40 \log 4/\omega_1$$
$$\log 4/\omega_1 = 0.9$$
$$4/\omega_1 = 8$$
$$\omega_1 = 0.5$$

(5) At low-frequencies, $G(j\omega)$ may be approximated as $G(j\omega) = K/j\omega$

At $\omega = \omega_1 = 0.5$,

$$G(j\omega_1)_{dB} = 36 = 20\log K/0.5$$
$$= 20 \log 2K$$
$$\log 2K = 1.8$$
$$2K = 63$$
$$K = 31.5$$

(6) To find ω_3, first determine $G(j\omega)_{dB}$ at $\omega = 8$.

$$0 = -40\log 4$$

$$x = -40\log 8$$

Subtracting the above two equations, we set

$$-x = 40\log 8/4$$

$$= 40\log 2 = 12dB$$

The slope between $\omega = 8$ and $\omega = \omega_3$ is $-20dB/dec$.

$$-12 = -20\log 8$$

$$-21 = -20\log \omega_3$$

Subtracting the above two equations, we set

$$9 = 20\log \omega_3/8$$

$$\log \omega_3/8 = 0.45$$

$$\omega_3/8 = 2.8/18$$

$$\omega_3 = 22.5.$$

Hence the transfer function is

$$G(j\omega) = 31.5(1 + j\omega/8)/[j\omega(1 + j\omega/0.5)(1 + j\omega/22.5)]$$

$$= 31.5(1 + 0.125j\omega)/[j\omega(1 + 2j\omega)(1 + 0.044j\omega)]$$

Example 11.13. *Determine the value of the error coefficient K_a from Fig. 11.36.*

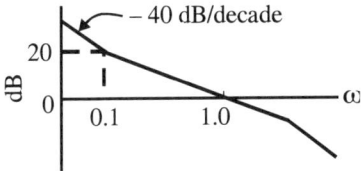

Fig. 11.36. G_{dB} plot for Example 11.13

Solution:

The initial slope is $-40dB/dec$. It is a Type 2 system. When the initial slope is extended, it will intersect the 0 dB-axis at $\omega = \omega_x$. Hence $K_a = \omega_x^2$.

$$20 = -40\log 0.1$$

$$0 = -40\log \omega_x$$

Subtracting the two equations, we get

$$20 = 40\log \omega_x/0.1$$

$$\log 10\omega_x = 0.5,$$

$$10\omega_x = 3.16$$
$$\omega_x = 0.316$$
$$K_a = \omega_x^2 = 0.1$$

Example 11.14. *Draw the Bode plot for the transportation lag system with*

$$G(s)H(s) = 10(1 + s)e^{-0.3s}/[s(1 + 0.2s)]$$

Solution:

The magnitude and phase of $e^{-0.3s}$ are:

$$|e^{-0.3s}| = 1,$$
$$\angle e^{-0.3s} = -0.3\omega$$

Hence magnitude plot is for

$$G(j\omega) = 10 (1 + j\omega)/[j\omega(1 + 0.2j\omega)]$$

This is a Type 1 system. Poles are at 0, −0.2 and a zero at −1.

(1) Starting point at $\omega = 0.1$ is 20log 10/0.1 = 40*dB*.

(2) Initial slope is −20*dB*/dec. up to $\omega = 0.2$. Then, it is −40*dB*/dec. up to $\omega = 1$. Thereafter, the slope is −20*dB*/dec.

(3) Final slope = −20 (2–1) = −20*dB*/dec.

(4) $\phi = \tan^{-1}\omega - 0.3\omega - 90° - \tan^{-1}0.2\omega$. Compute ϕ at $\omega = 0.1, 0.2, 1, 2, 5$ and 10 and plot.

The plotting of the magnitude and phase angle plots is left as an exercise to the reader.

11.23. SUMMARY

In this chapter, we have brought out the advantages of frequency response approach of system analysis. The frequency response of a system has been defined as the steady-state response of the system to a sinusoidal input signal of constant magnitude but varying frequency. The typical frequency response is the Bode plot. It consists of two plots: (i) magnitude plot and (ii) phase angle plot. We have also defined and explained the frequency response specifications such as peak resonance M_p, the resonant frequency ω_p, the band-width BW and the cut-off rate. The Bode plot has been explained for the various factors of the open-loop transfer function $G(j\omega)H(j\omega)$. The ease of obtaining asymptotic Bode plot has been explained with suitable illustrative examples. The relationship between the static error coefficients K_p, K_v, K_a and the Bode diagram has been clearly brought out. We have also explained the correlation between step response and frequency response. We have also discussed the Bode plot of non-minimum phase systems. We have discussed the polar plots for Type 1, 2 and 3 systems. We have illustrated the method of obtaining Lm-ϕ diagram of $G(j\omega)$ and obtaining the transfer function from the Bode plot.

REVIEW QUESTIONS

11.1. Define the frequency response of a system.

11.2. Mention four advantages of frequency response approach of system analysis.

11.3. What are the disadvantages of frequency response analysis?

11.4. What is the standard test signal used for determining the frequency response?

11.5. State the frequency domain specifications of a system.

11.6. What is an octave and a decade?

11.7. What are the four basic types of factors in a general transfer function?

11.8. Explain the Bode-plot.

11.9. Explain the relationship between system Type and initial slope and initial phase angle in a Bode-plot.

11.10. Explain the polar plot.

11.11. Sketch the polar plot of K, $j\omega$ and $(1/j\omega)$.

11.12. Sketch the polar plot of $(1 + j\omega T)$ and $1/(1 + j\omega T)$.

11.13. Explain Lm-ϕ diagram.

EXERCISE

11.1. What is meant by frequency response of a system? What are the advantages of frequency response approach?

11.2. Explain the frequency domain specifications.

11.3. Derive expressions for the resonance peak and resonant frequency of a second – order system when subjected to sinusoidal excitation.

11.4. Explain the correlation between resonance peak and bandwidth of a system and the time domain performance parameters?

11.5. What are minimum and non-minimum phase systems? Give suitable examples.

11.6. Sketch the polar plots of typical Type 0, 1 and 2 systems and explain the salient features of these plots.

PROBLEMS

11.1. Draw the Bode plot of the transfer function $G(s) = 1/(1 + s\tau)$.

11.2. For the function shown below, draw the Bode plot.

$G(s) = 10(1 + 2s)/[s(s + 1)(s^2 + 2s + 4)]$.

11.3. Draw the Bode plot for the following system

$G(s)H(s) = 25(s + 1)/[s(s + 4)(s + 7)]$.

11.4. Sketch the Bode plot for the unity feedback system with

$G(s) = (1 + 0.2s)(1 + 0.025s)/[s(1 + 0.001s)(1 + 0.005s)]$

11.5. A system has $G(s) = K(s + 10)/[s(s + 4)(s + 80)]$. The corresponding asymptotic magnitude Bode diagram passes through $|G(j\omega)| = 0.5$ at $\omega = 60$. Find K.

11.6. Draw the Bode plot for the function $G(s) = s^2/(1 + 0.02s)(1 + j0.28)$.

11.7. Draw the Bode plot for the fnction

$G(s)H(s) = (1 + 0.2s)(1 + 0.025s)/[s^3(1 + 0.1s)(1 + 0.005s)]$.

11.8. Draw the Bode plot for $G(j\omega)H(j\omega) = 5(1 + j0.1\omega)/\{j\omega(1 + j0.5\omega)[1 + j0.6\omega + (j\omega/50)^2]\}$.

11.9. Construct the asymptotic Bode plot for $G(s)H(s) = 36(s + 2)/[s^2(s + 4)(s + 6)]$.

11.10. Draw the asymptotic Bode plot for $G(j\omega)H(j\omega) = [1 + (j\omega/2) + (j\omega/2)^2]/[j\omega(1 + j\omega/0.5)(1 + j\omega4)]$.

11.11. Construct the asymptotic Bode plot for $G(s) = (1 + sT_a)(1 + sT_b)/[(1 + sT_1)(1 + sT_2)]$; $(T_1 > T_a > T_b > T_2)$.

11.12. Evaluate the transfer function for the plots shown in Fig. P. 11.12.

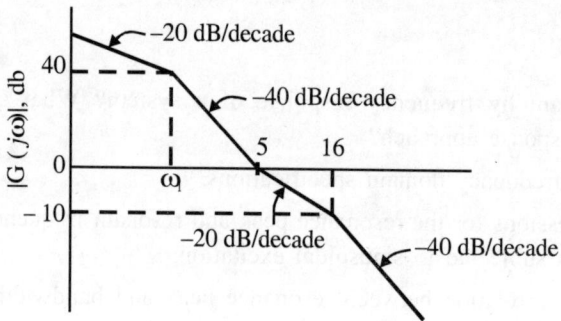

Fig. P. 11.12

11.13. Evaluate the transfer function for the plot shown in Fig. P. 11.13.

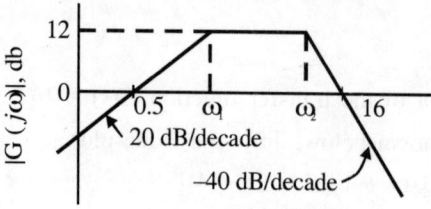

Fig. P. 11.13

11.14. Evaluate the error coefficients for the plot in Fig. P. 11.14 and obtain the transfer function.

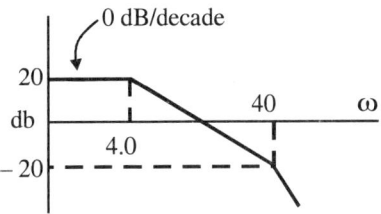

Fig. P . 11.14

11.15. Determine the error coefficient for the plot shown in Fig. P. 11.12.

11.16. Find the error coefficient for the plot shown in Fig. P. 11.13.

11.17. Find the error coefficient for the plot shown in Fig. P. 11.14.

11.18. A control system with unity feedback has the forward transfer function $G(s) = 1/[s (1 + 0.1s)](1 + 30s/625)$. Plot the polar plot and determine all the key points.

11.19. Sketch the polar plot for $G(s) = 1/[s^3(s + 1)]$ and determine all key points.

11.20. Prove that the polar plot of the function $G(s) = s/(1 + s)$ is a semicircle. Find its centre and radius.

Chapter 12

SYSTEM STABILITY IN FREQUENCY DOMAIN

12.1. INTRODUCTION

A system designer must be sure that the closed-loop system designed is stable. If the system is stable, it is further necessary to investigate the relative stability. The procedures so far considered for assuring the closed-loop system stability do not lead readily to practical design methods. Factoring the characteristic equation in order to determine the locations of poles is time consuming and impossible to carry out in terms of general parameter values. The Routh-Hurwitz criterion has similar limitations with the additional restriction that, although the number of poles in each half plane is indicated, the exact locations of these poles can be obtained only through extensive additional work. This method does not lead to a useful design procedure. The ideal design method should be relatively simple to apply, should directly lead to straight forward rules for system design, and should provide a sound intuitive understanding of changes in the closed-loop performance due to changes in system parameters. The Nyquist stability criterion satisfies these objectives and provides basis for classical design methods.

The Nyquist stability criterion is a frequency domain method and a graphical procedure for illustrating the system stability. It has the following features that make it desirable for the system analysis and design.

1. It provides the same information on the absolute stability of a system as the Routh-Hurwitz criterion.

2. It also indicates the relative stability of a stable system and is useful for determining suitable approaches to improve the relative stability of the system.

3. It gives information on the frequency-domain response of the system.

4. It can be used for the stability study of system with time delay.

This frequency domain stability criterion was developed by H. Nyquist in 1932 and is a fundamental approach to the investigation of stability of linear systems. The Nyquist stability criterion is based upon Cauchy's theorem in the Theory of Complex Variables. This theorem is concerned with the mapping of contours in the complex s-plane. This theorem can be understood without a formal proof.

The Nyquist stability criterion provides a graphical method for determining the closed-loop system stability from the frequency response curves of open-loop transfer function $G(j\omega)H(j\omega)$. Before embarking on the development of the Nyquist criterion, let us summarize the pole-zero relationships with respect to system function.

Let
$$G(s) = N_1/D_1 \text{ and } H(s) = N_2/D_2$$

Then ,
$$G(s)H(s) = N_1N_2/D_1D_2$$
$$F(s) = 1 + G(s)H(s) = (N_1N_2 + D_1D_2)/D_1D_2 \qquad \ldots (12.1)$$
$$C(s)/R(s) = N_1D_2/(N_1N_2 + D_1D_2)$$

It is seen from the above equations that

(i) The poles of $F(s)$ are the poles of $G(s)H(s)$

(ii) The poles of $C(s)/R(s)$ are the zeros of $F(s)$

For a stable system, the roots of the characteristic equation $F(s) = 1 + G(s)H(s) = (N_1N_2 + D_1D_2)/D_1D_2$ should not lie in the right-half s-plane or on the $j\omega$-axis. Or, none of the zeros of $F(s)$ can lie in the right-half s-plane or on the $j\omega$-axis. The Nyquist stability criterion relates the number of zeros and poles of $F(s)$ that lie in right-half s-plane of the polar plot of $G(s)H(s)$. We have assumed that $G(s)H(s)$ is represented as a ratio of two polynomials in s. For physically realizable systems, the degree of the numerator of $G(s)H(s)$ must be less than or equal to that of the denominator polynomial. This means that, as s approaches infinity, $G(s)H(s)$ approaches zero or a constant for physically realizable systems.

12.2. CONFORMAL MAPPING

We are concerned with the mapping of contours in the s-plane by the characteristic function $F(s)$. Since s is a complex variable, $F(s)$ is also complex. According to Cauchy's theorem, every point on the contour Γ_s is mapped into a corresponding point in the $F(s)$ plane. A closed path Γ_s in the s-plane is mapped into a closed path Γ_F in the $F(s)$ plane. Fig. 12.1(a) shows some of the zeros and poles of the characteristic equation $F(s)$. Consider an arbitrary closed path Γ_s which encloses the zero z_1. To any point s_1 on Γ_s, draw vectors from all zeros and poles. The lengths of these vectors are given by $s_1 + z_1$, $s_1 + p_1$ and $s_1 + p_2$. As the point s_1 is moved along the contour Γ_s in the prescribed counter-clockwise direction until it returns to the starting point, the vector $s_1 + z_1$ rotates through a net angle of 360° counterclockwise. All other vectors that are not enclosed by the contour Γ_s rotate through a net angel of 0°. The corresponding contour Γ_F in the $F(s)$ plane is shown in Fig. 12.1(b). The contour Γ_F encircles the origin once in the counterclockwise direction since only one zero is enclosed by Γ_s. If Z zeros are enclosed

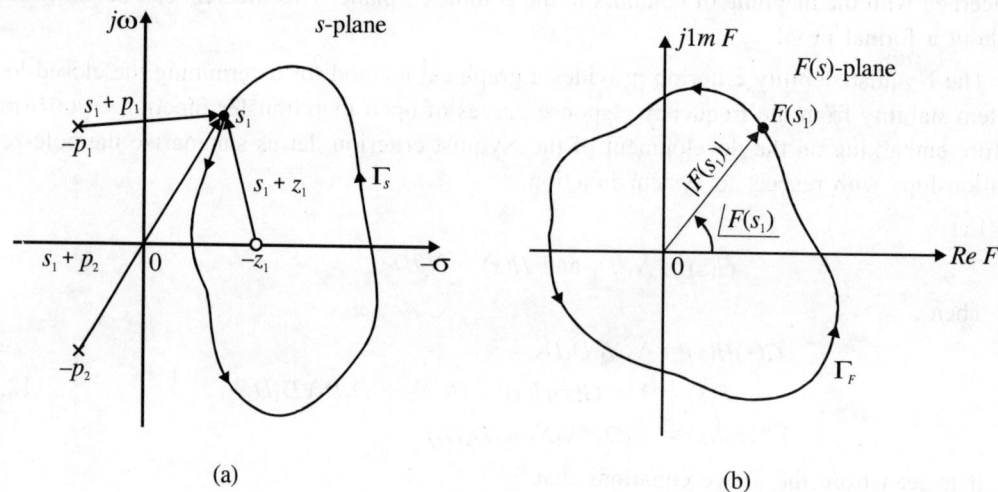

Fig. 12.1. (a) Arbitrary s-plane locus Γ_s (b) Corresponding $F(s)$ plane locus Γ_F

within Γ_s, the net angle is $2\pi Z$ radians. Similarly, if P poles are enclosed within Γ_s, the net angle is $2\pi P$ radians. The angular rotation of the pole is experienced by the characteristic equation in its denominator. Therefore, if Z zeros and P poles are enclosed by Γ_s, then the net angular rotation N is given by

$$2\pi N = 2\pi Z - 2\pi P$$

or

$$N = Z - P \qquad\qquad \ldots (12.2)$$

Thus, the number of encirclements of the origin by Γ_F in the $F(s)$–plane is equal to the number of zeros enclosed minus the number of poles enclosed.

The principle of argument can now be stated as:

If F(s) is a single-valued rational function that is analytic in a given region in the s-plane except at finite number of points (i.e., zeros and poles) and if a contour Γ_s in the s-plane encloses Z zeros and P poles of F(s) and does not pass through any zeros or poles of F(s), then the corresponding contour Γ_F in the F(s)-plane will encircle the origin of the F(s)-plane N = (Z – P) times in the same direction.

For the example shown in Fig. 12.1(a), the contour Γ_F in the $F(s)$–plane encircles the origin once since $N = Z - P = 1$. As a second example, consider the pole-zero patterns shown in Fig. 12.2(a).The contour Γ_s encloses three zeros and one pole. Hence $N = 3 - 1 = 2$. The contour Γ_F in the $F(s)$-plane shown in Fig. 12.2(b), encircles the origin twice in the counterclockwise direction.

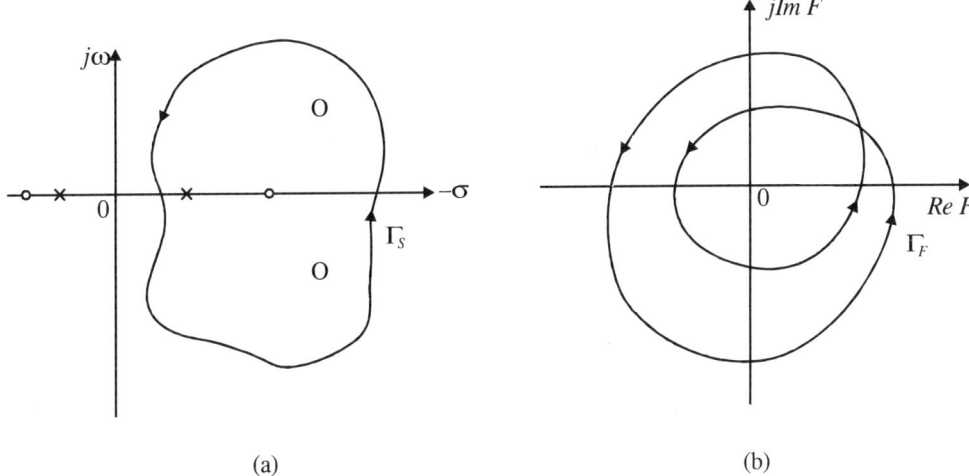

Fig. 12.2: (a) Locus Γ_s encircles 3 poles and one zero (b) Corresponding Γ_F

The following three cases arise:

Case 1: N is positive ($N > 0$). If Γ_s encloses more zeros than poles of $F(s)$ in a prescribed direction, N is positive. In this case, Γ_F will encircle the origin in the $F(s)$-plane N times in the same direction.

Case 2: N is Zero ($N = 0$). If Γ_s encloses as many zeros as poles, or, no zeros and poles of $F(s)$, then $N = 0$. The contour Γ_F will not encircle the origin in the $F(s)$-plane.

Case 3: N is negative ($N < 0$). If Γ_s encloses more poles than zeros of $F(s)$ in a prescribed direction, then N is negative. In this case, Γ_F will encircle the origin in the $F(s)$-plane N times in the opposite direction.

Now that we have illustrated the concept of mapping of contours with the help of the characteristic function $F(s)$ and developed the mathematical basis, we shall consider below the Nyquist stability criterion.

12.3. NYQUIST STABILITY CRITERION

In order to investigate the stability of control systems, we must consider the characteristic equation

$$F(s) = 1 + G(s)\ H(s) = 0 \qquad \qquad ...\ (12.3)$$

For a system to be stable, all the zeros of $F(s)$ must lie in the left-half s-plane. In other words, none of the zeros of $F(s)$ must lie in the right-half s-plane, It is therefore necessary to determine whether any of the zeros of $F(s)$ lies in the right-half s-plane. Hence, we choose a contour Γ_s in the s-plane which encloses the entire right-half s-plane, and we determine whether any zeros of $F(s)$ lies within Γ_s using Cauchy's theorem. That is, we plot Γ_F contour in the

$F(s)$–plane and determine the number of encirclements of the origin. The number of poles of $G(s)H(s)$ in the right-half s-plane is usually known since $G(s)H(s)$ is generally in the factored form. Therefore, the number of encirclements of the origin in the $F(s)$-plane is equal to

$$N = Z_R - P_R \qquad\qquad\qquad \text{... (12.4)}$$

where Z_R is the number of zeros of $F(s)$ in the right-half s-plane and P_R is the number of poles of $G(s)H(s)$ in the right half s-plane.

The contour Γ_s which encloses the entire right-half s-plane in the counterclockwise direction is shown in Fig. 12.3. This closed path is known as *Nyquist path*. Since the Nyquist path must not pass through any poles or zeros of $F(s)$, if there are poles or zeros on the $j\omega$–axis, they are avoided by taking a detour along a small semicircle with radius $r \to 0$ around them. Fig. 12.3 shows the Nyquist path when $G(s)H(s)$ has a pole or poles at origin. If $G(s)H(s)$ has no singularity at the origin, the small semicircle merges with the $j\omega$-axis.

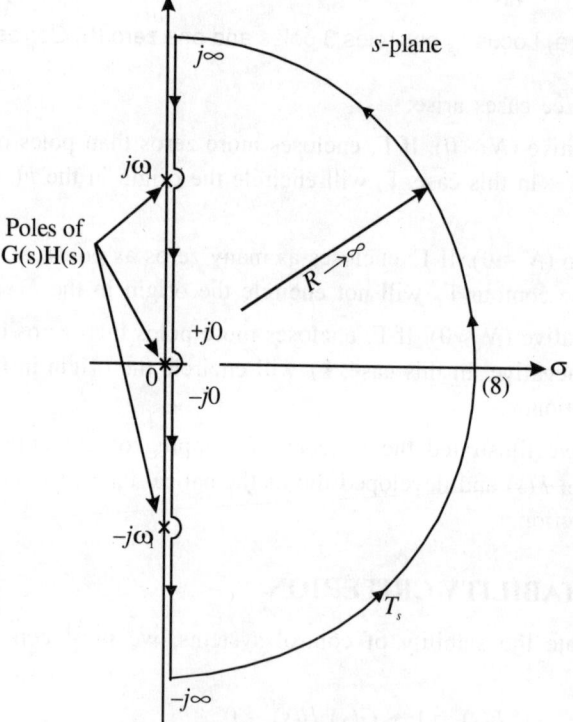

Fig. 12.3. Typical Nyquist path

For convenience, the Nyquist path is divided into a number of sections depending upon how many small semicircles are necessary on the $j\omega$-axis. For the case illustrated in Fig. 12.3, a total of four sections need to be defined.

Section I: From $s = + j\infty$ to $+ j0$ along the $j\omega$-axis.

Section II: From $+ j0$ to $- j0$ along the small semicircle around $s = 0$.

Section III: From $- j0$ to $- j\infty$ along the $j\omega$-axis. This is the mirror image of Section I.

Section IV: From $s = - j\infty$ to $+ j\infty$ along the semicircle of infinite radius.

In cases where $G(s)H(s)$ has no singularity (pole or zero) at the origin, there will be only three sections: Section I, III and IV. If the Nyquist path is specified, Γ_F may be plotted as Γ_s takes values along the Nyquist path. The Γ_F contour thus obtained is known as *Nyquist plot.* The stability of the system can be determined from Eq. (12.4). However, since $F(s) = 1 + G(s)H(s)$ is not generally available in factored form, it is difficult to obtain the Nyquist plot of $F(s)$, whereas the open-loop transfer function $G(s)H(s)$ is generally known and is typically available in factored form. Hence, it is simpler to construct the Nyquist plot of $G(s)H(s)$.

From Eq. (12.3),

$$G(s)H(s) = - 1 \qquad \qquad ... (12.5)$$

Then, the mapping of Γ_s in the s-plane will be through the function $G(s)H(s)$ into the $G(s)H(s)$-plane. In this case, the number of encirclements of the origin in the $F(s)$-plane becomes the number of encirclements of the $(-1, j0)$ point in the $G(s)H(s)$-plane. This is illustrated in Fig. 12.4. The $(-1, j0)$ point is called the *critical point.* With the Nyquist plot in the $G(s)H(s)$-plane, we can obtain the stability of the open-loop transfer function $G(s)H(s)$ (*the open-loop stability*), and the stability of closed-loop transfer function $C(s)/R(s)$ (*the closed-loop stability*). The closed-loop stability of a system can be obtained by applying the Nyquist stability criterion to the Nyquist plot in the $G(s)H(s)$ with the $(-1, j0)$ as the critical point, whereas the open-loop stability is obtained with the origin as the critical point. In the event of $G(s)H(s)$ has some poles in the right-half s-plane and $G(s)H(s)$ is not in the factored form, then the number of poles P_R of $G(s)H(s)$ in the right-half s-plane can obtained from the open-loop stability.

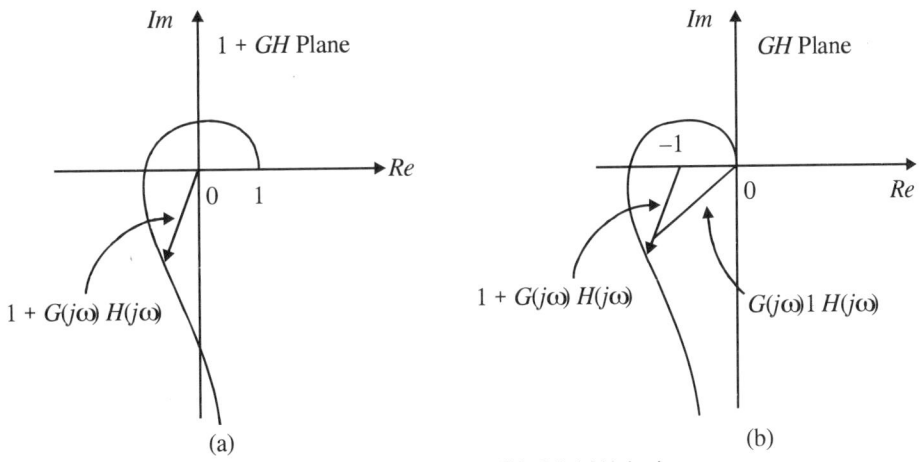

Fig. 12.4. (a) F(s) plane; (b) G(s) H(s) plane

The Nyquist stability criterion may be stated as follows:

A closed-loop system is stable if and only if the Nyquist plot of G(s)H(s) encircles the (-1, jo) point as many times as the number of poles of G(s)H(s) that are in the right-half s-plane, and the encirclements, if any, must be made in the clockwise direction.

Since $N = Z_R - P_R$, for the closed-loop system to be stable, Z_R should equal to zero. Hence $N = -P_R$. Thus the number of encirclements must be equal to P_R and in the clockwise direction.

$Z_R - P_R$ are zeros and poles in the right half s-plane.

In general, the open-loop transfer functions $G(s)H(s)$ of many physical systems do not have any poles in the right-half s-plane. Hence $P_R = 0$ and $N = Z_R$. In this case, the Nyquist stability criterion may be stated thus:

If G(s)H(s) has no poles in the right-half s-plane, for the closed-loop system to be stable, the Nyquist plot of G(s)H(s) must not encircle the critical point (−1, j0).

When the Nyquist plot of $G(s)H(s)$ passes through the critical point $(-1, j0)$, the number of encirclement N is indeterminate. This corresponds to the condition where the characteristic equation $F(s)$ has zeros on the $j\omega$-axis or the closed-loop transfer function $C(s)/R(s)$ has poles on the $j\omega$-axis. Simple imaginary zeros of $F(s)$ (or poles of $C(s)/R(s)$) mean that the closed-loop system is oscillatory in the steady-state and the system is considered to be unstable.

12.4. PROCEDURE FOR APPLYING NYQUIST STABILITY CRITERION

The closed-loop stability of a system is determined by the properties of the Nyquist plot of the open-loop transfer function $G(s)H(s)$. The procedure for determining the stability of closed-loop system is summarized below:

Step 1. Given the closed-loop transfer function of a system, determine the open-loop transfer function $G(s)H(s)$.

Step 2. According to pole-zero locations of $G(s)H(s)$, define the Nyquist path.

Step 3. Draw the polar plot corresponding to $G(s)H(s)$ as ω varies from $+\infty$ to 0 (Section I) and mark the direction.

Step 4. Draw its mirror image about the real axis. This corresponds to Section III of the Nyquist path. Mark the direction accordingly.

Step 5. To draw Section II of the Nyquist path,

    (i) Obtain the low-frequency approximation of $G(s)H(s)$.

    (ii) Substitute $s = \varepsilon e^{j\theta}$ in the resulting equation. As ε tends to zero, the magnitude tends to infinity.

    (iii) Allow θ to vary from $+90°$ to $-90°$ along the small semicircle and find the corresponding variation in the angle ϕ.

     (iv) Draw the plot for this section starting from $\omega = +0$ to $\omega = -0$ with infinite radius locus in the direction indicated by the variation in ϕ and mark the direction.

Step 6. To draw Section IV:

     (i) Obtain the high-frequency approximation of $G(s)H(s)$.

     (ii) Substitute $s = Re^{j\theta}$ in the resulting equation. As R tends to ∞, the magnitude tends to zero.

     (iii) Allow θ to vary from $-90°$ to $+90°$ along the infinite semicircle and find the corresponding variation in the angle ϕ.

     (iv) Complete the plot starting from $\omega = -\infty$ to $\omega = +\infty$ with small radius locus in the direction indicated by the variation in ϕ and mark the direction.

Step 7. Draw a vector from the critical point $(-1, j0)$ to any point on the Nyquist plot.

Step 8. Determine the number of encirclements N of the point $(-1, j0)$ as the vector is moved along the direction indicated on the plot.

Step 9. Since the number of poles of $G(s)H(s)$ is usually known, find Z_R from the relation $N = Z_R - P_R$.

Step 10. If Z_R is zero, the system is stable. Otherwise, it is unstable.

In examining the stability of linear systems using Nyquist stability criterion, the following three possibilities can occur.

    (a) *N is positive*: This is the case if $Z_R > P_R$. This means the closed-loop system has poles in the right-half s-plane. The system is *unstable*.

    (b) *N = 0*: There is no encirclement of the point $(-1, j0)$. This is the case where $Z_R = P_R$. The closed-loop system is stable if and only if $P_R = 0$. Otherwise, the system is unstable.

    (c) *N = Negative*: The encirclement of the point $(-1, j0)$ is clockwise in this case. This implies either $Z_R = 0$ or $Z_R < P_R$. If the number of encirclement of the point $(-1, j0)$ is equal to the number of poles P_R of $G(s)H(s)$ in the right-half s-plane, the closed-loop system is stable. Otherwise the system is unstable.

In the following section, the application of the Nyquist stability criterion is illustrated with examples.

12.5. APPLICATION OF NYQUIST STABILITY CRITERION

Several illustrative examples are considered in this section. The Nyquist plots can be obtained with the aid of the pole-zero diagrams and the Nyquist path. Both minimum and non-minimum phase systems are llustrated.

12.5.1. Minimum–Phase Systems

Example 12.1. *Draw the polar plot of the following Type 0 system with open-loop transfer function*

$$G(s)H(s) = K/[(1 + sT_a)(1 + sT_b)] \qquad \qquad ...(12.6)$$

with T_a and T_b positive. Investigate the system stability using Nyquist criterion.

Solution:

The pole-zero location and the Nyquist path are shown in Fig 12.5(a) and the Nyquist plot in 12.5(b) respectively. Since there are no poles or zeros at the origin, the path has three sections only.

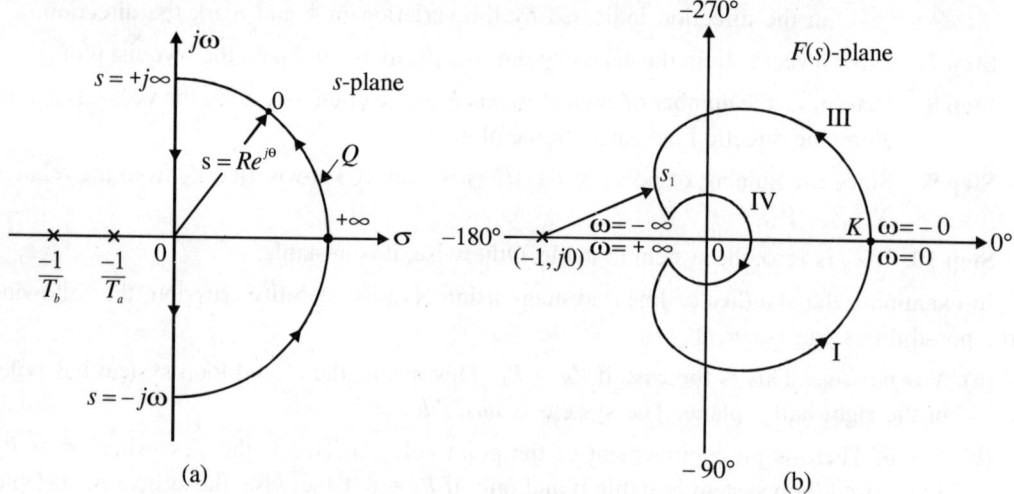

Fig. 12.5. (a) Pole-zero locations and Nyquist path (b) Nyquist plot of
$$G(s)H(s) = K/[(1+sT_a)(1+sT_b)]$$

(a) Polar plot corresponding to Sections I and III: First the polar plot for Section I is drawn as shown in Fig. 12.5. For Section III, the polar plot is the mirror image of that for Section I. This is also shown in Fig. 12.5.

(b) Polar plot corresponding to Section IV: For drawing the polar plot of Section IV, we obtain the high-frequency approximation of Eq. (12.6).

$$GH_{hF} = \lim_{s \to \infty} G(s)H(s) = K/(s^2 T_a T_b) \qquad \qquad ...(12.7)$$

Substituting $s = Re^{j\theta}$ in Eq. (12.7), we obtain

$$GH_{hf} = K/(T_a T_b\, R^2 e^{j\theta}) = K_1 e^{-j2\theta} = K_1 e^{j\phi}$$

As θ varies form $-90°$ to $+90°$ along the infinite radius, we tabulate the phase angle ϕ for various values θ in Table 12.1.

Table 12.1. θ *vs* φ for high frequency

θ	φ = −2θ
−90°	180°
−60°	120°
−45°	90°
−30°	60°
0°	0°
+30°	−60°
+45°	−90°
+60°	−120°
+90°	−180

Thus, Section IV is mapped into a small radius circular plot starting from $\omega = -\infty$ to $\omega = +\infty$ in the clockwise direction.

(c) System Stability: From Eq. (12.6), the value of $P_R = 0$. Therefore $N = Z_R - P_R = Z_R$. The number of encirclements N of the point $(-1, j0)$ is obtained by drawing a vector from this point to any point s_1 on the Nyquist plot and moving the *tip of the vector* once around the plot in indicated direction. The number of encirclements thus obtained is zero.

$$N = 0 = Z_R \qquad\qquad \dots (12.9)$$

Therefore, there are no zeros of $F(s)$ or no poles of $C(s)/R(s)$ in the right-half s-plane. Hence the system is stable.

Example 12.2. *Draw the polar plot of the Type 1 system with open-loop transfer function of the form*

$$G(s)H(s) = K/[s(1 + sT_a)(1 + sT_b)] \qquad\qquad \dots(12.10)$$

with T_i and T_b positive and investigate the system stability.

Solution:

The pole-zero locations and the Nyquist path are shown in Fig. 12.6(a) and the Nyquist plot in Fig. 12.6(b) respectively. Since there is a pole at the origin, it is bypassed by a small radius semicircle.

Thus the Nyquist path has 4 sections.

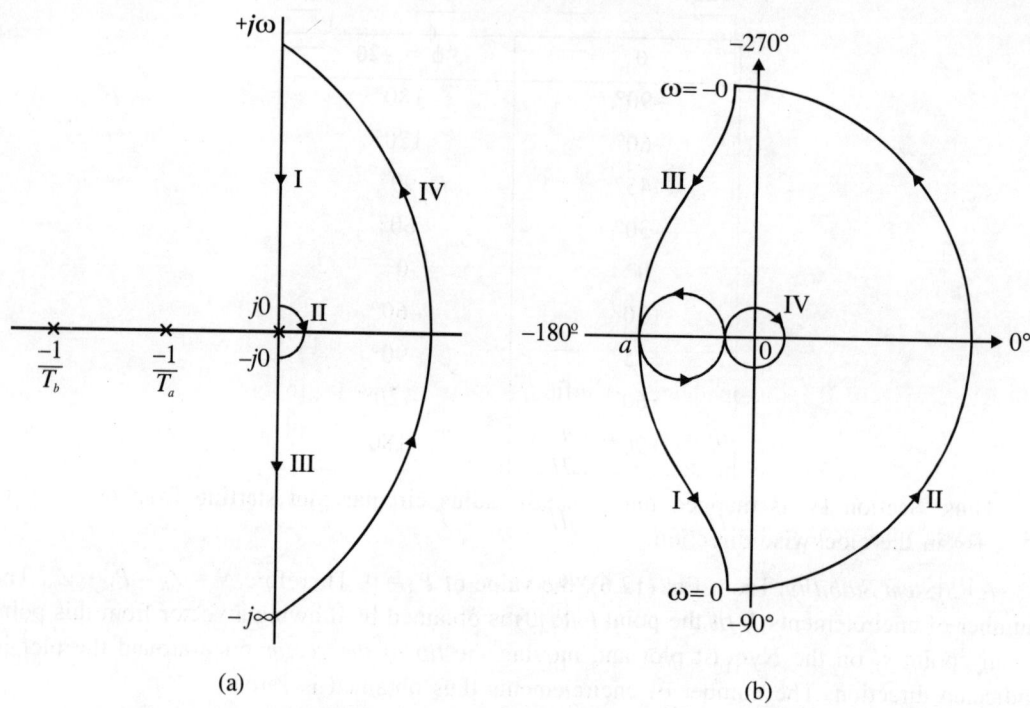

Fig. 12.6. (a) Pole-zero locations and Nyquist path (b) Nyquist plot of
$$G(s)H(s) = K/[s(1+sTa)(1+sTb)]$$

(a) Polar plots corresponding to Sections I and III: The polar plot for Section I is drawn first. Then, its mirror image about the real axis is drawn which corresponds to Section III.

(b) Polar plot corresponding to section II: To draw the polar plot of Section II, we first obtain the low-frequency approximation of $G(s)H(s)$.

$$GH_{lf} = \lim_{s \to 0} G(s)H(s) = K/s \qquad \qquad ...(12.11)$$

Substituting $s = \varepsilon e^{j\theta}$, we get

$$GH_{lf} = K/(\varepsilon e^{j\theta}) = K_1 e^{-j\theta} = K_1 e^{j\phi} \qquad \qquad ...(12.12)$$

As θ varies from $+90°$ to $-90°$ along the small radius semicircle, we tabulate in Table 12.2 the phase angle ϕ for various values of θ.

Table 12.2. θ vs φ for low frequency

θ	φ = −θ
+90°	−90°
+60°	−60°
+45°	−45°
+30°	−30°
0°	0°
−30°	+30°
−45°	+45°
−60°	+60°
−90°	+90°

Thus, Section II is mapped into an infinite radius semicircle from $\omega = +0$ to $\omega = -0$.

(c) Polar plot corresponding to Section IV: To draw the polar plot of Section IV, we obtain the high-frequency approximation of $G(s)H(s)$.

$$GH_{hf} = \lim_{s \to \infty} G(s)H(s) = K/(T_a T_b s^3) \qquad \qquad ...(12.13)$$

Substituting $s = Re^{j\theta}$ in Eq. (12.13), we get

$$GH_{hf} = K/(T_a T_b R^3 e^{j3\theta}) = K_1 e^{-j3\theta} = K_1 e^{j\phi} \qquad \qquad ...(12.14)$$

As θ varies from −90° to +90° along the infinite radius semicircle, we tabulate the values of φ in Table 12.3.

Table 12.3. θ vs φ for high frequency

θ	φ = −3θ
−90°	+270°
−60°	+180°
−45°	+135°
−30°	+90°
0°	0°
+30°	−90°
+45°	−135°
+60°	−180°
+90°	−270°

Thus, Section IV is mapped into a small radius circular plot from $\omega = -\infty$ to $\omega = +\infty$ in the clockwise direction. The complete Nyquist plot is shown in Fig. 12.6(b).

(d) System Stability: From Eq. (12.10), the value of $P_R = 0$. The stability of the system depends on the location of the point $(-1, j0)$ with respect to cross-over point a.

Case I: Let us assume that the point $(-1, j0)$ lies between the origin and a. Then the number of encirclements is $N = 2$. Therefore,

$$N = 2 = Z_R - P_R = Z_R - 0 = Z_R \qquad \qquad ...(12.15)$$

Since *two zeros* of the characteristic equation lies in the right-half s-plane, the system is *unstable*.

Case II: Let us now assume that the point $(-1, j0)$ lies between a and $-\infty$. The number of encirclements is zero. Hence,

$$N = 0 = Z_R - P_R = Z_R - 0 = Z_R \qquad \qquad ...(12.16)$$

In this case, the system is *stable*. However, the system may become unstable if the gain is increased sufficiently so that the Nyquist plot crosses the negative real axis to the left of $(-1, j0)$ point. Then, this corresponds to Case 1.

Example 12.3. *Draw polar plot of the Type 2 system with open-loop transfer function*

$$G(s)H(s) = K(1 + T_1 s)/[s^2(1 + T_a s)(1 + T_b s)(1 + T_c s)] \qquad ...(12.17)$$

where $T_1 > T_a > T_b > T_c$ *and* T_a, T_b, T_c *and* T_1 *are all positive. Investigate the system stability.*

Solution:

The pole-zero locations and the Nyquist path corresponding to Eq. (12.17) are shown in Fig. 12.7(a) and Nyquist plot is shown in Fig. 12.7(b). There are *two poles* at the origin; they are bypassed by detouring with a small radius semicircle. Thus, the Nyquist path has 4 sections.

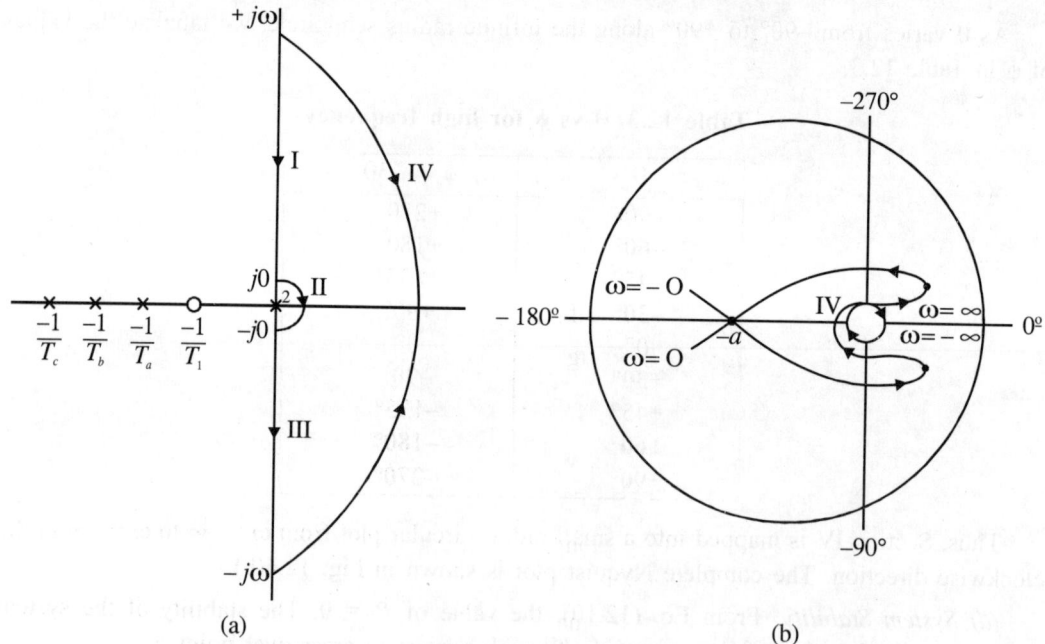

Fig. 12.7. (a) Pole-zero locations and Nyquist path (b) Nyquist plot of
$G(s) H(s) = K(1 + T_1 s)/[s^2(1 + T_a s)(1 + T_b s)(1 + T_c s)]$

(a) *Polar plots for Section I and III:* The polar plot for Section I is drawn first. Being a Type 2 system, the starting angle is $-180°$ and the magnitude is ∞. Since $T_a > T_1 > T_2 > T_3$, the zero occurs first. Therefore, the phase angle increases from $-180°$, then it decreases because of the three poles and finally ends up with $-360°$. The polar plot for Section III is the mirror image of that of Section I.

(b) *Polar Plot for Section II:* Let us obtain the low frequency approximation of $G(s)H(s)$.

$$GH_{lf} = \lim_{s \to 0} G(s)H(s) = K/s^2 \qquad \qquad ...(12.18)$$

Substituting $s = \varepsilon e^{j\theta}$, we get

$$GH_{lf} = K/(\varepsilon e^{j\theta})^2 = K_1 e^{-j2\theta} = K_1 e^{j\phi} \qquad \qquad ...(12.19)$$

As θ varies from $+90°$ to $-90°$ along the small radius semicircle, we tabulate the values of ϕ in Table 12.4.

Table 12.4. θ *vs* ϕ for low frequency

θ	$\phi = -2\theta$
$+90°$	$-180°$
$+60°$	$-120°$
$+45°$	$-90°$
$+30°$	$-60°$
$0°$	$0°$
$-30°$	$+60°$
$-45°$	$+90°$
$-60°$	$+120°$
$-90°$	$+180°$

Thus Section II is mapped into an infinite radius semicircle from $\omega = +0$ to $\omega = -0$.

(c) *Polar plot for Section IV:* The high frequency approximation of $G(s)H(s)$ is

$$GH_{hf} = \lim_{s \to \infty} G(s)H(s) = KT_1/(T_aT_bT_cs^4) \qquad \qquad ...(12.20)$$

Substituting $s = Re^{j\theta}$ in Eq. (12.20), we get

$$GH_{hf} = KT_1/[(T_aT_bT_c)(Re^{j\theta})^4] = K_1 e^{-j4\theta} = K_1 e^{j\phi} \qquad \qquad ...(12.21)$$

As θ varies from $-90°$ to $+90°$ along the infinite radius semicircle, the values of ϕ are tabulated in Table 12.5.

Thus, Section IV is mapped into a small radius circles staring from $\omega = -\infty$ to $\omega = +\infty$ in the clockwise direction. The complete Nyquist plot is shown in Fig. 12.7(b).

Table 12.5. θ *vs* φ for high frequency

θ	φ = −4θ
−90°	+360°
−60°	+240°
−45°	+180°
−30°	+120°
0°	0°
+30°	−120°
+45°	−180°
+60°	−240°
+90°	−360°

(d) System Stability: From Eq. (12.17), $P_R = 0$. The stability of the system depends on the location of the point $(-1, j0)$.

Case 1: Assume the point $(-1, j0)$ lies between the origin and the point $-a$. The number of encirclements is $N = 2$. Hence,

$$N = 2 = Z_R - P_R = Z_R.$$

The system is *unstable* since two zeros of the characteristic equation lie on the right half *s*-plane.

Case 2: Assume now the point $(-1, j0)$ lies between $-a$ and $-\infty$. The number of encirclements $N = 0$.

Hence,

$$N = 0 = Z_R - P_R = Z_R$$

Therefore, the system is *stable*. However, the system may become unstable if the gain is increased such that the Nyquist plot crosses the negative real axis to the left of $(-1, j0)$ point. This corresponds to Case 1.

Example 12.4. *Draw the polar plot of a system with the following open-loop transfer function*

$$G(s)H(s) = K(1 + T_1 s)^2 / [(1 + T_a s)(1 + T_b s)(1 + T_c s)(1 + T_d s)^2] \quad ...(12.22)$$

where $T_d < T_1 < T_c < T_b < T_a$, *and all* T_i *are positive. Investigate the system stability.*

Solution:

The pole-zero locations and the Nyquist path corresponding to Eq. (12.22) are shown in Fig. 12.8(a) and the Nyquist plot in Fig. 12.8(b) respectively. Since there are no poles at the origin, Section II need not be considered.

(a) *Polar plots for Sections I and III:* The polar plot for Section I is drawn first. Being a Type 0 system, the starting angle is 0° and the magnitude is K. First three poles due to T_a, T_b and T_c occur. Hence, the polar plot goes to the second quadrant through fourth and third in the clockwise direction. Next, we have two zeros due to T_1. Hence, plot turns towards negative real axis and enters the third quadrant. Now, the two poles due to T_d come into play. The polar plot again enters the second quadrant and terminates with $-270°$ phase angle at $\omega = +\infty$.

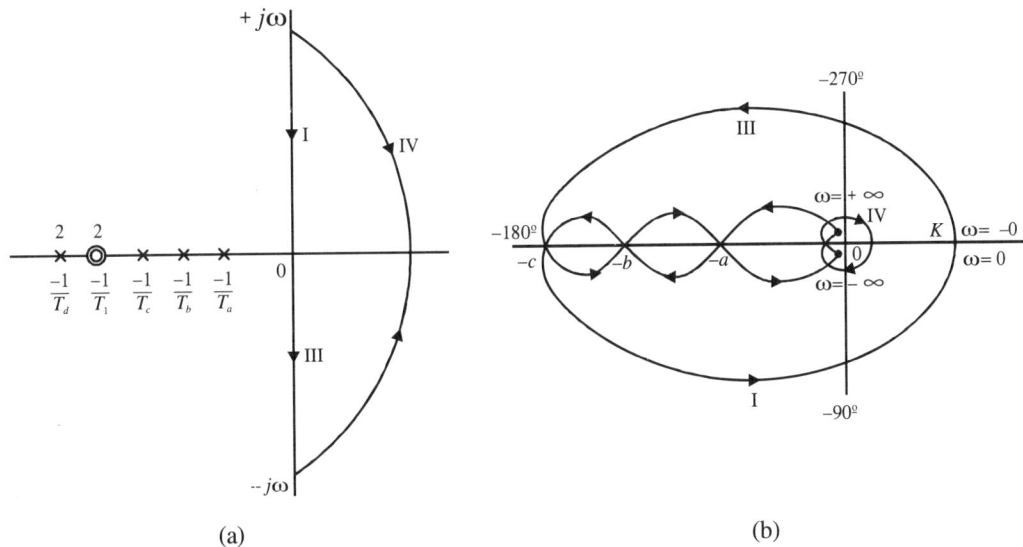

(a)   (b)

Fig. 12.8. (a) Pole-zero locations and Nyquist path (b) Nyquist pole of
$G(s)H(s) = K(1 + T_1 s)^2/[(1 + T_a s)(1 + T_b s)(1 + T_c s)(1 + T_d s)^2]$

The polar plot of Section III is the mirror image of that of Section I.

(b) *Polar plot for Section IV:* The high frequency approximation of $G(s)H(s)$ is

$$GH_{hf} = \lim_{s \to \infty} G(s)H(s) = KT_1^2/(T_a T_b T_c T_d s^3) \qquad \qquad ...(12.23)$$

Substituting $s = Re^{j\theta}$ in Eq. (12.23), we get

$$GH_{hf} = KT_1^2/[(T_a T_b T_c T_d)(Re^{j\theta})^3] = K_1 e^{-j3\theta} = K_1 e^{j\phi} \qquad \qquad ...(12.24)$$

The values of ϕ are tabulated in Table 12.6 as θ is varied from $-90°$ to $+90°$ along the infinite radius semicircle.

Table 12.6. θ *vs* φ for high frequency

θ	φ = −3θ
−90°	−270°
−60°	+180°
−45°	+135°
−30°	+90°
0°	0°
+30°	−90°
+45°	−135°
+60°	−180°
+90°	−270°

Section IV thus maps into a small radius circle starting from $\omega = -\infty$ to $\omega = +\infty$ in the clockwise direction. The complete Nyquist plot is shown in Fig. 12.8(b).

(c) System Stability: From Eq. (12.22), $P_R = 0$. The stability of the system depends on the location of $(-1, j0)$ point.

Case 1. Assume the point $(-1, j0)$ lies between the origin and point $-a$. The number of encirclements $N = 2$. Since $P_R = 0$, $Z_R = 2$. Hence the system is *unstable*.

Case 2. Let the point $(-1, j0)$ be between $-a$ and $-b$. The number of encirclements is $N = 0$. Since $P_R = 0$, $Z_R = 0$. The system is *stable*.

Case 3. Assume the point $(-1, j0)$ lies between $-b$ and $-c$. The number of encirclements $N = 2$. Since $P_R = 0$, $Z_R = 0$. The system is *unstable*.

Case 4. Let the critical point $(-1, j0)$ be between $-c$ and $-\infty$. The number of encirclements $N = 0$. Since $Z_R = 0$, the system is *stable*.

In this example, it can be seen that the system can become unstable not only by increasing the gain K but also by decreasing the gain. If K is increased sufficiently so that the critical point $(-1, j0)$ lies between the origin and $-a$ or between $-b$ and $-c$, the system is unstable. This corresponds to Case 1 or Case 3. One the other hand, the gain is decreased so that the critical point lies between $-a$ and $-b$ or $-c$ and $-\infty$, the system is stable. This corresponds to Case 2 or Case 4. Such systems are known as *conditionally stable systems*.

A conditionally stable system is one which is stable for a given range of values of gain K but becomes unstable if the gain is either increased or decreased sufficiently. Such systems place a greater restriction on the stability and system characteristics. The conditionally stable systems must be avoided, if possible, since a reduction in gain of the amplifier due to saturation and ageing may result into an unstable operation.

12.5.2. Non-Minimum Phase Systems

A non-minimum phase system is one which has at least a pole or zero in the right-half *s*-plane. We shall now consider a non-minimum phase system and apply the Nyquist criterion to find its stability.

Example 12.5. *The open-loop transfer function of a non-minimum phase system with a pole in the right-half s-plane is given by*

$$G(s)H(s) = K(1 + sT_1)/[s(sT_a - 1)] \qquad \qquad ...(12.25)$$

with T_1 and T_a positive. Draw the polar plot and investigate the system stability.

Solution:

The pole-zero locations and the Nyquist path corresponding to Eq. (12.25) are shown in Fig. 12.9(a) and the Nyquist plot in Fig. 12.9(b) respectively. Since there is a *pole at the origin*, the Nyquist path has 4 sections.

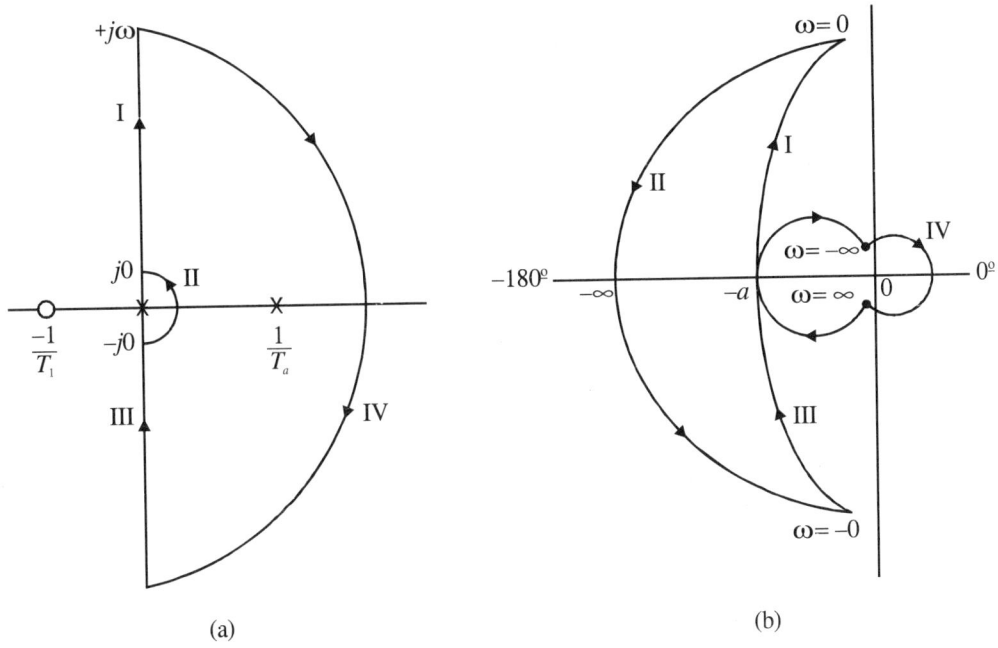

(a)   (b)

Fig. 12.9. (a) Pole-zero locations and Nyquist path (b) Nyquist plot

(a) Polar plot for Sections I and III: The polar plot for Section I is drawn first. This is a Type 1 systems and the initial phase for a minimum phase system is –90°. However, the presence of a pole-factor in the right-half s-plane adds another –180°. Hence, the initial phase for this

example is $-270°$. As ω is increased from $+0$ towards $+\infty$, the plot progresses in the third quadrant and ends up with a final phase of $-90°$ as shown in Fig. 12.9(c).

The polar plot for Section III is the mirror image of that for Section I.

(b) Polar Plot for Section II: The low frequency approximation of $G(s)H(s)$ is

$$GH_{lf} = \lim_{s \to 0} G(s)H(s) = K/(-s) \qquad \qquad ...(12.26)$$

Substituting $s = \varepsilon e^{j\theta}$ in Eq. (12.26), we get

$$GH_{lf} = K/(-\varepsilon\ e^{j\theta}) = K_1 e^{-j(180 + \theta)} = K_1 e^{j\phi}$$

As θ varies from $+90°$ to $-90°$ along the small radius semicircle, the variations of ϕ are tabulated in Table 12.7.

Table 12.7. θ *vs* ϕ for low frequency

θ	$\phi = -(180 + \theta)$
$+90°$	$-270°$
$+60°$	$-240°$
$+45°$	$-225°$
$+30°$	$-210°$
$0°$	$-180°$
$-30°$	$-150°$
$-45°$	$-135°$
$-60°$	$-120°$
$-90°$	$-90°$

Thus, Section II is mapped into an infinite radius semicircle starting from $\omega = +0$ to $\omega = -0$ in the counter-clockwise direction as shown in Fig. 12.9(c).

(c) Polar plot for Section IV: The high frequency approximation of $G(s)H(s)$ is

$$GH_{hf} = \lim_{s \to \infty} G(s)H(s) = KT_1/(T_a s) \qquad \qquad ...(12.28)$$

Substituting $s = R\ e^{j\theta}$ in Eq. (12.28), we obtain

$$GH_{hf} = KT_1/(T_a R\ e^{j\theta}) = K_1\ e^{-j\theta} = K_1\ e^{j\phi} \qquad \qquad ...(12.29)$$

As θ varies from $-90°$ to $+90°$ along the infinite radius semicircle, the values of ϕ are tabulated in Table 12.8.

Table 12.8. θ *vs* φ for high frequency

θ	φ = –θ
–90°	+90°
–60°	+60°
–45°	+45°
–30°	+30°
0°	0°
+30°	–30°
+45°	–45°
+60°	–60°
+90°	–90°

Section IV maps into a circle of small radius starting from $\omega = +\infty$ to $\omega = -\infty$ in the clockwise direction as shown in Fig. 12.9(b).

(d) System Stability: From Eq. (12.25), $P_R = 1$. This means the open-loop system with the given $G(s)H(s)$ is *unstable*. The number of encirclements of the origin is 1 in the clockwise direction because of the presence of a pole in the right-half *s*-plane. The closed-loop system stability depends on the location of the critical point $(-1, j0)$.

Case 1: Let the critical point $(-1, j0)$ be between the origin and the point $-a$. The number of encirclements of the point is $N = -1$ (clockwise). Since $P_R = 1$, $Z_R = 0$. Hence the system is *stable*. The region from origin to $-a$ is the *stable* region.

Case 2: Assume that the critical point $(-1, j0)$ lies between $-a$ and $-\infty$. The number of encirclement is $N = 1$ (counterclockwise). Since $P_R = 1$, $Z_R = 2$. The system is *unstable*. This region is the *unstable region*.

For low values of $K(0 < K < K_1)$, the system is *unstable*. For the range $K_1 < K < \infty$, the system is *stable* as this corresponds to Case 1.

12.5.3. Systems with Transportation Lag

The transfer function of transportation lag is e^{-Ts}. In the frequency domain, the transfer function is

$$G(j\omega)H(j\omega) = e^{-j\omega T} \qquad \text{...(12.30)}$$

The magnitude is *unity* and the phase angle is increasing in proportion to the frequency ω.

$$|G(j\omega)H(j\omega)| = 1 \qquad \text{...(12.31(a))}$$

$$\underline{G(j\omega)H(j\omega)} = -\omega T \qquad \text{...(12.31(b))}$$

The polar plot is a unit circle which is traced indefinitely as ω varies from zero to infinity as shown in Fig. 12.10.

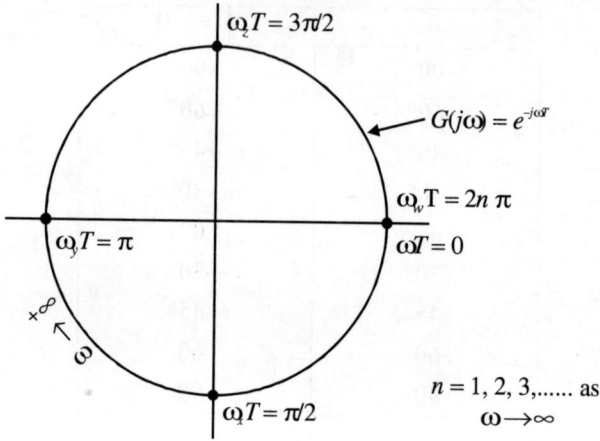

Fig. 12.10. Polar plot of transportation lag.

Example 12.6. *Draw the polar plot of the oepn-loop transfer function of a system with transportation lag given by*

$$G(s)H(s) = Ke^{-sT}/[s(1 + sT_1) (1 + sT_2)]$$ ...(12.32)

Investigate the system stability.

Solution:

The polar plot is similar to the system without transportation lag but with added phase lag due to the presence of the transportation lag. As $s \to \infty$, the polar plot tends to be a spiraling curve. The polar plots of the system without transportation lag and with transportation lag are shown in Fig. 12.11 (a) and (b) respectively.

The effects of transportation lag are:

(a) The polar plot is shifted clockwise due to the additional phase lag introduced by transportation lag. The polar plot is shifted close to the critical point $(-1, j0)$.

(b) If the dead time (the transportation lag) is increased sufficiently, the critical point $(-1, j0)$ will be encircled and the system becomes unstable.

(c) As $s \to \infty$, the magnitude goes on decreasing towards zero whereas the phase angle contributed by transportation lag increases indefinitely resulting in a spiraling curve.

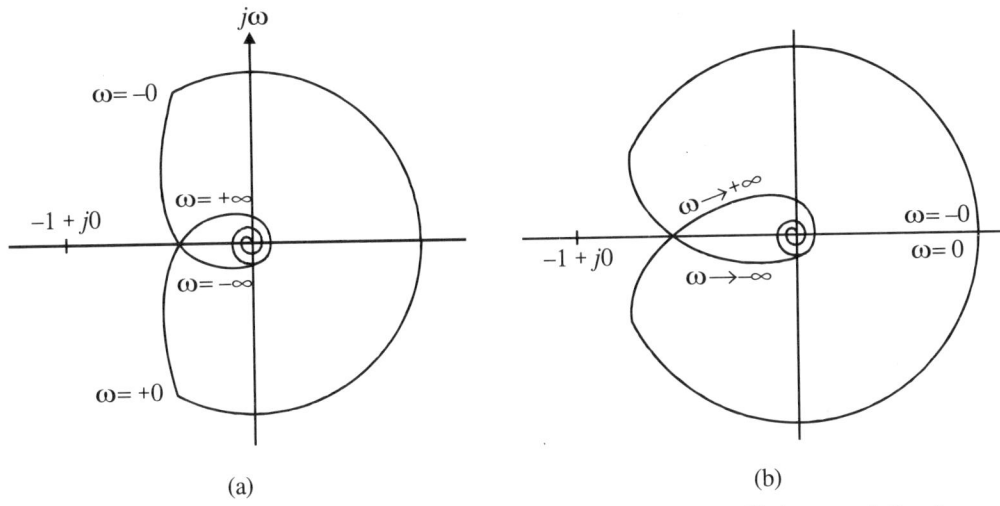

Fig. 12.11. (a) Pola plot without transportation lag (b) polar plot with transportation lag.

Thus the transportation lag tends to make a system less stable.

12.6. RELATIVE STABILITY OF SYSTEMS

We have demonstrated in the previous sections the application of the Nyquist stability criterion to obtain the absolute stability of a system. In this section we shall study how the Nyquist criterion can be used to define and ascertain the relative stability of a system.

The relative stability of a system is one of the criteria based on which a control system engineer selects a particular system for the process to be controlled from many available stable systems. To understand the concept of relative stability, let us consider the Nyquist plots and the corresponding step and frequency responses of a typical third-order system shown in Fig. 12.12 for four different values of system gain K.

Consider the case shown in Fig. 12.12 (a) which is for low gain K. The Nyquist plot of $G(s)H(s)$ intersects the negative real axis at a point, called phase-crossover point, quite far away from the critical point $(-1, j0)$ towards the $j\omega$-axis. The corresponding step response is quite well-behaved with low peak overshoot and settles down rapidly. In the corresponding frequency response, M_p is low. As K is increased, the phase-crossover point moves close to the $(-1, j0)$ point as seen from Fig. 12.12(b). The system is still *stable*. But the step response has a higher peak overshoot, and the frequency response shows a larger M_P. The phase curve $\phi(\omega)$ does not give as good an indication of relative stability as M_P, except that the slope of the phase curve gets steeper as the relative stability decreases. In Fig. 12.12(c), the Nyquist plot intersects the negative real axis at $(-1, j0)$ point. The system is *critically*, in this case, with constant amplitude oscillation as shown by the step response. In the frequency response, M_P becomes infinite. If the value of K is further increased, the Nyquist plot will enclose the critical

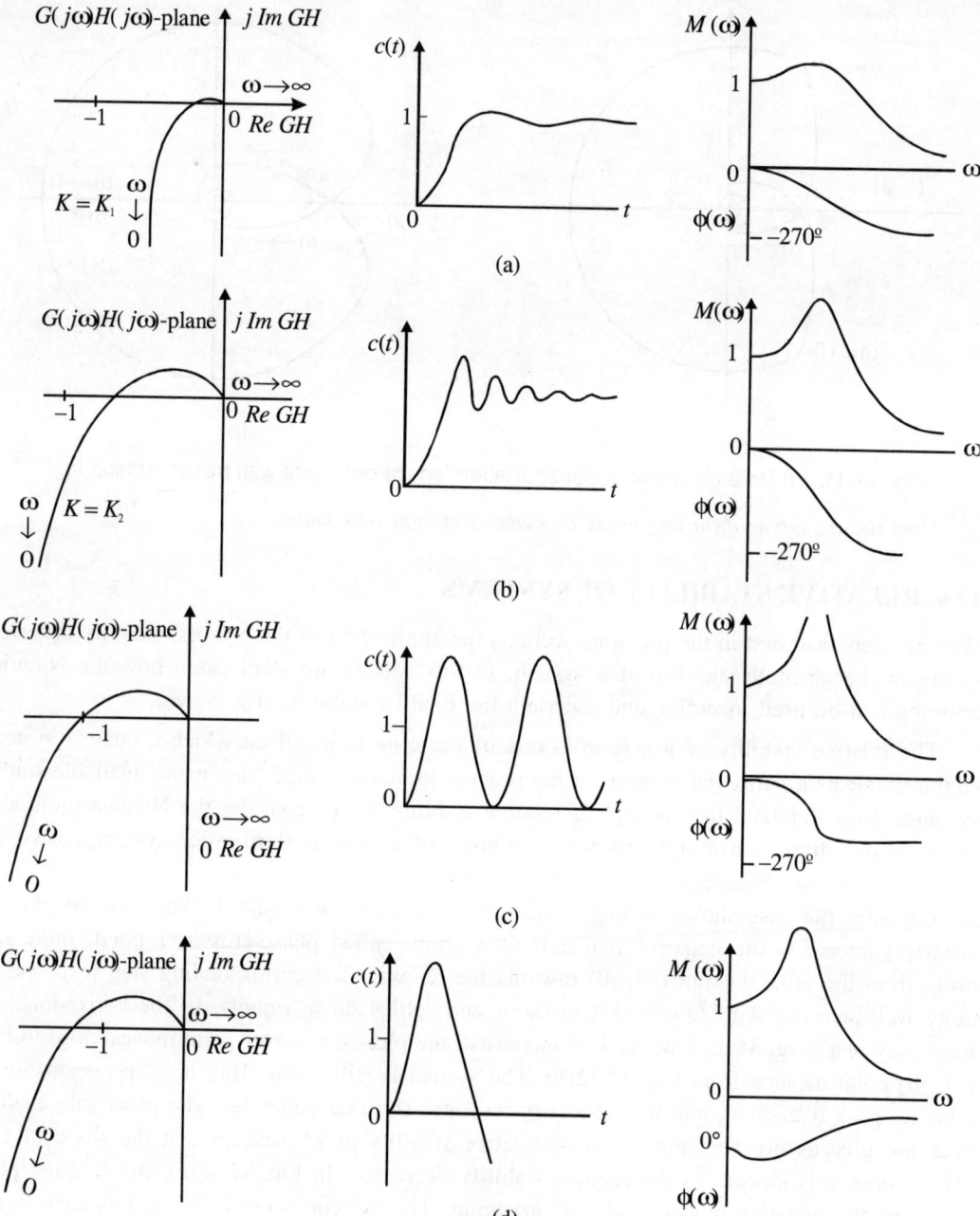

Fig. 12.12. Polar plot, time-response and frequency response of (a) Stable and well-damped system (b) Stable but oscillatory system (c) Marginally stable system (d) Unstable system

point $(-1, j0)$ and the system is *unstable*. The step response increases without bound. In this case, the amplitude response curve $M(\omega)$ ceases to have any significance but the only symptom of instability is that the slope of $\phi(\omega)$ curve is positive at and after the resonant frequency. Thus the relative stability of a system may be defined in terms of closeness of the $G(s)H(s)$ plot to the critical point $(-1, j0)$.

In this connection, we define the following quantities.

Phase Cross-over: This is the point at which the Nyquist plot of $G(s)H(s)$ crosses the negative real axis. The phase angle at this point is $-180°$. The frequency at which the phase crossover occurs is called the *phase crossover frequency* ω_{cp}.

Gain Margin: The gain margin is the factor or ratio a by which the gain of a system must be changed in order that the system becomes marginally stable.

Gain Crossover: This is the point on the Nyquist plot at which the magnitude of $G(s)H(s)$ is unity. The frequency at gain crossover is called the *gain crossover frequency* ω_{cg}.

Phase Margin: This is the amount of angle by which the phase angle ϕ of $G(s)H(s)$ must be changed in order that the phase angle becomes $180°$ and the system is *marginally stable*.

12.6.1. Gain Margin

The gain margin is a quantitative measure of the relative distance between the $G(s)H(s)$ plot and the $(-1, j0)$ point. Specifically, the gain margin is a measure of the closeness of the phase crossover point to the critical point $(-1, j0)$ in the $G(s)H(s)$-plane. The phase crossover frequency ω_{cp} is shown in Fig. 12.13. The magnitude of $G(j\omega)$ at $\omega = \omega_{cp}$ is denoted by $|G(j\omega_{cp})H(j\omega_{cp})|$. Then the gain margin of the closed-loop system with $G(s)H(s)$ as its open-loop transfer function is defined as

$$\text{Gain margin} = \text{G.M.} = 20 \log [1/|G(j\omega_{cp})H(j\omega_{cp})|] = -20 \log |G(j\omega_{cp})H(j\omega_{cp})|$$
$$\text{... (12.33)}$$

If $|G(j\omega_{cp})H(j\omega_{cp})|$ is less than 1, then the critical point $(-1, j0)$ is not encircled and the system is stable. In this case, the gain margin as defined in Eq. (12.33) is positive. Thus, *for stable systems, the gain margin will be positive.* The less the gain margin, the system is less stable. If the gain K is increased to the extend that the $G(j\omega)H(j\omega)$ plot passes through the critical point $(-1, j0)$, the system is *marginally stable* and the gain margin is 0 dB. Further increase in the gain K will result in the encirclement of the critical point $(-1, j0)$ and the system becomes *unstable*. In this case, $|G(j\omega_{cp})H(j\omega_{cp})|$ is greater than unity and the gain margin is negative. Thus, *if the gain margin is negative, the closed-loop system is unstable.* However, a word of caution is in order here. If $G(s)H(s)$ has poles or zeros in the right-half s-plane, the $G(j\omega)H(j\omega)$ plot will encircle the critical point $(-1, j0)$ in order for the closed-loop system to be stable. Under this condition, a stable system yields a negative gain margin, It is therefore necessary to determine the stability of a system first and then evaluate the gain margin. If the

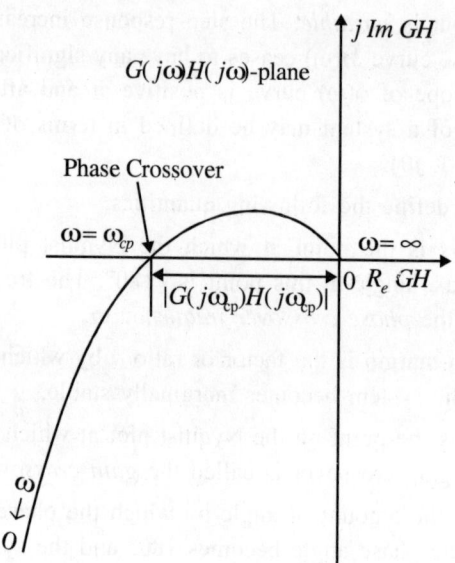

Fig. 12.13. Polar plot and ω_{cp}.

system is stable, the magnitude of gain margin simply denotes the stability margin and the sign has no significance.

The gain margin is defined as the additional gain in dB that can be introduced in the system so that the system becomes marginally stable.

12.6.2. Phase Margin

The gain margin is one of the many ways of representing the relative stability of a feedback control system. Generally, a system with a large gain margin should be relatively more stable than the one having a smaller gain margin. Unfortunately, the gain margin alone is not sufficient to evaluate the relative stability of all systems, especially when parameters other than the system gain K are variable. For example, consider the two systems represented by the polar plots of $G(j\omega)H(j\omega)$ shown in Fig. 12.14. Apparently, both the systems have the same gain margin. However, the locus A actually corresponds to a more state system than locus B because it is easier for locus B to pass through or encircle the critical point $(-1, j0)$ with any change in system parameters other than loop gain. In addition, system B has a much larger M_p than system A.

To supplement the gain margin, it is therefore necessary to define another quantity, the phase margin. The phase margin is defined as the additional angle in degrees to be introduced so that the $G(j\omega)H(j\omega)$ plot passes through the critical point $(-1, j0)$. In other words, the phase margin is the angle between the negative real axis and the phasor that passes through the gain crossover frequency ω_{cg} as shown in Fig. 12.15. The angle is always measured from the negative

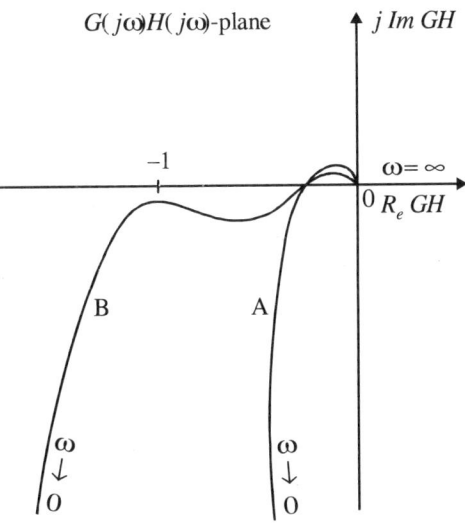

Fig. 12.14. Polar plots of two stable systems

real axis. The phase margin is negative if measured in the clockwise direction and positive if measured in the counter clockwise direction. The phase margin indicates the effect on stability of systems due to changes in system parameters whereas the gain on the closed-loop system stability.

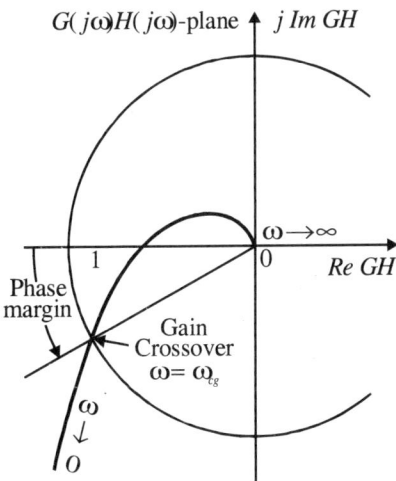

Fig. 12.15. Polar plot and ω_{cg}

$$\text{Phase margin} = \text{P.M.} = 180 + \phi° \qquad \qquad ...(12.34)$$

where ϕ is the phase angle of $G(j\omega)H(j\omega)$ at $\omega = \omega_{cg}$. Thus

$$\phi = \underline{|G(j\omega_{cg})H(j\omega_{cg})} \qquad \qquad ...(12.35)$$

where ω_{cg} is the gain crossover frequency. *For a system to be stable, the phase margin should be positive.*

The gain and phase margins of a control system are the measures of the closeness of the polar plot to the critical point $(-1, j0)$. Therefore, these margins are used as design criteria. Neither the gain margin alone nor the phase margin alone gives sufficient indication of the relative stability of systems. Both should be considered to determine the system stability. The gain and phase margins bound the behaviour of the closed-loop system near the resonant frequency. For minimum phase systems, both the margins must be positive for the systems to be stable, and for satisfactory performance, the phase margin should be between 30° and 60°, and the gain margin should be greater than 6 dB. With these values, minimum phase systems have guaranteed stability, even if the open-loop gain and time constants of the components vary to a certain extent.

Example 12.7. *Applying Nyquist stability criterion, find the gain margin and phase margin of the system with the open-loop transfer function*

$$G(s)H(s) = 10/[(s + 1)(s + 2)(s + 3)] \qquad \qquad ...(12.36)$$

Also find the maximum value of K for which the system is just stable.

Solution:

The frequency transfer function is given by

$$G(j\omega)H(j\omega) = 10/(-j\omega^3 - 6\omega^2 + j11\omega + 6) \qquad \qquad ...(12.37)$$

At the phase crossover frequency ω_{cp}, the imaginary part of $G(j\omega)H(j\omega)$ is zero. Hence ω_{cp} is obtained by equating the imaginary part to zero. Thus,

$$-j\omega^3 + j11\omega = 0$$

or

$$\omega_{cp} = \pm\sqrt{11} \qquad \qquad ...(12.38)$$

The magnitude of $G(j\omega)H(j\omega)$ at ω_{cp} is

$$|G(j\omega)H(j\omega)|\omega_{cp} = |10/(-6 \times 11 + 6) \qquad \qquad ...(12.39)$$

Therefore, the gain margin is

$$\text{G.M.} = -20 \log (1/6) = + 15.56 \text{ dB} \qquad \qquad ...(12.40)$$

This is the additional gain to be introduced so that the system becomes just *unstable.* Therefore, the maximum value of K is

$$K_{max} = 20 \log 10 + 15.56 = 35.56 \text{ dB}$$
$$= 60 \ (numeric) \qquad \qquad ...(12.41)$$

To find the phase margin, first the gain crossover frequency should be determined. At the gain crossover frequency ω_{cg}, $|G(j\omega)H(j\omega)| = 1$. Hence

$$|10/(-j\omega^3 - 6\omega^2 + j11\omega + 6)| = 1 \qquad \text{...(12.42)}$$

or,

$$(6 - 6\omega_{cg}^2)^2 + (11\omega_{cg} - \omega_{cg}^3)^2 = 100 \qquad \text{...(12.43)}$$

Expanding and simplifying, we get

$$\omega_{cg}^6 + 14\omega_{cg}^4 + 49\omega_{cg}^2 - 64 = 0$$

Solving the above equation by trial and error method, we obtain

$$\omega_{cg}^2 = 1, -7.5 \pm j2.78 \qquad \text{...(12.44)}$$

Since ω_{cg} can be only real, we have

$$\omega_{cg} = 1. \qquad \text{...(12.45)}$$

Hence, the phase angle of $G(j\omega)H(j\omega)$ at $\omega = \omega_{cg}$ is

$$\phi = -\tan^{-1}(\text{Im.Part/Real Part}) = -\tan\infty = -90° \qquad \text{...(12.46)}$$

Therefore, the phase margin is

$$\text{P.M.} = 180 + \phi = 90° \qquad \text{...(12.47)}$$

The analytical evaluation of ω_{cg} is a time consuming process. This is the limitation in applying the Nyquist criterion to find the phase margin. The order of the equation to be evaluated to obtain ω_{cg} is $2n$ where n is the order of the denominator. The phase margin can be obtained graphically by plotting the polar plot. This is also laborious as the magnitude and phase of $G(j\omega)H(j\omega)$ have to be computed for different values of ω for plotting the function. A simpler graphical method is to obtain the gain and phase margins from the Bode plot.

12.7. RELATIVE STABILITY FROM BODE PLOT

Compared to the polar plot, the Bode plot is easier to draw and the gain and phase margins can be easily evaluated from the Bode plot. The critical point $(-1, j0)$ in the $G(j\omega)H(j\omega)$-plane corresponds to the 0 *dB* and $-180°$ lines in the Bode plot. For a system to be stable, the slope of the magnitude curve at the gain crossover should preferably be -20 *dB*/dec.

The procedure to determine the gain and phase margins from the Bode plot are:

1. Draw the magnitude and phase angle curves.
2. From the phase angle curve, find the phase crossover frequency ω_{cp} at which $\phi = -180°$.
3. Read the gain G_{dB} at ω_{cp} from the magnitude curve.

4. The gain margin is given by

$$G.M. = -G_{dB} \qquad\qquad ...(12.48)$$

5. To find the phase margin, determine the gain crossover frequency ω_{cg} from the magnitude curve at which $G_{dB} = 0$.

6. Read the phase angle ϕ at ω_{cg} from the phase angle curve.

7. The phase margin is given by

$$P.M. = 180° + \phi \qquad\qquad ...(12.49)$$

Fig. 12.16 Illustrates the above procedure. The procedure is illustrated with an example.

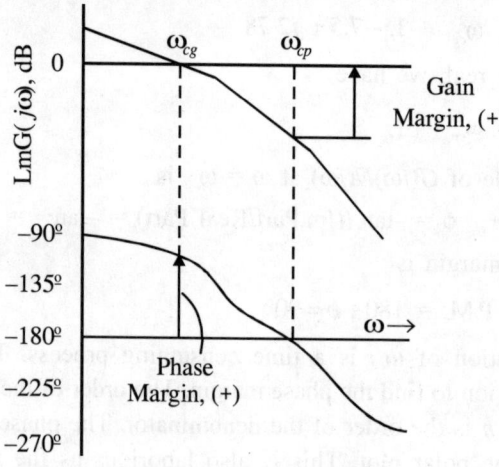

Fig. 12.16. Determination of GM and PM from Bode plot)

Example 12.8. *Determine the gain and phase margins of a unity feedback system with the open-loop transfer function*

$$G(s)H(s) = 10/[s(1 + 0.02s) (1 + 0.2s)] \qquad\qquad ...(12.50)$$

Solution:

The Bode plot corresponding to Eq. (12.50) is shown in Fig. 12.17. From the phase angle curve, the phase crossover frequency ω_{cp} is 15.9 rad/sec and the corresponding G_{dB} as read from the magnitude curve is $-14.8dB$. Hence the gain margin is $+14.8dB$. Next, the gain crossover frequency ω_{cg} is determined from the magnitude curve as 6.2 rad/sec. The phase angle ϕ read from the phase angle curve is $-148°$. Hence the phase margin is $32°$ ($=180° - 148°$). Since the gain margin and phase margin are both positive, the system is *stable*.

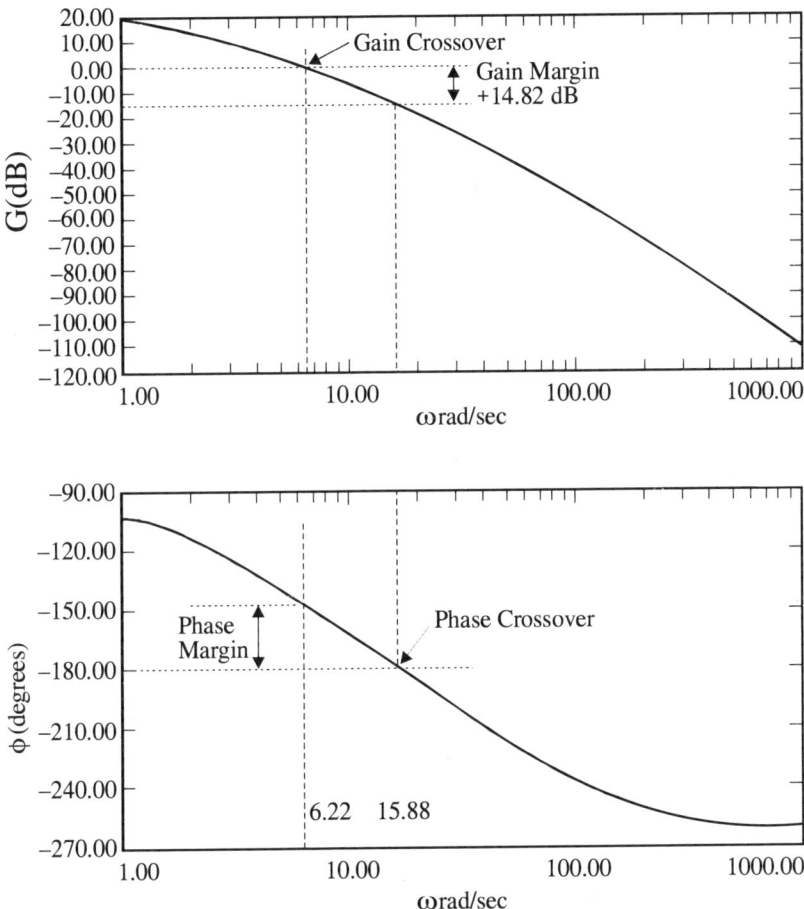

Fig. 12.17. Bode plot of 10/s(1 + 0.02s) (1 + 0.2s)

The Bode plot reveals some pertinent information. For instance, the gain can be adjusted (this raises or lowers the magnitude curve) to produce a desirable gain and phase margins. The slope of the low-frequency portion determines the system type and therefore the degree of steady-state accuracy. The system type and the gain together determine the error coefficients and hence the steady-state error. The gain crossover frequency ω_{cg} gives a qualitative indication of the speed of response of a system.

12.8. RELATIVE STABILITY FROM NICHOLS PLOT

The gain and phase margins are even better illustrated on the *log-magnitude versus phase plot* or Nichols plot. This plot is drawn by reading for each frequency, the values of G_{dB} and ϕ from the Bode plot. The resultant curve has frequency as a parameter. For the transfer function in Eq. (12.50), the plot is shown in Fig. 12.18. In this plot, *the phase crossover point is the intersection point of the locus with the –180°-axis. The gain crossover point is the intersection point of the locus with 0 dB-axis.* Hence the gain margin is simply the vertical distance in decibels measured from the phase crossover to the 0 dB-axis. *An upward measurement from the phase crossover is positive and a downward measurement is negative.* The phase margin is the horizontal distance in degrees measured from the –180° line to the gain crossover and is positive along the positive horizontal direction as indicated in Fig. 12.18. The gain and phase margins as determined from the Nichols plot are both positive and hence this represents a stable system. Changing the gain raises or lowers the curve without changing the angle characteristics. Increasing the gain raises the curve, thereby decreasing the gain and phase margins with the result that the relative stability is decreased. If gain is increased sufficiently so that the gain and phase margins become negative, this represents an unstable system. Decreasing the gain increases the relative stability. However, a large gain is desirable to reduce the steady-state error.

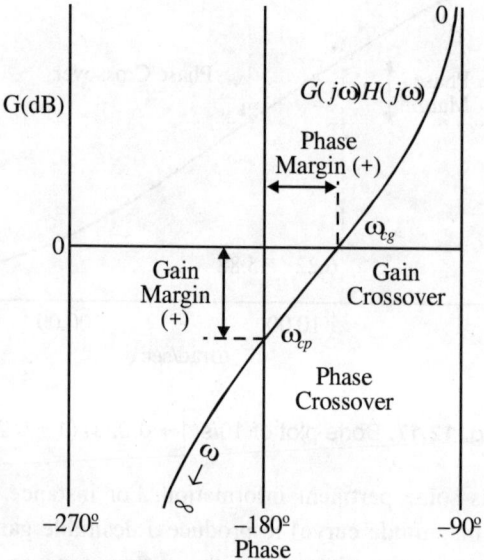

Fig. 12.18. Determination of GM and PM from Lm–ϕ diagram

12.9. RELATIVE STABILITY FROM ROUTH STABILITY CRITERIA

An alternative to the conventional graphical methods of estimating gain margin is to analytically determine the phase cross-over frequency and the gain margin from the imaginary and real parts of the open-loop transfer function (OLTF) respectively. For example, if the OLTF of a system is $G(j\omega)H(j\omega)$, we may write

$$G(j\omega)H(j\omega) = K[g_1(\omega) + jg_2(\omega)] \qquad \qquad ... (12.51)$$

where $g_1(\omega)$ and $g_2(\omega)$ are real and imaginary parts respectively of $G(j\omega)H(j\omega)$. The phase cross-over frequency ω_{cp} is then obtained from

$$g_2(\omega_{cp}) = 0 \qquad \qquad ... (12.52)$$

Therefore the gain margin in *dB* is

$$GM_{dB} = 20 \log |1/(K\, g_1(\omega)| \qquad \qquad ... (12.53)$$

Limitations: A limitation of the gain margin as in Eq. (12.53) is that it is not precisely defined in the case of all-pole second order systems which are stable for all positive values of gain [$\omega_{cp} = \infty$, $K\, g_1\, (\omega_{cp}) = 0$] and for systems which are globally unstable [$\omega_{cp} = 0$, $K\, g_1\, (\omega_{cp}) = \infty$]. A second limitation is that the solution of Eq. (12.52) becomes cumbersome for high order systems. In general, the figure of merit should provide the measure of gain margin of all systems without any limitation.

12.9.1. Stability Ratio

To overcome the above limitations, we define the gain margin as a *numeric* and designate this as the stability Ratio (S.R.). Thus *the stability ratio of a system is defined as the ratio of the marginal gain K_m to the actual system gain K* [57]-[63].

i.e.,

$$\text{S.R.} = K_m/K \qquad \qquad ... (12.54)$$

The proof of Eq. (12.54) follows:

Let the open-loop transfer function (OLTF) of a system is given by

$$G(s)H(s) = K\frac{b_m s^m + b_{m-1}s^{m-1} + ... + b_1 s + b_0}{a_n s^n + a_{n-1}s^{n-1} + ... + a_1 s + a_0} \qquad \qquad ... (12.55)$$

where $m < n$. Then, in the frequency domain

$$G(j\omega)H(j\omega) = K\frac{b_m (j\omega)^m + b_{m-1}(j\omega)^{m-1} + ... + b_1 (j\omega) + b_0}{a_n (j\omega)^n + a_{n-1}(j\omega)^{n-1} + ... + a_1 (j\omega) + a_0}$$

$$= K\frac{N_1(\omega) + jN_2(\omega)}{D_1(\omega) + jD_2(\omega)}$$

Rationalizing (and dropping ω), we get

$$G(j\omega)H(j\omega) = K\frac{(N_1 D_1 + N_2 D_2) - j(N_1 D_2 - N_2 D_1)}{D_1^2 + D_2^2}$$

$$= K\,[g_1(\omega) + jg_2(\omega)]$$

The phase cross-over frequency ω_{cp} is obtained by equating the imaginary part to zero. Thus,

$$g_2(\omega_{cp}) = (N_1 D_2 - N_2 D_1)$$

Therefore at $\omega = \omega_{cp}$, we have

$$G(j\omega_{cp})H(j\omega_{cp}) = K\frac{(N_1 D_1 + N_2 D_2)}{D_1^2 + D_2^2} \qquad \qquad \text{... (12.57)}$$

and the gain margin in terms of S.R. as a numeric is given by

$$\text{S.R.} = \frac{1}{G(j\omega_{cp})H(j\omega_{cp})} = \frac{D_1^2 + D_2^2}{K(N_1 D_1 + N_2 D_2)} = 1 \qquad \qquad \text{... (12.58)}$$

When $K = K_m$, the conventional gain margin is 0 dB and the S.R. = 1. Thus,

$$\text{S.R.} = \frac{D_1^2 + D_2^2}{K(N_1 D_1 + N_2 D_2)} = 1 \qquad \qquad \text{... (12.59)}$$

Or,

$$K_m = \frac{D_1^2 + D_2^2}{(N_1 D_1 + N_2 D_2)} \qquad \qquad \text{... (12.60)}$$

Substitution of K_m in Eq. (12.58) results in the S.R. being defined as

$$\text{S.R.} = K_m/K$$

This establishes a direct relationship between the gain margin of a system and the Routh stability criteria. Thus the gain margin in terms of S.R. can simply be evaluated from the knowledge of marginal gain K_m obtained from the s^1- row test function of the Routh array.

For any value of system gain K, the gain margin S.R. is immediately known and this circumvents the entire process of finding the phase cross-over frequency and the gain margin as in other methods.

In terms of S.R., the closed-loop system is stable, marginally stable or unstable according as the S.R. is greater than, equal to or less than 1. For a globally unstable system, the S.R. is zero and for a system which is absolutely stable, the S.R. is infinity. Thus the S.R. stands precisely defined in all cases whereas Eq. (12.53) does not.

12.9.2. System Gain for a Specified Gain Margin

The gain margin in terms of S.R. easily lends itself to determine the system gain K for a specified gain margin. The gain K may easily be obtained from the knowledge of the S.R. as under 20 $\log_{10}$ (S.R.) = specified Gain Margin (S.G.M).

Or,

$$S.R. = antilog\ (S.G.M.)/20 = K_m/K \qquad ...(12.61)$$

The value of K_m is obtained from the Routh stability criterion. Therefore K is given by

$$K = K_m/(S.R.) \qquad ...(12.62)$$

12.9.3. Phase Cross-over Frequency

The conventional method of obtaining the phase cross-over frequency is by graphical methods and this will yield only an approximate value of ω_{cp}. The exact value of ω_{cp} may be obtained by solving Eq. (12.52); however, the solution of Eq. (12.52) for high-order systems is tedious and time-consuming. Here, a simple method of computing ω_{cp} from the Routh array is highlighted.

From the s^1 – row test function of the Routh array, the marginal gain K_m is obtained. The roots of the auxiliary equation formed from the s^2-row elements with $s = j\omega$ and $K = K_m$ correspond to ω_{cp}. This yields the exact value of ω_{cp}. It may be noted that ω_{cp} is independent of the system gain K.

12.9.4. Gain Cross-over Frequency and Phase Margin

The procedure to obtain the phase margin of a given system is as follows:

Let the transfer function of the system be

$$G(s)H(s) = \frac{K\sum\limits_{i=0}^{m}b_i s^i}{\sum\limits_{j=0}^{n}a_i s^j} \qquad m \le n \qquad ...\ (12.63)$$

The characteristic equation is

$$F(s) = 1 + G(s)H(s) = 0$$

$$= \sum_{j=0}^{n}a_i s^j + K\sum_{i=0}^{m}b_i s^i = 0 \qquad ...(12.64)$$

Using the coefficients of Eq. (12.64), the Routh array is formulated and the Routh table is completed. From the s^1-row test function of Routh array, the marginal gain K_m of the system is determined. The auxiliary equation is formed from s^2-row elements with $s = j\omega$ and $K = K_m$. Then, its roots are obtained. This frequency is the *phase cross-over frequency* ω_{cp} and is computed from a second-order equation irrespective of the order of the numerator and denominator polynomial of $G(s)H(s)$.

For any value of K, let the gain cross-over frequency be denoted by ω_{cg}. This frequency ω_{cg} is computed using the empirical formula given in Eq. (12.65) or (12.66) as the case may be

(i) For Type 0 System:

$$\omega_{cg} = \omega_{cg} [(K - a_o)/K_m]^{0.5} \qquad \qquad ...(12.65)$$

(ii) For Type 1 and higher Types:

$$\omega_{cg} = \omega_{cg} [K/K_m]^{0.5} \qquad \qquad ...(12.66)$$

The phase angle ϕ of $G(s)H(s)$ is computed from Eq. (12.63) with $\omega = \omega_{cg}$. The phase margin is given by

$$\text{P.M.} = 180° + \phi \qquad \qquad ...(12.67)$$

where ϕ is negative.

Example 12.9. *Obtain the phase margin of the system with the OLTF*

$$G(s)H(s) = K(s+1)(s+5)/s^2(s+2)(s+10)(s+20)$$

Solution:

The Characteristic equation is

$$F(s) = s^2(s + 2)(s + 10)(s + 20) + K(s + 1)(s + 5) = 0$$
$$= s^5 + 32s^4 + 260s^3 + (400 + K)s^2 + 6Ks + 5K = 0 \qquad ...(12.69)$$

The Routh table corresponding to Eq. (12.69) is

Routh Table

s^5	1	260	6K	
s^4	32	400 + K	5K	
s^3	7920 – K	187K	(after multiplying by 32)	
s^2	3168000 + 1536K – K²	5K(7920 – K)	(after multiplying by 7920 – K)	
s^1	1452000K + 1908.5K² – K³		(after multiplying by 316800 + 1536K – K²)	
s^0	5K (7920 – K)			

The marginal gain K_m is obtained from the s^1-row test function.

$$K (1453000 + 1908.5K - K^2 = 0 \qquad \qquad ...(12.70)$$

$K_m = 0$ being a trivial case, the value of K_m is given by

$$K_m = 2491.3 \qquad \qquad ...(12.71)$$

The auxiliary equation formed with the elements of s^2-row with $K = K_m$ is

$$(3168000 + 1536 \times 2491.3 - 2491.3^2)s^2 + 5 \times 2491.3 (7920 - 2491.3) = 0 \qquad ...(12.72)$$

Substituting $s = j\omega$ and solving for ω, we get the phase cross-over frequency ω_{cp}.

$$\omega_{cp} = 9.26 \text{ rad/sec} \qquad \qquad ...(12.73)$$

The gain cross-over frequency for any value of K is found by using Eq. (12.66) as the given $G(s)H(s)$ is of Type 2 system. Hence, for $K = K_m/2$,

$$\omega_{cg} = 9.26 \ [(K_m/2 \ K_m)]^{0.5} = 6.55 \text{rad/s} \qquad \ldots (12.74)$$

The phase angle of $G(s)H(s)$ at ω_{cg} is

$$\phi = \tan^{-1}(\omega_{cg}) + \tan^{-1}(\omega_{cg}/5) - 180^{\circ} -$$
$$\tan^{-1}(\omega_{cg}/2) - \tan^{-1}(\omega_{cg}/10) - \tan^{-1}(\omega_{cg}/20)$$

$$= -170.42^{\circ}$$

The phase margin is

$$\text{P.M.} = 180^{\circ} + \phi = 180 - 170.42^{\circ}$$
$$= 9.58^{\circ}$$

The phase margin thus obtained is in close agreement with the exact value of 10.68°.

12.10. ADDITIONAL EXAMPLES

Example 12.10. *Consider a system with an open-loop transfer function*

$$G(s)H(s) = (4s + 1)/[s^2 \ (s + 1) \ (s + 2)]$$

Draw the Nyquist plot and find the system stability. Also calculate the G.M. and P.M. of the system.

Solution:

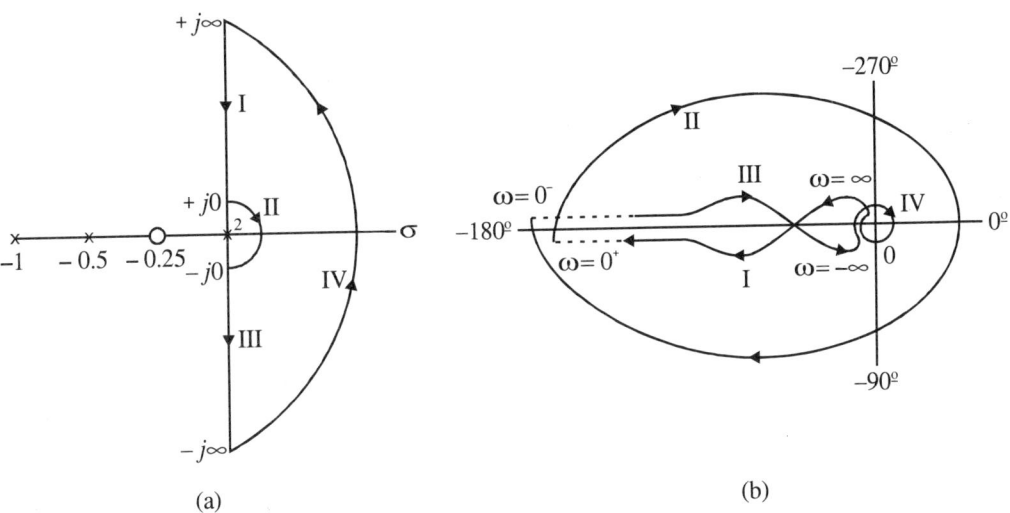

(a)   (b)

Fig. 12.19. (a) Pole-zero locations and Nyquist path; (b) Nyquist plot

(a) Polar plot corresponding to Sections I and III: The given system is Type 2 system and has a zero at $s = -0.25$. Hence, the polar plot corresponding to Section I starts at $-180°$ and at ∞ when $\omega = 0$; increases slightly due to the presence of zero, again decreases and ends at $-270°$ when $\omega = \infty$. Section III is the mirror image of Section I. They are shown in Fig. 12.19(b).

(b) Polar Plot corresponding to Section II: The low-frequency approximation of $G(s)H(s)$ is

$$GH_{lf} = \lim_{s \to 0} G(s)H(s) = 1/s^2$$

Substituting $s = \varepsilon\ e^{j\theta}$, we get

$$GH_{lf} = 1/(\varepsilon\ e^{j\theta})^2 = (1/\varepsilon^2)e^{-j2\theta} = K\ e^{j\phi}$$

As θ varies from $+90°$ to $-90°$ along the small-radius semicircle, we tabulate in Table 12.9 the phase angle ϕ for various values of θ.

Table 12.9. θ vs ϕ for low frequency

θ	$\phi = -2\theta$
$+90°$	$-180°$
$+60°$	$-120°$
$+30°$	$-60°$
$0°$	$0°$
$-30°$	$+60°$
$-60°$	$+120°$
$-90°$	$+180°$

Thus, Section II is mapped into an infinite radius semicircle from $\omega = +0$ to $\omega = -0$ as shown in Fig. 12.19(b).

(c) Polar plot corresponding to Section IV: The high-frequency approximation of $G(s)H(s)$ is

$$GH_{hf} = \lim\ G(s)H(s) = 2/s^3$$

Substituting $s = R\ e^{j\theta}$, we get

$$GH_{hf} = 4/(R\ e^{j\theta})^3 = (4/R^3)\ e^{-j3\theta} = K_1\ e^{j\phi}$$

As θ varies from $-90°$ to $+90°$ along the infinite radius semicircle, the values of ϕ are tabulated in Table 12.10.

Table 12.10. θ *vs* ϕ **for high frequency**

θ	$\phi = -3\theta$
$-90°$	$+270°$
$-60°$	$+180°$
$-45°$	$+135°$
$-30°$	$+90°$
$0°$	$0°$
$+30°$	$-90°$
$+45°$	$-135°$
$+60°$	$-180°$
$+90°$	$-270°$

Section IV maps into a circle of small radius starting from $\omega = +\infty$ to $\omega = -\infty$ in the clockwise direction. The complete Nyquist plot is shown in Fig. 12.19(b).

(d) *System Stability:* From the given $G(s)H(s)$ and Fig. 12.19(a), it can be seen that $P_R = 0$. The stability of the system depends upon the location of $(-1, j0)$ point. First we find the phase crossover frequency and then we find the magnitude of $G(s)H(s)$ at this frequency. This will locate the $(-1, j0)$ point.

At the phase crossover point, the imaginary part of $G(j\omega)H(j\omega)$ is zero. Hence, rationalizing we get,

$$= \left(\frac{(1 + 4j\omega)(1 - j\omega)(1 - 2j\omega)}{-\omega^2 (1 + j\omega)(1 + 2j\omega)(1 - j\omega)(1 - 2j\omega)} \right)$$

$$= [(1 + 10\omega^2 - j(8\omega^3 - \omega)]/[-\omega^2(1 + \omega^2)(1 + 4\omega^2)]$$

Equating the imaginary part to zero, we get

$$\omega_{cp} = 1/2\sqrt{2}$$

Hence magnitude of $G(j\omega)H(j\omega)$ at $\omega = 1/2\sqrt{2}$ is given by

$$|G(j\omega)H(j\omega)|_{\omega_{cp} = 1/2\sqrt{2}} = 10.67$$

Therefore the critical point $(-1, j0)$ is to the right of the crossover point. From the Nyquist plot, it is found that $N = 2$. Since $P_R = 0$, we have

$$N = Z_R = P_R$$
$$Z_R = 2$$

Therefore, the system is *unstable* since two zeros of the characteristic equation lie in the right half s-plane.

The conventional gain margin is given by

$$\text{G.M.} = 20 \log_{10} (1/|G(j\omega)H(j\omega)|_{\omega = 1/2\sqrt{2}}$$

$$= 20 \log_{10} (1/10.67)$$

$$= -20.56 \text{ dB}$$

The gain margin being negative, the system is *unstable*.

The phase margin of the system is given by

$$\text{P.M.} = 180° + \phi$$

where ϕ is the $\angle G(j\omega)H(j\omega)$ when its magnitude is unity.

First we find the frequency at which the magnitude is unity. This frequency is the gain crossover frequency.

$$|G(j\omega)H(j\omega)| = \{(1 + 16\ \omega^2)/[\omega^4 (1 + \omega^2) (1 + 4\omega^2)]\}^{1/2} = 1$$

Squaring both sides of the above equation

$$(1+16\ \omega^2)/[\ \omega^4 (1+ \omega^2) (1+4\ \omega^2)] = 1$$

or,

$$4\omega^8 + 5\omega^6 + \omega^4 - 16\omega^2 - 1 = 0$$

By trial-and-error method, we find the crossover frequency as $\omega_{cg} = 1.12$. The phase angle ϕ at $\omega_{cg} = 1.12$ is

$$\phi = \tan^{-1} 4.48 - 180° - \tan^{-1} 1.12 - \tan^{-1} 2.24$$

$$= 77.42° - 180° - 48.24° - 65.94°$$

$$= -216.76°$$

The phase margin is given by

$$\text{P.M.} = 180° + \phi = 180° - 216.76°$$

$$= -36.76°$$

Since the P.M. is negative, the system *is unstable*.

Example 12.11. *Applying Routh stability criterion, find the system stability, G.M. and P.M. of the system in Example 10.10.*

Solution:

The characteristic equation is

$$F(s) = 2s^4 + 3s^3 + 2s^2 + 4s + 1 = 0$$

Routh Table

s^4	2	2	1
s^3	3	4	
s^2	$-2/3$	1	
s^1	$-17/9$		
s^0	1		

There are two sign changes in the first column elements of the Routh table. Hence, there are *two zeros* of the characteristic equation in the right half s-plane and therefore the system is *unstable*.

To calculate the G.M., the open-loop transfer function is written as

$$G(s)H(s) = (4s + 1)/[s^2 (s + 1) (s + 2)]$$

The characteristic equation is

$$F(s) = 2s^4 + 3s^3 + s^2 + 4Ks + K = 0$$

Using the Routh stability criterion, we find K_m.

Routh Table

s^4	2	1	1
s^3	3	4K	
s^2	$(3 - 8K)/3$	K	
s^1	$\{[(3 - 8K)/3]4K - 3K\}/[(3 - 8K)/3]$		
s^0	K		

From the s^1-row test function, we obtain K_m as

$$K_m = 3/32$$

The gain margin with $K = 1$ is

$$\text{G.M.} = 20 \log_{10} (K_m/K)$$
$$= 20 \log_{10} 3/32$$
$$= -20.56 \text{ dB}$$

The gain margin being negative, the system is *unstable*.

To find the P.M., the gain crossover frequency is calculated first.

$$\omega_{cg} = \omega_{cp}\sqrt{K/K_m}$$

where ω_{cp} is the phase crossover frequency and is obtained from s^2-row function with $K = K_m$.

Thus,

$$[(3 - 8 \times 3/32)/3]s^2 + 3/32 = 0$$

Or,

$$2.75 \ s^2 + 9/32 = 0$$

$$\therefore \ \omega_{cp} = 0.3535$$

same as obtained in the previous example ($\omega = 1/2\sqrt{2} = 0.3535$).

$$\omega_{cg} = 0.3535 \times \sqrt{1/(3/32)}$$

$$= 1.15$$

The phase angle ϕ is obtained by

$$\phi = \tan^{-1} 4.6 - 180° - \tan^{-1} 1.15 - \tan^{-1} 2.3$$

$$= 77.7° - 180° - 49.1° - 66.5°$$

$$= -217.86°$$

The phase margin is given by

$$P.M. = 180° - 217.86°$$

$$= -37.86°$$

Example 12.12. *Find the stability of the system whose open loop transfer function is given by*

$$G(s)H(s) = (s + 2)/[(s + 1) \ (s - 1)]$$

Solution:

The pole-zero locations and the Nyquist path are shown in Fig. 12.20(a). As there is no pole or zero at the origin, the Nyquist path does not have Section II. There is a pole at $s = 1$ in the right half s-plane. Hence $P_R = 1$.

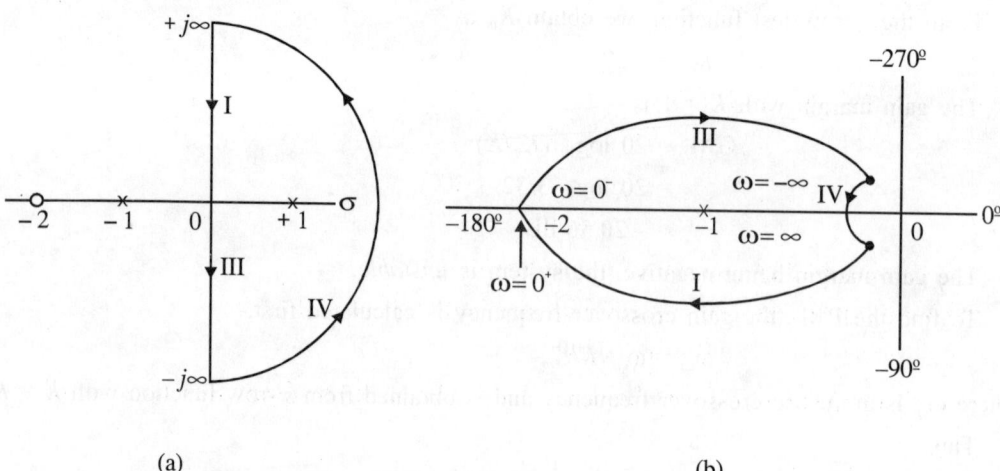

(a)   (b)

Fig. 12.20. (a) Pole-zero locations and Nyquist path (b) Nyquist plot

(a) Polar plot corresponding to Sections I and III: The low frequency approximation is

$$GH_{lf} = \lim_{s \to 0} G(j\omega)H(j\omega) = -2 = 2 \angle -180°$$

Hence the polar for Section I starts from the negative real axis with a magnitude of 2. As $\omega \to \infty$, the magnitude reduces to zero and the corresponding phase angle is –90°. The polar plot of Section III is the mirror image of Section I. They are shown in Fig. 12.20(b).

(b) Polar plot corresponding to Section IV: The system is a non-minimum system. The phase angle contributed by $(s + 1)$ and $(s - 1)$ are always +180°. Therefore the phase angle ϕ is given by

$$\phi = -180° + \tan\left(\frac{\omega}{2}\right) = -180° + \theta$$

As θ varies from $-90°$ to $+90°$ along the infinite -radius semicircle, the values of ϕ are tabulated in Table 12.11.

Table 12.11. θ *vs* ϕ for high frequency

θ	$\phi = -180° + \theta$
$-90°$	$-270°$
$-60°$	$-240°$
$-45°$	$-225°$
$-30°$	$-210°$
$0°$	$-180°$
$+30°$	$-150°$
$+45°$	$-135°$
$+60°$	$-120°$
$+90°$	$-90°$

Thus, Section IV is mapped into a small radius circular plot from $\omega = -\infty$ to $\omega = +\infty$ in the anticlockwise direction. The complete Nyquist plot is shown in Fig. 12.20(b).

(c) System Stability: The critical point $(-1, j0)$ is inside the Nyquist plot as shown in Fig. 12.20(c). If is found that $N = -1$. $P_R = 1$. Hence

$$\therefore \qquad -1 = Z_R - 1 = 0$$

Since the characteristic equation has no zeros in the right-half s-plane, the *closed-loop system is stable even though the open-loop transfer function G(s)H(s) is unstable.*

12.11. SUMMARY

The Nyquist stability criterion is a graphical frequency domain method of determining the system stability. In this chapter, we have first discussed the advantages of the Nyquist stability method. We then introduced the concept of conformal mapping as a preliminary for Nyquist stability criterion and derived the Nyquist stability criterion. Then we have defined the Nyquist path for Type 0, Type 1 and Type 2 systems.

The Nyquist path in general consists of four sections: (i) From $+j\infty$ to $j0^+$, (ii) From $j0^+$ to $j0^-$ along a semicircle of small radius, (iii) From $j0^-$ to $-j\infty$ and (iv) From $-j\infty$ to $+j\infty$ along a semicircle of infinite radius. We have discussed in detail the Nyquist stability criterion, the Nyquist-plot and the applications of Nyquist stability criterion to minimum and non-minimum phase systems as well as systems with transportation lag. We have defined the relative stability of a system in terms of gain – margin and phase – margin and studied the method of obtaining them from Nyquist plot, Bode plot and Nichols plot. Finally a method of finding the relative stability from Routh stability criterion, hitherto considered not possible, has been also dealt with in detail. All the methods have been adequately illustrated with suitable examples.

REVIEW QUESTIONS

12.1. Mention the special features of Nyquist plot.

12.2. State the principle of argument.

12.3. Explain the Nyquist path.

12.4. State the Nyquist stability criterion

12.5. Define phase crossover frequency and gain crossover frequency.

12.6. Define gain and phase margins.

12.7. Define relative stability. How do you find gain and phase margins from Bode plot?

12.8. What is meant by stability ratio? How is it superior to gain margin?

EXERCISE

12.1. What is meant by relative stability of a system?

12.2. Explain how the locations of the poles of closed-loop systems determine its stability.

12.3. Define gain and phase margin and indicate their physical significance.

12.4. Explain the significance of gain and phase margins in determining the relative stability of closed loop systems.

12.5. Define gain margin and phase margin with reference to Bode plot.

12.6. Explain clearly, with the necessary mathematical preliminaries, the Nyquist stability criterion.

12.7. Write a note on Nyquist path.

12.8. State and explain the Nyquist stability criterion.

12.9. Define gain margin and phase margin with reference to Nyquist plot.

12.10. State Nyquist stability criterion and explain the stability analysis procedure for a linear control system. How does this method of stability analysis compare with Routh stability criterion?

12.11. Explain the effects of adding a pole or zero to the loop transfer function of a control system on its closed loop stability.

12.12. Explain how gain margin and phase margin can be determined from the polar plot of the open-loop transfer function

PROBLEMS

12.1. Determine the gain and phase margin of a unity feedback control system with $G(j\omega)H(j\omega) = 10/j\omega(1 + j0.1\omega)(1 + j0.05\omega)$ by drawing the Bode plot. Determine the crossover frequencies.

12.2. The open-loop transfer function of a unity feedback control system is

$$G(s)H(s) = 50/s(1+0.05s)(1+0.25s)$$

Draw the Bode plot and evaluate the gain and phase margins.

12.3. The following experimental results were obtained from an open-loop frequency response test of a feedback control system.

ω: (rad./sec.)	3	4	5	6	8	10	12
Gain: (numeric)	0.95	0.66	0.48	0.362	0.230	0.150	0.130
ϕ	$-123°$	$-134°$	$-143°$	$-152°$	$-167°$	$-180°$	$-190°$

(i) Plot the Bode Plot.

(ii) Obtain the gain and phase margins.

12.4. Sketch the Bode plot for the following system with unity feedback

$$G(s)H(s) = 10(1+0.5s)/s(1+0.1s)(1+0.2s).$$

Find from the plot, the gain and phase margins.

12.5. Sketch the Bode plot for the unity feedback system with

$$G(s)H(s) = \frac{10(1+0.2s)(1+0.025s)}{s(1+0.001s)(1+0.005s)}$$

and find the gain margin and phase margin.

12.7. Draw the Bode diagram for the open-loop transfer function

$$G(s)H(s) = 5/s(1+0.6s)(1+0.1s)$$

Determine the gain and phase Margins.

12.7. The loop transfer function of a unity feedback control system is given by

$$G(s)H(s) = 1/s(1+s)(2+s)$$

Determine the phase margin of the system.

12.8. A system has the open-loop transfer function

$$G(s)H(s) = 2/s(s+1)(s+0.2)$$

Determine graphically the gain margin and phase margin.

12.9. The open-loop transfer function of a unity feedback system is

$$G(s)H(s) = K(1+s)/s(1+0.1s)(1+0.4s)$$

Using straight line approximation, draw the Bode plot and obtain the values of phase and gain margins.

12.10. The open-loop transfer function of a control system is given by

$$G(s)H(s) = 80/s(s+1)(s+4)$$

Draw the Bode plots and determine the gain and phase margins. Comment on the stability of the system.

12.11. Using the straight line approximation, draw the Bode plot for the open-loop transfer function

$$G(s)H(s) = 5/s(1+0.1s)(1+0.02s)$$

and determine the gain and phase margins.

12.12. The open loop transfer function of a unity feedback control system is $G(s)H(s) = K/s (1 + 0.1s) (1 + s)$. Draw Bode diagram and analyze the stability of the system for $K = 10$.

12.13. For the transfer function of a unity feedback system

$$G(s)H(s) = 1/s(1+0.04s)(1+s/50+s^2/2500)$$

(a) Draw the log-magnitude (asymptotic and exact) and phase diagrams.

(b) Find gain margin and phase margin.

(c) Also determine the respective crow-over frequencies,

12.14. Test the stability of the unity feedback system with

$$G(s)H(s) = 1/s^2(s+1)$$

using Nyquist criterion.

12.15. Draw the Nyquist plot and comment on stability of the system $G(s)H(s) = K/s^2(s + 3)$ $(s + 5)$.

12.16 Find the stability of the system whose open-loop transfer function is given by $G(s)H(s) = s/(s + 1) (s + 3)$.

If a pole is added at the origin, what will be the nature of stability?

12.17. Using Nyquist criterion, predict the closed loop stability of

$$G(s)H(s) = 10(s+1)/s^2(1+0.1s)(1+0.2s)$$

12.18. Sketch the Nyquist plot for a system with

$$G(s)H(s) = K(1+0.5s)(s+1)/(1+10s)(s-1)$$

Determine the range of values of K for which the system is stable.

12.19. Using Nyquist criterion, test the stability of the system whose open-loop transfer functions is given by

$$G(s)H(s) = 10(s+5)/s(s+2)(s+3)$$

12.20. A position control system has the forward loop transfer function

$$G(s) = 2/s(1+s)(1+0.1s)$$

Determine the gain and phase margins using Nyquist plot. Assme unity feedback.

12.21. Find the value of gain margin, phase margin and cross-over frequencies of the system whose open-loop transfer function is

$$G(s)H(s) = 1/s(1+0.2s)(1+0.5s)$$

Using Nyquist criterion.

12.22. Use Nyquist criterion determine closed loop stability for the system with open loop gain

$$G(s)H(s) = (1+4s)/s^2(1+s)(1+2s)$$

If not stable, how many closed loop poles lie on the r.h.s. plane?

12.23. Determine, using Nyquist stability criterion, the range of K for system stability for the openloop gain

$$G(s)H(s) = K/s(1+2s)(1+s)$$

12.24. The loop transfer function of a feedback system is

$$G(s)H(s) = K(s+1)/s^2(s+5)(s+10)$$

Using Nyquist stability criterion, determine the stability of the system.

12.25. By applying Nyquist stability criterion, find the value of K for which the system whose

$$G(s)H(s) = K(s+2)/(s+1)(s-1); \quad H(s) = 1$$

is stable.

12.26. Using Nyquist stability criterion, determine whether the system having the loop transfer function

$$G(s)H(s) = K/s^2(1-0.5s)$$

is stable or unstable for (i) $K = 10$ and (ii) $K = -10$

12.27. Show that the closed loop system with

$$G(s)H(s) = K/(1+sT_1)(1+sT_2)$$

is stable for all values of K.

12.28. Sketch the polar plot of the unity feedback system with open-loop transfer function

$$G(s)H(s) = 1/s(1+s)^2$$

and find the gain and phase margins.

12.29. A unity feedback control system has the forward transfer function

$$G(s) = K/(1+s)^3$$

(a) For $K = 4$, draw both the straight line and the corrected log-magnitude bode plot and the phase diagram.

(b) From the Bode plot, determine the phase and gain margins. Comment on the stability of the system.

(c) What is the maximum value of K for the system to be just stable?

12.30. Draw the Bode plot for a unity feedback system having the plant transfer function $G_p(s) = K_1/s(1+s)(1+0.2s)$ with a cascade compensator having a transfer function $G_c(s) = (1+14s)/(1+140s)$ and find the gain and phase margins.

Chapter 13

CLOSED-LOOP FREQUENCY RESPONSE

13.1. INTRODUCTION

In Chapter 11, we discussed three types of plotting the frequency response of the open-loop transfer function $G(j\omega)H(j\omega)$. In this chapter, we shall discuss the methods of obtaining the closed-loop frequency response from the open-loop frequency response. The closed-loop frequency response can be directly obtained if the closed-loop transfer function is given, and if its numerator and denominator are in factored form. Unfortunately, this is not the usual case. Only the open-loop transfer function is normally given and in the factored form. Therefore, the frequency response of the closed-loop system is obtained from the open-loop frequency response and the value of M_p can be predicted. Eventually we can design a closedloop system with a specified M_p.

13.2. POLAR PLOT AND CLOSED-LOOP FREQUENCY RESPONSE

Consider a simple control system with a forward transfer function of $G(j\omega)$ and unity feedback as shown in Fig. 13.1. The following equations describing the performance of the system are evident.

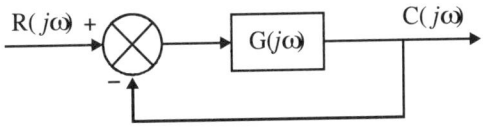

Fig. 13.1. Unity Feedback System

$$C(j\omega)/E(j\omega) = |G(j\omega)|e^{j\phi} = |A(j\omega)|e^{j\phi} \qquad \qquad ...(13.1)$$

$$\frac{C(j\omega)}{R(j\omega)} = \frac{G(j\omega)}{1+G(j\omega)} = M(j\omega) = M(\omega)e^{j\alpha(\omega)} \qquad \qquad ...(13.2)$$

$$\frac{E(j\omega)}{R(j\omega)} = \frac{1}{1+G(j\omega)} \qquad ...(13.3)$$

A typical polar plot of $G(j\omega)$ is shown in Fig 13.2. The directed line drawn from $(-1, j0)$ point to any point on the $G(j\omega)$ curve represents $B(j\omega) = 1 + G(j\omega)$. Hence

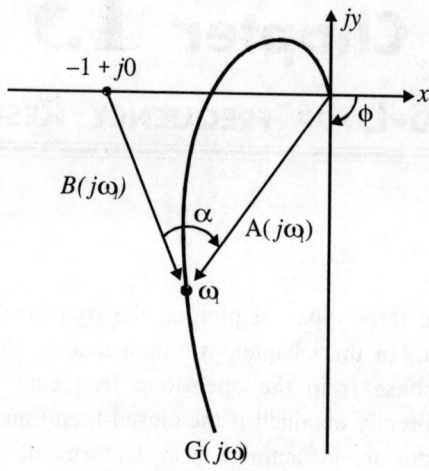

Fig. 13.2: Typical Polar Plot

$$B(j\omega) = 1+G(j\omega) = \left|B(j\omega)\right|e^{j\lambda(\omega)} \qquad ...(13.4)$$

The closed-loop frequency response is given by

$$\frac{C(j\omega)}{R(j\omega)} = \frac{\left|A(j\omega)\right|e^{j\phi}}{\left|B(j\omega)\right|e^{j\lambda}} = M(\omega)e^{j\alpha(\omega)} \qquad ...(13.5)$$

The values of $|A(j\omega)|$, $|B(j\omega)|$ and $\alpha(\omega)$ can be obtained directly from Fig 13.2. The angle $\alpha(\omega)$ is always negative since the magnitude of $\phi(\omega)$ is greater than the magnitude of $\lambda(\omega)$ Substituting Eq. (13.4) in Eq. (13.3), we get

$$\frac{E(j\omega)}{R(j\omega)} = \frac{1}{1+G(j\omega)} = \frac{1}{\left|B(j\omega)\right|e^{j\lambda(\omega)}} \qquad ...(13.6)$$

or

$$E(j\omega) = \frac{R(j\omega)}{\left|B(j\omega)\right|e^{j\lambda(\omega)}} \qquad ...(13.7)$$

where $|B(j\omega)|$ is the distance from $(-1, j0)$ point to the point on the $G(j\omega)$ locus. It is therefore clear from Eq. (13.7) that the greater the distance from $(-1, j0)$ point to a point on the $G(j\omega)$ locus for a given frequency, the smaller the steady-state error for a given sinusoidal input. Thus the closedloop response and the steady-state error for sinusoidal inputa can be determined from the polar plot of open-loop transfer function.

13.3. CONSTANT M(ω) AND $\alpha(\omega)$ LOCI OF C(jω)/R(jω)

The following information have, so far, been obtained from the open-loop transfer function and its polar plot of a given feedback control system.

1. Whether the given system is stable or not.
2. If the system is stable, how stable it is or its relative stability.
3. The system type.
4. The steady-state error.
5. A graphical method of obtaining $C(j\omega)/R(j\omega)$.

The loci of constant values of M drawn in the complex plane provide a rapid method of determining the values of M_p and ω_p, and also the value of the gain required to achieve a desired value of M_p. The loci of constant values of α may also be drawn in the complex plane. With these two loci, the plot of $C(j\omega)/R(j\omega)$ can be obtained rapidly. The M and α loci are developed for unity feedback system. However, we shall describe how these loci can be applied for non-unity feedback systems also.

13.3.1. Constant M(ω) Loci in the G(jω) Plane

The polar plot of the forward transfer function $G(j\omega)$ of a unity feedback system is shown in Fig. 13.2. From Eq.(13.5),

$$M(\omega) = \frac{|A(j\omega)|}{|B(j\omega)|} = \frac{|G(j\omega)|}{|1+G(j\omega)|} \qquad ...(13.8)$$

As shown in Fig. 13.2, if the value of M is M_a at $\omega = \omega_1$, to draw the constant M loci, we have to determine all the other points in the complex plane for which

$$\frac{|A(j\omega)|}{|B(j\omega)|} = M_a \qquad ...(13.9)$$

To derive the constant M loci, the transfer function is expressed in the rectangular coordinates. Thus,

$$G(j\omega) = x + jy \qquad ...(13.10)$$

Substituting Eq.(13.10) into Eq.(13.8), we obtain

$$M = \frac{|x + jy|}{|1 + x + jy|} = \left[\frac{x^2 + y^2}{(1+x)^2 + y^2}\right]^{1/2} \qquad ...(13.11)$$

or,

$$M^2 = \frac{x^2 + y^2}{(1+x)^2 + y^2} \qquad ...(13.12)$$

Rearranging,

$$x^2\left(M^2 - 1\right) + 2xM^2 + y^2\left(M^2 - 1\right) = -M^2 \qquad ...(13.13)$$

Dividing by $(M^2 - 1)$ and adding $M^4/(M^2 - 1)^2$ to both sides, we get

$$\left(x + \frac{M^2}{M^2 - 1}\right)^2 + y^2 = \frac{M^2}{\left(M^2 - 1\right)^2} \qquad ...(13.14)$$

Eq. (13.14) is the equation of a circle with center on the real axis. The coordinates (x_0, y_0) of the center are

$$(x_0, y_0) = \left(-\frac{M^2}{M^2 - 1}, 0\right) \qquad ...(13.15)$$

The radius of the circle is given by

$$r_0 = \left|\frac{M}{M^2 - 1}\right| \qquad ...(13.16)$$

For a given value of $M = M_a$, Eq. (13.14) describes a circle in the complex plane having a radius r_0 and its center at (x_0, y_0). This circle is called a constant M loci for $M = M_a$. For $M = 1$, Eq. (13.14) is invalid. From Eq. (13.13), for $M = 1$,

$$x = -1/2 \qquad ...(13.17)$$

This is the equation of straight line passing through the point $(-1/2, j0)$ in the $G(j\omega)$ plane and parallel to y-axis.

When M takes various values, Eq.(13.14) describes in the $G(j\omega)$ plane a family of circles called constant M loci or constant M circles. For different values of M, the coordinates of the centers and the radii are given in Table 13.1 and some of the loci are shown in Fig. 13.3. When

$M = 0$, the circle degenerates into a point at the origin. This is in agreement with the fact that the polar plots have zero magnitude at $\omega = \infty$. When $M = \infty$, the circle again degenerates into a point at critical point $(-1, j0)$. This is also in agreement with the fact that when the polar plot passes through the $(-1, j0)$ point, M_p is infinite and the system is marginally stable. From Fig. 13.3, it can be seen that the constant M loci in the $G(j\omega)$ plane are symmetrical with respect to the $M = 1$ line and the real axis. The circles to the right of the $M = 1$ locus correspond to values of M less than 1 and those to the left of $M = 1$ line are for M greater than 1.

Table 13.1. Constant M Circles

M	$Center\ x_0 = \dfrac{M^2}{M^2-1}, y_0 = 0$	$Radius\ r_0 = \left\|\dfrac{M}{M^2-1}\right\|$
0	0	0
0.3	0.01	0.33
0.5	0.33	0.67
0.7	0.96	1.37
1.0	∞	∞
1.05	−10.74	10.24
1.1	−5.76	5.24
1.15	−4.10	3.57
1.2	−3.27	2.73
1.25	−2.78	2.22
1.3	−2.45	1.88
1.35	−2.215	1.64
1.4	−2.04	1.46
1.5	−1.80	1.20
1.6	−1.64	1.30
1.7	−1.53	0.90
1.8	−1.46	0.80
1.9	−1.38	0.73
2.0	−1.33	0.67
2.5	−1.19	0.48
3.0	−1.13	0.38
4.0	−1.07	0.27
5.0	−1.04	0.21

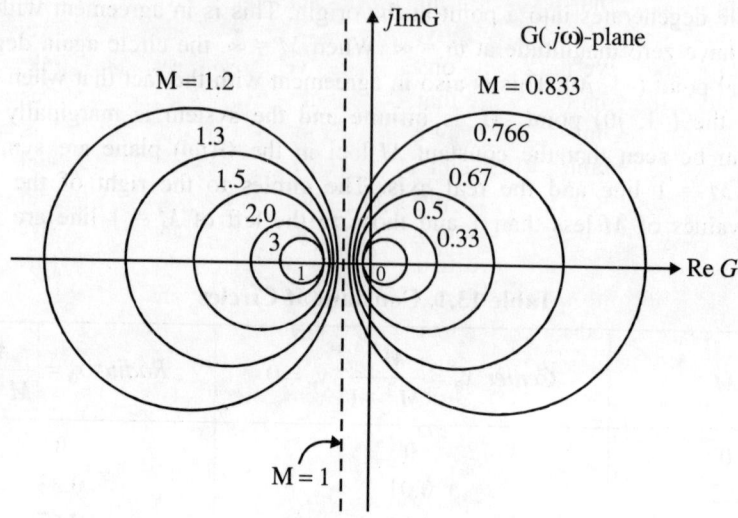

Fig. 13.3. M-Circles

13.3.2. Obtaining $M(\omega)$ from Constant M loci

The intersections of the polar plot with constant M loci give the values of $M(\omega)$ at the frequencies denoted on the plot. The constant M circle that is tangent to the plot gives the value of M_P. The resonant frequency ω_p is read off at the tangent point on the polar plot. If it is desired that the value of M_P should be less than a certain value M_c, the polar plot must not intersect the corresponding circle with $M = M_c$, and at the same time must not enclose the critical point $(-1, j_o)$.

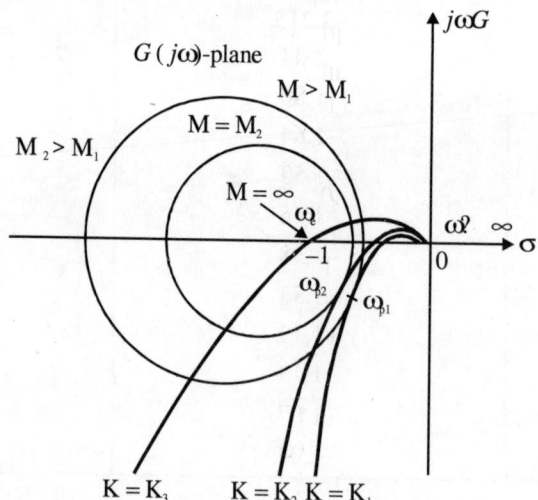

Fig. 13.4. Polar plots super imposed on M-circles

Fig. 13.4 shows the polar plots of a unity feedback control system for three values of gain K, together with constant M loci. For a given gain $K = K_1$, the intersects between the polar plot and the constant M loci give the points on the $M(\omega)$ versus ω curve. The peak resonance M_p is obtained by locating the smallest circle that is tangent to the polar plot. The resonant frequency ω_p is determined at the point of tangency and is denoted by ω_{p1}. If the gain is increased to K_2 and if the system is still stable, a constant M circle with a smaller radius which corresponds to a larger M is found tangent to the polar plot. Thus the peak resonance M_p is larger in this case. The resonant frequency is denoted by ω_{p2} which is closer to phase cross-over frequency ω_{cp} than ω_{p1}. When K is further increased to K_3 so that the polar plot passes through the critical point $(-1, j_o)$, M_p is infinite and $\omega_{p3} = \omega_{cg}$. The system is marginally stable. In all cases, the bandwidth BW of the closed-loop system is found at the intersection of the polar plot and the $M = 0.707$ circle. For values of K greater than K_3, the system is unstable. The $M(\omega)$ versus ω graph is plotted when enough data is obtained from the intersection points of polar plot and the constant M loci, as shown in Fig. 13.5.

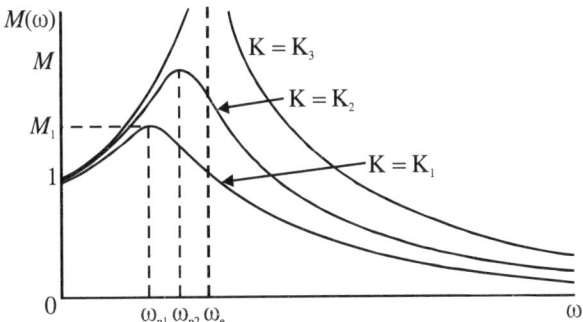

Fig. 13.5. Magnitude response M(ω) vs ω

13.3.3 Constant $\alpha(\omega)$ Loci in the $G(j\omega)$ plane

The constant $\alpha(\omega)$ loci for $C(\omega)/R(j\omega)$ can be determined by a method similar to that of $M(\omega)$ loci. Substitution of Eq.(13.10) into Eq.(13.2) yields,

$$M(\omega)e^{j\alpha(\omega)} = \frac{x + jy}{1 + x + jy} \qquad \qquad ...(13.18)$$

To obtain the $\alpha(\omega)$ loci, we have to determine all points in the $G(j\omega)$ plane which gives the same angel α. To derive the constant α loci, we obtain the phase angle α from Eq. (13.18). Thus

$$\alpha = \tan^{-1}(y/x) - \tan^{-1}y/(1 + x) = \theta_1 - \theta_2 \qquad ...(13.19)$$

or,
$$\tan \alpha = \tan(\theta_1 - \theta_2) = \frac{\tan\theta_1 - \tan\theta_2}{1 + \tan\theta_1 \tan\theta_2}$$

$$= \frac{(y/x) - y/(1 + x)}{1 + (y/x)[y/(1 + x)]} \qquad ...(13.20)$$

For a constant value of α, $\tan \alpha = N$ is also constant. Rearranging Eq.(13.20), we obtain

$$x^2 + x + y^2 - y/N = 0 \qquad ...(13.21)$$

Adding the term $(1/4) + (1/4N^2)$ to both sides of Eq. (13.21), we get

$$x^2 + x + 1/4 + y^2 - y/N + 1/4N^2 = 1/4 + 1/4N^2 = \frac{N^2 + 1}{4N^2} \qquad ...(13.22)$$

which can be written as

$$\left(x + 1/2\right)^2 + \left(y - 1/2N\right)^2 = \frac{N^2 + 1}{4N^2} \qquad ...(13.23)$$

When N assumes various values, Eq.(13.23) describes a family of circles with the coordinates of center (x_0, y_0) and the radius r_0 given by

$$(x_0, y_0) = (-1/2,\ 1/2N);\ r_0 = \sqrt{\left(\frac{N^2 + 1}{4N^2}\right)} \qquad ...(13.24)$$

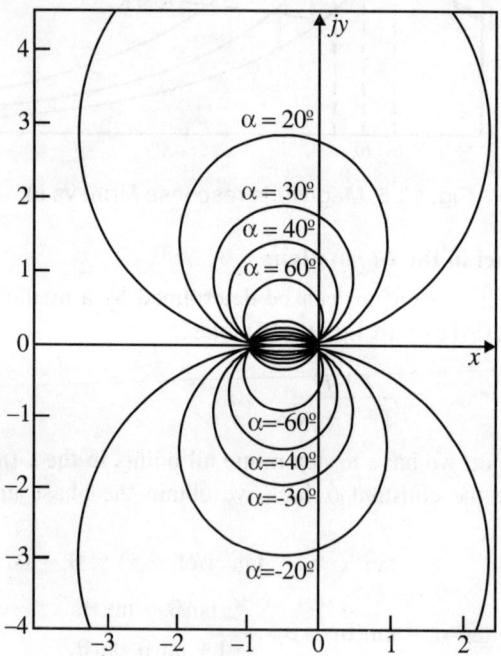

Fig. 13.6. α – loci

The tangent of angles in the first and third quadrant is positive. Therefore the coordinate y_0 is the same for an angle in the first quadrant and for the negative of its supplement which is in the third quadrant. As a result of this, constant α loci is only an arc of the circle. Thus $\alpha = 50°$ and $-130°$ are arcs of the same circle. Similarly angles α in the second and fourth quadrants have the same y_0 if they are negative supplement of each other. The constant α loci are shown in Fig. 13.6. The centers and radius of the constant N circles for various values of N are tabulated in Table 13.2.

Table 13.2. Constant N Circles

α	$N = \tan \alpha$	Center $x_0 = -1/2, \; y_0 = 1/2N$	Radius $r_0 = \sqrt{\dfrac{N^2+1}{4N^2}}$
$-90°$	$-\infty$	0	0.500
$-60°$	-1.732	-0.289	0.577
$-45°$	-1.000	-0.500	0.707
$-30°$	-0.577	-0.866	1.000
$-15°$	-0.268	-1.866	1.931
$-10°$	-0.176	-2.840	2.880
$0°$	0	∞	∞
$-15°$	0.268	1.866	1.931
$-30°$	0.577	0.866	1.000
$-45°$	1.000	0.500	0.707
$-60°$	1.732	0.289	0.577
$-90°$	∞	0	0.500

13.3.4. Obtaining Closed-Loop Frequency Response from Constant M and α loci

Fig. 13.7 shows a polar plot of $G(j\omega)$ with constant M and α loci superimposed. From this

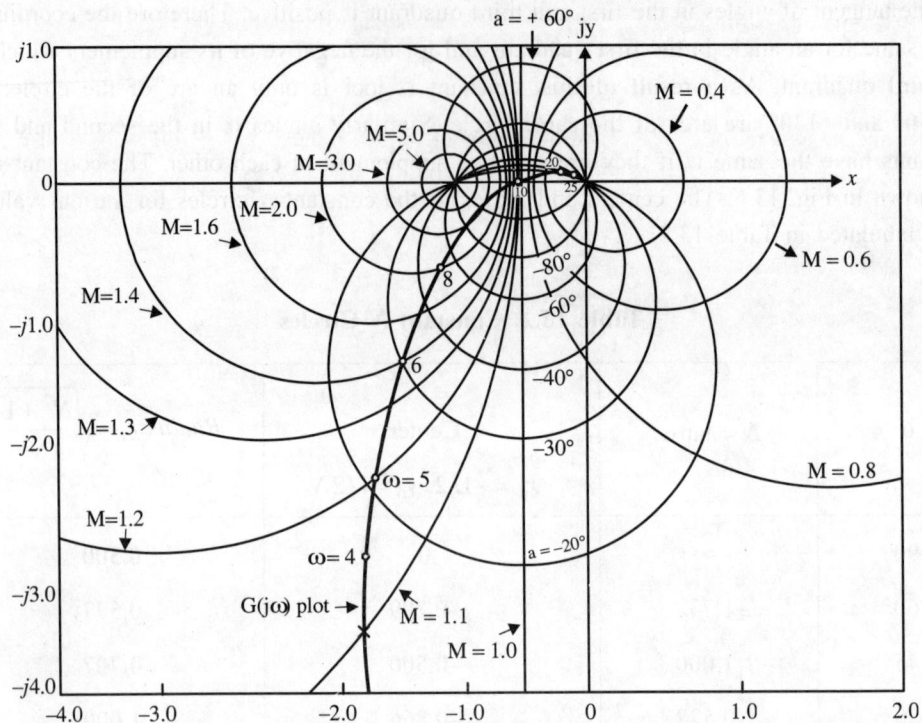

Fig. 13.7. Polar plot superimposed on M and α-loci

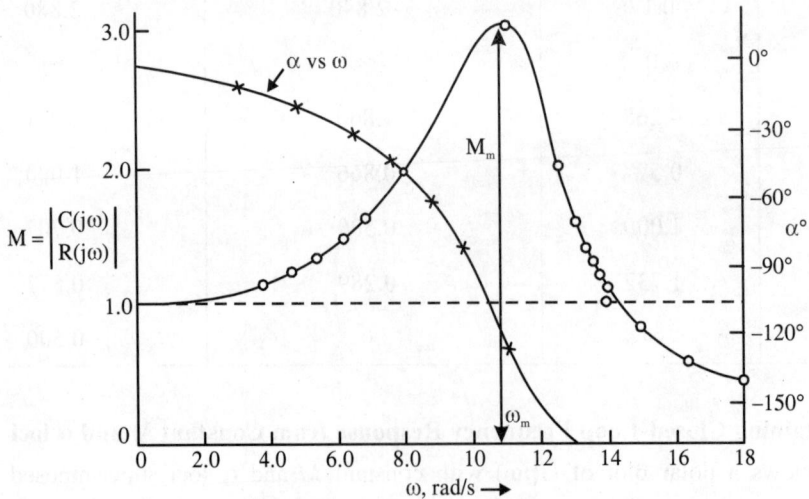

Fig. 13.8. Closed-loop response

figure, the magnitude M and the angel α of $C(j\omega)/R(j\omega)$ may be obtained for a number of values of frequency as indicated. The closed-loop response is plotted in Fig. 13.8.

13.4. NONUNITY FEEDBACK CONTROL SYSTEMS

To apply the M and α loci to a nonunity feedback system, the $C(j\omega)/R(j\omega)$ is written as

$$\frac{C(j\omega)}{R(j\omega)} = \frac{G(j\omega)}{1+G(j\omega)H(j\omega)} = \frac{1}{H(j\omega)} \cdot \frac{G(j\omega)H(j\omega)}{1+G(j\omega)H(j\omega)} \qquad ...(13.25)$$

or,

$$\frac{C(j\omega)}{R(j\omega)} = \frac{1}{H(j\omega)} \frac{G_0(j\omega)}{1+G_0(j\omega)} = \frac{1}{H(j\omega)} \frac{C_0(j\omega)}{R_0(j\omega)} \qquad ...(13.26)$$

where

$$G_0(j\omega) = G(j\omega)H(j\omega) \qquad ...(13.27)$$

The M and α loci can be applied to $G_0(j\omega)$ and $C_0(j\omega)/R_0(j\omega)$ is obtained. Then $C(j\omega)/R(j\omega)$ is obtained by multiplying $C_0(j\omega)/R_0(j\omega)$ by $1/H(j\omega)$. The phase α is obtained by the relation

$$\alpha = \lfloor C_0(j\omega) - \lfloor R_0(j\omega) - \lfloor H(j\omega) \qquad ...(13.28)$$

13.5. GAIN ADJUSTMENT FOR A SPECIFIED M_p OF A UNITY FEEDBACK SYSTEM

The tangent line drawn from the origin of $G(j\omega)$ plane to a given M circle plays an important part in gain adjustment. The tangent line along with a M circle is shown in Fig. 13.9. bc is the radius r_0 and 0b is the x-coordinate x_0, given by Eqs. (13.16) and (13.15) respectively. From Fig 13.9,

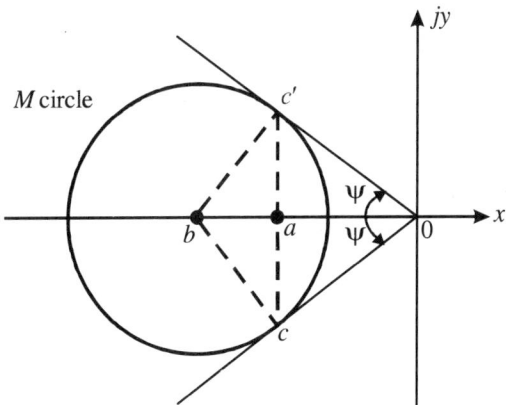

Fig. 13.9. M-circle and tangent lines

$$\sin \psi = \frac{bc}{0b} = \frac{M/\left(M^2-1\right)}{M^2/\left(M^2-1\right)} = 1/M \qquad \text{...(13.29)}$$

The point a in the figure is $(-1, j0)$ point as shown below :

$$oc^2 = ob^2 - bc^2;\ ac = oc \sin \psi;\ oa^2 = oc^2 - ac^2$$

$$oa^2 = oc^2 - oc^2 \sin^2 \psi$$

$$= \left(1 - \sin^2 \psi\right)\left(ob^2 - bc^2\right)$$

$$= \frac{M^2 - 1}{M^2} \cdot \frac{M^2\left(M^2-1\right)}{\left(M^2-1\right)^2} = 1$$

Hence $oa = 1$.

Adjusting the gain is the first and easiest step in setting the system for the desired performance. If satisfactory response cannot be achieved by gain adjustment alone, compensation techniques must be utilized.

In Fig. 13.10(a) is shown $G(j\omega)$ with its respective M_p circle in the $G(j\omega)$ plane. $G(j\omega)$ may be written as

$$G(j\omega) = x + jy = KG'(j\omega) = K(x' + jy') \qquad \text{...(13.30)}$$

Therefore

$$x' + jy' = \frac{x}{K} + j\frac{y}{K} \qquad \text{...(13.31)}$$

where $G'(j\omega) = G(j\omega)|K$ is the frequency-sensitive portion of $G'(j\omega)$ with unity gain. Changes in gain merely changes the amplitude and not the angle of $G(j\omega)$. Thus, in Fig. 13.10(a), a change of scale is made by dividing x and y coordinates by K so that the new coordinates are x' and y'. Then

1. $G(j\omega)$ plot becomes the $G'(j\omega)$ plot.

2. M_p circle becomes a circle which is simultaneously tangent to $G'(j\omega)$ and the line representing $\sin \psi = 1/M_p$.

3. The critical point $(-1, j0)$ becomes the $(-1/K, j0)$ point.

4. The radius r_0 becomes $r_0' = r_0/K$.

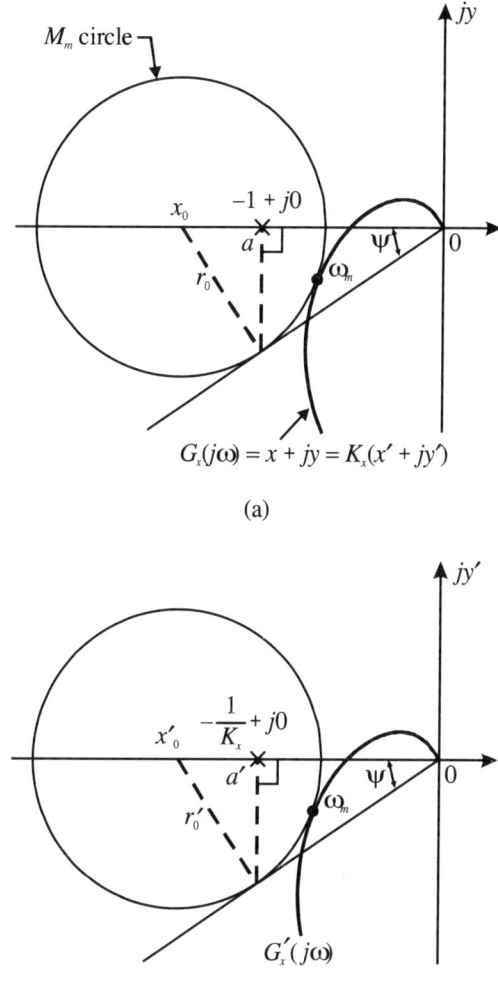

Fig. 13.10. (a) G(jω) and M_p circle (b) G'(jω) plots

The $G'(j\omega)$ plot is shown in Fig. 13.10(b). It is therefore possible to determine the required gain to achieve a desired M_p for a given system by the graphical procedure. The steps are:

Step 1: Given the original system transfer function

$$G(j\omega) = \frac{K(1+ j\omega T_1)(1+ j\omega T_2).....}{(j\omega)^l (1+ j\omega T_a)(1+ j\omega T_b).....} \qquad ...(13.32)$$

plot the frequency- sensitive portion $G'(j\omega) = G(j\omega)|K$.

Step 2: Draw the line representing the angle $\psi = \sin^{-1}(1/M_p)$.

Step 3: By trial and error find an M circle which is simultaneously tangent to the $G'(j\omega)$ plot and the line representing the angle ψ.

Step 4: Draw a line from the point of tangency on the ψ-angle line perpendicular to the negative real axis. Denote the point on the real axis as a'.

Step 5: For this circle to be the M_P circle, the point a' must be $(-1, j0)$ point. Therefore, the x' and y' coordinates must be multiplied by a gain factor K_1 such that this plot will become $G(j\omega)$ plot. The value K_1 is given by

$$K_1 = 1/0 \ a' \qquad \qquad \qquad \text{...(13.33)}$$

Step 6: The original gain K must therefore be changed by a factor

$$A = K_1/K \qquad \qquad \qquad \text{...(13.34)}$$

Example 13.1. *The transfer function of a system is given by*

$$G(j\omega) = \frac{1.47}{j\omega(1+j0.25\omega)(1+j0.1\omega)}$$

Determine the actual gain K_1 such that $M_p = 1.3$.

Solution:

The $G'(j\omega) = G(j\omega)/1.47$ is plotted on polar coordinates and by trial and error, an M circlec is found which is tangent to both $G'(j\omega)$ plot and ψ-line. The length oa' is found by drawing a perpendicular to the real axis from the point of tangency on the ψ-line. Thus o$a' = 0.34$.

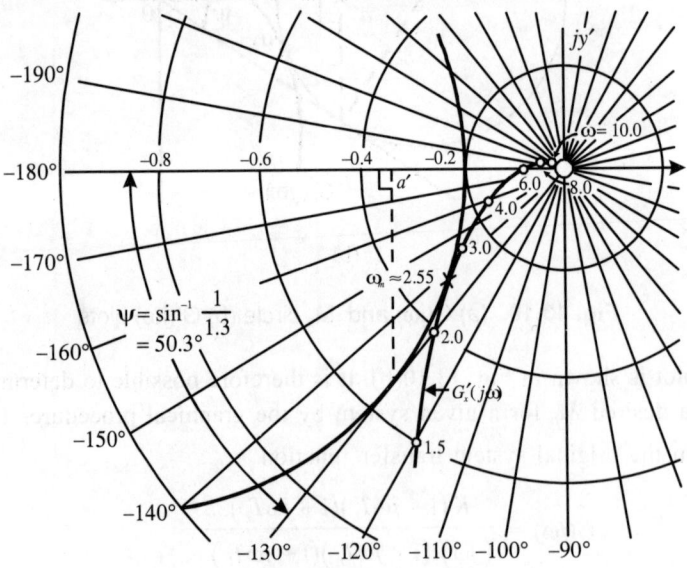

Fig. 13.11. G(jω) and G'(jω) plots

Therefore the required gain $K_1 = 1/0.34 = 2.94$ and the original gain must be increased by a factor of 2. The plots are shown in Fig. 13.11.

13.6. THE NICHOLS CHART

A major disadvantage of polar plots is that the plot does not retain its original shape when a change in gain is made. In design problems, frequently the loop gain has to be altered and also series or feedback controllers have to be added. This requires complete reconstruction of the

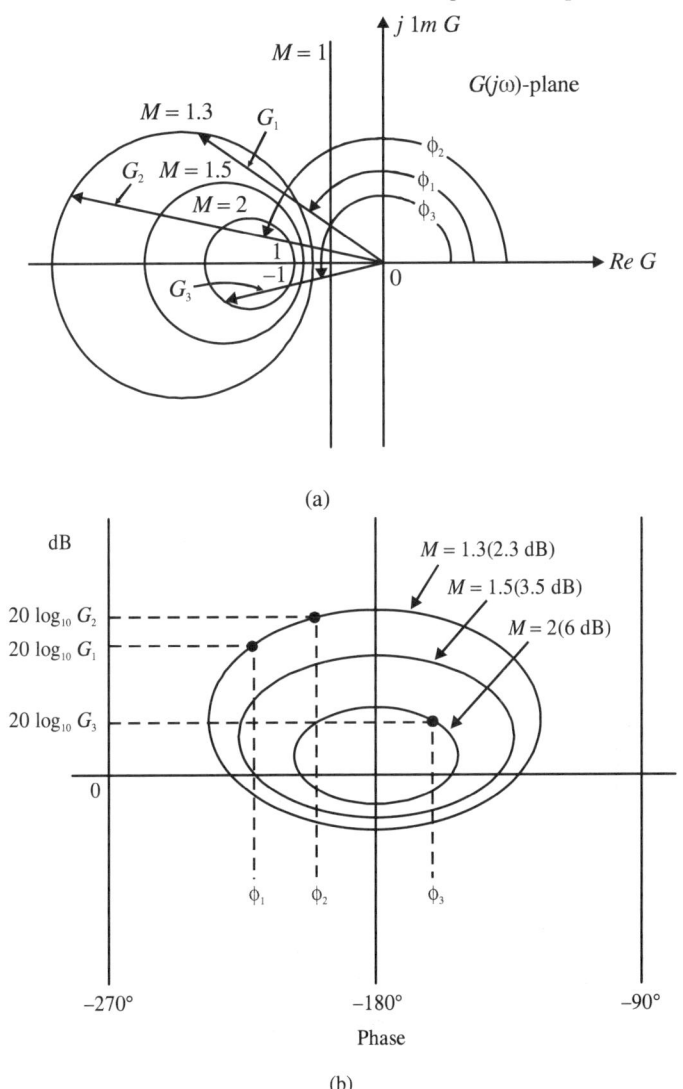

(a)

(b)

Fig. 13.12. (a) M-circles in G(jω) plane (b) M-circles in Lm-ϕ plane

resulting $G(j\omega)$. For design purposes, it is convenient to work in the Bode diagram or the log-magnitude versus phase plot. In the Bode diagram, the change in loop gain results in shifting the magnitude plot up or down. In the log-magnitude versus phase plot, the entire $G(j\omega)$ plot is shifted up or down when the gain is varied. Also, the Bode plot can be easily modified to accommodate any change made to $G(j\omega)$ in the form of added poles and zeros.

The constant M and constant α loci in the polar coordinates may be transferred to the log-magnitude versus phase coordinates easily. This is illustrated in Fig. 13.12. For a given point on a constant M loci in the $G(j\omega)$ plane, the corresponding point in the log-magnitude versus phase plane may be determined by drawing a vector directly from the origin of the $G(j\omega)$ plane to the particular point on the M loci. The length of this phasor expressed in dB and the phase angle in degrees give the corresponding point in the log-magnitude versus phase plane. Fig. 13.12 illustrates the process of locating three arbitrary point on the constant M loci in the log-magnitude versus phase plane.

In a similar manner, the constant α loci can also be transferred. The constant M and N loci in the log-magnitude versus phase coordinates were originated by Nichols and hence know as the Nichols chart. A typical Nichols chart is shown in Fig. 13.13. The $M = 1$ (0 dB) curve is asymptotic to $-90°$ and $-270°$. $M = \infty$ becomes a point at 0 dB and $-180°$. The curves for $M > 1$ are closed curves within the limits $-90°$ and $-270°$. To determine bandwidth, $M = 0.707$ locus should be used.

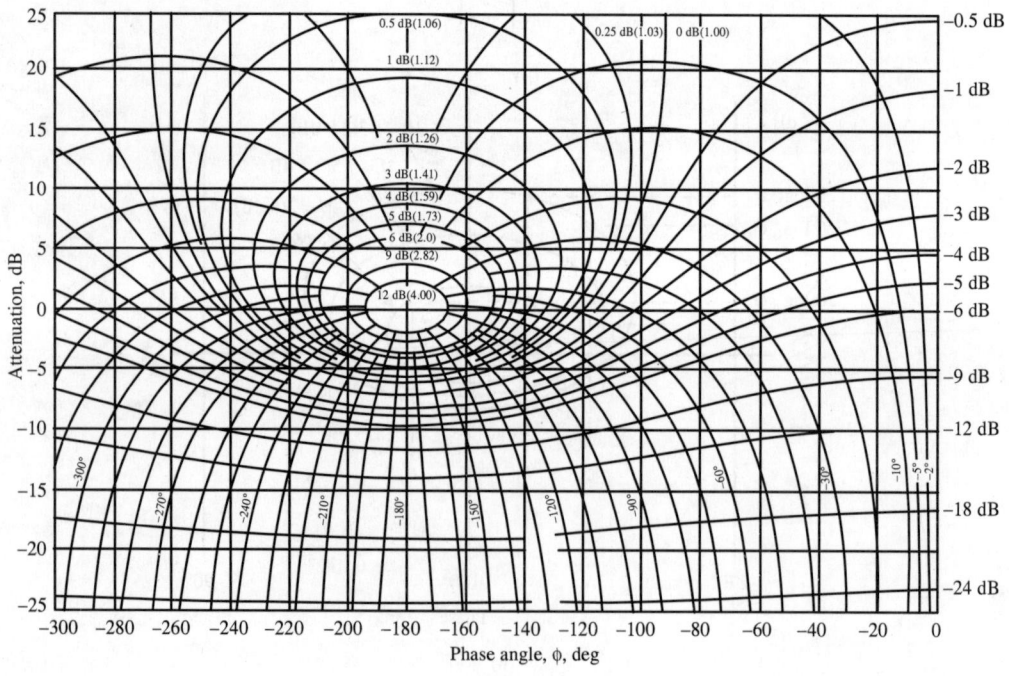

Fig. 13.13. Nichols chart

13.7. GAIN ADJUSTMENT USING NICHOLS CHART

By using Nichols chart, the gain can easily adjusted to achieve a desired M_p for a given system. This is illustrated by an example.

Example 13.2. *The transfer function of a system is given by*

$$G(j\omega) = \frac{2.04(1+ j2\omega/3)}{j\omega(1+ j\omega)(1+ j0.2\omega)(1+ j0.2\omega/3)} \qquad ...(13.35)$$

Determine the actual gain K_1 such that Mp = 1.26 (2dB)

Solution:

The log-magnitude-phase angle plot of $G(j\omega)$ is drawn on a transparent graph paper which has the same scales as the Nichols Chart. This graph paper is usually called an *overlay*. This graph is shown as solid curve, in Fig. 13.14 This is obtained by placing the overlay with the graph drawn on the Nichols charts with the two axes coinciding. This solid curve is tangent to $M = 1.12(1dB)$ curve at $\omega =1.1$. Hence

$$M_p = 1.12 \text{ and } \omega_p = 1.1.$$

The gain has to be changed to achieve an $M_p = 1.26$. The overlay is moved up or down until the graph on the overlay is tangent to $M = 1.26$ (2dB) curve. The dashed curve represents this. This is obtained by raising the overlay by a value equal to 4.5 dB from the original

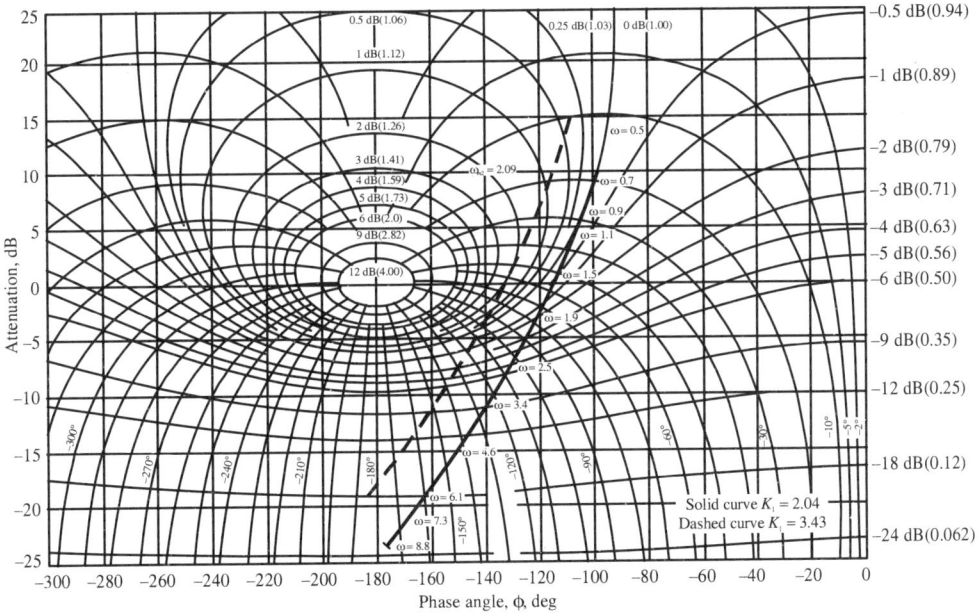

Fig. 13.14. Gain adjustment for a specified M_p

position. The gain therefore should be raised by 4.5 *dB* or the original gain should be increased by a factor of 1.706. Hence $K_1 = 3.43$. With this value of gain, $\omega_p = 2.20$.

When the gain has been adjusted for the desired M_p, the closed-loop frequency response can be easily obtained from Fig. 13.14. For example, with $M_p = 1.26$, the value of M at $\omega = 3.4$ is $-0.5dB$ (0.94) and α is $-110°$. The closed-loop frequency response obtained from Fig. 13.14 is shown in Fig. 13.15 for $K = 2.04$ and 3.43.

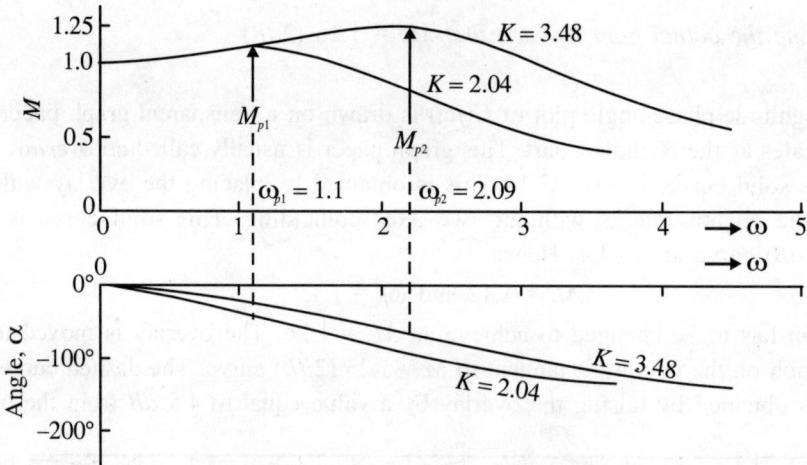

Fig. 13.15: Closed-loop response

13.8. ADDITIONAL PROBLEMS

Example 13.3. *The open-loop transfer function of a unity feedback system is*

$$G(j\omega) = K/j\omega(1 + j\omega)$$

Find the value of K so that $M_p = 1.4$

Solution:

The first step is to draw the polar plot of $G'(j\omega) = G(j\omega)/K$.

The value of ψ for $M_p = 1.4$ is given by

$$\psi = \sin^{-1}(1/M_p) = \sin^{-1}(1/1.4) = 45.6°$$

Draw the line *OP* making an angle of $\psi = 45.6°$ with negative real axis. Then draw a circle which is tangent to both $G'(j\omega)$ and *OP*. Then *PA* is drawn. *OA* is found to be 0.63. Hence $K = 1/0.63 = 1.58$. The graph is shown in Fig. 13.16.

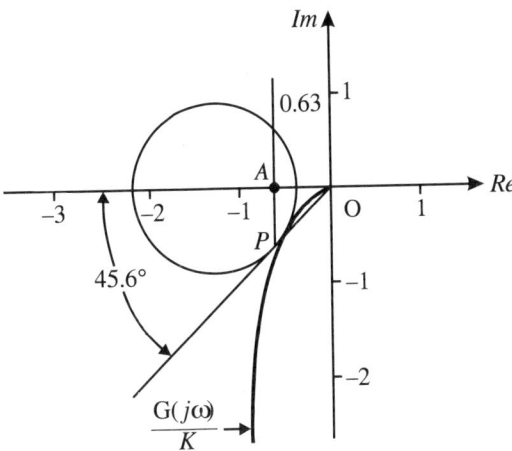

Fig. 13.16. Determination of gain for $M_p = 1.4$ in $G(j\omega)$ plane

Example 13.4: *Solve the problem in Example 13.3 using Nichols Chart.*

Solution:

The $G'(j\omega) = G(j\omega)/K$ locus and $M_p = 1.4$ locus are shown in Fig. 13.17. The $G'(j\omega)$ locus must be raised by $4dB$ in order to make it tangent to $M_p = 1.4$ locus. Hence $20 \log K = 4$, or $K = 1.58$.

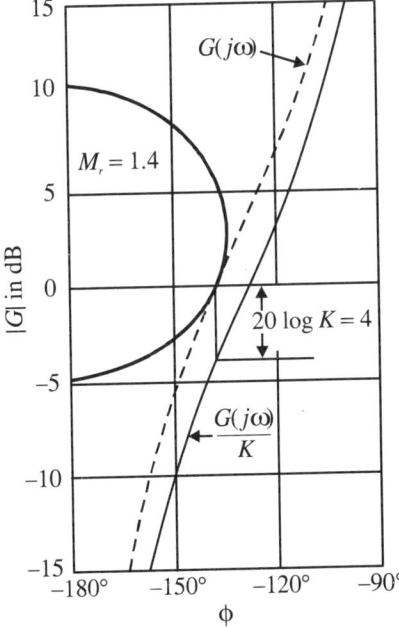

Fig. 13.17. Determination of gain for $M_p = 1.4$ from Lm–ϕ diagram

Example 13.5. *The open-loop transfer function of a unity feedback system is given by*

$$G(j\omega) = \frac{8.24K}{j\omega(1 + j\omega/49)(1 + j\omega/991)}$$

Using Nichols chart, find M_p, ω_p and BW for K = 1, 2.94 and 126.1.

Solution:

The three plots along with constant M loci are shown in Fig. 13.18. For $K = 1$, $M_p = 1$, $\omega_p = 0$, and $BW = 9.93$ rad/sec; for $K = 2.94$, $M_P = 1.0$, $\omega_p = 9$ and $BW = 35.1$ rad/sec and for $K = 126.1$, $M_p = \infty$, $\omega_p = 220$ and BW 338.3 rad/sec. For $K = 126.1$, since it passes through $(0dB, -180°)$ point, the system is marginally stable.

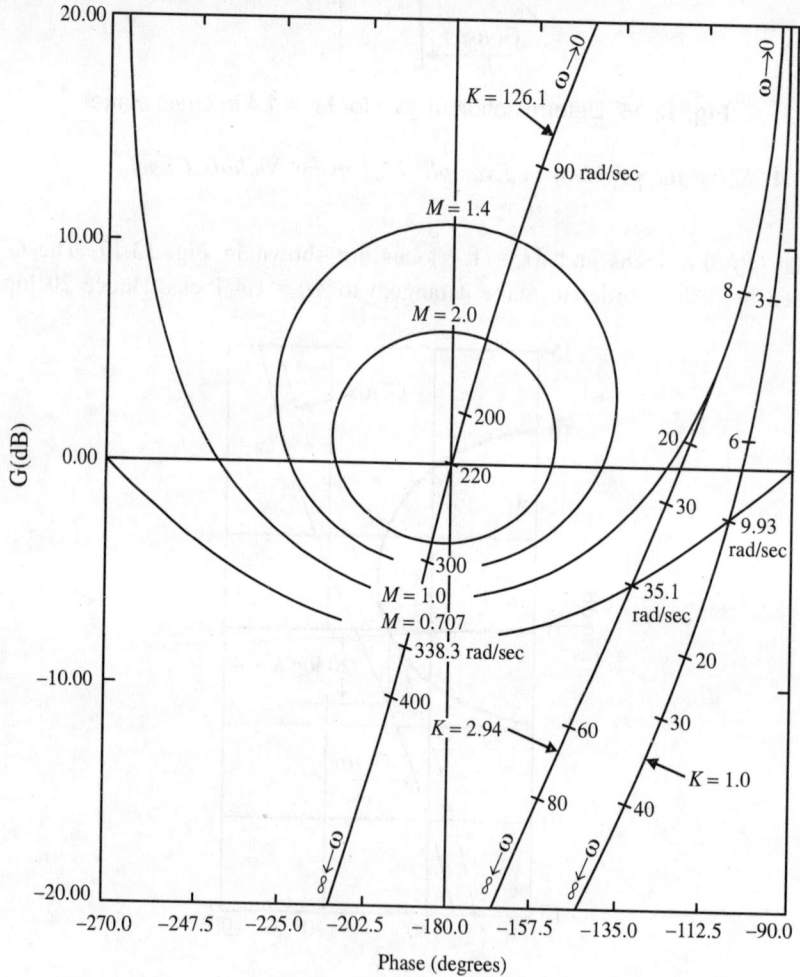

Fig. 13.18. G(jω) plots in Lm-φ diagram for K = 1, 2.94 and 126.1

Example 13.6. *A second order system has a maximum value of 9dB occurring at 0.88 rad/sec. Find its natural resonance frequency.*

Solution:

$$20 \log M_p = 9 \text{ or } M_p = 2.8$$

$$M_p = 2.8 = 1/\left[2\varsigma\sqrt{1-\varsigma^2}\right]$$

$$4\varsigma^2 - 4\varsigma^4 = (1/2.8)^2 = 0.128$$

$$\varsigma^2 = 0.026,$$

$$\varsigma = 0.16$$

$$\omega_n = \omega_p/\sqrt{1-2\varsigma^2} = 0.88/0.97 = 0.93 \text{ rad/sec}$$

Example 13.7. *Find the gain and phase margins for the system in Example 13.2 and comment on system stability*

Solution:

From Fg. 13.18,

For $K = 2.04$, $\omega_{cp} = 8.2,\ G_{dB} = -22$ dB, G.M. $= 22$ dB

$\omega_{cg} = 1.8,\ \phi = -128°,$ PM $= 52°$

System is stable since gain and phase margins are positive.

For $K = 3.43$ $\omega_{cp} = 8.2,\ G_{dB} = -17.5$ dB, G.M. $= 17.5$ dB

$\omega_{cg} = 2.6,\ \phi = -145°$ PM $= 35°$

System is stable since gain and phase margins are positive.

Example 13.8. *Find the phase margin for the system in Example 13.5 for three cases.*

Solution:

From Fg. 13.18,

(a) $K = 1.0$, $\phi_{\omega cg} = -100°,$ PM $= 80°$

(b) $K = 2.94$, $\phi_{\omega cg} = -118°,$ PM $= 62°$

(c) $K = 126.1$, $\phi_{\omega cg} = -180°,$ PM $= 0°$

13.9. SUMMARY

In this chapter, the M and α loci are developed which are used as graphical aids to obtain the closed-loop frequency response and in adjusting the gain to achieve a desired M_p. Usually the desired values of M_p are greater than 1 in order to obtain an underdamped response. The method of obtaining the closed-loop response and gain setting using polar plots and Nichols chart are suitably illustrated with examples.

REVIEW QUESTIONS

13.1. Explain constant M locus.

13.2. Give expressions for the radius and coordinates of the centre of M circles.

13.3. Explain constant α locus.

13.4. Give expressions for the radius and coordinates of the centre of N circles

13.5. What is a Nichols chart? Explain.

EXERCISE

13.1. Deduce that the locus of constant magnification and phase of closed loop system are circles. Find their centers and radius.

13.2. Explain how Nichol's chart is obtained from M and N circles.

13.3. What are constant M and constant N circles?

13.4. Discuss the use of Nichols chart in control systems.

13.5. Explain the use of constant M and constant N circles.

13.6. Explain the construction of Nichols Chart in gain phase plane and how it is used to predict system performance.

13.7. Show how the closed loop frequency response can be obtained graphically from the polar plot of a system.

PROBLEMS

13.1. The forward path transfer function of a unity feedback control system is $G(s) = K/s[(s + 10)]$. Determine the value of K so that the resonant peak of the closedloop system is 1.20.

13.2. A second-order system is given sinusoidal input and it is observed that the resonant peak of 1.4 occurs at a frequency of 2 rad/sec. Determine the natural frequency of oscillation and the damping factor of the system.

13.3. The openloop transfer function of a control system with unity feedback is $G(s) = K/s (1 - 0.1s) (1 + s)$. Determine the value of K so that the resonance peak M_p of the system is 1.4

13.4. A unity feedback control system has $G(s) = K/(1 + s)^2$. Find the value of K for an $M_P = 1.4$.

13.5. The openloop transfer function of a system is given by $G(s)H(s) = 40/s(s + 4) (s + 6)$. Find its M_P and ς.

Chapter **14**

DESIGN OF COMPENSATORS IN FREQUENCY DOMAIN

14.1. INTRODUCTION

In Chapter 8, we discussed the factors that determine the selection of compensators and their effects on the time response. The designer can determine the new values of M_P, ω_P and the system gain using frequency response method. However, the disadvantage of this method of design is that the closed-loop poles are not explicitly determined. Moreover, the correlation between frequency response parameters and the time response is only qualitative and further changes are required with the presence of real poles near the dominant complex poles.

In this chapter, we shall discuss the use of the lag, lead and lag-lead cascade compensators and the feedback compensator and their effects on the frequency response characteristics. The procedures for designing these compensators are also presented and illustrated with suitable examples.

14.2. NEED FOR COMPENSATORS

The performance of a closed-loop system is usually described by the maximum response M_P, the frequency ω_P at which the maximum response occurs, and the error coefficient. The value of M_P essentially describes the damping ratio ζ and therefore the amount of overshoot in the time response. For a specific value of M_P, the frequency ω_P determines the undamped natural resonance frequency ω_n which, in turn, determines the response time of the system. The error coefficient is important because it determines the steady-state error with appropriate standard input. The design of a control system in the frequency domain is usually based on selecting a value for M_P and, then, using the frequency response methods, to find the corresponding value of ω_P and the error coefficient. Once this is accomplished, the performance of the system is tested. If the desired performance specifications are not met with, compensators must be used. They reshape the frequency response plot of the system in order to meet all the performance specifications. Also, for those systems which are unstable for all values of gain, compensators must be used to stabilize the system. The compensators may be placed in cascade or in minor feedback loop.

The reasons for using compensators are:

(1) A given system is stable and its M_p and ω_p are satisfactory. But the *steady-state error is too large.*

The response is modified *by increasing the gain of the system* in order to reduce the steady-state error without appreciably altering the values of M_p and ω_p.

(2) A given system is stable but its *setting time is unsatisfactory.* This means M_p is satisfactory but ω_p is too low.

In order to increase ω_p, *the high frequency portion of the frequency response must be altered.*

(3) A given system is stable and has the desired M_p. But its *transient and steady-state responses are unsatisfactory.*

In this case, *the value of ω_p and the gain must be increased.* The portion of the frequency plot in the vicinity of the critical point $(-1, j0)$ or $(0dB, -180°)$ point must be altered to give the desired ω_p. The low frequency portion must be altered to obtain the desired increase in gain.

(4) A given system is *unstable* for all values of gain.

The *frequency response plot must be altered in the vicinity of $(-1, j0)$ point or $(0dB, -180°$ point to produce a stable system with desired M_p and ω_p.

Thus the objective of compensation is to reshape the frequency response of the original system by the use of a compensator in order to satisfy the performance specifications.

14.3. PHASE-LEAD COMPENSATOR

We have discussed the analysis and design of phase lead compensator in Chapter 8. In this section, we shall discuss the design procedure of phase lead compensators in the frequency domain. The circuit of a phase-lead compensator is shown in Fig. 14.1(a). The transfer function of a phase-lead compensator is given by

$$G_c(j\omega) = \frac{1 + j\omega aT}{1 + j\omega T} \qquad a > 1 \qquad \qquad ...(14.1)$$

Fig. 14.1(b) shows the polar plot of $G_c(j\omega)$ as ω is varied from 0 to ∞ for different values of *a. The angle ϕ_m between the tangent line drawn from the origin to the polar plot and the real axis is the maximum phase lead which the network can provide.* The frequency ω_m at the tangent point represents the frequency at which the maximum phase lead ϕ_m occurs. As the value of the *a* increases, the maximum phase lead ϕ_m also increases and the frequency ω_m decreases. As *a* approaches infinity, ϕ_m approaches 90°.

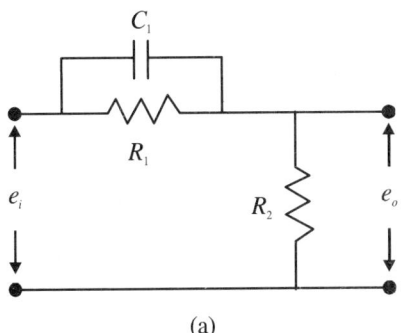

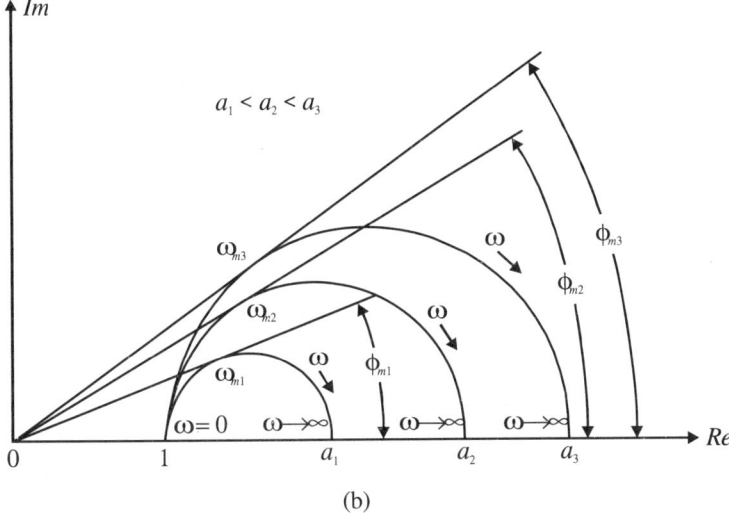

Fig. 14.1. (a) Phase-lead circuit (b) Polar plot *vs* values of *a*

The Bode plot of the phase-lead compensator is shown in Fig. 14.2. The compensator has a zero at $\omega = 1/aT$ and a pole at $\omega = 1/T$. Since $a > 1$, the zero occurs first and the angle contributed is positive. The phase-lead compensator is equivalent to a high-pass filter since the frequencies higher than $\omega = 1/T$ are passed without attenuation and the frequencies lower than $\omega = 1/T$ are attenuated. The lead angle is introduced in the frequency range from $\omega = 1/aT$ to $\omega = 1/T$.

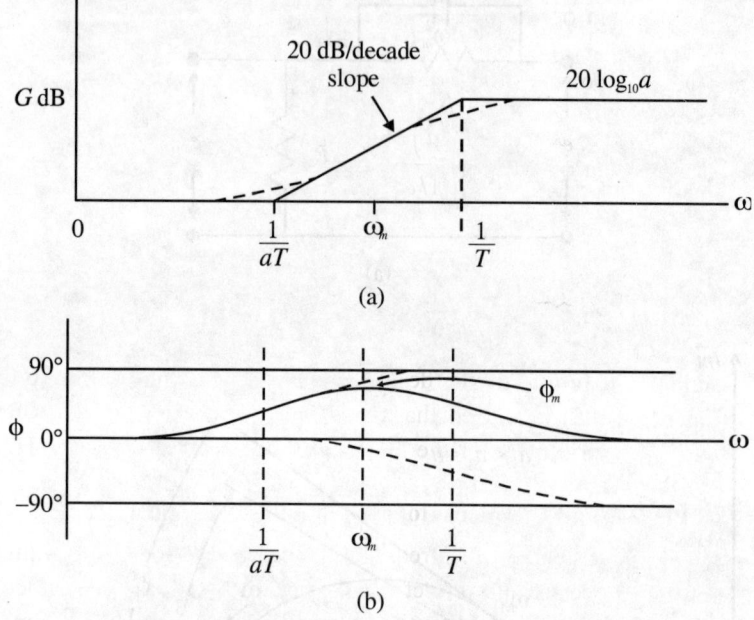

Fig. 14.2. Bode plot of the phase-lead compensator

14.3.1. Relationship between ϕ_m, ω_m, a and T

The phase angle of lead compensator is given by

$$\phi_c = \tan^{-1} a\omega t - \tan^{-1} \omega t \qquad \qquad ...(14.2)$$

The frequency ω_m at which maximum lead angle ϕ_m occurs is determined by setting the derivative of ϕ_c with respect to the frequency ω to zero. Therefore,

$$d\phi_c/d\omega = aT/(1 + a^2\omega^2T^2) - T/(1 + \omega^2T^2)0 \qquad \qquad ...(14.3)$$

or,

$$\omega_m^2 = 1/aT^2 \qquad \qquad ...(14.4)$$

Hence,

$$\omega_m = 1/T\sqrt{a} \qquad \qquad ...(14.5)$$

Let $\tan^{-1} a\omega_m T = A$ and $\tan^{-1}\omega_m T = B$. Then, from Eq. (14.2), we have

$$\tan\phi_m = \tan(A - B)$$

$$= \frac{\tan A - \tan B}{1 + \tan A \tan B} = \frac{a\omega_m T - \omega_m T}{1 + a\omega_m^2 T^2}$$

$$\tan \phi_m = \frac{\omega_m T (a-1)}{1 + a\omega_m^2 T^2} \qquad ...(14.6)$$

Substituting the value of ω_m from Eq. (14.5) in Eq. (14.6), we get

$$\tan \phi_m = \frac{(1/T\sqrt{a})T(a-1)}{1+a(1/aT^2)T^2} = \frac{(a-1)}{2\sqrt{a}} \qquad ...(14.7)$$

Therefore,

$$\sin \phi_m = \frac{(a-1)}{(a+1)} \qquad ...(14.8)$$

Eq. (14.5) and (14.8) are used in the design of phase-lead compensators to obtain the value of a and T. From Eq. (14.5) it is seen that the frequency ω_m at which maximum phase-lead angle occurs is the geometric mean of the two corner frequencies $1/aT$ and $1/T$.

14.3.2. Design of Phase-Lead Compensator by Bode Plot Method

To design linear control systems in the frequency domain, the Bode plot is preferred to other frequency domain plots because the effect of compensation is easily obtained by adding its magnitude and phase angle curves respectively to that of the original system. *The design of phase-lead compensator utilizes the maximum phase lead ϕ_m of the network.* The function of the phase-lead compensator is to increase the phase in the vicinity of the gain-crossover frequency while keeping the magnitude curve of the Bode plot relatively unchanged near that frequency. However, in phase-lead design, the gain-crossover frequency is increased because of the phase-lead network. Therefore, *the design is essentially finding a compromise between the increase in bandwidth and the desired phase-margin.*

The general procedure of the phase-lead compensator design is outlined below. The design specifications are assumed to be the steady-state error and the phase margin.

Step 1: Given the transfer function of the original system $G_p(j\omega)$, determine the value of the gain K according to the steady-state error requirement.

Step 2: Draw the Bode plot (magnitude and ϕ curves) of $G_p(j\omega)$ with K set to the above value.

Step 3: Find the phase margin of the original system from the Bode plot.

Step 4: If the specified phase margin is not met with, estimate the additional phase angle to be provided by the lead compensator. This is given by

$$\phi_m = \text{specified P.M} - \text{P.M of the original system} + \delta$$

where δ is *the safety margin* and is in the range of 5° to 10° and is used as a safety margin to account for the inevitable phase drop of the original system in the vicinity of the new gain crossover frequency.

Step 5: From Eq. (14.8), determine the value of a, $(a > 1)$.

Step 6: The maximum phase lead angle ϕ_m occurs at ω_m which is the geometric mean of the two corner frequencies $1/aT$ and $1/T$. To achieve the maximum phase margin with the value of a already determined, ϕ_m should occur at the new gain-crossover frequency ω'_{cg} which is not known.

The problem now is to locate the two corner frequencies so that ϕ_m occurs at ω'_{cg}. This is achieved graphically as explained below:

(a)   First, the zero-frequency attenuation of the lead compensator is calculated.

$$G_l = 20\log a$$

(b)   The geometric mean ω_m of the two corner frequencies should be so located at the frequency at which the magnitude of $G_p(j\omega)$ in *dB* is equal to $-G_1/2$. Find ω_m from the Bode plot.

Step 7: As ω_m is known, find T and aT.

Step 8: The transfer function of the phase-lead compensator is

$$G_c(j\omega) = \frac{1 + j\omega aT}{1 + j\omega T}$$

Step 9: Draw the Bode plot of the compensated system whose transfer function $G(j\omega)$ is given by

$$G(j\omega) = G_c(j\omega)G_p(j\omega)$$

Step 10: Check that all performance specifications are met with. If not, a new value of ϕ_m must be chosen and the steps from 5 onwards are repeated.

Step 11: If all performance specifications are satisfied, the design is complete.

14.3.3. DESIGN EXAMPLE

Example 14.1. *The open-loop transfer function of a process is given by*

$$G_p(j\omega) = K/[j\omega(1 + j\omega)(1 + j0.2\omega)]$$

Design a phase-lead compensator to meet the following specifications:

(a) $K_v = 3 \text{ sec}^{-1}$

(b) P.M. $\geq 45°$

Solution:

$$G_p(j\omega) = \frac{K}{j\omega(1 + j\omega)(1 + j0.2\omega)}$$

From the steady-state error condition specified in terms of K_v, we find

$$K_v = \lim_{s \to 0} sG_p(s) = K = 3$$

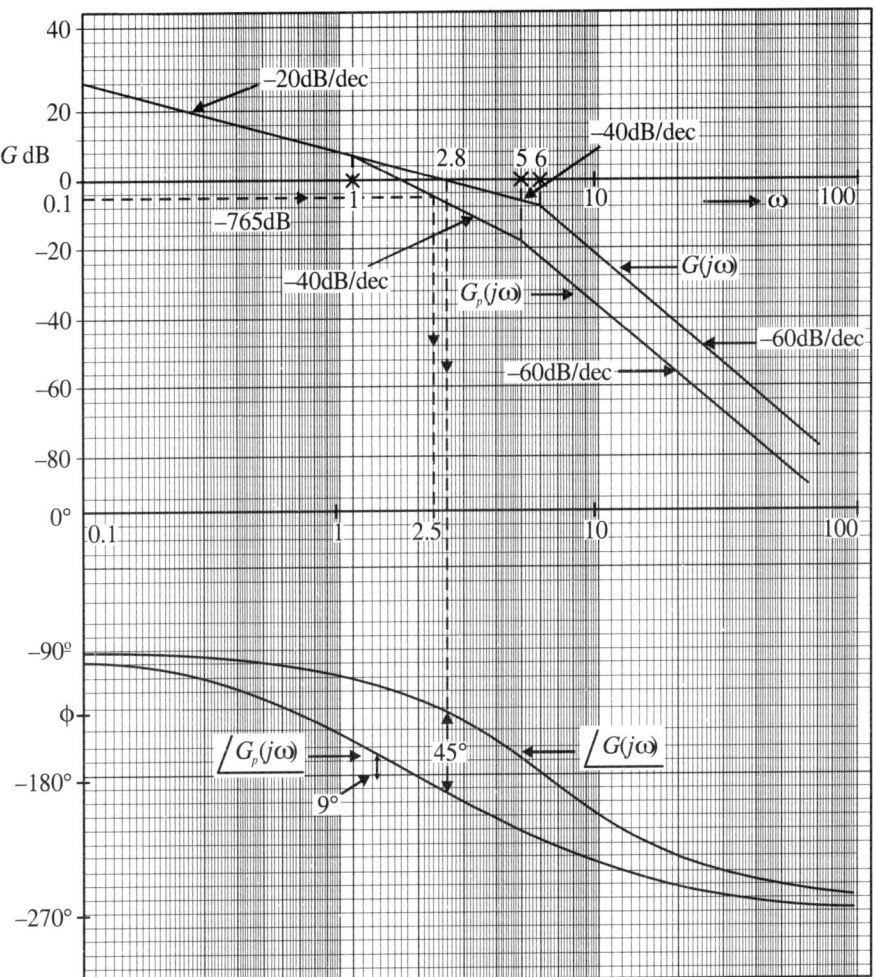

Fig. 14.3. Bode plot of $G_p(j\omega) = K/[j\omega(1 + j\omega)(1 + j0.2\omega)]$ for $K = 3$

The Bode plot with $K = 3$ is shown in Fig 14.3. The phase margin is found to be 9°. Therefore, $\phi_m = 45° - 9° + 9° = 45°$. The value of a found from Eq. (14.8) is 5.8. The zero-frequency attenuation of the lead compensator is 20 log 5.8 = 15.3dB. ω_m is to be located at the frequency at which $G_{dB} = -15.3/2 = -7.65dB$ and, from the Bode plot, it is found to be 2.5 rad./sec.

$$\omega_m = 2.5 \text{rad./sec}$$

$$T = 1/ \omega_m \sqrt{a} = 0.167$$

Assuming $a = 6$, $aT = 6 \times 0.167 = 1.0$

Therefore the transfer function of the phase-lead compensator is

$$G_c(j\omega) = \frac{1 + j\omega}{1 + j0/167\omega}$$

The over all open-loop transfer function $G(j)$ is

$$G(j\omega) = G_c(j\omega)G_p(j\omega)$$

$$= \frac{3}{j\omega(1 + j0.2\omega)(1 + j0.167\omega)}$$

The Bode plot of $G(j\omega)$ is shown in Fig. 14.3. It is found that the phase-margin is 45°.

14.3.4. Effects of Phase Lead Compensation

The general effects of phase lead compensation on the performance of control systems are:

1. The phase margin is improved due to increase in the phase of the open-loop transfer function in the vicinity of the gain-crossover frequency.

2. At the gain-crossover frequency, the slope of the magnitude curve of the open-loop transfer function is reduced which corresponds to an improvement in the gain and phase margins of the system. Thus, the relative stability of the system is improved.

3. The bandwidth is increased because of the increase in gain-crossover frequency. This gives rise to noise problem.

4. The overshoot of the step response is reduced.

5. The steady-state error is not affected.

In general, the phase lead compensation is ineffective if the original system is unstable because the additional phase lead required is excessive. Even though the large phase lead required may be achieved by cascading two or more phase lead compensators, the resulting system may still be unsatisfactory because a portion of the phase curve may still be below the $-180°$ line which corresponds to a conditionally stable system.

14.4. PHASE-LAG COMPENSATOR

The circuit of a phase-lag compensator is shown in Fig. 14.4(a). The transfer function of the phase-lag compensator is given by

$$G_c(j\omega) = \frac{1 + j\omega bT}{1 + j\omega T} \qquad (b < 1) \qquad\qquad ...(14.9)$$

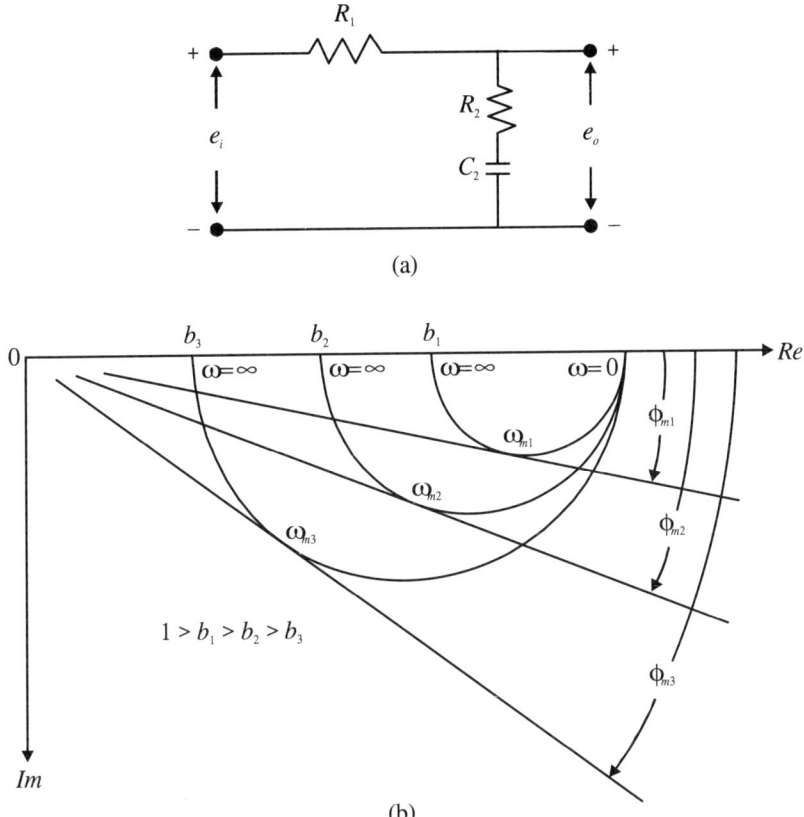

Fig. 14.4. (a) Phase-lag compensator circuit (b) Polar plot for various values of *b*

Fig. 14.4(b) shows the polar plot of $G_c(j\omega)$ as a function of ω for various values of *b*. *The angle ϕ_m between the tangent line drawn from the origin to the polar plot and the real axis is the maximum phase lag of the network.* The frequency ω_m at the tangent point is the frequency at which ϕ_m occurs. As the values of *b* decrease, ϕ_m becomes more negative. As *b* approaches zero, ϕ_m tends to $-90°$ and ω_m increases.

The Bode plot of the phase-lag compensator is shown in Fig. 14.5. The compensator has a pole at $\omega = 1/T$ and a zero at $\omega = 1/bT$. Since $b < 1$, the pole occurs first and the angle is negative. The lag compensator is equivalent to a low-pass filter since it passes the frequencies from 0 to $\omega = 1/T$ without attenuation. The lag angle is introduced in the frequency range from $\omega = 1/T$ to $\omega = 1/bT$.

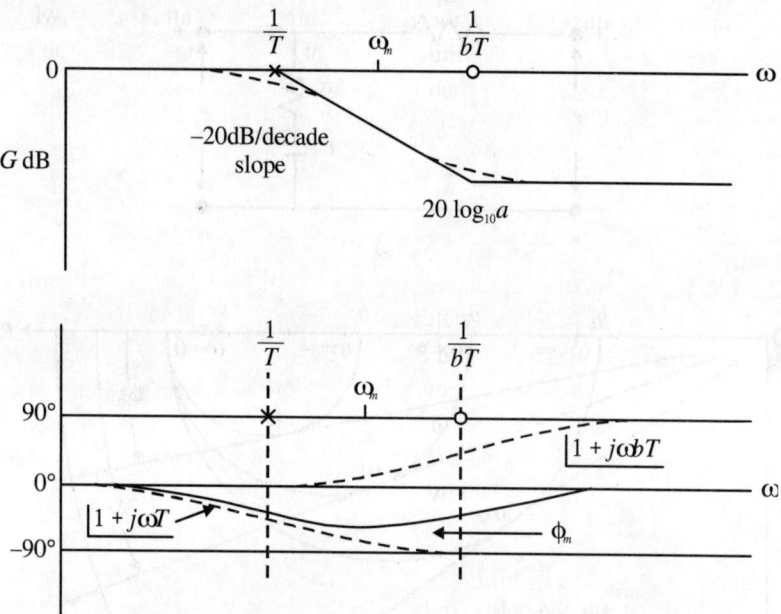

Fig. 14.5. Bode diagram of a Phase-lag compensator

14.14.1. Design of Phase-Lag Compensator by Bode Plot Method

The phase-lag compensator design utilizes the *attenuation characteristic of the network at the high frequencies* rather than the maximum phase-lag angle of the network. In the phase-lag compensation, the objective is to move the gain-crossover frequency to a lower value while keeping the phase curve of the Bode plot relatively unchanged at the gain-crossover frequency.

The general design procedure for phase-lag compensator using the Bode plot is outlined below.

Step 1: Given the open-loop transfer function of the process $G_p(j\omega)$, determine the open-loop gain K to satisfy the steady-state error requirement.

Step 2: Draw the Bode plot (magnitude and ϕ-curves) of $G_p(j\omega)$ with K set to the above value.

Step 3: Find the frequency ω'_{cg} where the required phase margin is found. The required phase margin is given by

$$\text{Required P.M.} = \text{Specified P.M.} + \delta \qquad\qquad ...(14.10)$$

where δ is the *safety margin* to account for the small negative phase angle of the lag compensator and is in the range of $5°$ to $10°$. The gain-crossover frequency of the compensated system must be located at ω'_{cg}.

Step 4: In order to bring the magnitude curve down to 0 dB at ω'_{cg}, the phase-lag compensator must provide the required amount of attenuation which is equal to the gain in dB of the magnitude curve at ω'_{cg}. If the gain of the uncompensated system at ω'_{cg} is $G_p(j\omega'_{cg})$ then

$$|G(j\omega'_{cg})| = -20\log b \qquad\qquad b < 1 \qquad\qquad ...(14.10)$$

Therefore,

$$b = 10^{-|G_p(j\omega'_{cg})/20|} \qquad\qquad b < 1 \qquad\qquad ...(14.11)$$

Step 5: Once the value of b is determined in Step 4, it is necessary only to select the proper value of T to complete the design. Since the phase curve of the original uncompensated system should be relatively unchanged near the new gain-crossover frequency ω'_{cg}, it is seen from Fig. 14.5, that the zero of the lag compensator should be placed far below ω'_{cg}. Since the phase lag of the lag compensator at a frequency a decade above $1/bT$ is in the order of 6°, it is the usual practice to locate $1/bT$ one decade below ω'_{cg}. Thus

$$1/bT = \omega'_{cg}/10 \qquad\qquad \text{rad./sec} \qquad\qquad ...(14.12)$$

or,

$$1/T = b\omega'_{cg}/10 \qquad\qquad \text{rad./sec} \qquad\qquad ...(14.13)$$

Step 6: The transfer function of the phase-lag compensator is

$$G_c(j\omega) = \frac{1 + j\omega bT}{1 + j\omega T} \qquad\qquad (b < 1) \qquad\qquad ...(14.14)$$

Step 7: Draw the Bode plot of the compensated system $G(j\omega)$ given by

$$G(j\omega) = G_c(j\omega)G_p(j\omega)$$

Step 8: Check that all performance specifications are met with. If not, a new value of ϕ_m must be chosen and the step from 5 onwards are repeated.

Step 9: If all performance specifications are met with, the design is complete.

14.4.2. Design Example

Example 14.2. *The open-loop transfer function of a process is given by*

$$G_p(j\omega) = \frac{K}{j\omega(1 + j\omega)(1 + j0.2\omega)}$$

Design a phase lag compensator to meet the following specifications:

(1) $K_v = 3 \text{ sec}^{-1}$

(2) PM $\geq 45°$

Solution:

From the first specification, $K = 3$. The Bode plot is drawn with $K = 3$ and shown in Fig 14.3. The phase-margin is found to be 9°.

The required phase margin is 50° (= 45° + 5°). The frequency at which this occurs is $\omega'_{cg} = 0.65$ rad./sec.

From the Bode plot, we find that $|G_{dB}|$ at ω'_{cg} is 12 dB.

Therefore, b is found from

$$b = 10^{-12/20} = 0.25 \qquad (b < 1)$$
$$1/bT = \omega'_{cg}/10$$

Or,

$$T = 10/b\omega'_{cg} = 10/(0.25 \times 0.65)$$
$$= 61.5$$
$$bT = 15.4$$

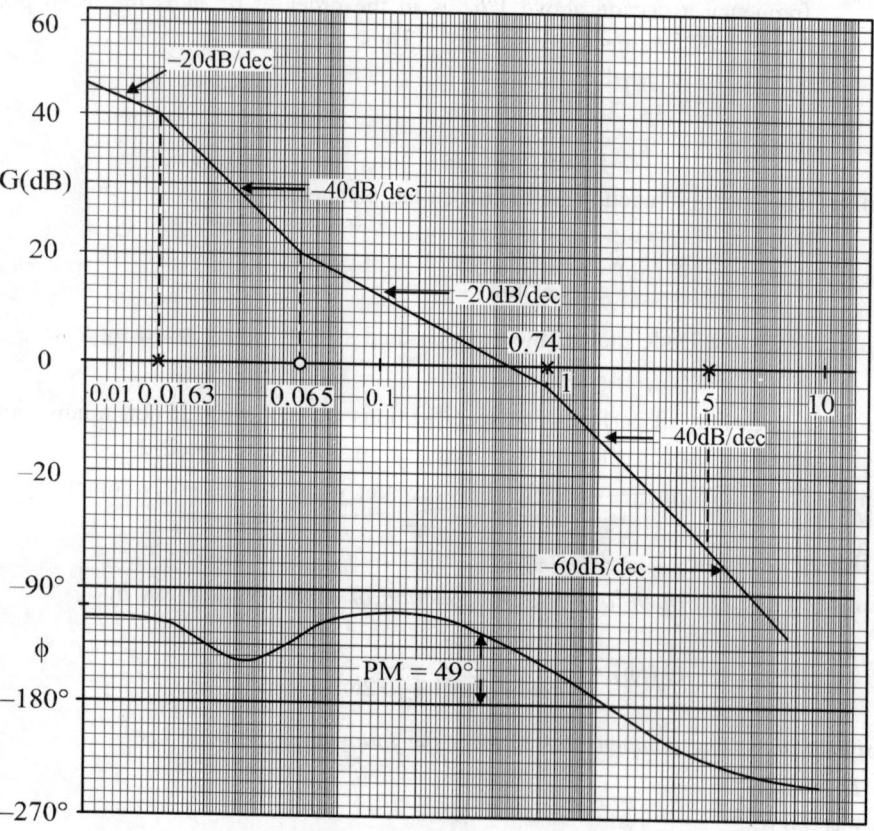

Fig. 14.6. Bode plot of compensated system of Example 14.2

Hence,

$$G_c(j\omega) = \frac{1 + j15.4\omega}{1 + j61.5\omega}$$

The transfer function of the compensated system is

$$G(j) = \frac{3(1 + j15.4\omega)}{j\omega(1 + j\omega)(1 + j0.2\omega)(1 + j61.5\omega)}$$

The Bode plot is shown in Fig. 14.6 and the phase margin is found to be 49°.

14.4.3. Effects of Phase Lag Compensation

The general effects of phase lag compensation on the system performance are:

1. For a specified relative stability, the steady-state error for a ramp input is reduced.

2. The bandwidth is decreased due to decrease in the gain–crossover frequency.

3. For a specified gain K, the gain and phase margins are increased and the resonance peak M_p is reduced due to attenuation at the gain crossover frequency.

4. The rise time is increased due to decrease in bandwidth.

14.5. LAG-LEAD COMPENSATOR

The lag compensator increases the gain of the system with a consequent reduction in the steady-state error whereas a lead compensator increases the resonance frequency ω_p and reduces the settling time. If both the steady-state error and the settling time are to be reduced, a lag-lead compensator should be used.

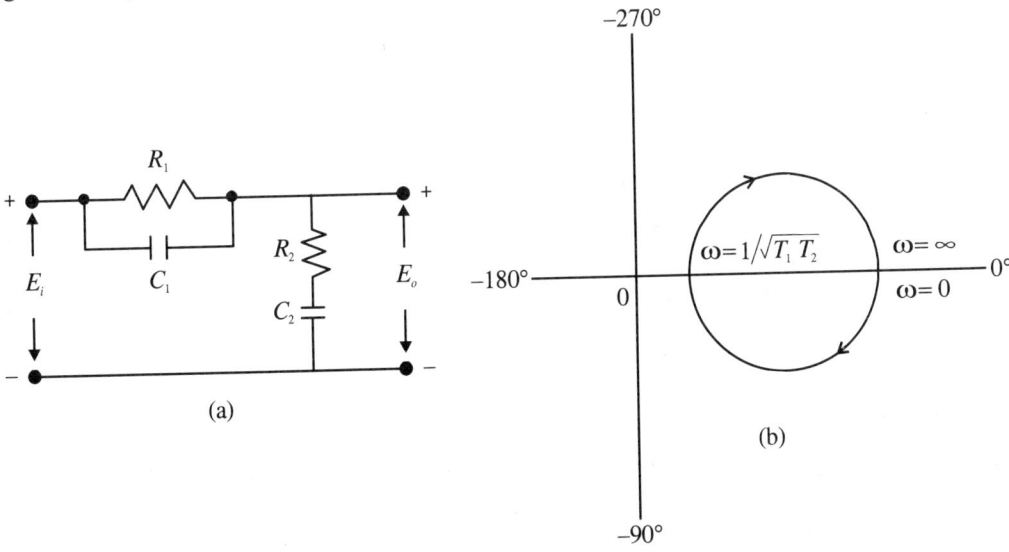

Fig. 14.7. (a) Lag-lead compensator circuit (b) Polar plot of the compensator

The circuit of a lag-lead compensator is shown in Fig. 14.7(a). The transfer function is given by

$$G_c(j\omega) = \frac{(1+ j\omega aT_1)}{(1+ j\omega T_1)} \frac{(1+ j\omega bT_2)}{(1+ j\omega T_2)}$$
$$\underbrace{\qquad\qquad}_{\text{(Lead compensator)}} \underbrace{\qquad\qquad}_{\text{(Lag compensator)}}$$

...(14.14)

where $ab = 1$, $a > 1$ and $b < 1$. Fig. 14.7 (b) shows the polar plot of the lag-lead compensator. The lag compensator is effective in the frequency range $0 \le \omega < 1/T_1T_2$ and the lead compensator in the frequency range $1/T_1T_2 < \omega \le \infty$.

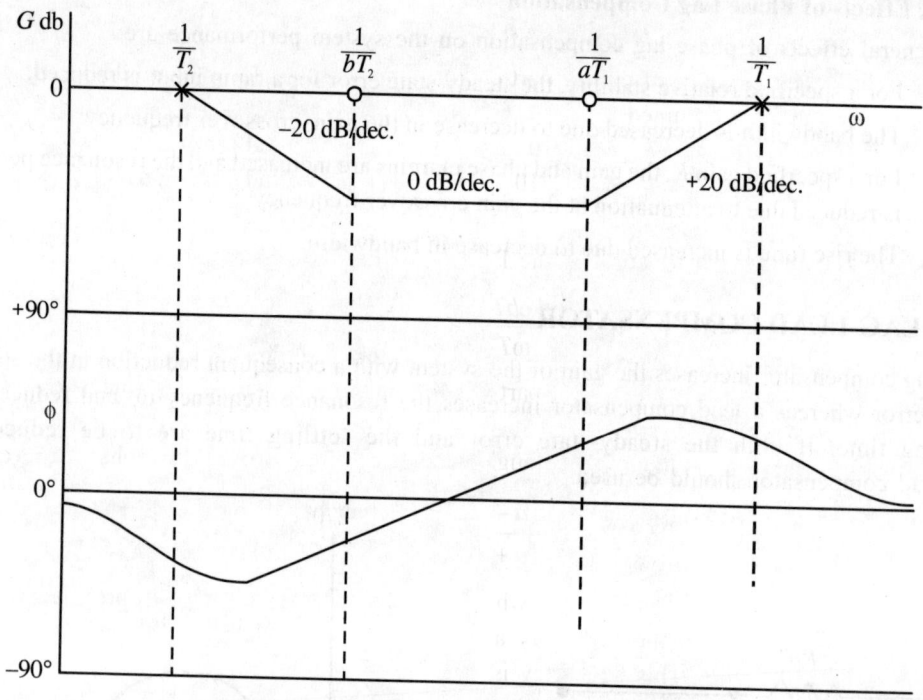

Fig. 14.8. Bode plot of lag-lead compensator

The Bode plot of the lag-lead compensator is shown in Fig. 14.8. The compensator has two poles and two zeros. The pole at $\omega = 1/T_2$ and the zero at $\omega = 1/bT_2$ are due to the lag network and the zero at $1/aT_1$ and the pole at $1/T_1$ are due to lead network.

14.5.1. Design of a lag-lead compensator by Bode Plot Method

There is no fixed procedure for the design of the lag-lead compensator. Usually, a trial and error method using the design techniques outlined for the phase-lag and phase-lead compensators is followed. First, the lag portion of the compensator is designed and then the lead part.

The design procedure for lag-lead compensator is outlined below:

Step 1: Given the open-loop transfer function $G_p(j\omega)$ of the process, determine the gain K to satisfy the steady-state error requirement.

Step 2: Draw the Bode plot (magnitude and ϕ curves) of $G_p(j\omega)$ with K set to the above value.

Step 3: To design the lag part of the compensator first, find the new gain-crossover frequency ω'_{cg} where the required phase margin is found.

Step 4: From the magnitude curve, find the magnitude in *dB* at ω'_{cg}. Let it be $G_p(j\omega'_{cg})$.

Step 5: The value of b is found from the Eq. (14.11).

$$b = 10^{-|G_p(j\omega'cg)|/20}$$

Step 6: Then, the zero of the lag part of the compensator is located one decade below ω'_{cg} and T_2 is obtained.

$$1/bT_2 = \omega'_{cg}/10 \qquad\qquad \text{rad./sec.}$$
$$1/T_2 = b\omega'_{cg}/10 \qquad\qquad \text{rad./sec.}$$

Step 7: The transfer function of the lag part of the compensator is

$$G_{c1} = \frac{1+ j\omega bT_2}{1+ j\omega T_2}$$

Step 8: To design the phase-lead part of the compensator, first determine

$$a = 1/b$$

Step 9: Find the maximum phase angle ϕ_m that could be provided by phase lead part.

$$\phi_m = \sin^{-1}\frac{(a-1)}{(a+1)}$$

Step 10: Find the frequency ω_m at which the phase angle ϕ_m is to be provided. For this purpose, find the frequency at which the magnitude of $G_p(j\omega)$ in *dB* is equal to $(20\log a)/2$. This frequency is ω_m.

Step 11: As ω_m is known, find T and aT.

Step 12: The transfer function of the lead part of the compensator is

$$G_{c2} = \frac{1+ j\omega aT_1}{1+ j\omega T_1}$$

Step 13: The transfer function of the compensated system is

$$G(j\omega) = G_{c1}(j\omega) \cdot G_{c2}(j\omega) \cdot G_p(j\omega)$$

Step 14: Draw the Bode plot of $G(j\omega)$. Check all performance specifications are met with. If not, repeat.

14.5.2. Design Example

Example 14.3. *The open-loop transfer function of a system is given by*

$$G_p(j\omega) = \frac{K}{j\omega(1+j\omega)(1+j0.2\omega)}$$

Design a lag-lead compensator to meet the following performance specifications:
1. $K_v = 3 \text{ sec}^{-1}$
2. Phase margin $\geq 45°$

Solution:

From the specification of K_v, we find the value of K as

$$K_v = \lim_{s \to 0} s\, G_p(s) = K = 3$$

The Bode plot of $G_p(j\omega)$ with $K = 3$ is shown in Fig. 14.9. To design the lag part of the compensator first, we find the required phase-margin ϕ_m. $\phi_m = 45° + 5° = 50°$. The frequency at which it occurs is $\omega'_{cg} = 0.65\text{rad./sec}$. The value of $|G_{pdB}|$ at $\omega'_{cg} = 12dB$. Hence b is

$$b = 10^{-12/20} = 0.25$$

Therefore $T = 61.5$ and $aT = 15.4$.

$$G_{c1}(j\omega) = \frac{1+j15.4\omega}{1+j61.5\omega}$$

To design the lead compensator, we obtain

$$a = 1/b = 4$$

Therefore,

$$\phi_m = \sin^{-1}\frac{(a-1)}{(a+1)} = 37°$$

This phase lead is provided at $\omega_m = 2.5\text{rad./sec}$. Assuming $a = 6$, $T = 1/(\omega_m\sqrt{a} = 0.67$ and $aT = 1.0$. Therefore the phase lead part of compensator is

$$G_{c2}(j\omega) = \frac{1+j\omega}{1+j\,0.167\omega}$$

The open-loop transfer function of the compensated system is

$$G(j\omega) = G_{c1}(\omega)G_{c2}(j\omega)G_p(j\omega)$$

$$= \frac{3(1+j15.4\omega)}{j\omega(1+j0.2\omega)(1+j0.167\omega)(1+j61.5\omega)}$$

The Bode plot of $G(j\omega)$ is shown in Fig. 14.9. The phase margin is found to be 51°.

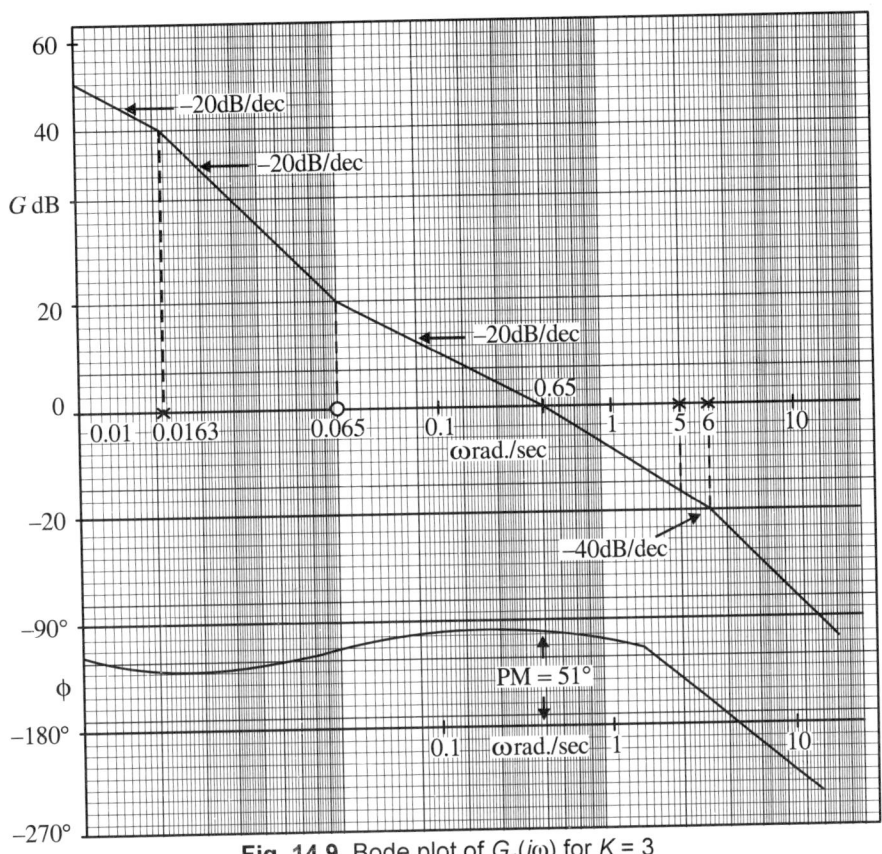

Fig. 14.9. Bode plot of $G_p(j\omega)$ for $K = 3$

14.6. COMPARISON OF LEAD, LAG AND LAG-LEAD COMPENSATION

1. Phase lead compensation achieves the desired result through its phase lead contribution whereas lag compensation accomplishes this through its high-frequency attenuation.

2. Lead compensators are used when large bandwidth and low settling time are required. However, if noise signals are present, lag compensator is preferred.

3. Lag compensators improve steady-state accuracy and reduces the bandwidth.

4. If both fast response and good steady-state accuracy are required. Lag-lead compensators are used.

5. Lead compensation requires an additional increase in gain to offset the inherent attenuation. Thus lead compensation requires larger gain than lag compensation.

14.7. Bridged-T Compensator

The bridged-T compensators are used for pole-zero cancellation of dominant complex poles of control systems. The reason for this approach is that, it is easier to design the compensator by canceling the undesirable complex poles of the process transfer function.

In this section, we shall discuss the properties of the bridged-T networks in the frequency domain and illustrate the design with a suitable example.

The circuit of the bridged-T compensators is shown in Fig. 8.18. The transfer function is of the form

$$G_c(s) = \frac{s^2 + 2\zeta_z\, \omega_{nz} s + \omega_n^2}{s^2 + 2\zeta_p\, \omega_{np} s + \omega_n^2}$$

with $\omega_{nz} = \omega_{np} = \omega_n$. The maximum attenuation provided at the resonant frequency ω_n is given by

$$G_c = |G_c(j\omega_n)| = \frac{\zeta_z}{\zeta_p} \qquad\qquad ...(14.15)$$

Substituting $\zeta_p = \dfrac{1+2\zeta_z^2}{2\zeta_z}$ Eq. (8.77) into the above equation, we get

$$\zeta_z^2 = G_c/[2(1-G_c)] \qquad\qquad ...(14.16)$$

Thus, given the amount of maximum attenuation required, the damping ratio of the zeros of the bridged-T network is given by Eq. (14.16). Once the location of the maximum attenuation is determined, the resonant frequency ω_n is known. Then using Eq. (8.77) the poles of the bridged-T network are determined.

The polar plot of the bridged-T compensator is shown in Fig. 14.10 as a function of ω. The maximum attenuation occurs at $\omega = \omega_n$ and the phase at this point is zero.

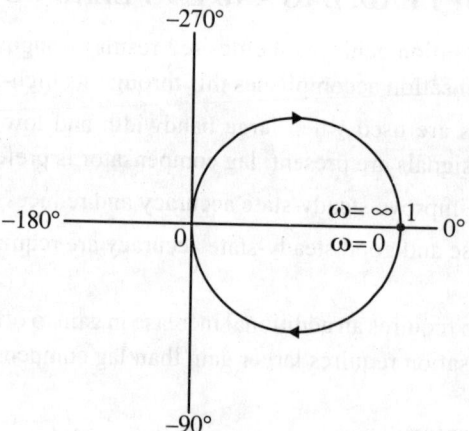

Fig. 14.10. Polar plot of the bridged-T compensator

The Bode plot is shown in Fig. 14.11. The magnitude plot of the bridged-T network typically has a "notch" at the resonant frequency ω_n. The phase plot is negative below ω_n and positive above ω_n and is zero at ω_n. The phase angle and the magnitude in the vicinity of the resonant frequency ω_n change steeply whereas the magnitude and phase angle contributed by the network do not affect the low-and high-frequency properties of the original system. This is the advantage of the bridged-T network over phase-lag and phase-lead networks. The attenuation of the magnitude curve and the positive phase characteristics are effectively used to improve the stability of a system. Because of the "notch" characteristics in the amplitude curve, the bridged-T networks are also known as *notch networks*.

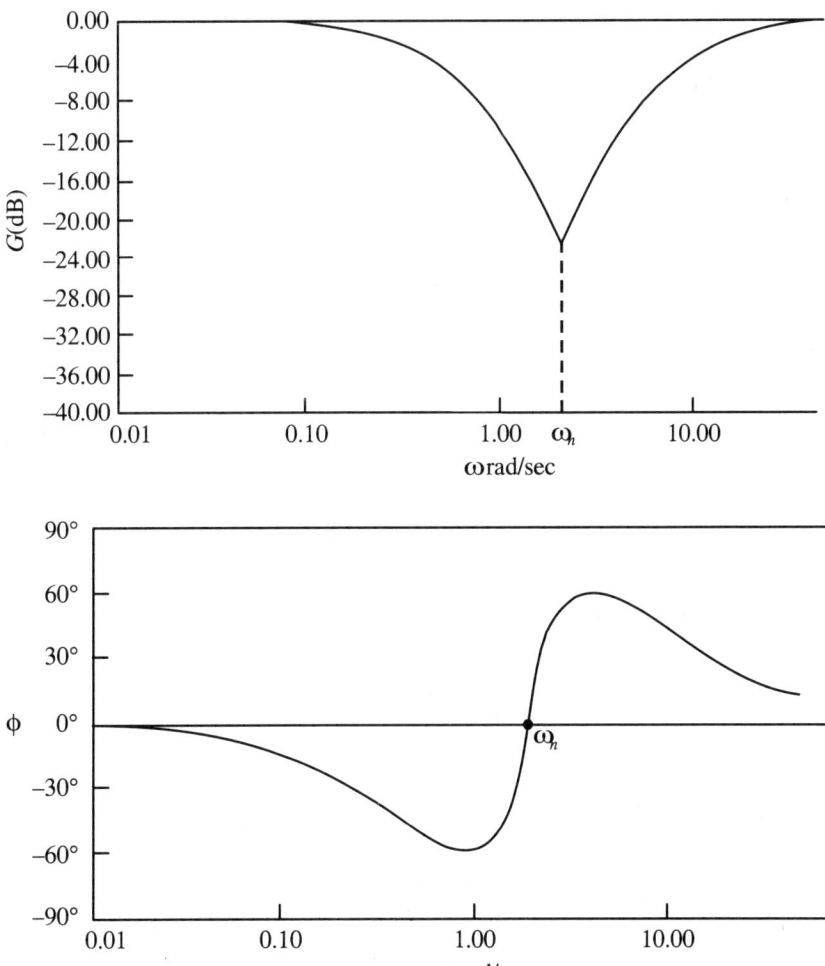

Fig. 14.11. Bode plot of the bridged-T compensator

14.7.1. Design Example

Example 14.4. *The transfer function of a controlled process is given by*

$$G_p(s) = \frac{1+10s}{s(1+0.2s+0.25s^2)}$$

Design a bridged-T compensator.

Solution:

The transfer function of the pole-zero cancellation bridged-T compensator as designed in Section 8.7 is

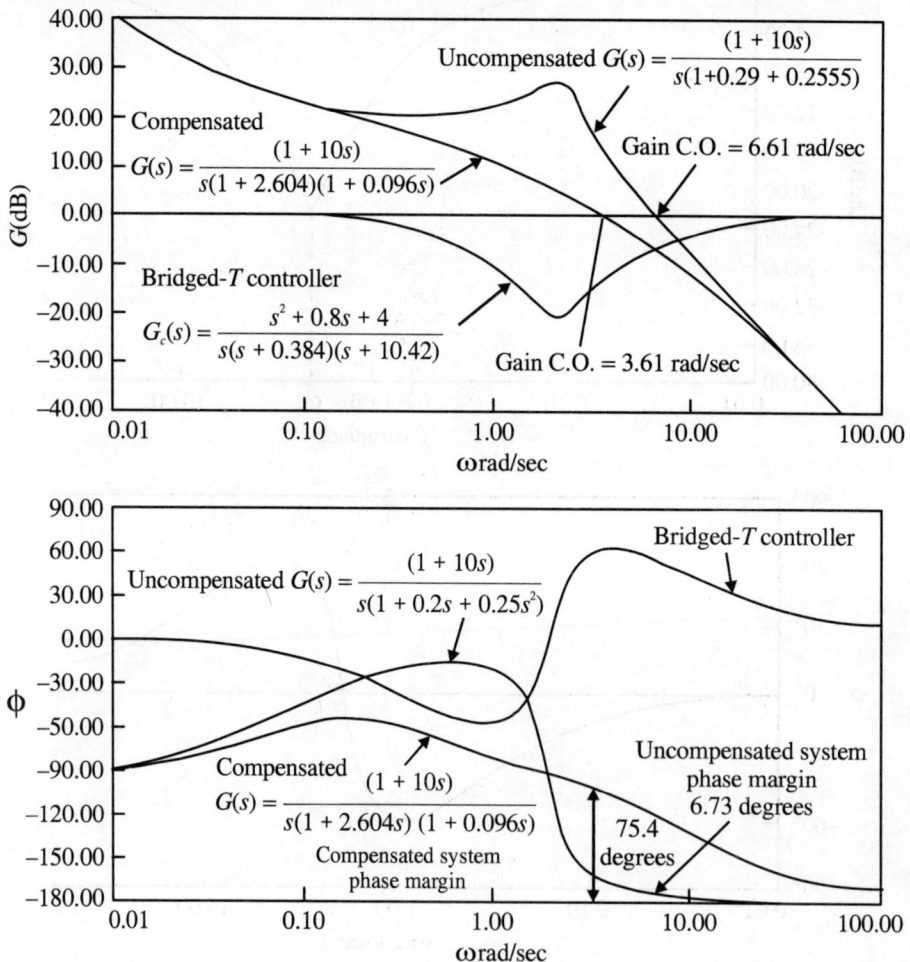

Fig. 14.12. Bode plots of the compensated and uncompensated systems of Example 14.4.

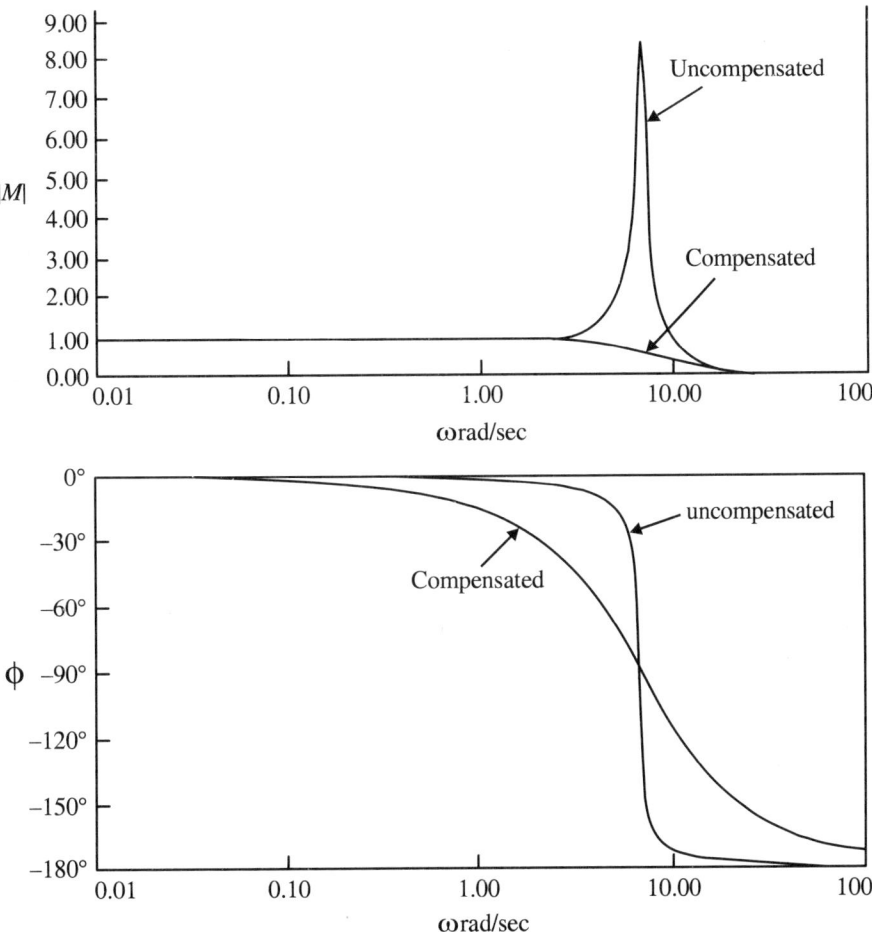

Fig. 14.13. Closed loop frequency response of compensated and uncompensated systems of Example 14.4.

$$G_p(s) = \frac{s^2 + 0.8s + 4}{(s + 0.384)(s + 10.42)}$$

The Bode plots of the compensated and the uncompensated systems are shown in Fig. 14.12. The closed-loop frequency responses are shown in Fig. 14.13.

The characteristics of the uncompensated system in the frequency domain are:

(i) Gain margin = Infinite

(ii) Phase margin = 6.73°

(iii) $M_p = 8.52$ (18.6dB)

The compensated system has the following performance characteristics:

(i)   Gain margin = Infinite

(ii) Phase margin = 75.4°

(iii) $M_p = 1.0$ (0dB)

From the Bode plot of Fig. 14.12, it is seen that the pole-zero cancellation provides an attenuation of –22dB at 2 rad./sec. This corresponds to the peak of 22dB of the magnitude of the uncompensated system at the same frequency.

14.8. ADDITIONAL EXAMPLES

Example 14.5. *The open-loop transfer function of a system is*

$$G_p(s) = \frac{2500K}{s(s+25)}$$

The performance specifications are:

1.   *The steady-state error due to a ramp input should be less than or equal to 1%*

2.   *Phase margin > 45*

Design a phase lead compensator to meet the above specifications.

Solution:

The value of K is found from the steady-state error specifications.

The steady-state error for unit ramp input is

$$e_{ss} = 1/K_v$$

$$K_v = \lim_{s \to 0} s G_p(s) = \lim_{s \to 0} s \frac{2500K}{s(s+25)} = 100K$$

Hence,

$$(1/100K) = 0.01$$

$$K = 1$$

The Bode plot of $G_p(j\omega)$ with $K = 1$ is shown in Fig. 14.14. The gain-crossover frequency $\omega_{cg} = 47$ rad./sec and the phase margin is 28°. Hence the additional lead angle required is

$$\phi_m = 45° - 28° + 8° = 25°$$

Therefore,

$$a = 2.46$$

The zero-frequency attenuation of the lead compensator is

$$G_{dB} = 20\log a = 20\log 2.46 = 7.82dB$$

Therefore, ω_m is placed at that frequency where

$$G_{pdB} = -7.82/2 = -3.91dB$$

This frequency is 60 rad./sec from the Bode plot. Therefore

$$1/T = \omega_m \sqrt{a} = 60\sqrt{2.46} = 94 \text{ rad./sec}$$

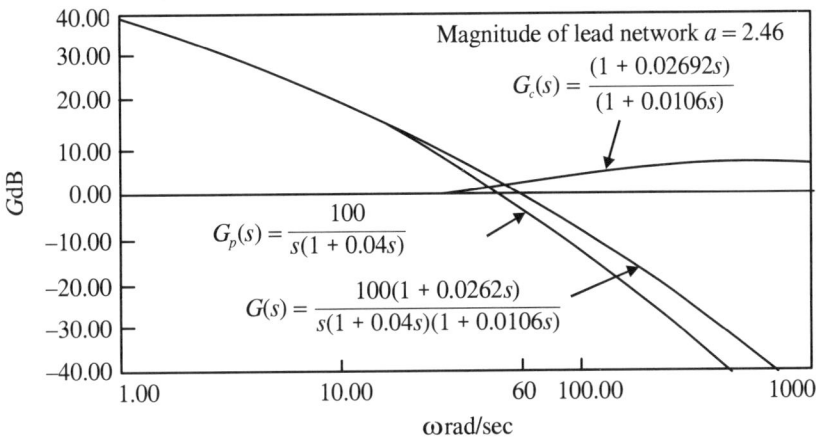

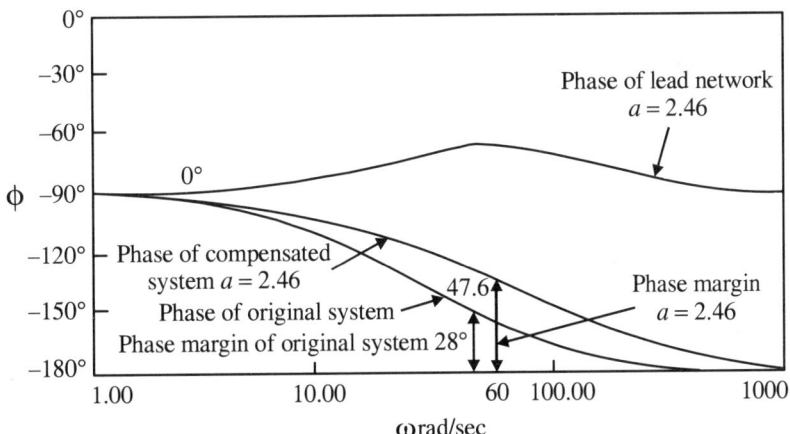

Fig. 14.14. Bode plot of $G_p(j\omega)$ with $K = 1$, $G_c(j\omega)$ and $G(j\omega)$

$$T = 0.0106$$

$$1/aT = 94/2.46 = 38.2 \text{ rad./sec}$$

$$aT = 1/38.2 = 0.0262$$

The transfer function of the compensator is

$$G_c(s) = \frac{1 + j0.0262\omega}{1 + j0.0106\omega}$$

The transfer function of the compensated system is

$$G(j\omega) = \frac{2500(1 + j0.0262\omega)}{j\omega(25 + j\omega)(1 + j0.0106\omega)}$$

$$= \frac{100(1+j0.0262\omega)}{j\omega(1+j0.04\omega)(1+j0.0106\omega)}$$

The Bode plot of $G(j\omega)$ is shown in Fig. 14.14. The phase margin is found to be 47.6°.

Example 14.6. *Design a phase lag compensator for the system considered in Example* 14.5.

Solution:

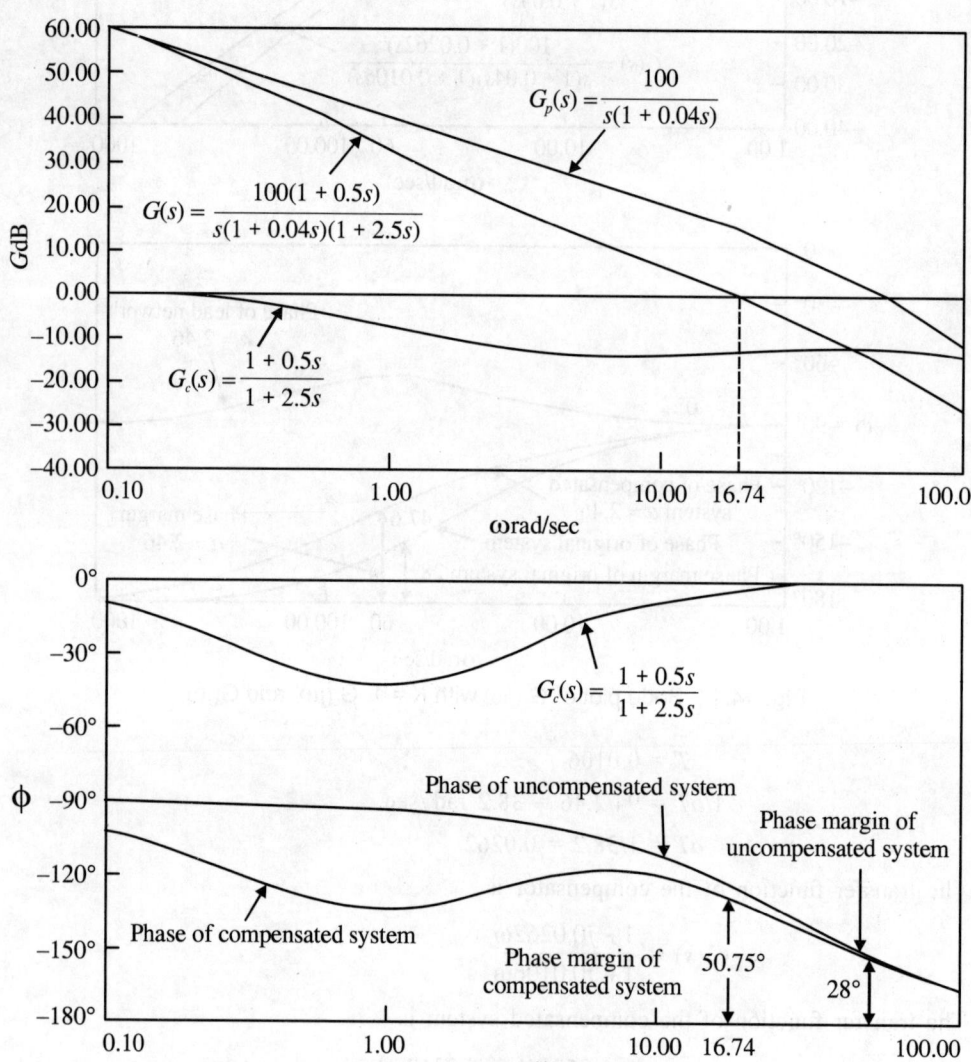

Fig. 14.15. Bode plot of Example 14.6

Since the specifications and the system are same, the value of $K = 1$. The Bode plot with $K = 1$ is drawn in Fig.14.15. The required phase margin is $50°(45° + 5°)$ and occurs at $\omega'_{cg} = 20$ rad./sec. The G_{dB} at 20 rad./sec is $14dB$. Hence

$$a = 10^{-14/20} = 0.2$$

$$1/aT = \omega'_{cg}/10 = 2 \text{ rad./sec}$$

$$aT = 0.5$$

$$1/T = 2 \times a = 0.4 \text{ rad./sec}$$

$$T = 2.5$$

The transfer function of the lag compensator is

$$G_c = \frac{1 + j0.5\omega}{1 + j2.5\omega}$$

The open-loop transfer function of the compensated system is

$$G(j\omega) = \frac{100(1 + j0.5\omega)}{j\omega(1 + j0.04\omega)(1 + j2.5\omega)}$$

The Bode plot of $G(j\omega)$ is shown in Fig. 14.15. The phase margin is found to be $50.75°$.

Example 14.7 *The open-loop transfer function of a system is*

$$G_p(s) = \frac{K}{s(1 + 0.1s)(1 + 0.2s)}$$

Design a lag-lead compensator to meet the following specifications:

1. $K_v = 100 \text{ sec}^{-1}$
2. P.M. $\geq 30°$

Solution:

The system gain K is obtained from K_v as

$$K_v = \lim_{s \to 0} sG_p(s) = K = 100$$

The Bode plot with $K = 100$ is shown in Fig. 14.16. The required phase margin of 30° occurs at $\omega'_{cg} = 5$ rad./sec and $|G_{dB}|$ at this point is $22dB$. Hence

$$b = 10^{-22/20} = 0.08$$

$$1/bT_2 = \omega'_{cg}/10 = 0.5 \text{rad./sec}$$

$$bT_2 = 2$$

$$1/T_2 = 0.5 \times b = 0.04 \text{rad./sec}$$

$$T_2 = 25$$

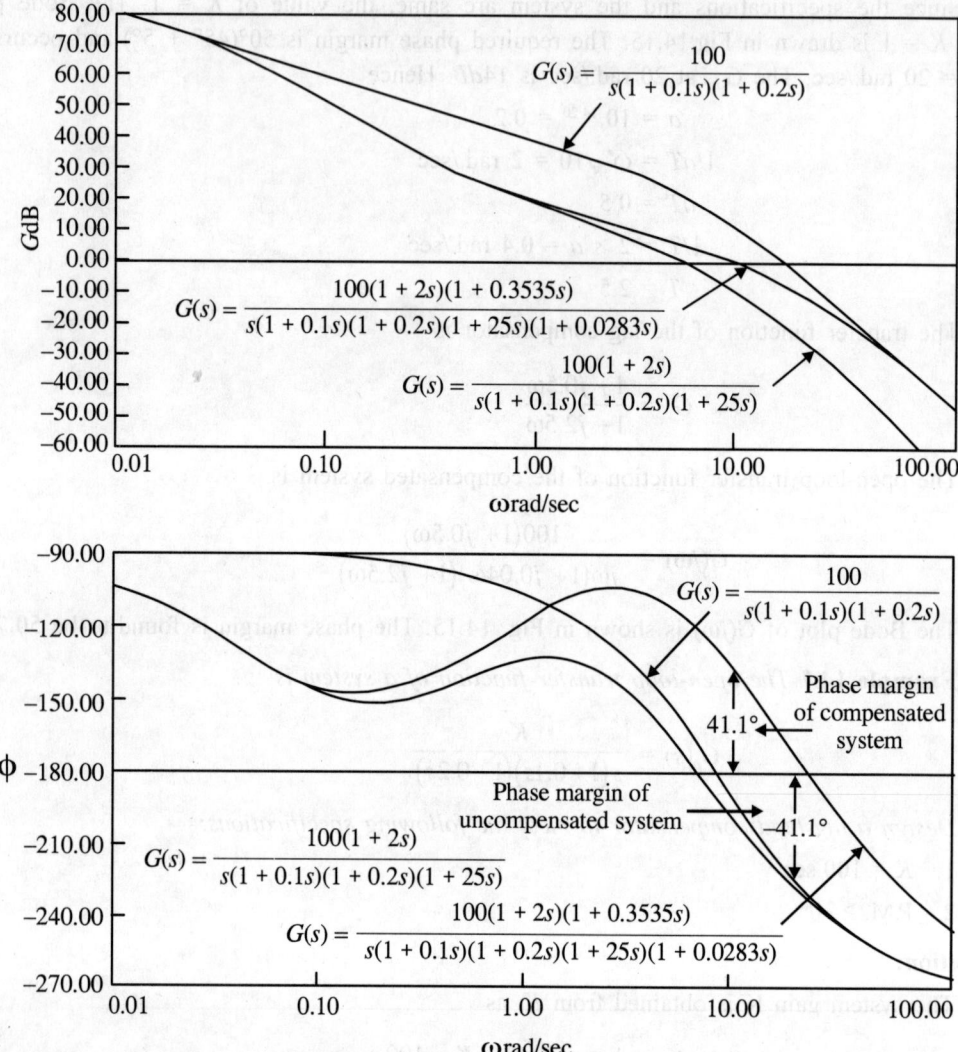

Fig. 14.16. Bode plot of Example 14.7

The transfer function of the lag compensator is

$$G_{c1}(j\omega) = \frac{1+j2\omega}{1+j25\omega}$$

The value of a for phase lead compensator is

$$a = 1/b = 1/0.08 = 12.5$$

$$\sin \phi_m = \frac{(a-1)}{(a+1)} = \frac{11.5}{13.5} = 0.85$$

$$\phi_m = 58°$$

The zero-frequency attenuation of the phase lead compensator is

$$20\log a = 20\log 12.5 = 21.9 dB$$

The new gain-crossover frequency ω_m at $G_{dB} = 10.95$ is 10 rad./sec.

$$1/T_1 = \omega_m \sqrt{a} = 10.95 \sqrt{12.5} = 35.35 \text{ rad/sec.}$$

$$T_1 = 0.028$$

$$1/aT_1 = 35.35/12.5 = 2.83 \text{ rad./sec}$$

$$aT_1 = 0.354$$

The transfer function of the lag-lead compensator is

$$G_c(j\omega) = \frac{(1+j0.354\omega)(1+j2\omega)}{(1+j0.028\omega)(1+j25\omega)}$$

The open-loop transfer function of the compensated system is

$$G(j\omega) = \frac{100(1+j0.354\omega)(1+j2\omega)}{j\omega(1+j0.1\omega)(1+j0.2\omega)(1+j0.028\omega)(1+j25\omega)}$$

The Bode plot of $G(j\omega)$ is shown in Fig. 14.16 and the phase margin is found to be 41°.

14.9. SUMMARY

In this chapter, we have discussed the need for compensators. They are to reshape the frequency response of the original system by using suitable compensators in order to satisfy the specified system performance. We have then explained the design of lag, lead, lag-lead and bridged-T compensators. We have been primarily concerned only with the transfer function of compensators. In addition, we must satisfy additional design constraints such as cost, size, weight and reliability. Under normal operating conditions, the system designed may meet the performance specifications but may deviate considerably from the specifications due to environmental changes. It is therefore necessary to provide automatic or manual gain control to compensate for such environmental changes, for nonlinear effects which were not taken into account in the design and also to compensate for manufacturing tolerances of system components. We have adequately illustrated the design procedure by designing all types of compensators discussed here.

REVIEW QUESTIONS

14.1. State the need for using compensators.

14.2. Mention the different types of compensators.

14.3. Write down the transfer function of a phase-lag, phase-lead and phase lag-lead compensator.

14.4. Mention the effects of phase-lag compensator.

14.5. Mention the effects of phase-lead compensator.

14.6. Compare phase-lag, phase lead and phase-lag-lead compensator.

EXERCISE

14.1. Explain the need for compensators in control systems.

14.2. Mention compensators that are used in the design of control systems and derive the transfer function of any one.

14.3. Derive the transfer function of a phase-lead compensator and explain its frequency characteristics.

14.4. Enumerate the design steps involved in the phase lead compensation.

14.5. Derive the transfer function of a phase-lag network.

14.6. What are the steps involved in designing a phase-lag compensator?

14.7. Derve the transfer function of a phase lag-lead compensator.

14.8. Explain the procedure for lag-lead compensator design.

14.9. Compare lead, lag and lag-lead compensation.

14.10. How are lag and lead compensators realized?

14.11. Show the realization of lag-lead network.

14.12. Explain the effects and limitations of phase-lag compensation.

14.13. Explain the effects and limitations of phase-lead compensators.

14.14. What are the merits of lag-lead compensators?

PROBLEMS

14.1. A unity feedback system has open-loop gain $G_p(s) = K/s^2$. It is desired to compensate the system so that the setting time is less than 4 sec. and peak overshoot for step input is less than 20%. Design a compensator and draw the Bode plots for the uncompensated and compensated systems.

14.2. A unity feedback control system has $G(s) = K\, G_c(s)/(1 + s)^2$. Design a compensator $G_c(s)$ that will increase the step error coefficient by a factor of 8 with $M_p = 1.4$

14.3. A unity feedback system with $G_p(s) = 4/s(2s + 1)$ is to be compensated to have a phase margin of 40° without sacrificing the velocity error constant. Design a suitable lag network and compute the component values for a suitable impedance level.

14.4. A unity feedback system has an open-loop transfer function $G_p(s) = 1/s^2$. Design a suitable compensating network such that a phase margin of 45° is achieved without sacrificing velocity error constant. Sketch the Bode plot of uncompensated and compensated system.

14.5. A unity feedback system with the open-loop transfer function $G_p(s) = K/s(1 + 0.3s)$ $(1 + 0.5s)$ is required to have (i) velocity error constant $K_v \geq 10$ sec^{-1}, and (ii) phase margin $\geq 28°$, Design a suitable phase lag compensator to meet the above specifications.

14.6. The system whose open-loop gain is $G_p(s) = 2/s(s + 0.5)$ is to have a phase margin of 40° without sacrificing the velocity error coefficient. Design a compensator. Assume unity feedback.

14.7. A unity feedback system has open-loop transfer function. $G_p(S) = K/s(s + 1)(0.2s + 1)$. Design a phase lag compensator to achieve velocity error coefficient of 8 and phase margin of 40°.

14.8. Design a compensator for the system with $G_p(s) = 250/s (1 + 0.1s)$ to have a phase margin of 30°.

Chapter 15

STATE–VARIABLE ANALYSIS

15.1. INTRODUCTION

One of the most important tasks in the analysis and design of control systems is the mathematical modeling of systems. The two common methods of modeling linear systems are the transfer function approach and the state variable approach. In the preceding chapters, we have described in detail the transfer function approach. The transfer function approach is convenient in the frequency domain analysis and design of systems, but suffers a major disadvantage in that all the initial conditions of the system lose their natural significance. The system design in the frequency domain involves a trial and error procedure. It is difficult to obtain optimal solutions for system design using the classical approach.

In contrast to the transfer function approach to the analysis and design of control systems, the state-variable approach decomposes a complex system into a set of first-order systems which can be normalized to have minimum interaction and which can be solved individually. It also provides a unified approach that is extensively used in modern control theory.

In this chapter, the basic concepts of state variables and state equations are introduced. A brief review of the standard matrix operations used in this chapter is presented in Appendix B. Development of state equations from the original integro-differential equations or transfer functions is discussed. Transformations of the state equation in several canonical forms are developed.

The general solution for linear time-invariant systems in state equation form is presented. The transition matrix concept is introduced and used in describing the general solution for forced linear time invariant systems. Finally the controllability and observability of linear systems are defined and their applications are illustrated.

15.2. ADVANTAGES OF STATE-VARIABLE ANALYSIS

The advantages of state-variable analysis over the classical transfer function approach are:

1. It decomposes an n-th order system into n first-order equations which are simpler to handle.

2. It enables us to include initial conditions in the design.

3. It is a unified approach applicable to time-invariant, time variant, non-linear and multiple-input multiple-output systems.

4. The use of vector-matrix notation greatly simplifies the mathematical representation of systems.

5. It can be used to design optimal systems for the given performance indices.

6. It is suited for computer solutions.

7. The state variables need not be physically measurable or observable quantities.

15.3. DEFINITIONS

We shall first define the following terms frequently used with the state-variable approach.

(i) **State Variables:** Let the set of variables, $x_1(t)$, $x_2(t)$, ... , $x_n(t)$ describe the dynamic characteristics of a system. Let us define these variables as state variables. At any time $t = t_o$, the state variables, $x_1(t_o)$, $x_2(t_o)$, ..., $x_n(t_o)$ define the initial states of the systems at the selected initial time t_o. If the inputs of the system for $t \geq t_o$ and the initial states are specified, the state variables will completely define the future behaviour of the system. Thus, *the state variables are defined as a minimal set of variables which determine the state of a dynamic system.*

(ii) **State:** The state of a dynamic system is the minimal set of state variables such that the knowledge of these variables at any time $t = t_o$, together with the information of the input excitation subsequently applied, completely determines the behaviour of the system at any time $t > t_o$. Thus *the state of a system refers to the past, present and future conditions of the system.*

(iii) **State Vector:** If the behaviour of a dynamic system is determined by a set of n variables, then these variables are the components of the n-dimensional vector $x(t)$, called *state vector*. The state vector uniquely determines the system behaviour for any $t > t_o$, when all inputs to the system are specified for $t > t_o$.

(iv) **State Space:** State space is defined as the n-dimensional space whose co-ordinate axes consist of the x_1-axis, x_2-axis, ..., x_n-axis. Any state can be uniquely represented by a point in the state space.

(v) **State Trajectory:** State trajectory is defined as the path traversed in the state space by the state vector $x(t)$ as it changes with the passage of time. State space and state trajectory in the two-dimensional case are known as *the phase plane* and *phase trajectory* respectively.

(vi) **State Equations:** The system equations expressed in terms of state variables are called state equations. The state equations of a system are a set of n first-order differential equations, where n is the number of independent states.

15.4. STATE-VARIABLE REPRESENTATION OF SYSTEMS

The first step in representing a system in the state-variable form is to select the system variables that are to represent the state of the system. However, there is no unique way of making this selection. The three common methods for expressing the system states are (*i*) *physical*, (*ii*) *phase* and (*iii*) *canonical state variables*. We shall discuss these representations in the following sections.

15.4.1. Physical Variable Method

The selection of state variables for the physical variable method is usually based on the *energy-storage elements of the system*. Some common energy-storage elements that exist in physical systems are listed in Table 15.1 along with the energy equations. The physical variables in these energy equations may be selected as the state variables of the system.

Table 15.1. Energy Storage Elements

Elements	Energy	Physical variable
Capacitor C	$\frac{1}{2}\,CV^2$	Voltage V
Inductor L	$\frac{1}{2}\,LI^2$	Current I
Mass M	$\frac{1}{2}\,Mv^2$	Translational Velocity v
Inertia J	$\frac{1}{2}\,J\omega^2$	Rotational velocity ω
Spring K	$\frac{1}{2}\,Kx^2$	Displacement x

Only independent physical variables are chosen as the state variables. Independent state variables are those state variables which cannot be expressed in terms of the remaining assigned state variables. In some systems, it may be necessary to identify more state variables than just the energy-storage variables. For example, in a mechanical system, if we select the velocity as a state variable, the position may also be of interest. In such cases, the position which is the integral of the velocity must also be assigned as a state variable.

The following examples illustrate the physical variable method of system representation.

Example 15.1. *Consider a series resistor-inductor circuit shown in Fig. 15.1. Derive its state equations.*

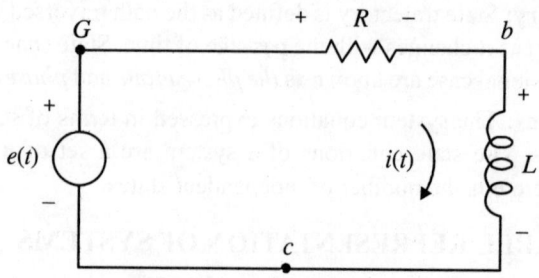

Fig. 15.1. Series RL circuit

Solution:

This circuit has only one energy storage element, the inductor L. Thus, only one state variable is required. From Table 15.1, it is seen that the corresponding state variable is the current i. The mesh equation of the circuit is

$$R\,i + L\,di/dt = e \qquad\qquad\qquad ...\,(15.1)$$

Assigning the state $x_1 = i$ and letting $u = e$, Eq. (15.1) can be rewritten as

$$Rx_1 + L\dot{x}_1 = u$$

Or,

$$\dot{x}_1 = (-R/L)x_1 + (1/L)u \qquad\qquad ...\,(15.2)$$

The letter u is the standard notation for the input forcing function and is known as *control variable*. Eq. (15.2) is called the *state equation of the system*. There is only one state equation since the system is of first order ($n = 1$). Eq. (15.2) may be written in the vector-matrix form:

$$[\dot{x}_1] = [-(R/L)]\,[x_1] + [1/L]\,[u] \qquad ...\,(15.3)$$

Eq. (15.3) is called the *state variable representation* of the system.

Example 15.2. *Consider a series resistor-inductor-capacitor circuit shown in Fig. 15.2(a). Derive its state equation.*

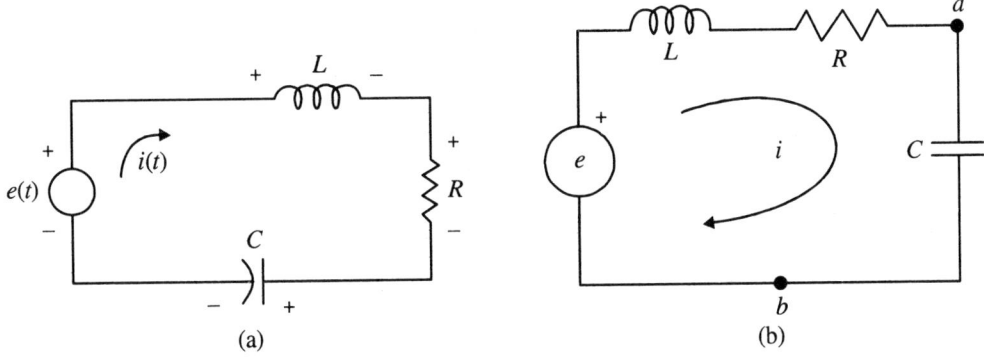

Fig. 15.2. (a) Series RLC circuit for mesh analysis (b) Series RLC circuit for node analysis

Solution:

There are two energy-storage elements, L and C. From Table 15.1, the two state variables are current i through the inductor and the voltage v_c across the capacitor. Thus, two equations are required.

The number of loop equations to be written is equal to the number of state variables representing currents in inductors and the number of node equations to be written is equal to the number of state variables representing voltage across capacitors. The loop and node equations are written in terms of inductor branch currents and capacitor branch voltages respectively. From these equations, one should determine which of the assigned physical variables are independent. Fig. 15.2(a) is redrawn in Fig. 15.2(b) with node b as the reference node. The node equation for node a and the loop equation are respectively:

Node Equation:

$$v_c = (1/C)\int i \; dt$$

or,

$$Cdv_c/dt = i \qquad \qquad \qquad \qquad \text{... (15.4)}$$

Mesh Equation:

$$Ldi/dt + Ri + v_c = e \qquad \qquad \qquad \qquad \text{... (15.5)}$$

Let

$$x_1 = v_c$$

and

$$x_2 = i$$

Eq. (15.4) and (15.5) are written in terms of the state variables x_1, and x_2 with $u = e$ as

$$C\dot{x}_1 = x_2$$

$$L\dot{x}_2 + Rx_2 + x_1 = u \qquad \qquad \qquad \qquad \text{... (15.7)}$$

Eq. (15.7) is written in the form:

$$\dot{x}_1 = (1/C) x_2$$

$$\dot{x}_2 = -(1/L) x_1 - (R/L) x_2 + (1/L) u \qquad \qquad \text{... (15.8)}$$

Eq. (15.8) represents the state equations of the electrical system containing two independent state variables. They are two first-order linear differential equations and $n = 2$. They are the minimum number of state equations required to represent the performance of the system. Eq. (15.8) is expressed in the matrix form as:

$$\begin{pmatrix} \dot{x}_1 \\ \dot{x}_2 \end{pmatrix} = \begin{pmatrix} 0 & 1/C \\ -1/L & -R/L \end{pmatrix} \begin{pmatrix} x_1 \\ x_2 \end{pmatrix} + \begin{pmatrix} 0 \\ 1/L \end{pmatrix} u \qquad \text{... (15.9)}$$

or, in more compact form,

$$\dot{x} = A x + B u \qquad \qquad \qquad \qquad \text{... (15.10)}$$

where $\dot{x}$ is a 2×1 column vector, A is an 2×2 plant coefficient matrix, x is a 2×1 state vector, B is a 2×1 control matrix and u is a one-dimensional control vector.

15.4.2 State Variable Representation of nth-order system with a single Forcing Function

Let us consider the following n^{th} order system

$$y^{(n)} + a_1 \, y^{(n-1)} + a_2 \, y^{(n-2)} + \ldots + a_{n-1} \, y' + a_n \, y = u \qquad (15.11)$$

The future behavior of the system is completely determined by the initial conditions together with the input $u(t)$ for $t \geq 0$.

Let us assume $y(t), \, \dot{y}(t), \, \ldots, \, y^{(n-1)}(t)$ be the set of state variables. Let us now define

$$
\begin{aligned}
x_1 &= y \\
x_2 &= \dot{x}_1 = \dot{y} \\
x_3 &= \dot{x}_2 = \ddot{y} \\
& \quad \cdot \qquad \cdot \qquad \cdot \\
& \quad \cdot \qquad \cdot \qquad \cdot \\
& \quad \cdot \qquad \cdot \qquad \cdot \\
x_n &= \dot{x}_{n-1} = y^{(n-1)}
\end{aligned}
\qquad \ldots (15.12)
$$

Then, we have

$$\dot{x}_n \;=\; y^{(n)} = -a_n x_1 - a_{n-1} x_2 - \ldots - a_1 x_n + u \qquad \ldots (15.13)$$

In the matrix form, Eq.(15.12) and (15.13) is written as

$$\dot{x} \;=\; A_c x + B_c u \qquad \ldots (15.14)$$

where

$$
x = \begin{pmatrix} x_1 \\ x_2 \\ x_3 \\ \cdot \\ \cdot \\ \cdot \\ \cdot \\ x_n \end{pmatrix} ; \;
A_c = \begin{pmatrix}
0 & 1 & 0 & \cdot & \cdot & \cdot & 0 \\
0 & 0 & 1 & 0 & \cdot & \cdot & 0 \\
0 & 0 & 0 & 1 & 0 & \cdot & 0 \\
\cdot & & & & & \cdot \cdot \cdot & 0 \\
\cdot & & & & & \cdot \cdot \cdot & 0 \\
\cdot & & & & & \cdot \cdot \cdot & 0 \\
0 & 0 & & & 0 & 0 \; 0 & 1 \\
-a_n & -a_{n-1} & -a_{n-2} & \cdot & \cdot & \cdot & -a_1
\end{pmatrix} ; \;
B_c = \begin{pmatrix} 0 \\ 0 \\ 0 \\ 0 \\ 0 \\ \cdot \\ 0 \\ 0 \\ 1 \end{pmatrix}
$$

$$\ldots (15.15)$$

The output equation is

$$y = [1\,0\,0\cdots0]\begin{pmatrix} x_1 \\ x_2 \\ x_3 \\ \cdot \\ \cdot \\ \cdot \\ x_n \end{pmatrix} \qquad \cdots (15.16)$$

In matric form

$$y = C_c\,x \qquad \cdots (15.17)$$

where,

$$C_e = [1\ 0\ \ 0\ \ 0\ .\ .\ .\ .\ .\ 0] \qquad \cdots (15.18)$$

This form of representation of a system by Eq.(15.14) and (15.17) and defined by Eq.(15.15) and (15.16) is called *phase-variable canonical form* and the method is called *phase-variable method*. Eq.(15.14) is the *state equation* and Eq. (15.17) is the *output equation*. The matrix A_c is $n \times n$ coefficient matrix, $B_c = n \times 1$ control matrix and C_e is $1 \times n$ output matrix. The matrix A_c is said to be in *companion form* and is known as *companion matrix*.

Example 15.3. *Consider the system defined by*

$$\dddot{y} + 6\ddot{y} + 11\dot{y} + 6y = 6u \qquad \cdots (15.19)$$

where y is the output and u is the input. Obtain the state variable representation.

Solution:

Let us define the state variables as

$$x_1 = y$$

$$x_2 = \dot{x}_1 = \dot{y}$$

$$x_3 = \dot{x}_2 = \ddot{y}$$

Then, we obtain

$$\dot{x}_3 = -6x_1 - 11x_2 - 6x_3 + 6u$$

These equations are combined to yield

$$\begin{pmatrix} \dot{x}_1 \\ \dot{x}_2 \\ \dot{x}_3 \end{pmatrix} = \begin{pmatrix} 0 & 1 & 0 \\ 0 & 0 & 1 \\ -6 & -11 & -6 \end{pmatrix}\begin{pmatrix} x_1 \\ x_2 \\ x_3 \end{pmatrix} + \begin{pmatrix} 0 \\ 0 \\ 6 \end{pmatrix}u \qquad \cdots (15.20)$$

The output equation is given by

$$y = [1 \quad 0 \quad 0] \begin{pmatrix} x_1 \\ x_2 \\ x_3 \end{pmatrix} \qquad \text{... (15.21)}$$

In the matrix form, we may write

$$\dot{x} = A_c x + B_c u \qquad \text{... (15.22)}$$

where

$$A_c = \begin{pmatrix} 0 & 1 & 0 \\ 0 & 0 & 1 \\ -6 & -11 & -6 \end{pmatrix}; \ B_c \begin{pmatrix} 0 \\ 0 \\ 6 \end{pmatrix}; \ C_c = [1 \quad 0 \quad 0] \qquad \text{... (15.23)}$$

Fig. 15.3 shows the block-diagram representation of the present state equation and output equation. The coefficients of feedback block are identical with the negatives of the coefficients of the original differential equation, Eq. (15.19).

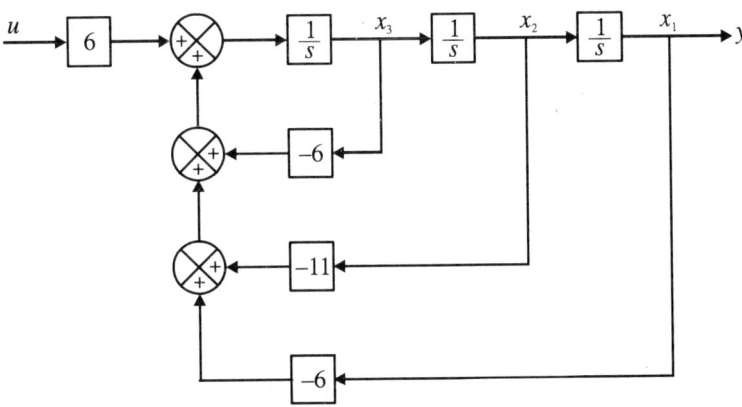

Fig. 15.3. Black diagram represetation of state and output equations

15.4.3. State Variable Representation of nth-order System with Forcing Functions involving Derivative Terms

The differential equation of the system involving derivatives of the forcing function is

$$y^{(n)} + a_1 y^{(n-1)} + a_2 y^{(n-2)} + \dots + a_{n-1} \dot{y} + a_n y = b_{n-1} u^{(n)} + \dots + b_{n-1} \dot{u} + b_n u \qquad \text{... (15.24)}$$

In this case, the set of n variables y, $\dot{y}$, ..., $y^{(n-1)}$ do not qualify as set of state variables because the n first-order equations with $x_1 = y$ may not yield a unique solution. We should define the state variables such that they will eliminate the derivatives of u in the state equation. We therefore define the following set of n variables as a set of n state variables.

$$x_1 = y - \beta_o u$$
$$x_2 = \dot{y} - \beta_o \dot{u} - \beta_1 u = \dot{x}_1 - \beta_1 u$$
$$x_3 = \ddot{y} - \beta_o \ddot{u} - \beta_1 \dot{u} - \beta_2 u = \dot{x}_2 - \beta_2 u$$

. . . .
$$\qquad \qquad \qquad \qquad \qquad \qquad \qquad \qquad \qquad \qquad \text{... (15.25)}$$

$$x_n = y^{(n-1)} - \beta_o u^{(n-1)} - \beta_1 u^{(n-2)} - .... - \beta_{n-2} \dot{u} - \beta_{n-1} u = \dot{x}_{n-1} - \beta_{n-1} u$$

where

$$\beta_o = b_o$$
$$\beta_1 = b_1 - a_1 \beta_o$$
$$\beta_2 = b_2 - a_1 \beta_1 - a_2 \beta_o$$

. . . . .
$$\beta_n = b_n - a_1 \beta_{n-1} - a_2 \beta_{n-2} .... - a_{n-1} \beta_1 - a_n \beta_o \qquad \qquad \text{... (15.26)}$$

Then, we obtain

$$\dot{x}_n = -a_n x_1 - a_{n-1} x_2 - ..... - a_2 x_{n-1} - a_1 x_n + \beta_n u \qquad \qquad \text{... (15.27)}$$

The existence and uniqueness of the solution of the state equation is now guaranteed. With this choice of state variables, the following state and output equations of the system are obtained.

$$\dot{x} = A x + B u \qquad \qquad \qquad \qquad \qquad \qquad \qquad \text{... (15.28)}$$
$$y = C x + D u \qquad \qquad \qquad \qquad \qquad \qquad \qquad \text{... (15.29)}$$

where

$$A = \begin{pmatrix} 0 & 1 & 0 & . & . & . & 0 \\ 0 & 0 & 1 & 0 & . & . & 0 \\ 0 & 0 & 0 & 1 & 0 & . & 0 \\ . & & & . & . & . & 0 \\ . & & & . & . & . & 0 \\ . & & & . & . & . & 0 \\ 0 & 0 & 0 & 0 & 0 & 0 & 1 \\ -a_n & -a_{n-1} & -a_{n-2} & . & . & . & -a_1 \end{pmatrix} ; B = \begin{pmatrix} \beta_1 \\ \beta_2 \\ . \\ . \\ . \\ . \\ . \\ \beta_n \end{pmatrix} \qquad \text{... (15.30a)}$$

$$C = \begin{bmatrix} 1 & 0 & 0 & 0 & . & . & . & 0 \end{bmatrix}; D = \beta_o = b_o \qquad \qquad \text{... (15.30b)}$$

The initial condition $x(0)$ is determined from Eq. (15.25). In this representation, the matrix A is essentially the same as in the system of Eq. (15.11). The derivatives of the forcing function affect only the elements of the B matrix.

Example 15.4. *The differential equation of a system is given by*

$$\dddot{y} + 18\ddot{y} + 192\dot{y} + 640y = 160\,\dot{u} + 640u$$

Obtain the state variable representation of the system.

Solution:

From Eq. (15.25),

$$x_1 = y - \beta_0 u$$

$$x_2 = \dot{y} - \beta_0\,\dot{u} - \beta_1 u = \dot{x}_1 - \beta_1 u$$

$$x_3 = \ddot{y} - \beta_0\ddot{u} - \beta_1\dot{u} - \beta_2 u = \dot{x}_2 - \beta_2 u$$

From Eq.(15.26), we have

$$\beta_0 = b_0 = 0$$

$$\beta_1 = b_1 - a_1\beta_0 = 0$$

$$\beta_2 = b_2 - a_1\beta_1 - a_2\beta_0 = 160$$

$$\beta_3 = b_3 - a_1\beta_2 - a_2\beta_1 - a_3\beta_0 = -2240$$

The state equation of the system is

$$\begin{pmatrix} \dot{x}_1 \\ \dot{x}_2 \\ \dot{x}_3 \end{pmatrix} = \begin{pmatrix} 0 & 1 & 0 \\ 0 & 0 & 1 \\ -640 & -192 & -18 \end{pmatrix} \begin{pmatrix} x_1 \\ x_2 \\ x_3 \end{pmatrix} + \begin{pmatrix} 0 \\ 160 \\ -2240 \end{pmatrix} u \qquad \text{... (15.31)}$$

The output equation is

$$y = \begin{bmatrix} 1 & 0 & 0 \end{bmatrix} \begin{pmatrix} x_1 \\ x_2 \\ x_3 \end{pmatrix} \qquad \text{... (15.32)}$$

15.4.4. State Variable Representation of Multiple-Input Multiple-Output (MIMO) Systems

Let us consider the MIMO system shown in Fig. 15.4. In this system, $x_1, x_2, ..., x_n$ represent the n state variables; $u_1, u_2, ...u_r$ denote the r input variables and $y_1, y_2, ... y_m$ are the m output variables. The system equations are:

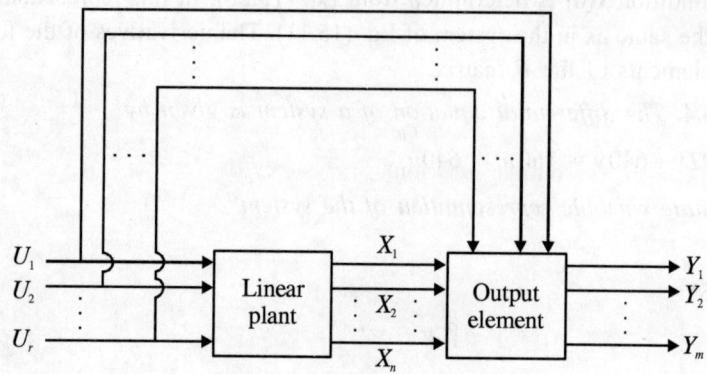

Fig. 15.4. MIMO system

$$\dot{x}_1 = a_{11}x_1 + a_{12}x_2 + \ldots + a_{1n}x_n + b_{11}u_1 + b_{12}u_2 + \ldots + b_{1r}u_r$$

$$\dot{x}_2 = a_{21}x_1 + a_{22}x_2 + \ldots + a_{2n}x_n + b_{21}u_1 + b_{22}u_2 + \ldots + b_{2r}u_r$$

$$\dot{x}_n = a_{n1}x_1 + a_{n2}x_2 + \ldots + a_{nn}x_n + b_{n1}u_1 + b_{n2}u_2 + \ldots + b_{nr}u_r \qquad \ldots (15.33)$$

where a_{ij} and b_{ij} are constants or functions of t.

In the matrix from, the state equation is written as

$$\dot{x} = Ax + Bu \qquad \ldots (15.34)$$

where $x = n \times 1$ state vector, u is $n \times 1$ input or control vector.

$$A = \begin{pmatrix} a_{11} & a_{12} & \cdot & \cdot & \cdot & a_{1n} \\ a_{21} & a_{22} & \cdot & \cdot & \cdot & a_{2n} \\ \cdot & \cdot & \cdot & & & \cdot \\ \cdot & \cdot & & \cdot & & \cdot \\ \cdot & \cdot & & & \cdot & \cdot \\ \cdot & \cdot & & & & \cdot \\ -a_{n1} & -a_{n2} & -a_{n3} & \cdot & \cdot & -a_{nn} \end{pmatrix} ; B = \begin{pmatrix} b_{11} & b_{12} & \cdot & \cdot & b_{1n} \\ b_{21} & b_{22} & \cdot & \cdot & b_{2n} \\ \cdot & \cdot & & & \cdot \\ \cdot & \cdot & \cdot & & \cdot \\ \cdot & \cdot & & \cdot & \cdot \\ \cdot & \cdot & & & \cdot \\ b_{n1} & b_{n2} & \cdot & \cdot & b_{nn} \end{pmatrix} \qquad \ldots (15.35)$$

The Output equations are:

$$y_1 = c_{11}x_1 + c_{12}x_2 + \ldots + c_{1n}x_n$$

$$y_2 = c_{21}x_1 + c_{22}x_2 + \ldots + c_{2n}x_n$$

$$\cdot \quad \cdot \quad \cdot \quad \cdot \quad \cdot \quad \cdot$$

$$\cdot \quad \cdot \quad \cdot \quad \cdot \quad \cdot \quad \cdot$$

$$y_n = c_{n1}x_1 + c_{n2}x_2 + \ldots + c_{nn}x_n \qquad \ldots (15.36)$$

In the matrix form, the output equation is

$$y = Cx + Du \qquad \ldots (15.37)$$

where y is $n \times 1$ output vector.

$$C = \begin{pmatrix} c_{11} & c_{12} & \cdot & \cdot & \cdot & \cdot & c_{1n} \\ c_{21} & c_{22} & \cdot & \cdot & \cdot & \cdot & c_{2n} \\ \cdot & \cdot & \cdot & \cdot & \cdot & \cdot & \cdot \\ \cdot & \cdot & \cdot & \cdot & \cdot & \cdot & \cdot \\ \cdot & \cdot & \cdot & \cdot & \cdot & \cdot & \cdot \\ \cdot & \cdot & \cdot & \cdot & \cdot & \cdot & \cdot \\ -c_{n1} & -c_{n2} & -c_{n3} & \cdot & \cdot & \cdot & -c_{nn} \end{pmatrix} ; D = \begin{pmatrix} d_{11} & d_{12} & \cdot & \cdot & \cdot & d_{1n} \\ d_{21} & d_{22} & \cdot & \cdot & \cdot & d_{2n} \\ \cdot & \cdot & \cdot & \cdot & \cdot & \cdot \\ \cdot & \cdot & \cdot & \cdot & \cdot & \cdot \\ \cdot & \cdot & \cdot & \cdot & \cdot & \cdot \\ d_{n1} & d_{n2} & \cdot & \cdot & \cdot & d_{nn} \end{pmatrix} \qquad \ldots (15.37b)$$

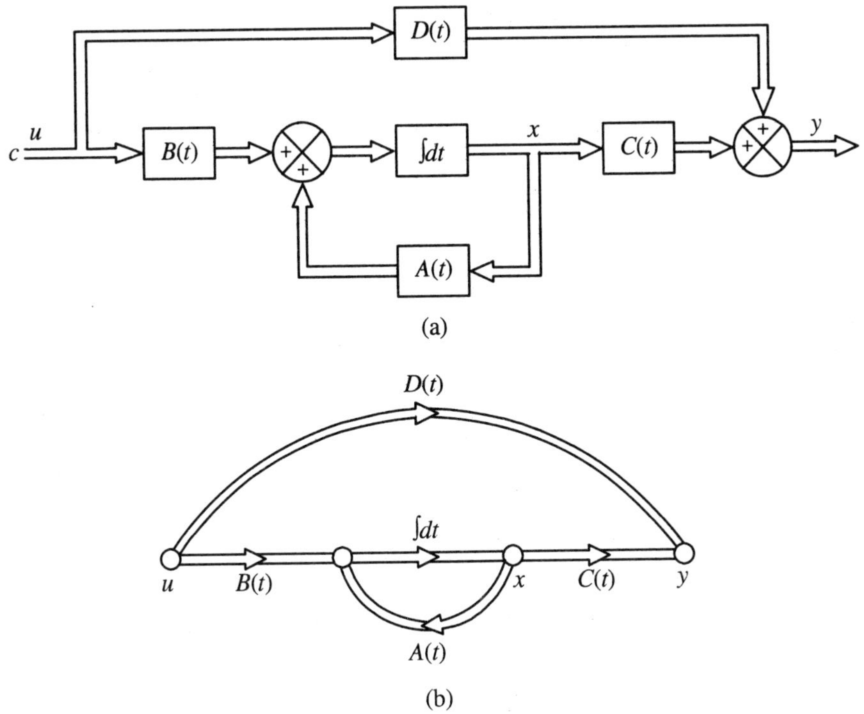

(a)

(b)

Fig. 15.5. (a) Black diagram and (b) signal flow graph represnetation of MIMO system

The matrices A, B, C and D completely characterize the system dynamics. A block diagram and a signal flow graph representation are shown in Fig. 15.5. Double arrows in the diagram indicate the vector quantities.

15.5. TRANSFORMATION TO PHASE VARIABLE CANONICAL FORM

When the state variables are functions of the dependent variables and their derivatives, they are called *phase variables*. When the coefficient matrices A and B are given by Eq. (15.14) is called the *phase variable canonical form*. A linear time-invariant system, represented in the phase variable canonical form, has certain unique properties with regard to controllability and pole-placement design through state feedback. When the state equation is not in the phase variable canonical form, it may be transformed to that form by a linear transformation discussed below.

Let the state equation of a linear time-invariant system by given by

$$\dot{x} = Ax + Bu \qquad \qquad \text{... (15.38)}$$

If the matrix

$$S = [B\, AB\, A^2B \cdots A^{n-1}B] \qquad \qquad \text{... (15.39)}$$

is nonsingular, then there exists a linear transformation

$$z = Tx \qquad \qquad \text{... (15.40)}$$

or,

$$x = T^{-1}z \qquad \qquad \text{... (15.41)}$$

which transforms Eq. (15.38) in the phase variable canonical form

$$\dot{z} = A_c z + B_c u \qquad \qquad \text{... (15.42)}$$

where A_c and B_c are in the form given in Eq. (15.15). The transformation matrix T is given by

$$T = \begin{pmatrix} T_1 \\ T_1 A \\ \cdot \\ \cdot \\ T_1 A^{n-1} \end{pmatrix}$$

where

$$T_1 = [0\ 0\ 0\ \cdot\ \cdot\ \cdot\ 1]S^{-1}$$

The proof follows:

Let

$$
x = \begin{pmatrix} x_1 \\ x_2 \\ \cdot \\ \cdot \\ x_n \end{pmatrix}; \quad z = \begin{pmatrix} z_1 \\ z_2 \\ \cdot \\ \cdot \\ z_n \end{pmatrix} \qquad \text{... (15.43)}
$$

$$
T = \begin{pmatrix} t_{11} & t_{12} & \cdot & \cdot & t_{1n} \\ t_{21} & t_{22} & \cdot & \cdot & t_{2n} \\ \cdot & \cdot & \cdot & \cdot & \cdot \\ t_{n1} & t_{n12} & \cdot & \cdot & t_{nn} \end{pmatrix} = \begin{pmatrix} T_1 \\ T_2 \\ \cdot \\ T_n \end{pmatrix} \qquad \text{... (15.44)}
$$

where

$$
T_i = \begin{bmatrix} t_{-i1} & t_{-i2} & \cdot & \cdot & t_{in} \end{bmatrix}, \quad i = 1, 2, \dots, n \qquad \text{... (15.45)}
$$

From Eq. (15.40), z_1 is given by

$$
z_1 = T_1 x \qquad \text{... (15.46)}
$$

Taking the time derivative on both sides of Eq. (15.46) we get

$$
\dot{z}_1 = z_2 = T_1 \dot{x} = T_1 A x + T_1 B u \qquad \text{... (15.47)}
$$

Since z is a function of x only, $T_1 B = 0$. Therefore

$$
\dot{z}_1 = z_2 = T_1 A x \qquad \text{... (15.48)}
$$

Taking the time derivative of Eq. (15.48), it can be shown that

$$
\dot{z}_2 = z_3 = T_1 A^2 x \qquad \text{... (15.49)}
$$

with $T_1 A B = 0$. Repeating the above procedure, we obtain

$$
\dot{z}_{n-1} = z_n = T_1 A^{n-1} x \qquad \text{... (15.50)}
$$

with $T_1 A^{n-2} B = 0$. Hence, Eq. (15.40) may be written as

$$
z = Tx = \begin{pmatrix} T_1 \\ T_1 A \\ \cdot \\ \cdot \\ T_1 A^{n-1} \end{pmatrix} \qquad \text{... (15.51)}
$$

The transformation matrix T is given by

$$T = \begin{pmatrix} T_1 \\ T_1 A \\ \cdot \\ \cdot \\ T_1 A^{n-1} \end{pmatrix} \qquad \qquad \ldots (15.52)$$

and T_1 should satisfy the condition

$$T_1 B = T_1 AB = T_1 A^2 B = \cdots = T_1 A^{n-2} B = 0 \qquad \ldots (15.53)$$

Taking the derivative of Eq. (15.40), we obtain

$$\dot{z} = T\dot{x} = TAx + TBu \qquad \ldots (15.54a)$$

Substituting Eq. (15.41) in Eq. (15.54b), we get

$$\dot{z} = TAT^{-1}z + TBu \qquad \ldots (15.54b)$$

Comparing Eq. (15.42) and (15.54b), we get

$$A_c = TAT^{-1} \qquad \ldots (15.55)$$
$$B_c = TB \qquad \ldots (15.56)$$

Then, from Eq. (15.52),

$$TB = \begin{pmatrix} T_1 B \\ T_1 AB \\ T_1 A^2 B \\ \cdot \\ \cdot \\ \cdot \\ T_1 A^{n-1} B \end{pmatrix} = \begin{pmatrix} 0 \\ 0 \\ 0 \\ \cdot \\ \cdot \\ \cdot \\ 1 \end{pmatrix} \qquad \ldots (15.57)$$

Eq. (15.57) proves that Eq. (15.38) has been transformed to the phase-variable canonical form with the transformation matrix T given by Eq. (15.52). Since T_1 is $l \times n$ matrix, Eq. (15.57) may be written as

$$T_1[B\ AB \ldots A^{n-1}B] = [0\ 0\ 0\ \ldots\ 1] \qquad \ldots (15.58)$$

Therefore,

$$T_1 = [0\ 0\ 0 \ldots 1]\ [B\ AB \ldots A^{n-1}B]^{-1}$$
$$= [0\ 0\ 0 \ldots 1]\ S^{-1} \qquad (15.59)$$

provided the matrix S is nonsingular. This is the condition of *complete state controllability*.

Example 15.5. *Consider a linear system described by Eq.* (15.38) *with*

$$A = \begin{pmatrix} 1 & -1 \\ 0 & -1 \end{pmatrix}; B = \begin{pmatrix} 1 \\ 1 \end{pmatrix}$$

Obtain its phase-variable canonical form.

Solution:

The matrix S is

$$S = [B \ AB] = \begin{pmatrix} 1 & 0 \\ 1 & -1 \end{pmatrix}$$

It is nonsingular. Therefore the row-matrix T_1 is

$$T_1 = [0 \ 1] \ S^{-1} = [0 \ 1] \begin{pmatrix} 1 & 0 \\ 1 & -1 \end{pmatrix} = [1 \ -1]$$

Hence, the transformation matrix T is given by

$$T = \begin{pmatrix} T_1 \\ T_1 A \end{pmatrix} = \begin{pmatrix} 1 & -1 \\ 1 & 0 \end{pmatrix}$$

Its inverse is

$$T^{-1} = \begin{pmatrix} 0 & 1 \\ -1 & 1 \end{pmatrix}$$

Thus,

$$A_c = TAT^{-1} = \begin{pmatrix} 1 & -1 \\ 1 & 0 \end{pmatrix} \begin{pmatrix} 1 & -1 \\ 0 & -1 \end{pmatrix} \begin{pmatrix} 0 & 1 \\ -1 & 1 \end{pmatrix} = \begin{pmatrix} 0 & 1 \\ 1 & 0 \end{pmatrix}$$

$$B_c = TB = \begin{pmatrix} 1 & -1 \\ 1 & 0 \end{pmatrix} \begin{pmatrix} 1 \\ 1 \end{pmatrix} = \begin{pmatrix} 0 \\ 1 \end{pmatrix}$$

15.6. TRANSFER FUNCTION AND STATE EQUATION

The transfer function of a linear time-invariant system is defined as the Laplace transform of the impulse response. It is also defined as the ratio of the Laplace transform of the output to the Laplace transform of the input with *all initial conditions of the system set to zero*. By using the transfer function, the output $y(t)$ of a system can be obtained for any input $u(t)$.

In the study of linear systems, the characteristic equation plays an important role as the response of the system is primarily determined by it. It can be defined from the basis of differential equation, the transfer function or the state equations.

Consider a linear time-invariant system described by the differential equation

$$(d^n y/dt^n) + a_n(d^{n-1}y/dt^{n-1}) + a_2(dt/dy) + a_1 y = b_{n+1}(d^n u/dt^n) + \ldots + b_2(du/dt) + b_1 u \ldots \quad (15.60)$$

Assuming zero initial conditions, the transfer function is

$$T(s) = \frac{Y(s)}{U(s)} = \frac{b_{n+1}s^n + b_n s^{n-1} + b_{n-1}s^{n-2} + \cdots + b_2 s + b_1}{s^n + a_n s^{n-1} + a_{n-1}s^{n-1} + \cdots + a_2 s + a_1} \qquad \ldots \quad (15.61)$$

The characteristic equation $F(s)$ of the system is obtained by equating the denominator of the transfer function to zero. Thus,

$$F(s) = s^n + a_n \, s^{n-1} + \cdots + a_2 s + a_1 = 0 \qquad \ldots \quad (15.62)$$

15.6.1. Characteristic Equation from State Equations

We shall now derive the transfer function of a *single-input single-output* (SISO) system from the state and output equation and then obtain the characteristic equation.

Consider the system with the transfer function given in Eq. (15.61). The state variable representation of the system is

$$\dot{x} = A x + B u \qquad \ldots \quad (15.63)$$
$$y = C x + D u \qquad \ldots \quad (15.64)$$

where x is the state vector, u is the input and y is the output.

Taking the Laplace transforms of the above equations with all initial conditions set to zero, we obtain

$$sX(s) = AX(s) + BU(s) \qquad \ldots \quad (15.65)$$

$$Y(s) = CX(s) + DU(s) \qquad \ldots \quad (15.66)$$

Substituting the value of $X(s) = (sI - A)^{-1} BU(s)$ obtained from Eq. (15.65) in Eq.(15.66), we get

$$Y(s) = [C(sI - A)^{-1} B + D]U(s) \qquad \ldots \quad (15.67)$$

Or,

$$G(s) = \frac{Y(s)}{U(s)} = [C(sI - A)^{-1} B + D] \qquad \ldots \quad (15.68)$$

$$= \frac{C[adj(sI - A)]B + |(sI - A)| D}{|sI - A|} \qquad \ldots \quad (15.69)$$

where *adj M* is the adjoint of the matrix M. Hence, the characteristic equation is

$$|sI - A| = 0 \qquad \ldots \quad (15.70)$$

15.6.2. Eigenvalues

The roots of the above characteristic equation are often referred to as the *eigenvalues* of the matrix A. If the state equations are expressed in the phase-variable canonical form, the coefficients of the characteristic equation are readily given by the elements of the last row of the A matrix. In other words, if A is of the form given in Eq. (15.15), then the characteristic equation is readily given by Eq. (15.62).

An important property of the characteristic equation and the eigenvalues is that they are invariant under a linear transform.

Consider that the matrix A is transformed by a linear transformation $x = Tz$ so that

$$A_1 = T^{-1}AT \qquad \ldots (15.71)$$

The new characteristic equation is

$$|sI - A_1| = |T^{-1}|\,|sI - A|\,|T| \qquad \ldots (15.72)$$

Since the determinant of a product is the product of determinants, Eq. (15.72) can be written as

$$|sI - A_1| = |T^{-1}|\,|sI - A|\,|T|$$
$$= |T^{-1}|\,|T|\,|sI - A| = |sI - A| \qquad \ldots (15.73)$$

Since the characteristic equations are same, the eigenvalues are also the same.

15.6.3. Eigenvectors

The $n \times 1$ nonzero column vector v_i that satisfies the matrix equation

$$(\lambda_i I - A)\,v_i = 0 \qquad \ldots (15.74$$

where λ_i is the i-th eigenvalue of the matrix A and is known as the *eigenvector* of A associated with the eigenvalue λ_i. The eigenvectors are determined by similarity transformation as illustrated in the following section.

15.7. SIMILARITY TRANSFORMATION

A general method of converting the state equation to the desirable form in which the A matrix is diagonal, is called *linear similarity transformation*. Since the state vectors are not unique, the intention is to transform the state vector x to a new state vector z by means of a constant, square, nonsingular transformation matrix T so that

$$x = Tz \qquad \ldots (15.75)$$

Since T is a constant matrix, the time derivative of the above equation yields

$$\dot{x} = T\dot{z} \qquad \ldots (15.76)$$

Substituting Eq. (15.75) and (15.76) into the state equation $\dot{x} = Ax + Bu$, we obtain

$$T\dot{z} = AT\,z + Bu \qquad \ldots (15.77)$$

Premultiplying by T^{-1}, we have

$$\dot{z} = T^{-1}AT\,z + T^{-1}Bu = A_d z + B_d u \qquad \qquad ... (15.78)$$

The corresponding output equation is

$$y = CTz + Du = C_d z + D_d u \qquad \qquad ... (15.79$$

If the matrix T is selected such that $T^{-1}AT$ is diagonal, then the matrix T is known as *modal matrix*. Thus,

$$T^{-1}AT = A_d = \begin{pmatrix} \lambda_1 & & & & & \\ & \lambda_2 & & & \mathbf{0} & \\ & & \lambda_3 & & & \\ & & & \cdot & & \\ \mathbf{0} & & & & \cdot & \\ & & & & & \lambda_n \end{pmatrix} \qquad \qquad ... (15.80)$$

This assumes that the eigenvalues of the matrix A are distinct. The matrix A_d is in diagonal form. Eq. (15.78) is the state equation and is known as the *canonical form*.

15.7.1. Obtaining Modal Matrix

In general, there are several methods of determining the modal matrix T. We shall now discuss four methods of obtaining T for the case where A has distinct eigenvalues.

Method 1: Where the matrix A has distinct eigen values, $\lambda_1, \lambda_2, ..., \lambda_n$ and it is in companion form $(A = A_c)$, the modal matrix T is given by *Vandermonde matrix* which is easily obtained. The *Vandermonde matrix* is defined by

$$T = \begin{pmatrix} 1 & 1 & 1 & \cdot & \cdot & \cdot & 1 \\ \lambda_1 & \lambda_2 & \lambda_3 & \cdot & \cdot & \cdot & \lambda_n \\ \lambda_1^2 & \lambda_2^2 & \lambda_3^2 & \cdot & \cdot & \cdot & \lambda_n^2 \\ \cdot & \cdot & \cdot & \cdot & \cdot & \cdot & \cdot \\ \cdot & \cdot & \cdot & \cdot & \cdot & \cdot & \cdot \\ \cdot & \cdot & \cdot & \cdot & \cdot & \cdot & \cdot \\ \lambda_1^{n-1} & \lambda_2^{n-1} & \lambda_3^{n-1} & \cdot & \cdot & \cdot & \lambda_n^{n-1} \end{pmatrix} \qquad \qquad ... (15.81)$$

Example 15.6. *Transfer the state variables in the following equation into canonical or modal form.*

$$\dot{x} = \begin{pmatrix} 0 & 1 & 0 \\ 0 & 0 & 1 \\ -24 & -26 & -9 \end{pmatrix} x + \begin{pmatrix} 1 \\ 0 \\ 2 \end{pmatrix} u \qquad \dots (15.82)$$

$$y = \begin{bmatrix} 3 & 3 & 1 \end{bmatrix} x \qquad \dots (15.83)$$

Solution:

Method 1: The characteristic equation $|\lambda I - A_c| = s^3 + 9s^2 + 26s + 24 = (s + 2)(s + 3)(s + 4)$ $= 0$. The roots are $\lambda_1 = -2$, $\lambda_2 = -3$ and $\lambda_3 = -4$. Since they are distinct and the matrix A is in the companion form, the Vandermonde matrix can be used as the modal matrix. From Eq. (15.81), we get

$$T = \begin{pmatrix} 1 & 1 & 1 \\ -2 & -3 & -4 \\ 4 & 9 & 16 \end{pmatrix} \qquad \dots (15.84)$$

The inverse of T is given by

$$T^{-1} = \frac{1}{2} \begin{pmatrix} 12 & 7 & 1 \\ -16 & -12 & -2 \\ 6 & 5 & 1 \end{pmatrix} \qquad \dots (15.85)$$

$$T^{-1}A_c T = \begin{pmatrix} -2 & 0 & 0 \\ 0 & -3 & 0 \\ 0 & 0 & -4 \end{pmatrix} = A_d \qquad \dots (15.86)$$

$$T^{-1}B_c = \begin{pmatrix} 7 \\ -10 \\ 4 \end{pmatrix} \qquad \dots (15.87)$$

$$CT = \begin{bmatrix} 1 & 3 & 7 \end{bmatrix} \qquad \dots (15.88)$$

Hence, the new matrix system equations are:

$$\dot{z} = \begin{pmatrix} -2 & 0 & 0 \\ 0 & -3 & 0 \\ 0 & 0 & -4 \end{pmatrix} z + \begin{pmatrix} 7 \\ -10 \\ 4 \end{pmatrix} u \qquad \dots (15.89)$$

$$y = \begin{bmatrix} 1 & 3 & 7 \end{bmatrix} z \qquad \qquad \text{... (15.90)}$$

Method 2: When the A matrix is in the general form and not in the companion form, the eigenvectors can by used to form the modal matrix T.

Thus,

$$T = [v_1 \ v_2 \ \cdots \ v_n]$$

where v_i are the i-th eigenvector of A associated with the eigenvalue λ_i, Now, premultiplying Eq. (15.80) by T,

$$AT = TA_d = \begin{pmatrix} \lambda_1 v_{11} & \lambda_2 v_{12} & \lambda_3 v_{13} & \cdot & \cdot & \cdot & \lambda_n v_{1n} \\ \lambda_1 v_{21} & \lambda_2 v_{22} & \lambda_3 v_{23} & \cdot & \cdot & \cdot & \lambda_n v_{2n} \\ \cdot & \cdot & \cdot & \cdot & \cdot & \cdot & \cdot \\ \cdot & \cdot & \cdot & \cdot & \cdot & \cdot & \cdot \\ \lambda_1 v_{n1} & \lambda_2 v_{n2} & \lambda_3 v_{n3} & \cdot & \cdot & \cdot & \lambda_n v_{nn} \end{pmatrix} \qquad \text{... (15.91)}$$

The necessary condition is that T is not singular. This is satisfied if the v_i are linearly independent. Eq. (15.91) may be written as

$$Av_i = \lambda_i v_i \qquad \qquad \text{... (15.92)}$$

Or

$$[\lambda_i I - A]v_i = 0 \qquad \qquad \text{... (15.93)}$$

Using the matrix property,

$$[M] \, adj[M] = |M| I \qquad \qquad \text{... (15.94)}$$

and letting

$$M = \lambda_i I - A \qquad \qquad \text{... (15.95)}$$

we have

$$[\lambda_i I - A] \, adj[\lambda_i I - A] = |\lambda_i I - A| \, I = 0 \qquad \qquad \text{... (15.96)}$$

where $(\lambda_i I - A)$ is the characteristic polynomial. Comparing Eq. (15.93) and (15.95), we obtain

$$v_i = adj[\lambda_i I - A] \qquad \qquad \text{... (15.97)}$$

Since Eq. (15.93) is a homogeneous equation of rank $(n - 1)$, when the eigenvalues are distinct, each value of λ_i yields only one eigenvector v_i. When the eigenvalue λ_i has multiplicity r, the rank deficiency α of Eq. (15.93) is $1 < \alpha < r$. Rank deficiency denotes that the rank of Eq. (15.93) is $(n - \alpha)$. Then the linearly independent eigenvectors v_i which satisfy Eq. (15.93) are α in number.

Example 15.7 *Obtain the modal matrix for the following equation using adj $[\lambda_i I - A]$.*

$$\dot{x} = \begin{pmatrix} -9 & 1 & 0 \\ -26 & 0 & 1 \\ -24 & 0 & 0 \end{pmatrix} x + \begin{pmatrix} 2 \\ 5 \\ 0 \end{pmatrix} u \qquad \ldots (15.97)$$

$$y = [1 \quad 2 \quad -1]x \qquad \ldots (15.98)$$

Solution:

The roots of the characteristic equation $|\lambda I - A| = 0$ are $\lambda_1 = -2$, $\lambda_2 = -3$ and $\lambda_3 = -4$. The eigenvalues are distinct but A is not in the companion form. Hence, the Vandermonde matrix is not the modal matrix. Using the adjoint method, we get

$$adj\ [\lambda I - A] = adj \begin{pmatrix} \lambda+9 & -1 & 0 \\ 26 & -\lambda & -1 \\ 24 & 0 & \lambda \end{pmatrix}$$

$$= \begin{pmatrix} \lambda^2 & \lambda & 1 \\ 26\lambda - 24 & \lambda^2 + 9\lambda & \lambda+9 \\ -24\lambda & -24 & \lambda^2 + 9\lambda + 26 \end{pmatrix}$$

For $\lambda_1 = -2$:

$$adj[-2I - A] = \begin{pmatrix} 4 & -2 & 1 \\ 28 & -14 & 7 \\ 48 & -24 & 12 \end{pmatrix}; v_1 \begin{pmatrix} 1 \\ 7 \\ 12 \end{pmatrix}$$

For $\lambda_2 = -3$:

$$adj[-3I - A] = \begin{pmatrix} 9 & -3 & 1 \\ 54 & -18 & 6 \\ 72 & -24 & 8 \end{pmatrix}; v_2 \begin{pmatrix} 1 \\ 6 \\ 8 \end{pmatrix}$$

For $\lambda_3 = -4$:

$$adj[-4I - A] = \begin{pmatrix} 16 & -4 & 1 \\ 80 & -20 & 5 \\ 96 & -24 & 6 \end{pmatrix}; v_2 \begin{pmatrix} 1 \\ 5 \\ 6 \end{pmatrix}$$

It may be noted that in each case, the columns of $adj\ [\lambda_i I - A]$ is linearly related. In other words, they are proportional. The v_i may be multiplied by a constant and are selected to contain the smallest integers. Frequently, the leading term is reduced to 1. In practice, it is necessary to calculate only one column of the adjoint matrix.

The Modal matrix so formed is

$$T = \begin{bmatrix} v_1 & v_2 & v_3 \end{bmatrix} = \begin{pmatrix} 1 & 1 & 1 \\ 7 & 6 & 5 \\ 12 & 8 & 6 \end{pmatrix}$$

$$T^{-1} = -\frac{1}{2} \begin{pmatrix} -4 & 2 & -4 \\ 18 & -6 & 2 \\ -16 & 4 & -1 \end{pmatrix}$$

$$T^{-1}B = \begin{pmatrix} -1 \\ -3 \\ 6 \end{pmatrix}; \ CT = (3 \ \ 5 \ \ 5)$$

The system matrix equations in terms of the new state vector z are

$$\dot{z} = \begin{pmatrix} -2 & 0 & 0 \\ 0 & -3 & 0 \\ 0 & 0 & -4 \end{pmatrix} z + \begin{pmatrix} -1 \\ -3 \\ 6 \end{pmatrix} u = A_d x + B_d u \qquad \qquad ... \ (15.99)$$

$$y = [3 \ 5 \ 5] \ z = C_d z \qquad \qquad ... \ (15.100)$$

Eigenvalues appear along the diagonal of A_d in the order they are assigned, i.e., λ_1, λ_2, and λ_3. The systems in Examples (15.6) and (15.7) have the same eigenvalues but require different modal matrix because of the fact that the A matrices are in different forms.

Method 3: Another method of evaluating the n elements of v_i is to form a set of n equations from the matrix equation (15.92). This method is illustrated by the following example.

Example 15.8. *For the system in Example 15.7, obtain the eigenvectors using* $Av_i = \lambda_i v_i$.

Solution:

Substituting A-matrix from Eq. (15.97), we get

$$\dot{x} = \begin{pmatrix} -9 & 1 & 0 \\ -26 & 0 & 1 \\ -24 & 0 & 0 \end{pmatrix} \begin{pmatrix} v_{11} \\ v_{21} \\ v_{31} \end{pmatrix} = \lambda_i \begin{pmatrix} v_{1i} \\ v_{2i} \\ v_{3i} \end{pmatrix}$$

Or,

$$-9\ v_{1i} + v_{2i} = \lambda_i v_{1i}$$
$$-26\ v_{1i} + v_{3i} = \lambda_i v_{2i}$$
$$-24\ v_{11} = \lambda_i v_{3i}$$

For each value of λ_i, the corresponding elements are equated to form three equations and then solved. For $\lambda_1 = -2$, we have

$$-9v_{11} + v_{21} = -\ 2v_{1i}$$
$$-26v_{11} + v_{31} = -2v_{2i}$$
$$-24v_{11} = -2v_{3i}$$

Of these three equations, only two are independent because the insertion of v_{21} from the first equation and v_{31} from the third equation into the second equation yields the identity $v_{11} = v_{11}$. This merely confirms that the rank deficiency $\alpha = 1$ and the rank of $[\lambda_1 I - A] = n - \alpha = 2$. The procedure, therefore, is to arbitrarily select one of the elements as unity. If we let $v_{11} = 1$, we get

$$v_{21} = 7 \text{ and } v_{31} = 12$$

Hence,

$$v_1 = \begin{pmatrix} v_{11} \\ v_{21} \\ v_{31} \end{pmatrix} = \begin{pmatrix} v_{11} \\ 7v_{11} \\ 2v_{11} \end{pmatrix} = \begin{pmatrix} 1 \\ 7 \\ 12 \end{pmatrix}$$

The result is same as in the previous example. Repeating the procedure for λ_2 and λ_3, we can obtain v_2 and v_3. It is left as an exercise.

Method 4: The fourth method of selecting the eigenvectors is from the matrix equation $[\lambda_i I - A]v_i = 0$. The eigenvectors, in this case, is said to lie in the null space of the matrix $[\lambda_i I - A]$.

The procedure for computing these eigenvectors is:

1. Transform $[\lambda_i I - A\ ]$ into Hermite Normal Form (HNF) by elementary row operations.

2. Rearrange the rows of this matrix, if required, so that the leading unit elements appear on the principal diagonal.

3. Identify the column of this matrix which has zero on the principal diagonal. After replacing the zero by -1, this column vector is the basis vector for the eigenvector space.

This method is illustrated with the following example.

Example 15.9. *For the system in Example 15.7, determine the modal matrix by Method 4.*

Solution:

From Eq. (15.97), we get

$$[\lambda_i I - A] = \begin{pmatrix} \lambda_i + 9 & -1 & 0 \\ 26 & \lambda_i & -1 \\ 24 & 0 & \lambda_i \end{pmatrix}$$

For $\lambda_1 = -2$:

$$[-2I - A] = \begin{pmatrix} 7 & -1 & 0 \\ 26 & -2 & -1 \\ 24 & 0 & -2 \end{pmatrix}$$

Its H.N.F. is

$$\begin{pmatrix} 1 & 0 & -1/12 \\ 0 & 1 & -7/12 \\ 0 & 0 & 0 \end{pmatrix}$$

The eigenvector, therefore, lies in the space defined by the third column after inserting -1 on the principal diagonal. Thus,

$$v_1 \in \text{span} \left\{ \begin{pmatrix} -1/12 \\ -7/12 \\ -1 \end{pmatrix} \right\} \qquad \ldots (15.101)$$

For $\lambda_2 = -3$:

$$[-3I - A] = \begin{pmatrix} 6 & -1 & 0 \\ 26 & -3 & -1 \\ 24 & 0 & -3 \end{pmatrix}$$

Its H.N.F. is

$$\begin{pmatrix} 1 & 0 & -1/8 \\ 0 & 1 & -6/8 \\ 0 & 0 & 0 \end{pmatrix}$$

Therefore,

$$v_2 \in \text{span} \left\{ \begin{pmatrix} -1/8 \\ -6/8 \\ -1 \end{pmatrix} \right\}$$

... (15.102)

For $\lambda_3 = -4$:

$$[-4I - A] = \begin{pmatrix} 5 & -1 & 0 \\ 26 & -4 & -1 \\ 24 & 0 & -4 \end{pmatrix}$$

Its H.N.F. is

$$\begin{pmatrix} 1 & 0 & -1/6 \\ 0 & 1 & -5/6 \\ 0 & 0 & 0 \end{pmatrix}$$

Therefore,

$$v_3 \in \text{span} \left\{ \begin{pmatrix} -1/6 \\ -5/6 \\ -1 \end{pmatrix} \right\}$$

... (15.103)

The eigenvectors are selected from the one-dimensional spaces given in Eq. (15.101) to (15.103). These vectors are linearly independent. For convenience, the basic vectors are multiplied by a suitable negative constant so that the first element is 1. Therefore, one selection of the Modal matrix is

$$T = \begin{pmatrix} 1 & 1 & 1 \\ 7 & 6 & 5 \\ 12 & 8 & 6 \end{pmatrix}$$

This is same as in Example (15.7)

When the matrix A has multiple eigenvalues, the number of independent eigenvectors associated with the eigenvalue may be equal to the multiplicity r of the eigenvalue. This property is determined using *Method 4*. The number of columns of the modified HNF of $[\lambda_i I - A]$ containing zeros on the principal axis may be equal to r. When this occurs, the independent eigenvectors are used to form the modal matrix T resulting in $A_d = T^{-1}AT$ being a diagonal matrix.

15.7.2. Motivation for Diagonalization

There are some advantages derived by diagonolizing the A matrix. They are summarized below:

1. Usually there is coupling between the various states of the system. It is therefore difficult to obtain the states. If the matrix A is diagonalized, the eigenvalues are located on the main diagonal of A provided the eigenvalues are all distinct, and it can be solved independently of other states.

2. Diagonalization simplifies the procedure for finding the state transition matrix.

3. The state transition matrix is also diagonal with elements given by $e^{\lambda_i t}$.

4. This form is also useful in study of controllability and observability of a system.

15.8. JORDAN CANONICAL FORM

When the A matrix has repeated eigenvalues, it cannot be diagonalized unless it is symmetric and has real elements. However, there exists a similarity transformation T which transform the A matrix to almost a diagonal matrix. The transformed matrix A_J is called the Jordan canonical form, where A_J is given by

$$A_J = \begin{pmatrix} \lambda_1 & 1 & 0 & 0 & 0 & 0 \\ 0 & \lambda_1 & 1 & 0 & 0 & 0 \\ 0 & 0 & \lambda_1 & 1 & 0 & 0 \\ 0 & 0 & 0 & \lambda_2 & 1 & 0 \\ 0 & 0 & 0 & 0 & \lambda_2 & 0 \\ 0 & 0 & 0 & 0 & 0 & \lambda_3 \end{pmatrix} \qquad \ldots (15.104)$$

in which λ_1 is repeated thrice and λ_2 twice.

The general properties of Jordan canonical form are:

1. The eigenvalues form the main diagonal elements.

2. All elements below the main diagonal are zero.

3. Some of the elements immediately above the repeated eigenvalues on the main diagonal are 1's as illustrated in Eq. (15.104).

4. The 1's, together with the repeated eigenvalues form typical blocks of dimension $r \times r$ where r is multiplicity of the eigenvalues. These blocks are known as *Jordan blocks* indicated by broken lines in Eq. (15.104).

5. When the nonsymmetrical A matrix has repeated eigenvalues, its eigenvectors are not linearly independent.

6. The number of Jordan blocks is equal to the number of independent eigenvectors v. Only one linearly independent eigenvector is associated with each Jordan block.

7. The number of 1's above the main diagonal is equal to $(n - v)$.

The transformation matrix T is determined as detailed below:

Assume that the matrix A has k distinct eigenvalues among n eigenvalues. First the eigenvectors corresponding to the distinct eigenvalues are determined by *Method 4* from

$$[\lambda_i I - A]v_i = 0 \qquad \text{... (15.105)}$$

where λ_i is the i-th distinct eigenvalue $(i = 1, 2,\ldots., k)$,

If λ_j is of multiplicity of r, the corresponding Jordan block of dimension $r \times r$ is

$$\begin{pmatrix} \lambda_j & 1 & & & \mathbf{0} \\ & \lambda_j & 1 & & \\ & & \lambda_j & 1 & \\ & & & \ddots & \ddots \\ \mathbf{0} & & & & \cdot & 1 \\ & & & & & \lambda_j \end{pmatrix} \qquad \text{... (15.106)}$$

Then the following transformation must hold:

$$[v_{k+1}\, v_{k+2}\, v_{k+3} \,\ldots\, v_{k+r}] \begin{pmatrix} \lambda_j & 1 & & & \mathbf{0} \\ & \lambda_j & 1 & & \\ & & \lambda_j & 1 & \\ & & & \ddots & \ddots \\ \mathbf{0} & & & & \cdot & 1 \\ & & & & & \lambda_j \end{pmatrix} = A[v_{k+1}\, v_{k+2}\, v_{k+3} \,\ldots\, v_{k+r}] \qquad \text{... (15.107)}$$

From the above equation, we have

$$\lambda_j v_{k+1} = Av_{k+1}$$
$$v_{k+1} + \lambda_j v_{k+2} = Av_{k+2}$$
$$v_{k+2} + \lambda_j v_{k+3} = Av_{k+3} \qquad \text{... (15.108)}$$
$$\cdot$$
$$\cdot$$
$$v_{k+r-1} + \lambda_j v_{k+r} = Av_{k+r}$$

where v_{k+1} is the eigenvector associated with λ_j and the remaining $(r - 1)$ vectors, $v_{k+2},\ldots., v_{k+r}$ are auxiliary vectors. Rearranging Eq. (15.108), the r vectors defined above can be determined from the following r vector equations.

$$[\lambda_j I - A]v_{k+1} = 0$$
$$[\lambda_j I - A]v_{k+2} = -v_{k+1}$$
$$[\lambda_j I - A]v_{k+3} = -v_{k+2} \qquad \qquad \qquad \dots \text{(15.109)}$$
$$\cdot$$
$$\cdot$$
$$[\lambda_j I - A]\, v_{k+r} = -v_{k+r-1}$$

Example 15.10. *Obtain the Jordan canonical form for the following A matrix:*

$$A = \begin{pmatrix} 0 & 6 & -5 \\ 1 & 0 & 2 \\ 3 & 2 & 4 \end{pmatrix}$$

Solution:

The characteristic equation is

$$|\lambda\, I - A| = \begin{vmatrix} \lambda & -6 & 5 \\ -1 & \lambda & -2 \\ -3 & -2 & \lambda - 4 \end{vmatrix} = \lambda^3 - 4\lambda^2 + 5\lambda - 2 = 0$$

Or,

$$= (\lambda - 2)(\lambda - 1)^2 = 0$$

Therefore, the matrix A has a simple eigenvalue at $\lambda_1 = 2$ and a double eigenvalue at $\lambda_2 = 1$.

The eigenvector v_1 associated with $\lambda_1 = 2$ is determined from

$$[\lambda_1 I - A]v_1 = 0$$

Or,

$$\begin{pmatrix} 2 & -6 & 5 \\ -1 & 2 & -2 \\ -3 & -2 & -2 \end{pmatrix} \begin{pmatrix} v_{11} \\ v_{21} \\ v_{31} \end{pmatrix} = 0$$

Selecting $v_{11} = 1$ arbitrarily, and solving for v_{21} and v_{31}, we get

$$v_{21} = -0.5 \text{ and } v_{23} = -1$$

Hence,

$$v_1 = \begin{pmatrix} 1 \\ -1/2 \\ -1 \end{pmatrix}$$

For $\lambda_2 = 1$, the eigenvectors v_2 and v_3 are obtained from

$$[\lambda_2 I - A]v_2 = 0 \qquad \qquad ... \ (15.110)$$

$$[\lambda_3 I - A]v_3 = 0 \qquad \qquad ... \ (15.111)$$

Eq. (15.110) in the expanded form is

$$\begin{pmatrix} 1 & -6 & 5 \\ -1 & 1 & -2 \\ -3 & -2 & -3 \end{pmatrix} \begin{pmatrix} v_{12} \\ v_{22} \\ v_{32} \end{pmatrix} = 0$$

Selecting arbitrarily $v_{12} = 1$ and solving for v_{22} and v_{23}, we get

$$v_{22} = -3/7 \text{ and } v_{23} = -5/7.$$

$$\therefore \quad v_2 = \begin{pmatrix} -1 \\ -3/7 \\ -5/7 \end{pmatrix}$$

Eq. (15.111) in its expended form is

$$\begin{pmatrix} 1 & -6 & -6 \\ -1 & 1 & -2 \\ -3 & -2 & -3 \end{pmatrix} \begin{pmatrix} v_{13} \\ v_{23} \\ v_{33} \end{pmatrix} = \begin{pmatrix} -1 \\ -3/7 \\ -5/7 \end{pmatrix}$$

Setting arbitrarily $v_{13} = 1$ and solving for v_{23} and v_{33}, we get

$$v_{23} = -22/49 \text{ and } v_{33} = -46/49$$

Therefore

$$v_3 = \begin{pmatrix} 1 \\ -22/49 \\ -46/49 \end{pmatrix}$$

Thus, the transformation matrix T is given by

$$T = [v_1 \quad v_2 \quad v_3] = \begin{pmatrix} 1 & 1 & 1 \\ -1/2 & -3/7 & -22/49 \\ -1 & -5/7 & -46/49 \end{pmatrix}$$

Therefore, the Jordan canonical form A_j is

$$A_j = T^{-1}AT = \begin{pmatrix} 2 & | & 0 & 0 \\ \hline 0 & | & 1 & 1 \\ 0 & | & 1 & 1 \end{pmatrix}$$

There are two Jordan blocks and the number of 1's in the upper diagonal is one.

15.9. TRANSFORMATION OF *A* MATRIX WITH COMPLEX EIGENVALUES

The diagonalization of A matrix, as described in the above sections, *decouples* the system modes. Since each state equation has the form

$$\dot{z}_i = \lambda_1 z_i + b_i u \qquad \qquad ... (15.112)$$

The state variable z_i can be solved independently of all the other states.

When the eigenvalues of the A matrix are complex, the modal matrix T which produces the diagonal matrix A_d contains complex numbers.

Consider the differential equation:

$$\ddot{x} - 2\sigma\dot{x} + \left(\sigma^2 + \omega_d^2\right)x = u(t) \qquad \qquad ... (15.113)$$

where $\omega_n^2 = \sigma^2 + \omega_d^2$.

Let $x_1 = x$ and $x_2 = \dot{x}_1$. The state and output equations are:

$$\dot{x} = \begin{pmatrix} 0 & 1 \\ -\omega_n^2 & 2\sigma \end{pmatrix} x + \begin{pmatrix} 0 \\ 1 \end{pmatrix} u \qquad \qquad ... (15.114)$$

$$y = \begin{bmatrix} 1 & 0 \end{bmatrix} x \qquad \qquad ... (15.115)$$

Since the eigenvalues are distinct and A is in the companion form, the modal matrix T is the Vandermonde matrix.

$$T = \begin{pmatrix} 1 & 1 \\ \sigma + j\omega_d & \sigma - j\omega_d \end{pmatrix} = \begin{bmatrix} v_1 & v_2 \end{bmatrix} \qquad \qquad ... (15.116)$$

When the transformation $x = Tz$ is used, we obtain

$$\dot{z} = \begin{pmatrix} \sigma + j\omega_d & 0 \\ 0 & \sigma - j\omega_d \end{pmatrix} z + \begin{pmatrix} 1/j2\omega_d \\ -1/j2\omega_d \end{pmatrix} u \qquad \qquad ... (15.117)$$

$$y = \begin{bmatrix} 1 & 1 \end{bmatrix} z \qquad \qquad ... (15.118)$$

Eq. (15.117) is undesirable though modes are decoupled. It gives rise to complex states which has no physical interpretation. This may be avoided by another transformation $z = Q\omega$ where Q is given by

$$Q = \begin{pmatrix} 1/2 & -j/2 \\ 1/2 & j/2 \end{pmatrix} \qquad \qquad \dots (15.119)$$

The new state and output equations are

$$\dot\omega = Q^{-1}A_d Q\omega + Q^{-1}\ T^{-1}\ B - u = A_d'\omega + B_d'u \qquad \dots (15.120)$$
$$y = CTQ\omega + Du \qquad \dots (15.121)$$

The above equations yield

$$\dot\omega = \begin{pmatrix} \sigma & \omega_d \\ -\omega_d & \sigma \end{pmatrix}\omega + \begin{pmatrix} 0 \\ 1/\omega_d \end{pmatrix}u \qquad \dots (15.122)$$

$$y = [1 \quad 0]\ \omega \qquad \dots (15.123)$$

The above matrices contain only real quantities. But the two modes are not isolated. This is an advantage because the pair of complex-conjugate eigenvalues λ_1 and λ_2 jointly contributes to one transient mode of the form of damped sinusoid.

The two transformations can be considered as one transformation given by

$$x = TQ\omega = \begin{pmatrix} 1 & 0 \\ \sigma & \omega_d \end{pmatrix} = T_m^T\omega \qquad \dots (15.124)$$

Comparing Eq. (15.124) and (15.116), the modified modal matrix is

$$T_m = [Re(v_1) \quad Im\ (v_1)] \qquad \dots (15.125)$$

Therefore, T_m can be obtained directly from T without using the Q transformation.

In general, when A contains both complex and real eigenvalues, the modified block diagonal matrix A_d' and the resulting B_d' and C_d' have the form

$$A_d' = \begin{pmatrix} \sigma_1 & \omega_1 & 0 & 0 & 0 & 0 \\ -\omega_1 & \sigma_1 & 0 & 0 & 0 & 0 \\ 0 & 0 & \sigma_2 & \omega_2 & 0 & 0 \\ 0 & 0 & -\omega_2 & \sigma_2 & 0 & 0 \\ & & & & \lambda5 & \\ & & & & & \ddots \\ & & & & & & \lambda_n \end{pmatrix}; B_d' = T_m^{-1}B \qquad \dots (15.126)$$

$$C_d' = CT_m \qquad \dots (15.127)$$

With this modified matrix $A_d{}'$, the two oscillating mode produced by the two sets of complex eigenvalues $\sigma_1 \pm j\omega_1$ and $\sigma_2 \pm j\omega_2$ are uncoupled. The column vectors contained in T for the conjugate eigenvalues are also conjugate. Thus

$$T_m = [v_1 \ v_1{}^* \ v_3 \ v_3{}^* \ v_5 \ . \ . \ . \ v_n] \qquad \text{... (15.128)}$$

The modified modal matrix T_m can be obtained directly by using

$$T_m = [Re(v_1) \ Im(v_1) \ Re \ (v_3) \ Im \ (v_3) \ v_5 \ . \ . \ . \ v_n] \qquad \text{... (15.129)}$$

15.10. STATE DIAGRAM

The state diagram is a signal flow graph to portray the state equations and differential equations. It includes the initial conditions of the states. Since the state diagram in the Laplace transform satisfies the rules of the signal flow graph discussed in Chapter 3, it can be used to obtain state equations, computer simulations, transfer functions, and the state transition equations. The state diagram is constructed following all the rules of the signal flow graph. Hence, the state diagram may be used for solving linear systems either analytically or by computers.

Consider the integral operation given by

$$x_1(t) = \int_{t_0}^{t} ax_2(\tau)d\tau \qquad a \leq 1 \qquad \text{... (15.130)}$$

Taking the Laplace transform on both sides,

$$X_1(s) = a(X_2(s)/s) + (x_1(t_0)/s) \qquad \tau \geq t_o \qquad \text{... (15.131)}$$

Eq. (15.130) is an algebraic equation and can be represented by signal flow graph shown in Fig. 15.6(a) and (b). These form the basic elements of the state diagram.

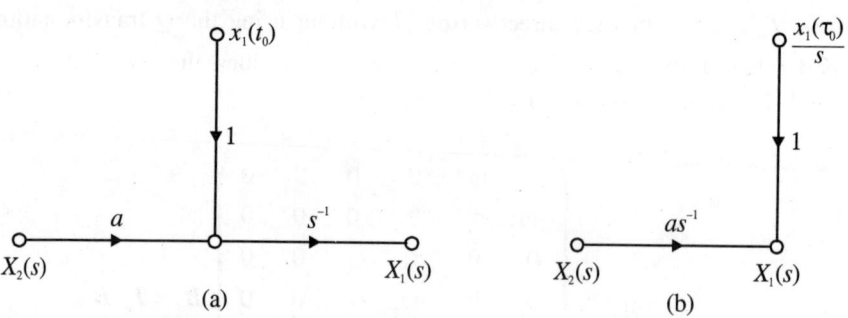

Fig. 15.6: Two signal flow graphs representation of Eq. (15.131)

15.10.1. Importance of State Diagrams

The important advantages of the state diagrams are:

1. A state diagram can be constructed directly from the differential equation that describes

the system. The state variables and the state equations can determined from the state diagram thus constructed.

2. A state diagram can be constructed from the transfer function of the system. This is known as the *decomposition of transfer functions.*

3. The transfer function of a system can be obtained from the state diagram.

4. The state equations and the output equations can be derived from the state diagram.

5. The state diagram can be used to program the system on analog computers.

6. The state diagram can be used for digital computer simulation.

7. The state transition equation in the Laplace transform may be obtained from the state diagram using the Mason's gain formula for signal flow graph.

We shall discuss the details of these techniques in the following sections.

15.10.2. State Diagram from Differential Equation

When a linear system is described by a high-order differential equation, a state diagram can be constructed from this equation.

Let us consider the following differential equation,

$$d^n y/dt^n + a_n d^{n-1}y/dt^{n-1} + \ldots + a_2 dy/dt + a_1 y = u \qquad \ldots (15.132)$$

Rearranging the terms to give the highest derivative, we have

$$d^n y/dt^n = -a_n d^{n-1}y/dt^{n-1} - \ldots -a_2 dy/dt - a_1 y + u \qquad \ldots (15.133)$$

Let us denote

$$d^i y/dt^i = y^{(i)}, \; i = 1, 2, \ldots, n \qquad \ldots (15.134)$$

The procedure for drawing the state diagram is:

1. The variables u, $y^{(n)}$, $\ldots$, $y^{(1)}$ and y are represented by nodes arranged as shown in Fig. 15.7(a). In terms of Laplace transform, these variables are $U(s)$, $s^n Y(s)$, $s^{n-1} Y(s)$, $\ldots$, $Y(s)$ respectively.

2. Connect these nodes by branches to portray Eq. (15.133) resulting in Fig. 15.7(b). Since the variables $y^{(i)}$ and $y^{(i-1)}$ are related through integration with respect to time, they are interconnected by a branch with gain of $1/s$, and the elements of Fig. 15.6 may be used. Thus, the complete state diagram is drawn as shown in Fig. 15.7(c).

3. The output variables of the integrators are defined as the state variables $x_1, x_2, \ldots, x_n$. This turn out to be the natural choice of state variables once the state diagram is drawn.

When the differential equation has the derivatives of the input on the right side as in Eq. (15.24), it is more convenient to obtain first the transfer function from the differential equation and then obtain the state diagram through transfer function decomposition.

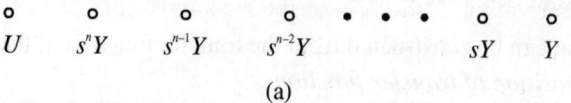

(a)

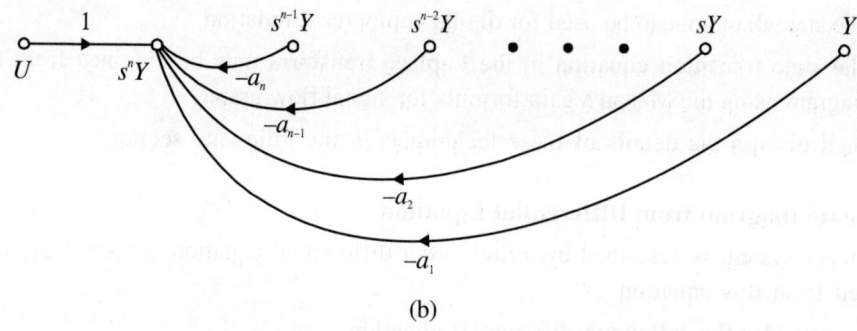

(b)

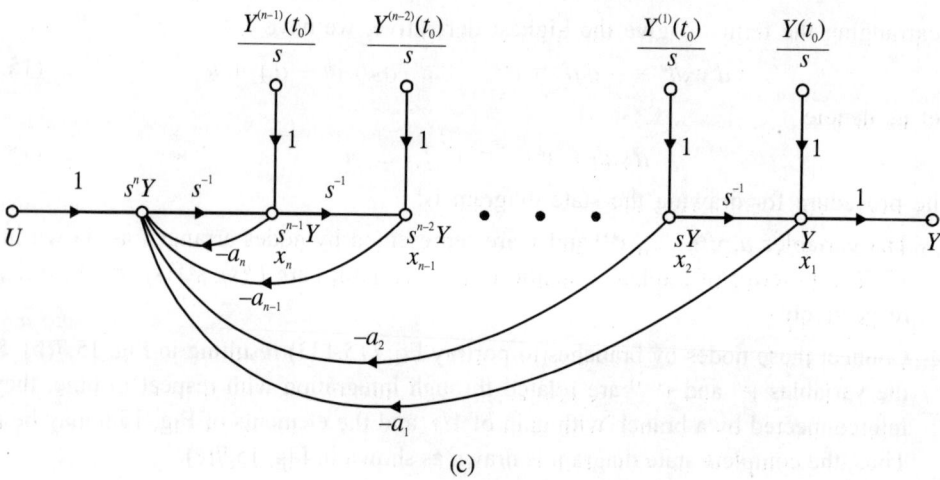

(c)

Fig. 15.7. (a) Arrangement of nodes (b) Nodes connected with branches
(c) Complete state diagram

Example 15.11. *Draw the state diagram for the second-order differential equation given*
by

$$\ddot{y} + 3\dot{y} + 2y = u.$$

Solution:

Equating the highest derivative to the rest of the terms, we get

$$\ddot{y} = -3\dot{y} - 2y + u.$$

Following the procedure outlined above, the state diagram of the system is drawn. It is shown in Fig. 15.8.

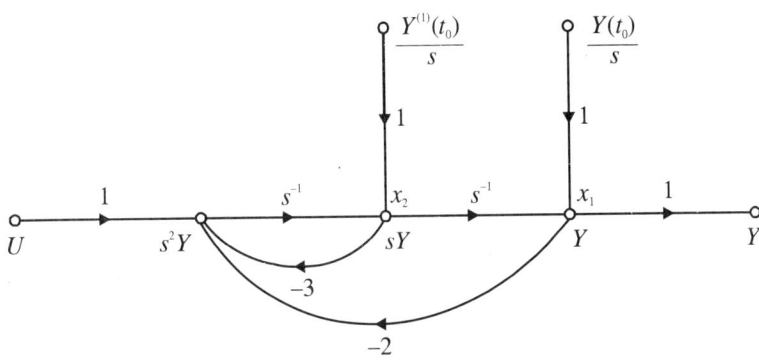

Fig. 15.8. State diagram

15.10.3. Transfer Function from State Diagram

The transfer function between an input and an output is obtained from the state diagram, after settling all initial states and other inputs to zero, by applying the Mason's gain formula.

This is illustrated by an example.

Example 15.12. *Obtain the transfer function corresponding to the state diagram shown in Fig. 15.8.*

Solution:

By setting $x_1(t_o)$ and $x_2(t_o)$ to zero and applying the Mason's gain formula, the transfer function $Y(s)/U(s)$ is

$$Y(\text{s})/U(\text{s}) = \frac{s^2}{1 - (-3s^{-1} - 2s^{-2})} = 1/(s^2 + 3s + 2)$$

15.10.4. State Equations from State Diagram

The state equations and the output equations can be obtained directly from the state diagram by the use of Mason's gain formula. The state equation contains the first-order time derivative of the state variable x_i on the left side and the state variables and the input variables on the right side. It contains no Laplace operator s or the initial state variables. Therefore, to obtain the state equations from the state diagram, eliminate the initial states and the branches with the gain 1/s from the state diagram. Then, applying the gain formula between the nodes, the state equations are written directly. This is illustrated with an example.

Example 15.13. *Obtain the state equations for the state diagram shown in Fig. 15.8.*

Solution:

The state diagram of Fig. 15.8 is simplified by eliminating the initial states and the branches with gain 1/s. The resulting state diagram is shown in Fig. 15.9. Using $\dot{x}_1$ and $\dot{x}_2$ as output nodes and x_1, x_2 and u as input nodes, the state equations are written directly by the application of the gain formula.

$$\dot{x}_1 = x_2$$

$$\dot{x}_2 = -2x_1 - 3x_2 + u$$

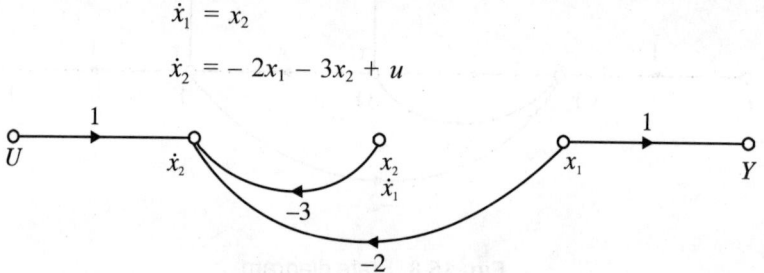

Fig. 15.9. Simplified state diagram

15.11. STATE DIAGRAM FROM TRANSFER FUNCTIONS (DECOMPOSITION OF TRANSFER FUNCTIONS)

The process of obtaining the state diagram or the state equations from the transfer function is called the decomposition of the transfer function. In general, the three basic ways of decomposition are:

    (i) Direct decomposition

    (ii) Cascaded decomposition

    (iii) Parallel decomposition.

Each has its own merits and is best suited for a particular situation. These methods are discussed in the following sections.

15.11.1. Direct Decomposition

The direct decomposition method is applied when the transfer function is not in the factored form.

Consider the transfer function

$$Y(s)/U(s) = (a_0 s^2 + a_1 s + a_2)/(b_0 s^2 + b_1 s + b_2) \qquad\qquad ...\ (15.135)$$

The procedure to obtain the state diagram and the state equations by direct decomposition is outlined below:

(i) Rewrite the transfer function such that it has only the negative powers of s. This is achieved by dividing the numerator and denominator of the transfer function by its highest power of s. For the transfer function in Eq. (15.135) we divide the numerator and the denominator by s^2.

(ii) Multiply the numerator and the denominator of the modified transfer function by a dummy variable $X(s)$. Now, Eq. (15.135) becomes

$$Y(s)/U(s) = \frac{(a_0 + a_1 s^{-1} + a_2 s^{-2})\, X(s)}{(b_0 + b_1 s^{-1} + b_2 s^{-2})\, X(s)} \qquad \dots (15.136)$$

(iii) Equate the numerators and the denominators of Eq. (15.136) to each other respectively. Thus,

$$Y(s) = (a_0 + a_1 s^{-1} + a_2 s^{-2})\, X(s) \qquad \dots (15.137)$$

$$U(s) = (b_0 + b_1 s^{-1} + b_2 s^{-2})\, X(s) \qquad \dots (15.138)$$

(iv) The state diagram is constructed from equations which are in the proper *cause-and–effect* relation. Eq. (15.137) satisfies this prerequisite.

However, Eq. (15.138) does not satisfy this condition since the input appears on the left side of the equation and, hence must be rearranged. To achieve this, divide both sides of Eq. (15.138) by b_0 and then write X(s) in terms of other terms. Thus,

$$X(s) = (1/b_0)\, U(s) - (b_1/b_0)s^{-1}\, X(s) - (b_2/b_0)s^{-2} X(s) \qquad \dots (15.139)$$

(v) Using Eq. (15.137) and (15.139), the state diagram is drawn as shown in Fig. 15.10. The outputs of the integrators are taken to be the state variables.

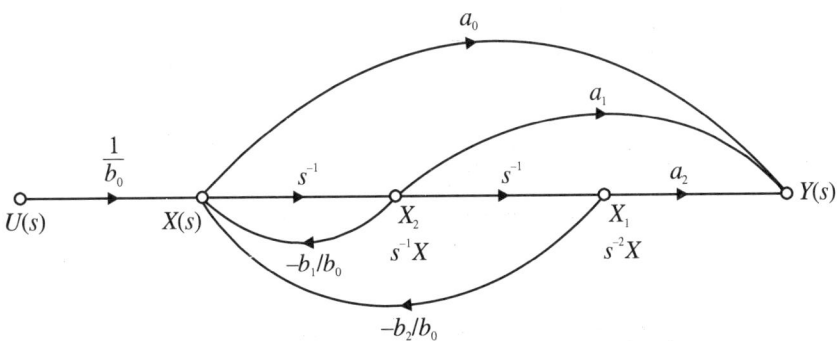

Fig. 15.10. State diagram of direct decomposition

As explained in the previous section, the state equations in matrix form are written directly from the state diagram

$$\begin{pmatrix} \dot{x}_1 \\ \dot{x}_2 \end{pmatrix} = \begin{pmatrix} 0 & 1 \\ -\dfrac{b_2}{b_0} & -\dfrac{b_1}{b_0} \end{pmatrix} \begin{pmatrix} x_1 \\ x_2 \end{pmatrix} + \begin{pmatrix} 0 \\ 1/b_0 \end{pmatrix} u \qquad \text{... (15.140)}$$

The output equation is obtained from Fig. 15.10 by applying the Mason's gain formula with $y(t)$ as the output node and $x_1(t)$, $x_2(t)$ and $u(t)$ as input nodes. Therefore,

$$y = \left(a_2 - \frac{a_0 b_2}{b_0} \right) x_1 + \left(a_1 - \frac{a_0 b_1}{b_0} \right) x_2 + \frac{a_0}{b_0} u \qquad \text{... (15.141)}$$

15.11.2. Cascade Decomposition

When the transfer function is in the factored form, the cascade decomposition is applied. Let the transfer function of Eq. (15.135) be written in the factored form as

$$Y(s)/U(s) = \frac{a_0}{b_0} \left(\frac{s + z_1}{s + p_1} \right) \left(\frac{s + z_2}{s + p_2} \right) \qquad \text{... (15.142)}$$

where z_1, z_2, p_1 and p_2 are real constants. Then, it can be treated as the product of two first-order functions. The state diagram for each first-order function is realized by using direct decomposition and then cascading them with a branch of unity gain. The complete state diagram is shown in Fig. 15.11. The outputs of the integrators are assigned as state variables. The state equations in matrix form are

$$\begin{pmatrix} \dot{x}_1 \\ \dot{x}_2 \end{pmatrix} = \begin{pmatrix} -p_2 & z_1 - p_1 \\ 0 & -p_1 \end{pmatrix} \begin{pmatrix} x_1 \\ x_2 \end{pmatrix} = \begin{pmatrix} a_0/b_0 \\ a_0/b_0 \end{pmatrix} u \qquad \text{... (15.143)}$$

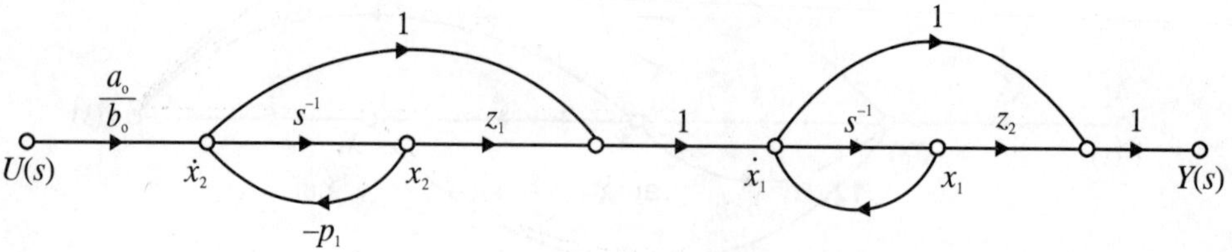

Fig. 15.11: State diagram of cascade decomposition

The output equation is

$$y = (z_2 - p_2) x_1 + (z_1 - p_1) x_2 + (a_0/b_0) u \qquad \text{... (15.144a)}$$

Or

$$y = \left(z_2 - p_2 \ \ z_1 - p_1\right)\begin{pmatrix} x_1 \\ x_2 \end{pmatrix}. \qquad \ldots (15.144b)$$

The advantage of cascade decomposition is that the poles and zeros of the transfer function appear as isolated branch gains on the state diagram. This facilitates the study of effects on the system when zeros and poles are varied.

15.11.3. Parallel Decomposition

When the denominator of the transfer function alone is in the factored form, the parallel decomposition is used.

Consider the second-order transfer function

$$Y(s)/U(s) = P(s)/[(s + p_1) \ (s + p_2)] \qquad \ldots (15.145)$$

where $P(s)$ is a polynomial of order less than 2. Assuming p_1 and p_2 are distinct, Eq. (15.145) may be written in the partial fraction form

$$Y(s)/U(s) = K_1/(s + p_1) + K_2/(s + p_2) \qquad \ldots (15.146)$$

where K_1 and K_2 are constants. The state diagram for each of the first order term is drawn by direct decomposition method and then they are connected in parallel as shown in Fig. 15.12.

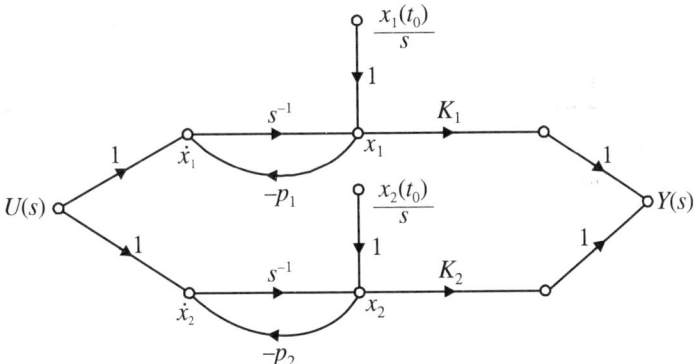

Fig. 15.12. State diagram of parallel decomposition

The state equations of the system in matrix form are written directly from the state diagram as

$$\begin{pmatrix} \dot{x}_1 \\ \dot{x}_2 \end{pmatrix} = \begin{pmatrix} -p_1 & 0 \\ 0 & -p_2 \end{pmatrix}\begin{pmatrix} x_1 \\ x_2 \end{pmatrix} = \begin{pmatrix} 1 \\ 1 \end{pmatrix} u \qquad \ldots (15.147)$$

The output equation is

$$y = [K_1 \ K_2]\begin{pmatrix} x_1 \\ x_2 \end{pmatrix} \qquad \qquad ... \ (15.148)$$

The parallel decomposition for transfer functions with simple poles yields the A matrix always in the diagonal form. Therefore, it may be used for the diagonalization of the A matrix. When the transfer function has repeated poles, the resulting A matrix due to parallel decomposition is in the Jordan form. This is illustrated in the following example.

Example 15.14. *Obtain the state diagram for the following transfer function:*

$$Y(s)/U(s) = (2s^2 + 6s + 5)/[(s + 1)^2(s + 2)]$$

Solution:

In partial fraction expansion form,

$$Y(s)/U(s) = 1/(s + 1)^2 + 1/(s + 1) \ + 1/(s + 2)$$

The state diagram is drawn for each term on the right-side of the above equation and then they are connected in parallel as shown in Fig. 15.13.

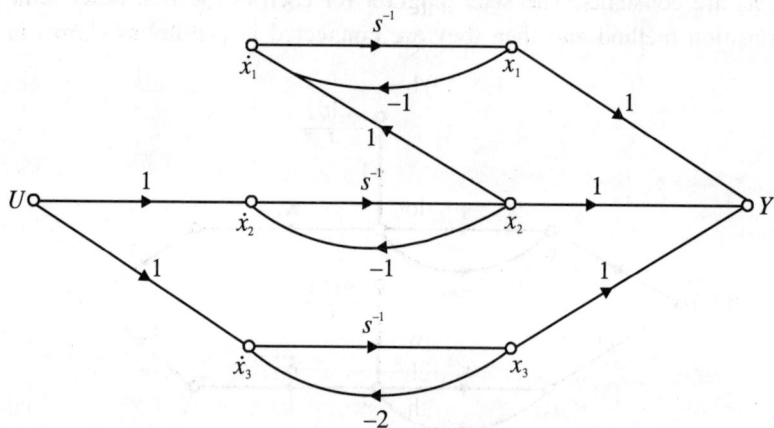

Fig. 15.13. State diagram for Example 15.14

The state equations of the system in matrix from are:

$$\begin{pmatrix} \dot{x}_1 \\ \dot{x}_2 \\ \dot{x}_3 \end{pmatrix} = \begin{pmatrix} -1 & 1 & 0 \\ 0 & -1 & 0 \\ 0 & 0 & -2 \end{pmatrix} \begin{pmatrix} x_1 \\ x_2 \\ x_3 \end{pmatrix} = \begin{pmatrix} 0 \\ 1 \\ 1 \end{pmatrix} u \qquad \qquad ... \ (15.147)$$

The output equation is

$$y = [1 \quad 1 \quad 1] \begin{pmatrix} x_1 \\ x_2 \\ x_3 \end{pmatrix}$$

The A matrix is in the Jordan canonical form.

15.12. SOLUTION OF HOMOGENEOUS STATE EQUATION (STATE TRANSITION MATRIX)

Once the state equations are expressed in the form

$$\dot{x} = A\,x + B\,u \qquad \qquad \text{... (15.149)}$$

the next step is to solve these equations. In general, the solution of this equation is of the form given in Eq. (15.150)

$$x(t) = \phi(t)x(0) + g(t) \qquad \qquad \text{... (15.150)}$$

The solution consists of two distinct parts:

(i) A part which is due to the initial states of the system.

(ii) Another part which represents the system response due to the driving function. When the driving function is zero, Eq. (15.149) becomes Eq. (15.151) and is called the *homogeneous equation*

$$\dot{x} = Ax \qquad \qquad \text{... (15.151)}$$

The solution of the homogeneous equation becomes

$$x(t) = \phi(t)x(0) \qquad \qquad \text{... (15.152)}$$

The square matrix $\phi(t)$ in Eq. (15.152) transforms the initial states of the system to a new state $x(t)$ at a later time t. Hence it is known as *state transition matrix* or simply *transition matrix*. It states precisely how the system changes from one state to the next state in a time interval with no driving function applied to the system. It contains all the information about the free motions of the system. Eq. (15.151) is called the *homogeneous state equation.*

The state transition matrix in Eq. (15.152) may be evaluated by employing any one of the following methods:

1. By Laplace transform method
2. By Series summation method
3. By Diagonal Matrix method
4. By using Cayley-Hamilton theorem
5. By Numerical method

15.12.1. Laplace Transform Method

Before we solve the state equations, let us first consider the scalar case:

$$\dot{x} = ax \qquad \qquad \text{... (15.153)}$$

Taking the Laplace transform of both sides of Eq. (15.153),

$$sX(s) - x(0) = aX(s) \qquad \qquad \text{... (15.154)}$$

Solving for $X(s)$, we have,

$$X(s) = x(0)/(s - a) = (s - a)^{-1} x(0) \qquad \qquad \text{... (15.155)}$$

By taking the inverse Laplace transform, we obtain the solution

$$x(t) = e^{at}x(0) \qquad \qquad \text{... (15.156)}$$

The above approach to the solution of the homogeneous scalar differential equation can be extended to solve the homogeneous state equation

$$\dot{x} = A\,x$$

Taking the Laplace transform of both sides of the above equation

$$sX(s) - x(0) = AX(s) \qquad \qquad \text{... (15.157)}$$

Hence

$$[sI - A]\,X(s) = x(0) \qquad \qquad \text{... (15.158)}$$

Premultiplying both sides of Eq. (15.158) by $[sI - A]^{-1}$, we get

$$X(s) = [sI - A]^{-1}\,x(0) \qquad \qquad \text{... (15.159)}$$

The solution is given by the inverse Laplace transform of $X(s)$.

$$x(t) = \mathcal{L}^{-1}[sI - A]^{-1}x(0) \qquad \qquad \text{... (15.160)}$$

Expanding $[sI - A]^{-1}$ in an infinite series, we have

$$[sI - A]^{-1} = (I/s) + (A/s^2) + (A^2/s^3) + ... \qquad \qquad \text{... (15.161)}$$

Hence the inverse transform of Eq. (15.161) is

$$\mathcal{L}^{-1}[sI - A]^{-1} = I + At + A^2\,t^2/2! + A^3\,t^3/3!+... \qquad \qquad \text{... (15.162)}$$

Therefore, the solution of the homogeneous state equation is

$$x(t) = e^{At}x(0) = \phi(t)\,x(0) \qquad \qquad \text{... (15.163)}$$

e^{At} is called *the matrix exponential* and provides a convenient means for finding the closed solution of the infinite series. Thus the state transition matrix is given by the matrix exponential.

15.12.2. Properties of Matrix Exponential

Since the matrix exponential is very important in the state-variable analysis of linear systems, we shall now examine its properties.

(*i*) The matrix exponential of an $n \times n$ matrix A is

$$e^{At} = \sum_{k=0}^{\infty} A^k t^k / k! \qquad \qquad ... (15.163)$$

It converges absolutely for all finite t.

(*ii*) Because of the convergence property, the series can be differentiated term by term to give

$$de^{At}/dt = A + A^2 t + A^3 t^2/2! + ... + A^k t^{k-1}/(k-1)! + ...$$

$$= A [I + At + A^2 t^2/2! + ... + A^{k-1} t^{k-1}/(k-1)! + ...] \quad ... (15.164)$$

$$= A e^{At} \qquad \qquad ... (15.165)$$

(*iii*) $e^{A(t+s)} = e^{At} . e^{As}$ ... (15.166)

(*iv*) e^{At} is nonsingular and its inverse e^{-At} always exists.

(*v*) $e^{(A+B)t} = e^{At} . e^{B}$ if $AB = BA$ ... (15.167)

    $e^{(A+B)t} \neq e^{At} . e^{B}$ if $AB \neq BA$ ... (15.168)

15.12.3. Series Summation Method

Let us first consider the solution of the scalar differential equation.

$$\dot{x} = ax \qquad \qquad ... (15.169)$$

Let us assume a solution of the form

$$x(t) = b_0 + b_1 t + b_2 t^2 + \cdots + b_k t^k + \cdots \qquad ... (15.170)$$

By substituting Eq. (15.170) into Eq. (15.169), we get

$$b_1 + 2b_2 t + 3b_3 t^2 + ... + kb_k t^{k-1} + ...$$
$$= a(b_0 + b_1 t + b_2 t^2 + ... + b_k t^k + ...) \qquad ... (15.171)$$

If the assumed solution were to be the true solution, Eq. (15.171) must hold for any t. Hence equating the coefficients of the equal power of t, we obtain

$$b_1 = ab_0$$
$$b_2 = (1/2)ab_1 = a^2 b_0/2!$$
$$b_3 = (1/3)ab_2 = (1/6)a^2 b_1 = a^3 b_0/3!$$

$$\cdot$$

$$b_k = a^k b_0/k!$$

The value of b_0 is obtained by substituting $t = 0$ in Eq. (15.170) Thus

$$x(0) = b_0$$

Hence, the solution $x(t)$ given by Eq. (15.170) may be written as

$$x(t) = (1 + at + a^2 t^2/2! + a^3 t^3/3! + ... + + a^k t^k/k! ...)x(0)$$
$$= e^{at}x(0) \qquad \qquad ... (15.172)$$

Let us now consider the homogeneous state equation

$$\dot{x} = Ax \qquad \qquad \dots (15.173)$$

By analogy with the scalar case, we assume that the solution is in the form of a vector power series in t

$$x(t) = b_0 + b_1 t + b_2 t^2 + \dots + b_k t^l + \dots \qquad \dots (15.174)$$

By substituting this assumed solution into Eq. (15.173), we get

$$b_1 + 2b_2 t + 3b_3 t^2 + \dots + kb_k t^{k-1} + \dots$$
$$= A (b_0 + b_1 t + b_2 t^2 + \dots + b_k t^k + \dots) \qquad \dots (15.175)$$

If the assumed solution is to be the true solution, Eq. (15.175) must hold for all t. Then, equating the coefficients of equal powers of t, we obtain

$$b_1 = Ab_0$$
$$b_2 = (1/2) Ab_1 = A^2 b_0/2!$$
$$b_3 = (1/3) Ab_2 = (1/6) A^2 b_1 = A^3 b_0/3!$$

$$\cdot$$
$$\cdot$$

$$b_k = A^k b_0/k!$$

By substituting $t = 0$ in Eq. (15.174), we get

$$x(0) = b_0$$

Thus the solution $x(t)$ of Eq. (15.173) can be written as

$$x(t) = (I + At + A^2 t^2/2! + \dots + A^k t^k/k! + \dots) x(0) \qquad \dots (15.176)$$

The expression in the right-hand side of Eq. (15.176) is an n x n exponential matrix. Therefore,

$$x(t) = e^{At} x(0) = \phi(t)x(0) \qquad \dots (15.177)$$

15.12.4. Diagonal Matrix Method

Consider the homogeneous state equation

$$\dot{x} = A x \qquad \qquad \dots (15.178)$$

Assume a similarity transformation

$$x = Tz \qquad \qquad \dots (15.179)$$

Substituting Eq. (15.179) into Eq. (15.178) and then premultiplying by T^{-1}, we obtain

$$\dot{z} = TAT^{-1} = A_d z \qquad \qquad \dots (15.180)$$

If T is the modal matrix, A_d is the diagonal matrix. Taking the Laplace transform and rearranging, we get

$$Z(s) = [sI - A_d]^{-1} z(0) \qquad \dots (15.181)$$

Since this equation is similar in form to Eq. (15.159), the solution is given by

$$z(t) = e^{A_d t} z(0) = \phi(t) z(0) \qquad \qquad \text{... (15.182)}$$

The advantage of this expression is that, if the eigenvalues $\lambda_1, \lambda_2, ..., \lambda_n$ of the A matrix are distinct and since A_d is in the diagonal form, then the matrix exponential is also diagonal. Thus,

$$\phi(t) = e^{A_d t} = \begin{pmatrix} e^{\lambda_d t} & & & \\ & e^{\lambda_d t} & \mathbf{0} & \\ & & \cdot & \\ \mathbf{0} & & & \cdot \\ & & & e^{\lambda_d t} \end{pmatrix} \qquad \qquad \text{... (15.183)}$$

If there is a multiplicity of eigenvalues, for example, if the eigenvalues are $\lambda_1, \lambda_1, \lambda_1, \lambda_4, ..., \lambda_n$, then the matrix exponential is given by

$$\phi(t) = e^{A_d t} = \begin{pmatrix} e^{\lambda_1 t} & & & & \\ & te^{\lambda_1 t} & \mathbf{0} & & \\ & & t^2 e^{\lambda_1 t} & & \\ \mathbf{0} & & & e^{\lambda_4 t} & \\ & & & & \cdot \\ & & & & e^{\lambda_n t} \end{pmatrix} \qquad \qquad \text{... (15.184)}$$

15.12.5. Method using Cayley-Hamilton Theorem

Cayley-Hamilton theorem states that an $n \times n$ matrix A satisfies his own characteristic equation. This enables us to write all powers of A in terms of a polynomial in A of degree equal to or less than $(n - 1)$. Thus,

$$\phi(t) = e^{At} = \alpha_0(t)\, I + \alpha_1(t)A + ... + \alpha_{n-1}(t)A^{n-1} \qquad \qquad \text{... (15.185)}$$

If A has n distinct eigenvalues, then to determine the coefficients, $\alpha_0(t), \alpha_1(t), ..., \alpha_{n-1}(t)$, we substitute each of the eigenvalues into the scalar version of Eq. (15.185) and obtain n simultaneous equations, Solution of these equations yields the coefficients and the substitution of these values into Eq. (15.185) results in the evaluation of e^{At} or the state transition matrix $\phi(t)$.

Example 15.15. *Obtain the state transition matrix $\phi(t)$ of the following system by using Cayley-Hamilton Theorem.*

$$\begin{bmatrix} \dot{x}_1 \\ \dot{x}_2 \end{bmatrix} = \begin{pmatrix} 0 & 1 \\ -2 & -3 \end{pmatrix} \begin{pmatrix} x_1 \\ x_2 \end{pmatrix}$$

Solution:

For this systems

$$A = \begin{pmatrix} 0 & 1 \\ -2 & -3 \end{pmatrix}$$

The characteristic equation is

$$F(\lambda) = \lambda^2 + 3\lambda + 2 = 0$$

The eigenvalues are $\lambda_1 = -1$ and $\lambda_2 = -2$.

Using Cayley-Hamilton Theorem, we have the transition matrix $\phi(t)$ as

$$\phi(t) = e^{At} = \alpha_0(t)I + \alpha_1(t)A$$

The scalar equations are:

For $\lambda_1 = -1$: $\qquad e^{\lambda_1 t} = e^{-t} = \alpha_0(t) - \alpha_1(t)$

For $\lambda_2 = -2$: $\qquad e^{\lambda_2 t} = e^{-2t} = \alpha_0(t) - 2\alpha_1(t)$

Solving the two simultaneous equations for $\alpha_0(t)$ and $\alpha_1(t)$, we get

$$\alpha_0(t) = 2e^{-t} - e^{-2t}$$

$$\alpha_1(t) = e^{-t} - e^{-2t}$$

Therefore the state transition matrix is

$$\phi(t) = e^{At} = (2e^{-t} - e^{-2t})I + (e^{-t} - e^{-2t})\begin{pmatrix} 0 & 1 \\ -2 & -3 \end{pmatrix}$$

$$= \begin{pmatrix} 2e^{-t} - e^{-2t} & e^{-t} - e^{-2t} \\ -2e^{-t} + 2e^{-2t} & -e^{-t} + 2e^{-2t} \end{pmatrix}$$

15.12.6. Numerical Method

The state transition matrix is given by the infinite series

$$\phi(t) = e^{At} = I + At + A^2 t^2/2! + \dots + A^k t^k/k! + \dots \qquad \dots (15.186)$$

From Eq. (15.186) it can be seen that each term can be derived from its previous term, Eq. (15.186) can therefore be written as

$$\phi(t) = e^{At} = \sum_{k=0}^{\infty} A^k t^k / k! \qquad \dots (15.187)$$

Eq. (15.187) provides a convenient algorithm to evaluate e^{At} using a digital computer. The computations may be carried out to any desired accuracy. The advantage of this method is that it is *simple, easy to program and does not require the evaluation of the eigenvalues of the plant matrix A*.

15.13. PROPERTIES OF STATE TRANSITION MATRIX

Since the state transition matrix satisfies the homogeneous state equation, it represents the free response of the system or the response due to initial conditions only. Thus, the state transition matrix $\phi(t)$ completely defines the transition of the states from the initial states $x(0)$ at $t = 0$ to a subsequent state $x(t)$ at any time t when the inputs are zero.

The state transition matrix has the following properties.

(i) $\phi(0) = I$, the *identity matrix.*

Proof: Since $\phi(t) = I + At + A^2\, t^2/2! + \dots$

$$\phi(0) = I$$

(ii) $\phi(t)^{-1} = \phi(-t)$

Proof: $\phi(t) = e^{At}$

Post-multiplying both sides by e^{-At}, we get

$$\phi(t)e^{-At} = e^{At}e^{-At} = I$$

Pre-multiplying both sides of the above equation by $\phi(t)^{-1}$, we obtain

$$e^{-At} = \phi(t)^{-1} = \phi(-t) \qquad \text{since } e^{-At} = \phi(-t).$$

When this property is applied to Eq. (15.162), we get

$$x(0) = \phi(-t)\, x(t) \qquad \dots (15.190)$$

Eq. (15.190) implies that the state transition process may be considered as bilateral in time. This means that the transition in time takes place in either direction.

(*iii*) $\phi(t_2 - t_1)\, \phi(t_1 - t_0) = \phi(t_2 - t_0)$ $\qquad \dots (15.191)$

By definition,

$$\phi(t) = e^{At}.$$

Hence,

$$\phi(t_2 - t_1) = e^{A(t_2 - t_1)}$$

$$\phi(t_1 - t_0) = e^{A(t_1 - t_0)}$$

Therefore,

$$\phi(t_2 - t_1)\, \phi(t_1 - t_0) = e^{A(t_2 - t_1)}e^{A(t_1 - t_0)}$$

$$= e^{A(t_2 - t_1 + t_1 - t_0)}$$

$$= e^{A(t_2 - t_0)}$$

$$= \phi(t_2 - t_0)$$

This property implies that a state transition process may be broken up into a number of parts.

(iv) $\phi^n(t) = \phi(nt)$ ... (15.192)

$$\phi^n(t) = [\phi(t)]^n$$
$$= (e^{At})^n$$
$$= (e^{nAt})$$
$$= \phi(nt)$$

(v) $\phi(t)$ is the solution of the equation

$$\phi(t) = A\phi(t) \ , \ \phi(0) = I \qquad \qquad ... (15.193)$$

Taking the Laplace transform, we get

$$s\phi(s) - \phi(0) = A\phi(s)$$

Or,

$$\phi(s) = [sI - A]^{-1}\phi(0)$$

Substituting $\phi(0) = I$,

$$\phi(0) = [sI - A]^{-1}I = [sI - A]^{-1}$$

Taking the inverse Laplace transform,

$$\phi(t) = \mathcal{L}^{-1} [sI - A]^{-1}$$

which is the definition of the state transition matrix.

15.14. SOLUTION OF NON-HOMOGENEOUS STATE EQUATION (STATE TRANSITION EQUATION)

Consider the non-homogeneous state equation

$$\dot{x}(t) = A \, x(t) + B \, u(t) \qquad \qquad ... (15.194)$$

where

$$x = n - \text{dimensional vector}$$
$$u = r - \text{dimensional vector}$$
$$A = n \times n \text{ plant matrix}$$
$$B = n \times r \text{ input matrix}$$

Eq. (15.194) may be written as

$$\dot{x}(t) - A \, x(t) = B \, u(t) \qquad \qquad ... (15.195)$$

Premultiplying both sides of the above equation by e^{-At}, we get

$$e^{-At}[\dot{x}(t) - Ax(t)] = d[e^{-At}x(t)]/dt = e^{-At}Bu(t) \qquad \qquad ... (15.196)$$

Integrating the above equation, we obtain

$$e^{-At}x(t) = \int_0^t e^{-A\tau} Bu(\tau)d\tau + x(0) \qquad \qquad ... (15.197)$$

The lower limit is the time at which $u(t)$ is applied. Multiplying Eq. (16.197) by e^{At} and rearranging the terms, we get

$$x(t) = e^{At}x(0) + \int_0^t e^{-A(t-\tau)} Bu(\tau)d\tau$$

$$= \phi(t)x(0) + \int_0^t \phi(t-\tau)Bu(\tau)d\tau \qquad \qquad \dots (15.198)$$

Generalizing for initial conditions at $t = t_0$, the solution of the non-homogeneous equation is

$$x(t) = \phi(t-t_0)x(t_0) + \int_0^t \phi(t-\tau)Bu(\tau)d\tau \qquad \qquad \dots (15.199)$$

Eq. (15.199) is called the *state transition equation* which is the solution of linear nonhomogeneous equation. It describes the change of state relative to the initial conditions $x(t_0)$ and the input $u(t)$. Eq. (16.199) may be written as

$$x(t) = x_{zi}(t) + x_{zs}(t) \qquad \qquad \dots (15.200)$$

where x_{zi} is the *zero-input response* and x_{zs} is the *zero-state response* $[x(t_0) = 0]$.

Example 15.16. *Solve the state equation*

$$\begin{pmatrix} \dot{x}_1 \\ \dot{x}_2 \end{pmatrix} = \begin{pmatrix} 0 & 1 \\ -2 & -3 \end{pmatrix}\begin{pmatrix} x_1 \\ x_2 \end{pmatrix} + \begin{pmatrix} 0 \\ 1 \end{pmatrix}u(t)$$

where $u(t)$ is the unit step input function occurring at $t = 0$.

Solution:

For this system,

$$A = \begin{pmatrix} 0 & 1 \\ -2 & -3 \end{pmatrix}$$

The state transition matrix $\phi(t) = e^{At}$ was obtained in Example 15.15 as

$$\phi(t) = e^{At} = \begin{pmatrix} 2e^{-t}-e^{-2t} & e^{-t}-e^{-2t} \\ -2e^{-t}+2e^{-2t} & -e^{-t}+2e^{-2t} \end{pmatrix}$$

The response to the unit-step input is

$$x(t) = \phi(t)x(0) + \int_0^t \begin{pmatrix} 2e^{-(t-\tau)}-e^{-2(t-\tau)} & e^{-(t-\tau)}-e^{-2(t-\tau)} \\ -2e^{-(t-\tau)}+2e^{-2(t-\tau)} & -e^{-(t-\tau)}+2e^{-2(t-\tau)} \end{pmatrix}\begin{pmatrix} 0 \\ 1 \end{pmatrix}d\tau$$

Or,

$$\begin{pmatrix} \dot{x}_1(t) \\ \dot{x}_2(t) \end{pmatrix} = \begin{pmatrix} 2e^{-t} - e^{-2t} & e^{-t} - e^{-2t} \\ -2e^{-t} + 2e^{-2t} & -e^{-t} + 2e^{-2t} \end{pmatrix} \begin{pmatrix} x_1 \\ x_2 \end{pmatrix} + \begin{pmatrix} \frac{1}{2}(1 - 2e^{-t} + e^{-2t}) \\ e^{-t} - e^{-2t} \end{pmatrix}$$

15.14.1. Laplace Transform Method

The nonhomogeneous state equation

$$\dot{x}(t) = Ax(t) + Bu(t)$$

may also be solved by the Laplace transform approach. The Laplace transform of the above equation yields

$$sX(s) - x(0) = AX(s) + BU(s) \qquad \ldots (15.201)$$

Or,

$$[sI - A]X(s) = x(0) + BU(s) \qquad \ldots (15.202)$$

Premultiplying both sides of Eq. (15.202) by $[sI - A]^{-1}$

$$X(s) = [sI - A]^{-1}x(0) + [sI - A]^{-1} BU(s) \qquad \ldots (15.203)$$

Using Eq. (15.162), the above equation may be written as

$$X(s) = \mathcal{L}^{-1}[e^{At}]x(0) + \mathcal{L}^{-1} [e^{At}]BU(s) \qquad \ldots (15.204)$$

The inverse Laplace transform of Eq. (15.204) is obtained by the use of convolution integral as

$$x(t) = e^{At}x(0) + \int_0^t e^{A(t-\tau)} Bu(\tau)\, d\tau \qquad \ldots (15.205)$$

$$= \phi(t)x(0) + \int_0^t \phi(t-\tau)Bu(\tau)\, d\tau \qquad \ldots (15.206)$$

The state transition equation given in Eq. (15.206) is useful only when the initial time is defined as $t = 0$. If, however the initial time is t_0, then the state transition equation is obtained as

$$x(t) = \phi(t-t_0)\, x(0) + \int_0^t \phi(t-\tau)Bu(\tau)\, d\tau \qquad \ldots (15.207)$$

15.14.2. State Transition Equation from state Diagram

The Laplace transform method of solving the state equations involves the inverse of the matrix $[sI - A]$. The state transition equation may be obtained using the state diagram and the Mason's gain formula.

The state transition equation in the Laplace transform, setting the initial time to be t_o is given by

$$X(s) = [sI - A]^{-1} x(t_o) + [sI - A]^{-1} BU(s) \qquad \qquad ... (15.208)$$

The above equation can be written directly from the state diagram using the gain formula, with $X_i(s)$, $(i = 1,2,3, ..., n)$ as output nodes, and $x_i(t_o)$, $(i = 1,2,..., n)$ and $U_j(s)$, $(j = 1,2,...,r)$ as the input nodes. The state diagram method of obtaining the state transition equation is illustrated by the following example.

Example 15.17. *Draw the state diagram for the system described by the state equation*

$$\begin{pmatrix} \dot{x}_1 \\ \dot{x}_2 \end{pmatrix} = \begin{pmatrix} 0 & 1 \\ -2 & -3 \end{pmatrix} \begin{pmatrix} x_1 \\ x_2 \end{pmatrix} + \begin{pmatrix} 0 \\ 1 \end{pmatrix} u(t)$$

when $u(t) = 1$ *for* $t \geq t_0$ *and* t_0 *is the initial time. Also obtain the state transition equation.*

Solution:

The state diagram is shown in Fig. 15.14. The output of the integrators are assigned as state variables. Considering $X_1(s)$ and $X_2(s)$ as output nodes with $x_1(t_o)$, $x_2(t_o)$ and $U(s)$ as input nodes, and applying the gain formula, we obtain

$$X_1(s) = (1/\Delta)\{[s^{-1}(1 + 3s^{-1}) x_1(t_o)] + s^{-2}[x_2(t_o) + U(s)]\} \qquad ... (15.209)$$

$$X_2(s) = (1/\Delta)\{[-2s^{-2}x_1(t_o)] + s^{-1}[x_2(t_o) + U(s)]\} \qquad ... (15.210)$$

where

$$\Delta = 1 + 3s^{-1} + 2s^{-2}$$

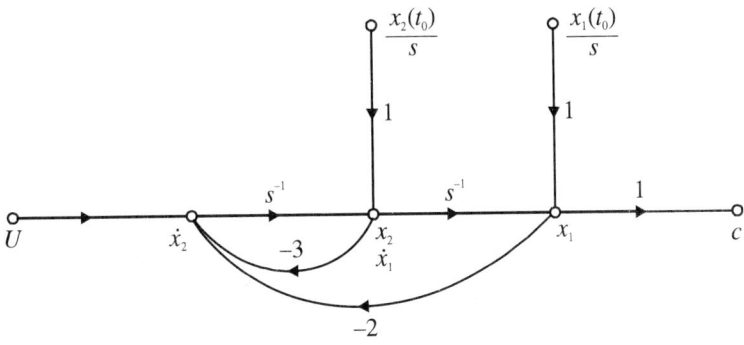

Fig. 15.14. State diagram for Example 15.17

After simplification, Eq. (15.209) and (15.210) in matrix form are:

$$\begin{pmatrix} X_1(s) \\ X_2(s) \end{pmatrix} = \frac{1}{(s+1)(s+2)} \begin{pmatrix} s+3 & 1 \\ -2 & s \end{pmatrix} \begin{pmatrix} x_1(t_o) \\ x_2(t_o) \end{pmatrix} + \begin{pmatrix} 1/(s+1)(s+2) \\ s/(s+1)(s+2) \end{pmatrix} U(s)$$

$$... (15.211)$$

The state transition equation for $t \geq t_o$ is given by the inverse Laplace transform of Eq.(15.211). Thus,

$$\begin{pmatrix} x_1(t) \\ x_2(t) \end{pmatrix} = \begin{pmatrix} 2e^{(t-t_0)} - e^{-2(t-t_0)} & e^{-(t-t_0)} - e^{-2(t-t_0)} \\ -2e^{(t-t_0)} + 2e^{-2(t-t_0)} & -e^{-(t-t_0)} + 2e^{-2(t-t_0)} \end{pmatrix} \begin{pmatrix} x_1(t_0) \\ x_2(t_0) \end{pmatrix}$$

$$+ \begin{pmatrix} \dfrac{1}{2}u(t-t_0) - e^{-(t-t_0)} + \dfrac{1}{2}e^{-2(t-t_0)} \\ e^{-(t-t_0)} - e^{-2(t-t_0)} \end{pmatrix} \quad t \geq t_0 \qquad ... (15.212)$$

15.15. TRANSFER MATRIX

In this section, we shall extend the transfer function concept of single-input single-output systems to multiple-input multiple-output (MIMO) systems.

Consider a MIMO system with r inputs and m outputs. The r inputs may considered to be a component of vector called *input vector*. Similarly the m outputs are the components of a *output vector*. The matrix which relates the Laplace transform of the output vector to the Laplace transform of the input vector is known as *transfer matrix* between the output vector and the input vector.

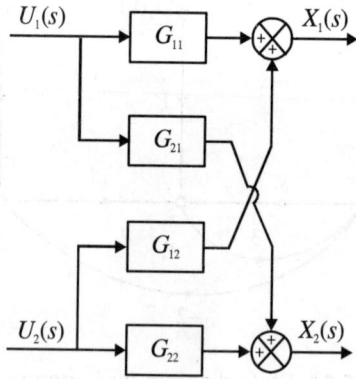

Fig. 15.15. MIMO system with two inputs and two outputs

Fig. 15.15 shows a MIMO systems with two inputs and two outputs. The relationship between the outputs and inputs is given by

$$X_1(s) = G_{11}(s)\, U_1(s) + G_{12}(s)\, U_2(s) \qquad \qquad \dots (15.213)$$

$$X_2(s) = G_{21}(s)\, U_1(s) + G_{22}(s)\, U_2(s) \qquad \qquad \dots (15.214)$$

where $G_{ij}(s)$ is the transfer function relating the *i*-th output to the *j*-th input. In vector-matrix form, Eqs. (15.213) and (15.214) can be written as

$$\begin{bmatrix} X_1(s) \\ X_2(s) \end{bmatrix} = \begin{bmatrix} G_{11}(s) & G_{12}(s) \\ G_{12}(s) & G_{22}(s) \end{bmatrix} \begin{bmatrix} U_1(s) \\ U_2(s) \end{bmatrix} \qquad \qquad \dots (15.215)$$

A MIMO system is also known as *multivariable system*. In general, if such a system has *r* inputs and *m* outputs, then the Laplace transform of the *i*-th output is related to the Laplace transform of the *r* inputs by

$$X_i(s) = G_{i1}(s)U_1(s) + G_{i2}(s)U_2(s) + \cdots G_{ir}(s)U_r(s) \quad (i = 1, 2, \dots, r) \qquad \dots (15.216)$$

It may be noted that in defining $G_{ij}(s)$, only the *j*-th input is considered and the other inputs are assumed to be zero. In matrix form the relationship between the Laplace transform of the output vector and the Laplace transform of the input vector is

$$\begin{bmatrix} X_1(s) \\ X_2(s) \\ \vdots \\ X_m(s) \end{bmatrix} = \begin{bmatrix} G_{11}(s) & G_{12}(s) & \cdots & G_{1r}(s) \\ G_{21}(s) & G_{22}(s) & \cdots & G_{2r}(s) \\ \vdots & & \vdots & \\ G_{m1}(s) & G_{m2}(s) & \cdots & G_{mr}(s) \end{bmatrix} \begin{bmatrix} U_1(s) \\ U_2(s) \\ \vdots \\ U_r(s) \end{bmatrix} \qquad \dots (15.217)$$

or,

$$X(s) = G(s)\, U(s) \qquad \qquad \dots (15.218)$$

Eq.(15.127) shows the interaction between the *r* inputs and *m* outputs. In Eq.(15.128), $X(s)$ is the Laplace transformed output vector, $U(s)$ is the Laplace transformed input vector and $G(s)$ is the transfer matrix between $X(s)$ and $U(s)$.

By following the same steps as used in the derivation of Eq.(15.68), we obtain the transfer matrix for MIMO system as

$$G(s) = C[sI - A]^{-1} B + D \qquad \qquad \dots (15.219)$$

Thus, Eq.(15.68) is a special case of Eq.(15.219).

15.15.1. Transfer Matrix of Closed-Loop Systems.

Let us consider the MIMO system shown in Fig. 15.16. The transfer matrix of the feed-forward path is $G(s)$ and that of the feedback path is $H(s)$. The transfer matrix between the feedback signal vector $B(s)$ and the error vector $E(s)$ is obtained as follows:

$$B(s) = H(s)\,Y(s)$$

$$= H(s)\,G(s)\,E(s) \qquad\qquad ... (15.220)$$

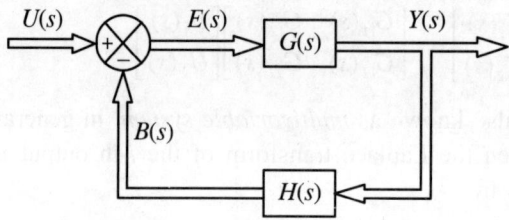

Fig. 15.16. Block diagram of a MIMO system

Hence,

$$\frac{B(s)}{E(s)} = H(s)\,G(s) \qquad\qquad ... (15.221)$$

Thus, *the transfer matrix of the cascaded elements is the product of the transfer matrices of the individual elements.*

The transfer matrix of the closed-loop system is obtained as follows

$$Y(s) = G(s)\,[U(s) - B(s)]$$

$$= G(s)\,[U(s) - H(s)Y(s)] \qquad\qquad ... (15.222)$$

Therefore,

$$[I + G(s)H(s)]\,Y(s) = G(s)U(s) \qquad\qquad ... (15.223)$$

Premultiplying both sides of Eq.(15.223) by $[I + G(s)\,H(s)]^{-1}$

$$Y(s) = [I + G(s)H(s)]^{-1}\,G(s)U(s) \qquad\qquad ... (15.224)$$

Then, the closed-loop transfer matrix is given by

$$M(s) = \frac{Y(s)}{U(s)} = [I + G(s)\,H(s)]^{-1}G(s) \qquad\qquad ... (15.225)$$

15.15.2. Noninteraction in MIMO systems

In practice, many process control systems have multiple inputs and multiple outputs. It is often desired that changes in one input affect only one output since it is easier to maintain each output value at a desired constant value in the absence of external disturbance.

Consider the transfer matrix $G_p(s)$ of dimension $n \times n$ of a plant and design a series compensator (dimension $n \times n$) such that the n inputs and n outputs are uncoupled. If noninteraction or uncoupling between the n inputs and n outputs are desired, then the closed-loop transfer matrix must be diagonal.

$$M(s) = \frac{Y(s)}{U(s)} = \begin{bmatrix} G_{11}(s) & & & 0 \\ & G_{22}(s) & & \\ & & \ddots & \\ 0 & & & G_{nn}(s) \end{bmatrix} \qquad \text{... (15.226)}$$

If we consider the feedback matrix $H(s)$ to be the identity matrix, then Eq.(15.225) becomes,

$$M(s) = \frac{Y(s)}{U(s)} = [I + G(s)]^{-1} G(s) \qquad \text{... (15.227)}$$

where

$$G(s) = G_p(s) G_c(s) \qquad \text{... (15.228)}$$

From Eq.(15.227) we get

$$[I + G(s)] M(s) = G(s) \qquad \text{... (15.229)}$$

or,

$$G(s) [I - M(s)] = M(s) \qquad \text{... (15.230)}$$

Post multiplying Eq.(15.230) by $[I - M(s)]^{-1}$, we obtain

$$G(s) = M(s) [I - M(s)]^{-1} \qquad \text{... (15.231)}$$

Since $M(s)$ is a diagonal matrix, $[I - M(s)]$ is also a diagonal matrix. Thus, $G(s)$ is the product of two diagonal matrices and hence it is also diagonal. Thus, to achieve noninteraction, $G(s)$ should be made diagonal, provided the feedback matrix $H(s)$ is the identity matrix.

15.16. CONTROLLABILITY AND OBSERVABILITY

The concepts of controllability and observability was first introduced by Kalman in 1960. They play an important role in the design of systems that have an optimum performance. The optimal control is based on the optimization of some specific performance criterion. The conditions on controllability and observability often govern the existence of a solution to an optimal control problem. Further, they establish complete equivalence between the state variable and transfer function representations. The concept of observability relates to the condition of observing or estimating the state variables from the output variables which are generally measurable. A study of controllability and observability provides a basis for consideration of the optimal control problem and pole-placement design through state feedback.

15.16.1. Definitions of Controllability and Observability

Controllability: A system is said to be completely state controllable if, for any initial time t_0, each initial state $x(t_0)$ can be transferred to any final state $x(t_f)$ in a finite time, $t_f > t_0$, by means of an unconstrained control input vector $u(t)$. An unconstrained control vector has no limit on the amplitude of $u(t)$. This definition implies that $u(t)$ is able to affect each state variable in $x(t)$ given by

$$x(t) = \phi(t - t_0)x(t_0) + \int_{t_0}^{t_f} \phi(t - \tau)Bu(\tau) \, d\tau \qquad \dots (15.232)$$

Observability: A system is said to be completely observable if every initial state $x(t_0)$ can be exactly determined from the measurements of the output $y(t)$ over the finite interval of time $t_0 \le t \le t_f$. This definition implies that every state of $x(t)$ affects the output $y(t)$ given by

$$y(t) = Cx(t) = C\phi(t - t_0)x(t_0) + \int_{t_0}^{t} \phi(t - \tau) Bu(\tau) \, d\tau \qquad \dots (15.233)$$

where the initial state $x(t_0)$ is the result of control inputs prior to t_0.

15.16.2. Concepts of Controllability and Observability

The concepts of controllability and observability can be illustrated with reference to the block diagram shown in Fig.15.17. By the proper selection of the state variables, it is possible to divide a system into four subdivisions shown in Fig. 15.17.

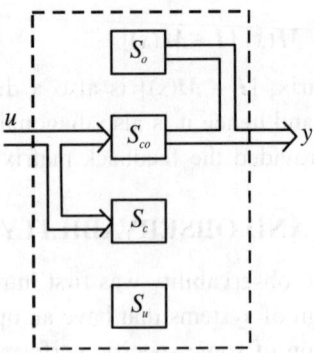

Fig. 15.17. Four sub-divisions of MIMO system

S_{co} = Completely controllable and completely observable subsystem

S_o = Completely observable out uncontrollable subsystem

S_c = Completely controllable but unobservable subsystem

S_u = Uncontrollable and unobservable system

An inspection of Fig.15.17 readily reveals that only the completely controllable and completely observable subsystem S_{co} satisfies the definition of a transfer function matrix. Thus

$$Y(s) = G(s)U(s) \qquad \text{... (15.234)}$$

If the entire system is completely controllable and completely observable, then the state variable and transfer function matrix representations of a system are equivalent and accurately represent the system. In such cases, the resulting transfer function has all the information characterizing the dynamic performance of the system. As there is no unique method of selecting the variables to represent a system, the transfer function matrix is completely and uniquely specified once the state variable representation of the system is known.

15.16.3. Determination of Controllability and Observability

The determination of controllability and observability of a system by subdividing the system into its four possible subdividions is difficult. A simpler method is developed by using Eq.(15.232) with $t_o = 0$ and specifying that the final state vector $x(t_f) = 0$. Then, Eq.(15.232) becomes,

$$0 = e^{At}x(0) + \int_0^t e^{A(t_f - \tau)} \, Bu(\tau) \, d\tau$$

or,

$$x(0) = -\int_0^{t_f} e^{-A\tau} Bu(\tau) \, d\tau \qquad \text{... (15.235)}$$

From Eq.(15.185), we have

$$e^{A\tau} = \sum_{k=0}^{n-1} \alpha_k(\tau) A^k \qquad \text{... (15.236)}$$

Substituting Eq.(15.236) into Eq.(15.235) we obtain

$$x(0) = -\sum_{k=0}^{n-1} A^k B \int_0^{t_f} \alpha_k(\tau) u(\tau) \, d\tau \qquad \text{... (15.237)}$$

Assuming that the input $u(t)$ is of dimension r, the integral in the above equation can be evaluated. The result is

$$\int_0^{t_f} \alpha_k(\tau) u(\tau) \, d\tau = \beta_k \qquad \text{... (15.238)}$$

Eq.(15.237) can be expressed as

$$x(0) = -\sum_{k=0}^{n-1} A^k B \beta_k = -[B \ AB \cdots\cdots A^{n-1}B]\begin{bmatrix} \beta_0 \\ \beta_1 \\ \vdots \\ \beta_{n-1} \end{bmatrix} \qquad \text{... (15.239)}$$

According to the definition of complete controllability, each mode $v_i e^{\lambda_i t}$ must be directly affected by the input $u(t)$. This requires that

$$\text{Rank } M_c = \text{Rank } [B \ AB \cdots A^{n-1}B] = n \qquad \text{... (15.240)}$$

where M_c is the controllability matrix, and of dimension $n \times nr$.

For a single-input system, the matrix B is a vector and M_c is of dimension $n \times n$. Since the matrix A and B are involved in Eq.(15.240), we state that the matrix pair (A, B) is controllable which implies that M_c is of rank n. When the matrix pair (A, B) is not completely controllable, the uncontrollable modes can be determined by transforming the state equation so that A is in the diagonal form. This approach is described below.

A simple method to determine controllability of a system when the system has distinct eigenvalues is to use the transformation $x = Tz$ to transform the A matrix into diagonal form so that the modes are decoupled. The resulting state equation is

$$\dot{z} = T^{-1}ATz + T^{-1}Bu = A_d z + B_d u \qquad \text{... (15.241)}$$

Each transformed state z_i represents a mode and can be directly affected by the input $u(t)$ only if $B_d = T^{-1}B$ has no zero row.

Hence, *when eigenvalues are distinct, a system is completely controllable if B_d has no zero row. B_d is called the mode-controllability matrix.* A zero row in B_d indicates that the corresponding mode is uncontrollable. The eigenvalues associated with uncontrollable modes are called *input-decoupling zeros*.

Similarly, the condition for observability is derived from the homogeneous state equations

$$x = Ax \qquad \text{... (15.242)}$$

$$y = Cx \qquad \text{... (15.243)}$$

Using Eq.(15.236) we get,

$$y(t) = Ce^{At}x(0) = \sum_{k=0}^{n-1} \alpha_k(t)CA^k x(0) \qquad \text{... (15.244)}$$

For observability, the output $y(t)$ must be affected by each state x_i. This imposes restrictions on CA^k. It can be shown that the system is completely observable if the observability matrix M_o has the property

$$\text{Rank } \boldsymbol{M}_o = \text{Rank } [\boldsymbol{C}^T \ \boldsymbol{A}^T \boldsymbol{C}^T \ (\boldsymbol{A}^T)^2 \boldsymbol{C}^T \cdots (\boldsymbol{A}^T)^{n-1} \boldsymbol{C}^T] = n \qquad \dots (15.245)$$

For MIMO systems, $\boldsymbol{M}_o$ is of dimension $n \times nr$. For single-input single-output system, the matrix $\boldsymbol{C}^T$ is a column matrix and the observability matrix $\boldsymbol{M}_o$ has dimension $n \times n$. When the matrix pair $(\boldsymbol{A}, \boldsymbol{C})$ is not completely observable, the unobservable modes can be determined by transforming the state and output equations so that the plant matrix $\boldsymbol{A}$ is in the diagonal form. This approach is described below.

A simple method to determine observability of a system when the system has distinct eigenvalues is to transform the $\boldsymbol{A}$ matrix into diagonal form. The output equation is then

$$y(t) = \boldsymbol{Cx} = \boldsymbol{CTz} = \boldsymbol{C}_d \boldsymbol{z} \qquad \dots (15.246)$$

If a column of $\boldsymbol{C}_d$ has all zeros, then one made is not coupled to any of the outputs and the system is unobservable. Hence, *when the eigenvalues are distinct, a system is completely observable if $\boldsymbol{C}_d$ has no zero column.* $\boldsymbol{C}_d$ is called the *mode-observable matrix*. The eigenvalues associated with an unobservable mode are called *output-decoupling zeros.*

15.17. CONTROLLABILITY, OBSERVABILITY AND TRANSFER FUNCTIONS

In the classical analysis of control systems, transfer functions are used to model the linear time-invariant systems. Although controllability and observability are concepts of modern control theory, they are closely related to the properties of transfer functions. The relationship among them can be stated as follows.

If the input-output transfer function of a linear time-invariant system has pole-zero cancellation, the system will be either not state controllable or not observable. If there is no pole-zero cancellation, the system can always be represented by dynamic equations as completely controllable and observable system.

Consider an n-th order system with a single-input and single-output and distinct eigenvalues given by

$$\dot{x}(t) = \boldsymbol{Ax}(t) + \boldsymbol{Bu}(t) \qquad \dots (15.247)$$

$$y(t) = \boldsymbol{Cx}(t) \qquad \dots (15.248)$$

The $n \times n$ Vandermonde matrix $\boldsymbol{T}$ to diagonalize $\boldsymbol{A}$ is

$$\boldsymbol{T} = \begin{bmatrix} 1 & 1 & 1 & & 1 \\ \lambda_1 & \lambda_2 & \lambda_3 & \cdots & \lambda_n \\ \lambda_1^2 & \lambda_2^2 & \lambda_3^2 & \cdots & \lambda_n^2 \\ \vdots & \vdots & \vdots & & \vdots \\ \lambda_1^{n-1} & \lambda_2^{n-1} & \lambda_3^{n-1} & & \lambda_n^{n-1} \end{bmatrix} \qquad \dots (15.249)$$

The new state equation in the canonical (diagonal) form is

$$\dot{z}(t) = A_d z(t) + B_d u(t) \qquad \ldots (15.250)$$

where $A_d = T^{-1} A T$. The transformed output equation is

$$y(t) = C T z(t) = C_d z(t) \qquad \ldots (15.251)$$

Since A_d is a diagonal matrix, the i-th equation of Eq.(15.250) is

$$\dot{z}_i(t) = \lambda_i y_i(t) + \gamma_i u(t) \qquad \ldots (15.252)$$

where λ_i is the i-th eigenvalue of A and λ_i is the i-th element of B_d. In the present case, B_d is an $n \times 1$ matrix. Assuming zero initial conditions and taking Laplace transform of Eq.(15.252),

$$Z_i(s) = \frac{\gamma_i}{s + \lambda_i} U(s) \qquad \ldots (15.253)$$

The Laplace transform of Eq.(15.251) is

$$Y(s) = C_d Z(s) = C\, T\, Z(s) \qquad \ldots (15.254)$$

If we assume that

$$C = [c_1 \quad c_2 \ldots c_n] \qquad \ldots (15.255)$$

then,

$$C_d = C\, T = [C_{d1} \quad C_{d2} \ldots C_{dn}] \qquad \ldots (15.256)$$

where

$$C_{di} = c_1 + c_2 \lambda_i + c_3 \lambda_i^2 + \ldots + c_n \lambda_i^{n-1} \qquad \ldots (15.257)$$

for $i = 1,2,3,\ldots,n$. Eq.(15.254) is written as

$$Y(s) = [C_{d1} \quad C_{d2} \ldots C_{dn}]\, Z(s) \qquad \ldots (15.258)$$

$$= [C_{d1}\ C_{d2} \ldots\ C_{dn}] \begin{bmatrix} \dfrac{\gamma_1}{s + \lambda_1} \\[6pt] \dfrac{\gamma_2}{s + \lambda_2} \\[4pt] \vdots \\[4pt] \dfrac{\gamma_n}{s + \lambda_n} \end{bmatrix} U(s) \qquad \ldots (15.259)$$

$$= \sum_{i=1}^{n} \frac{C_{di}\gamma_i}{s + \lambda_i} U(s) \qquad \ldots (15.260)$$

For the n-th order system with distinct eigenvalues, assume that the input-output transfer function is of the form

$$\frac{Y(s)}{U(s)} = K\frac{(s+a_1)(s+a_2)\ \ldots\ldots\ (s+a_m)}{(s+\lambda_1)(s+\lambda_2)\ \ldots\ (s+\lambda_n)} \quad n > m \qquad \ldots (15.261)$$

Eq.(15.261) in partial-fraction expanded form is

$$\frac{Y(s)}{U(s)} = \sum_{i=1}^{n}\frac{\sigma_i}{s+\lambda_i} \qquad \ldots (15.262)$$

where σ_i denotes the residue of $Y(s)/U(s)$ at $s = -\lambda_i$.

For the system described by Eq. (15.250) is to be state controllable, all the rows of B_d must be non-zero. This means that $\gamma_i \neq 0$ for $i = 1,2,\ \ldots.n$. If $Y(s)/U(s)$ has one or more pairs of identical pole and zero, (for example, $a_i = \gamma_i$ in Eq.(15.261) then σ_i is zero. From Eqs.(15.259) and (15.262), we see that, in general,

$$\sigma_i = C_{di}\gamma_i \qquad \ldots (15.263)$$

Hence, if $\sigma_i = 0$ then γ_i will be zero if $C_{di} \neq 0$ and the state z_i is not controllable.

For observability, C_d must not have zero columns. In the present case $C_{di} \neq 0$, for $i = 1,2,\ldots,\ n$. However, from Eq.(15.263),

$$C_{di} = \sigma_i / \gamma_i \qquad \ldots (15.264)$$

When there is pole-zero cancellation at $a_i = \lambda_i$, then $\sigma_i = 0$.

Thus, $C_{di} = 0$ if $\gamma_i \neq 0$.

Example 15.18. *The state and output equations of a system are*

$$\dot{x} = \begin{bmatrix} -2 & 0 \\ -1 & -1 \end{bmatrix}x + \begin{bmatrix} 1 \\ 1 \end{bmatrix}u,\ y = \begin{bmatrix} 0 & 1 \end{bmatrix}x$$

Determine (a) the eigenvalues from the state equation, (b) the transfer function $Y(s)/U(s)$, (c) controllability and observability and (d) the transform T which converts the A matrix into canonical form.

Solution:

(a) To determine the eigenvalues:

$$|sI - A| = \begin{vmatrix} s+2 & 0 \\ 1 & s+1 \end{vmatrix} = (s+2)(s+1)$$

Hence, the eigenvalues are $\lambda_1 = -2$, $\lambda_2 = -1$

(b) To obtain the transfer function:

$$\phi(s) = [sI - A]^{-1} = \frac{\begin{bmatrix} s+1 & 0 \\ -1 & s+2 \end{bmatrix}}{(s+1)(s+2)}$$

$$G(s) = C\phi(s)B = \begin{bmatrix} 0 & 1 \end{bmatrix} \begin{bmatrix} s+1 & 0 \\ -1 & s+2 \end{bmatrix} \begin{bmatrix} 1 \\ 1 \end{bmatrix} \frac{1}{(s+1)(s+2)}$$

$$= \frac{s+1}{(s+1)(s+2)} = \frac{1}{s+2}$$

Thus, a pole-zero cancellation occurs. Only the mode with $\lambda_1 = -2$ is controllable.

(c) To determine controllability and observabilitys

$$\text{Rank } M_c = \text{Rank } [B \ AB] = \text{Rank} \begin{bmatrix} 1 & -2 \\ 1 & -2 \end{bmatrix} = 1$$

The matrix [B AB] is singular. Hence, the system is not completely controllable.

$$\text{Rank } M_o = \text{Rank } [C \ A^T C] = \text{Rank} \begin{bmatrix} 0 & -1 \\ 1 & -1 \end{bmatrix} = 2$$

The matrix M_o is nonsingular. Thus, the system is completely observable.

(d) T may be obtained by the Vandermonde matrix

$$T = \begin{bmatrix} 1 & 1 \\ \lambda_1 & \lambda_2 \end{bmatrix} = \begin{bmatrix} 1 & 1 \\ -2 & -1 \end{bmatrix}$$

15.18. LINEAR TIME-VARYING SYSTEMS

An advantage of the state-variable approach to control system analysis is that it can be extended to linear time-varying systems. Most of the results obtained in the previous section can be applied to linear time-varying systems by changing the transition matrix $\phi(t)$ to $\phi(t, t_o)$. The transition matrix $\phi(t, t_o)$ of a time-varying system depends on both t and t_o. Hence, we cannot set the initial time always equal to zero. In general $\phi(t, t_o)$ cannot be expressed as a matrix exponential.

15.18.1. Solution of State Equations

Consider the state equation of a linear time-varying system given by

$$\dot{x}(t) = A(t)x(t) \qquad\qquad \text{... (15.265)}$$

where $x(t)$ is n-dimension vector and $A(t) = n \times n$ matrix whose elements are piecewise continuous functions of t in the interval $t_o \leq t \leq t_f$.

The solution of Eq. (15.265) is given by

$$x(t) = \phi(t, t_0)\, x(t_0) \qquad\qquad \text{... (15.266)}$$

where $\phi(t, t_o)$ is the $n \times n$ nonsingular matrix satisfying the following matrix differential equation:

$$\dot{\phi}(t, t_0) = A(t)\, \phi(t_0, t_0),$$

$$\phi\,(t_0, t_0) = I \qquad\qquad \text{... (15.267)}$$

The fact that Eq. (15.266) is a solution of Eq. (15.267) can be verified easily since

$$x(t_0) = \phi(t_0, t_0)\, x(t_0) = I\, x(t_0)$$

and

$$\dot{x}(t) = \frac{d}{dt}[\phi(t, t_0)\, x(t_0)]$$

$$= \dot{\phi}\,(t, t_0)\, x(t_0)$$

$$= A(t)\, \phi(t, t_0)\, x(t_0) = A(t)\, x(t)$$

From Eq.(15.266), it can be seen that the solution of Eq.(15.265) is simply a transformation of the initial state. The matrix $\phi(t, t_o)$ is the state transition matrix for the time-varying system described by Eq.(15.265).

15.18.2. State Transition Matrix

The state transition matrix for a time-varying system is given by a matrix exponential *if and only if $A(t)$ and $\int_0^t A(\tau)d\tau$* are commutable. Thus,

$$\phi(t, t_0) = \exp\left[\int_{t_0}^t A(\tau)d\tau\right] \text{ if and only if } A(t) \text{ and } \int_{t_0}^t A(\tau)\, d\tau \text{ are commutable.}$$

If $A(t)$ is a constant matrix or a diagonal matrix, $A(t)$ and $\int_{t_0}^t A(\tau)\, d\tau$ are commutable.

$\phi(t, t_o)$ may be computed numerically using the series expansion for $\phi(t, t_o)$.

$$\phi(t, t_o) = I + \int_{t_0}^t A(\tau)d\tau + \int_{t_0}^t A(\tau_1)\left[\int_{t_0}^t A(\tau_2)d\tau_2\right]d\tau_1 + \cdots \qquad \text{... (15.268)}$$

Thus, in general $\phi(t, t_o)$ will not be in a closed form.

15.18.3. Properties of $\phi(t, t_o)$

The state transition matrix of a time-varying system has the following properties:

1.
$$\phi(t_2, t_1)\, \phi(t_1, t_0) = \phi(t_2, t_0)$$

Proof:

Since

$$x(t_1) = \phi(t_1, t_0)\, x(t_0)$$

$$x(t_2) = \phi(t_2, t_0)\, x(t_0)$$

and

$$x(t_2) = \phi(t_2, t_1)\, x(t_1)$$

Therefore we have,

$$x(t_2) = \phi(t_2, t_1)\, x(t_1) = \phi(t_2, t_0)\, x(t_0)$$

Substituting for $x(t_1)$ from Eq. (i) in Eq. (iv), we get

$$\phi(t_2, t_1)\, \phi(t_1, t_0)\, x(t_0) = \phi(t_2, t_0)\, x(t_0)$$

Therefore,

$$\phi(t_2, t_1)\phi(t_1, t_0) = \phi(t_2, t_0)$$

2.
$$\phi(t_1, t_0) = \phi^{-1}(t_0, t_1)$$

Proof: From the first property, we have

$$\phi(t_1, t_0) = \phi^{-1}(t_2, t_1)\, \phi(t_2, t_0)$$

If we let $t_2 = t_o$ in the above equation, then

$$\phi(t_1, t_0) = \phi^{-1}(t_0, t_1)\, \phi(t_0, t_0)$$

$$= \phi^{-1}(t_0, t_1)$$

15.18.4. General Solution of State Equations

Let us consider the state equation given by

$$\dot{x}(t) = A(t)\, x(t) + B(t)u \qquad\qquad \dots (15.271)$$

where

$x(t)$ = n-dimensional vector

u = r-dimensional vector

$A(t)$ = $n \times n$ plant matrix

$B(t) = n \times r$ control matrix

The elements of $A(t)$ and $B(t)$ are assumed to be piecewise continuous functions of t in the interval $t_o \leq t \leq t_f$.

To obtain the solution of Eq.(15.271), let

$$x(t) = \phi(t, t_0) \, \xi(t) \qquad \qquad \text{... (15.272)}$$

where $\phi(t, t_o)$ is the unique matrix satisfying the following equation:

$$\dot{\phi}(t, t_0) = A(t) \, \phi(t, t_0) \qquad (\phi(t_0, t_0) = I) \qquad \text{... (15.273)}$$

Then, from Eq.(15.272),

$$\dot{x}(t) = \dot{\phi}(t, t_0) \, \xi(t) + \phi(t, t_0) \, \dot{\xi}(t)$$

$$= A(t) \, \phi(t, t_0) \, \xi(t) + \phi(t, t_0) \, \dot{\xi}(t)$$

$$= A(t) \, \phi(t, t_0) \, \xi(t) + B(t) \, u(t) \qquad \text{... (15.274)}$$

Therefore,

$$\phi(t, t_0) \dot{\xi}(t) = B(t) \, u(t)$$

or,

$$\dot{\xi}(t) = \phi^{-1}(t, t_0) \, B(t) \, u(t) \qquad \text{... (15.275)}$$

Integrating Eq.(15.275) with respect to t,

$$\dot{\xi}(t) = \xi(t_0) + \int_{t_0}^{t} \phi^{-1}(\tau, \tau_0) \, B(\tau) \, u(\tau) \, d\tau$$

Since

$$\xi(t_0) = \phi^{-1}(t_0, t_0) \, x(t_0) = x(t_0)$$

$$x(t) = \phi(t, t_0) \, x(t_0) + \phi(t, t_0) \int_{t_0}^{t} \phi^{-1}(\tau, \tau_0) \, B(\tau) \, u(\tau) \, d\tau$$

$$= \phi(t, t_0) \, x(t_0) + \int_{t_0}^{t} \phi(t, \tau) \, B(\tau) \, u(\tau) \, d\tau \qquad \text{... (15.276)}$$

Example 15.19. *Obtain* $\phi(t, 0)$ *for the time-varying system*

$$\begin{bmatrix} \dot{x}_1 \\ \dot{x}_2 \end{bmatrix} = \begin{bmatrix} 0 & 1 \\ 0 & t \end{bmatrix} \begin{bmatrix} x_1 \\ x_2 \end{bmatrix}$$

Solution:

$\phi(t, 0)$ may be computed from Eq. (15.268).

$$\int_0^t A(\tau)d\tau = \int_0^t \begin{bmatrix} 0 & 1 \\ 0 & \tau \end{bmatrix} d\tau = \begin{bmatrix} 0 & t \\ 0 & t^2/2 \end{bmatrix}$$

$$\int_0^t \begin{bmatrix} 0 & 1 \\ 0 & \tau \end{bmatrix} \left\{ \int_0^t \begin{bmatrix} 0 & 1 \\ 0 & \tau_2 \end{bmatrix} d\tau_2 \right\} d\tau_1 = \int_0^t \begin{bmatrix} 0 & 1 \\ 0 & \tau_1 \end{bmatrix} \begin{bmatrix} 0 & \tau_1 \\ 0 & \tau_1^2/2 \end{bmatrix} d\tau_1 = \begin{bmatrix} 0 & t^3/6 \\ 0 & t^4/8 \end{bmatrix}$$

Therefore,

$$\phi(t, 0) = \begin{bmatrix} 1 & 0 \\ 0 & 1 \end{bmatrix} + \begin{bmatrix} 0 & t \\ 0 & t^2/2 \end{bmatrix} + \begin{bmatrix} 0 & t^3/6 \\ 0 & t^4/8 \end{bmatrix} + \cdots$$

$$= \begin{bmatrix} 1 & t + t^3/6 + \cdots \cdots \\ 0 & 1 + t^2/2 + t^4/8 + \cdots \cdots \end{bmatrix}$$

15.19. STATE–FEEDBACK CONTROL

In modern control theory, a majority of design techniques is based on the state feedback configuration. The control is achieved by feeding back the state variables through constant gains instead of using controllers with fixed configuration.

The PD control and the rate-feedback control are special cases of state feedback control scheme. In the case of rate feedback control, consider the second-order process described by

$$G_p(s) = \omega_n^2/s(s + 2\zeta\omega_n) \qquad\qquad ...(15.277)$$

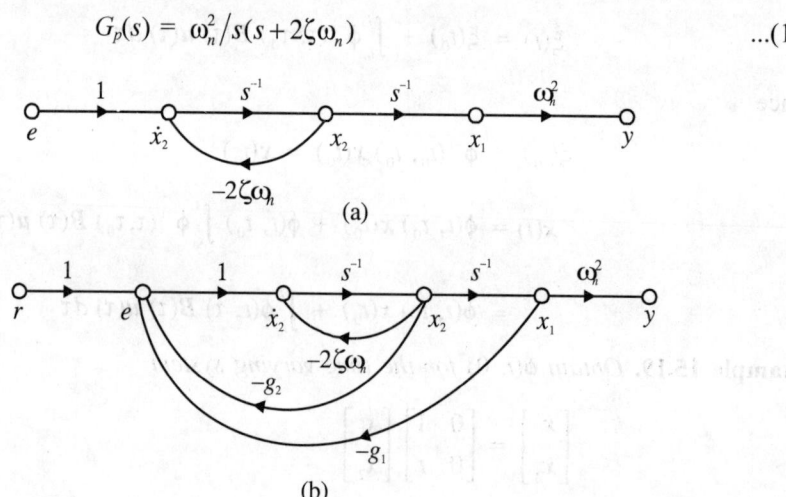

(a)

(b)

Fig. 15.18. (a) State diagram of direct decomposition
(b) Complete state diagram with x_1 and x_2 feedback control

The process is decomposed by direct decomposition and is represented by the state diagram of Fig.15.18(a). If the states x_1 and x_2 are physically accessible, they may be fed back through constant gains g_1 and g_2 respectively to form the control as shown in Fig. 15.18(b). The closed loop transfer function of the system is

$$\frac{C(s)}{R(s)} = \frac{\omega_n^2}{s^2 + (2\zeta\omega_n + g_2)\,s + g_1} \qquad \qquad ...(15.278)$$

Comparing this transfer function with Eq.(8.87), we notice that, if $g_1 = \omega_n^2$ and $g_2 = K_t\omega_n^2$, they are identical. If the system is to have zero steady state error, then g_1 should equal ω_n^2. The value of g_2 is selected to satisfy the damping requirements.

In case of PD control, the closed-loop transfer function is described in Eq. (8.8). Comparing this with Eq. (15.278), the characteristic equations would be identical if $g_2 = K_D\omega_n^2$ and $g_1 = \omega_n^2 K_p$.

15.20. POLE PLACEMENT DESIGN

The design methods discussed so far are all characterized by the property that the poles are selected on what can be achieved with fixed controller configuration and the practical range of controller parameters. The fixed configuration controllers would not be able to independently control three or more poles of a system since there are only two free parameters in these controllers. Under certain conditions, the poles may be placed arbitrarily. This is an entirely new design philosophy.

Let us investigate the condition required for arbitrary pole placement in an n-th order system. Consider a linear process described by the state equation

$$\dot{x}(t) = Ax(t) + Bu(t) \qquad \qquad ...(15.279)$$

where $x(t)$ is the $n \times 1$ state vector and $u(t)$ is the scalar control input. The state feedback control is

$$u(t) = -Gx(t) + \upsilon(t) \qquad \qquad ...(15.280)$$

where G is the $1 \times n$ constant gain feedback matrix. Hence, Eq.(15.279) becomes

$$\dot{x}(t) = (A - BG)\,x(t) + B\,\upsilon(t) \qquad \qquad ...(15.281)$$

If the pair $[A, B]$ is completely controllable, then a matrix G exists which can give an arbitrary set of eigenvalues of $(A - BG)$. The n roots of the characteristic equation

$$|\lambda I - A + BG| = 0 \qquad \qquad ...(15.282)$$

can be arbitrarily placed.

If a system is state controllable, it can always be represented in the phase variable canonical form as in Eq.(15.283). Conversely, if a system is represented in a phase-variable canonical form, it is always state controllable.

$$A = \begin{bmatrix} 0 & 1 & 0 & \cdots & 0 \\ 0 & 0 & 1 & \cdots & 0 \\ \cdot & \cdot & \cdot & & \cdot \\ \cdot & \cdot & \cdot & \cdots & \cdot \\ \cdot & \cdot & \cdot & & \cdot \\ 0 & 0 & 0 & & 1 \\ -a_n & -a_{n-1} & -a_{n-2} & \cdots & a_1 \end{bmatrix} ; \quad B = \begin{bmatrix} 0 \\ 0 \\ 0 \\ \cdot \\ \cdot \\ 1 \end{bmatrix} \qquad ...(15.283)$$

The feedback matrix G is given by

$$G = [g_1, g_2, g_3 \cdots g_n] \qquad ...(15.284)$$

Hence,

$$A - BG = \begin{bmatrix} 0 & 1 & 0 & \cdots & 0 \\ 0 & 0 & 1 & \cdots & 0 \\ 0 & 0 & 0 & \cdots & 0 \\ \cdot & & \cdot & & \cdot \\ 0 & 0 & 0 & \cdots & 1 \\ -(a_n + g_1) & -(a_{n-1} + g_2) & \cdot & \cdots & -(a_1 + g_n) \end{bmatrix} \qquad ...(15.285)$$

Then the eigenvalues of $(A - BG)$ is found from the characteristic equation

$$|\lambda I - A + BG| = \lambda^n + (a_1 + g_n) \lambda^{n-1} + \ldots + (a_n + g_1) = 0 \qquad ...(15.286)$$

Thus it can be seen that the eigenvalues may be selected by proper choice of $g_1, g_2 \ldots, g_n$.

Example 15.20. *The forward transfer function of a linear process is given by*

$$C(s)/E(s) = 20/s^2(s + 1) \qquad ...(15.287)$$

It is desired to have zero steady state error for a unit step input and two of the closed-loop poles must be s = −1 ± j1. Find the G matrix.

Solution:

The closed loop transfer function is

$$\frac{C(s)}{R(s)} = \frac{20}{s^3 + (1 + g_3)s^2 + g_2 s + g_1} \qquad ...(15.288)$$

The steady state specification requires the value of g_1 to be 20. The process is completely state controllable as the transfer function does not have identical poles and zeros.

The characteristic equation is

$$F(s) = s^3 + (1 + g_3)s^2 + g_2 s + 20 = (s + a)(s + 1 - j1)(s + 1 + j1) \qquad \text{...(15.289)}$$

Equating the coefficients of corresponding terms,

$$2a = 20; \ a = 10$$
$$g_2 = 2a + 2 = 22$$
$$g_3 = a + 1 = 11$$

The real pole is at $s = -10$.

15.21. GENERAL METHOD OF DETERMINING *G* MATRIX

In general, the feedback matrix *G* may be found as long as the system is completely controllable even though it is not represented in phase variable canonical form. The general approach is to determine *G* based on *A* and *B* matrices and the openloop and closed loop characteristic equations with a single input. Let

$$\Delta_0(s) = |sI - A| = \text{openloop characteristic equation} \qquad \text{...(15.290)}$$

Let

$$\Delta_c(s) = |sI - A + BG| = \text{Closedloop characteristic equation} \qquad \text{...(15.291)}$$

Let

$$\Delta(s) = 1 + G(sI - A)^{-1}B \qquad \text{...(15.292)}$$

$$= 1 + G\frac{adj(sI - A)B}{\Delta_0(s)} \qquad \text{...(15.293)}$$

$$= \frac{\Delta_0(s) + Gk(s)}{\Delta_0(s)} \qquad \text{...(15.294)}$$

where

$$k(s) = adj(sI - A)B \qquad \text{...(15.295)}$$

Therefore,

$$Gk(s) = \Delta(s)\Delta_0(s) - \Delta_0(s)$$
$$= \Delta_c(s) - \Delta_0(s) \qquad \text{...(15.296)}$$

If the process is completely controllable and $k(s)$, $\Delta_c(s)$ and $\Delta_0(s)$ are known, *G* matrix can be found from Eq.(15.296).

15.22. ADDITIONAL EXAMPLES

Example 15.21. *Obtain the state equations for the electrical network shown in Fig. 15.19, selecting (a) q_1, q_2, and q_3 as variables and (b) v_1 and v_2 as outputs.*

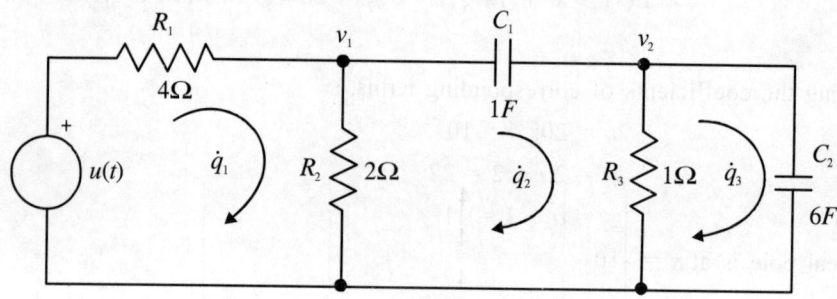

Fig. 15.19. Electrical network

Solution:

(a) *With q1, q2 and q3 as outputs*

The three loop equations are:

(i) $R_1 \dot{q}_1 + R_2(\dot{q}_1 - \dot{q}_2) = u(t)$

or, $6\dot{q}_1 - 2\dot{q}_2 = u(t)$ ...(15.297)

(ii) $R_2(\dot{q}_2 - \dot{q}_1) + \dfrac{1}{C_1} q_2 + R_3(\dot{q}_2 - \dot{q}_3) = 0$

or, $-2\dot{q}_1 + 3\dot{q}_2 - \dot{q}_3 = -q_2$ ...(15.298)

(iii) $R_3(\dot{q}_1 - \dot{q}_2) + \dfrac{1}{C_2} q_3 = 0$

or, $-\dot{q}_1 + \dot{q}_3 = \dfrac{-q_3}{6}$ ...(15.299)

Writing the equations (15.297), (15.298) and (15.299) in matrix form,

$$\begin{bmatrix} 6 & -2 & 0 \\ -2 & 3 & -1 \\ 0 & -1 & 1 \end{bmatrix} \begin{bmatrix} \dot{q}_1 \\ \dot{q}_2 \\ \dot{q}_3 \end{bmatrix} = \begin{bmatrix} 0 & 0 & 0 \\ 0 & -1 & 0 \\ 0 & 0 & -1/6 \end{bmatrix} \begin{bmatrix} q_1 \\ q_2 \\ q_3 \end{bmatrix} + \begin{bmatrix} 1 \\ 0 \\ 0 \end{bmatrix} u(t)$$

i.e.,

$$T\dot{q} = Aq + Bu$$

or,

$$\dot{q} = T^{-1}Aq + T^{-1}Bu$$

$$T^{-1} = \begin{bmatrix} 6 & -2 & 0 \\ -2 & 3 & -1 \\ 0 & -1 & 1 \end{bmatrix}^{-1} = \begin{bmatrix} 1/4 & 1/4 & 1/4 \\ 1/4 & 3/4 & 3/4 \\ 1/4 & 3/4 & 7/4 \end{bmatrix}$$

Hence

$$\begin{bmatrix} \dot{q}_1 \\ \dot{q}_2 \\ \dot{q}_3 \end{bmatrix} = \begin{bmatrix} 0 & -1/4 & -1/24 \\ 0 & -3/4 & -1/8 \\ 0 & -3/4 & -7/24 \end{bmatrix} \begin{bmatrix} q_1 \\ q_2 \\ q_3 \end{bmatrix} + \begin{bmatrix} 1/4 \\ 1/4 \\ 1/4 \end{bmatrix} u(t) \qquad \text{...(15.300)}$$

(b) *With* v_1 *and* v_2 *as outputs*

$$v_1 = u(t) - R_1 \dot{q}_1$$

$$v_1 = u(t) - 4\dot{q}_1 \qquad \text{...(15.301)}$$

Substituting $\dot{q}_1$ obtained from Eq. 15.300 in Eq. 15.301, we get

$$v_1 = u(t) - 4\left[-(1/4)q_2 - (1/24)q_3 + (1/4)\,u(t)\right]$$

$$= q_2 + \frac{q_3}{6} \qquad \text{...(15.302)}$$

$$v_2 = v_1 - q_2 = q_3/6 \qquad \text{...(15.303)}$$

Therefore,

$$\begin{bmatrix} v_1 \\ v_2 \end{bmatrix} = \begin{bmatrix} 0 & 1 & 1/6 \\ 0 & 0 & 1/6 \end{bmatrix} \begin{bmatrix} q_1 \\ q_2 \\ q_3 \end{bmatrix}$$

Example 15.22. *The differential equation describing a system is*

$$\dddot{x} + 3\ddot{x} + 4\dot{x} + 3x = u_1(t) + 3u_2(t) + 4u_3(t)$$

$$y_1 = 4\dot{x} + 3u_1(t)$$

$$y_2 = \ddot{x} + 4u_2(t) + u_3(t)$$

Obtain its state variable representation.

Solution:

Let

$$x_1 = x$$

$$x_2 = \dot{x}_1 = \dot{x}$$

$$x_3 = \dot{x}_2 = \ddot{x}$$

Then

$$\dot{x}_3 = \dddot{x}$$

$$\dot{x}_3 = u_1(t) + 3u_2(t) + 4u_3(t) - 3x_3 - 4x_2 + 3x_1$$

$$y_1 = 4x_2 + 3u_1(t)$$

$$y_2 = x_3 + 4u_2(t) + u_3(t)$$

Therefore,

$$\begin{bmatrix} \dot{x}_1 \\ \dot{x}_2 \\ \dot{x}_3 \end{bmatrix} = \begin{bmatrix} 0 & 1 & 0 \\ 0 & 0 & 1 \\ -3 & -4 & -3 \end{bmatrix} \begin{bmatrix} x_1 \\ x_2 \\ x_3 \end{bmatrix} + \begin{bmatrix} 0 & 0 & 0 \\ 0 & 0 & 0 \\ 1 & 3 & 4 \end{bmatrix} \begin{bmatrix} u_1(t) \\ u_2(t) \\ u_3(t) \end{bmatrix}$$

$$\begin{bmatrix} y_1 \\ y_2 \end{bmatrix} = \begin{bmatrix} 0 & 4 & 0 \\ 0 & 0 & 1 \end{bmatrix} \begin{bmatrix} x_1 \\ x_2 \\ x_3 \end{bmatrix} + \begin{bmatrix} 3 & 0 & 0 \\ 0 & 4 & 1 \end{bmatrix} \begin{bmatrix} u_1(t) \\ u_2(t) \\ u_3(t) \end{bmatrix}$$

Example 15.23: *Evaluate the state transition matrix of a system with*

$$A = \begin{bmatrix} 0 & 1 \\ -2 & -3 \end{bmatrix}$$

by (a) Laplace transform Method (b) Series summation method and (c) Diagonal matrix method.

Solution:

(a) *Laplace transform method*

$$\phi(t) = \mathcal{L}^{-1}[sI - A]^{-1}$$

$$[sI - A]^{-1} = \frac{1}{|sI - A|} \begin{bmatrix} s+3 & 1 \\ -2 & s \end{bmatrix}$$

$$= \frac{1}{s^2 + 3s + 2} \begin{bmatrix} s+3 & 1 \\ -2 & s \end{bmatrix}$$

Therefore,

$$\phi(t) = \begin{bmatrix} 2e^{-t} - e^{-2t} & e^{-t} - e^{-2t} \\ -2e^{-t} + 2e^{-2t} & -e^{-t} + 2e^{-2t} \end{bmatrix}$$

(b) *Series Summation Method*

$$\phi(t) = \mathbf{I} + \mathbf{A}t + \frac{\mathbf{A}^2 t^2}{2!} + \frac{\mathbf{A}^3 t^3}{3!} + \cdots$$

$$\mathbf{A}^2 = \begin{bmatrix} 0 & 1 \\ -2 & -3 \end{bmatrix} \begin{bmatrix} 0 & 1 \\ -2 & -3 \end{bmatrix} = \begin{bmatrix} -2 & -3 \\ 6 & 9 \end{bmatrix}$$

$$\mathbf{A}^3 = \begin{bmatrix} -2 & -3 \\ 6 & 9 \end{bmatrix} \begin{bmatrix} 0 & 1 \\ -2 & -3 \end{bmatrix} = \begin{bmatrix} 6 & 9 \\ -18 & -27 \end{bmatrix}$$

Hence

$$\phi(t) = \begin{bmatrix} 1 & 0 \\ 0 & 1 \end{bmatrix} + \begin{bmatrix} 0 & 1 \\ -2 & -3 \end{bmatrix} t + \begin{bmatrix} -2 & -3 \\ 6 & 9 \end{bmatrix} \frac{t^2}{2} + \begin{bmatrix} 6 & 9 \\ -18 & -27 \end{bmatrix} \frac{t^3}{6} + \cdots$$

$$= \begin{bmatrix} 1 - t^2 + t^3 + \cdots & t - \dfrac{3t^2}{2} + \dfrac{3t^3}{2} + \cdots \\ -2t + 3t^2 - 3t^3 + \cdots & 1 - 3t + \dfrac{9t^2}{2} - \dfrac{9t^3}{2} + \cdots \end{bmatrix}$$

(c) *Diagonal matrix Method.*

From $| s\mathbf{I} - \mathbf{A} | = 0$, the eigenvalues are: $\lambda_1 = -2$, $\lambda_2 = -1$. The Vandermonde matrix $\mathbf{T}$ is

$$\mathbf{T} = \begin{bmatrix} 1 & 1 \\ \lambda_1 & \lambda_2 \end{bmatrix} = \begin{bmatrix} 1 & 1 \\ -2 & -1 \end{bmatrix}$$

$$\mathbf{T}^{-1} = \begin{bmatrix} -1 & -1 \\ 2 & 1 \end{bmatrix}$$

Hence

$$\phi(t) = Te^{\lambda t} T^{-1} = \begin{bmatrix} 1 & 1 \\ -2 & -1 \end{bmatrix} \begin{bmatrix} e^{-2t} & 0 \\ 0 & e^{-t} \end{bmatrix} \begin{bmatrix} -1 & -1 \\ 2 & 1 \end{bmatrix}$$

$$= \begin{bmatrix} 2e^{-t} - e^{-2t} & e^{-t} - e^{-2t} \\ -2e^{-t} + 2e^{-2t} & -e^{-t} + 2e^{-2t} \end{bmatrix}$$

Example 15.24: *By partial fraction expansion method, obtain the state variable representation of the system described by the differential equation*

$$\dddot{y} + 6\ddot{y} + 11\dot{y} + 6y = 6u$$

and draw the block diagram representation.

Solution:

From the given differential equation, the transfer function is

$$\frac{Y(s)}{U(s)} = \frac{6}{s^3 + 6s^2 + 11s + 6} = \frac{6}{(s+1)(s+2)(s+3)}$$

In the partial-fraction expansion form,

$$\frac{Y(s)}{U(s)} = \frac{3}{s+1} + \frac{-6}{s+2} + \frac{3}{s+3}$$

or

$$Y(s) = X_1(s) + X_2(s) + X_3(s)$$

where

$$X_1(s) = \frac{3}{s+1} U(s); \quad \dot{x}_1(t) = -x_1 + 3u$$

$$X_2(s) = \frac{-6}{s+2} U(s); \quad \dot{x}_2(t) = -2x_2 - 6u$$

$$X_3(s) = \frac{3}{s+3} U(s); \quad \dot{x}_3(t) = -3x_3 + 3u$$

The output $y(t)$ is given by

$$y(t) = x_1(t) + x_2(t) + x_3(t)$$

In the matrix form, we obtain

$$\begin{bmatrix} \dot{x}_1 \\ \dot{x}_2 \\ \dot{x}_3 \end{bmatrix} = \begin{bmatrix} -1 & 0 & 0 \\ 0 & -2 & 0 \\ 0 & 0 & -3 \end{bmatrix} \begin{bmatrix} x_1 \\ x_2 \\ x_3 \end{bmatrix} + \begin{bmatrix} 3 \\ -6 \\ 3 \end{bmatrix} u$$

$$y = \begin{bmatrix} 1 & 1 & 1 \end{bmatrix} \begin{bmatrix} x_1 \\ x_2 \\ x_3 \end{bmatrix}$$

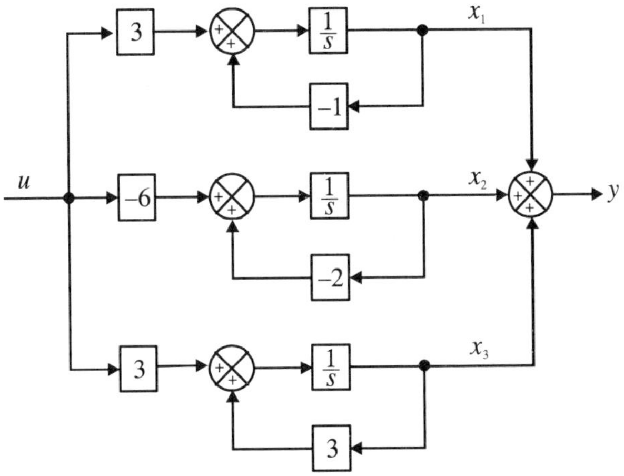

Fig. 15.20. Block diagram representation of example 15.24

The block diagram representation is shown in Fig. 15.20. It may be noted that the transfer functions in the feedback block are identical with the eigenvalues of the system. The residues of the poles of the transfer function, i.e. the coefficients in the partial fraction expansions of $Y(s)/U(s)$, appear in the feed forward blocks.

Example 15.25. *For the MIMO system shown in Fig. 15.21 determine the transfer matrix of the series compensator such that the closed-loop transfer matrix is*

$$M(s) = \begin{bmatrix} \dfrac{1}{s+1} & 0 \\ 0 & \dfrac{1}{5s+1} \end{bmatrix}$$

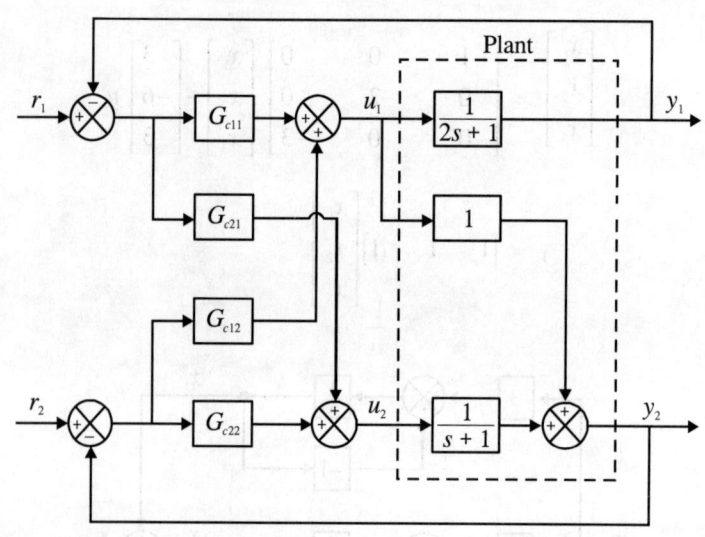

Fig. 15.21. MIMO system for example 15.25

Solution:

Since

$$G(s) = M(s)[I-M(s)]^{-1}$$

$$= \begin{bmatrix} \dfrac{1}{s+1} & 0 \\ 0 & \dfrac{1}{5s+1} \end{bmatrix} \begin{bmatrix} \dfrac{s+1}{s} & 0 \\ 0 & \dfrac{5s+1}{5s} \end{bmatrix} = \begin{bmatrix} \dfrac{1}{s} & 0 \\ 0 & \dfrac{1}{5s} \end{bmatrix}$$

$$\begin{bmatrix} Y_1(s) \\ Y_2(s) \end{bmatrix} = \begin{bmatrix} \dfrac{1}{2s+1} & 0 \\ 1 & \dfrac{1}{s+1} \end{bmatrix} \begin{bmatrix} U_1(s) \\ U_2(s) \end{bmatrix}$$

and

$$\begin{bmatrix} U_1(s) \\ U_2(s) \end{bmatrix} = \begin{bmatrix} G_{c11}(s) & G_{c12}(s) \\ G_{c21}(s) & G_{c22}(s) \end{bmatrix} \begin{bmatrix} R_1(s) - Y_1(s) \\ R_2(s) - Y_2(s) \end{bmatrix}$$

we obtain

$$
\begin{bmatrix} Y_1(s) \\ Y_2(s) \end{bmatrix} = \begin{bmatrix} \dfrac{1}{2s+1} & 0 \\ 1 & \dfrac{1}{s+1} \end{bmatrix} \begin{bmatrix} G_{c11}(s) & G_{c12}(s) \\ G_{c21}(s) & G_{c22}(s) \end{bmatrix} \begin{bmatrix} R_1(s) - Y_1(s) \\ R_2(s) - Y_2(s) \end{bmatrix}
$$

$$
= \begin{bmatrix} \dfrac{1}{s} & 0 \\ 0 & \dfrac{1}{5s} \end{bmatrix} \begin{bmatrix} R_1(s) - Y_1(s) \\ R_2(s) - Y_2(s) \end{bmatrix}
$$

Therefore,

$$
G_c(s) = \begin{bmatrix} G_{c11}(s) & G_{c12}(s) \\ G_{c21}(s) & G_{c22}(s) \end{bmatrix}
$$

$$
= \begin{bmatrix} \dfrac{1}{2s+1} & 0 \\ 1 & \dfrac{1}{s+1} \end{bmatrix}^{-1} \begin{bmatrix} \dfrac{1}{s} & 0 \\ 0 & \dfrac{1}{5s} \end{bmatrix}
$$

$$
= \begin{bmatrix} \dfrac{2s+1}{s} & 0 \\ -\dfrac{(s+1)(2s+1)}{s} & \dfrac{(s+1)}{s} \end{bmatrix} \qquad ...(15.304)
$$

Eq.(15.304) is the transfer matrix of the series compensator $G_{c11}(s) = 2 + \dfrac{1}{s}$ and $G_{c22}(s)$

$= \dfrac{1}{5} + \dfrac{1}{5s}$ are proportional and integral controllers and $G_{c21}(s) = (-3 - \dfrac{1}{s} - 2s)$ is a proportional-

integral-derivative (PID) controller.

Example 15.26. *Show that* $\Delta_0(s)\Delta(s) = \Delta_c(s)$

Solution:

$$sI - A + BG = (sI - A)[I + (sI - A)^{-1} BG]$$

Taking determinant on both sides,

$$\Delta_c(s) = |sI - A + BG| = \Delta_0(s)|I + (sI - A)^{-1} BG|$$

$$= \Delta_0(s)\left|I + BG(sI - A)^{-1}\right|$$

$$= \Delta_0(s)\left|I + G(sI - A)^{-1}B\right|$$

$$= \Delta_0(s)\,\Delta(s)$$

Example 15.27. *A linear process is described by*

$$\dot{x}(t) = \begin{bmatrix} 1 & 0 & 0 \\ -1 & 0 & 2 \\ 0 & -1 & 1 \end{bmatrix} x(t) + \begin{bmatrix} 1 \\ 0 \\ 0 \end{bmatrix} u(t)$$

The desired closed loop pole locations are –1, –1 and –2. Find the G-matrix

Solution:

From Eq. (15.295), $k(s) = adj\ (sI - A)B = \begin{bmatrix} s^2 - s + 2 \\ -(s-1) \\ 1 \end{bmatrix}$

$$\Delta_0(s) = |sI - A| = s^3 - 2s^2 + 3s - 2$$

For poles at –1, –1 and –2,

$$\Delta_c = s^3 + 4s^2 + 5s + 2$$

From Eq (15.296)

i.e., $Gk(s) = \Delta_c(s) - \Delta_o(s)$

$$(g_1\ g_2\ g_3)\begin{bmatrix} s^2 - s + 2 \\ -(s-1) \\ 1 \end{bmatrix} = \begin{bmatrix} 6s^2 + 2s + 4 \end{bmatrix}$$

$$g_1(s^2 - s + 2) - g_2(s - 1) + g_3 = g_1 s^2 - (g_1 + g_2)s + 2g_1 + g_2 + g_3$$

$$= 6s^2 + 2s + 4$$

Hence,

$$g_1 = 6,\ g_2 = -8 \text{ and } g_3 = 0$$

15.23. SUMMARY

In this chapter, the basic concepts of state and state variables have been introduced. The methods of representing a system in state-variable form have been presented. Derivations of state and output equations of systems and their transformation into several canonical forms have been discussed. The methods of solving the state and output equations have been presented. The main advantage of the modern control theory is that these methods are applicable to MIMO, time-varying and nonlinear systems. The concept of controllability and observability have been explained and illustrated. Examples have been extensively used to illustrate the approach.

REVIEW QUESTIONS

15.1. Define state variable, state and state vector.

15.2. Define state space, state trajectory and state equation.

15.3. Mention four advantages of state space variable representation.

15.4. State the three methods of expressing the system states.

15.5. What is meant by MIMO system?

15.6. Define eigenvalues and eigen vectors.

15.7. Mention 4 properties of Jordon Canonical form.

15.8. What is a state diagram?

15.9. State four advantages of state diagrams.

15.10. What is meant by decomposition of transfer function?

15.11. What are the types of decomposition?

15.12. Explain state transition matrix.

15.13. State the different methods used to solve state equations.

15.14. What is a modal matrix?

15.15. State Cayley-Hamilton theorem.

15.16. Mention the properties of a state transition matrix.

15.17. Explain controllability and observability.

15.18. State the properties of the state transition matrix of a time-varying system.

EXERCISE

15.1. State the advantages of state-variable approach to system analysis.

15.2. Define State variable, state, state vector, state space, state trajectory and state equation.

15.3. Illustrate the physical variable method of system representation with a suitable example.

15.4. Explain eigen value and eigen vector.

15.5. With a suitable example, explain the phase variable method of system representation.

15.6. Describe, with an illustrative example, the canonical variable method of system representation.

15.7. What is meant by similarity transformation? What are its advantages?

15.8. Write down Jordan canonical form of a system. State its general properties.

15.9. What is a state diagram? Bring out its importance.

15.10. Distinguish between direct, cascade and parallel decomposition.

15.11. Outline the steps involved in direct decomposition.

15.12. What is state transition matrix? Enumerate the methods of evaluating state transition matrix.

15.13. Discuss the properties of matrix exponential.

15.14. State the Cayley-Hamilton theorem. Explain how it is useful in evaluating state transition matrix.

15.15. Discuss the properties of state transition matrix.

15.16. Explain the method of obtaining state transition equation from the state diagram with an illustrative example.

15.17. Explain the concept of controllability and observability.

PROBLEMS

15.1. Obtain the state space equations for the mechanical translational system shown in Fig. 3.9.

15.2. Obtain the state space equations for the mechanical translational system shown in Fig. 3.11.

15.3. Obtain the state space equations for the electrical network shown in Fig. 3.5 with i_1, i_2 and v_c as variables.

15.4. Obtain the state space equations for the electrical network shown in Fig. 15.19 with i_1, i_2 and v_{c_2} and v_{c_3} as state variables.

15.5. Obtain the state space equations for the d.c. generator shown in Fig. 5.19.

15.6. obtain the state space equations for the armature controlled d.c servomotor shown in Fig. 5.23.

15.7. A system is defined by

$$2y'''' + 8y''' + 6y'' + 12y' + 10y = 4u$$

where y is the output and u is the input. Obtain the state variable representation.

15.8. The transfer function of a system is given by

$$\frac{Y(s)}{U(s)} = -\frac{5}{s^5 + 10s^4 + 6s^3 + 5s^3 + 2s + 9}$$

Obtain its state variable representation.

15.9. Obtain the state variable representation of the transfer function

$$\frac{Y(s)}{U(s)} = \frac{10s^2 + 5s + 2}{4s^4 + 5s^3 + 3s^2 + 9s + 6}$$

15.10. Obtain the state variable representation of the following transfer function

$$\frac{Y(s)}{U(s)} = \frac{s^3 + 4s^2 + 2s + 6}{2s^3 + 3s^2 + 6s + 5}$$

15.11. Obtain the eigenvalues, eigen vectors and modal matrix for the A matrix given below:

$$A = \begin{bmatrix} 0 & 1 & 0 \\ 3 & 0 & 2 \\ -12 & -7 & -6 \end{bmatrix}$$

by methods 2 and 3.

15.12. A system is represented by the following

$$A = \begin{bmatrix} -3 & 1 \\ -2 & 0 \end{bmatrix}, B = \begin{bmatrix} 1 \\ 0 \end{bmatrix}, x(0) = \begin{bmatrix} 1 \\ -1 \end{bmatrix}, C = \begin{bmatrix} 0 & 1 \end{bmatrix}$$

(a) Find e^{At} by Laplace transform method. (b) Find its response for a unit step input.

15.13. Solve problem 15.12 by series summation method.

15.14. Solve problem 15.12 by Caryley-Hamilton method.

15.15. Solve problem 15.12 by numerical method.

15.16. Check whether the system is state controllable with the following A and B Matrix.

(a)
$$A = \begin{bmatrix} 0 & 6 \\ -1 & -5 \end{bmatrix}, B = \begin{bmatrix} 3 & 6 \\ -1 & -2 \end{bmatrix}$$

(b)
$$A = \begin{bmatrix} 0 & 1 & 0 \\ 3 & 0 & 2 \\ -12 & -7 & -6 \end{bmatrix}, B = \begin{bmatrix} 1 \\ 4 \\ 6 \end{bmatrix}$$

15.17. Check whether the system described below is observable.

(a)
$$A = \begin{bmatrix} -5 & 4 \\ -6 & 5 \end{bmatrix}, B = \begin{bmatrix} 1 \\ 1 \end{bmatrix}, C = \begin{bmatrix} -2 & 3 \end{bmatrix}$$

(b)
$$A = \begin{bmatrix} 0 & 1 & 0 \\ 3 & 0 & 2 \\ -12 & -7 & -6 \end{bmatrix}, B = \begin{bmatrix} 1 \\ 4 \\ 6 \end{bmatrix}, C = \begin{bmatrix} 1 & 1 & 2 \end{bmatrix}$$

15.18. Obtain A, B and C matrices for $Y(s)/U(s) = 10(s + 1)/s(s + 2)(s + 3)$

15.19. Obtain A, B and C matrices for $Y(s)/U(s) = (s + 2)/(s + 1)(s^2 + 2s + 2)$

15.20. Given

$$A = \begin{bmatrix} 0 & 1 \\ -2 & -3 \end{bmatrix}, \ B = \begin{bmatrix} 1 \\ 1 \end{bmatrix}, \ C = \begin{bmatrix} 1 & 2 \end{bmatrix}$$

(a) Check for controllability and observability

(b) Repeat (a) for the state feedback

$$U = r(t) \ GX(t)$$

where $G = [g_1 \ g_2]$.

Chapter 16

MATLAB APPLICATION

16.1. INTRODUCTION

MATLAB is a software package extensively used in control system analysis and design. It has several built-in functions for numerical computations and graphics. It can handle *linear, non-linear, continuous and discrete systems*. It also provides easy extensibility. The user can create his/her own functions in MATLAB language. They can be used as any other built-in function.

Matrix is the main building block of the MATLAB. Complex matrix is the only data type used. It need not be declared. Scalars, vectors and matrices are automatically dealt with as special cases of the basic data type. MATLAB offers several toolboxes. Some of them are:

1. Control systems
2. Simulink
3. System identification
4. Robust control
5. Signal processing
6. Neural Networks

16.2. BASIC STRUCTURE OF MATLAB

MATLAB works through three basic windows. They are:

1. Command window
2. Graphics window
3. Edit Window

Command window: This is the main window. It is characterized by the prompt "»". All commands are typed in this window.

Graphics window: The output of all graphic commands, typed in this windows, are flushed to the graphics window. The user can also create his/her own graphics window.

Edit window: This is the window in which the user writes, edits, creates and saves his/her own programs in files, known as *M-files*.

16.3. FILE TYPES

MATLAB offers three basic type of files for storing information. They are:
1. M-files
2. Mat-files
3. Mex-files

16.3.1. M-files

These files are standard ASCII text-files. Each file has .m extension. The two types of M-files are:
1. Script files
2. Function files

Script file: This file consists of a set of valid commands in it. It is executed by typing the name of the file without .m extension. This file works on global variable, that is, variables that are currently available in the workspace. On the execution of the script file, the results are left in the workspace. This file is useful when one has to use a set of commands frequently.

Function file: This file is the same as script file except that the variables in this file are all local. The function file is defined with a function name with well-defined list of inputs and outputs. The syntax of the function definition line is

Function[output variables] = function name[input variables]

The function name is the same as the file name. Examples of function files are given below:

Function definition	File name
Function [a, b] = square (x)	Square.m
Function [] = angle (p, q)	angle.m
Funciton area (x)	area.m

The function name is typed in *lower case*. Multiple output variables are enclosed in the square bracket. Single variables need not be enclosed within the square bracket. When no output is required, the square bracket and equal sign is dropped.

16.3.2. Mat-files

These are the files generated by MATLAB when data is saved using "save" command. They are binary data files with a .*mat* extension. Mat files are loaded into the MATLAB using "load" command.

16.3.3. Mex files

These files are used to call FORTRAN and C programs. Mex-files are MATLAB files with .*mex* extension.

16.4. MATLAB SYSTEM COMMANDS

Some of the important system commands that are frequently used while working with MATLAB are described below:

1.	» help	Lists all topics on which help is available
2.	» help *command*	Provides information on the specified command
3.	» expo	Executes the demonstration programs
4.	» pwd	Shows the present (or current) working directory
5.	» dir	Displays contents of the current directory
6.	» cd	Changes the current directory
7.	» print	Prints the content of the active window
8.	» who	Displays variables currently in the workspace
9.	» whos	Displays variables currently in the workspace with their size
10.	» what	Shows MATLAB files and MATLAB data files in the current directory
12.	» load *file name*	Loads the specified file
13.	» quit	Quits MATLAB
14.	» exit	Exits MATLAB

16.5. MATLAB CONTROL SYSTEM COMMANDS

Some of the important commands frequently used in the analysis of control system are described below:

conv	:	Convolution of two polynomials
poly	:	Roots to polynomial conversion
residue	:	Partial fraction expansion
ss2tf	:	State-space to transfer function conversion
tf2ss	:	Transfer function to state-space conversion
eig	:	Eigen values and eigen vectors
roots	:	Polynomial roots
impulse	:	Impulse response
initial	:	Initial condition response
margin	:	Gain and phase margin
nichols	:	Nichols plot
ngrid	:	Draw gridlines for Nichols plot
nyquist	:	Nyquist plot
pzmap	:	Pole-zero map
rlocus	:	Root locus

16.6. MATLAB PARTIAL FRACTION EXPANSION

We shall discuss the application of the MATLAB to find the partial fraction expansion of a given transfer function with an example.

Example 16.1. *The transfer function* $T(s) = 10/[(s^2 + 6s + 25)(s + 2)]$. *Find the partial fraction expansion using MATLAB.*

Solution:

$$T(s) = 10/[(s^2 + 6s + 25)(s + 2)]$$
$$= A_1/(s + 3 - j4) + A_2/(s + 3 + j4) + A_3/(s + 2)$$

By conventional method, we obtain the values of A_1, A_2 and A_3 as

$$A_1 = 0.303 \angle 166°$$
$$A_2 = 0.303 \angle -166°$$
$$A_3 = 0.59 \angle 0°$$

```
% MATLAB program for partial fraction expansion
%
% Transfer function T(s) = 10/[(s**2 + 6*s + 25)(s + 2)]
%
num = [10];                         % Form the numerator polynomial
p(1) = -3+j*4, p(2) = -3-j*4, p(3) = -2;  % Poles of F(s)
den = poly(p);                      % Form denominator polynomial
[r1, p1, k1] = residue (num, den);   % Find the residue
r1                                  % List the residues in complex form
abs (r1)                            % Magnitude of residues
angle (r1)*180/pi                   % Angles of residues indegrees
end;
```

The results obtained after the execution are shown below:

```
r1 =
-0.2941 + 0.0735i                   % Complex value of A₁
-0.2941 - 0.0735i                   % Complex value of A₂
0.5882                              % Complex value of A₃


ans =
0.3032                              % Absolute value of A₁
0.3032                              % Absolute value of A₂
0.5882                              % Absolute value of A₃


ans =
165.9638                            %Phase angle of A₁ in degrees
-165.9638                           %Phase angle of A₂ in degrees
0                                   %Phase angle of A₃ in degrees
```

Example 16.2. *The transfer function of a system is* $T(s) = (7s + 2)/[(s + 1)(s + 3)(s + 4)]$. *Find the partial fraction expansion using MATLAB.*

Solution:

$$T(s) = (7s + 2)/(s^3 + 8s^2 + 19s + 12)$$
$$= A_1/(s + 1) + A_2/(s + 3) + A_3/(s + 4)$$

% MATLAB program for Partial Fraction expansion
% Transfer function $T(s) = (7s + 2)/[(s + 1)(s + 3)(s + 4)]$
%

num = [7 2];	**% Form the numerator polynomial**
den = [1 8 19 12];	**% Form denominator polynomial**
[r1,p1,k1] = residue (num, den);	**% Find the residue**
r1	**% List the residues in complex form**
abs(r1)	**% Magnitude of residues**
angle(r1)*180/pi	**% Angles of residues**
end;	

The results obtained after the execution are shown below:

r1 =	
−8.667	**% Residue of** $p = -4$
9.5000	**% Residue of** $p = -3$
0.8333	**% Residue of** $p = -1$

ans =	
8.6667	**% Absolute value of** A_1
9.5000	**% Absolute value of** A_2
0.8333	**% Absolute value of** A_3

ans =	
180	**% Angle of** $p = -4$
0	**% Angle of** $p = -3$
180	**% Angle of** $p = -1$

Example 16.3. *The transfer function of a system is* $T(s) = 1/[(s + 1)^3(s + 2)]$. *Find the partial fraction expansion using MATLAB.*

Solution:

$$T(s) = 1/[(s + 1)^3 (s + 2)]$$
$$= B_1/(s + 1)^3 + B_2/(s + 1)^2 + B_3(s + 1) + A/(s + 2)$$

METALAB program for
% Partial fraction expansion
% Transfer function $T(s) = 1/[(s + 1)**3(s + 2)]$
%

num[1];	Form the numerator polynomial
den = [1 5 9 7 2];	% Form the denominator polynomial
[r1, p1, k1] = residue (num, den);	% Find the residue
r1	% List the residues in complex form
abs(r1)	% Magnitude of residues
angle(r1)*180/pi	% angles of residues
end;	

The results obtained after the execution are shown below:

r1 =

−1.0000	% B_1
1.0000	% B_2
−1.0000	% B_3
1.0000	% A

ans =

1.0000	% Absolute value of B_1
1.0000	% Absolute value of B_2
1.0000	% Absolute value of B_3
1.0000	% Absolute value of A

ans =

180	% Angle of B_1
0	% Angle of B_2
180	% Angle of B_3
0	% Angle of A

Example 16.4. *The transfer function of a system is* $T(s) = (s^2 + 5s + 9)/[(s + 3)(s^2 + 2s + 6)]$. *Find the partial fraction expansion using MATLAB.*
Solution:

Transfer function $T(s) = (s**2 + 5s + 9)/[(s + 3)(s**2 + 2s + 6)]$
$$= A/(s + 3) + C/(s + 1 + j\sqrt5) + C^*/(s + 1 - j\sqrt5)$$

```
% METALAB program for Partial Fraction expansion
% Transfer function  T(s) = (s**2 + 5s + 9)/[(s + 3)(s**2 + 2s + 6)]
%
num = [1 5 9];                           % Form the numerator polynomial
den = [1 5 12 18];                       % Form the denominator polynomial
[r1, p1, k1] = residue (num, den);       % Find the residue
r1                                       % List the residues in complex form
abs(r1)                                  % Magnitude of residues
angle(r1)*180/pi                         % Angles of residues
end;
```

The results obtained after the execution are shown below:

```
r1 =
0.3333
0.3333 – 0.3727i
0.3333 – 0.3727i


ans =
0.3333                                   % residue A
0.5000                                   % residue C
0.5000                                   % residue C*


ans =
0                                        % Angle of A
–48.1897                                 % Angle of C
48.1897                                  % Angle of C*
```

16.7. TIME-RESPONSE OF SYSTEMS

In this section, we shall write MATLAB programs to find the impulse and step response of systems of different orders. This is illustrated with suitable examples.

16.7.1. Impulse Response of a First-order System

The impulse response of the first-order system with transfer function $F(s) = 1/(s + 1)$ is e^{-t}. The MATLAB program and the resultant response are illustrated with an example.

Example 16.5. *Write a MATLAB program to find the impulse response of the system with transfer function $T(s) = 1/(s + 1)$.*

Solution:
The given transfer function is $T(s) = 1/(s + 1)$.

```
% Impulse response of T(s) = 1/(s + 1)
%
num = [1];                              % Form the numerator polynomial
den = [1 1];                            % Form the denominator polynomial
y = impulse (num, den);                 % Find the impulse response
plot (y);                               % Plot the response
xlabel ('time');                        % Label x axis as time
ylabel ('Impulse response');            % Lable y axis as impulse response
title ('Impulse response of 1/(s + 1)'); % Label title as Impulse response of 1/(s + 1)
grid;                                   % Show the grid
end;
```

The impulse response of the first-order system with transfer function $T(s) = 1/(s + 1)$ is shown in Fig 16.1 obtained by executing the MATLAB program.

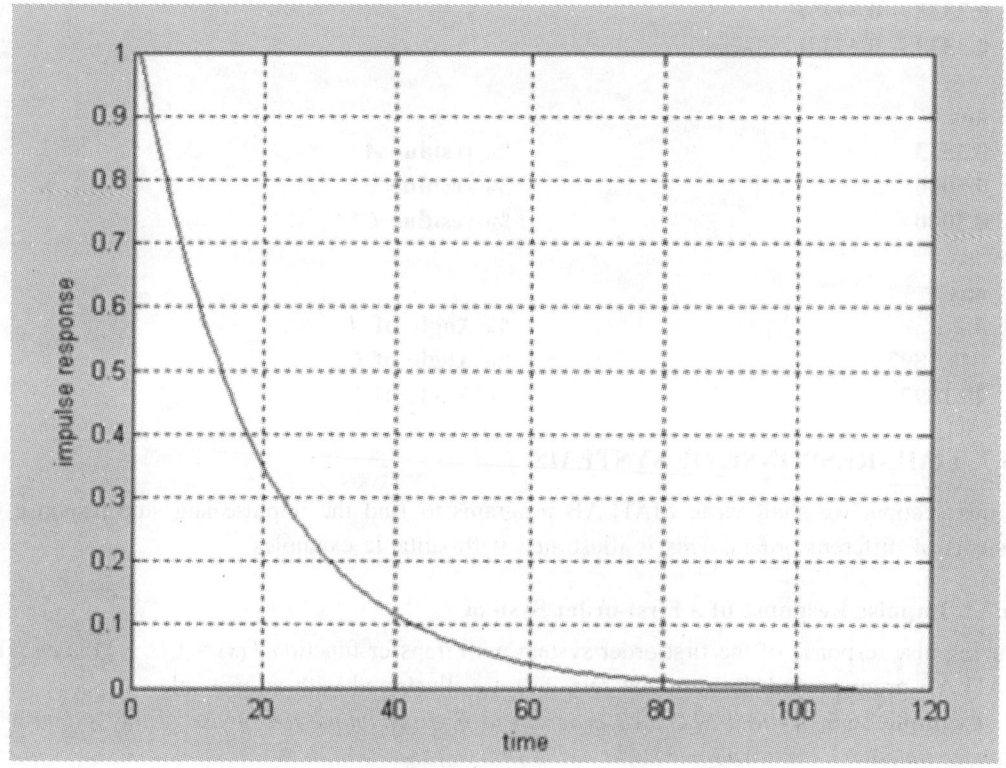

Fig. 16.1. Impulse response of $T(s) = 1/(s + 1)$

16.7.2. Step Response of a First-order System

The step response of the first-order system with transfer function $T(s) = 1/(s + 1)$ is $(1-e^{-t})$. The MATLAB program and the resultant response are illustrated with an example.

Example 16.6. Write a MATLAB program to find the step response of a system with the transfer function $T(s) = 1/(s + 1)$.

Solution:

The given transfer function is $T(s) = 1/(s + 1)$. Its step response is $(1-e^{-t})$.

We shall verify this with the MATLAB program.

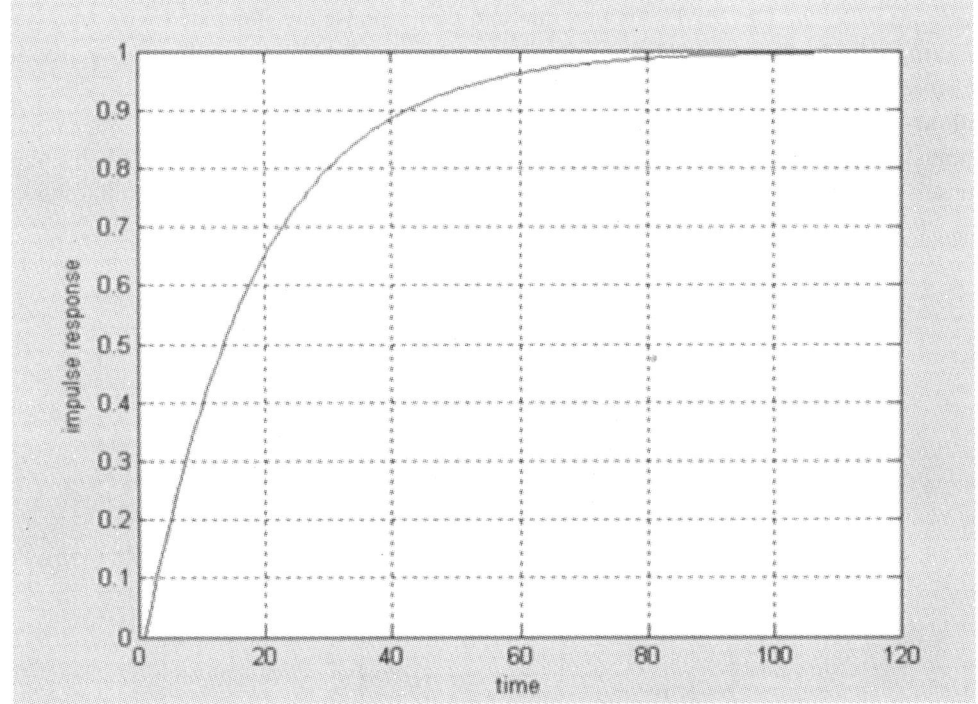

Fig. 16.2. Step response of $T(s) = 1/(s + 1)$

```
% Step response of 1/(s + 1)                 
%                          
num = [1];                        % Form the numerator polynomial
den = [1 1];                      % Form the denominator polynomial
y = step (num, den);              % Find step response
plot (y);                         % Plot the response
xlabel ('time');                  x label x axis as time
```

ylabel ('step response');   **% Label *y* axis as Step Response**
title ('Fig. 16.2 Step response of 1/(*s* + 1)');   **% Label Title**
grid;   **% Show the grid**
end;

The step response of the first-order system with transfer function $T(s) = 1/(s+1)$ is shown in Fig. 16.2 obtained by executing the MATLAB program.

16.7.3. Impulse Response of an Un-damped Second-order System

The impulse response of the second-order un-damped system with transfer function $T(s) = 5/(s**2 + 25)$ is *sin 5t*. The MATLAB program and the response are illustrated with an example.

Example 16.7. *Write a MATLAB program to find the impulse response of a second-order un-damped system with $T(s) = 5/(s^2 + 25)$.*

Solution:

The transfer function of an un-damped second order system is $T(s) = 5/(s^2 + 25)$. Its impulse response is *sin 5t*. We shall verify this with the MATLAB program and the results obtained.

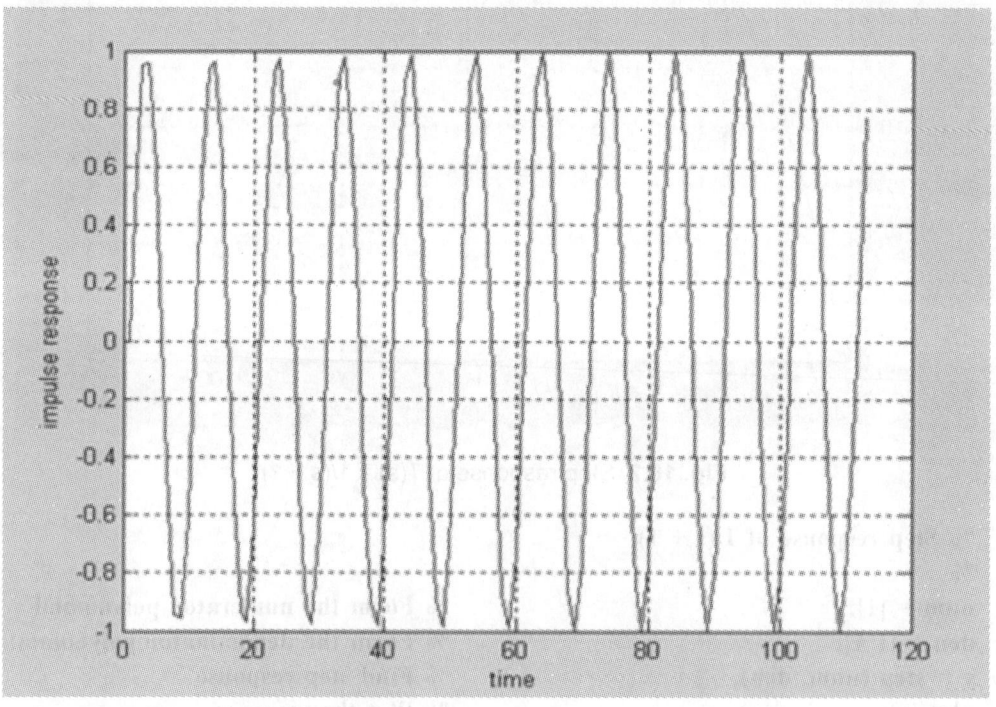

Fig. 16.3. Impulse response of second-order system

```
% MATLAB program for Second-order un-damped system–impulse response
% Transfer Function T(s) = 5/(s**2 + 25)
% zeta = 0
%
num = [5];                        % Form the numerator polynomial
den = [1 0 25];                   % Form the denominator polynomial
y = impulse (num, den);           % Find the impulse response
plot(y);                          % Plot the response
xlabel ('time');                  % Label x axis as time
ylabel ('impulse response');      % Label y axis as impulse response
title ('Fig. 16.3 Impulse response of second-order un-damped systems')
grid;                             % Show the grid
end;
```

The impulse response of a second-order un-damped system with $T(s) = 5/(s**2 + 25)$ is shown in Fig.16.3 obtained by executing the above MATLAB program.

16.7.4. Step Response of an Un-damped Second-order System

The step response of an un-damped second-order system with transfer function $T(s) = 25/(s**2 + 25)$ is shown in Fig. 16.4. The MATLAB program and the resultant response are given below:

Example 16.8. *Write a MATLAB program to find the step response of a second-order un-damped system with* $T(s) = 5/(s^2 + 25)$

Solution:
The transfer function of an un-damped second order system is $T(s) = 25/(s^2 + 25)$. Its impulse response is $(1 - \cos 5t)$.

We shall verify with the MATLAB program.

```
% MATLAB PROGRAM Second-order un-damped system–step response
% transfer Function T(s) = 25/(s² + 25).
% zeta = 0
%
num [25];                         % Form the numerator polynomial
den [10 25];                      % Form the denominator polynomial
y = step (num, den);             % Find the step response
plot (y);                        % Plot the response
xlabel ('time');                 % Label x axis
ylabel ('step response');        % Label y axis
title ('Fig. 16.4 Step response of 25/(s² + 25)'); % Label title
```

grid;   **% Show the grid**
end;

The step response of second-order un-damped system with $T(s) = 25/(s**2 + 25)$ is shown in Fig. 16.4. obtained by executing the MATLAB program.

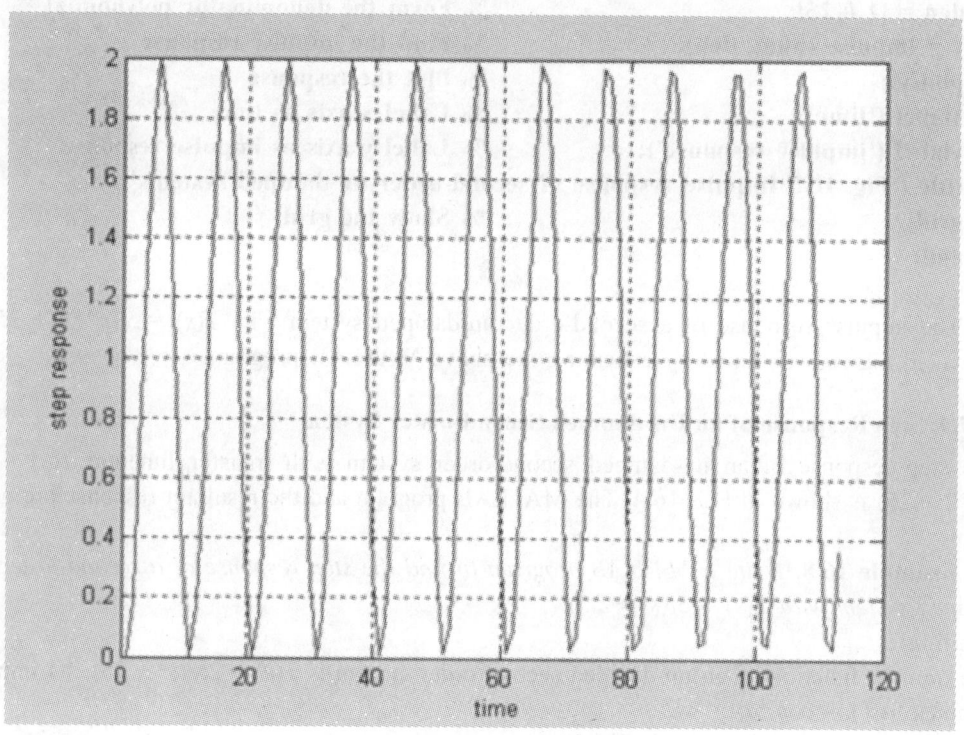

Fig. 16.4. Step response of $25/(s**2 + 25)$

16.7.5. Impulse Response of an Under-damped Second-order System

The impulse response of an under-damped second-order system with transfer function $T(s) = 25/(s**2 + 3s + 25)$ is shown in Fig. 16.5. The MATLAB program and the resultant response is given below:

Example 16.9. *Write a MATLAB program to find the impulse response of a second-order under-damped system with transfer function $T(s) = 25/(s^2 + 3s + 25)$.*

Solution:

The transfer function of an under-damped second order system is $T(s) = 25/(s^2 + 3s + 25)$. The MATLAB program and the results are given below:

% **MATLAB program for Second-order under-damped system–impulse response**
% **Transfer Function** $T(s) = 25/(s^2 + 3s + 25)$.
% **zeta = 0.3**
%

num [25];	% **Form Nr. polynomial**
den [1 3 25];	% **Form Dr. polynomial**
y = impulse (num, den);	% **Find the impulse response**
plot (y);	% **Plot the response**
xlabel ('time');	% **Label x axis**
ylabel ('impulse response');	% **Label y axis**

title ('Fig. 16.5 Impulse response of second-order system, zeta = 0.3')
grid;   % **Show the grid**
end;

The impulse response of the second-order system with a zeta = 0.3 is shown in Fig. 16.5 obtained by executing the MATLAB program.

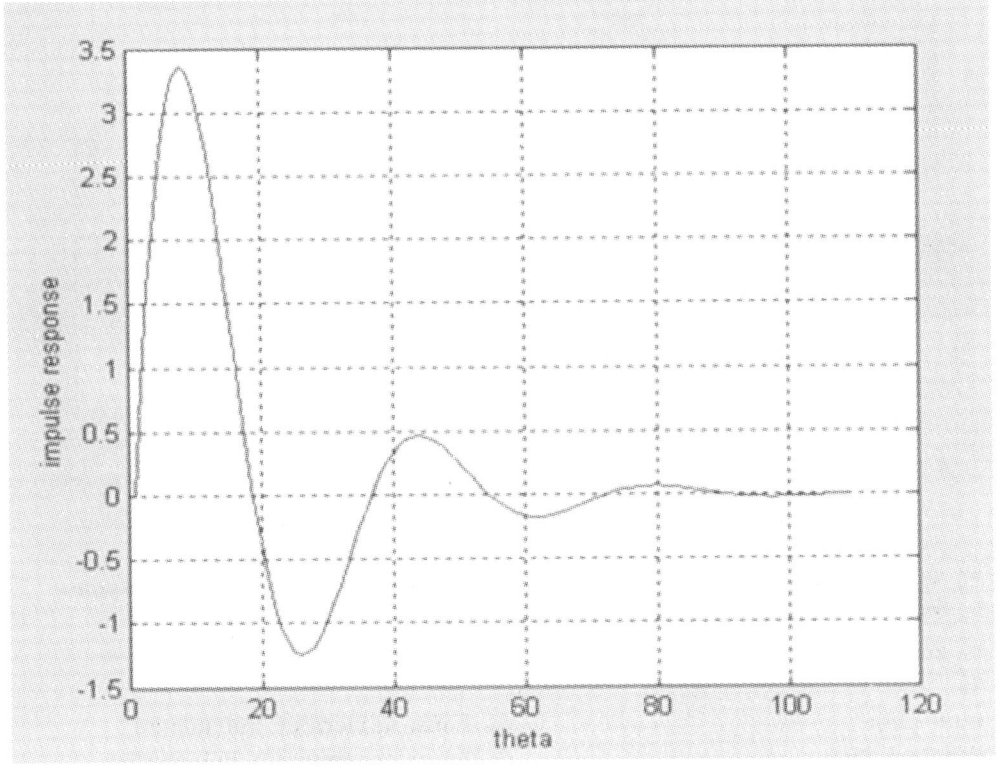

Fig. 16.5. Impulse response of second-order system (zeta = 0.3)

16.7.6. Step Response of an Under-damped Second-order System

The MATLAB program to obtain the step response of an under-damped second-order system with transfer function $T(s) = 25/(s**2 + 3s + 25)$ is given below:

Example 16.10. *Write a MATLAB program to find the step response of a second-order under-damped system with transfer function* $T(s) = 25/(s^2 + 3s + 25)$.

Solution:

The transfer function of an under-damped second order system is $T(s) = 25/(s^2 + 3s + 25)$. Its step response is $5.24\ e^{-1.5t} \sin 4.77t$. The MATLAB program and the response are given below:

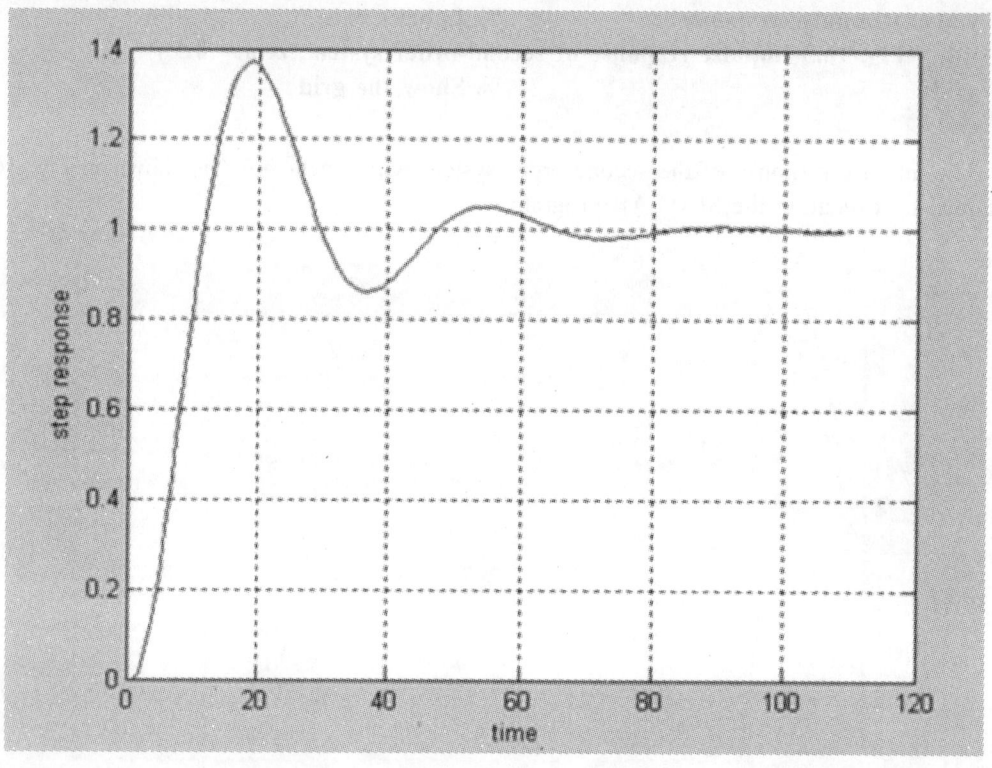

Fig. 16.6: Step response of second-order system, zeta = 0.3

% MATLAB program for Second-order under-damped system - step response
% Transfer Function T(s) = 25/(s² + 3s + 25).
% zeta = 0.3
%
num [25];   **% Form numerator Polynomial**
den [1 3 25];   **% Form denomuiator Polynomial**
y = step (num, den);   **% Find step response**

```
plot (y);                           % Plot the response
xlabel ('time');                    % Label x axis
ylabel ('step response');           % Label y axis
title ('Fig. 16.6 Step response of second-order system, zeta = 0.3')
grid;                               % Show grid
end;
```

Fig. 16.6 shows the step response of an under-damped second-order system obtained by executing the MATLAB program.

16.7.7. Impulse Response of a Critically Damped Second-order System

The MATLAB program to obtain the impulse response of a critically damped second-order system with transfer function $T(s) = 25/(s**2 + 10s + 25)$ is given below:

Example 16.11. *Write a MATLAB program to find the impulse response of a second-order critically damped system with transfer function $T(s) = 25/s^2 + 10s + 29$).*

Solution:

The transfer function of a critically damped second order system is $T(s) = 25/(s^2 + 10s + 25)$. The MATLAB program and the response are given below:

```
% MATLAB program for Second-order critically damped system - impulse response
% T(s) = 25/(s² + 10s + 25).
% zeta = 1.0
%
num [25];                           % Form Nr. polynomial
den [1 10 25];                      % Form Dr. polynomial
y = impulse (num, den);             % Find response
plot (y);                           % Plot the response
xlabel ('time');                    % Label x axis
ylabel ('impulse response');        % Label y axis
title ('Fig. 16.7 Impulse response of second-order critically damped system');
grid;                               % Show grid
end;
```

Fig. 16.7 shows the impulse response of a critically damped second-order system obtained by executing the MATLAB program.

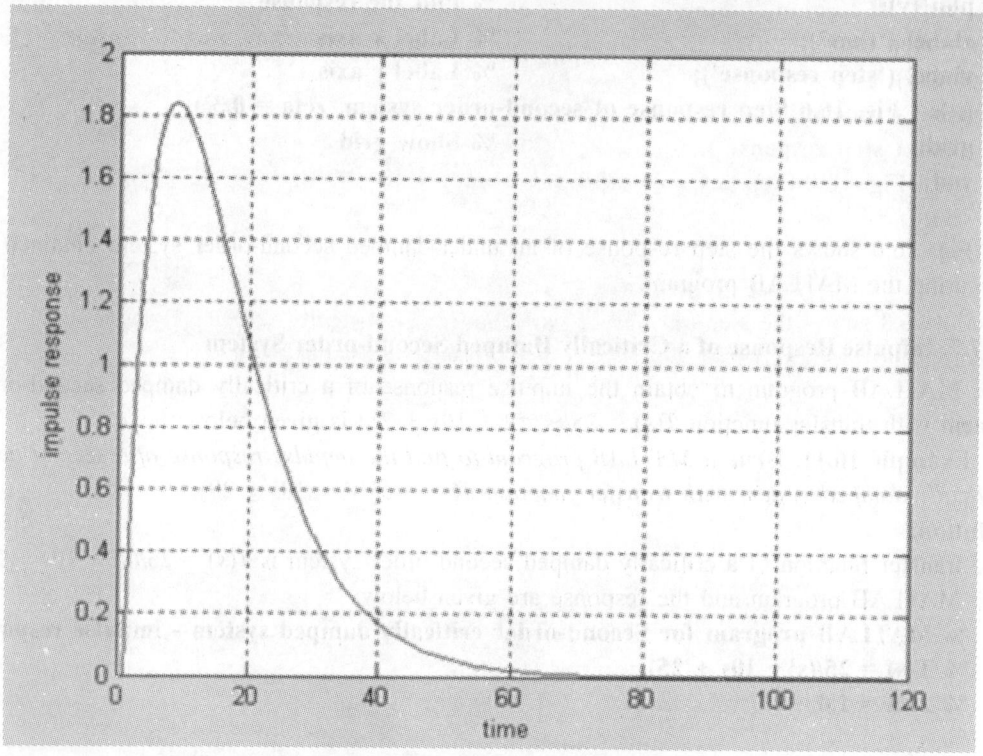

Fig. 16.7. Impulse response of second-order critically damped system

16.7.8. Step Response of a Critically Damped Second-order System

The step response of a critically damped second-order system with transfer function $T(s) = 25/(s{**}2 + 10s + 25)$ is shown in Fig. 16.8. The MATLAB program and the resultant response are given below:

Example 16.12. *Write a MATLAB program to find the step response of a second-order critically damped system with transfer function* $T(s) = 25/(s^2 + 10s + 25)$.

Solution:

The transfer function of a critically damped second order system is $T(s) = 25/(s^2 + 10s + 25)$. The MATLAB program and the response are given below:

```
% MATLAB program for Second-order critically damped system–step response
% Transfer Function T(s) = 25/(s² + 10s + 25).
% zeta = 1
%
num [25];
```

```
den [1 10 25];
y = step (num, den);
plot (y);
xlabel ('time');
ylabel ('step response');
title ('Fig. 16.8 Step response of second-order critically damped system');
grid;
end;
```

Fig. 16.8 shows the step response of a second-order critically damped system obtained by executing the MATLAB program.

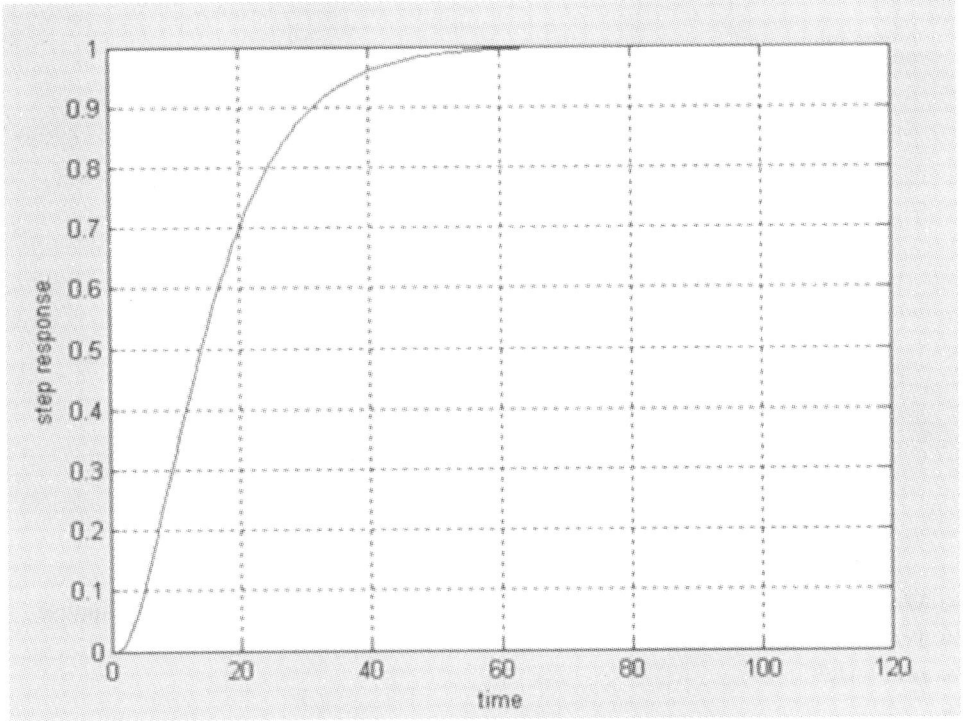

Fig. 16.8. Step response of second-order critically damped system

16.7.9. Impulse Response of an Over-damped Second-order System

The impulse response of an over-damped second-order system with transfer function $T(s) = 25/(s**2 + 15s + 25)$ is shown in Fig. 16.9. The MATLAB program and the resultant response are given below:

Example 16.13. *Write a MATLAB program to find the impulse response of a second-order over-damped system with transfer function* $T(s) = 25/(s^2 + 15s + 25)$.

Solution:

The Transfer function of an over-damped second order system is $T(s) = 25/(s^2 + 15s + 25)$. The MATLAB program and the response are given below:

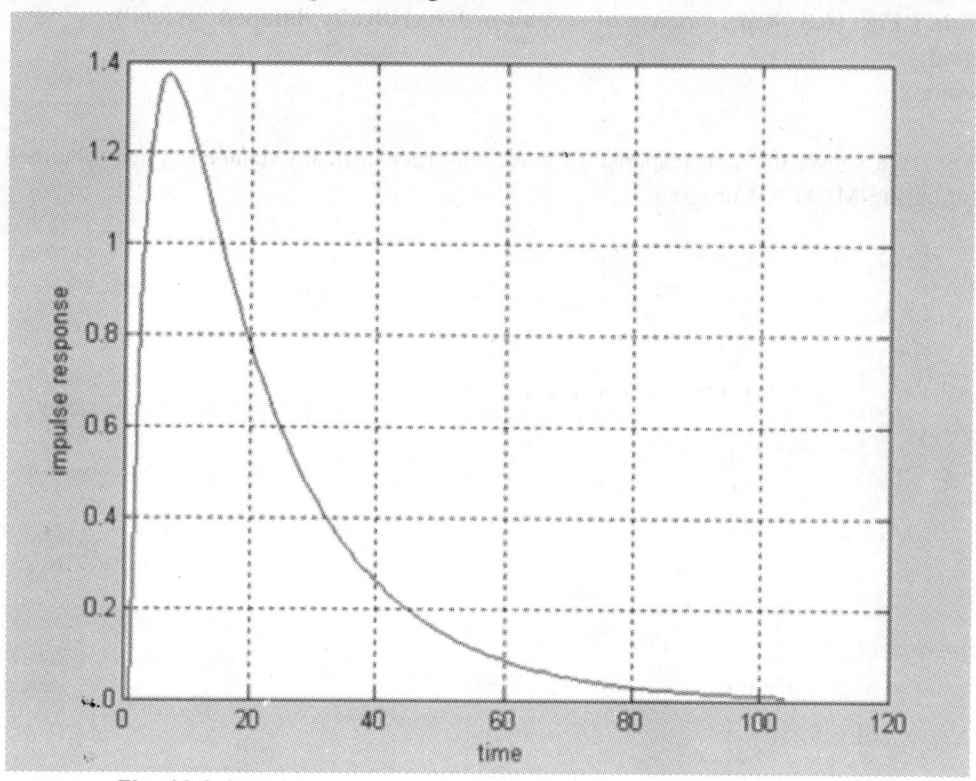

Fig. 16.9. Impulse response of a second-order over-damped system

% MATLAB program for second-order over-damped system-impulse response
% Transfer Function $T(s) = 25/(s^2 + 15s + 25)$.
% zeta = 1.5
%
num [25];
den [1 15 25];
y = impulse (num, den);
plot (y);
xlabel ('time');
ylabel ('impulse response');

title ('**Fig. 16.9 impulse response of second-order over-damped system**');
grid;
end;
Fig. 16.9 gives the impulse response of a second-order over-damped system obtained by executing the MATLAB program.

16.7.10. Step Response of an Over-damped Second-order System

The step response of an over-damped second-order system with transfer function $T(s) = 25/(s**2 + 15s + 25)$ is shown in Fig. 16. 10. The MATLAB program and the resultant response are given below:

Example 16.14. *Write a MATLAB program to find the step response of a second-order over-damped system with transfer function* $T(s) = 25/(s^2 + 15s + 25)$.

Solution:
The Transfer function of a over-damped second order system is $T(s) = 25/(s^2 + 15s + 25)$. The MATLAB program and the response are given below:

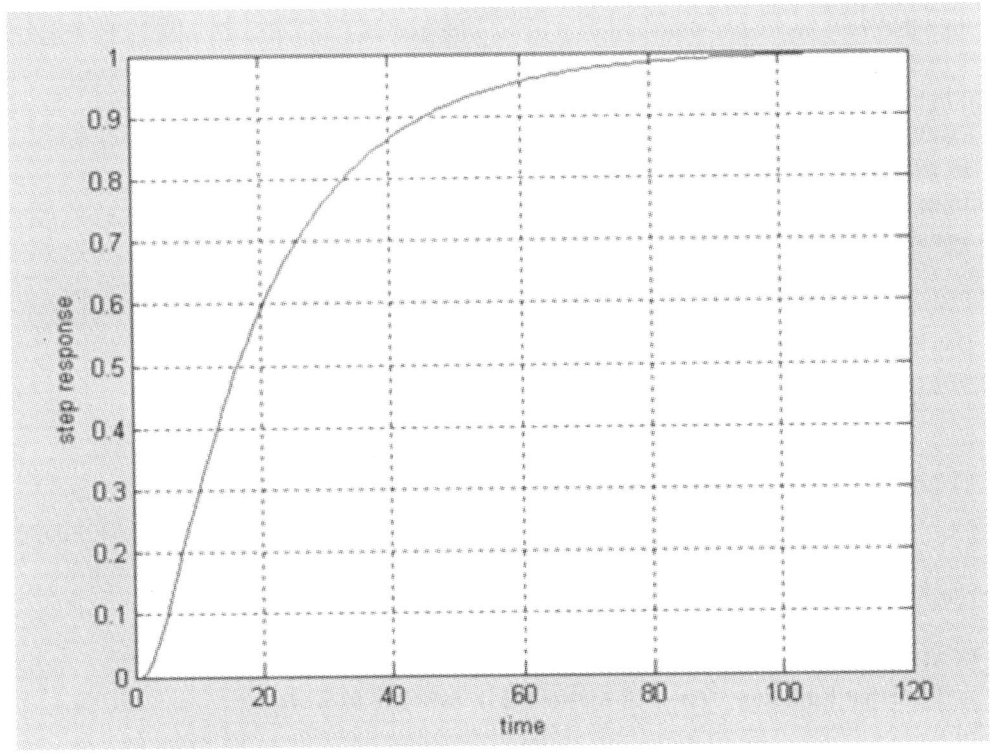

Fig. 16.10. Step response of a second-order over-damped system

```
% MATLAB program for Second-order over-damped system-step response
% Transfer Function T(s) = 25/(s² + 15s + 25).
% zeta = 1.5
%
num [25];
den [1 15 25];
y = step (num, den);
plot (y);
xlabel ('time');
ylabel ('step response');
title ('Fig. 16.10 Step response of second-order over-damped system');
grid;
end;
```

Fig. 16.10 shows the step response of a second-order over-damped system obtained by executing the MATLAB program.

16.7.11. Impulse Response of a Third-order System

The impulse response of a third-order system with transfer function $T(s) = (s + 3)/(s**3 + 5s**2 + 8s + 25)$ is shown in Fig. 16.11. The MATLAB program and the resultant response are given below:

Example 16.15. *Write a MATLAB program to find the impulse response of a third-order system with transfer function* $T(s) = (s + 3)/(s^3 + 5s^2 + 8s + 4)$.

Solution:

The transfer function of a third-order system is

$$T(s) = (s + 3)/(s**3 + 5s**2 + 8s + 4)$$
$$= (s + 3)/(s + 1)(s + 2)^2$$

The impulse response $C(s)$ is

$$C(s) = \frac{2}{s+1} - \frac{1}{(s+2)^2} - \frac{2}{s+2}$$

$$\therefore \quad c(t) = 2e^{-t} - t\, e^{-2t} - 2e^{-2t}$$

Since t is involved in the calculation of $c(t)$, it is included as a parameter in the expression for y.

The MATLAB program and the impulse response are given below:

```
% MATLAB program for Third-order system-impulse response
% Transfer Function T(s) = (s + 3)/(s**3 + 5s**2 + 8s + 4)
%
num = [1 3];
den = [1 5 8 4];
```

```
y = impulse (num, den, time);
plot (y);
x label('time');
y label('impulse response');
title ('Fig. 16.11 impulse response of third-order system');
grid;
end;
```

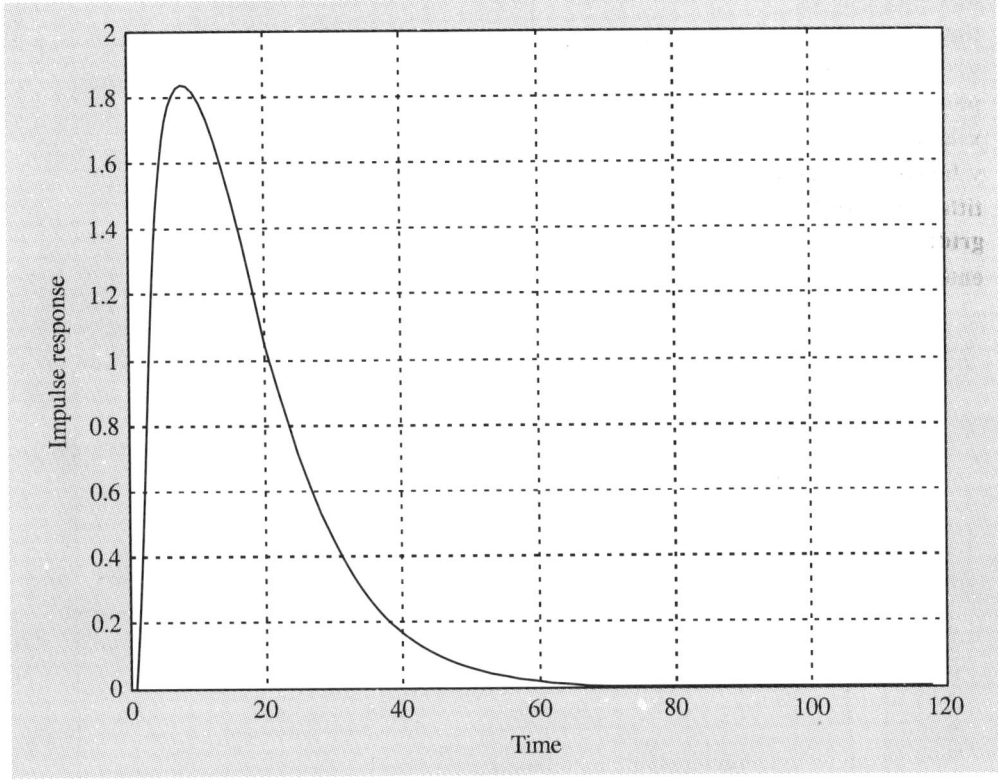

Fig. 16.11. Impulse response of third-order system

Fig. 16.11 shows the impulse response of the third-order system.

16.7.12. Step Response of a Third-order System

The step response of a third-order system with transfer function $T(s) = (s + 3)/(s**3 + 5s**2 + 8s + 4)$ is shown in Fig. 16.12. The MATLAB program and the resultant response are given below:

Example 16.16. *Write a MATLAB program to find the step response of a third-order system with transfer function* $T(s) = (s + 3)/(s^3 + 5s^2 + 8s + 4)$.

Solution:

The transfer function of a third-order system is $T(s) = (s + 3)/(s**3 + 5s**2 + 8s + 4)$. The MATLAB program and the response are given below:

```
% MATLAB program for Third-order system-step response
% Transfer Function T(s) = (s + 3)/(s**3 + 5s**2 + 8s + 4)
%
num = [1 3];
den = [1 5 8 4];
time = (0: .2: 30);
y = step (num, den, time);
plot (y);
x label('time');
y label('step response');
title ('Fig. 16.12 step response of third-order system');
grid;
end;
```

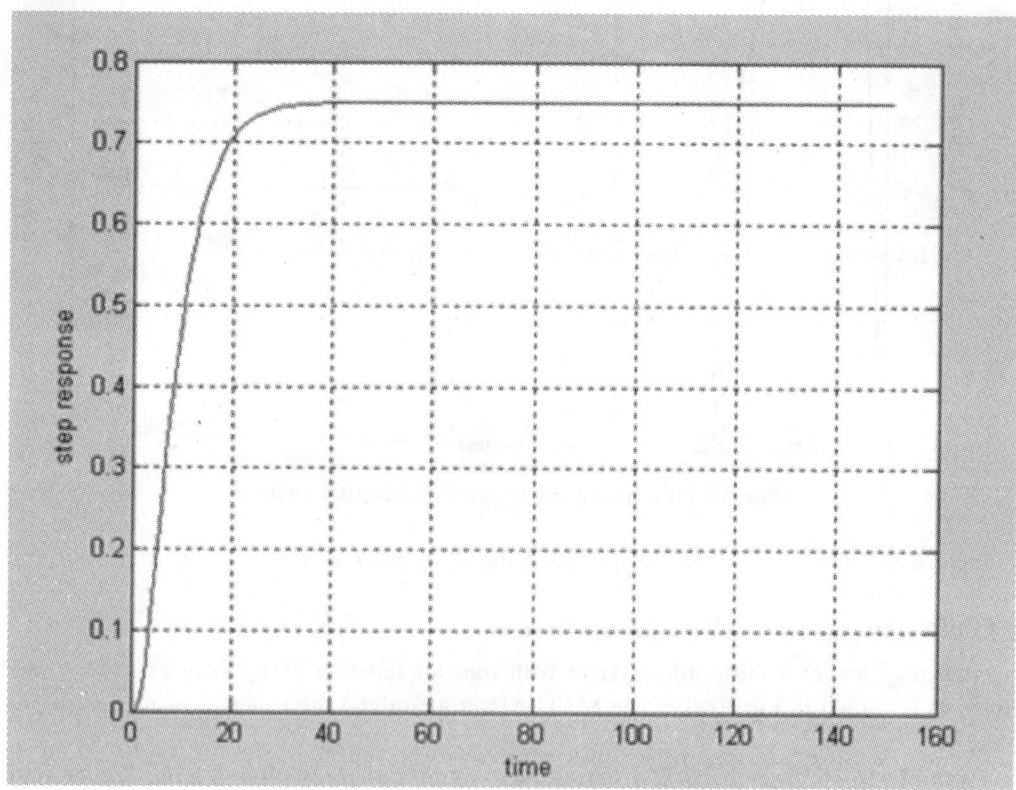

Fig. 16.12. Step response of third-order system

Fig. 16.12 shows the step response of the third-order system obtained by executing the MATLAB program.

16.8. FREQUENCY RESPONSE (BODE PLOTS) OF SYSTEMS

In this section, we shall write MATLAB programs to find the frequency response (Bode plots) of systems of different orders. The command we use for this purpose is "bode". This is illustrated with suitable examples.

16.8.1. Frequency Response of a Lag Compensator

Example 16.17. *Write a MATLAB program to find the frequency response of a lag compensator with transfer function* $T(s) = (1 + 0.1s)/(1 + s)$.
Solution:
The frequency response (Bode plots) of a lag compensator with transfer function $T(s) = (1 + 0.1s)/(1 + s)$ is shown in Fig. 16.13. The MATLAB program and the resultant response are given below:

```
% MATLAB program for Frequency Response–Bode plots
% T(s) = (1 + 0.1s)/(1 + s)
% Lag compensator
%
num = [0.1 1];
den = [1 1];
w = logspace (–2, 3, 200);
[mag, phase] = bode (num, den, w);
subplot (2, 1, 1)
semilogx (w, 20*log 10(mag))
axis ([0.01, 1000, –60, 60]);
title ('Fig. 16.13 Bode plot of Lag compansator');
xlabel ('omega (rad/sec)');
ylabel ('dB');
grid;
subplot (2, 1, 2);
semilogx (w, phase)
axis ([0.01, 1000, –60, 0]);
xlabel ('omega(rad/sec)');
ylabel ('PHI(deg)');
grid;
end;
```

Fig. 16.13 shows the Bode plot of a lag compencator obtained by executing MATLAB program.

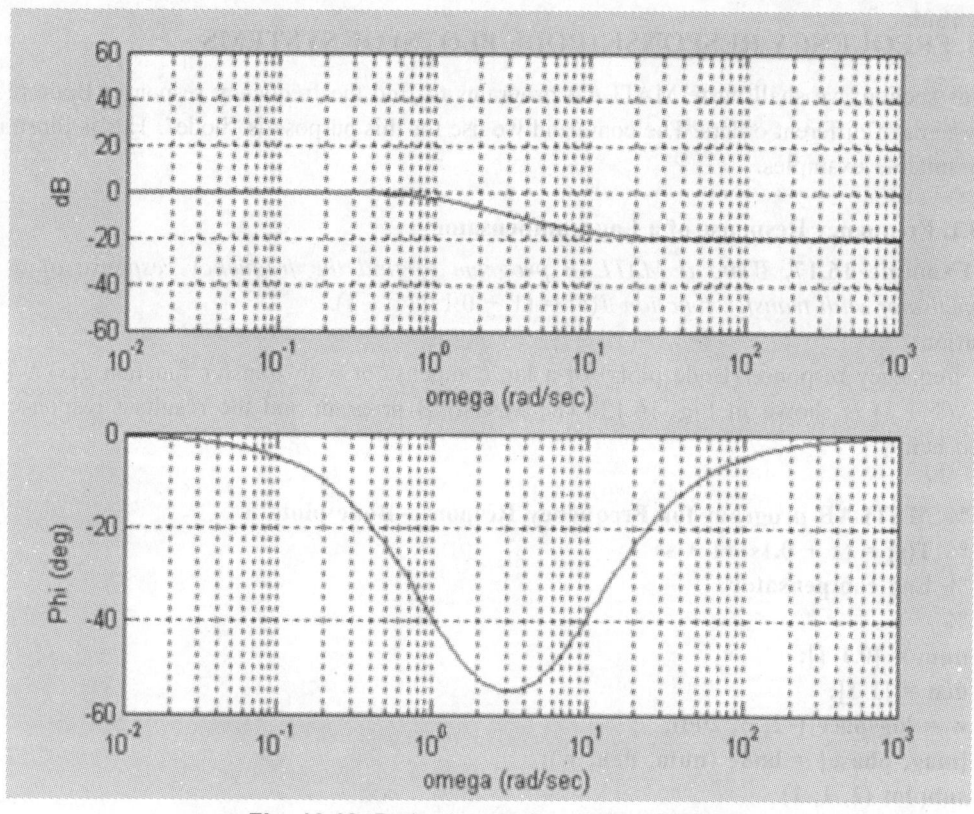

Fig. 16.13. Bode plot of a Lag Compensator

16.8.2. Frequency Response of a Lead Compensator

The frequency response (Bode plots) of a lead compensator with transfer function $T(s) = (1 + s)/(1 + 0.1s)$ is shown in Fig. 16.14. The MATLAB program and the resultant response are given below:

Example 16.18. *Write a MATLAB program to find the frequency response of a lead compensator with transfer function* $T(s) = (1 + s)/(1 + 0.1s)$.

Solution:

% MATLAB program for Frequency Response–Bode plots
% Transfer function $T(s) = (1 + s)/(1 + 0.1s)$
% Lead compensator
%
num = [1 1];
den = [.1 1];

```
w = logspace (-2, 3, 200);
[mag, phase] = bode (num, den, w);
subplot (2, 1, 1)
semilogx (w, 20*log10(mag))
axis ([0.1, 100, -40, 40]);
title ('Fig. 16.14 Bode plot of lead compensator');
xlabel ('omega (rad/sec)');
ylabel ('dB');
grid;
subplot (2, 1, 2);
semilogx (w, phase)
axis ([0.1, 100, 0, 90]);
xlabel ('omega (rad/sec)');
ylabel ('Phi (deg)');
grid;
end;
```

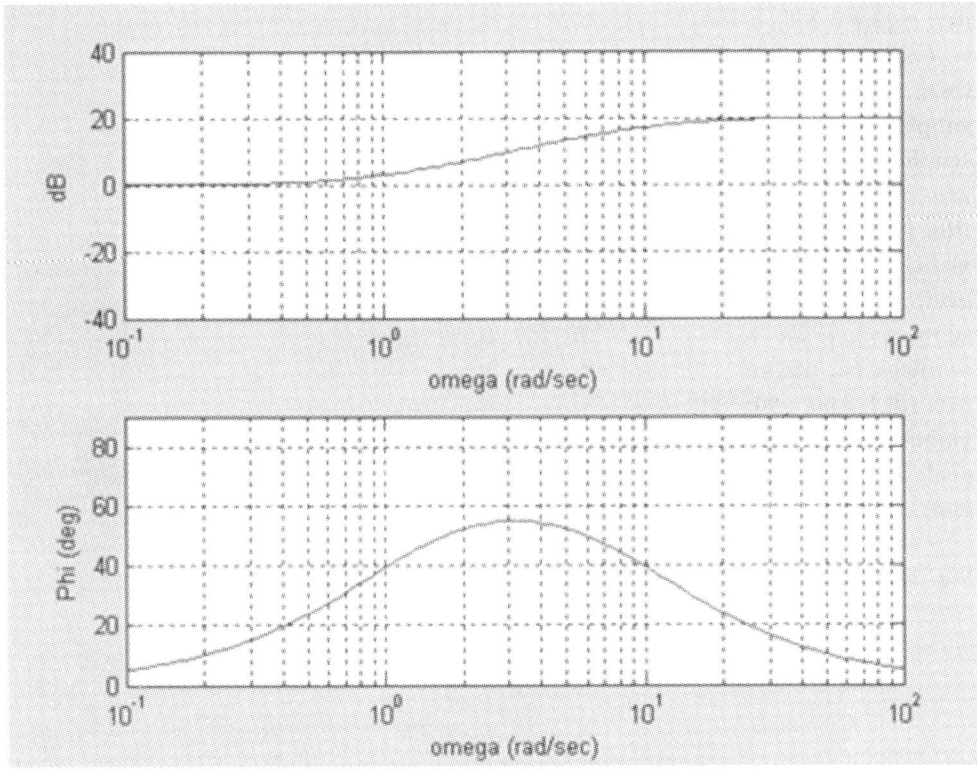

Fig. 16.14. Bode plot of a lead compensator

Fig. 16.14 shows the frequency response of the lead compensator with $T(s) = (1 + s)/(1 + 0.1s)$.

16.8.3. Frequency Response of a lag-lead Compensator

The frequency response (Bode plots) of a lag-lead compensator with transfer function $T(s) = (1 + 0.1s)(1 + 22s)/(1 + s)(1 + 2.2s$) is shown in Fig. 16.15. The MATLAB program and the resultant response are given below:

Example 16.19. *Write a MATLAB program to find the frequency response of a lag-lead compensator with transfer function $T(s) = (1 + 0.1s)(1 + 22s)/(1 + s)(1 + 2.2s)$.*

Solution:

```
% MATLAB program for Frequency Response–Bode plots
% Transfer function T(s) = (1 + 0.1s)(1 + 22s)/(1 + s)(1 + 2.2s)
% Lag-Lead compensator
%
num = [2.2 22.1 1];
den = [2.2 3.2 1];
w = logspace (–2, 3, 200);
[mag, phase] = bode (num, den, w);
subplot (2, 1, 1)
semilogx (w, 20*log10(mag))
axis ([0.1, 100, –40, 40]);
title ('Fig. 16.15 Bode plot of lag-lead compensator');
xlabel ('omega (rad/sec)');
grid;
subplot (2, 1, 2);
semilogx (w, phase)
axis ([0.1, 100, –90, 90]);
ylabel ('phi (deg)');
grid;
end;
```

Fig. 16.15 shows the frequency response of a lag-lead compensator.

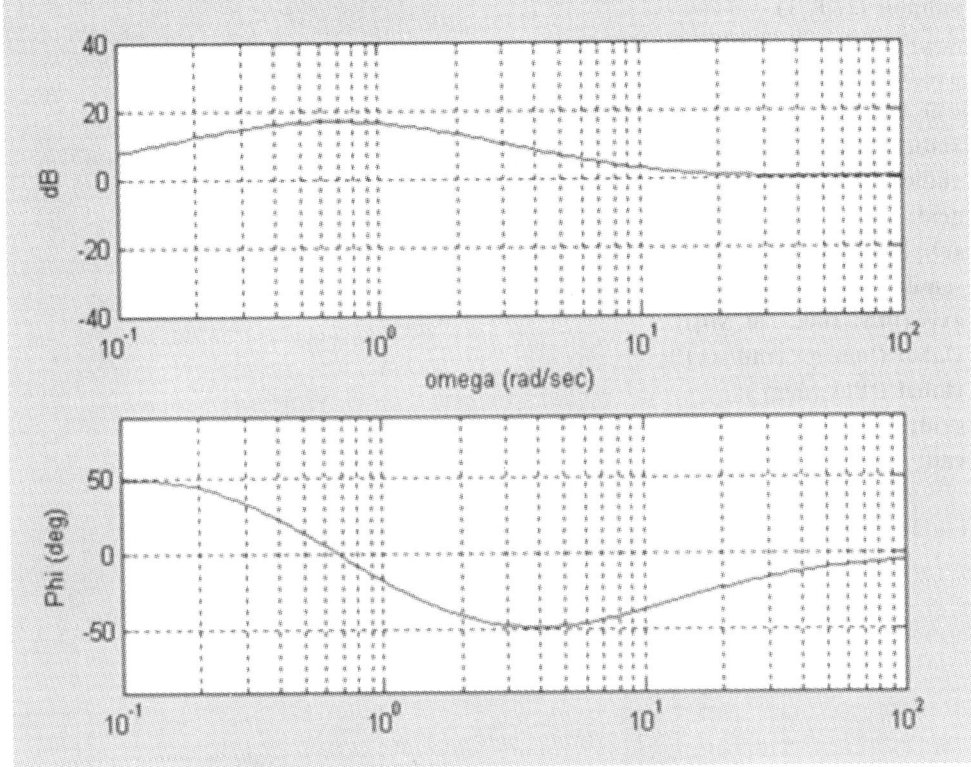

Fig. 16.15. Bode plot of a lag-lead compensator

16.8.4. Frequency Response of a Bridged-T Compensator

Example 16.20. *Write a MATLAB program to find the frequency response of a bridged-T compensator with transfer function* $T(s) = (s**2 + s + 10)/(s**2 + 15s + 25)$.

Solution:

The frequency response (Bode plots) of a bridged-T compensator with transfer function $T(s) = (s^2 + s + 10)/(s^2 + 15s + 25)$ is shown in Fig. 16.16. The MATLAB program and the resultant response are given below:

```
% MATLAB program Frequency Response–Bode plots-Bridged-T compensator
% Transfer function T(s) = (s**2 + s + 10)/(s**2 + 15s + 25)
%
num = [1 1 10];
den = [1 15 25];
w = logspace (–2, 3, 200);
[mag, phase] = bode (num, den, w);
```

```
subplot (2, 1, 1)
semilogx (w, 20*log10(mag))
axis ([0.1, 1008, –40, 40]);
title ('Fig. 16.16 Bode plot of Bridged-T compensator');
xlabel ('omega (rad/sec)');
ylabel ('dB');
grid;
subplot (2, 1, 2);
semilogx (w, phase)
axis ([0.1, 100, –90, 90]);
xlabel ('omega (rad/sec)');
ylabel ('Phi (deg)');
grid;
end;
```

Fig 16.16 shows the Bode plots of a bridged-T compensator.

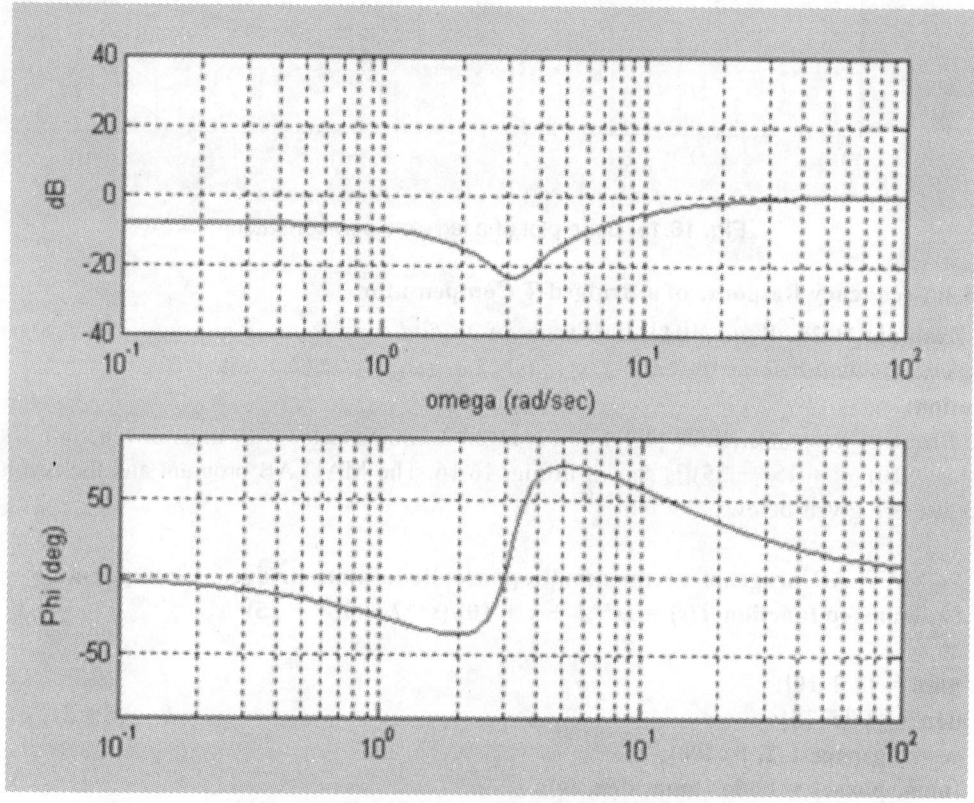

Fig. 16.16. Bode plot of a bridged-T compensator

16.8.5. Frequency Response of Second-order Systems

In this section, we shall write MATLAB programs to find the frequency response of a second-order system for various values of damping constant zeta.

Example 16.21. *Write a MATLAB program to find the frequency response of an un-damped second-order system with transfer function T(s) = 25/(s² + 25).*

Solution:

Let the transfer function be $GH(s) = 25/(s^2 + 25)$. The MATLAB program and the response are given below:

```
% MATLAB program for Frequency Response–Bode plots
% GH(s) = 25/(s**2 + 25)
% Second-order system–un-dampeddelta = 0
%
num = [25];
den = [1 0 25];
w = logspace(–2, 3, 200);
[mag, phase] = bode (num, den, w);
subplot (2, 1, 1)
semilogx (w, 20*log10(mag))
axis ([0.1, 100, –40, 40]);
title ('Fig. 16.17 Bode plot of Second-order un-damped system');
xlabel ('omega (rad/sec)');
ylabel ('dB');
grid;
subplot (2, 1, 2);
semilogx (w, phase)
axis ([0.1, 100, –270, 60]);
ylabel ('Phi (deg)');
grid;
end;
```

Fig. 16.17 shows the Bode plots of a second-order un-damped system.

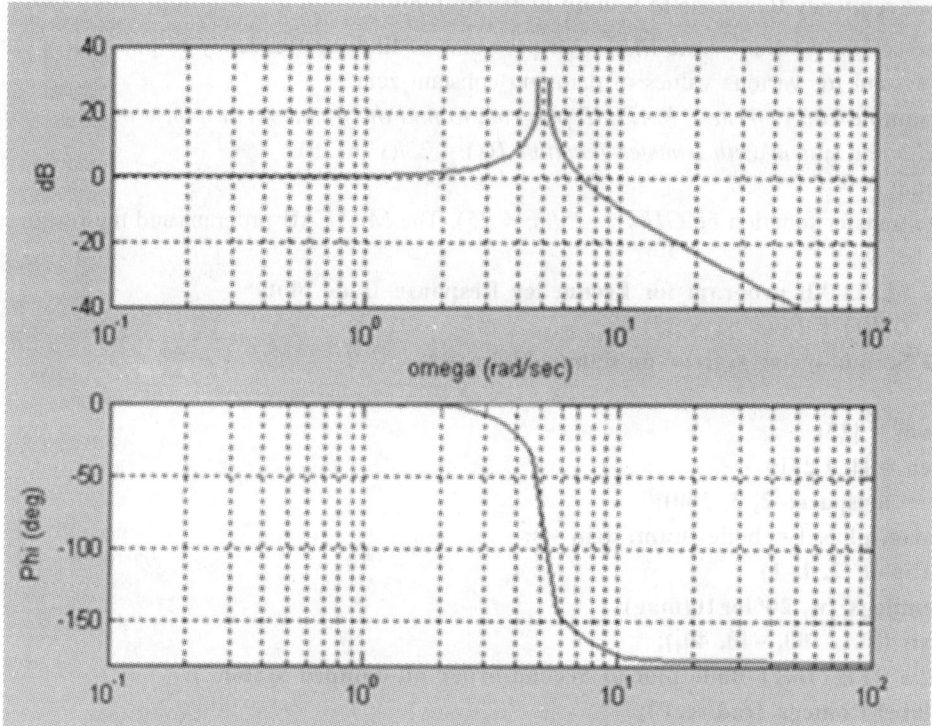

Fig. 16.17. Bode plot of a Second-order undamped system

Example 16.22. *Write a MATLAB program to find the frequency response of an under-damped second-order system with G(s)H(s)* = $25/(s^2 + 3s + 25)$.

Solution:

Let the transfer function be $GH(s) = 25/(s^2 + 3s +25)$ with the value of ζ as 0.3 to represent under-damped case. The result is shown in Fig. 16.18.

```
% MATLAB program for Frequency Response – Bode plots-Second-order system
% G(s)H(s) = 25/(s² + 3s +25)
% ζ = 0.3
%
num = [25];
den = [1 3 25];
w = logspace(–2, 3, 200);
[mag, phase] = bode (num, den, w);
subplot (2, 1, 1)
semilogx (w, 20*log10(mag))
```

```
axis ([0.1, 100, -40, 40]);
title ('Fig. 16.18 Bode plot of Second-order system, ζ = 0.3');
xlabel ('omega (rad/sec)');
ylabel ('dB');
grid;
subplot (2, 1, 2);
semilogx (w, phase)
axis ([0.1, 100, -180, 180]);
xlabel('omega (rad/sec)');
ylabel ('Phi (deg)');
grid;
end;
```

The Bode plot of the under-damped second-order system is shown in Fig. 16.18.

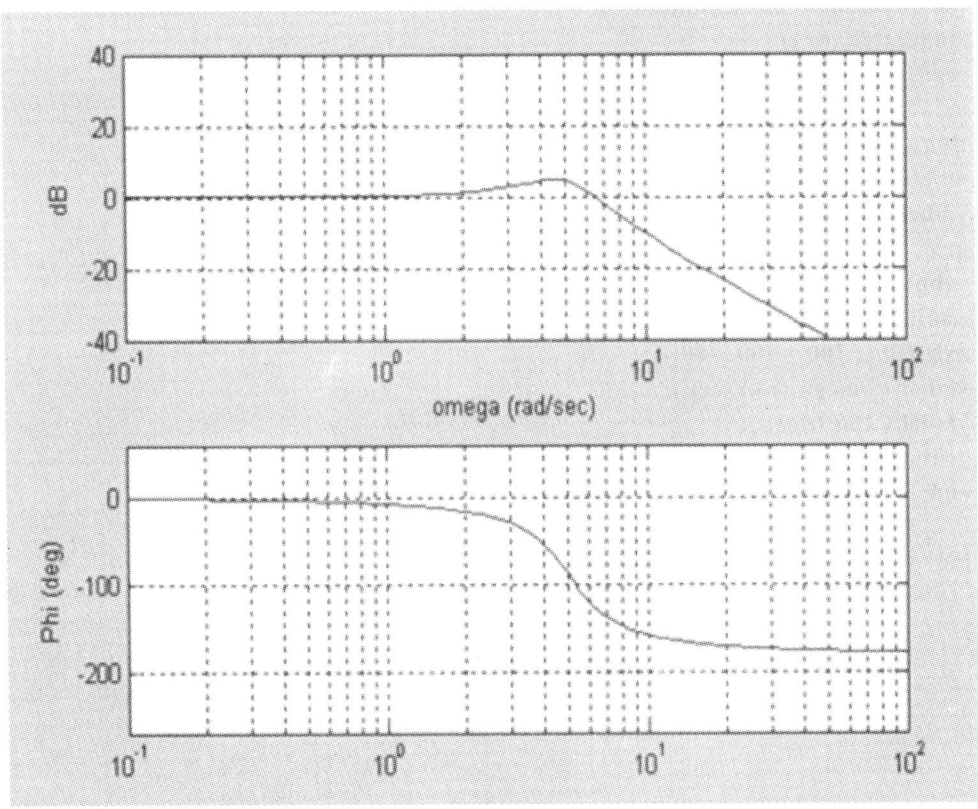

Fig. 16.18. Bode plot of a Second-order under-damped system

Example 16.23. *Write a MATLAB program to find the frequency response of a critically damped second-order system* with $G(s) = 25/(s^2 + 7.07s + 25)$.

Solution:

Let the openloop transfer function be $GH(s) = 25/(s^2 + 7.07s + 25)$ with the value of $\zeta = 0.707$ to represent critically damped case. The result is shown in Fig. 16.19.

```
% Frequency Response–Bode plots-Second-order system
% Transfer function GH(s) = 25/(s**2 + 7.07s +25)
% Critically damped, z = 0.707
%
num = [25];
den = [1 7.07 25];
w = logspace(–2, 3, 200);
[mag, phase] = bode (num, den, w);
subplot (2, 1, 1)
semilogx (w, 20*log10(mag))
axis ([0.01, 1000, –40, 40]);
title ('Fig. 16.19 Bode plot of Second-order system, zeta = 0.707');
xlabel ('omega (rad/sec)');
ylabel ('dB');
grid;
subplot (2, 1, 2);
semilogx (w, phase)
axis ([0.1, 100, –180, –60]);
xlabel ('omega (rad/sec)');
ylabel ('Phi (deg)');
grid;
end;
```

Fig. 16.19 shows the Bode plot of a critically damped second-order system.

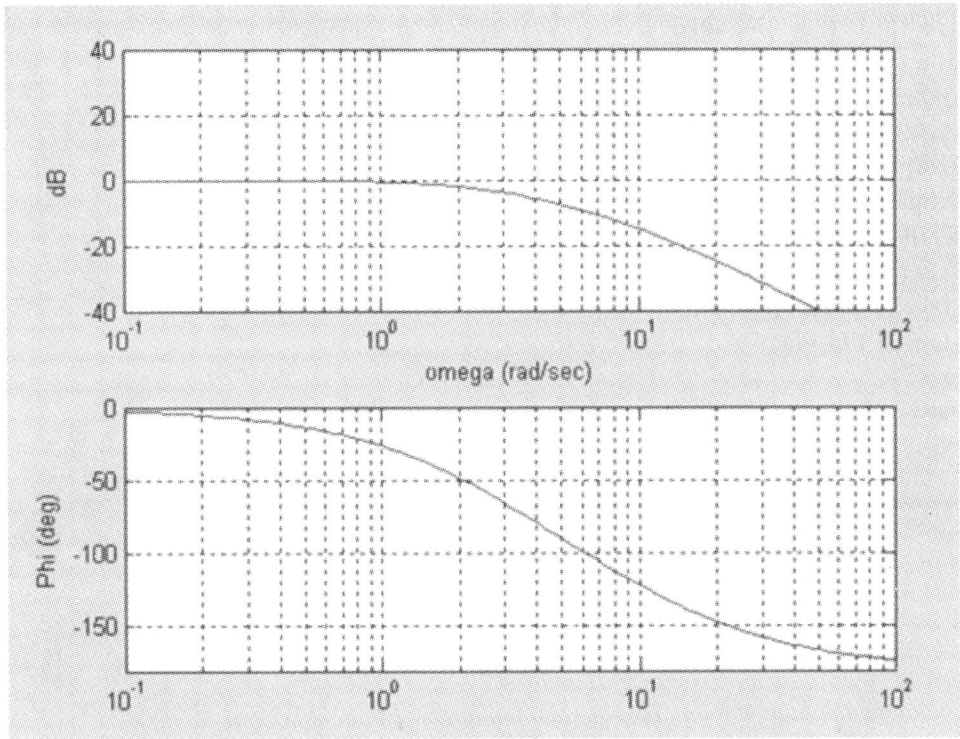

Fig. 16.19. Bode plot of a Second-order critically damped system

Example 16.24. *Write a MATLAB program to find the frequency response of an over-damped second-order system with G(s)H(s) = 25/(s² + 12s + 25).*

Solution:

$GH(s) = 25/(s^2 + 12s + 25)$ with the value of zeta = 1.2 to represent over-damped case. The MATLAB program and the response are given below:

```
% MATLAB program for Frequency Response–Bode plots-Second-order system
% Transfer function G(s)H(s) = 25/(s² + 12s +25)
% Over-damped, delta = 1.2
%
num = [25];
den = [1 12 25];
w = logspace(–2, 3, 200);
[mag, phase] = bode (num, den, w);
subplot (2, 1, 1)
```

```
semilogx (w, 20*log10(mag))
axis ([0.1, 100, –40, 40]);
title ('Fig. 16.20 Bode plot of Second-order system, delta = 1.2');
xlabel ('omega (rad/sec)');
ylabel ('dB');
grid;
subplot (2, 1, 2);
semilogx (w, phase)
axis ([0.1, 100, –180, 0]);
xlabel ('omega (rad/sec)');
ylabel ('Phi (deg)');
grid;
end;
```

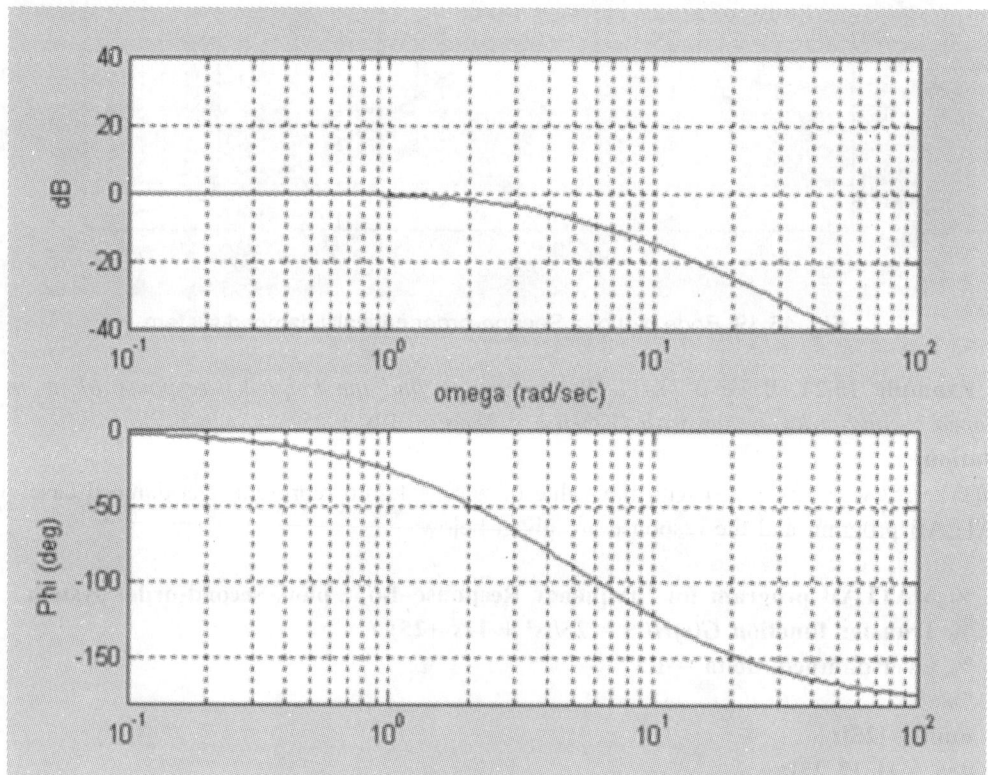

Fig. 16.20. Bode plot of an over-damped second-order system

Fig 16.20 shows the Bode plot of an over-damped second-order system.

16.8.6. Frequency Response of Third-order Systems

In this section, we shall write MATLAB programs to find the frequency response of a third-order system with $GH(s) = (s + 3)/(s**3 + 5s**2 + 8s + 4)$.

Example 16.25. *Write a MATLAB program to find the frequency response of a third-order system with* $G(s)H(s) = (s + 3)/(s**3 + 5s**2 + 8s + 4)$.

Solution:

```
% Frequency Response–Bode plots-Third-order system
% G(s)H(s) = (s + 3)/(s**3 + 5s**2 + 8s + 4)
% Over-damped, delta = 1.2
%
num = [1 3];
den = [1 5 8 4];
w = logspace(–2, 3, 200);
[mag, phase] = bode (num, den, w);
subplot (2, 1, 1)
semilogx (w, 20*log10(mag))
axis ([0.01, 1000, –60, 60]);
title ('Fig. 16.21 Bode plot of third-order system');
xlabel ('omega (rad/sec)');
ylabel ('dB');
grid;
subplot (2, 1, 2);
semilogx (w, phase)
axis ([0.01, 1000, –300, 0]);
xlabel ('omega (rad/sec)');
ylabel ('Phi (deg)');
grid;
end;
```

Fig 16.21 Shows the Bode plot of a third-order system with $GH(s) = (s + 3)/(s**3 + 5s**2 + 8s + 4)$

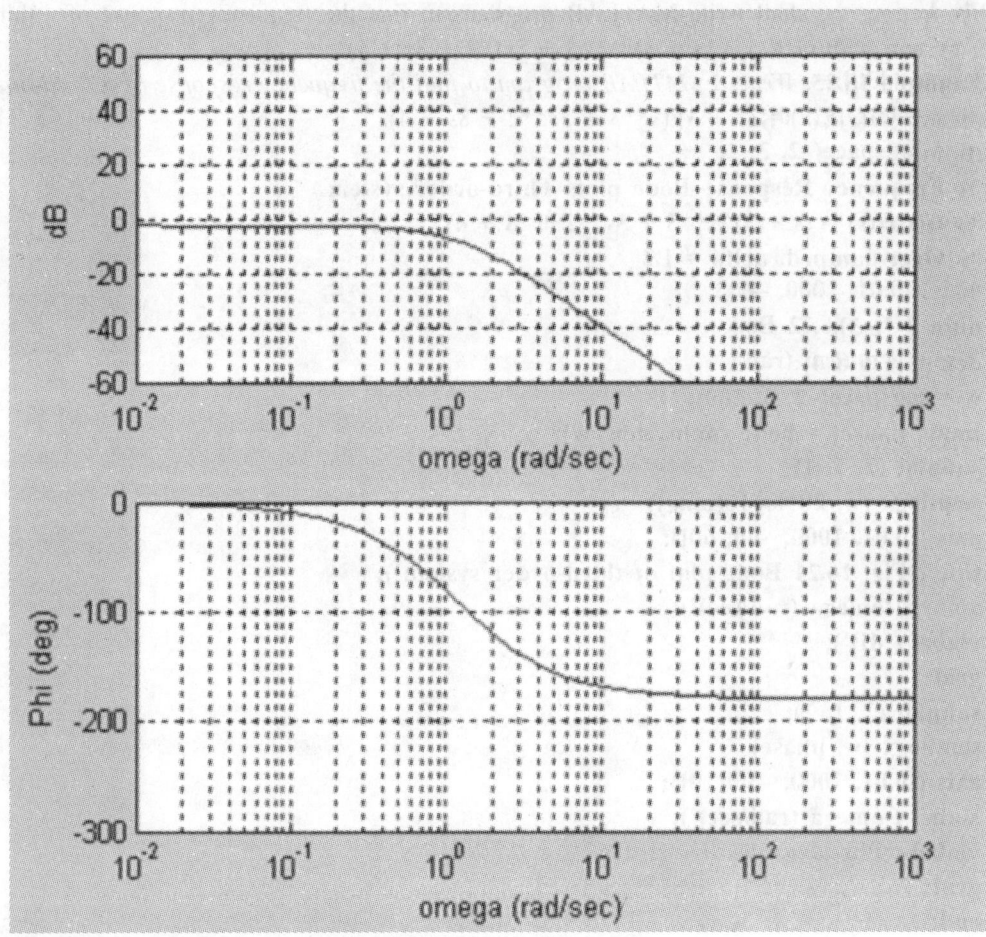

Fig. 16.21. Bode plot of a third-order system

16.9. SYSTEM DESIGN USING COMPENSATORS

In this section, we shall write MATLAB program to synthesize a system with compensators.

16.9.1. Lag compensation

Example 16.26. *Write a MATLAB program for the system with $G_p(s) = K/[s(1 + 0.1s)$ $(1 + 0.2s)]$ to have $K_v = 30$ sec^{-1} and phase margin $\geq 40°$ using lag compensator.*

Solution:

$K = K_v = 30$

% **MATLAB program for Frequency Response–Bode plots**
% **Lag Compensation** $G_p(s) = 30/[s(1 + 0.1s)(1 + 0.2s)]$
%
```
num = [30];
den = [.02 .3 1 0];
w = logspace(-2, 3, 200);
[mag, phase] = bode (num, den, w);
subplot (2, 1, 1)
semilogx (w, 20*log10(mag))
axis ([0.01, 1000, -60, 60]);
title ('Fig. 16.22 Bode plot of uncompensated system');
xlabel ('omega (rad/sec)');
ylabel ('dB');
grid;
subplot (2, 1, 2);
semilogx (w, phase)
```

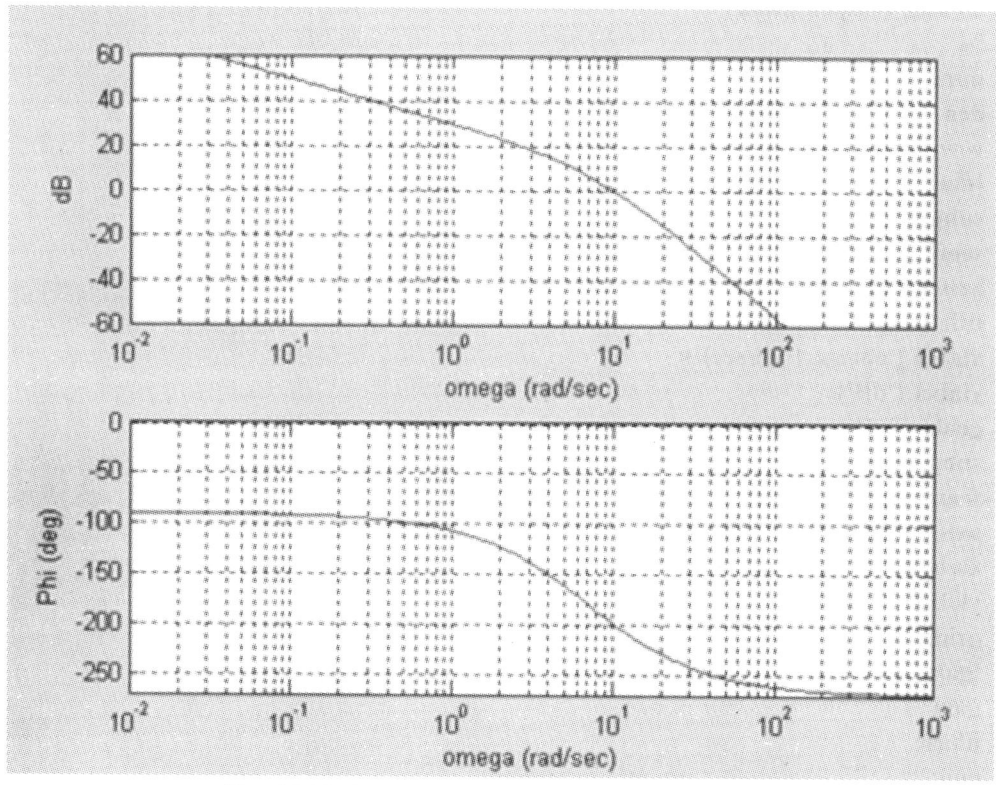

Fig. 16.22. Bode plot of an uncompensated system

```
axis ([0.01, 1000, -270, 0]);
xlabel ('omega (rad/sec)');
ylabel ('Phi (deg)');
grid;
[gain, phase] = margin (num, den);
gain
phase
end;
```

Uncompensated system Phase margin = −17.24° at ω_g = 10 rad/sec.

If the gain crossover frequency is shifted to ω = 3 rad./sec, the phase margin will be about 40°.

Select b = 0.1 and bT = 3.33. Then G_c = (1 + bTs)/(1 + Ts) = (1 + 3.33s)/(1 + 33.3s).

```
% MATLAB programe for Frequency Response–Bode plots
% G(s) = 30(1 + 3.33s)/[s(1 + 0.1s)(1 + 0.2s)(s + 33.3s)]
% Lag compensation
%
num = [100 30];
den = [.67 10.02 33.6 1 0];
w = logspace(-2, 3, 200);
[mag, phase] = bode (num, den, w);
subplot (2, 1, 1)
semilogx (w, 20* log 10(mag))
axis ([0.01, 1000, -75, 60]);
title ('Fig. 16.23 Bode plot of compensated system');
xlabel ('omega (rad/sec)');
ylabel ('dB');
grid;
subplot (2, 1, 2);
semilogx (w, phase)
axis ([0.01, 1000, -270, 0]);
xlabel ('omega (rad/sec)');
ylabel ('Phi (deg)');
grid;
[gain, phase] = margin (num, den);
gain
phase
end;
```

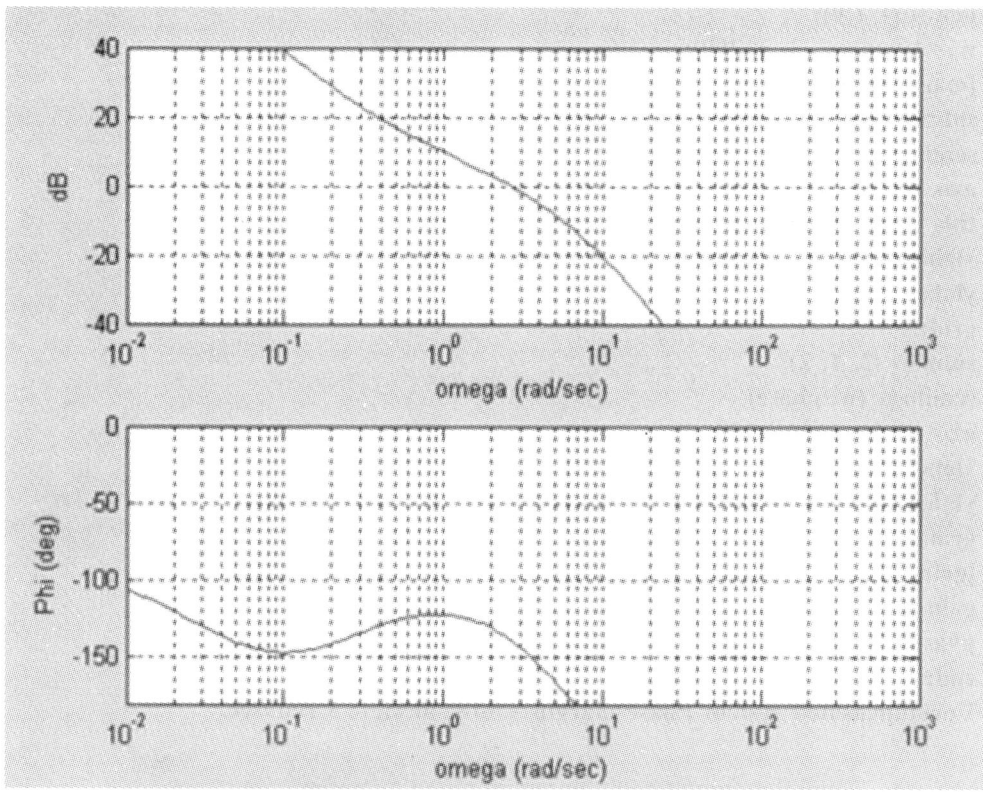

Fig. 16.23. Bode plot of a compensated system

Compensated system Phase margin = + 42°

16.9.2. Lead compensation

Example 16.27. *Write a MATLAB program for the system with* $G_p(s) = K/[s(s + 1)]$ *to satisfy the following specifications using lag compensator:*
 (i) Phase Margin $\geq 45°$
 (ii) $K_v = 10$ sec-1

Solution:
   $K = K_v = 10$
   % **MATLAB program for Frequency Response–Bode plots**
   % **Transfer function** $G_p(s) = 10/[s(s + 1)]$
   % **Lead compensation**
   %
   num = [10];

```
den = [1 1 0];
w = logspace(-2, 3, 200);
[mag, phase] = bode (num, den, w);
subplot (2, 1, 1)
semilogx (w, 20*log10(mag))
axis ([0.01, 1000, -60, 60]);
title ('Fig. 16.24 Bode plot of uncompensated system');
xlabel ('omega (rad/sec)');
ylabel ('dB');
grid;
subplot (2, 1, 2);
semilogx (w, phase)
axis ([0.01, 1000, -270, 0]);
xlabel ('omega (rad/sec)');
ylabel ('Phi (deg)');
grid;
[gain, phase] = margin (num, den);
gain
phase
end;
```

Uncompensated system Phase margin = $-18°$ at $\omega_g = 3$ rad./sec.

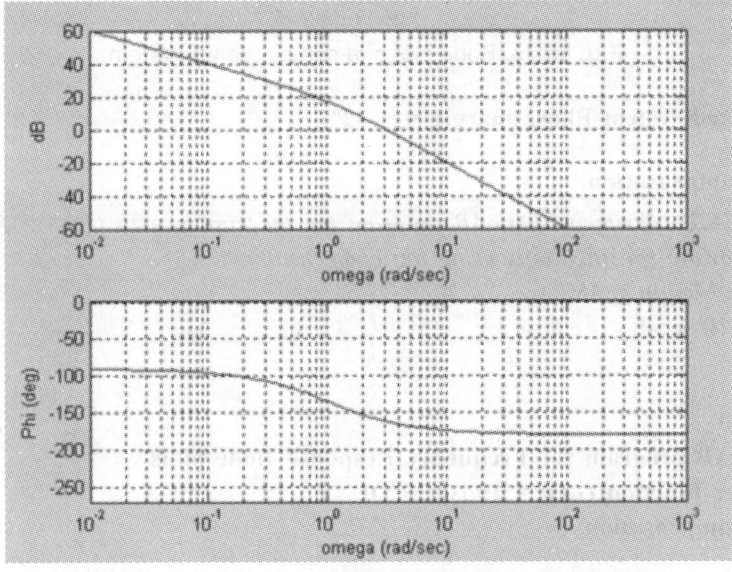

Fig. 16.24. Bode plot of uncompensated system

The transcription of the lead compensator $G_c(s) = (1 + 0.416s)/(1 + 0.139s)$. The MATLAB program and the response of the compensated system are given below:

```
% Frequency Response–Bode plots
% Transfer function G(s) = 10(1 + 0.416s)/[s(s + 1)(1 + 0.139s)]
% Lead compensation
%
num = [4.16 10];
den = [.139 1.139 1 0];
w = logspace(–2, 3, 200);
[mag, phase] = bode (num, den, w);
subplot (2, 1, 1)
semilogx (w, 20*log10(mag))
axis ([0.01, 1000, –60, 60]);
title ('Fig. 16.25 Bode plot of compensated system');
xlabel ('omega (rad/sec)');
ylabel ('dB');
grid;
subplot (2, 1, 2);
semilogx (w, phase)
axis ([0.01, 1000, –300, 0]);
xlabel ('omega (rad/sec)');
ylabel ('Phi (deg)');
grid;
[gain, phase] = margin (num, den);
gain
phase
end;
Compensated system phase margin = + 42°
```

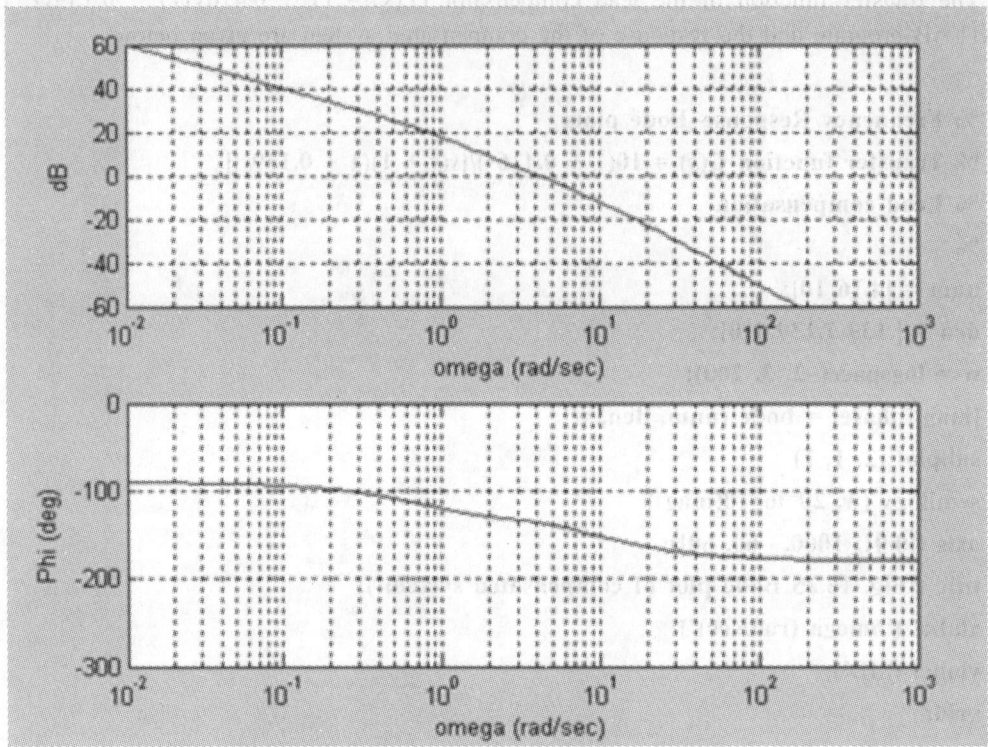

Fig. 16.25. Bode plot of compensated system

16.9.3. Lag-lead compensation

Example 16.28. *Write a MATLAB program for the system with $G_p(s) = K/[s(1 + 0.1s)(1 + 0.2s)]$ to satisfy the following speciications using lag-lead compensator.*

   (i) Phase Margin $\geq 30°$

   (ii) $K_v = 100$ sec-1

Solution:

   $K = K_v = 100$

   %Frequency Response–Bode plots

   %$G_p(s) = 100/[s(1 + 0.1s)(1 + 0.2s)]$

   %

   num = [100];

   den = [.02 .3 1 0];

   w = logspace(–2, 3, 200);

   [mag, phase] = bode (num, den, w);

   subplot (2, 1, 1)

```
semilogx (w, 20*log10(mag))
axis ([0.01, 1000, –60, 60]);
title ('Fig. 16.26 Bode plot of uncompensated system');
xlabel ('omega (rad/sec)');
ylabel ('dB');
grid;
subplot (2, 1, 2);
semilogx (w, phase)
axis ([0.01, 1000, –300, 0]);
xlabel ('omega (rad/sec)');
ylabel ('Phi (deg)');
grid;
[gain, phase] = margin (num, den);
gain
phase
end;
Uncompensated phase margin = –40°
```

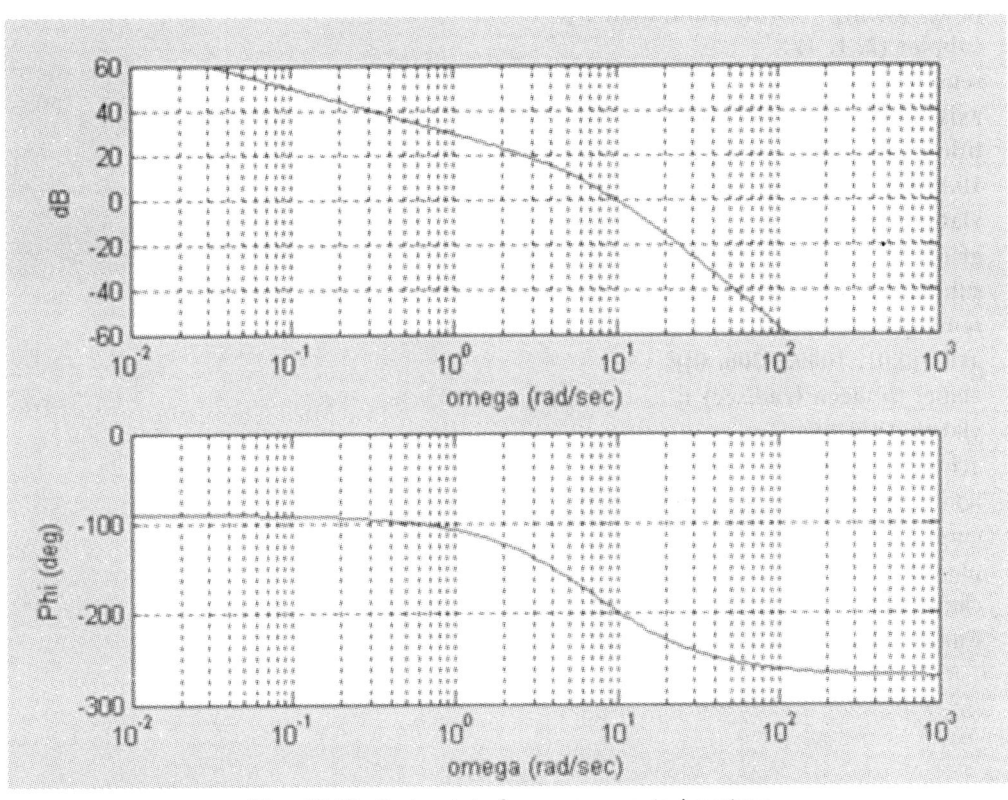

Fig. 16.26. Bode plot of uncompensated system

The transcription of the lag-lead compensator is $G_c(s) = (1 + 0.35s)(1 + 2s)/[(1 + 0.028s)(1 + 25s)]$. The MATLAB program and the response are given below:

```
%  MATLAB program for Frequency Response–Bode plots
%  Transfer function G(s) = 100 (1 + 0.35s)(1 + 2s)/[s(1 + 0.1s)(1 + 0.2s)(1 + 0.028s)
%  (1 + 25s)]
%  Lag-Lead compensation
%
%  Numerator = 70s**2 + 235s + 100
%  Denominator = 0.014s**5 + 0.71s**4 + 8.22s** 3 + 0.328s** 2 + s
num = [70 235 100];
[z1] = roots (num);
z1
den = [.014 .71 8.22 .328 1 0];
[p1] = roots (den);
p1
w = logspace(–2, 3, 200);
[mag, phase] = bode (num, den, w);
subplot (2, 1, 1)
semilogx (w, 20*log 10(mag))
axis ([0.01, 1000, –60, 80]);
title ('Fig. 16.27 Bode plot of compensated system');
xlabel ('omega (rad/sec)');
ylabel ('dB');
grid;
subplot (2, 1, 2);
semilogx (w, phase)
axis ([0.01, 1000, –300, 0]);
xlabel ('omega (rad/sec)');
ylabel ('Phi (deg)');
grid;
[gain, phase] = margin (num, den);
gain
phase
end;
Compensated system Phase margin = 30°
```

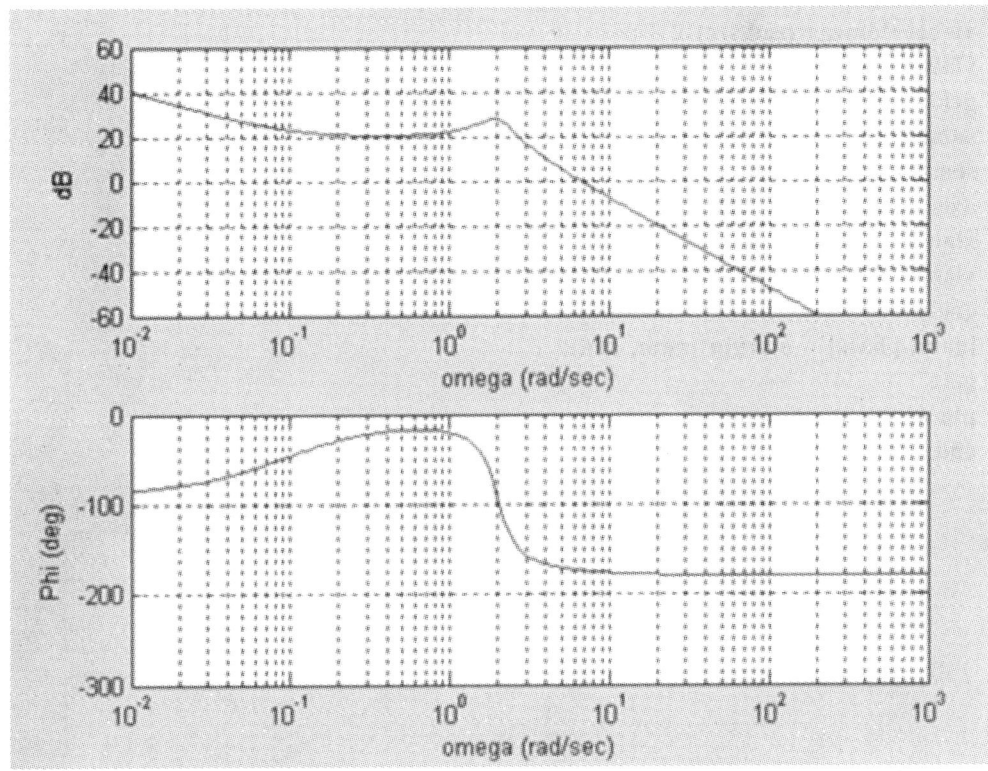

Fig. 16.27. Bode plot of compensated system

16.9.4. Bridge-T compensation

Example 16.29. *Write a MATLAB program for the system with* $G_p(s) = (1 + 10s)/[s(1 + 0.2s + 0.25s^2)]$ *with pole-zero-cancellation type bridged-T compensator.*

Solution:

```
% MATLAB program for Frequency Response–Bode plots
% Transfer function Gp(s) = (1 + 10s)/[s(1 + 0.2s + 0.25s**2)]
%
num = [10 1];
den = [.25 .2 1 0];
w = logspace(–2, 3, 200);
[mag, phase] = bode (num, den, w);
subplot (2, 1, 1)
semilogx (w, 20*log10(mag))
axis ([0.01, 1000, –60, 60]);
```

```
title ('Fig. 16.28 Bode plot of uncompensated system');
xlabel ('omega (rad/sec)');
ylabel ('dB');
grid;
subplot (2, 1, 2);
semilogx (w, phase)
axis ([0.01, 1000, -300, 0]);
xlabel ('omega (rad/sec)');
ylabel ('Phi (deg)');
grid;
[gain, phase] = margin (num, den);
gain
phase
end;
```

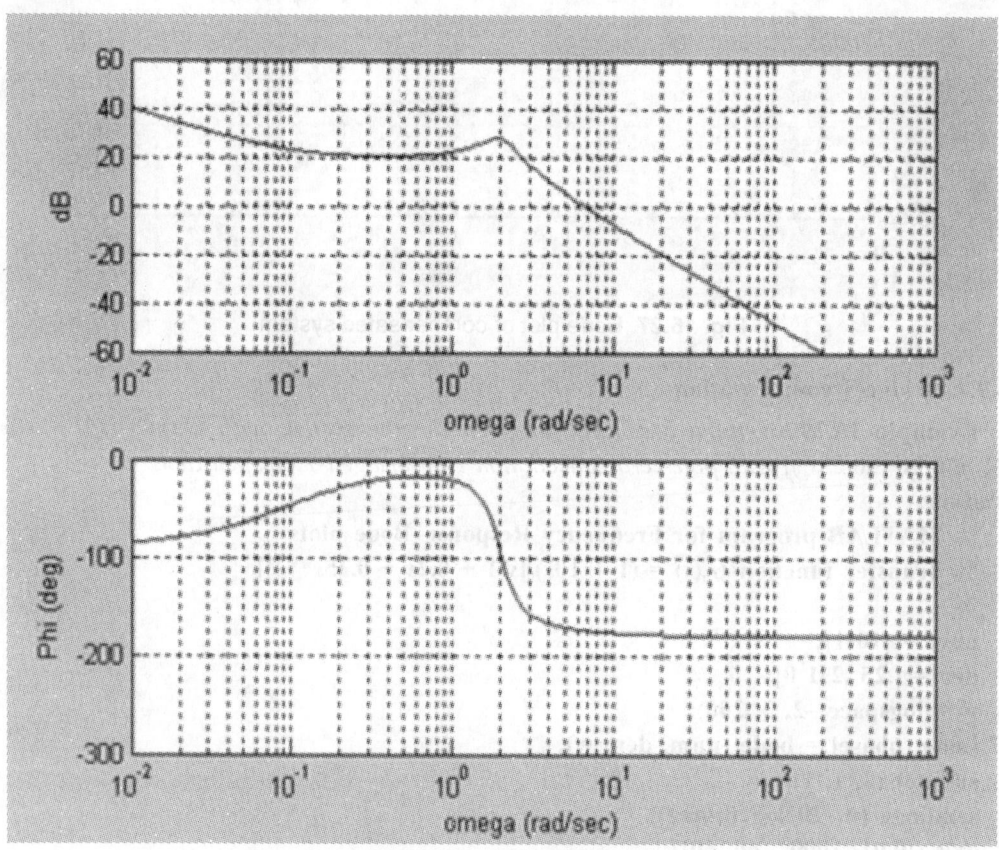

Fig. 16.28. Bode plot of uncompensated system

Uncompensated system: phase margin = 6.7°

The transfer function of the bridged-T compensator for pole-zero cancellation is $G_c(s)$ = $4(1 + 0.2s + 0.25s^2)/(s^2 + 10.8s + 4)$. The MATLAB program and the response are given below:

```
% MATLAB program for Frequency Response–Bode plots
% Transfer function G(s) = (40s + 4)/[s(s + 0.384)(s + 10.42)]
% Bridged-T compensation
%
num = [40 4];
den = [1 10.8 4 0];
w = logspace(–2, 3, 200);
[mag, phase] = bode (num, den, w);
subplot (2, 1, 1)
semilogx (w, 20*log10(mag))
axis ([0.01, 1000, –60, 60]);
title ('Fig. 16.29 Bode plot of compensated system');
xlabel ('omega (rad/sec)');
ylabel ('dB');
grid;
subplot (2, 1, 2);
semi logx (w, phase)
axis ([0.01, 1000, –300, 0]);
xlabel ('omega (rad/sec)');
ylabel ('Phi (deg)');
grid;
[gain, phase] = margin (num, den);
gain
phase
end;
```

Compensated System: Phase margin = 75.4°

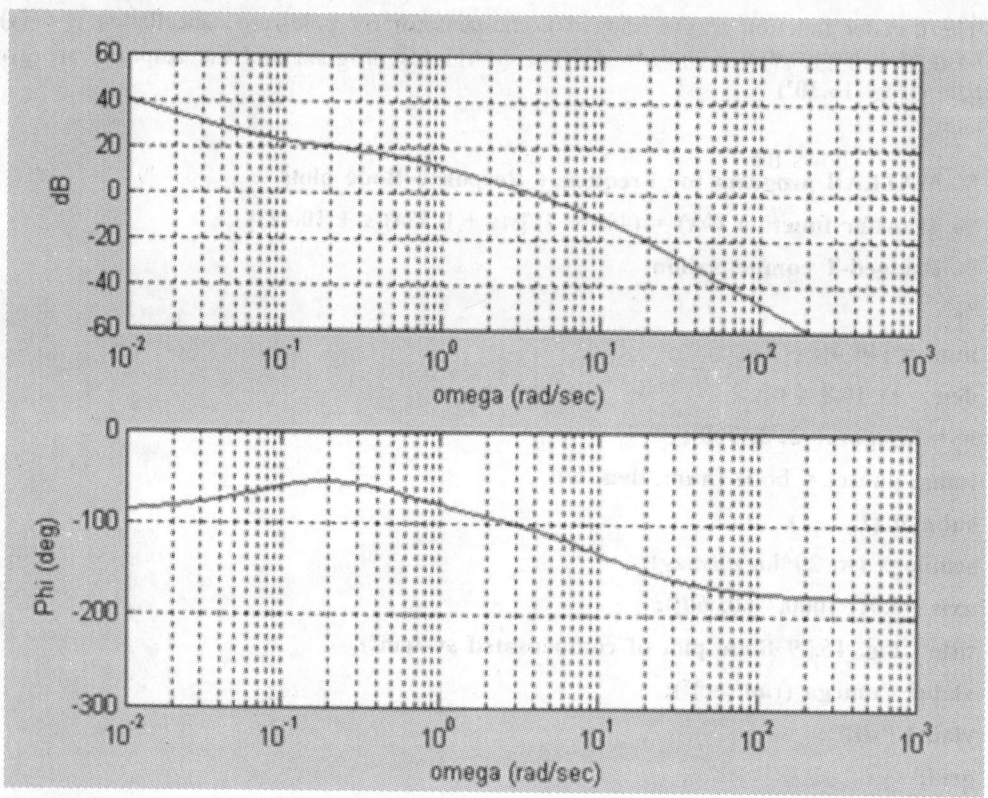

Fig. 16.29. Bode plot of compensated system

16.10. NYQUIST PLOT

In this section, we shall write MATLAB programs to obtain the Nyquist plot for various order of system.

For this purpose, we use the command "nyquist" as described in the following examples.

Example 16.30. *Write a MATLAB program for the system with $G(s)$ $H(s) = 10/(s + 1)$ to obtain the Nyquist plot.*

Solution:

The MATLAB program and the result are shown below:

% MATLAB program for Nyquist plot–Type zero system–First order
% $G(s)H(s) = 10/(s + 1)$
%

num = [10];
den = [1 1];
nyquist (tf(num, den));
title ('Fig. 16.30') Nyquist diagram of $G(s)H(s) = 10/(s + 1)$;
end;
Fig. 16.30 shows the Nyquist diagram of $G(s)H(s) = 10/(s + 1)$

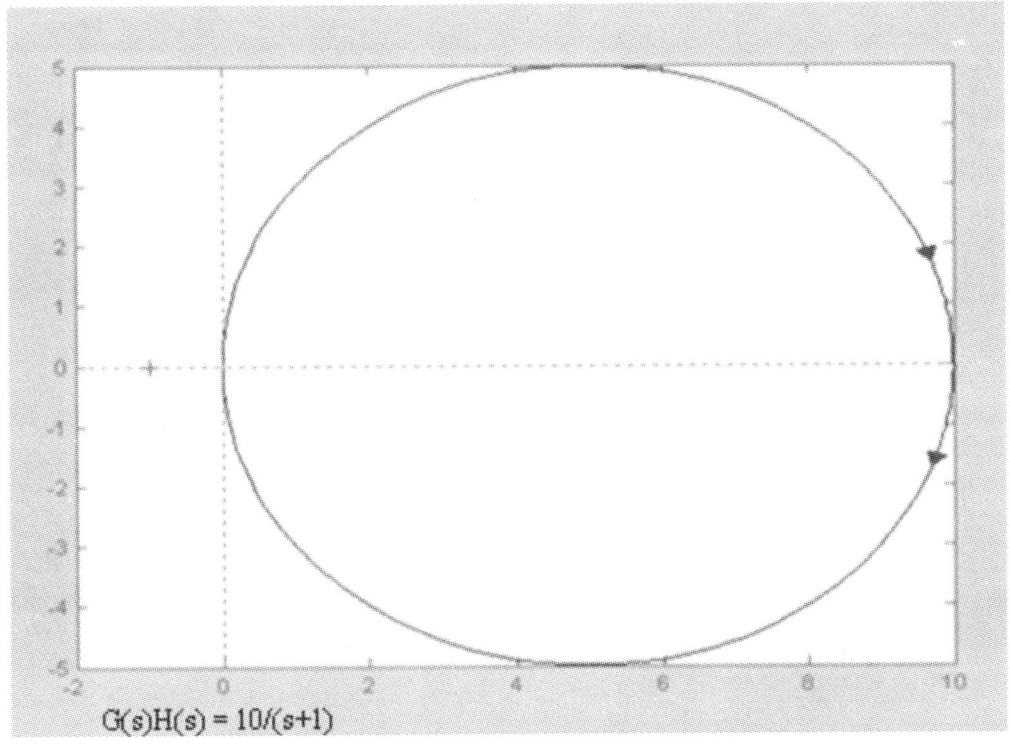

$$G(s)H(s) = 10/(s+1)$$

Fig. 16.30: Nyquist diagram of $GH\ (j\omega) = 10/(1 + j\omega)$

Example 16.31. *Write a MATLAB program for the system with $G(s)H(s) = 10/[(s + 1)$
$(s + 2)]$ to obtain the Nyquist plot.*

Solution:

The MATLAB Program and the result are shown below:

% MATLAB program for Nyquist plot–Type zero system–Second order
% $G(j\omega)H(j\omega) = 10/[(s + 1)(S + 2)]$
%
num = [10];

den = [1 3 2];
nyquist (tf(num, den));
title ('Fig.16.31 nyquist diagram of $G(s)H(s) = 10/[(s + 1)(s + 2)'$];
end;

Fig 16.31 shows the nyquist diagram of $G(s)H(s) = 10/[(s + 1)(s + 2)]$

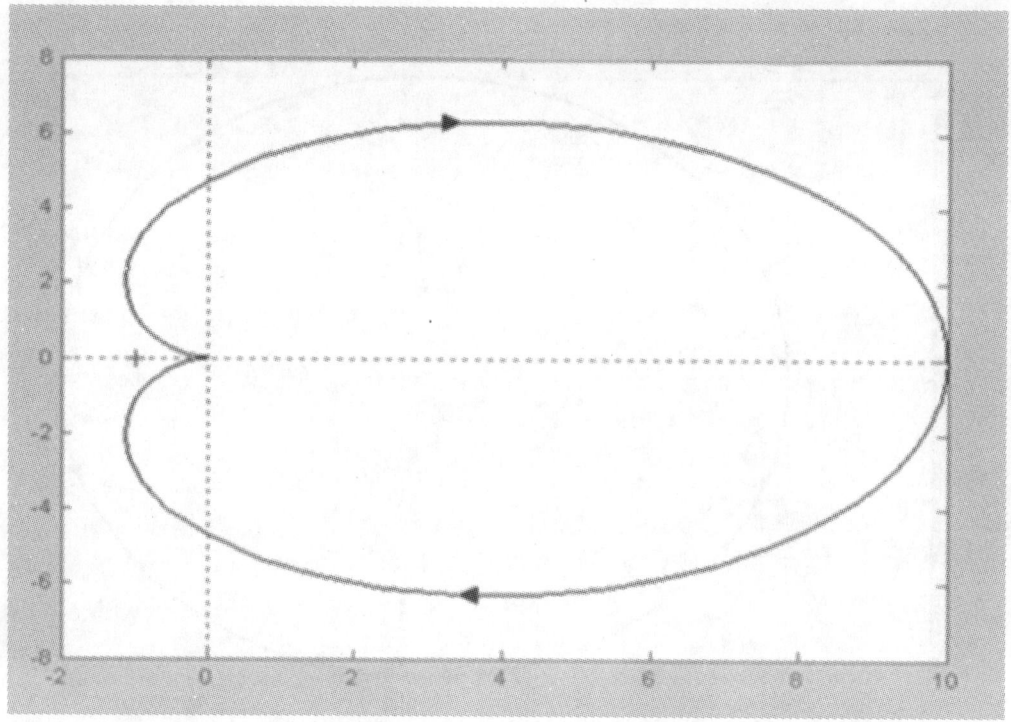

Fig. 16.31: Nyquist diagram of $G(s)H(s) = 10/(s + 1)$

Example 16.32. *Write a MATLAB program for the system with* $G(s)H(s) = 10/[(s + 1)$
$(s + 2)(s + 3)]$ *to obtain the Nyquist plot.*
Solution:
   The MATLAB program and the result are shown below:
   % MATLAB program for nyquist plot–Type zero system–third-order
   % $G(s)H(s) = 10/[(s + 1) (s + 2) (s + 3)]$
   %
   num = [10];
   den = [1 6 11 6];
   nyquist (tf(num, den));

title ('Fig.16.32 Nyquist diagram of *G*(*s*)*H*(*s*) = 10/[(*s* + 1) (*s* + 2) (*s* + 3)];
end;

Fig 16.32 shows the Nyquist diagram of $G(s)H(s) = 10/[(s + 1)(s + 2)(s + 3)]$

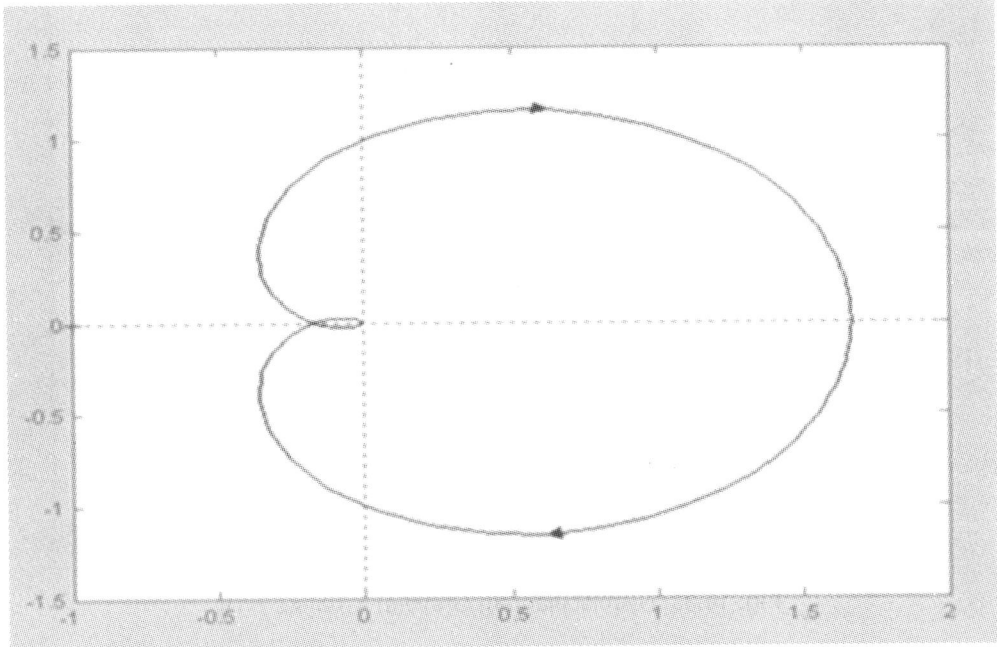

Fig. 16.32. Nyquist diagram of $G(j\omega)H(j\omega) = 10/[(1 + s)(2 + s)(3 + s)]$

Example 16.33. *Write a MATLAB program for the system with* $G(s)H(s) = 10/[s(s + 1)]$ *to obtain the Nyquist plot.*

Solution:

The MATLAB program and the result are shown below:

```
% MATLAB program for nyquist plot–Type one second-order system
% G(s)H(s) =10/[(s)(1 + s)]
%
num = [10];
den = [1 1 0];
nyquist (tf(num, den));
title ('Fig.16.33 Nyquist diagram of G(s)H(s) = 10/[(s)(1 + s)]');
end;
```

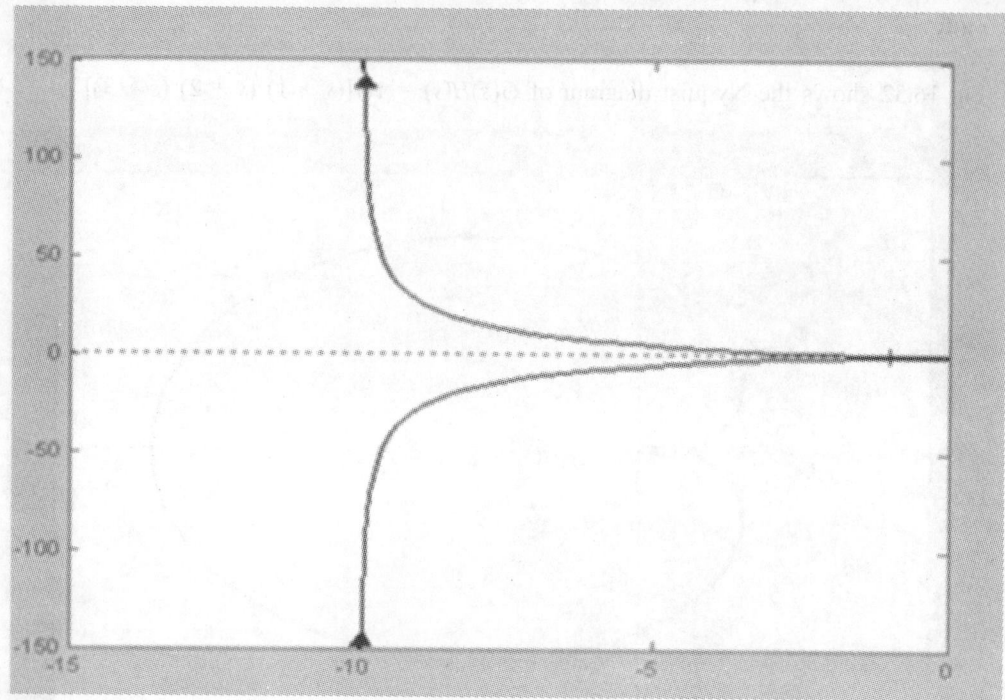

Fig. 16.33: Nyquist diagram $G(s)H(s) = 10/[s(s + 1)]$

Example 16.34. *Write a MATLAB program for the system with* $G(s)H(s) = 10/[s(s + 1)$ $(s + 2]$ *to obtain the Nyquist plot.*

Solution:

The MATLAB program and the result are shown below:

```
% MATLAB program for nyquist plot–Type one third-order system
% Transfer Function G(s)H(s) = 10/[s(1 + s)(2 + s)]
%
num = [10];
den = [1 3 2 0];
nyquist (tf(num, den));
title ('Fig.16.34 Nyquist diagram of G(s)H(s) = 10/[(s)(1 + s)(2 + s)]');
end;
```

Fig. 16.34 Shows the Nyquist plot for Example 16.34

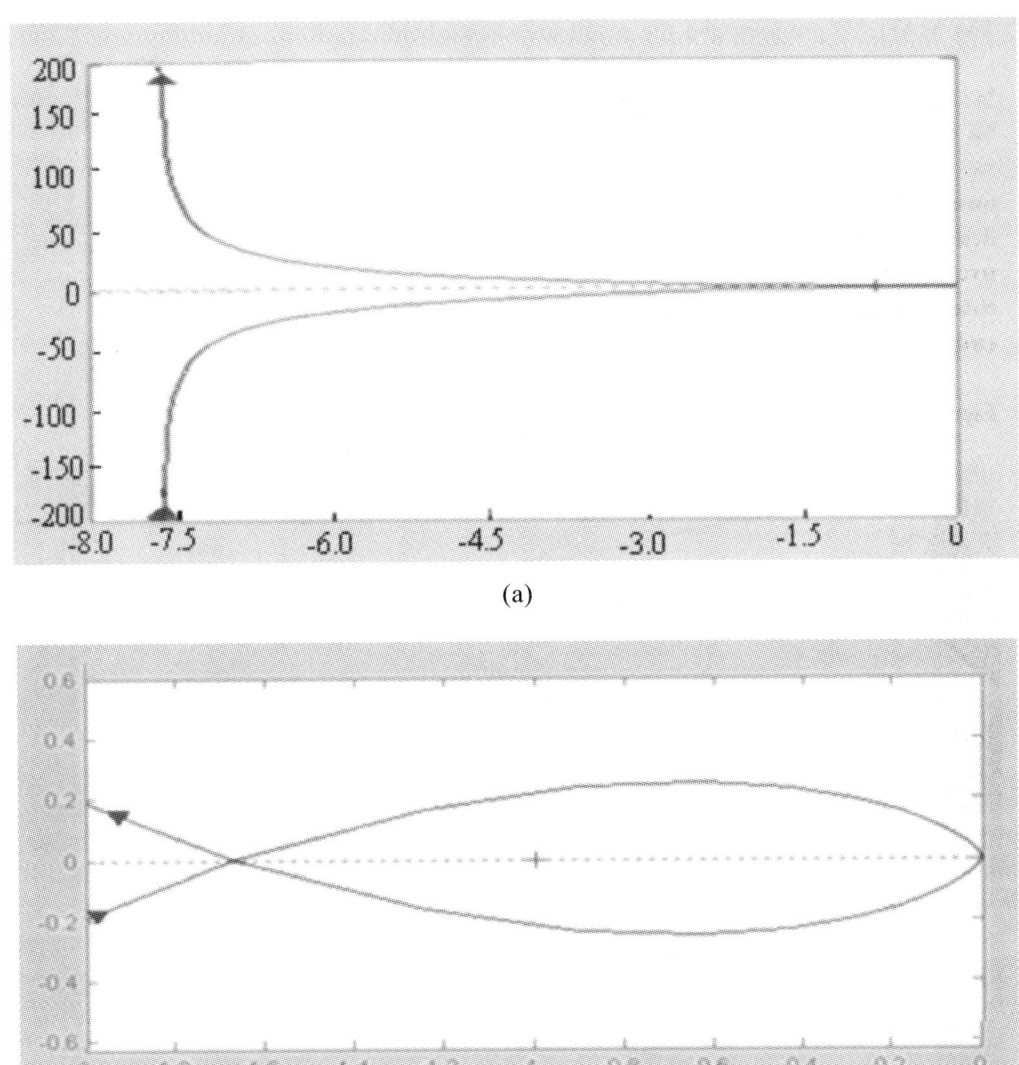

(a)

(b)

Fig. 16.34. (a) Nyquist diagram of $G(s)H(s) = 10/[s(1 + s)(2 + s)]$;
(b) Expanded view at high frequencies.

Example 16.35. *Write a MATLAB program for the system with* $G(s)H(s) = 10/[s^2(s + 1)]$ *to obtain the Nyquist plot.*

Solution:
The MATLAB program and the result are shown below:

% MATLAB program for nyquist plot–Type two third-order system
% G(jω)H(jω) = 10/[(s)²(1 + s)]
%
num = [10];
den = [1 1 0 0];
nyquist (tf(num, den));
title ('Fig.16.35 Nyquist diagram of G(s)H(s) = 10/[s²(1 + s)]');
end;

Fig. 16.35 shows the Nyquist diagram $G(s)H(s) = 10/[s^2(s + 1)]'$);.

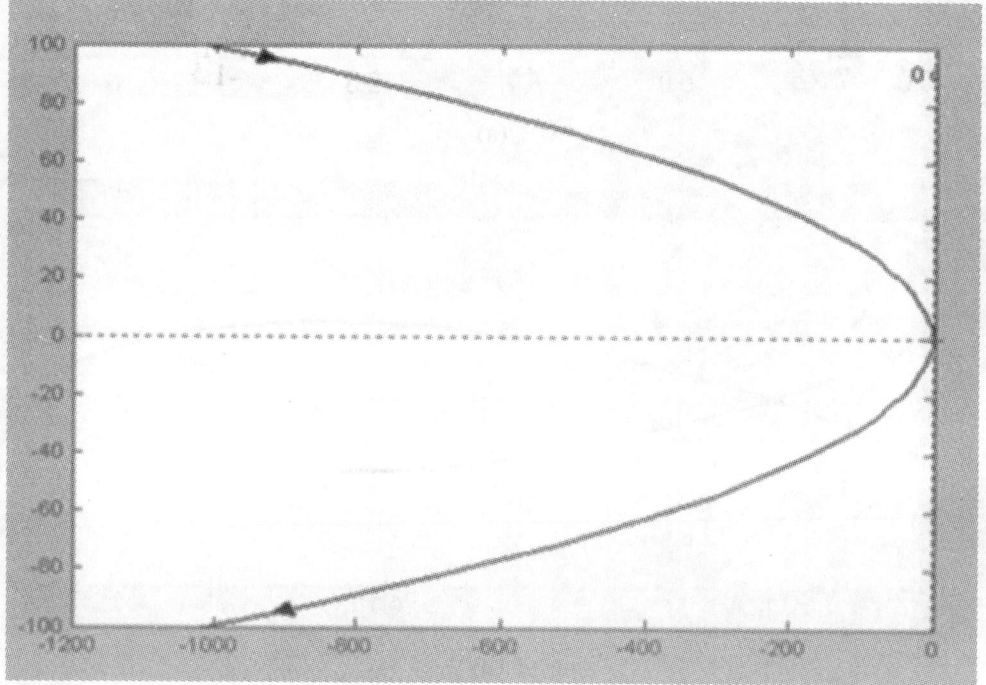

Fig. 16.35. Nyquist diagram of $G(j\omega)H(j\omega) = 10/[s^2(1 + s)]$

Example 16.36. *Write a MATLAB program for the system with* $G(s)H(s) = 100/[s^2(s + 1)(s + 2)]$ *to obtain the Nyquist plot.*
Solution:
The MATLAB program and the result are shown below:

% **MATLAB program for Nyquist plot–Type two fourth-order system**
% $G(j\omega)H(j\omega) = 100/[(s)^2(1 + s)(2 + s)]$
%
num = [100];
den = [1 3 2 0 0];
nyquist (tf(num, den));
title ('Fig.16.36 Nyquist diagram of $G(s)H(s) = 100/[(s)^2(1 + s)(2 + s)]$');
end;

Fig. 16.36 shows the Nyquist plot of $G(s)H(s) = 100/[(s)^2(1 + s)(2 + s)]$');.

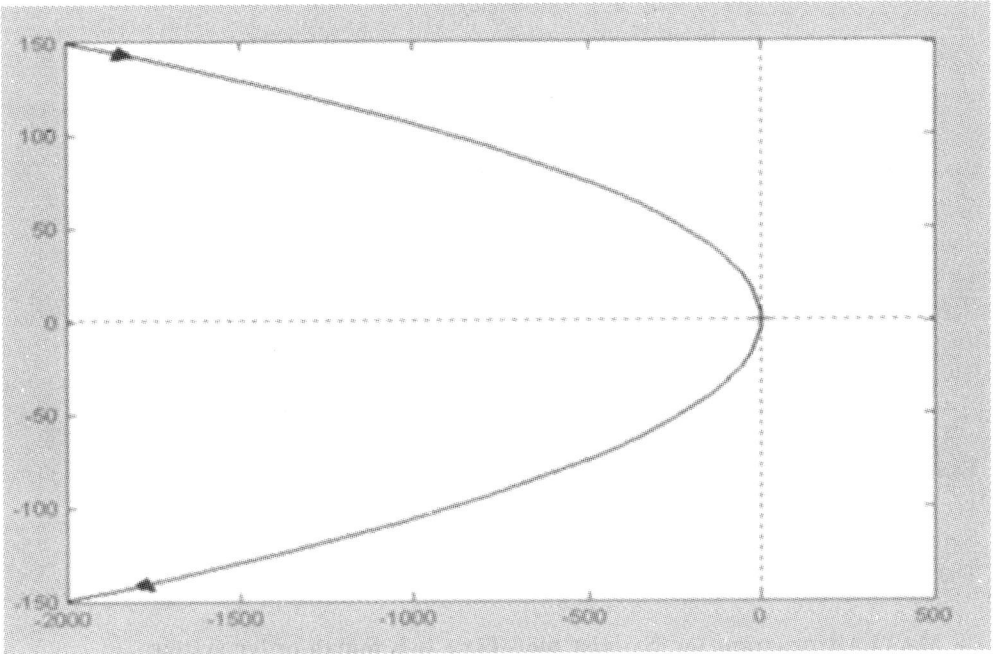

Fig.16.36. (a) Nyquist plot of $100/[(s)^2(1 + s)(2 + s)]$;

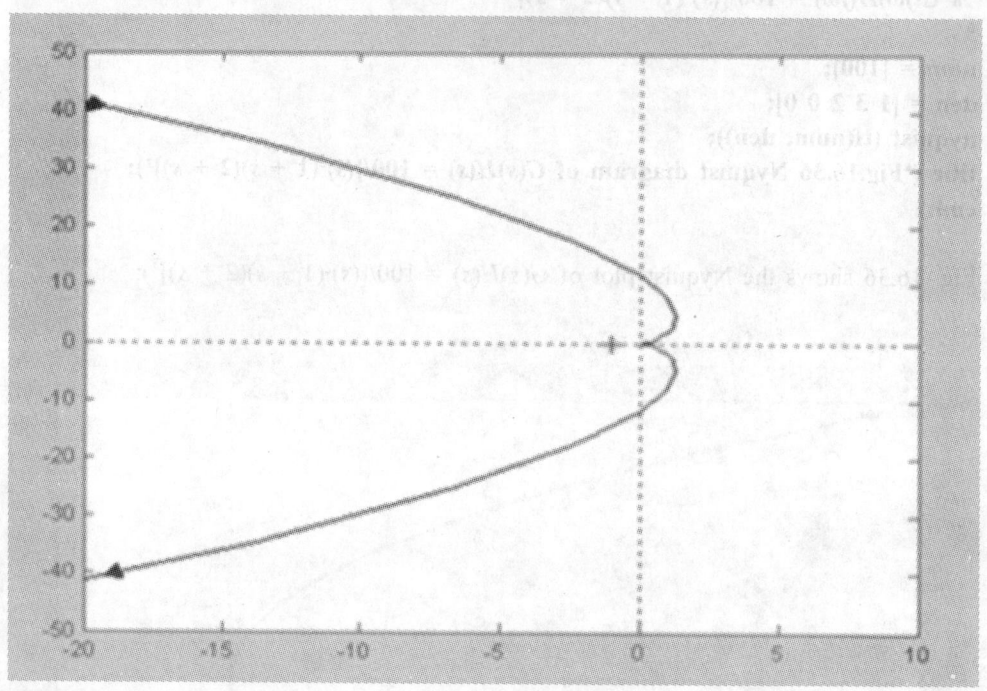

Fig.16.36. (b) Expanded view at high frequencies

Example 16.37. *Write a MATLAB program for the system with* $G(s)H(s) = (4s + 1)/[s^2(s + 1)(2s + 1)]$ *to obtain the Nyquist plot.*
Solution:
The MATLAB program and the result are shown below:

```
% MATLAB program for Nyquist plot–Type two fourth-order system
% G(jω)H(jω) = (1 + 4s)/[(s)²(1 + s)(1 + 2s)]
%
num = [4 1];
den = [2 3 1 0 0];
nyquist (tf(num, den));
title ('Fig.16.37 Nyquist diagram of G(s)H(s) = (1 + 4s)/[(s)²(1 + s)(1 + 2s)]');
end;
```

Fig. 16.37 shows the Nyquist diagram for Example 16.37.

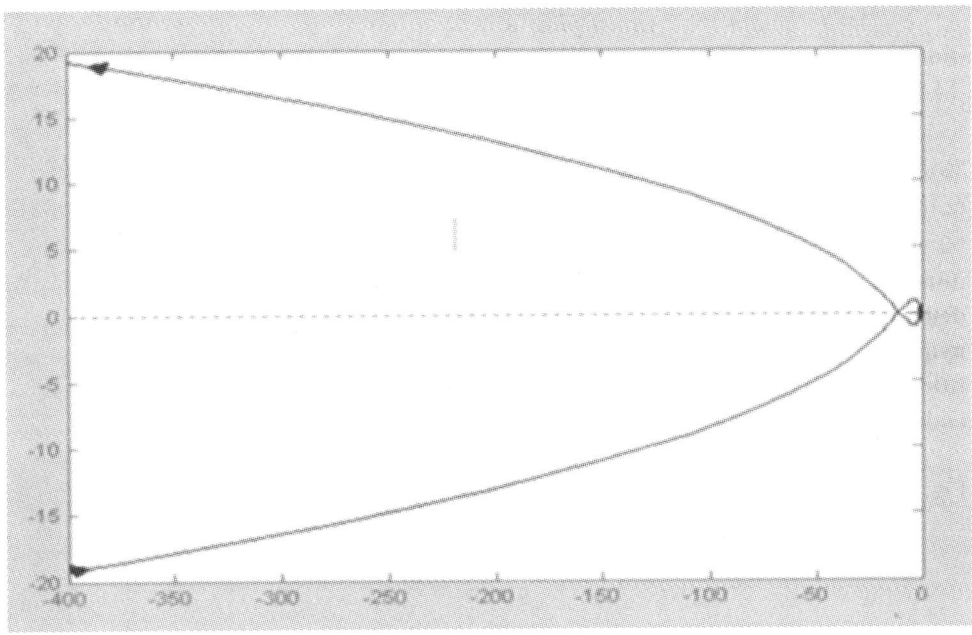

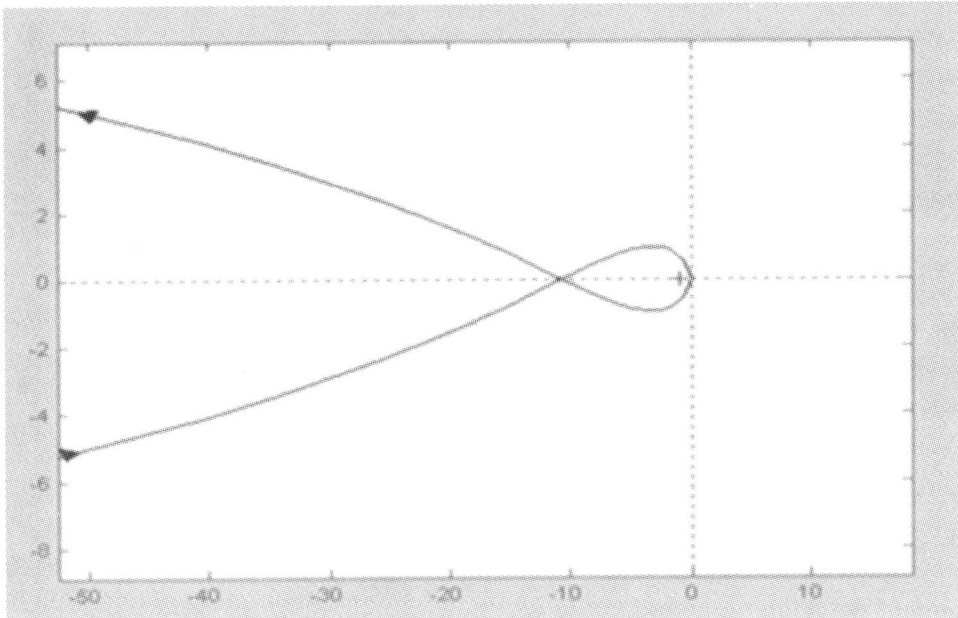

Fig. 16.37. (a) Nyquist diagram of $G(s)H(s) = (1 + 4s)/[(s)^2(1 + s)(1 + s)]$;
(b) Expanded view at high frequencies

Example 16.38. *Write a MATLAB program for the system with* $G(s)H(s) = (s + 2)/[(s + 1)(s - 1)]$ *to obtain the Nyquist plot.*

Solution:

The MATLAB program and the result are shown below:

% MATLAB program for Nyquist plot–Non-minimum phase system

% $G(j\omega)H(j\omega) = (s + 2)/(s + 1)(s - 1)]$

%

num = [1 2];

den = [1 0 –1];

nyquist (tf(num, den));

title ('Fig.16.38 Nyquist diagram of $G(j\omega)H(j\omega) = (s + 2)/(s + 1)(s - 1)$;

end;

Fig. 16.38 shows the Nyquist plot $G(j\omega)H(j\omega) = (s + 2)/(s + 1)(s - 1)$.

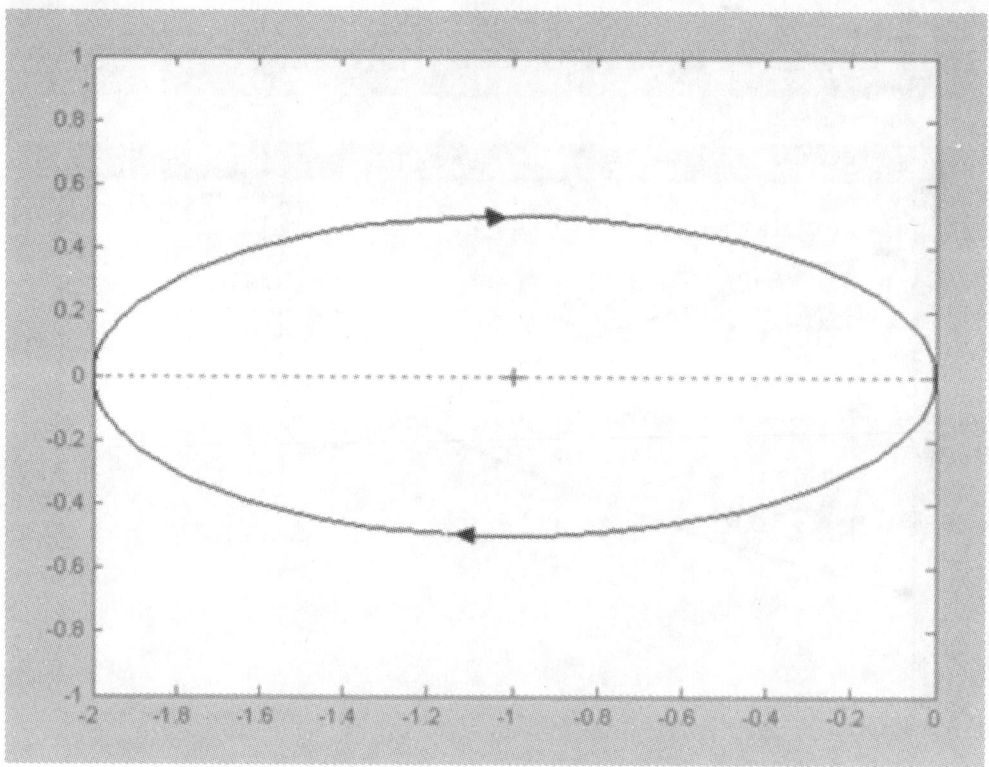

Fig. 16.38. Nyquist plot of $G(j\omega)H(j\omega) = (s + 2)/(s + 1)(s - 1)$.

Example 16.39. *Write a MATLAB program for the system with* $G(s)H(s) = 10^6$ $(0.5 + s)(0.1 + s)/[s^3(40 + s)(80 + s)]$ *to obtain the Nyquist diagram.*

Solution:

The MATLAB Program and the result are shown below:

```
% MATLAB program for nyquist plot–Type three-fifth-order system
% G(jω)H(jω) = 10⁶(0.5 + s)(0.1 + s)/[s³(40 + s)(80 + s)]
%
num = [1000000 600000 50000];
den = [1 120 3200 0 0 0];
nyquist (tf(num, den));
title ('Fig.16.39 Nyquist diagram of G(s)H(s)');
end;
```

Fig. 16.39 Nyquist diagram for Example 16.39.

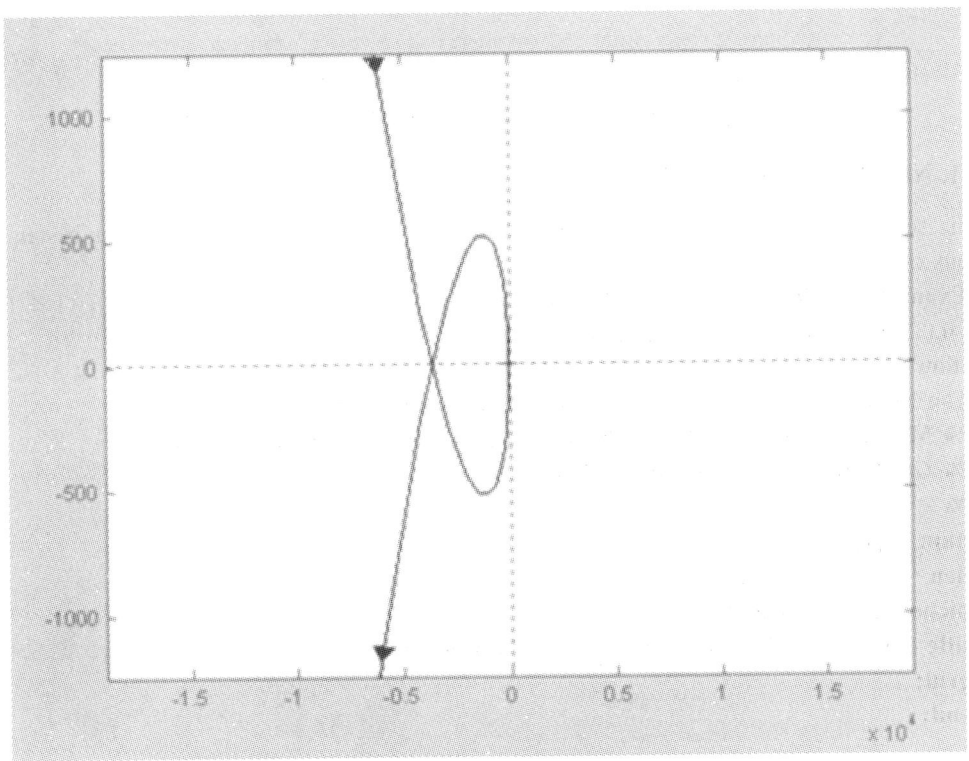

Fig. 16.39. (a) Nyquist diagram for Example 16.39

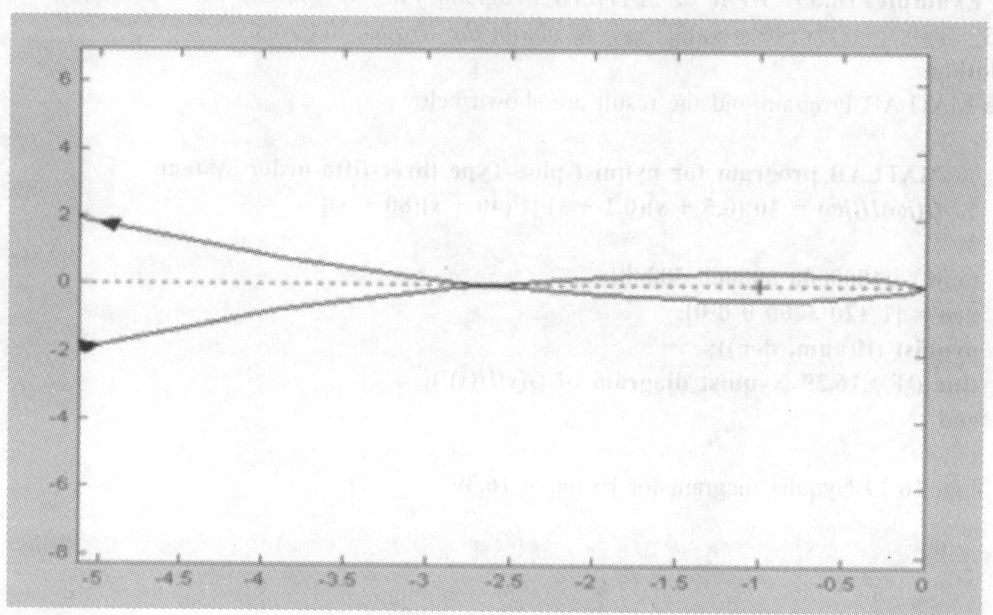

Fig. 16.39. (b) Expanded view at high frequencies

16.11. NICHOLS DIAGRAM

In this section, we shall write MATLAB program to obtain Nichols diagram for systems of various orders.

Example 16.40. *Write a MTLAB program for the system with* $G(s)H(s) = 1/[s(1 + 0.2s)(1 + 0.02s)]$ *to obtain the Nichols diagram.*

Solution:

The MATLAB program and the result are shown below:

```
% MATLAB program for Nichols diagram–Type one third-order system
% G(s)H(s) = 1/[s(1 + 0.2s)(1 + 0.02s)]
%
num = [1];
den = [.004 .22 1 0];
nichols ((num), (den));
title ('Fig. 16.40 Nichols diagram for Example 16.40');
grid;
end;
```

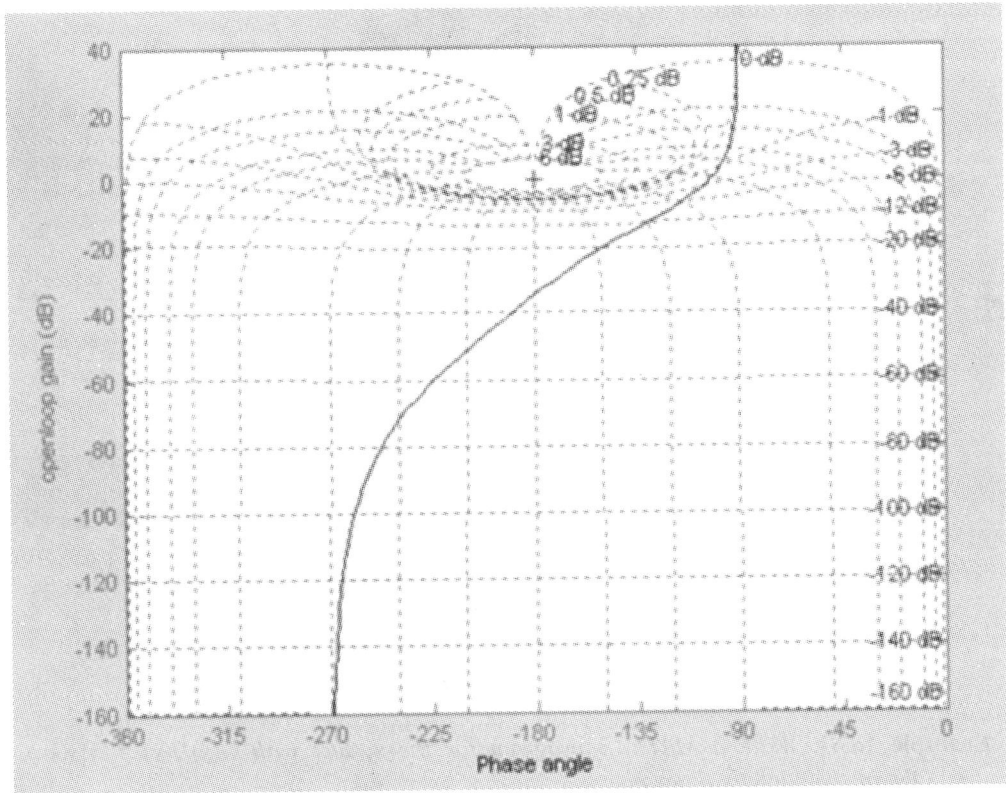

Fig. 16.40. Nichols diagram for $G(s)H(s) = 1/[s(1 + 0.2s)(1 + 0.02s)]$

The Nichols diagram for $K = 3.16$ is shown in Fig. 16.40. The Nichols diagram for $K = 10$ is shown in Fig. 16.41.

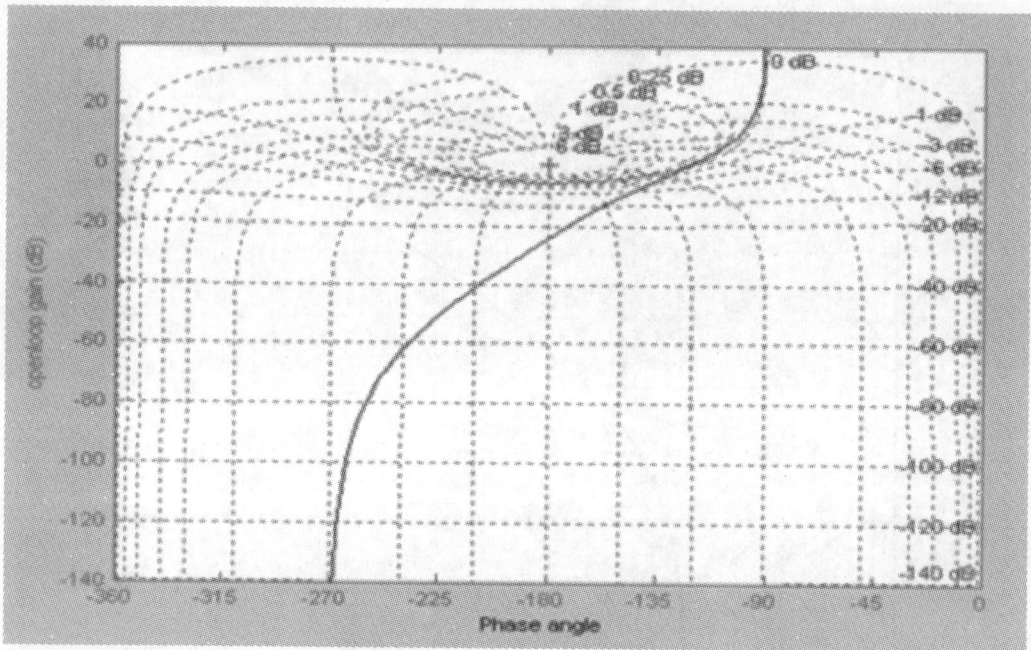

Fig. 16.41. Nichols diagram for Example 16.40 with K = 3.16

Example 16.41. *Write a MATLAB program for the system with* $G(s)H(s) = 1[s(s + 2)$ $(s + 5)]$. *Obtain the Nichols diagram.*

Solution:

The MATLAB program and the result are shown below:

```
% MATLAB program for Nichols diagram–Type-one Third-order system
% G(s)H(s) = 10/[s(s + 2)(s + 5)]
%
num = [10];
den = [1 7 10 0];
nichols ((num), (den));
title ('Fig. 16.42 Nichols diagram for Example 16.41');
grid;
end;
```

Fig.16.42 shows the Nichols diagram $G(s)H(s) = 10[s(s + 2)(s + 3)]$

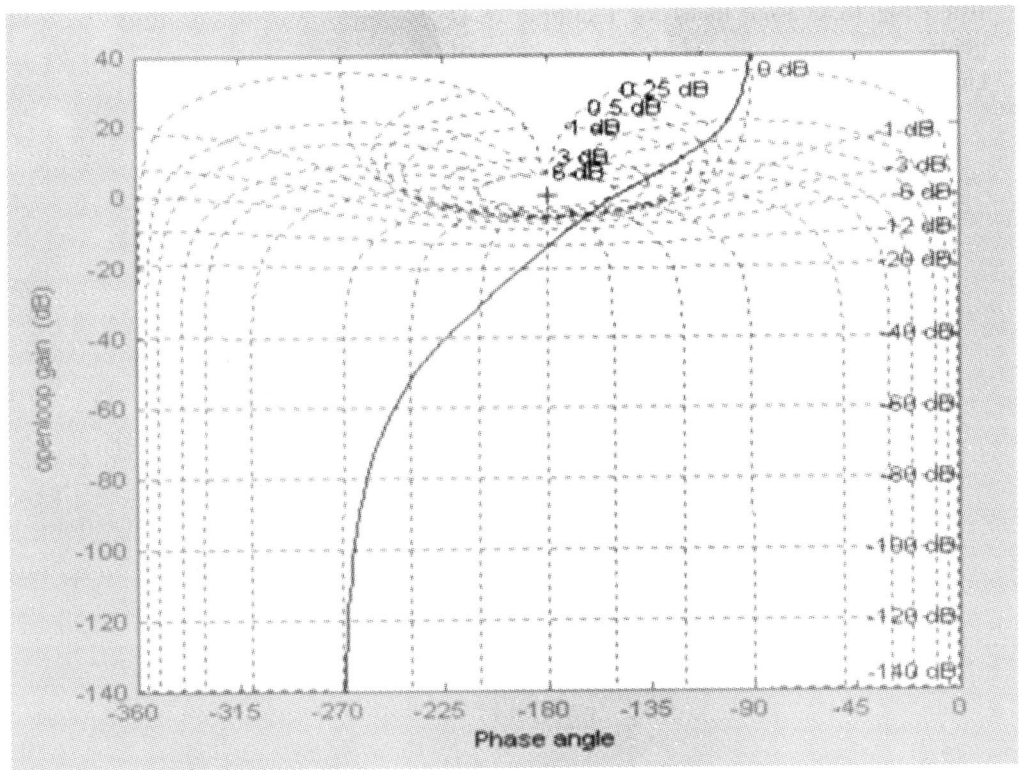

Fig. 16.42. Nichols diagram of $G(s)H(s) = 10/[s(s + 2)(s + 5)]$

16.12. ROOT LOCUS PLOT

In this section, we shall write MATLAB program to obtain the root locus of systems with various $G(s)H(s)$.

Example 16.42. *Write a MATLAB program for a system with* $G(s)H(s) = 10/[s(s + 1)]$ *to determine the root locus.*

Solution:

The MATLAB program and the root locus are shown below:

```
% MATLAB program for Root Locus Plot-Type1-second-order system
% G(s)H(s) = 10/[s(s + 2)]
%
num = [10];
den = [1 2 0];
rlocus (num, den)
```

```
axis ([-2, 0], [-0.8, 0.8]);
title ('Fig 16.43 Root locus for Example 16.42');
grid;
end;
```

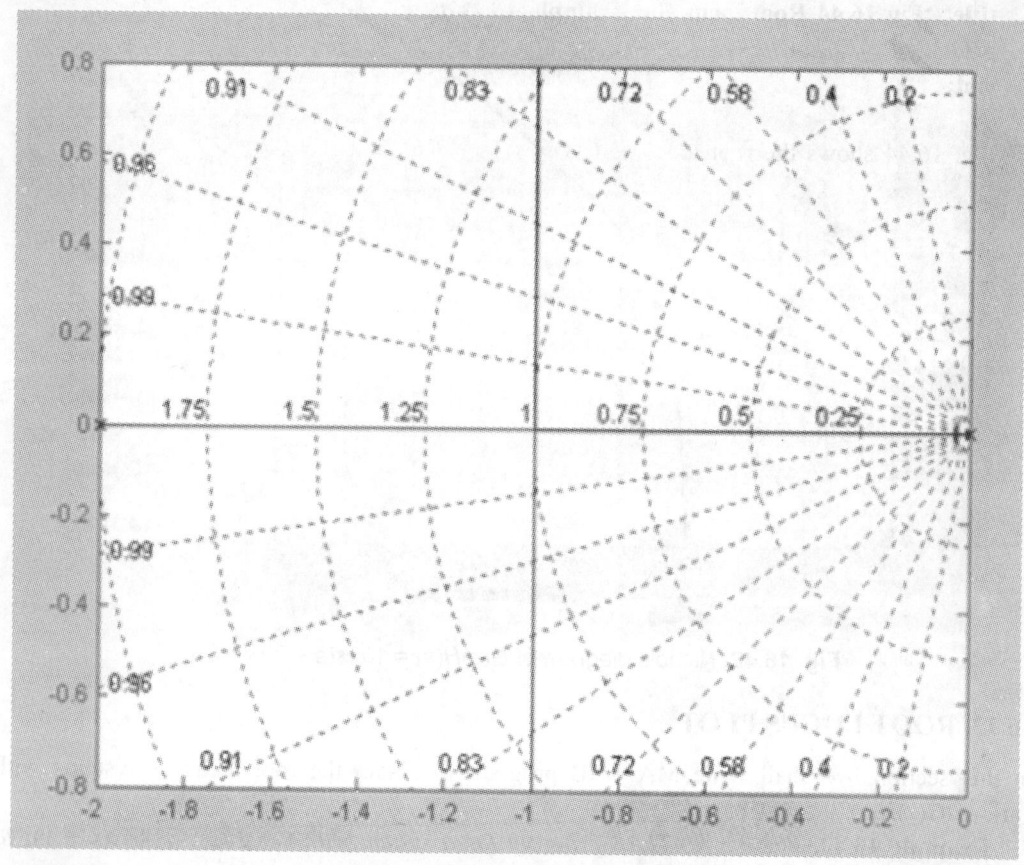

Fig. 16.43. Root locus for Example 16.42

Fig. 16.43 shows the root locus of $G(s)H(s) = 10/[s(s + 1)]$

Example 16.43. *Write a MATLAB program for a system with* $G(s)H(s) = 10/[s(s + 2)$
$(s + 3)]$.

Solution:

The MATLAB program and the root locus are shown below:

% MATLAB program for Root Locus Plot-Type1-Third-order system

```
% GH(s) = 10/[s(s + 2)(s + 3)]%
num = [10];
den = [1 5 6 0];
r locus (num, den)
axis ([-10, 2], [-6, 6]);
title ('Fig 16.44 Root locus for Example 16.43');
grid;
end;
```

Fig. 16.44 shows the root locus of $G(s)H(s) = 10/[s(s + 2)(s + 3)]$.

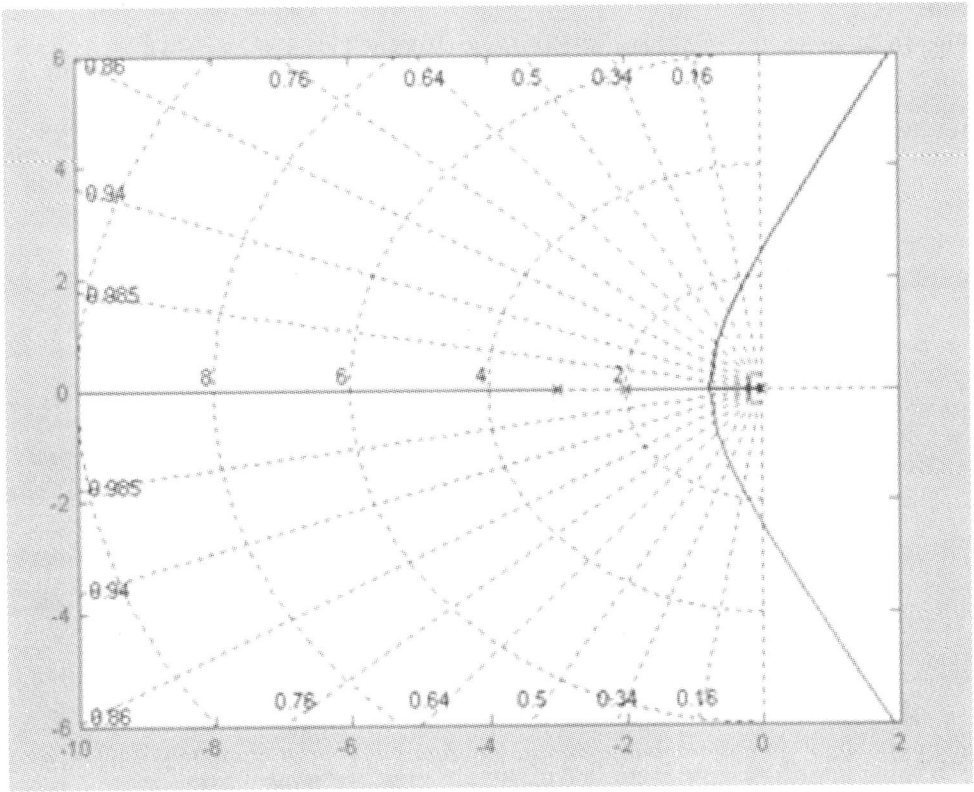

Fig. 16.44. Root locus of Example 16.43

Example 16.44. *Write A MATLAB program for a system with $G(s)H(s) = 10(s + 1)/[s(s + 2)(s + 3)]$.*

Solution:

The MATLAB program and the root locus are given below:

```
% MATLAB program for Root Locus Plot-Type1-Third-order system
% G(s) = 10(s + 1)/[s(s + 2)(s + 3)]
%
num = [10 10];
den = [1 5 6 0];
rlocus (num, den)
axis ([-3, 0], [-8, 8]);
title ('Fig 16.45 Root locus for Example 16.44');
grid;
end;
```

Fig. 16.44 shows the root locus of $G(s)H(s) = 10(s + 1)/[s(s + 2)(s + 3)]$

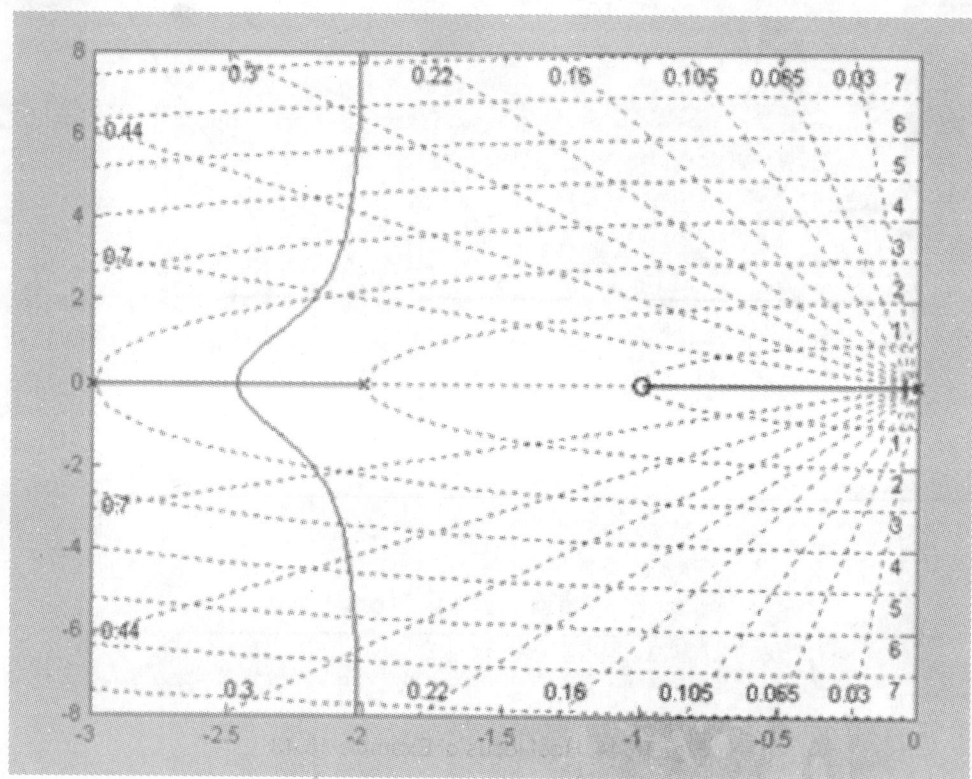

Fig. 16.45. Root locus for Example 16.44

Example 16.45. Write a MATLAB program for a system with $G(s)H(s) = 2600/[s(s^2 + 50s + 2600)]$, $H(s) = 25/(s + 25)$.

Solution:

The MATLAB program and the root locus are given below:

```
% MATLAB program for Root Locus Plot-Type1-Third-order system
% G(s) = 2600 /[s(s**2 + 50s + 2600)], H(s) = 25/(s + 25)
%
num = [65000];
den = [1 75 3850 65000 0];
rlocus (num, den)
axis ([-100, 80], [-100, 100]);
title ('Fig 16.46 Root locus for Example 16.45');
grid;
end;
```

Fig. 16.46 shows the root locus of $G(s)H(s) = 65000/[s^4 + 75s^3 + 3850s^2 + 65000s]$.

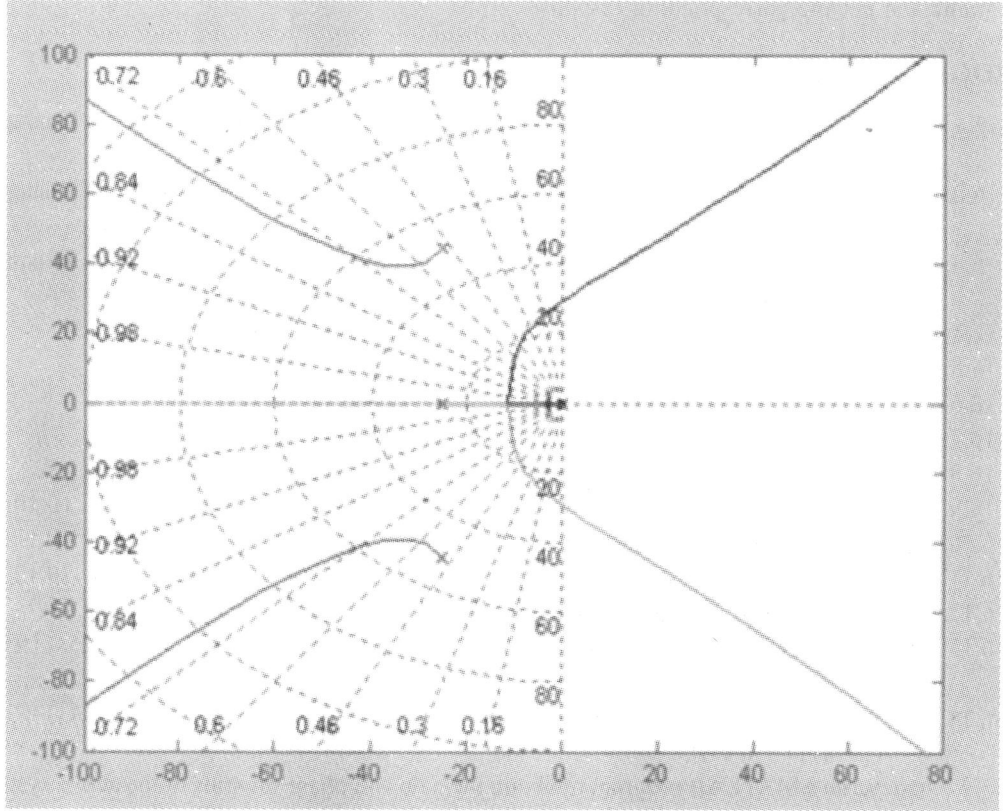

Fig. 16.46. Root locus for Example 16.45.

16.14. SUMMARY

In this chapter, we have first introduced MATLAB and its features followed by definition of system commands and MATLAB commands. We have then developed MATLAB programs to determine (i) the partial fraction expansion for real poles, repeated poles and complex poles. We have then written MATLAB programs to determine impulse and step responses of first-order systems. We have developed MATLAB programs to determine impulse and step responses of un-damped, under-damped, critically damped and over-damped second-order systems. We have also determined the impulse and step responses of a third-order system. We have proceeded to write MATLAB programs to draw Bode plots for lag, lead, lag-lead and bridged-T compensators. We have then determined the frequency response of un-damped, under-damped, critically damped and over-damped second-order systems and also a third-order system. We have designed systems by using Bode plots for given gain and phase margins using compensators. We have also developed MATLAB programs to determine the Nyquist plots, Nichols diagrams and root locus for various systems. All the executed programs are given in detail with explanations wherever required. Hence, it would be easier for the reader to write his/her own programs and get the corresponding response.

PROGRAMMING EXERCISE

16.1. Write a MATLAB program for partial fraction expansion of the problem in Example 2.16.

16.2. Write a MATLAB program for partial fraction expansion of the problem in Example 2.17.

16.3. Write a MATLAB program for partial fraction expansion of the problem in Example 2.18.

16.4. Write a MATLAB program to obtain impulse response of Eq. (6.77) of $K = 2.94$.

16.5. Write a MATLAB program to obtain step response of Eq. (6.77) for $K = 2.94$.

16.6. Write a MATLAB program to obtain impulse response of Eq. (6.81) of $K = 1$.

16.7. Write a MATLAB program to obtain step response of Eq. (6.81) for $K = 2$.

16.8. Write a MATLAB program to obtain step response of the problem in Example 6.5.

16.9. Write a MATLAB program to obtain the root locus of the problem in Example 10.10.

16.10. Write a MATLAB program to obtain the root locus of the problem in Example 10.11.

16.11. Write a MATLAB program to obtain the root locus of the problem in Example 10.12.

16.12. Write a MATLAB program to obtain the Bode plot of the problem in Example 12.9.

16.13. Write a MATLAB program to obtain the Bode plot of $G(s)H(s) = 31.5 (1 + 0.125s)/[s(1 + 2s) (1 + 0.44s)]$.

16.14. (a) Write a MATLAB program to obtain the gain and phase margins of $G(s)H(s) = 3/[s (1 + s) (1 + 0.2s)]$.

(b) Write a MATLAB program to obtain the gain and phase margins of the above system with a lead compensator with $G_c(s) = (1 + s)/(1 + 0.167s)$ and compare.

16.15. (a) Write a MATLAB program to obtain the gain and phase margin of $G(s)H(s) = 3/[s(1 + s)(1 + 0.2s)]$.

(b) Write a MATLAB program to obtain the gain and phase margins of the above system with a lag compensator with $G_c(s) = (1 + 15.4s)/(1 + 61.5s)$ and compare.

16.17. (a) Write a MATLAB program to obtain the gain and phase margins of $G(s)H(s) = 3/[s(1 + s)(1 + 0.2s)]$.

(b) Write a MATLAB program to obtain the gain and phase margins of the above system with a lag-lead compensator with $G_c(s) = (1 + 15.4s)(1 + s)/(1 + 61.5s)(1 + 0.167s)$ and compare.

16.18. Write a MATLAB program to obtain the Nyquist plot of the problem in Example 13.7.

16.19. Write a MATLAB program to obtain the Nyquist plot of the problem in Example 13.8.

16.20. Write a MATLAB program to obtain the Nyquist plot of the problem in Example 13.9.

16.21. Write a MATLAB program to obtain the Nichols plot of the problem in Example 14.1.

16.22. Write a MATLAB program to obtain the Nichols plot of the problem in Example 14.2.

Appendix A

SOME LAPLACE TRANSFORM PAIRS USEFUL FOR CONTROL SYSTEMS ANALYSIS

	$F(s)$	$f(t) \quad t \geq 0$
1.	1	unit impulse $\delta(t)$
2.	$\dfrac{1}{s}$	unit step $u(t)$
3.	$\dfrac{1}{s^2}$	t
4.	$\dfrac{1}{s+a}$	e^{-at}
5.	$\dfrac{1}{(s+a)^2}$	te^{-at}
6.	$\dfrac{\omega}{s^2+\omega^2}$	$\sin \omega t$
7.	$\dfrac{s}{s^2+\omega^2}$	$\cos \omega t$
8.	$\dfrac{n!}{s^{n+1}}$	$t^n \quad (n = 1,2,3, ...)$
9.	$\dfrac{n!}{(s+a)^{n+1}}$	$t^n e^{-et} \quad (n = 1,2,3, ...)$
10.	$\dfrac{1}{(s+a)(s+b)}$	$\dfrac{1}{b-a}(e^{-at} - ae^{-bt})$

11.	$\dfrac{s}{(s+a)(s+b)}$	$\dfrac{1}{b-a}(be^{-bt}-ae^{-at})$
12.	$\dfrac{1}{s(s+a)(s+b)}$	$\dfrac{1}{ab}\left[1+\dfrac{1}{a-b}(be^{-at}-ae^{-bt})\right]$
13.	$\dfrac{\omega}{(s+a)^2+\omega^2}$	$e^{-at}\sin\omega t$
14.	$\dfrac{s+a}{(s+a)^2+\omega^2}$	$e^{-at}\cos\omega t$
15.	$\dfrac{1}{s^2(s+a)}$	$\dfrac{1}{a^2}(at-1+e^{-at})$
16.	$\dfrac{\omega_n^2}{s^2+2\zeta\omega_n S+\omega_n^2}$	$\dfrac{\omega_a}{\sqrt{1-\zeta^2}}e^{-\zeta\omega_n t}\sin\omega_n\sqrt{1-\zeta^2}\,t$
17.	$\dfrac{s}{s^2+2\zeta\omega_n S+\omega_n^2}$	$\dfrac{-1}{\sqrt{1-\zeta^2}}e^{-\zeta\omega_n t}\sin(\omega_n\sqrt{1-\zeta^2}\,t-\phi)$ $\phi=\tan^{-1}\dfrac{\sqrt{1-\zeta^2}}{\zeta}$
18.	$\dfrac{\omega_n^2}{s(s^2+2\zeta\omega_n S+\omega_n^2)}$	$1-\dfrac{-1}{\sqrt{1-\zeta^2}}e^{-\zeta\omega_n t}\sin(\omega_n\sqrt{1-\zeta^2}\,t-\phi)$ $\phi=\tan^{-1}\dfrac{\sqrt{1-\zeta^2}}{\zeta}$

Appendix B

INTRODUCTORY MATRIX ALGEBRA

In Appendix B, we shall define matrices and discuss the elementary matrix operations required for the analysis of linear time-invariant systems.

A. DEFINITIONS

(*i*) *Matrix*: A matrix is an ordered rectangular array of elements. The elements may be real or complex numbers, functions or operations. In general, a rectangular matrix consists of m rows and n columns. It is said to be of *order $m \times n$*. The matrix $\mathbf{A}$ with m rows and n columns is shown in Eq. (B.1).

$$\mathbf{A} = \begin{bmatrix} a_{11} & a_{12} & a_{13} & \cdots & a_{1n} \\ a_{21} & a_{22} & a_{23} & \cdots & a_{2n} \\ \cdot & \cdot & \cdot & \cdots & \cdot \\ \cdot & \cdot & \cdot & \cdots & \cdot \\ a_{m1} & a_{m2} & a_{m3} & \cdots & a_{mn} \end{bmatrix} \tag{B.1}$$

The matrix $\mathbf{A}$ has mn elements. Each element in the matrix is known by its location. Thus, an element in *i-th* row and *j-th* column is denoted by a_{ij}.

(*ii*) *Square matrix*: In Eq. (B.1), if $m = n$, then the matrix has equal number of rows and columns. Such a matrix is known as a square matrix. Its order is n. A 3×3 square matrix is shown in Eq. (B.2).

$$\mathbf{B} = \begin{bmatrix} a_{11} & a_{12} & a_{13} \\ a_{21} & a_{22} & a_{23} \\ a_{31} & a_{32} & a_{33} \end{bmatrix} \tag{B.2}$$

(*iii*) **Column Vector:** In Eq. (B.1), if $n = 1$, then the matrix has m columns and one row. Such a matrix is known as a *column vector*. A 3×1 column vector is shown in Eq. (B.3).

$$\mathbf{C} = \begin{bmatrix} a_{11} \\ a_{21} \\ a_{31} \end{bmatrix} \tag{B.3}$$

(*iv*) **Row vector:** In Eq. (B.1), if $m = 1$, then the matrix has one row and m columns. Such a matrix is known as a **row** vector. A 3×1 row vector is shown in Eq. (B.4).

$$\mathbf{C} = [a_{11} \ a_{12} \ a_{13}] \tag{B.4}$$

(*v*) **Diagonal Matrix:** This is a *square matrix* with elements on the *main diagonal only*. All other elements are zeros. An $n \times n$ diagonal matrix is shown in Eq. (B.5).

$$\mathbf{D} = \begin{bmatrix} a_{11} & 0 & 0 & \cdots & 0 \\ 0 & a_{22} & 0 & \cdots & 0 \\ \cdot & \cdot & \cdot & \cdots & \cdot \\ \cdot & \cdot & \cdot & \cdots & \cdot \\ 0 & 0 & 0 & \cdots & a_{nn} \end{bmatrix} \tag{B.5}$$

Thus, the diagonal matrix $\mathbf{D}$ has a_{ii} ($i = 1$ to n) present and $a_{ij} = 0$ for $i \neq j$.

(*vi*) **Unit** or **Identity Matrix:** A Unit or Identity matrix $\mathbf{I}$ is a diagonal matrix with all elements on the main diagonal being unity. All other elements are zeros. An $n \times n$ unit matrix is shown in Eq. (B.6).

$$\mathbf{I} = \begin{bmatrix} 1 & 0 & 0 & \cdots & 0 \\ 0 & 1 & 0 & \cdots & 0 \\ \cdot & \cdot & 1 & \cdots & \cdot \\ \cdot & \cdot & \cdot & \cdots & \cdot \\ 0 & 0 & 0 & \cdots & 1 \end{bmatrix} \tag{B.6}$$

Thus, the identity matrix $\mathbf{I}$ has a_{ii} ($i = 1$ to n) equal to 1 and $a_{ij} = 0$ for $i \neq j$.

(*vii*) **Transpose of a matrix:** The transpose of a matrix $\mathbf{A}$ is a matrix formed by interchanging the rows and columns of the matrix. Let the matrix $\mathbf{A}$ is given by

$$\mathbf{A} = \begin{bmatrix} 1 & -2 & 3 \\ 4 & 6 & -2 \\ 8 & 4 & 1 \end{bmatrix} \tag{B.7}$$

Its transpose denoted by $\mathbf{A}^T$ is given by

$$\mathbf{A}^T = \begin{bmatrix} 1 & 4 & 8 \\ -2 & 6 & 4 \\ 3 & -2 & 1 \end{bmatrix} \tag{B.8}$$

The transpose of the transpose of a matrix is the matrix itself. Thus,

$$(\mathbf{A}^T)^T = \mathbf{A} \tag{B.9}$$

(*viii*) *Symmetric Matrix*: The symmetric matrix is a square matrix if its transpose ie equal to the matrix itself. Thus, a symmetric matrix $\mathbf{S}$ is characterized by

$$\mathbf{S}^T = \mathbf{S} \tag{B.10}$$

$$\mathbf{S} = \begin{bmatrix} 1 & 2 & 3 \\ 2 & 3 & 4 \\ 3 & 4 & 5 \end{bmatrix} \tag{B.11}$$

A symmetric matrix $\mathbf{S}$ is shown in Eq. (B.10).

(*ix*) *Determinant of a matrix*: The determinant exists *only for square matrix*. The method of determining the determinant of a matrix is explained below:

Let the matrix $\mathbf{A}$ is given by

$$\mathbf{A} = \begin{bmatrix} 3 & -2 & 1 \\ -2 & 6 & 4 \\ 1 & 4 & 8 \end{bmatrix} \tag{B.12}$$

The determinant of $\mathbf{A}$ is given by

$$\text{Det}(\mathbf{A}) = |\mathbf{A}| = 3 \begin{vmatrix} 6 & 4 \\ 4 & 8 \end{vmatrix} - (-2) \begin{vmatrix} -2 & 4 \\ 1 & 8 \end{vmatrix} + 1 \begin{vmatrix} -2 & 6 \\ 1 & 4 \end{vmatrix} \qquad - (-2)$$

$$= 3 \times (48 - 16) + 2 \, (-16 - 4 \,) + 1 \times (-8 - 6)$$

$$= 96 - 40 - 14 = \mathbf{42}$$

(*x*) *Minor*: The minor M_{ij} of an $n \times n$ matrix is the determinant of $(n - 1) \times (n - 1)$ matrix formed by deleting the *i-th* row and *j-th* column of the original matrix.

$$M_{11} = \begin{bmatrix} a_{22} & a_{23} & \cdot & \cdot & \cdot & a_{2n} \\ \cdot & \cdot & \cdot & \cdot & \cdot & \cdot \\ a_{m2} & a_{m3} & \cdot & \cdot & \cdot & a_{mn} \end{bmatrix} \tag{B.13}$$

(*xi*) *Cofactor*: The cofactor A_{ij} of element a_{ij} of the matrix $\mathbf{A}$ is defined as given below:

$$A_{ij} = (-1)^{(i+j)} M_{ij} \tag{B.14}$$

(xii) Adjoint matrix: The adjoint matrix of a square matrix **A** is determined by replacing each element a_{ij} of matrix **A** by its cofactor A_{ij} and then transposing. If the matrix **A** is as defined in Eq. (B.12), then the adjoint matrix is given by

$$Adj\ A = \begin{bmatrix} +A_{11} & -A_{12} & +A_{13} \\ -A_{21} & +A_{22} & -A_{23} \\ +A_{31} & -A_{23} & +A_{33} \end{bmatrix}^T = \begin{bmatrix} +A_{11} & -A_{21} & +A_{31} \\ -A_{12} & +A_{22} & -A_{32} \\ +A_{13} & -A_{23} & +A_{33} \end{bmatrix} \tag{B.15}$$

where

$$A_{11} = \begin{vmatrix} 6 & 4 \\ 4 & 8 \end{vmatrix} = 32 \quad A_{12} = \begin{vmatrix} -2 & 4 \\ 1 & 8 \end{vmatrix} = 20 \quad A_{13} = \begin{vmatrix} -2 & 6 \\ 1 & 4 \end{vmatrix} = -14$$

$$A_{21} = \begin{vmatrix} -2 & 1 \\ 4 & 8 \end{vmatrix} = 20 \quad A_{22} = \begin{vmatrix} 3 & 1 \\ 1 & 8 \end{vmatrix} = 23 \quad A_{23} = \begin{vmatrix} 3 & -2 \\ 1 & 4 \end{vmatrix} = -14$$

$$A_{31} = \begin{vmatrix} -2 & 1 \\ 6 & 4 \end{vmatrix} = -14 \quad A_{32} = \begin{vmatrix} 3 & 1 \\ -2 & 4 \end{vmatrix} = -23 \quad A_{33} = \begin{vmatrix} 3 & -2 \\ -2 & 6 \end{vmatrix} = 14$$

$$Adj\ A = \begin{pmatrix} 32 & 20 & -14 \\ 20 & 23 & -23 \\ -14 & -14 & 14 \end{pmatrix} \tag{B.16}$$

(xiii) Singular Matrix: A singular matrix is a square matrix whose determinant is *zero*.

(xiv) Non-singular Matrix: A non-singular matrix is a square matrix whose determinant is *not zero*.

(xv) Rank of a Matrix: The rank of the matrix **A** is *r* (i) if there exists an *r* × *r* submatrix of the matrix **A** which is *non-singular* and (ii) all other *p* × *p* submatrices are singular (*p* ≥ *r* + 1).

(xvi) Equality of Matrices: Two matrices **A** and **B** are equal if they have same number of rows and columns and the corresponding elements are equal. That is, if **A** is an *m* × *n* matrix, then **B** is also an *m* × *n* matrix and $a_{ij} = b_{ij}$ for $i = 1$ to m and $j = 1$ to n.

(xvii) Matrix Addition: Addition of two matrices **A** and **B** results in a new matrix **C** with its elements c_{ij} equal to the sum of the corresponding elements a_{ij} and b_{ij}. Thus,

$$\mathbf{C} = \mathbf{A} + \mathbf{B} \text{ and } c_{ij} = a_{ij} + b_{ij}.$$

The two matrices should have the same number of rows and columns.

The transpose of matrix **C** is the sum of the transpose of matrices **A** and **B**. Thus,

$$C^T = (A+B)^T = A^T + B^T$$

(*xviii*) *Matrix Subtraction*: Subtraction of two matrices **A** and **B** results in a new matrix **C** with its elements c_{ij} equal to the difference of the corresponding elements a_{ij} and b_{ij}. Thus,

$$C = A - B \text{ and } c_{ij} = a_{ij} - b_{ij}.$$

The two matrices should have the same number of rows and columns.

The transpose of matrix **C** is the difference of the transpose of matrices **A** and **B**. Thus,

$$C^T = (A - B)^T = A^T - B^T$$

(*xix*) *Rank of a Matrix*: The rank of a matrix **A** is r if there exists an $r \times r$ submatrix that is *non-singular* and all other $p \times p$ submatrices are singular ($p \geq r + 1$).

(*xx*) *Equality of Matrices*: Two matrices **A** and **B** are said to be equal if they have same number of rows and columns and the corresponding elements are equal. Thus, if **A** is an $m \times n$ matrix, then **B** should also be an $m \times n$ matrix and $a_{ij} = b_{ij}$ for $i = 1$ to m and $j = 1$ to n.

(*xxi*) *Addition of Matrices*: Addition of two matrices is possible *only if* they have same number of rows and columns. Addition of matrix **A** and **B** results in a new matrix **C** with its element c_{ij} equal to the sum of the corresponding elements. Thus,

$$C = A + B$$

$$c_{ij} = a_{ij} + b_{ij} \quad (i = 1 \text{ to } m \text{ and } j = 1 \text{ to } n)$$

The transpose of **C** is the sum of transpose of **A** and **B**. Thus,

$$C^T = (A + B)^T = A^T + B^T$$

(*xxii*) *Subtraction of Matrices*: Subtraction of two matrices is possible *only if* they have same number of rows and columns. Subtraction of matrix **A** and **B** results in a new matrix **C** with its element c_{ij} equal to the difference of the corresponding elements. Thus,

$$C = A - B$$

$$c_{ij} = a_{ij} - b_{ij} \quad (i = 1 \text{ to } m \text{ and } j = 1 \text{ to } n)$$

The transpose of **C** is the difference of transpose of **A** and **B**. Thus,

$$C^T = (A - B)^T = A^T - B^T$$

B. MATRIX MULTIPLICATION

(*i*) *Multiplication by a Scalar*: If a matrix **A** is multiplied by a scalar, it results in a matrix with all elements of **A** multiplied by the scalar. Thus, **A** given in Eq. (B.1) is multiplied by k, the resulting matrix is given by

$$kA = \begin{pmatrix} ka_{11} & ka_{12} & ka_{13} & \cdot & \cdot & ka_{1n} \\ ka_{21} & ka_{22} & ka_{23} & \cdot & \cdot & ka_{2n} \\ \cdot & \cdot & \cdot & \cdot & \cdot & \cdot \\ ka_{m1} & ka_{m2} & ka_{m3} & \cdot & \cdot & ka_{mn} \end{pmatrix} \quad (B.17)$$

(ii) Conformable Matrices: Two matrices are said to be conformable if the number of columns of the first matrix equal to the number of rows of the second matrix.

(iii) Multiplication of Matrices: If matrix **A** and matrix **B** are conformable, they can be multiplied and it results into a product matrix **C** given by

$$\mathbf{C} = \mathbf{AB}$$

and its element c_{ij} is given by

$$c_{ij} = \sum_{k=1}^{n} a_{ik} b_{kj}$$

If A is of order $m \times n$ and **B** is of order $n \times p$, then the order of **C** is $m \times p$. The element c_{ij} is obtained by multiplying the elements of *i-th row* of **A** with elements of *j-th column* of **B** and summing up them. Thus,

$$\begin{pmatrix} c_{11} & c_{12} & c_{13} \\ c_{21} & c_{22} & c_{23} \\ c_{31} & c_{32} & c_{33} \end{pmatrix} = \begin{pmatrix} a_{11} & a_{12} & a_{13} \\ a_{21} & a_{22} & a_{23} \\ a_{31} & a_{32} & a_{33} \end{pmatrix} + \begin{pmatrix} b_{11} & b_{12} & b_{13} \\ b_{21} & b_{22} & b_{23} \\ b_{31} & b_{32} & b_{33} \end{pmatrix} \tag{B.18}$$

where

$$c_{11} = a_{11} b_{11} + a_{12} b_{21} + a_{13} b_{31}$$
$$c_{12} = a_{11} b_{12} + a_{12} b_{22} + a_{13} b_{32}$$

$$\cdot$$

$$\cdot$$

In general, the multiplication of matrices is *not commutative*. Thus,

$$\mathbf{AB} \ne \mathbf{BA}$$

(iv) Transpose of the product Matrix: The transpose of the product matrix is the product of their transpose in the reverse order. Thus,

$$\mathbf{C}^{\mathrm{T}} = (\mathbf{AB})^{\mathrm{T}} = \mathbf{B}^{\mathrm{T}} \mathbf{A}^{\mathrm{T}}$$

(v) Multiplication by Identity Matrix: If any matrix is multiplied by the identity matrix **I**, the result is the original matrix itself. Thus,

$$\mathbf{AI} = \mathbf{A}$$

(vi) Matrix Inversion: Only square matrices have inverse matrix. The inverse of matrix **A** is denoted by $\mathbf{A}^{-1}$. It is determined by the relation

$$\mathbf{A}^{-1} = Adj\ \mathbf{A}\ /\ \det \mathbf{A}$$

The inverse of Eq. (B.12) is given by

$$\mathbf{A}^{-1} = Adj \; \mathbf{A} \, / \, \det \mathbf{A} = \begin{pmatrix} 32 & 20 & -14 \\ 20 & 23 & -23 \\ -14 & -14 & 14 \end{pmatrix} \quad \text{(B.19)}$$

The multiplication of matrix **A** by its inverse will result in the identity matrix, Thus,

$$\mathbf{A}^{-1}\mathbf{A} = \mathbf{A}\,\mathbf{A}^{-1} = \mathbf{I}$$

(*vii*) *Matrix form of Simultaneous Linear Equation*: The simultaneous linear equation can be represented in a compact form using matrix multiplication.

Let the simultaneous linear equation be

$$\begin{aligned}
a_{11}\, x_1 + a_{12}\, x_2 + \ldots + a_{13}\, x_n &= c_1 \\
a_{21}\, x_1 + a_{22}\, x_2 + \ldots + a_{23}\, x_n &= c_2 \\
a_{31}\, x_1 + a_{32}\, x_2 + \ldots + a_{33}\, x_n &= c_3
\end{aligned} \quad \text{(B.20)}$$

Using the matrix multiplication rule, Eq. (B.20) cab be represented in compact form as

$$\mathbf{Ax} = \mathbf{c} \quad \text{(B.21)}$$

Where

$$\mathbf{A} = \begin{pmatrix} a_{11} & a_{12} & \cdots & a_{1n} \\ a_{21} & a_{22} & \cdots & a_{2n} \\ \cdot & \cdot & \cdots & \cdot \\ \cdot & \cdot & \cdots & \cdot \\ \cdot & \cdot & \cdots & \cdot \\ a_{m1} & a_{m2} & \cdots & a_{mn} \end{pmatrix}, \quad \mathbf{x} = \begin{pmatrix} x_1 \\ x_2 \\ \cdot \\ \cdot \\ \cdot \\ x_n \end{pmatrix}, \quad \mathbf{c} = \begin{pmatrix} c_1 \\ c_2 \\ \cdot \\ \cdot \\ \cdot \\ c_3 \end{pmatrix} \quad \text{(B.22)}$$

The solution of **x** is given by

$$\mathbf{x} = \mathbf{A}^{-1}\mathbf{c}$$

C. MATRIX CALCULUS

(*i*) *Derivative of m × n Matrix*: The derivative of *m* × *n* matrix **A**(t) is defined as the derivative of each of the *mn* elements of the matrix **A**(t). Thus,

$$d\mathbf{A}(t)/dt = da_{ij}(t)/dt = \begin{pmatrix} da_{11}(t)/dt & da_{12}(t)/dt & \cdots & da_{1n}(t)/dt \\ da_{21}(t)/dt & da_{22}(t)/dt & \cdots & da_{2n}(t)/dt \\ \cdot & \cdot & \cdots & \cdot \\ \cdot & \cdot & \cdots & \cdot \\ da_{m1}(t)/dt & da_{m2}(t)/dt & \cdots & da_{mn}(t)/dt \end{pmatrix} \quad \text{(B.23)}$$

(ii) Integral of m × n Matrix: The integral *of m × n matrix* **A**(t) is defined as the integral of each of the *mn* elements of the matrix **A**(t). Thus,

$$\int \mathbf{A}(t)dt = \int a_{ij}(t)dt = \begin{pmatrix} \int a_{11}(t)/dt & \int a_{12}(t)/dt & \cdots & \int a_{1n}(t)/dt \\ \int a_{21}(t)/dt & \int a_{22}(t)/dt & \cdots & \int a_{2n}(t)/dt \\ \cdot & \cdot & \cdots & \cdot \\ \cdot & \cdot & \cdots & \cdot \\ \int a_{m1}(t)/dt & \int a_{m2}(t)/dt & \cdots & \int a_{mn}(t)/dt \end{pmatrix} \tag{B.24}$$

D. MATRIX EXPONENTIAL

(i) Matrix Exponential: The matrix exponential of a square matrix **A** is function of time defined as a power series. Thus,

$$e^{\mathbf{A}t} = \mathbf{I} + \mathbf{A}t + (1/2!)\ \mathbf{A}^2t^2 + (1/3!)\mathbf{A}^3t^3 + \ldots$$

(ii) Differentiation of Matrix Exponential: If **A** is a matrix with constant coefficient, then its time derivative is given by

$$d(e^{\mathbf{A}t})/dt = \mathbf{A} + \mathbf{A}^2t + \mathbf{A}^3t^2 + \ldots$$
$$= (\mathbf{I} + \mathbf{A}t + (1/2!)\ \mathbf{A}^2\ t^2 + (1/3!)\ \mathbf{A}^3\ t^3 + \ldots)\ \mathbf{A} = e^{\mathbf{A}t}\mathbf{A}$$
$$= \mathbf{A}(\mathbf{I} + \mathbf{A}t + (1/2!)\ \mathbf{A}^2t^2 + (1/3!)\ \mathbf{A}^3t^3 + \ldots) = \mathbf{A}e^{\mathbf{A}t}$$

(iii) Differentiation of $e^{\mathbf{A}t}\ x$: The differentiation of $e^{\mathbf{A}t}x$ is given by

$$d(e^{\mathbf{A}t}\mathbf{x})/dt = [d(e^{\mathbf{A}t})/dt]\ \mathbf{x} + e^{\mathbf{A}t}\ (d\mathbf{x}/dt) = e^{\mathbf{A}t}\mathbf{A}\mathbf{x} + e^{\mathbf{A}t}\ \mathbf{x}'$$

(iv) Product of Matrix Exponential: The product of matrix exponential is given by

$$e^{\mathbf{A}t_1} \cdot e^{\mathbf{A}t_2} = e^{\mathbf{A}t_2} \cdot e^{\mathbf{A}t_1} = e^{\mathbf{A}(t_1 + t_2)}$$

It may be noted that

$$e^{\mathbf{A}t} \cdot e^{-\mathbf{A}t} = \mathbf{I}$$

Appendix C

DETERMINATION OF ROOTS OF FUNCTIONS

In the analysis of control systems we have very often to determine the roots of the characteristic equations. We discuss in this appendix one method of obtaining the roots.

Reminder Theorem Method: The remainder theorem states that if in a function a value is substituted, the resulting value is the remainder of the function for that value. This implies that if there is no remainder then, the value is a root of the function. This property we will utilize to determine the roots of a function.

Rule 1: *If all the coefficients of the function F(s) are positive, add up odd coefficients and even coefficients. If they are equal, then (–1) is a root and (s + 1) is a factor.*

Rule 2: *If one or more coefficients of the function F(s) are negative, add up the positive coefficients and negative coefficients. If they are equal, then (1) is a root and (s – 1) is a factor.*

These rules we will utilize in determining the roots of a function.

Example C.1 *Find the roots of the cubic equation $F(s) = s^3 + 4s^2 + 6s + 4 = 0$.*

Solution:

All coefficients are positive. Rule 1 is not satisfied. Hence, try (–2) as a root. Let us substitute this value in F(s) and find the remainder. Thus,

$$F(-2) = (-2)^3 + 4(-2)^2 + 6(-2) + 4 = 0$$

Hence –2 is a root and (s + 2) is a factor of the function. To find the other roots, divide F(s) by (s + 2) and determine the quotient which is in this case a quadratic function for which the roots can easily be evaluated. Thus,

$$F(s) = (s + 2)(s^2 + 2s + 2) = (s + 2)(s + 1 + j1)(s + 1 - j1)$$

Therefore, the roots are $-2, -1 \pm j1$.

Example C.2 *Find the roots of $F(s) = s^4 + 9s^3 + 30s^2 + 42s + 20 = 0$.*

Solution:

All coefficients are positive. Rule 1 is satisfied. Hence, (-1) is a root. Let us substitute this value in $F(s)$ and find the remainder. Thus,

$$F(-1) = (-1)^4 + 9(-1)^3 + 30(-1)^2 + 42(-1) + 20 = 0$$

Hence (-1) is a root and $(s+1)$ is a factor of the function. To find the other roots, divide $F(s)$ by $(s + 1)$ and determine the quotient. The quotient is

$$F_1(s) = F(s)/(s + 1) = s^3 + 8s^2 + 22s + 20$$

All coefficients are positive. Rule 1 is not satisfied. Hence, try (-2) as a root. Let us substitute this value in $F(s)$ and find the remainder. Thus,

$$F_1(-2) = (-2)^3 + 8(-2)^2 + 22(-2) + 20 = 0$$

Hence (-2) is a root and $(s + 2)$ is a factor of the function. To find the other roots, divide $F_1(s)$ by $(s + 2)$ and determine the quotient which is a quadratic function for which the roots can easily be evaluated. Thus,

$$F_2(s) = F_1(s)/(s + 2) = s^2 + 6s + 10 = (s + 3 + j1)(s + 3 - j1)$$

Therefore, the roots are $-1, -2, -3 \pm j1$.

Example C.3 *Find the roots of* $F(\omega^2) = \omega^6 + 14\omega^4 + 49\omega^2 - 64 = 0$

Solution:

All coefficients are not positive. Rule 2 is satisfied. Hence, try (1) as a root. Let us substitute this value in $F(s)$ and find the remainder. Thus,

$$F(1) = 1 + 14 + 49 - 64 = 0$$

Hence (1) is a root and $(\omega^2 - 1)$ is a factor of the function. To find the other roots, divide $F(s)$ by $(\omega^2 - 1)$ and determine the quotient. The quotient is

$$F_1(\omega^2) = F_1(\omega^2)/(\omega^2 - 1) = \omega^4 + 15w^2 + 64$$

$$\omega^2 = (-15 \pm \sqrt{225 - 256})/2 = -7.5 \pm j2.78$$

Therefore, the roots are $\pm 1, \sqrt{-7.5 \pm j2.78}$.

Example C.4. *Find the roots of* $F(\omega^2) = 4\omega^8 + 5\omega^6 + \omega^4 - 16\omega^2 - 1 = 0$

Solution:

$$F(\omega^2) = 4\omega^8 + 5\omega^6 + \omega^4 - 16\omega^2 - 1 = 0$$

All coefficients are not positive. Rule 2 is not satisfied. Hence, try (1) as a root. Let us substitute this value in $F(s)$ and find the remainder. Thus,

$$F(1) = 4 + 5 + 1 - 16 - 1 = -7.$$

Since the remainder is negative, increase the value by a small amount because the order of the function is high. Let us try (1.2). Thus,

$$F(1.2) = 4 \times (1.2)^4 + 5(1.2)^3 + (1.2)^2 - 16(1.2) - 1$$

$$= 8.2944 + 10.843 + 1.44 - 19.8 - 1$$
$$= -0.2226$$

Since the remainder is still negative, increase the value slightly to (-1.25). The remainder is

$$F(1.25) = 4 \times (1.25)^4 + 5 \times (1.25)^3 + (1.25)^2 - 16 \times (1.25) - 1$$
$$= 9.7656 + 9.7656 + 1.5625 - 20 - 1$$
$$= +0.09$$

Assuming the remainder is negligible, the root is 1.25 and $(\omega^2 - 1.25)$ is a factor.

$$F_1(s) = F(s)/(\omega^2 - 1.25) = (4\omega^6 + 10\omega^4 + 13.5\omega^2 + 0.875)$$
$$F_1(s) = 4\omega^6 + 10\omega^4 + 13.5\omega^2 + 0.875$$
$$= 4(\omega^2)^3 + 10(\omega^2)^2 + 13.5\omega^2 + 0.875$$

Since all coefficients are positive and the constant term is small, assume $\omega^2 = -0.1$. We get

$$F_1(-0.1) = -0.004 + 0.1 - 1.35 + 0.875 = -0.379$$

This is negative. Hence decrease the value to -0.075.

$$F_1(-0.075) = 4 \times (-0.075)^3 + 10 \times (-0.075)^2 + 13.5 \times (0.075) + 0.875$$
$$= -0.00169 + 0.05625 - 1.0125 + 0.875 = -0.083$$

Decrease the value to (-0.068). Then,

$$F_1(-0.07) = 4 \times (-0.07)^3 + 10 \times (-0.07)^2 + 13.5 \times (-0.07) + 0.875$$
$$= -0.001372 + 0.049 - 0.945 + 0.875 = -0.025372$$

Try (-0.068). We get

$$F_1(-0.068) = 4 \times (0.068)^3 + 10 \times (-0.068)^2 + 13.5 \times (-0.068) + 0.875$$
$$= -0.0012576 + 0.04624 - 0.918 + 0.875 = -0.002$$

The remainder is negligible. Hence $(\omega^2 + 0.068)$ is a factor.

Dividing $F_1(s)$ by $(\omega^2 + 0.068)$, we get

$$F_2(s) = F_1(s)/(\omega^2 + 0.068) = 4\omega^4 + 9.752\omega^2 + 12.84$$

The roots are:

$$s_{1,2} = (-9.752 \pm \sqrt{(9.752)^2 - 4 \times 4 \times 12.84})/8$$
$$= (-9.752 \pm \sqrt{(9.752)^2 - 4 \times 4 \times 12.84})/8$$
$$= (-9.752 \pm \sqrt{95.1 - 207.4})/8$$
$$= (-9.752 \pm \sqrt{112.3})/8$$
$$= -1.219 \pm j\,10.59$$

Therefore the roots are $\pm\sqrt{1.25}, \pm\sqrt{(-0.068)}, \pm\sqrt{-1.219 \pm j10.59}$

Appendix D

GREEK ALPHABETS

In the analysis of control systems, we use the Greek alphabets freely and hence in Appendix D they are tabulated.

Table D.1 Greek Alphabets

Pronunciation	Upper case	Lower case
Alpha	A	α
Beta	B	β
Gamma	Γ	γ
Delta	Δ	δ
Epsilon	E	ε
Zeta	Z	ζ
Eta	H	η
Theta	Θ	θ
Iota	I	ι
Kappa	K	κ
Lambda	Λ	λ
Mu	M	μ
Nu	N	ν
Xi	Ξ	ξ
Omicron	O	o
Pi	Π	π

Pronunciation	Upper case	Lower case
Rho	P	ρ
Sigma	Σ	σ
Tan	T	τ
Upsilon	Y	υ
Phi	Φ	φ
Chi	X	χ
Psi	Ψ	ψ
Omega	Ω	ω

ANSWERS

Chapter 1

1.1. $(3s + 1)/(3s + 101)$

1.2. (a) $1/(1 + G_1G_2) = 1/21$

   (b) $V_o = (20r + 2n)/21$

   (c) $G_1 = 210$ and $V_o = (420r + 2n)/421$

1.3. (a) $S_{G_3}^M = 1/(1 + G_2G_3); S_{G_4}^M = G_3/(1 + G_2G_3)$

   (b) G_3

1.4. $K(G_2G_3 - G_1G_4)/(G_3 + KG_4)(G_1 + KG_2)$

1.5. $(10s + 1)(2s + 1)/(20s^2 + 12s + 1 + K)$

1.6. (a) $M_1 = M_2 = 100$

   (b) 0.01; 0.1

Chapter 2

2.2. (a) $5/s$

   (b) $1/s^2 + 4/(s + 2) + 2/(s^2 + 4)$

   (c) $5/s + s/(s^2 + 25)$

   (d) $a(s^2 - a^2)$

   (e) $s/s^2 - b^2)$

   (f) $\omega/[(s + a)^2 + \omega^2]$

   (g) $A[as + (a^2 - b^2)^2]/[s^4 + 2(a^2 + b^2)s^2 + (a^2 - b^2)^2]$

2.3. (a) $1/(s - a)$

   (b) (i) $1/s$;

      (ii) $\omega/(s^2 + \omega^2); s/(s^2 + \omega^2)$;

      (iii) $(s + a)/[(s + a)^2 + \omega^2]; \omega/[(s + a)^2 + \omega^2]$;

      (iv) $1/(s - 1)$; (v) $n!/s^{n+1}$

2.4. (a) $3 + 3e^{-3t} - 6e^{-t}$

   (b) $2 - 2e^{-t} \sin 2t$

   (c) $0.2 + 0.224\ e^{-t} \sin (2t - 63.43°)$

   (d) $0.33\ e^{-3t} + e^{-t} \sin (-\sqrt{5}t + 41.8°)$

2.5. (a) $1/s^2 - (3/4s) + 2/[3(s + 1)]$

(b) $1/s + 6/(s + 3)^2 - 10/(s + 3) + 9/(s + 4)$

(c) $1/2s - 25/[12(s + 1)] + 10/[3(s + 2)] - 5/[2(s + 3)] + 5/[6(s + 4)] - 1/[12(s + 5)]$

2.6. (a) $5/12 - (5/3) e^{-6t} + (5/4)e^{-8t}$

(b) $\frac{1}{2} - (25/12)e^{-t} + (10/3)e^{-2t} - 2.5e^{-3t} + (5/6)e^{-4t} - (1/12)e^{-5t}$

(c) $100\ e^{-4t} - 170e^{-5t}$

(d) $[-2 + 8t - 6t^2 + (4/3)t^3 - (1/12)t^4]e^{-t}$

(e) $10e^{-t} (2 - t + t^2/2 - e^{-t})$

(f) $15(e^{-4t} - 2e^{-5t}) + 10t((e^{-3t} + 2e^{-4t}) - 5t^2e^{-4t}$

(g) $5e^{-3t} \sin t$

(h) $0.6e^{-3t}\sin (4t - 104°) + 0.59e^{-2t}$

(i) $0.75 (1 - \cos 2t)$

(j) $(2/9) (1 - \cos 3t + 1.5\sin 3t)$

(k) $3e^{-3t} - e^{-4t} (3 + 2t + 2t^2)$

(l) $-(e^{-2t} - 3e^{-3t} + e^{-4t})$

(m) $10e^{-t} (t - \sin t)$

(n) $1 - 1.3e^{-3t} + 0.029e^{-2t}\sin(3t + 164.7°)$

(o) $1 - 1.047e^{-t} + 0.0673e^{-t}\sin(t + 135°)$

2.7. $x(t) = 2.5(1 - 10e^{-t} + 3.5e^{-4t});\ x(\infty) = 2.5$

2.8. $x(t) = 10$

2.9. $i(t) = -9.5e^{-2t} + (295/34)e^{-4t} + 70\cos t + 60\sin t$

2.10. (a) $x(t) = 2.125 - (24.5t - 1.875\ ) e^{-4t}$

(b) $x(\infty) = 2.125$

2.11. $\theta_g = 0.4 (1 - e^{-3t} - 3t\ e^{-3t})$

2.12. $f(0) = 10 \sin \theta;\ f'(0) = 30 \cos \theta - 20 \sin \theta$

2.13. $f(0) = 0;\ f'(0) = 2;\ f''(0) = 5$

2.14. (a) $x(0) = 0;\ x(\infty) = 3$

(b) 2, 2

(c) 0, 0.2

(d) 1, 0

2.15. (a) 1, 0

(b) 6, 0

(c) 0, 0.5

2.16. $f(0) = 0;\ K = 12$

2.17. (a) $0.5\ e^{-2t} (1 - e^{-2t})$

(b) $(1/81) (e^{-9t} + 9t - 1)$

(c) $(2/9)(3t^2 - \sin 3t)$

(d) $(1 - 4t\ e^{-4t} - e^{-4t})/16$

Chapter 3

3.1. $E_o(s)/E_i(s) = s/(s + R/L)$; $E_o(s)/E_i(s) = s/L$

3.2. (*a*) $E_o(s)/E_i(s) = (R/L)/[s + (R + R_L)/L]$

    (*b*) $E_o(s)/E_i(s) = (s + R_L/L)/[s + (R + R_L)/L]$

    (*c*) $E_o(s)/E_i(s) = (s + 1/RC)/[s + (C + C_1)/RCC_1]$

3.4. $E_o(s)/E_i(s) = 1/[(1 + R_1C_1s)(1 + R_2C_2s)]$

3.5. $E_o(s)/E_i(s) = (1 + 2 R_1Cs + R_1R_2C^2s^2)/[1 + (2R_1 + R_2) Cs + R_1R_2C^2s^2]$

3.6. $X(s)/F(s) = 1/(Ms^2 + Bs + K)$

3.7. $X(s)/F(s) = (1/M)/[(s^2 + (K/B)s + K/M]$

3.8. (*a*) $X(s)/F(s) = 1/(Ms^2 + Bs + K)$

    (*b*) $X(s)/F(s) = K/[s(MBs^2 + MKs + BK)$

3.9. $Y(s)/X(s) = (Bs + K)/(Ms^2 + Bs + K)$

3.10. (*a*) $X(s)/F(s) = K_2/[M_1M_2s^4 + M_2B_1s^3 + (M_1K_2 + M_2K_1 + M_2K_2)s^2 + B_1K_2s + K_1K_2]$

    (*b*) $X(s)/F(s) = (B_2s + K_2)/[M_1M_2s^4 + M_2B_1s^3 + (M_1K_2 + M_2K_1 + M_2K_2)s^2 +$

$$B_1K_2s + K_1K_2]$$

3.11. $X_2(s)/F(s) = K_1/(as^4 + bs^3 + cs^2 + ds + e)$

    $a = M_1M_2 + JM/r^2$

    $b = BM_2$

    $c = K_1(M_1 + M_2 + J/r^2)$

    $d = B(K_1 + K_2)$

    $e = K_1K_2$

3.12. $E_o(s)/E_i(s) = -0.9999/(s + 0.5015)$

3.13. $\Theta(s)/T(s) = K_1K_2/[(as^4 + bs^3 + cs^2 + ds + e)(J_3s^2 + B_3s + K_2 + K_3)$

    $a = J_1J_2$

    $b = J_1B_2 + J_2B_1$

    $c = J_1(K_1 + K_2) + J_2K_1$

    $d = B_1(K_1 + K_2) + B_2K_2$

    $e = K_1K_2$

3.14. $X(s)/F(s) = K_{12}K_{23}/(A.B - K_{12}^2) C$

    $A = M_1s^2 + B_1s + K_1 + K_{12}$

    $B = M_2s^2 + B_2s + K_2 + K_{12}$

    $C = M_3s^3 + K_{23}$

3.15. $X(s)/F(s) = (MBs^2 + MKs + BK)/MBKs^2$; $Y(s)/F(s) = (Ms + B)/MBs^2$; $Z(s)/F(s) = 1/Ms^2$

3.16. (*a*) $E(s)/R(s) = (M + MG_1G_2H - G_1G_2)/(1 + G_1G_2H)$

    (*b*) $C(s)/R(s) = G_1G_2/(1 + G_1G_2H)$

    (*c*) $X(s)/R(s) = G_1/(1 + G_1G_2H)$

3.17. (*a*) $C(s)/R(s) = G_1G_2G_3/(1 + G_2G_3H)$

    (*b*) $E(s)/R(s) = G_1/(1 + G_2G_3H)$

    (*c*) $C(s)/D(s) = G_3G_4/(1 + G_2G_3H)$

    (*d*) $E(s)/D(s) = -G_3G_4H(1 + G_2G_3H)$

3.18. $C(s)/R(s) = G_1G_2G_3/(1 + G_2H_1 + G_2G_3H_2)$

3.19. $C(s)/R(s) = G_1G_2G_3G_4/(1 - G_3 G_4H_1 + G_2G_3H_2 + G_1G_2G_3G_4H_3)$

3.20. $X_1(s) = (1 - a_{22}) R_1(s)/\Delta + a_{12} R_2(s)/\Delta$

   $X_2(s) = (1 - a_{11}) R_2(s)/\Delta + a_{21} R_1(s)/\Delta$

   $\Delta = 1 - a_{11} - a_{22} + a_{11} a_{22} - a_{12} a_{21}$

3.21. $C(s)/R(s) = [A.B + C.D]/[1 - E + F]$

   $A = G_1G_2G_3G_4$

   $B = (1 - L_3 - L_4)$

   $C = G_5G_6G_7G_8$

   $D = (1 - L_1 - L_2)$

   $E = (L_1 + L_2 + L_3 + L_4)$

   $F = (L_1L_3 + L_1L_4 + L_2L_3 + L_3L_4)$

3.22. $C(s)/R(s) = (P_1 + P_2 \Delta_2 + P_3)/\Delta$

   $P_1 = G_1G_2G_3G_4G_5G_6$

   $P_2 = G_1G_2G_6G_7$

   $P_3 = G_1G_2G_3G_4 G_8$

   $\Delta_2 = 1 - L_5$

   $$\Delta = 1 - \sum_{i=1}^{8} L_i + L_3L_4 + L_4L_5 + L_5L_7$$

   $L_1 = -G_2G_3G_4G_5H_2$

   $L_2 = -G_5 G_6H_1$

   $L_3 = -G_8H_1$

   $L_4 = -G_2G_7H_2$

   $L_5 = -G_4H_4$

   $L_6 = -G_1G_2G_3G_4G_5G_6H_3$

   $L_7 = -G_1G_2G_6G_7H_3$

   $L_8 = -G_1G_2G_3G_4G_8H_3$

3.23. $E_2(s)/E_1(s) = Y_1Z_3Y_3Z_4/\Delta$

   $\Delta = (1 + Y_1Z_2 + Y_3Z_4 + Y_1Z_2Y_3Z_4)$

3.24.

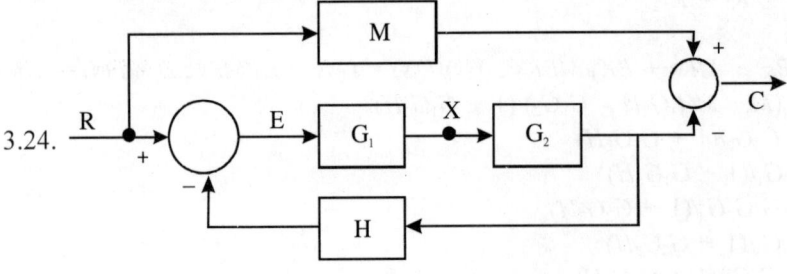

Fig. (a)

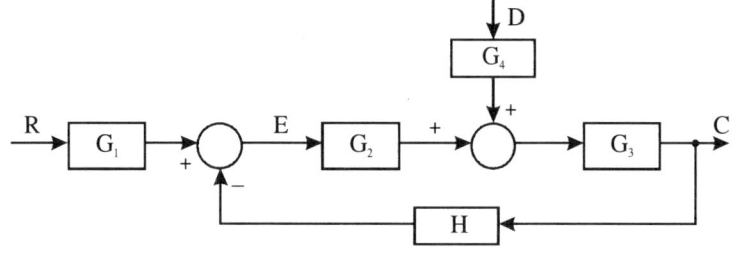

Fig. (b)

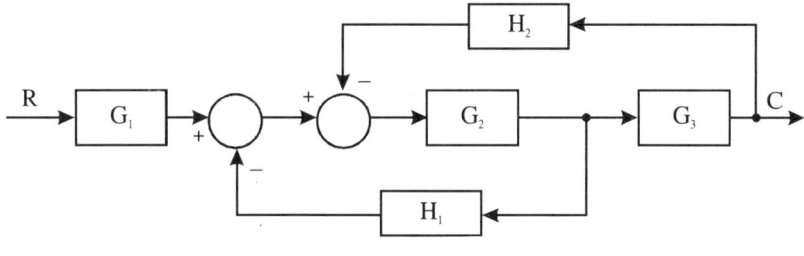

Fig. (c)

Chapter 4

4.1.

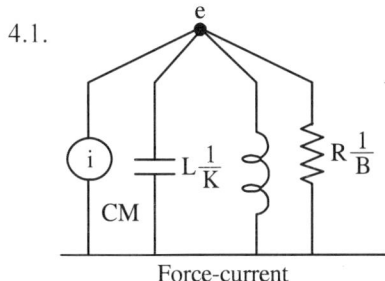

Force-current

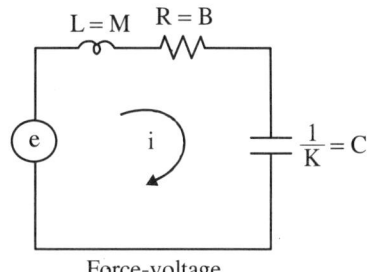

Force-voltage

4.2.

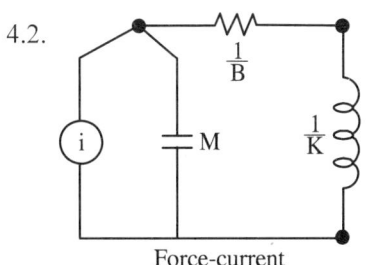

Force-current

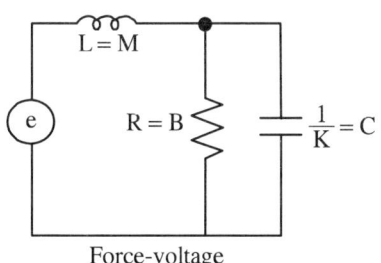

Force-voltage

4.3. (a)

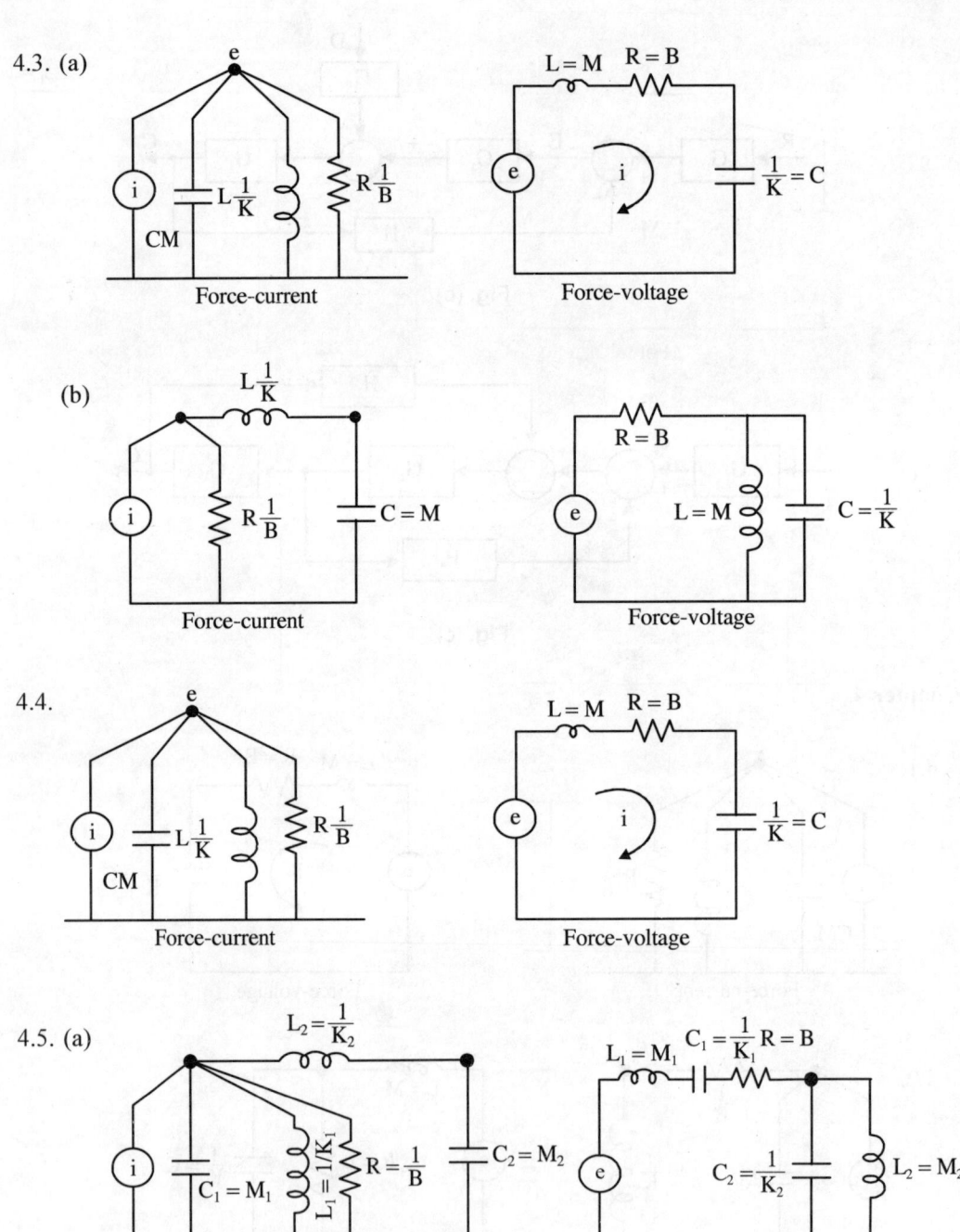

Force-current   Force-voltage

(b)

Force-current   Force-voltage

4.4.

Force-current   Force-voltage

4.5. (a)

Force-current   Force-voltage

(b)

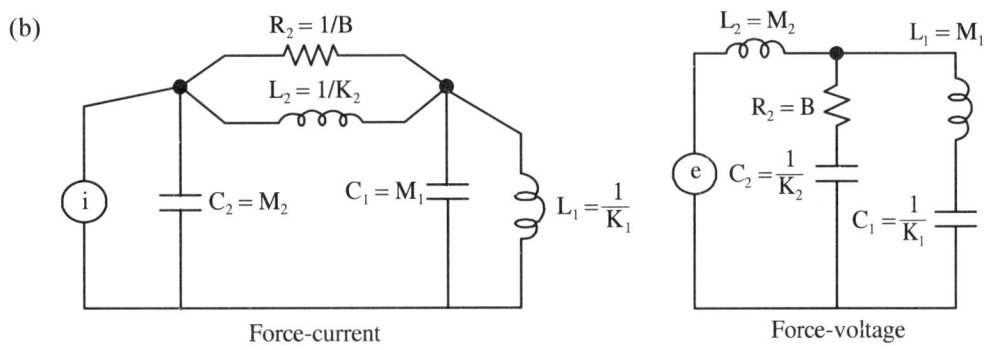

Force-current   Force-voltage

4.6.

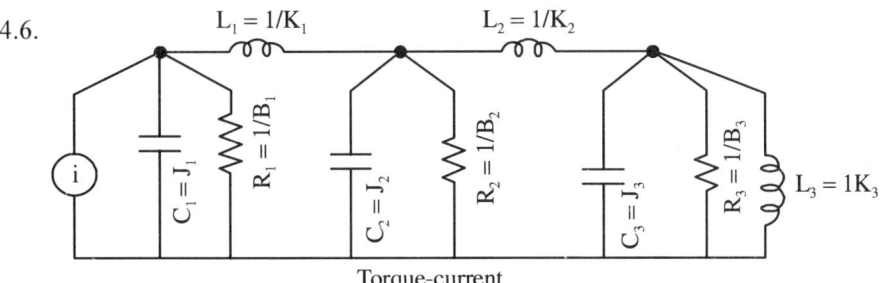

Torque-current

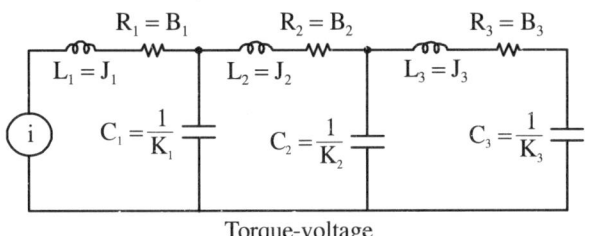

Torque-voltage

4.7.

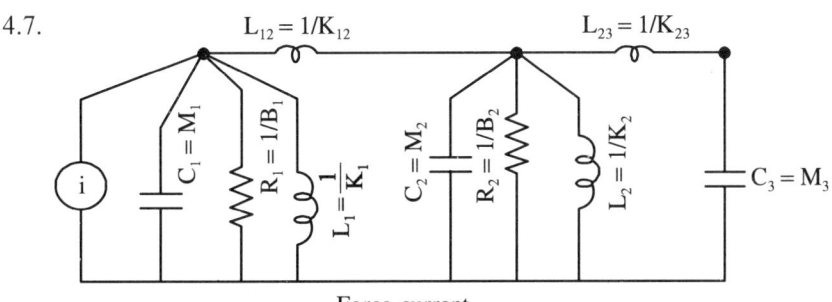

Force-current

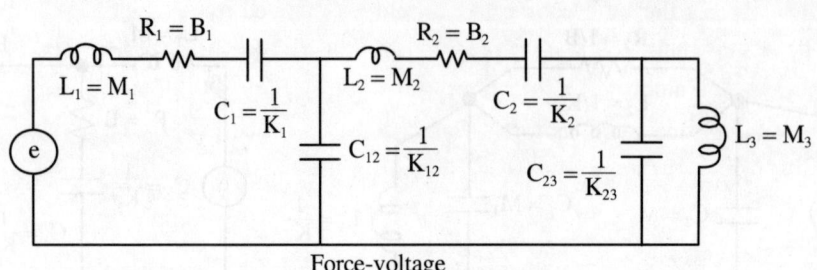

Force-voltage

4.8. $\dfrac{v_2(s)}{F(s)} = \dfrac{B_2 s}{M_1 s + B_1 + B_2)(M_2 s^2 + B_2 s + K) - s B_2^2}$

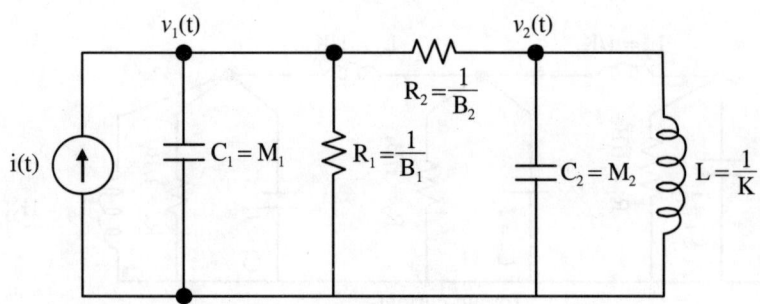

4.9. (a) $e_o = -(2e_1 + 5e_2)$

   (b) $e_o = -0.5e_1$

   (c) $e_o = -\int (10e_1 + e_2 + 2e_3)dt$

   (d) $E_o(s) = (-R_2/R_1)E_1(s)/(1 + R_2 C_2 s)$

   (e) $e_o = -\int (e_1 + 5e_2 + 10e_3)dt$

   (f) $e_o = -(2e_1 + e_2 + 4e_3 + 10e_4)$

   (g) $e_o = -e_1/3$

   (h) $E_o(s) = -2\,E_1(s)/(s + a)$

4.10. [Refer Fig. 4.23(d)]

   (a) $f(t) = 16,\ K = 2,\ B = 1.4,\ IC_1 = -3,\ IC_2 = -2$

   (b) $f(t) = 10,\ K = 2,\ B = 2.5,\ IC_1 = 0,\ IC_2 = -3$

   (c) $f(t) = 10,\ K = 1,\ B = 0.25,\ IC_1 = -2,\ IC_2 = 1$

   (d) $f(t) = f(t)/M,\ K = K/M,\ B = B/M,\ IC_1 = IC_2 = 0$

4.11. (a) and (b) similar to Fig. 4.28 with appropriate changes.

4.12. Same as 4.11 except the integrator gains should be changed by a factor $(1/\alpha)$

4.13. Same as Eq. (4.30) with $k_2/RC = 2$, $k_1 = 0.7$ and include $k_3 = 0.1$ in series with 1 μF capacitor across A1 amplifier

4.14. Input impedance is due to capacitor C_i. Feedback impedance is due to a resistor R_f in series with C_f. $C_1 = 1\mu F$, $C_f = 0.2\mu F$ $R_f = 1\mu$

Chapter 5

5.1. (*a*) 1.5%

    (*b*) 2%

5.2. (*a*) 0.028

    (*b*) 0.1 *V/deg.*

    (*c*) 1.59 V/*deg.*

    (*d*) 10 V/turn

5.3. (*a*) 4 *V*/turn

    (*b*) 20 ± 0.4 *V*

    (*c*) (*i*) 20*V*; (ii) 13.33 *V*

5.4. (*a*) 1*V*/deg.

    (*b*) 10 *V*

    (*c*) 10.42 *V*

5.5. 1.3 *V*/deg.

5.6. (*a*) 84.57 *V*

    (*b*) 88.63 *V*

5.7. $E_a(s)/\Theta(s) = (K_T/JL)/[s(s^2 + as + b)$

    $a = B/J + R/L;\ b = (BR + K_bK_T)/JL$

5.8. $E_o(s)/\Theta(s) = (K_bR_Ls/L)/[s + (R_a + R_b)/L]$

5.9. (*a*) 0.3 Nm/*V*

    (*b*) 12 rad./sec

5.10. (*a*) T = 0.67 sec.

    (*b*) (*i*) 100/(s + 6); (*ii*) 100/[s(s + 6)]

5.11. $\Omega(s)/E(s) = 50/(s + 0.625)$

5.12. (*a*) 700 rad./sec.; 300 rad./se

    (*b*) 800 Nm

5.13. $Hs/(Js^2 + Bs + K)$

5.14. (a) $H/(Js^2 + Bs + K)$

(b) $H/[s(Js + B)]$

5.15. (a) $C = 0.0051 \ F$

(b) $T_s = 0.2$ sec

5.16. (a) $(1 + 0.2s^2)/[G(1 = 0.2s + GK_a)]$

(b) $K_a = 74$

(c) $c(\infty) = 15$

Chapter 6

6.1. $c(t) = (1/T)e^{-t/T}$

(i) $T = 1$, $c(1) = 0.368$; $c(2) = 0.135$; $c(3) = 0.0498$, $c(0) = 1$, $c'(0) = -1$.

(ii) $T = 2$, $c(1) = 0.607$; $c(2) = 0.368$; $c(3) = 0.223$, $c(0) = 0.5$, $c'(0) = -0.25$

(iii) $T = 3$, $c(1) = 0.717$; $c(2) = 0.\ 513$; $c(3) = 0.368$, $c(0) = 0.333$, $c'(0) = -1/9$

6.2.

T	$c \ (1)$	$c \ (2)$	$c \ (3)$	$c' \ (0)$	t_s
1	0.632	0.865	0.9592	1	4
2	0.393	0.632	0.777	2	8
3	0.283	0.487	0.632	3	12

6.3. (a) $c \ (t) = 1 - 3e^{-2}t + 2e^{-3}t$; $c \ (\infty) = 1$;

Transient part $= -3e^{-2} + 2e^{-3t}$;

Transient part at $t = \infty$ is zero.

$\omega_n = \sqrt{K/J} = \sqrt{6}$; $\zeta = 1.02$

(b) $c \ (t) = 1 - 1.414e^{-t} \sin (t + 45°)$; $c \ (\infty) = 1$;

Transient part $= -1.414e^{-t} \sin (t + 45°)$;

Transient part at $t = \infty$ is zero.

$\omega_n = \sqrt{2}$; $\zeta = 0.707$

(c) $c(t) = 3 \sin 3t$; $c \ (\infty)$ is *indeterminate* since the response is oscillatory.

Transient part $= 3 \sin t$;

Transient part at $t = \infty$ is indeterminate.

$\omega_n = 3$; $\zeta = 0$.

6.4. $\omega_n = 3$, $\zeta = 0.5$, $\omega_d = 2.6$, $T = 0.667$

6.5. $\omega_n = 3$, $\omega_d = 2.6$, $t_r = 1.08$, $t_p = 1.21$, $t_s = 2.7$,

$M_o = 0.163$, $t_u = 2.42$, N = 1.12.

6.6. (a) $c(t) = 1 - 1.09 \ e^{-2t} \sin (\sqrt{21} \ t + 0.37)$

$\omega_n = 5$; $\zeta = 0.4$, $M_o = 0.254$, $t_p = 0.685$, $t_s = 2$

  (*b*) 0.06 ra*d*.

6.7. $T(s) = 2/(s^2 + 4)$

6.8. (*a*) $Y(s)/R(s) = (s + 8)/[(s + 1)\ (s + 2)\ (s + 4)]$

  (*b*) $2.33\ e^{-t} - 3\ e^{-2t} + 0.667\ e^{-4t}$

6.9. (*a*) $T(s) = 3(s^2 + 2s + 5)/[(s + 3)\ (s^2 + 2s + 2)]$

  (*b*) $c\ (t) = 2.5 - 0.4\ e^{-3t} - 2.12\ e^{-t}(cos\ t - 8.13°)$

6.10. $c\ (t) = - 4 + e^{-t} + e^{t} + 2\ cos\ t$

6.11. (*a*) $c\ (t) = 1 + 2\ e^{-3t} - 3\ e^{-3t}$

  (*b*) 2 sec.

  (*c*) Overdamped.

6.12. (*a*) $c(t) = 1 - 1.414\ e^{-t} sin\ (t + 45°)$

  (*b*) 4 *sec*.

  (*c*) Un*d*er-damped.

6.13. (*a*) 3

  (*b*) 0.1

  (*c*) 2.98 ra*d*./*sec*

  (*d*) 3.33 *sec*

  (*e*) 70%

6.14. (*a*) 0.1 rad.

  (*b*) $\omega_n = 5,\ \zeta = 0.5$

  (*c*) Under-damped

  (*d*) 0.3535

6.15. $\omega_n = 2,\ \zeta = 0.5$

6.16. (*a*) $- 0.5$ ra*d*.

  (*b*) 0.1 rad.

6.17. (*a*) 0.09 rad.

  (*b*) 0.01 rad.

6.18. (*a*) 10

  (*b*) 0.2

  (*c*) 9.8

  (*d*) 50%

6.19. (*a*) 5

  (*b*) 0.6

  (*c*) 4

  (*d*) 10%

6.20. (*a*) 20%

  (*b*) 6.25

  (*c*) 0

6.21. (*a*) 100 ra*d*./*sec*

  (*b*) 0.83

6.22. $c(t) = 1 - 5e^{-2t} + 5e^{-3t} - e^{-5t}$

6.23. (a) $\omega_n = 8.94$ rad./sec., $\zeta = 0.11$

    (b) $a = 0.132$

6.24. 9

6.25. $a = 0.2$, $b = 1.6$

6.26. $e^{-t} - e^{-3t}$

6.27. $t_r = 2.42$ sec., $t_p = 3.63$ sec., $t_s = 8$ sec., $M_o = 16.32\%$

6.28. $M_p = 2.15$, $\omega_p = 0.489$

6.29. $c(t) = 0.5 + 0.6\,e^{-1.65t}\,\sin\,(1.51t - 1.34)$

Chapter 7

7.1. (a) ∞, ∞, -2

    (b) ∞, -50, 0

7.2. ∞, 10, 0

7.3. $0.1\,a_1$

7.4. (i) ∞, ∞, 2;

    (ii) Type 2, order 3,

    (iii) 3

7.5. ∞, 5, 0

7.6. ∞, 20, 0; $e_{ss} = \infty$

7.7. Second-order, 1.5, 0, 0

7.8. $K = 25$, unstable

7.9. $K = 20$

7.10. $K = 20$

7.11. $K = 200$

7.12. $K_v = 7.41$

7.13. $0.004784 + 0.004\,t$

7.14. $6 + 4t$

7.15. 1.89

7.16. $125,000/(s^3 + 87.5\,s^2 + 5372.9s + 125,000)$

7.17. (a) 2

    (b) 0, 0

    (c) 2

7.18. $(52s + 64)/(s^3 + 7s^2 + 52s + 64)$; $\omega_n = 4$

Chapter 8

8.1. $K_D = 2.32$

8.2. $K_1 = 26.53$, $K_2 = 1$, $K_3 = 0.85$, $\zeta\omega_n = 1.2$

8.3. $(s + 10)/(s + 100)$, $a = 10$, $T = 0.1$

8.4. $(1 + 1.6s)/(1 + 20s)$, $a = 0.08$, $T = 20$

8.5. $(s + 10)(s + 0.5)/[(s + 100)(s + 0.05)]$, $a = 10$, $b = 0.1$

8.6. $(s + 7.5)/(s + 75)$, $a = 10$

8.7. $(s^2 + s = 4)/[(s + 1.22)(s + 3.28)]$

Chapter 9

9.1. (a), (b) and (e) stable (c), (d), (f) and (g) unstable with 2 roots in RHP

9.2. (a) For all values of K, the system is unstable.

    (b) $1 < K < \infty$

    (c) $0 < K < 61.6$

    (d) $0 < K < 4$

    (e) $0 < K < 1.56$

9.3. (a) $0 < K < 35.5$

    (b) $0 < K < 150$

    (c) $-20 < K < \infty$

    (d) $-10 < K < 126$

    (e) $-6 < K < 60$

9.4. 150

9.5. $T = 0.021$, $\omega_n = 9.8$

9.6. $K \geq 72$

9.7. $0 < K \leq 48$

9.8. 2 roots

9.9. $0.53 < K < \infty$

9.10. One root

9.11. $-200 < K \leq 666.25$, $K > 666.25$

9.12. $0 < K \leq 6$

9.13. $0 < K < \infty$

9.14. 2 roots

9.15. $-1.2 < K \leq 11.95$

9.16. $0 < K < \infty$

9.17. No root in RHP; two roots on jω-axis.

9.18. $0 < K \leq 61.69$

9.19. Two roots between 0 and -2.

9.20. $0 < K \leq 60$; $0 < a \leq \infty$

9.21. $0 < T \leq 0.125$, $K \geq 4 - 16\ T$

Chapter 10

10.1. (a) $-1, -3$

    (b) -4

    (c) $-0.424, -3.6$

(*d*) 0, – 3.

10.2. (*a*) 1.414 ∠–98.13°

    (*b*) 2.24 ∠–153.4°

    (*c*) 2.28 ∠–232.1°

10.3. 123.1°

10.4. 193°

10.5. 116.6°

10.6. – 198.4 °

10.7. 29.5°

10.8. (*a*) Between – 1 and – 2

    (*b*) Between 0 and – 2.

10.9. – 1.57 and – 4.4

10.10. $2(s + 2)/(s^2 + 4s + 5)$

10.11. $2(s + 1)(s + 2)/ (s^2 + 4s + 5)$

10.12. (*b*) $K = 136$, $\omega_n = \sqrt{17}$

    (*c*) $K = 25.4$

10.13. (*b*) $K = 36.1$

10.14. (*b*) $K \geq 12$

10.15. $0 < K \leq 48$

10.16. (*b*) $1 < K \leq \infty$

    (*c*) $K = 1$, $\omega = 1$.

10.17. Stable

Chapter 11

11.5. K = 2965.6

11.12. $\dfrac{50(1+ j0.2\omega)}{j\omega(1+ j2\omega)(1+ j0.0625\omega)}$

11.13. $\dfrac{j3.17\omega}{(1+ j0.5\omega)(1+ j0.125\omega)}$

11.14. $\dfrac{10}{(1+ j0.25\omega)^2 (1+ j0.025\omega)}$

11.15. $K_v = 50$

11.16. $K_v = 2$

11.17. $K_p = 10$

11.18. $V_x = 0.148$

    Phase crossover frequency = 12.5 rad./sec.

    $j\omega$-axis crossover frequency = 30.4 rad./sec.

$$|G(j\omega)|_{\omega=0} = \infty; \quad |G(j\omega)|_{\omega=0} = -90°$$
$$|G(j\omega)|_{\omega=\infty} = 0; \quad |G(j\omega)|_{\omega=\infty} = -360°$$

11.19. $V_x = 1$

$$|G(j\omega)|_{\omega=0} = \infty; \quad |G(j\omega)|_{\omega=0} = -270°$$
$$|G(j\omega)|_{\omega=\infty} = 0; \quad |G(j\omega)|_{\omega=\infty} = -360°$$

11.20. Centre at $\left(\dfrac{1}{2}, 0\right)$; radius $= 1/2$.

Chapter 12

(Note: Answers may differ as they are obtained from plots)

12.1. $\omega_{cg} = 10$ rad./sec., $\omega_{cp} = 14$ rad./sec.,
$GM = + 7$ dB, $PM = + 18.4°$

12.2. $\omega_{cg} = 12$ rad./sec., $\omega_{cp} = 9$ rad./sec.,
$GM = - 4$ dB, $PM = - 12.6°$

12.3. $\omega_{cg} = 2.85$ rad./sec., $\omega_{cp} = 10$ rad./sec.,
$GM = +16.5$ dB, $PM = + 60°$

12.4. $\omega_{cg} = 15.5$ rad./sec., $\omega_{cp} = \infty$ rad./sec.,
$PM = + 50°$

12.5. $\omega_{cg} = 10,000$ rad./sec., $\omega_{cp} = \infty$ rad./sec.,
$PM = + 96.6°$

12.6. $\omega_{cg} = 2.9$ rad./sec., $\omega_{cp} = 4.4$ rad./sec.,
$GM = + 8$ dB, $PM = + 18°$

12.7. $\omega_{cg} = 0.55$ rad./sec., $\omega_{cp} = 1.45$ rad./sec.,
$GM = + 12$ dB, $PM = + 45°$

12.8. $\omega_{cg} = 1.2$ rad./sec., $\omega_{cp} = 0.47$ rad./sec.,
$GM = -18$ dB, $PM = - 37°$

12.9. $\omega_{cg} = 1.5$ rad./sec., $\omega_{cp} = \infty$ rad./sec.,
$PM = + 106°$

12.10. $\omega_{cg} = 4.2$ rad./sec., $\omega_{cp} = 2$ rad./sec.,
$GM = - 14$ dB, $PM = - 24°$

12.11. $\omega_{cg} = 4.5$ rad./sec., $\omega_{cp} = 20$ rad./sec.,
$GM = + 19$ dB, $PM = + 60°$

12.12. $\omega_{cg} = 1.0$ rad./sec., $\omega_{cp} = 4.3$ rad./sec.,
$GM = + 26$ dB, $PM = + 40°$

12.13. $\omega_{cg} = 1.0$ rad./sec., $\omega_{cp} = 29$ rad./sec.,
$GM = + 32$ dB, $PM = + 87°$

12.14. Z_R = 1, unstable
12.15. Z_R = 1, unstable
12.16. Z_R = 0, stable
12.17. $0 < K \leq 10.53$
12.18. $K > 6$
12.19. Z_R = 0, stable
12.20. ω_{cg} = 3.16 rad./sec., ω_{cp} = 1.24 rad./sec.,
    $GM = + 14.8$ dB, $PM = + 32°$
12.21. ω_{cg} = 0.9 rad./sec., ω_{cp} = 3.16 rad./sec.,
    $GM = + 16.7$ dB, $PM = + 55.6°$
12.22. $GM = - 20.6$ dB, $PM = - 36.7°$
12.23. $GM = + 15.6$ dB, $PM = = 10.7°$
12.24. $GM = + 20.7$ dB, $PM = + 30°$ for $K = 1$
12.25. $K \geq 0.5$
12.26. Uns*t*able for $K = \pm 10$
12.28. $GM = + 6$ dB, $PM = + 22°$
12.29. $GM = +6$ dB, $PM = + 27°$, $K_{max} = 8$

Chap*t*er 13
13.1. $K = 111.3$
13.2. $\zeta = 0.39$, $\omega_n = 2.4$
13.3. $K = 1.5$
13.4. $K = 5.6$
13.5. $\zeta = 0.866$, $M_p = 1.155$

Chap*t*er 14
14.1. $G_C(J\omega) = (1 + j0.5\omega)/(1 + j0.083\omega)$, $a = 5.8$, $K' = K$
14.2. $G_C(J\omega) = (1 + j0.5\omega)/(1 + j0.83\omega)$, $a = 5.8$, $K' = 8$
14.3. $G_C(J\omega) = (1 + j2\omega)/(1 + j20\omega)$, $b = 0.1$, $K' = 4$
14.4. $G_C(J\omega) = (1 + j0.5\omega)/(1 + j0.083)$, $a = 5.8$, $K' = K$
14.5. $G_C(J\omega) = (1 + j6.67)/(1 + j33.3\omega)$, $K' = 10$
14.6. $G_C(J\omega) = (1 + j2\omega)/(1 + j20\omega)$, $b = 0.1$, $K' = 4$
14.7. $G_C(J\omega) = (1 + j13.3\omega)/(1 + j111.1\omega)$, $K' = 8$
14.8. $G_C(J\omega) = (1 + j0.6\omega)/(1 + j4.5\omega)$, $K' = 250$

Chap*t*er 15

15.1. $\begin{bmatrix} \dot{x}_1 \\ \dot{x}_2 \end{bmatrix} = \begin{bmatrix} 0 & 1 \\ -\dfrac{K}{M} & -\dfrac{B}{M} \end{bmatrix} \begin{bmatrix} x_1 \\ x_2 \end{bmatrix} + \begin{bmatrix} 0 \\ \dfrac{1}{M} \end{bmatrix} f$

15.2.
$$\begin{bmatrix} \dot{x}_1 \\ \dot{x}_2 \\ \dot{x}_3 \end{bmatrix} = \begin{bmatrix} 0 & 1 & 0 \\ -\dfrac{K_2}{M_1} & 0 & -\dfrac{K_1}{M_1} \\ 0 & 1 & -\dfrac{K_1}{B} \end{bmatrix} \begin{bmatrix} x_1 \\ x_2 \\ x_3 \end{bmatrix} + \begin{bmatrix} 0 \\ \dfrac{1}{M_1} \\ 0 \end{bmatrix} f$$

15.3.
$$\begin{bmatrix} i_1^* \\ i_2^* \\ i_C^* \end{bmatrix} = \begin{bmatrix} -40 & 0 & -5 \\ -20 & 0 & -5 \\ 0 & 1000 & 0 \end{bmatrix} \begin{bmatrix} i_1 \\ i_2 \\ i_3 \end{bmatrix} + \begin{bmatrix} 10 \\ 5 \\ 0 \end{bmatrix} v$$

15.4.
$$\begin{bmatrix} i_1 \\ i_2 \\ \dot{v}_{c_2} \\ \dot{v}_{c_3} \end{bmatrix} = \begin{bmatrix} -\dfrac{R_1 + R_2}{L_1} & \dfrac{R_2}{L_1} & -\dfrac{1}{L_1} & 0 \\ \dfrac{R_2}{L_3} & -\dfrac{R_2}{L_3} & \dfrac{1}{L_3} & -\dfrac{1}{L_3} \\ \dfrac{1}{C_2} & -\dfrac{1}{C_2} & 0 & 0 \\ 0 & -\dfrac{1}{C_3} & 0 & 0 \end{bmatrix} \begin{bmatrix} i_1 \\ i_2 \\ v_{c_2} \\ v_{c_3} \end{bmatrix} + \begin{bmatrix} \dfrac{1}{L_1} & 0 \\ 0 & \dfrac{1}{L_3} \\ 0 & 0 \\ 0 & 0 \end{bmatrix} \begin{bmatrix} v_1 \\ v_2 \end{bmatrix}$$

15.5.
$$\begin{bmatrix} \dot{x}_1 \\ \dot{x}_2 \end{bmatrix} = \begin{bmatrix} 0 & 1 \\ -\dfrac{1}{T_a T_f} & -\dfrac{T_a + T_f}{T_a T_f} \end{bmatrix} \begin{bmatrix} x_1 \\ x_2 \end{bmatrix} + \begin{bmatrix} 0 \\ \dfrac{K_g R_L}{L_a L_f} \end{bmatrix} E_f; \; x_1^2 = E_L, \; \dot{x}_1 = \dot{E}_L$$

15.6.
$$\begin{bmatrix} \dot{x}_1 \\ \dot{x}_2 \\ \dot{x}_3 \end{bmatrix} = \begin{bmatrix} 0 & 1 & 0 \\ 0 & 0 & 1 \\ 0 & -a & -b \end{bmatrix} \begin{bmatrix} x_1 \\ x_2 \\ x_3 \end{bmatrix} + \begin{bmatrix} 0 \\ 0 \\ \dfrac{K_T}{L_a J} \end{bmatrix} E_a$$

$$a = (R_a B + K_T K_b)/L_a J \quad b = -(T_a + T_b)/T_a T_b$$

15.7.
$$\begin{bmatrix} \dot{x}_1 \\ \dot{x}_2 \\ \dot{x}_3 \\ \dot{x}_4 \end{bmatrix} = \begin{bmatrix} 0 & 1 & 0 & 0 \\ 0 & 0 & 1 & 0 \\ 0 & 0 & 0 & 1 \\ -5 & -6 & -3 & -4 \end{bmatrix} \begin{bmatrix} x_1 \\ x_2 \\ x_3 \\ x_4 \end{bmatrix} + \begin{bmatrix} 0 \\ 0 \\ 0 \\ 2 \end{bmatrix} u$$

$$y = \begin{bmatrix} 1 & 0 & 0 & 0 \end{bmatrix} \begin{bmatrix} x_1 \\ x_2 \\ x_3 \\ x_4 \end{bmatrix}$$

15.8.
$$\begin{bmatrix} \dot{x}_1 \\ \dot{x}_2 \\ \dot{x}_3 \\ \dot{x}_4 \\ \dot{x}_5 \end{bmatrix} = \begin{bmatrix} 0 & 1 & 0 & 0 & 0 \\ 0 & 0 & 1 & 0 & 0 \\ 0 & 0 & 0 & 1 & 0 \\ 0 & 0 & 0 & 0 & 1 \\ -9 & -2 & -5 & -6 & -10 \end{bmatrix} \begin{bmatrix} x_1 \\ x_2 \\ x_3 \\ x_4 \\ x_5 \end{bmatrix} + \begin{bmatrix} 0 \\ 0 \\ 0 \\ 0 \\ 5 \end{bmatrix} u$$

$$y = \begin{bmatrix} 1 & 0 & 0 & 0 & 0 \end{bmatrix} \begin{bmatrix} x_1 \\ x_2 \\ x_3 \\ x_4 \\ x_5 \end{bmatrix}$$

15.9.
$$\begin{bmatrix} \dot{x}_1 \\ \dot{x}_2 \\ \dot{x}_3 \\ \dot{x}_4 \end{bmatrix} = \begin{bmatrix} 0 & 1 & 0 & 0 \\ 0 & 0 & 1 & 0 \\ 0 & 0 & 0 & 1 \\ -1.5 & -2.25 & -0.75 & -1.25 \end{bmatrix} \begin{bmatrix} x_1 \\ x_2 \\ x_3 \\ x_4 \end{bmatrix} + \begin{bmatrix} 0 \\ 2.5 \\ -1.875 \\ 0.969 \end{bmatrix}$$

15.10.
$$\begin{bmatrix} \dot{x}_1 \\ \dot{x}_2 \\ \dot{x}_3 \end{bmatrix} = \begin{bmatrix} 0 & 1 & 0 \\ 0 & 0 & 1 \\ -2.5 & -3 & -1.5 \end{bmatrix} \begin{bmatrix} x_1 \\ x_2 \\ x_3 \end{bmatrix} + \begin{bmatrix} 1.25 \\ -2.375 \\ 1.5625 \end{bmatrix} u$$

15.11. Eigenvalues = $-1, -2, -3$
Eigenvector = $(1, -1, -1), (2, -4, -1), (1, -3, -3)$

15.12. $e^{-At} = \begin{pmatrix} -e^{-t}+2e^{=2t} & e^{-t}-e^{-2t} \\ 2(-e^{-t}+e^{-2t}) & 2e^{-t}-e^{-2t} \end{pmatrix}$

$x(t) = \begin{pmatrix} 2e^{-t}-e^{-2t} \\ 2e^{-2t}-2e^{-1}-1 \end{pmatrix}$

15.13. (a) Not completely state controllable
(b) Completely state controllable

15.14. (a) Observable
(b) Not observable

15.15. $A = \begin{bmatrix} 0 & 0 & 0 \\ 0 & -2 & 0 \\ 0 & 0 & -3 \end{bmatrix}; B = \begin{bmatrix} 1 \\ 1 \\ 1 \end{bmatrix}; C[5/3 \quad 5 \quad -20/3]$

15.16. $A = \begin{bmatrix} -1 & 0 & 0 \\ 0 & -1-j1 & 0 \\ 0 & 0 & -1+j1 \end{bmatrix}; B = \begin{bmatrix} 1 \\ 1 \\ 1 \end{bmatrix}; C[10-3.536(1-j1) \quad -3.536(-1+j1)]$

15.17. $A - BG = \begin{bmatrix} -g_1 & 1-g_2 \\ -2-g_1 & -3-g_2 \end{bmatrix}$

Observability matrix is

$= [C^T(A-BG)^T C^T] = \begin{bmatrix} 1 & -3g_1-4 \\ 2 & -3g_2-5 \end{bmatrix}$

$\begin{vmatrix} 1 & -3g_1-4 \\ 2 & -3g_2-5 \end{vmatrix} = 6g_1 - 3g_2 + 3$

If g_1 and g_2 are chosen such that the above determinant is zero, the system would be not observable.

REFERENCES

1. Kuo, B.C., Linear Networks and systems, McGraw Hill, N.Y., 1967

2. Hilderbrand, F.B. Methods of Applied Mathematics, PH, N.J. 1965

3. Cannon, R. H. Jr. Dynamics of Physical Systems, McGraw Hill, N.Y. 1967

4. Truxal, J. G., Control System Synthesis, McGraw Hill, N.Y. 1955

5. Kennedy, E.J., Operational Amplifier Circuits, HRW, Texas., 1988

6. Wait, J.V. et al: Introduction to Operational Amplifier Theory and applications, McGraw Hill, N. Y., 1992

7. Gabel, R.A. and Roberts, R.A., Signals and Linear Systems, John Wiley, N.Y., 1987

8. Borgan, W.L., Modern Control Theory, PH, N. J., 1985

9. Gantmacher, F.A. Matrix Theory, Vol. II, Chelsea, N.Y., 1964

10. Evans, W.R., Control System Dynamics, McGraw Hill, N.Y. 1954

11. Kuo, B.C., Automatic Control Systems, PH, N.J. 1991

12. Ogata, K. Modern Control Engineering, PHI, New Delhi, 2001

13. Bode, H.W., Network Analysis and Feedback Design, VNR, N.Y. 1945

14. Cheng, D. K. Analysis of Linear Systems, AW, MA, 1959

15. Dorf, R.C., Modern Control Systems, AW, MA, 1992

16. Doyle, J.C. et al, Feedback Control Theory, Macmillan, 1992

17. Kailath, T., Linear Systems, PH, N.J. 1980

18. Ogata, K. State Space Analysis of Control Systems, PH, N.J. 1967

19. Stefani, R. T. et al, Design of Feedback Systems, SCP, 1994

20. Bode, H. W., Feedback – The History of an Idea, in selected papers in Mathematical Trends in Control Theory, Dover, N. Y. 1964

21. Fuller, A.T. The early Development of Control Theory, Trans. ASME J., of Dynamic Systems, Measurements and Control, Sci. Amer., June & Sept., 1976

22. Mayr, O., The Origins of Feedback Control, Sci. Amer., Oct. 1970

23. Close, C.M. and Frederik, D.K. Modeling and Analysis of Dynamic Systems, HM, MA, 1978

24. Luenberger, D.G., Introduction to Dynamic Systems, Wiley, N.Y., 1979

25. Perkins, W. R. and Cruz, J. B. Engineering of Dynamic Systems, Wiley, N.Y. 1969

26. Van Valkenberg, M. E., Network Analysis, PH, N. J. 1974

27. Dertpuzos, M.L. et al., Systems, Network and Computation, McGraw Hill, N.Y. 1973

28. Mason, S. J., Feedback Theory : Some Properties of Signal Flow Graphs, Proc. IRE. Sept. 1953

29. Mason, S. J., Feedback Theory : Further Properties of Signal Flow Graphs, Proc. IRE. July, 1956

30. Chestnut, H and Mayor, R.W., Servomechanisms and Regulating System Design, Wiley, N.Y., 1959

31. James, H.M. et al., Theory of Servomechanisms, McGraw Hill, N.Y., 1972

32. Savant, C.J. Jr., Basic Feedback Control System Design, McGraw Hill, N.Y. 1986

33. Cruz, J.B. r., Feedback Systems, McGraw Hill, N.Y., 1972

34. Friedland, B., Control Systems Design : An Introduction to State Space Methods, McGraw Hill, N.Y., 1986

35. Dorf, R.C., Time Domain Analysis and Design of Control Systems, AW, MA, 1965

36. Schultz, D.G. and Melsa, J. L., State Functions and Linear Control Systems, McGraw Hill, N.Y., 1967

37. D'Azzo, J. J., and Houpis, C.H. Linear Control system Analysis and Design, McGraw Hill. N.Y., 1995

38. Jackson, A.S., Analog Computation, McGraw Hill, N.Y., 1960

39. Korn, G.A. and Korn, T.M., Electronic Analog Computers, McGraw Hill, N.Y., 1952

40. Decarlo, R.A., Linear Systems : A State variable Approach with Numerical Implementation, PH. N.J., 1989

41. Gupta, S.C., Transforms and State variable methods in Linear Systems, Wiley, N.Y., 1966

42. Kalman, R.E., Mathematical Description of Linear Dynamical Systems, SIAM J, Control series, 1963

43. Ogata, K., State Space Analysis of Control Systems, PH, N. J., 1967

44. Timothy, L. K. and Bona, B. E., State Space Analysis : An Introduction, McGraw Hill, N.Y., 1968

45. Zadeh, L.A. and Desoer, C.A., Linear System Theory : The state Space Approach, McGraw Hill, N.Y., 1963

46. Bellman, R., Introduction to Matrix Analysis, McGraw Hill, N.Y., 1960

47. Davis, S.A. and Ledgerwood, B.K., Electromechanical Components for Servomechaninsms, McGraw Hill, N.Y., 1961

48. Cochin, I., Analysis and Design of Dynamic Systems, Harper-Adrow, N.Y., 1980

49. Rugh, W.J., Linear System Theory, PH, N.J., 1993

50. Ward, J.R., and Strum, R.D, State variable Analysis, PH, N.J., 1970

51. Wiberg, D.M., State Space and Linear Systems, Schaum's Outline Series, McGraw Hill, N.Y., 1971

52. Russo et al., State Variables for Engineers, Krieger, Fla., 1990

53. Thomson, W.T., Laplace Transformation, P.H., N.J., 1966

54. Aseltine, J.A., Transform Methods in Linear System Analysis, McGraw Hill, N.Y. 1958

55. Krishnamurthi, V., Implications of Routh Stability Criteria, IEEE. Trans. AC-25, 1980

56. Krishnamurthi, V., Correlation between Routh's Stability Criteria and Relative Stability of Linear Systems, IEEE. Trans. AC-17, 1972

57. Krishnamurthi V., Gain Margin of conditionally Stable Systems from Routh's Stability Criterion, IEEE. Trans. AC-17, 1972

58. Krishnamurthi V. and Sastri, G.V.K.R., Relative Stability using Simplified Routh Approximation method, JIETE, Vol. 33, No. 3, May 1987

59. Krishnamurthi V., Routh's Stability Criteria and Relative Stability of Linear Systems, JIE, Vol. 52, No. 9, ET-5, 1972

60. Krishanmurthi V. and Gopal Reddy, P., Relative Stability from Strum's Sequence, vol. 31, No. 5, 1985

61. Krishnamurthi, V. and Sastri, G.V.K.R., Relative Stability using Simplified Routh approximation Method, JIETE., Vol. 35, No. 3, 1987

62. Krishnamurthi, V., A Generalized Method of Stability Analysis, IASTED, Intl. Conf. Computer Aided Design and Applications, Paris, 1985

63. Krishnamurthi, V., Analog and Digital Computational Methods, CBS Publications, New Delhi, 1984

64. Burns, R.A. and Saunders, R.M., Analysis of Feedback Control Systems, McGraw Hill, N.Y., 1955

65. Chu, Y., Correlation between Frequency and Transient Response of Feedback Control Systems, Trans. AIEE, Vo.2, pt-II. May 1953

66. Levine, W.S.(Eds.), The Control Hand Book, IEEE Press, 1996

67. Rainville, E.D., The Laplace Transform: An Introduction, Macmillan, London, 1983

68. Wolovich, W.A., Automatic Control Systems: Basic Analysis and Design, Saunders College Publishing, M.A., 1994

69. Brown, G.S., and Campbell, D.P., Principles of Servomechanisms, Wiley, N.Y., 1948

70. James, H.M. et al., Theory of Servomechanisms, McGraw-Hill, N.Y. 1947

71. Ahrent, W.R., Servomechanism Practice, McGraw-Hill, N.Y., 1954

72. Ahrent, W.R. and Taplin, J.F., Automatic Feedback Control, McGraw-Hill, N.Y., 1951

73. Aizerman, M.A., Theory of Automatic Control, A-W, MA, 1963

74. Atkinson, P., Feedback Control Theory for Engineers, Heinmann, London, 1968

75. Bower, J.L. and Schultheiss, P.M., Introduction to design of Servomechanisms, Wiley, N.Y., 1958

76. Chen, C.F. and Hass, I.J., Elements of Control System Analysis, PH, N.J., 1969

77. Clark, R.N., Introduction to Automatic Control System, Wiley, N.Y., 1962

78. Del Toro, V. and Parker, S., Principles of Control Systems, McGraw-Hill, N.Y., 1960

79. Eckman, D.P., Automatic Process Control, Wiley, N.Y., 1958

80. Elgerd, O.L., Control System Theory, McGraw-Hill, N.Y., 1967

81. Gibson, J.E., and Tuteur, F.B., Control System Components, McGraw-Hill, N.Y., 1958

82. Horowitz, I. M., Synthesis of Feedback System, AP. N.Y., 1963

83. Murphy, G.J., Basic Automatic Control Theory, Van Nostrand, Princeton, 1957

84. Pitman, R. J.G., Automatic Control Systems Explained, Macmillian, London, 1966

85 Raven, P., Automatic Control Engineering, McGraw-Hill, N.Y., 1960

86. Smith, O., J.M., Feedback Control Systems, McGraw-Hill, N.Y., 1958

87. Taylor, P.L., Servomechanisms, Longmans, London, 1960

88. Thaler, G.J. and Brown, R.G., Analysis and Design of Feedback Control Systems. McGraw-Hill, N.Y., 1964

89. Tou, J. T., Modern Control Theory, McGraw-Hill, N.Y.,1964

90. Evans, W.R., Control System Dynamics, McGraw-Hill, N.Y., 1954

91. Evans, W.R., Graphical Analysis of Control Systems, Trans. AIEE, Vol. 67, 1948

92. Evans, W.R., Control System Synthesis by Root Locus method, Trans. AIEE, vol. 67. 1950

93. Gardener, M.F. and Barnes, J.L., Transient in Linear Systems, Wiley, N.Y., 1942

94. Churchill, R.V., Introduction to Complex variables and Applications, McGraw-Hill, N.Y., 1948

95. Watkins, B.O., Introduction of Control Systems, Macmillan, London, 1969.

INDEX